THE BIOLOGY OF

MOSQUITOES

THE BIOLOGY OF

MOSQUITOES

VOLUME 1

DEVELOPMENT, NUTRITION AND REPRODUCTION

A. N. CLEMENTS

London School of Hygiene and Tropical Medicine

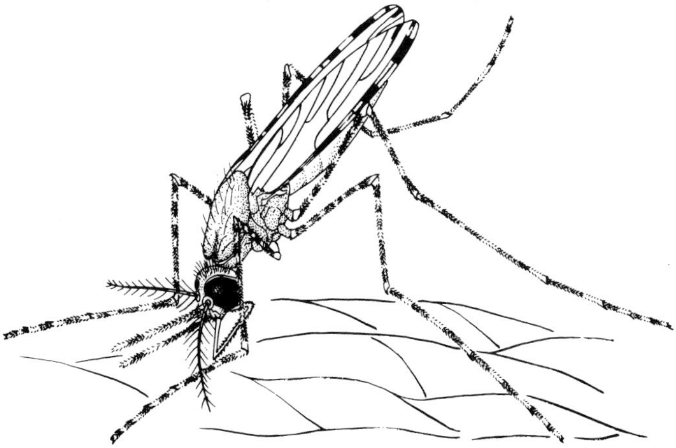

CHAPMAN & HALL

London · Glasgow · New York · Tokyo · Melbourne · Madras

Published by Chapman & Hall, 2–6 Boundary Row, London SE1 8HN

Chapman & Hall, 2–6 Boundary Row, London SE1 8HN, UK

Blackie Academic & Professional, Wester Cleddens Road, Bishopbriggs, Glasgow G64 2NZ, UK

Chapman & Hall, 29 West 35th Street, New York NY10001, USA

Chapman & Hall Japan, Thomson Publishing Japan, Hirakawacho Nemoto Building, 6F, 1–7–11 Hirakawa-cho, Chiyoda-ku, Tokyo 102, Japan

Chapman & Hall Australia, Thomas Nelson Australia, 102 Dodds Street, South Melbourne, Victoria 3205, Australia

Chapman & Hall India, R. Seshadri, 32 Second Main Road, CIT East, Madras 600 035, India

First edition 1992

Typset in Goudy Old Style 10½/12½pt by Falcon Typographic Art Ltd, Fife, Scotland
Printed in Great Britain at the University Press, Cambridge

ISBN 0 412 40180 0

A catalogue record for this book is available from the British Library

Library of Congress Cataloging-in-Publication data
Clements, A. N. (Alan Neville)
The biology of mosquitoes / A. N. Clements.
p. cm.
Rev. ed. of: The physiology of mosquitoes. 1963.
Includes bibliographical references and index.
Contents: v. 1. Development, nutrition, and reproduction.
ISBN 0–412–40180–0 (v. 1)
1. Mosquitoes. I. Clements, A. N. (Alan Neville). Physiology of mosquitoes. II. Title.
QL536.C556 1992
595.77'1 – dc20 91–5096 CIP

Printed on permanent acid-free text paper, manufactured in accordance with the proposed ANSI/NISO Z 39.48-199X and ANSI Z 39.48-1984

Contents

Acknowledgements ix
Preface xi
Introduction Mosquitoes: life cycle, biology, disease transmission xiii

1 Aspects of genetics **1**
1.1 The chromosomes 1
1.2 Molecular characteristics of the genome 22
1.3 Meiosis 29
1.4 Sex determination, sexual differentiation and sex-ratio regulation 29
1.5 Cytoplasmic incompatibility 37

2 Embryology **46**
2.1 Larval embryogenesis 46
2.2 Experimental embryology 56

3 The egg shell **63**
3.1 Structure 63
3.2 Sclerotization 69
3.3 Water relations 70
3.4 Hatching 71

4 Larval feeding **74**
4.1 Food sources and feeding modes 74
4.2 Structure of the mouthparts and pharynx 80
4.3 Feeding mechanisms 87
4.4 Feeding rates 97

5 Larval nutrition, excretion and respiration **100**
5.1 Structure of the alimentary canal 100
5.2 Digestion 106
5.3 Nutritional requirements 109
5.4 Metabolic reserves 117
5.5 Excretion 117
5.6 Larval and pupal respiration 118

6 Osmotic and ionic regulation **124**
6.1 Chemical properties of larval habitats 124
6.2 Regulation of the haemolymph 126
6.3 Water balance 131

6.4 Mechanisms of ion uptake 135
6.5 Mechanisms of ionic and osmotic regulation 138
6.6 Ultrastructure of the organs associated with ion and water transport 143

7 Growth and development 150
7.1 Larval growth 150
7.2 Cytogenetic aspects of growth and development 158
7.3 Cuticle ultrastructure, deposition and ecdysis 160
7.4 Endogenous developmental rhythms 167

8 Metamorphosis 171
8.1 The imaginal disks 171
8.2 Sexual differentiation 175
8.3 Development and metamorphosis of the central nervous system and compound eyes 183
8.4 Metamorphosis of the alimentary canal 188
8.5 Metamorphosis of other organs 190
8.6 Hormones and metamorphosis 191
8.7 Maturation 193

9 The circulatory system 195
9.1 Anatomy 195
9.2 Origin and characteristics of the heartbeat 197
9.3 Haemolymph 198
9.4 Immune responses 202

10 The endocrine system and hormones 206
10.1 The endocrine organs 206
10.2 The stomatogastric nervous system 212
10.3 Hormones 215

11 Adult food and feeding mechanisms 220
11.1 Food 220
11.2 Structure of the mouthparts and associated organs 224
11.3 Feeding mechanisms 235
11.4 Sensory reception and regulation of food intake 241

12 The adult salivary glands and their secretions 251
12.1 Structure of the salivary glands 251
12.2 The constituents, actions and secretion of saliva 255
12.3 Skin reactions of mammalian hosts 259

13 Structure of the adult alimentary canal 263
13.1 The foregut 263
13.2 The midgut 264
13.3 The hindgut and Malpighian tubules 269

14 Adult digestion	**272**
14.1 Enzymes of the digestive tract	272
14.2 The peritrophic membrane	278
14.3 Digestion of the blood meal	287
14.4 Regulation of enzyme synthesis	288
15 Adult energy metabolism	**292**
15.1 Synthesis of reserves	292
15.2 Utilization of reserves	300
16 Adult diuresis, excretion and defaecation	**304**
16.1 Diuresis after emergence and after feeding	305
16.2 Mechanisms of tubular fluid production	310
16.3 Regulation of diuresis	320
16.4 Excretion and defaecation after emergence	322
16.5 Composition of excreta and faeces after blood feeding	323
16.6 Regulation of excretion and defaecation after blood feeding	325
17 Structure of the gonads and gonoducts	**327**
17.1 Structure of the testes and male genital ducts	327
17.2 Structure of the ovaries and oviducts	328
18 Spermatogenesis and the structure of spermatozoa	**333**
18.1 Spermatogenesis	333
18.2 Spermatozoa	335
19 Oogenesis	**340**
19.1 The formation of ovarian follicles	340
19.2 The ovarian cycle	342
19.3 Effects of external factors	346
19.4 rRNA synthesis in oocyte and nurse cells	348
19.5 Follicle growth and development	349
19.6 Chorion formation	353
20 Vitellogenesis	**360**
20.1 The nature of yolk	360
20.2 Expression of the vitellogenin genes	363
20.3 Ultrastructural changes in the trophocytes	370
20.4 Incorporation of vitellogenin	374
21 Hormonal regulation of ovarian development in anautogenous mosquitoes	**380**
21.1 The previtellogenic phase	381
21.2 The initiation of vitellogenic development	385
21.3 The promotion of vitellogenic development	387
21.4 Stimuli from the blood meal	392
21.5 20-Hydroxyecdysone	395

21.6 Juvenile hormone 400
21.7 Antagonistic and synergistic effects of hormones 401
21.8 Oostasis 404
21.9 The second ovarian cycle 405

22 Nutrition and fertility of anautogenous mosquitoes 408
22.1 Factors affecting fecundity and fertility 408
22.2 Nutritional requirements for oogenesis 416
22.3 Correlation of fertility with food intake 421

23 Autogeny 424
23.1 The phenomenon of autogeny 424
23.2 Hormonal regulation of autogenous ovarian development 432
23.3 The genetic basis of autogeny 440

References 442
Species index 497
Subject index 505

Acknowledgements

One of the pleasures of writing this book has been experiencing the helpfulness of many individuals, who have generously provided copies of published papers, manuscripts and micrographs, and willingly shared their knowledge. I have sought the comments and advice of specialists as extensively as possible and am very grateful to all who have enlightened me, too many to acknowledge fully here. However, I should like to name those individuals who scrutinized and commented on particular chapters or sections of Volume 1, contributing greatly to its final form. They are, I hope without any omission, D.T. Anderson, A. Ralph Barr, Klaus W. Beyenbach, Mary Bownes, P.F.Billingsley, Timothy J. Bradley, H.Briegel, G.M.Coast, D.A.Craig, J.M.Crampton, C.F.Curtis, R.H.Dadd, C.M.Dye, Morton S. Fuchs, C.A.Green, J.D.Gillett, Henry H. Hagedorn, H.Hecker, W.R.Horsfall, Anne Hudson, Jack Colvard Jones, M.D.R.Jones, Thomas J. Kelly, J.R.Larsen, Arden O. Lea, J.D.Lines, Susan McIver, Richard W. Merritt, George F. O'Meara, W.B.Owen, Larry G. Pappas, Werner Peters, John E. Phillips, Alexander S. Raikhel, K.Sander, Andrew Spielman, David W. Stanley-Samuelson, R.H.Stobbart, D.J.Tritton, Annelise Wandall, G.B.White, R.J.Wood and K.A.Wright. Ron Page's advice on word processing was invaluable. The figures of mosquitoes in the Introduction were drawn by Catherine Constable from photographs taken by Ludwig Gomulski.

I express my appreciation to the following publishers for permission to reproduce illustrations: Academic Press; Akademie Verlag; American Association for the Advancement of Science; American Mosquito Control Association; American Physiological Society; Blackwell Scientific Publications; Company of Biologists; CSIRO; Elsevier Science Publishers; Entomological Society of America; Entomological Society of Canada; Humana Press; National Research Council of Canada; Natural History Museum, London; New Jersey Mosquito Control Association; Pergamon Press; Royal Entomological Society; Royal Society; Smithsonian Institution; Springer-Verlag; Taylor & Francis; Wiley-Liss, a Division of John Wiley and Sons; and the World Health Organization.

Preface

This is the first of two volumes arising from the rewriting of *The Physiology of Mosquitoes*, published in 1963. That book addressed many aspects of mosquito biology, so the new title is felt to be appropriate, but as before the subject is treated from the viewpoint of a physiologist. The expansion of the work into two volumes reflects the extent and diversity of mosquito research, an activity which is driven partly by the medical importance of mosquitoes but very largely by the suitability of certain species as experimental animals. The accessibility of wild populations of many species is also valuable. Some observations that were made first on mosquitoes introduced concepts that were later shown to apply to all insects, or even to all animals. A notable example is receptor-mediated endocytosis, the universal mechanism for uptake of key macromolecules by animal cells. In recent years research in mosquito biology has become increasingly technical, and it can be a struggle to decipher the language. With the exception of a few passages, I have tried to describe recent advances in a way that can be readily understood by all informed biologists. For readers who know little or nothing of mosquitoes, this volume starts with a thumb-nail sketch of their biology and medical importance. Medawar* observed that it is a strength of many single-author books that the subject is presented as a whole, with recent developments placed in the context of earlier knowledge. I have not approached mosquito biology historically, but I have included much important information from earlier years that today is forgotten or overlooked. It is a disadvantage of some single-author books that they are a long time appearing; the development time of this one has matched the longer generation time of the periodical cicada *Magicicada septendecim* (L.).

Alan Clements

* P.Medawar, *The Times Higher Education Supplement*, 18 March 1983.

Introduction

Mosquitoes: life cycle, biology, disease transmission

The mosquitoes, or Culicidae, are a family of about three and a half thousand species within the order Diptera, the two-winged flies. They are one of the more primitive families of Diptera, being more closely related to midges, gnats and crane flies, for example, than to houseflies and blowflies. Mosquitoes are found throughout the world except in places that are permanently frozen. Three quarters of all mosquito species live in the humid tropics and subtropics, where the warm moist climate is favourable for rapid development and adult survival, and the diversity of habitats permitted the evolution of many species. Although the arctic zone has fewer than a dozen species, the greatest concentrations of mosquitoes are found in the arctic tundra, where dense populations of adults, which sometimes blacken the sky, emerge from larvae that have bred in summer-melt pools overlying permafrost.

Mosquitoes are classified into three subfamilies, the largest and most diversified of which is divided into a number of tribes:

Family	Culicidae:
Subfamily	Toxorhynchitinae
	Anophelinae
	Culicinae:
Tribes	Culicini, Aedini, Sabethini, Mansoniini, etc.

Like other true flies, culicids exhibit 'complete metamorphosis', i.e. the juvenile form passes through both larval and pupal stages. The larvae are anatomically different from the adults, live in a different habitat and feed on a different type of food. Transformation to the adult takes place during the non-feeding pupal stage.

THE EGGS

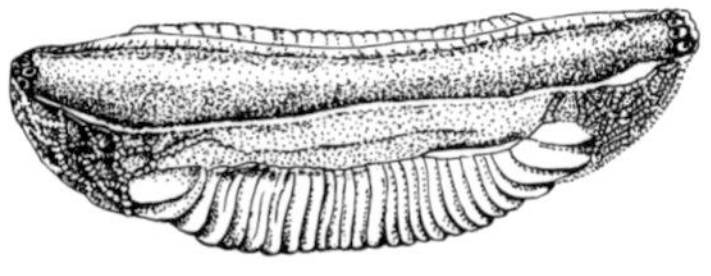

Egg of *Anopheles gambiae*.

Female mosquitoes lay some 50 to 500 eggs at one time, depositing them on water or on sites that will be flooded. Each egg is protected by an egg shell, which in many species is elaborately sculpted (Section 3.1). Spermatozoa stored by the inseminated female fertilize the oocytes as they are ovulated, and embryonic development starts almost immediately after the eggs have been laid.

Within one to two days to a week or more, depending on temperature, the embryo develops into a fully formed larva. In most species the larva hatches once it is formed, and can survive for a few days at most in the absence of water. Mosquitoes of the tribe Aedini have water-proofed egg shells capable of resisting desiccation, and fully-formed but unhatched aedine larvae can survive for months or even years in the absence of free water. Aedine species lay their eggs in places that may not be flooded for days, weeks or months (Section 3.3). A fall of rain that inundates oviposition sites, or a high tide flooding a salt marsh, stimulates hatching and can lead to an apparent population explosion.

THE LARVAE

Mosquito larvae are legless, but they retain a well-formed head and so do not appear maggot-like. The larval habitats are small or shallow bodies of water with little or no water movement – typically shallow pools, sheltered stream edges, marshes and water-filled treeholes, leaf axils or man-made containers. The habitats range in size from animal footprints to marshes and rice fields. Most species live in fresh water but a few are adapted for a life in brackish or saline water in salt marshes, rock pools or inland saline pools. All aquatic animals have problems of water and salt balance, whether they live in fresh or salt water, and the means by which mosquito larvae have solved these problems has been the subject of much investigation. In addition to the usual internal organs of ion regulation, mosquito larvae have four external balloon-like anal papillae, which are capable of ion uptake from very dilute solution (Chapter 6). When it hatches from the egg the young mosquito larva is fully adapted for living in water, and two features determine its manner of life: use of atmospheric oxygen for respiration and use of water-borne particles as food.

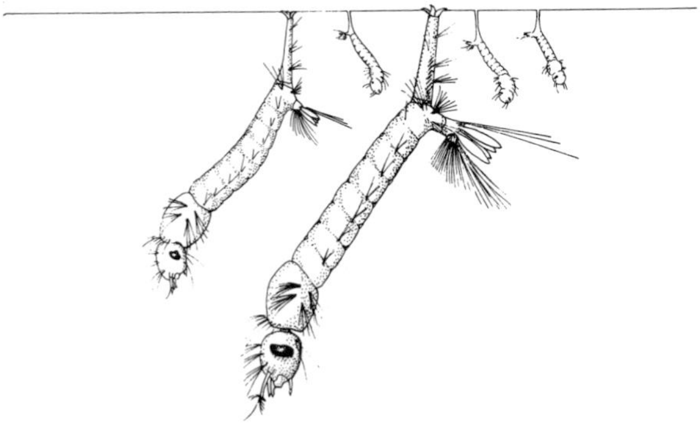

Culex pipiens larvae.

The air-breathing habit requires mosquito larvae either to live more or less permanently at the air/water interface, as most anopheline and some culicine larvae do, or to make frequent visits to the water surface. The only functional respiratory openings are a pair of spiracles, near the end of the abdomen, from which air-filled tracheae extend to all parts of the body. The spiracles of culicine and toxorhynchitine larvae are situated at the end of a tube or siphon, and the larvae hang downwards from the surface membrane by their siphons with their spiracles open to the air. The spiracles of anopheline larvae are flush with the dorsal surface of the last abdominal segment, and the larvae lie horizontally below the surface membrane, their spiracles opening through it. Larvae of two culicine genera (*Mansonia* and *Coquillettidia*) are able to remain permanently submerged. They live with their respiratory siphons, which have modified saw-like tips, forced into the air-filled tissues that fill the stems and roots of certain aquatic plants (Section 5.6).

The characteristic food resource of mosquito larvae is 'particulate matter'. This includes aquatic microorganisms, such as bacteria, diatoms and algae, and also, as an important component, particles of detritus that are largely derived from decayed plant tissues. Such particles provide food for diverse aquatic invertebrates, which filter them from the water by a variety of mechanisms. Mosquito larvae, which live mainly in still water, are exceptional in not relying on natural water currents to bring the particles to them. Through the regular beating of their 'mouth brushes' mosquito larvae generate water currents which flow towards the head, and in a manner that is not well understood they separate particles of a certain size from the water. Anopheline larvae typically feed at the water surface, in a particle-rich layer just below the surface membrane. Culicine larvae feed on particles suspended in the water column, and many supplement this feeding mode by abrading with their mouthparts the layers of organic matter that cover submerged surfaces, so generating new particles. Toxorhynchitine larvae are predatory on small invertebrates, as are a very few species in the other two subfamilies (Chapter 4).

For organisms the size of mosquito larvae, the aquatic medium is very different from our perception of it. The 'viscous force' of a fluid is proportional to its relative velocity across a solid object and inversely proportional to the size of the object, so for small organisms that move at low velocities water is highly viscous, even syrup-like. This is particularly significant for the current-generating and particle-capturing actions of mosquito larvae (Section 4.3.1).

The organs that compose the larval body serve larval functions and are mostly very different in structure from the adult organs. Within these larval organs there remain groups of undifferentiated cells that will eventually form adult organs. They include the 'imaginal disks', pocketed invaginations of the epidermis that are nascent adult appendages. The growing mosquito larva moults four times. On the first three occasions that it leaves its cast cuticle the larva appears very much as before. During the period of the fourth moult the imaginal disks develop rapidly, changing the form of the insect crudely to that of an adult, and the organism that leaves the fourth larval skin is a pupa. The rapid growth rates of many tropical species permit the exploitation of transient water bodies (Chapters 7 and 8).

THE PUPAE

The pupa remains an aquatic organism. That it has assumed the form of an adult is largely concealed because the head and thorax, with their elongate appendages, are cemented together in the form of a cephalothorax. The abdomen, which now terminates in two large paddles, has retained the strong larval musculature and is an effective organ of propulsion. An air bubble, which is enclosed between the appendages, provides buoyancy, and the pupa floats at the water surface with the top of its thorax in contact with the surface membrane and its abdomen hanging down. The new form and posture preclude use of

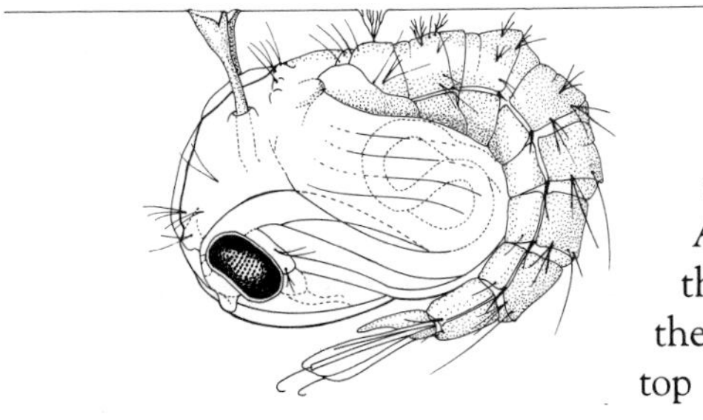

Pupa of *Anopheles gambiae*.

the terminal abdominal spiracles for respiration. That function is taken over by the mesothoracic spiracles, which open within large 'respiratory trumpets'. As the pupa floats at the air/water interface the hydrophobic rims of the respiratory trumpets protrude through the surface membrane.

During the pupal stage certain larval organs are destroyed, e.g. the alimentary canal, while replacement adult organs are constructed from undifferentiated embryonic cells. Other organs, including the heart and fat body, are carried over to the adult stage. These final stages of metamorphosis can be completed within one to two days if the temperature is sufficiently high (Chapter 8). When the adult is fully formed within the pupal cuticle, the insect rests at the water surface and starts to swallow air. The consequent increase in internal pressure forces a split along the midline of the pupal thoracic cuticle, and the adult slowly expands out of the pupal cuticle and steps on to the water surface (Section 7.3).

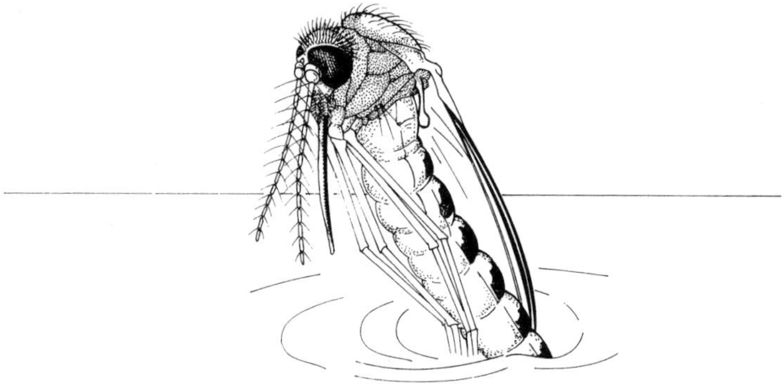

Culex pipiens emerging.

THE ADULTS

Like many of the more primitive Diptera, adult mosquitoes have an elongate body and long wings and legs, which provide an aerodynamically stable form. The hind wings are modified as small oscillating sense organs, or halteres, which assist flight control. Like other Diptera, mosquitoes are fluid feeders. Exceptionally among the more primitive Diptera, their mouthparts have evolved into an elongate composite proboscis, half as long as the body, suitable for probing nectaries and, in the case of the female, adapted for piercing skin and imbibing blood from peripheral blood vessels. The outer sheath-like part of the proboscis, the labium, encloses the remaining mouthparts, which have the form of needle-like stylets. Of these, the female's mandibles and maxillae, which are flattened and toothed, can be driven through tissue by the muscles at their bases, making a channel for other styletized mouthparts which contain canals for the delivery of saliva and the removal of blood (Chapter 11). Both males and females use the sugar in plant juices as a source of energy, usually obtaining it from nectaries but sometimes from other sources such as rotting fruit and honeydew. Anopheline and culicine females have a requirement for protein, from which to develop large batches of eggs, and they engorge on vertebrate blood for that purpose. Toxorhynchitine females feed only on plant juices.

Body odour and carbon dioxide, carried on the wind, stimulate sense receptors on the antennae and palps of female mosquitoes, alerting them to the presence of a host. The females respond by flying upwind, which takes them towards the host. Close to the host, visual stimuli and the convection currents of warm moist air that rise from the host provide additional cues. The females of some species are able to detect individual vertebrate hosts at a distance. *Anopheles melas* responded to one calf at a distance of 14 m, and to two calves at over 36 m. The females of all blood-feeding species show a degree of specificity in their choice of host, whether mammal, bird or cold-blooded vertebrate. Some species are highly specific, feeding predominantly on one or a few host species only, others are less specific. Individual human beings differ in their attractiveness to mosquitoes; the cause of the difference has not been elucidated (Volume 2).

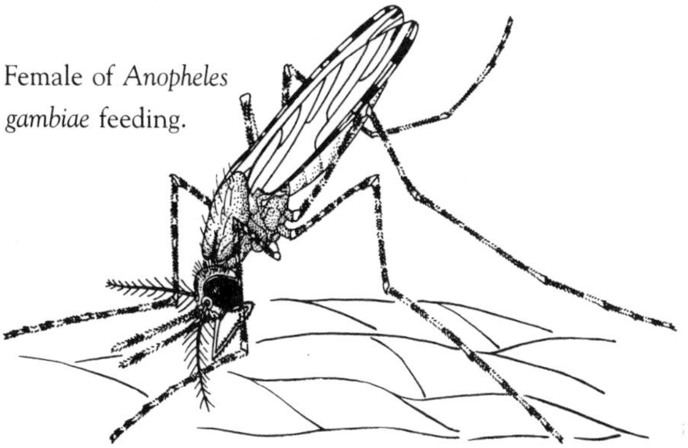
Female of *Anopheles gambiae* feeding.

Once landed on an appropriate part of the host, the female drives the styletized components of her mouthparts into its skin. The saliva that is injected as the mouthparts penetrate contains a substance that prevents haemostasis, the aggregation of blood platelets that is the host's first defence against the laceration of small blood vessels (Chapter 12). The saliva is also the source of immunogens that are responsible for the characteristic skin reactions to mosquito bites. Sooner or later the probing stylets pierce a blood vessel, and the presence of blood is identified from its content of ADP and ATP. If the female is undisturbed, feeding continues until abdominal stretch receptors signal repletion. Within a few minutes gorging mosquitoes can imbibe up to four times their own weight of blood (Chapter 11). This provides the protein needed for egg production, but also inflicts upon the mosquito a water load, which renders flight difficult, and potentially toxic amounts of sodium and potassium. The adult excretory system is capable of rapid elimination of water and salts, and diuresis commences while the female is still feeding (Chapter 16).

Digestion of blood proteins yields amino acids which are reconstituted in the mosquito's fat body as proteinaceous yolk. This is transported to the ovaries and incorporated into the oocytes, which are matured in a number that matches the provision of yolk (Chapters 14, 20, 22). It is a feature of mosquito biology that eggs are not matured continuously but in batches, following the periodic blood meals. This phenomenon has been a boon for reproductive physiologists, enabling them to probe the complex process of hormone secretion that regulates the conversion of blood protein into eggs (Chapter 21). A small percentage of species are able to develop one batch of eggs from protein and lipid reserves carried over from the larval stage, i.e. without a blood meal. Species with this capability are termed autogenous (Chapter 23). Male mosquitoes can be readily distinguished by their large and elaborate antennae, in which rings of fibrils encircle the shaft. These antennae resonate in response to a pure tone of a certain pitch. Female mosquitoes in flight produce a familiar whining sound, the pitch of which reflects the wing beat frequency of the species. That sound activates the antennae of conspecific males, and provides directional indicators which the massive sense organs at the base of the antennae can resolve (Volume 2).

The role of adult male mosquitoes is insemination of females, and when not resting the males are either feeding or exhibiting a behaviour pattern that is likely to bring them into contact with females. One conspicuous manifestation of male behaviour is swarming – the localized assembly of from two or three individuals to many thousands of individuals of a single species. Any conspecific female that enters a swarm will be seized immediately by a male. Mating also occurs outside swarms. Inseminated females store sufficient sperm in their spermathecae to fertilize a number of egg batches. A factor called matrone that is transferred in semen renders females unreceptive to males and refractory to further copulation, but its effectiveness may not persist throughout the life of the female.

The behavioural activities of adult mosquitoes – emergence, mating, feeding, oviposition – take place at particular times of day and night, which vary between species. Adults of *Anopheles gambiae*, for example, emerge during the late afternoon and, once mature, mate during a twenty-minute period at dusk. The females take blood meals principally during the four hours after midnight. The timing is not governed directly by light and dark but by endogenous or so-called circadian rhythms, which are reset daily by the change from light to dark at sunset (Section 7.4 and Volume 2).

When a female has matured a batch of eggs she takes to the wing and responds to stimuli from suitable oviposition sites. For most mosquitoes the oviposition site is a water body with particular characteristics; odour, taste, flow and shade are known to influence different species. The eggs may be dropped individually to float on the water surface, as by females of *Anopheles*, or packed together to form a floating egg raft, as by *Culex*. Aedine species deposit their eggs on moist surfaces, often at the edge of a body of water or on an area of soil that will be flooded. By whatever means, the female finds the appropriate habitat, and the larvae hatch into conditions for which they are adapted (Volume 2). The passage of eggs from the ovarioles into the oviducts leaves permanent structural changes which form a record of a female's reproductive history, and which also provide a remarkably accurate guide to her age (Volume 2).

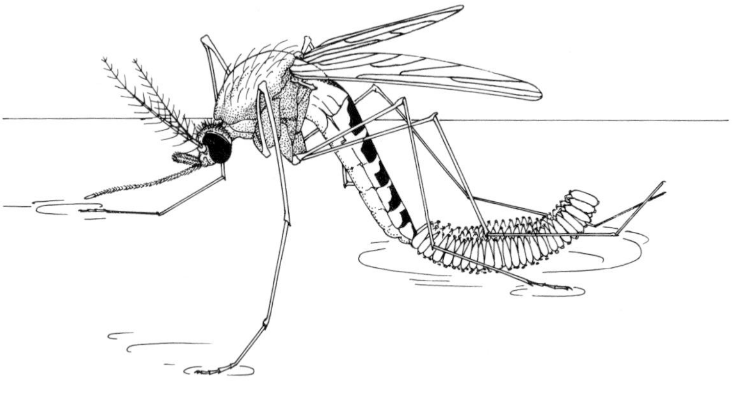

Female of *Culex pipiens* ovipositing.

In tropical regions the life span of adult mosquitoes ranges from a few days to several weeks. In temperate regions it is frequently longer, and in species that overwinter as adults the life span of females may approach one year. The females' behavioural responses and physiological processes follow a pattern, the gonotrophic cycle, which starts with response to the vertebrate host and feeding, continues with the digestion of blood and formation of a batch of mature oocytes, and ends with oviposition. Within an hour of completing one gonotrophic cycle a female may commence another. At warmer temperatures, tropical *Anopheles* oviposit regularly every two or three days.

PATHOGENS, PARASITES AND DISEASE TRANSMISSION

Mosquitoes are hosts to a variety of pathogens and parasites, including viruses, bacteria, fungi, protoctistans and nematodes. Some of these organisms alternate a parasitic phase with a free-living phase; others are entirely parasitic, and many of these alternate between their mosquito hosts and other invertebrate or vertebrate hosts. The blood-sucking habit renders adult mosquitoes prone to acquire pathogens and parasites from one vertebrate host and to pass them on to another, but even so, many aspects of a mosquito's ecology and physiology must be appropriate for it to acquire, harbour and transmit a particular organism. Mosquitoes are protected to a degree against invasive organisms by structural barriers and by their immune system (Section 9.4).

The role of blood-sucking arthropods as agents of human and animal diseases was established in the last quarter of the 19th century. The first firm evidence that a human parasite underwent obligatory development in an insect was obtained in China in 1877 by Manson, who discovered that mosquitoes were intermediate hosts of the nematode *Wuchereria bancrofti*, and that they acquired the parasite by feeding on infected human beings. He observed the growth and development of the nematode larvae in the mosquito, but believed that re-infection of human beings occurred when they drank water in which infected mosquitoes had died. In 1880 in Algeria, Laveran observed, in drops of human blood, malaria parasites undergoing what we now know to be the exflagellation of male gametocytes. In India, Ross time and again observed exflagellation and the production of male gametes within the stomachs of mosquitoes that had fed on malaria patients, but obtained no evidence of further development until, in 1897, he found oocysts in the stomach wall of bloodfed 'dapple-winged' mosquitoes (*Anopheles*). Later he worked out the life cycle of the causal organism of bird malaria, and established the 'bird – culicine mosquito – bird' cycle. Mosquitoes were proposed as active agents in transmission of yellow fever pathogen by Nott in 1848, Beauperthuy in 1853 and Finlay in 1881. Finlay's attempts to demonstrate transmission through the bite of infected mosquitoes failed because he did not allow sufficient time for the infective agent to replicate in the mosquitoes. In 1900, a Commission headed by Reed and working in Cuba, first achieved transmission of the yellow fever pathogen from an infected to an uninfected person. The yellow fever story is characterized by the heroism of many people who exposed themselves to the disease organism in the course of attempts, over many years, to establish its manner of transmission. The advance of knowledge during a few momentous decades enabled a rational approach to be made to the prevention of a number of major diseases. It is salutary to survey the extent of mosquito-borne disease today.

About half of the 500 and more viruses listed in the international catalogue of viruses have been isolated from mosquitoes, but only a proportion of these fall into the category of arboviruses, i.e. viruses that replicate in both vertebrate and invertebrate hosts. (This definition excludes the few, such as myxoma virus, that are transmitted mechanically because they are sufficiently resistant to survive exposure on arthropod mouthparts.) After ingestion in a bloodmeal, arboviruses initially infect and replicate in the cells of the mosquito's midgut epithelium. Later they are found in other tissues, including the salivary glands. Virus particles produced by replication in the salivary glands can pass with saliva into a vertebrate host.

Almost one hundred arboviruses have been recorded as causing clinical symptoms in man. A number of these cause severe morbidity and death, but there is space here to consider only yellow fever and dengue. Yellow fever, in its rural (or sylvan) epidemiological form, is maintained in forest primates, with canopy-dwelling mosquitoes providing the invertebrate host and vector. Man becomes infected when he enters an enzootic area and is bitten by an infected mosquito. In the New World, human infections follow contact with infected *Haemagogus* mosquitoes from the forest canopy. In East Africa, *Aedes africanus*, a rain-forest canopy species that will accept human hosts, and *Ae simpsoni*, which breeds in vegetation around human dwellings, are involved in the transfer to man. If an infected person travels to a town where *Aedes aegypti* is present, an outbreak of urban yellow fever can result, involving the human population and *Ae aegypti*. Yellow fever is endemic in tropical Africa, and since it was carried to the New World has become endemic in Central and South America. Little progress has been made in reducing the risk of rural yellow fever in the enzootic areas of Africa and America, and, although the disease can be controlled by immunization, outbreaks of urban yellow fever still occur where *Ae aegypti* is present. An outbreak in Ethiopia in 1960–1962 resulted in an estimated 115 000 deaths. In 1986–1987 serious outbreaks occurred in West African cities.

The diseases dengue fever and dengue haemorrhagic fever, caused by dengue viruses, are increasing in importance. The vectors are four man-biting species of *Aedes* – *aegypti*, *albopictus*, *scutellaris* and *polynesiensis* – which breed efficiently in urban environments. Dengue was once largely restricted to India, south-east Asia and the southern Pacific, but its range increased dramatically with tropical urbanization, and it is now present in Africa and the Americas. Since its first description from the Philippines in 1953, dengue haemorrhagic fever has become one of the leading causes of childhood illness and mortality in southeast Asia, and it has been reported increasingly in the Americas over the past decade.

In the life cycle of filarioid nematodes of the family Onchocercidae, a phase of parasitism in a vertebrate, in which the blood or lymphatic system is often invaded, alternates with a phase of parasitism in adult insects. This family includes *Wuchereria bancrofti* and *Brugia malayi* which are the causative agents of human lymphatic filariasis, a disease that is characterized by periodic inflammation of lymph nodes and lymph vessels, accompanied by fever; it can culminate in elephantiasis. It occurs in humid tropical regions of Africa and Asia and in numerous Pacific islands, and retains a foothold in the Americas. Due to population increase in the areas where the disease is endemic, there is now far more filariasis in the world than at the time of Manson's breakthrough in 1877. In rural situations, Bancroftian filariasis (*Wuchereria bancrofti*) is mainly transmitted by the species of *Anopheles* that transmit malaria, and where antimalarial house spraying is maintained its incidence is greatly reduced. In urban situations, Bancroftian filariasis is becoming increasingly prevalent due to transmission by *Culex quinquefasciatus*, a mosquito that is closely associated with man and breeds in polluted water. In the Pacific region, where *Aedes polynesiensis* is the vector, control is difficult. Brugian filariasis (*Brugia malayi*) occurs only in China and southeast Asia, where its prevalence has been much reduced by control of the vectors, species of *Mansonia*, by removal from the larval breeding places of the plants that provide their air supply.

The haemosporidian genus *Plasmodium* includes parasites of reptiles, birds and mammals. Four species are pathogenic to human beings and cause human malaria. Vertebrates constitute the 'intermediate host', mosquitoes the 'definitive host' within which fertilization occurs shortly after the ingestion of blood containing gametocytes. The zygotes resulting from the fusion of gametes in the mosquito stomach become motile ookinetes and migrate to the outside of the stomach wall where they round up and secrete a protective coat, becoming oocysts. Successive divisions of the sporoplasm within the oocysts result in the formation of many haploid, spindle-shaped sporozites. Rupture of the oocysts releases sporozites which migrate throughout the mosquito. Some enter the salivary glands, and may be injected with saliva into the next host on which the mosquito feeds.

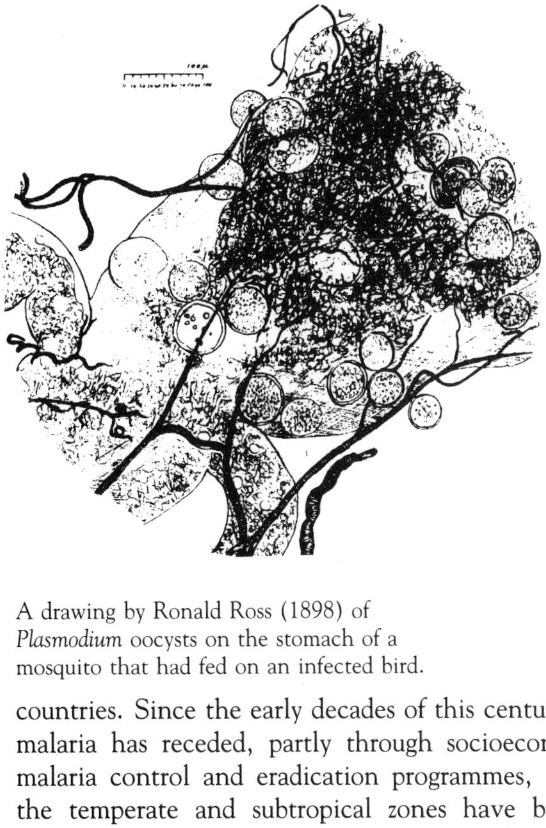

A drawing by Ronald Ross (1898) of *Plasmodium* oocysts on the stomach of a mosquito that had fed on an infected bird.

For the first thirty years of this century, drainage of mosquito breeding sites constituted the principal form of attack on malaria. This proved outstandingly successful in some places, a failure in many others. In the 1930s, pyrethrum spraying demonstrated the potential of insecticides. With the discovery of DDT, an active, safe and persistent insecticide, the spraying of house interiors proved an effective method of controlling many species of *Anopheles*, and led to hope of eradicating malaria from many countries. Since the early decades of this century the geographical range of human malaria has receded, partly through socioeconomic progress and partly through malaria control and eradication programmes, with the result that huge areas in the temperate and subtropical zones have been rendered free of the disease. Unfortunately, for a number of reasons, not least the difficulty and cost of sustaining intensive country-wide attack on the vectors, and also insecticide resistance, there has been a resurgence of malaria in countries such as India and Sri Lanka in which it was for a brief time almost under control.

Despite advances in control methods and drug treatment, malaria is the most

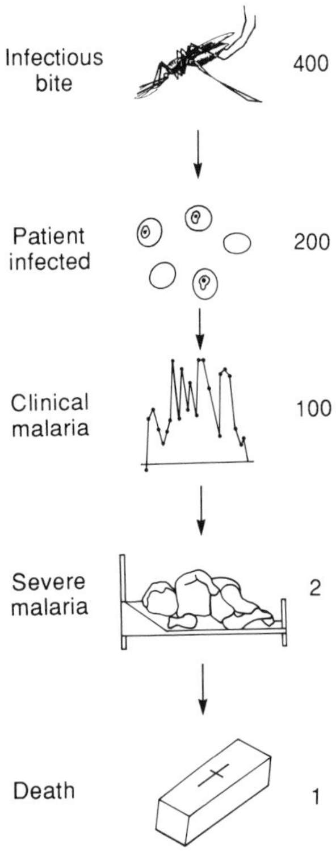

Illustrative values of the numbers of African children who enter upon different categories of malaria for each that dies of the disease. The likelihood that inoculation of sporozoites will lead to a clinical infection varies greatly from area to area, being low in areas of high malaria endemicity. (After Greenwood *et al.*, 1991.)

widespread of all tropical diseases and one of the most lethal. It remains endemic in most tropical countries and in a number of countries in the northern subtropical zone. Two-fifths of the world's population remain exposed to malaria. Estimates of morbidity and mortality range from 270 million infected with one million deaths to 400 million infected with two million deaths. In some areas of sub-Saharan Africa malaria is believed to account for one in ten infant deaths and for a quarter of deaths of children one to four years of age. Here it is *Anopheles gambiae* and *Anopheles arabiensis* – highly efficient vectors due to their longevity and their association with man, and impossible to eradicate – that are responsible for the continued high incidence of the disease. In a number of tropical countries the impact of malaria is increasing, particularly at the frontiers of economic development, where deforestation, mining and increased irrigation for agricultural purposes lead to the migration of large numbers of non-immune workers. Today, control of *Anopheles*, not eradication, is attempted, employing either environmental management or insecticides, or both, depending on the biology of the vector species and what can be afforded.

With all mosquito-borne diseases, the local epidemiology of the disease is strongly influenced by the behaviour and ecology of the local species of vectors. Whatever the future holds for the prevention and treatment of these diseases, knowledge of mosquito biology will remain a keystone in the fight against them.

1

Aspects of genetics

Mosquitoes have proved excellent subjects for genetical research, for a number of reasons. (a) Many species have short life cycles and are easy to rear. (b) The small chromosome complement simplifies chromosome recognition, linkage studies and mapping. (c) The excellent polytene chromosome preparations that can be made from the larval salivary glands and ovarian nurse cells of a number of species have provided opportunities for many types of investigation, for example into population genetics. (d) Nurse cell polytene chromosomes (not available in *Drosophila*) can indicate the karyotype of individual wild females for which other data such as habitat, blood-meal type and parasite load might be obtainable. (e) In a few species, notably *Aedes aegypti*, substantial numbers of mutants are available for genetic investigations.

This chapter is concerned with the more functional aspects of mosquito genetics. It contains information on mosquito chromosomes and the molecular characteristics of the genome. It also contains descriptions of the mechanisms of meiosis and of sex determination and an account of cytoplasmic incompatibility. Other aspects of mosquito genetics are described elsewhere in this volume, e.g. cytogenetic aspects of growth (Section 7.2) and the genetic regulation of autogeny (Section 23.3).

Genetic mapping is of fundamental importance for the interpretation and application of genetic data. Three types of map are commonly used by mosquito geneticists, and construction of a fourth type commenced recently. (i) Physical or polytene-chromosome maps: photomicrographs or drawings of polytene chromosomes, which illustrate specific banding patterns (Figure 1.2), and upon which the locations of various types of genetic entity can be determined by *in situ* hybridization (Section 1.1.3.c). (ii) Genetic or linkage maps: plots of the relative positions of genes and other genetic entities, and of the distances between them (Figure 1.3, Section 1.1.4). (iii) Restriction maps: linear arrays of sites on DNA cleaved by restriction enzymes. (iv) Whole genome or molecular maps; this type of mapping has just started, for one or two mosquito species. The integration of data from the different types of maps is necessary for the fullest understanding of genetic constitutions.

1.1 THE CHROMOSOMES

An important characteristic of mosquito chromosomes, shared with those of other Diptera, is the more or less permanent synapsing of homologous chromosomes, a phenomenon called somatic pairing. Homologous chromosomes remain tightly synapsed throughout their lengths, and not only at their centromeres, during interphase and into the prophase of the next mitotic cycle. Somatic pairing is distinct in the polytene chromosomes, which appear in the haploid number.

1.1.1 Mitotic karyotypes

The chromosomes can be clearly seen during the mitotic metaphase when all chromatin is condensed, making it possible to observe the number and morphology of the chromosomes of a complement. Brain cells from late 4th instar larvae are commonly used as the source of metaphase chromosomes. The appearance of chromosomes is affected by the distribution of heterochromatic and euchromatic regions. The heterochromatic regions are composed of non-coding repetitive sequences of nucleotides (Sections 1.2.1, 1.2.2) that remain condensed during interphase and that stain differently from the remainder of the chromosome. The euchromatic

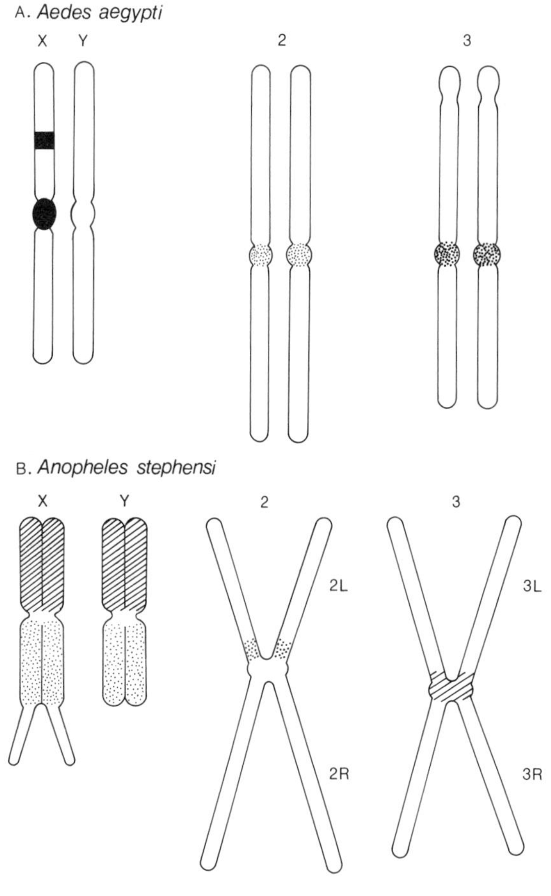

Figure 1.1 Idiograms of culicine and anopheline mitotic karyotypes. (A) *Aedes aegypti* male. The black and stippled regions represent C-bands, of different intensities, produced by Giemsa staining. The intercalary band is sometimes present on the Y chromosome also. (After Newton *et al.*, 1978b.) (B) *Anopheles stephensi* male. The cross-hatched areas on the sex chromosomes and chromosome-3 indicate regions with which satellite I hybridized (see Section 1.2.2); the stippled areas represent regions seen to be heterochromatic under phase contrast illumination. (After Redfern, 1981a.)

regions show the normal cycle of chromosome coiling and have normal staining properties. In mitotic chromosomes, where the heterochromatic regions are more highly condensed than the euchromatic regions, blocks of heterochromatin are distinguishable by phase contrast microscopy. Heterochromatin is classified into two types: (a) constitutive heterochromatin, which is always present in certain locations on both chromosomes of a homologous pair in all cells of a species, and (b) facultative heterochromatin, which varies in its state in different cell types and at different times, and which may differ between homologous chromosomes.

Two related terms that are sometimes used are 'heteropycnotic', indicating differences of coiling and appearance, and 'isopycnotic', indicating similarities of coiling and appearance to the major parts of the chromosomes. Heteropycnotic regions have affinities for certain stains and fluorochromes, e.g. Giemsa, quinacrine HCl and Hoechst 33258. Exposure of metaphase chromosomes to heat or dilute acid or alkali, followed by treatment with a Giemsa stain, consistently stains certain regions, producing so-called C-bands. The heterochromatic pericentromeric region stains in this way. Treatment with trypsin or urea, followed by Giemsa staining, produces G-bands. Staining techniques can be used to distinguish the different chromosomes of a species and to distinguish between species. Methods for preparing and staining mitotic and meiotic chromosomes have been described by French *et al.* (1962), Newton *et al.* (1974), Marchi *et al.* (1980), Gatti *et al.* (1982), and King and Pasteur (1985).

The somatic cells of almost all mosquitoes contain three pairs of chromosomes, but there is one known exception; a species of the anopheline genus *Chagasia* has four pairs. From comparisons of the karyotypes, C-banding patterns and nuclear DNA content of species of Tipulidae, Dixidae, Chaoboridae and Culicidae, Rao and Rai (1987b) concluded that the Culicidae arose from a chaoborid *Monochlonyx*-like ancestor. The *Chagasia* and *Toxorhynchites* karyotypes were considered to be primitive, and the Anophelinae and Culicinae were believed to have evolved along separate lines from a common ancestral stock.

(a) Culicines

The karyotypes of culicine (and toxorhynchitine) genera are remarkably uniform. A typical karyotype consists of three pairs of homomorphic chromosomes. Two of the pairs are of about equal length and one pair is shorter. By convention, the shortest pair is called chromosome-1, the longest pair chromosome-2, and the pair of intermediate length chromosome-3. Typically, all three pairs are metacentric, i.e. the centromere, which attaches to a spindle fibre during mitosis, is centrally placed on the chromosome with the result that the chromosomes are V-shaped at anaphase. In some species one or more of the pairs may be slightly submetacentric, with one chromosome arm a little longer than the other (Kitzmiller, 1967). Sex is inherited by a single pair of alleles or of chromosomal segments on chromosomes which carry many other alleles. For many years it was believed that the sex chromosomes were indistinguishable from the autosomes, but in a number of species it has proved possible to distinguish the sex chromosomes by staining.

The karyotypes of most species of *Aedes* are so similar that they cannot be used to distinguish species, but some can be distinguished by their chromosome staining characteristics (Motara and Rai, 1978; Rao and Rai, 1987b). The chromosomes of *Ae aegypti* have been well characterized. Chromosomes 1 and 2 are metacentric. Chromosome-3 is usually considered submetacentric, and during prophase a secondary constriction is apparent in the distal third of the longer arm of one or both homologues (Figure 1.1A) (McDonald and Rai, 1970). Appropriate Giemsa staining revealed heterochromatic C-bands which in females were present in the centromeric regions of all six chromosomes. In males pericentromeric C-bands were present in five chromosomes only. The unstained chromosome was associated with linkage group I which includes the sex locus, therefore the sex chromosomes of females are XX and those of the males XY, the Y being the chromosome without centromeric heterochromatin. The centromeric staining was most intense in the X chromosome, somewhat less intense in chromosome-3 and weak in chromosome-2. Specific silver-staining of centromeres showed that the centromeric regions that had affinity for Giemsa were more extensive than the centromeres themselves. Giemsa staining also revealed an intercalary (not terminally situated) C-band in one arm of the X chromosome in most strains of *Ae aegypti* and in the Y chromosome of some geographic strains (Figure 1.1A). This variability in the presence or absence of an intercalary C-band on the Y chromosome has led to the suggestion that it represents facultative heterochromatin. Treatment with fluorochromes revealed a narrow band of responsiveness within the broader X-borne intercalary band and another in an adjacent euchromatic segment; a further type of responsiveness was apparent in the band on the Y chromosome. The intercalary band therefore is heterogeneous (Newton *et al.*, 1974, 1978; Motara and Rai, 1977; Wallace and Newton, 1987; Wood and Newton, 1991). The presence or absence of an intercalary C-band is sufficient to distinguish X from Y chromosomes in a number of species of *Aedes* and *Culex* (Motara and Rai, 1978; Motara, 1982).

In *Culiseta longiareolata* a small but significant difference between the lengths of one arm of the longest pair of chromosomes was observed in males but not in females. The longer chromosome of the pair was designated Y and the shorter X. Staining for C-bands revealed a structural homology between all chromosome pairs of the homogametic sex but differences in the heterogametic sex, two telomeric bands, often fused, being present on one arm of the Y chromosome but absent from the homologous arm of the X. This telomeric heterochromatic region may account for the difference in length between X and Y (Mezzanotte *et al.*, 1979).

(b) Anophelines

All species of *Anopheles* that have been examined have one pair of heteromorphic sex chromosomes

(X and Y) and two pairs of autosomes numbered 2 and 3. Usually one pair of autosomes is metacentric and the other submetacentric. The sex chromosomes are generally shorter than the autosomes. In females the two sex chromosomes are of equal length and are homologous (XX). In males the sex chromosomes are usually of different lengths and only parts are homologous (XY). Accordingly the females are said to be homogametic since they produce only one kind of gamete, whereas the males are heterogametic and produce two kinds of gametes in equal numbers.

Evidence from coiling characteristics and from the polytene complement suggests that at least half of the X chromosome is heterochromatic. The X chromosomes of a number of species of *Anopheles* exhibit one or other of two types of structure, categories which do not always follow species relationships. In one type, found for example in *An maculipennis*, one of the chromosomal arms is euchromatic while the other is completely or mostly heterochromatic. The euchromatic arm is usually the shorter, with a length of 1–2 μm. The length of the heterochromatic arm varies, and determines whether the X chromosome is subtelocentric, submetacentric, or metacentric. The other type of X chromosome is telocentric or subtelocentric. The short arm, which is no more than 0.5 μm long and almost undetectable, is presumed to be heterochromatic. The long arm, which is 2.5–3 μm long, includes both euchromatic and heterochromatic regions. *An gambiae* has the second type of X chromosome (Coluzzi and Kitzmiller, 1975).

In most species of *Anopheles* the entire Y chromosome appears heterochromatic, and it often resembles the heterochromatic region of the X. In *An stephensi*, in which the X chromosome includes two similar sized blocks of heterochromatin separated by the centromere, with one block terminating in a euchromatic region, the Y appears identical except for the absence of the euchromatic constituent (Figure 1.1B) (Redfern, 1981a).

The extensive heterochromatin of anopheline sex chromosomes is susceptible to staining; for example, in *An stephensi*, Q-, G- and C-positive regions can be distinguished (Marchi *et al.*, 1980). The distribution of satellite DNA in these chromosomes is described in Section 1.2.2. The application of fluorescence banding techniques revealed intraspecific polymorphism of sex chromosome heterochromatin in *An gambiae* and *An arabiensis*, wild populations being polymorphic for X chromosome variants (Bonaccorsi *et al.*, 1980). *Anopheles labranchiae* and *An atroparvus* are homosequential species, having identical polytene banding patterns. Their mitotic karyotypes were distinguishable through differences in the Q-banding patterns of the heterochromatic arms of the sex chromosomes, particularly when these patterns had been altered and made more distinctive by treatment of the chromosome preparations with restriction endonucleases. It was concluded that chromosome divergence between the two species had involved changes, and probably rearrangements, in the heterochromatin DNA of the sex chromosomes, whereas the structural organization of the autosomes remained essentially unaltered (Marchi and Mezzanotte, 1990). The heterochromatic banding patterns of the sex chromosomes have been useful for identification of members of the *An balabacensis* complex, in which it is difficult to make good salivary gland chromosome preparations (Baimai, 1988).

The X chromosomes of *Anopheles* carry genes for eye colour and for certain enzymes. In the males these genes are present on the single X chromosome only, and are therefore expressed as a single allele (Adak *et al.* (1988) and Section 1.4.1). The Y chromosomes of mosquitoes are frequently described as competely heterochromatic, and considered genetically inert. However, there is evidence that in *An culicifacies* the presence of the Y is positively male-determining (Section 1.4.1). In *An atroparvus* the long arms of the X and Y chromosomes are of equal length and have an identical C-, Q- and G-banding pattern (Tiepolo *et al.*, 1975). The short arm of the X chromosome is half the length of the long arm and is euchromatic; the short arm of the Y is distinctly shorter than that of the X. Autoradiographic studies showed that the long arms of the X and Y chromosomes were late replicating, whereas the short arms were not and were therefore presumably genetically active (Fraccaro *et al.*, 1976). These authors stated that the short arm of the Y chromosome was euchromatic, possibly coming to that view from their evidence that it was not late replicating. However, no part of the Y chromosome is visible in the polytene complement (Farci *et al.*, 1973).

Crosses between *An atroparvus*, which is stenogamous, and the closely related *An labranchiae*, which is eurygamous, revealed that when the male offspring carried the Y chromosome of *atroparvus*, copulation occurred spontaneously in small cages, whereas when they carried the Y chromosome of *labranchiae* it did not. Fraccaro *et al.* (1977) proposed that in these species mating behaviour was determined by the Y chromosome, postulating the presence of one or more genes on the short arm.

Chagasia bathana has a diploid chromosome number of eight. In the female mitotic karyotype all four pairs of chromosomes are homomorphic; in the male one pair is heteromorphic and consists of a long (X) and a short (Y) chromosome. The sex chromosomes are telocentric; the autosomes are subtelocentric, submetacentric and metacentric (Kreutzer, 1978).

1.1.2 Endoreduplication

Within certain mosquito tissues the phenomenon of endoreduplication is observed, i.e. the chromosomes replicate repeatedly without any mitosis-like event. In some cases only the euchromatin and centromeric heterochromatin are replicated, and the sister chromatids produced by endoreduplication remain synapsed forming banded polytene chromosomes (Section 1.1.3). In other cases the products are endopolyploid (the term 'polyneme' was once used); the chromosomes are fully replicated, and are not banded but have the reticulate appearance of interphase chromosomes (Nagl, 1978; Cave, 1982). In mosquitoes, the endopolyploid nuclei with the highest degrees of ploidy are found in the epithelial cells of the larval fore- and hindgut. Nuclei with lower degrees of ploidy occur in the larval epidermis, tracheae and neurilemma.

In most organisms cells containing endopolyploid nuclei are terminally differentiated, but mosquitoes are highly unusual in undergoing somatic reduction. At metamorphosis the giant chromosomes of the fore- and hindgut separate into 48, 96 or rarely 192 chromosomes, after which repeated mitoses and cell division occur. During each anaphase, chromatids move to the two poles in the normal manner, and the completion of each mitotic cycle sees a halving of the chromosome number until the diploid condition is restored (Section 7.2).

1.1.3 Polytene chromosomes

(a) Occurrence

The formation of polytene chromosomes by endoreduplication is described in the previous section. Cells containing polytene chromosomes are believed to be terminally differentiated. They occur in glandular tissues of Collembola and Diptera, but in few other organisms. In mosquitoes polytene chromosomes occur in the nuclei of larval tissues, such as the salivary glands and midgut epithelium, which are destined to be histolysed in the pupal stage (Section 7.2). Polytene chromosomes are also found in Malpighian tube cells, which persist from the larval to the adult stage, and in the ovarian nurse cells of adult females. Readable squash preparations can be made of polytene chromosomes in the salivary glands of late 4th instar larvae and the nurse cells of semigravid females of some species of mosquitoes (Figure 1.2). With other species the polytene chromosomes are not amenable to cytological manipulation or make only poor preparations (possibly because the chromosomes are poorly synapsed or are difficult to spread due to ectopic connections between non-homologous regions). The genus *Anopheles* has been most extensively studied. Excellent polytene chromosome preparations can be made from a number of species, notably those of the subgenus *Cellia*

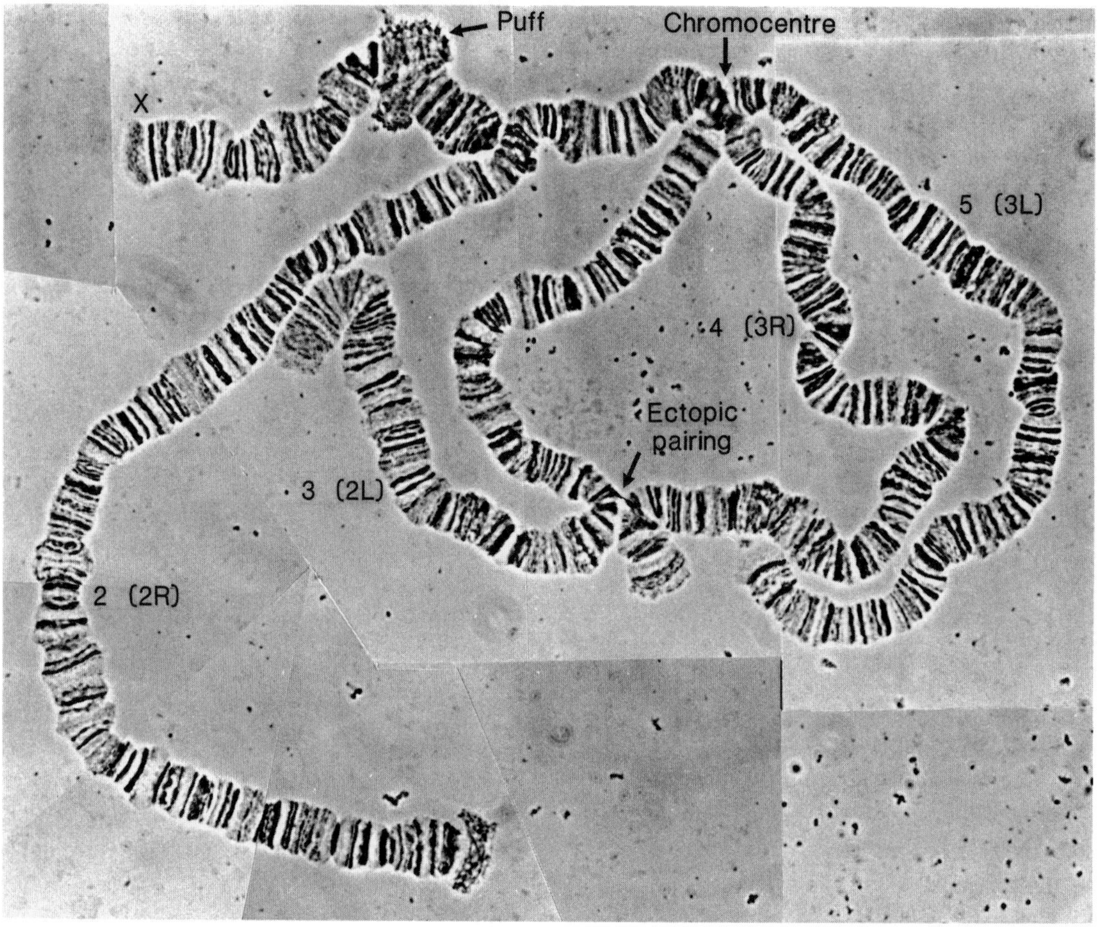

Figure 1.2 Polytene chromosomes from an ovarian nurse cell of *Anopheles* (*Cellia*) *listeri* De Meillon. During interphase the chromosomes are joined at their centromeres forming the chromocentre, a heterochromatic region which is less conspicuous in *An listeri* than in some other *Anopheles* species. The X chromosome has become detached from the chromocentre in this spread. Ectopic pairing is the connection by a DNA strand of ostensibly non-homologous parts of two different chromosomes. The arms of the autosomes are numbered (in parentheses) according to the right/left convention. They are also numbered 2 to 5 under the convention that uses the arm numbers designated for *An gambiae* as the standard for homologous arms in other species of the subgenus *Cellia*, whichever autosome they are on (Section 1.1.4.) In all species of the Paramyzomyia series (which includes *listeri*) and all species of the Pyretophorus series (which includes *gambiae*) the arrangement of the autosomal arms, when designated according to the 2R, 2L, 3R, 3L convention, is identical to that found in *An gambiae*. (Micrograph by courtesy of Dr C.A. Green and Dr R.H.Hunt. Retouched.)

(except the Neomyzomyia series). Among medically important species, *subpictus*, *superpictus*, *stephensi* and members of the *gambiae* complex provide very readable chromosomes; *balabacensis* does not. With *An gambiae* s.l., *An stephensi* and *An quadrimaculatus*, the nurse cell polytene chromosomes are easier to prepare and have a better morphology than those from the larval salivary glands. In contrast, good preparations could be made from the larval salivary glands of *An farauti* but organized polytene chromosomes could not be seen in the nurse cells (Bryan and Coluzzi, 1971; Kitzmiller, 1977; Redfern, 1981a; Green, 1982a, b; Kaiser and Seawright, 1987).

With some exceptions, it has not proved possible to make readable polytene chromosome preparations from culicines. However, excellent preparations have been made from the larval salivary glands of *Orthopodomyia pulcripalpis* (Munstermann *et al.*, 1985), *Wyeomyia smithii* (Moeur and Istock, 1982) and *Sabethes cyaneus* (Munstermann and Marchi, 1986). Less satisfactory preparations have been made from the larval salivary glands of *Culex* species (Dennhöfer, 1968, 1974; Kanda, 1970; Chaudhry, 1981), *Aedes aegypti* (Sharma *et al.*, 1979), and *Toxorhynchites brevipalpis* (White, 1980), and from the larval Malpighian tubules of *Ae vittatus* (Sharma *et al.*, 1986).

Methods for making polytene chromosome preparations have been described by French *et al.* (1962), Green (1972), Hunt (1973), Green and Hunt (1980), and Graziosi *et al.* (1990).

(b) Structure

Much that has been discovered of the structure and function of the polytene chromosomes of *Drosophila* must also be true of the polytene chromosomes of mosquitoes. Nuclei that contain polytene chromosomes are in permanent interphase. Repeated replication cycles give rise to hundreds of daughter chromosomes which stay synapsed. In *Anopheles stephensi* the mean levels of ploidy, estimated from nuclear volume measurements, were of the order of 330n in the larval salivary glands and 620n in the ovarian nurse cells at 24 h post-bloodmeal (Redfern, 1981a). α heterochromatin is not replicated during the formation of polytene chromosomes which, apart from their centromeres, consist of euchromatin and β heterochromatin (Section 1.2.2). The chromosomes are relatively uncoiled and so are elongate. There is variation in the degree of coiling, however, with alternating regions of condensed and dispersed DNA called chromomeres and interchromomeres, or bands and interbands respectively. Because each band and interband has characteristic dimensions and appearance, a reproducible banding pattern is seen. Genes are present in both bands and interbands. Individual bands often decondense and expand to form **puffs** when the genes they contain become active, and recondense when the genes become quiescent. Thus puffs are sites of synthetic activity, and their size is thought to result from the uncoiling of deoxyribonucleoprotein and the accumulation of RNA and non-histone proteins (Richards, 1985).

Because of tight somatic pairing the maternal and paternal homologues are closely linked; consequently polytene chromosomes always appear in the haploid number. For this reason the polytene karyotypes of the Culicinae consist of six arms, unlike preparations of the mitotic chromosomes in which twelve arms are visible. The polytene karyotype of *Chagasia bathana* ($2n = 8$) includes six autosomal arms and two unsynapsed and poorly banded sex chromosomes (White, 1980).

In *Anopheles* the heterochromatic region of the X chromosome and the whole of the Y chromosome are under-replicated during the endoreduplication process, and the euchromatic region of the X chromosome is the only part of the sex chromosomes to be polytenized. All α heterochromatin, including that of the Y chromosome, melts into a diffuse 'chromocentre', from which the five polytenized arms project. In strains of *An albimanus* carrying an artificial translocation whereby the Y chromosome was connected to an autosome, the Y was visible in squash preparations as a small block of heterochromatin attached to the free end of a polytene autosomal arm (Rabbani and Kitzmiller, 1975; Kaiser *et al.*, 1979).

In most species of *Anopheles*, including *gambiae*, *albimanus* and *atroparvus*, the centromeres of all chromosomes are associated with a chromocentre on the nuclear membrane, so that all polytenized arms radiate from a common point. In other species the arms appear to extend from separate associations of the centromeres with the nuclear membrane. In *An stephensi*, and presumably other species also, the chromosome arms are tenuously linked to their centromeres by fine chromatin threads which are readily broken when the nucleus is squashed. For this reason the natural conformation of the polytenized arms is rarely seen in squash preparations. Despite the fusion of centromeres, mosquito chromocentres are much smaller than those of *Drosophila melanogaster* (Redfern, 1981a).

A part of the X chromosome of *Anopheles* is structurally continuous with the nucleolus; in *An atroparvus* and *An subpictus* this is the distal end of the euchromatic arm of the X (Kreuzer, 1970; Chaudhry, 1986). In *An atroparvus* this region incorporated [^{3}H]thymidine and was identified as the nucleolus organizer (Farci *et al.*, 1973), which consists of loops of DNA containing multiple copies of the ribosomal RNA genes. *In situ* hybridization studies located the nucleolus organizer of *An quadrimaculatus* on the heterochromatic arms of the X and Y chromosomes (Section 1.2.3). In *An albimanus* a heterochromatic strand connects the nucleolus to the chromocentre (Seawright *et al.*, 1985).

It has been a matter of considerable interest whether the banding patterns of the larval salivary glands and of the adult nurse cells, which contain germ-line nuclei, are identical. Working with *An superpictus*, Coluzzi *et al.* (1970) found it difficult to homologize most bands because, with few exceptions, their expression, i.e. state of puffing, did not correspond. Redfern (1981b) made a detailed investigation of this question with *An stephensi*. He concluded that the banding patterns of chromosomes from the two tissues were essentially homologous, and that the organization or fine structure of homologous bands and interbands was mostly the same, but that there were clear differences of structure in several regions which might be attributable to differences in band condensation. Numerous dissimilarities were observed between the banding patterns of the salivary gland and nurse cell polytene chromosomes of *An quadrimaculatus*, but these were partly due to differences in puff distribution. Homologies were most distinct at the ends of the arms (Kaiser and Seawright, 1987). Differences in transcriptional activity between the two types of polytene chromosome are described in the next section.

Patterns of puffing in larval salivary gland and nurse cell chromosomes were found to be stable over periods of several hours. Analysis of one region of a nurse cell chromosome during the period 12–36 h post-bloodmeal failed to detect any consistent changes in puff morphology or to detect the appearance of any new puffs (Redfern, 1981b).

(c) Physical maps

Photomicrographs or drawings of complete sets of polytene chromosomes are the basis of physical maps. Banding sequences may change by rearrangements within or between chromosome arms, the most frequent mechanisms of change being translocations and inversions. Differences of banding pattern, and particularly differences due to paracentric inversions, have proved to be of great value in resolving inter-relationships between species, and as diagnostic tools in distinguishing members of sibling species complexes (Section 1.1.3.e). However, there are several known cases of homosequential species, i.e. biologically distinct species which possess apparently identical banding patterns, e.g. *Anopheles atroparvus* and *labranchiae*; *An maculipennis*, *melanoon* and *messeae*.

Physical and genetic maps are co-linear, and the information contents of these different maps can be integrated by use of *in situ* hybridization. At the simplest level, DNA clones are linked to a visible marker and then hybridized with spread chromosomes, permitting resolution of location to individual bands or to regions within the larger bands (Redfern, 1981a). Thus, the acetylcholinesterase gene of *An stephensi* was shown to be located on the boundary between zones 7B and 7C, near the free end of chromosome 2R (Malcolm and Hall, 1990; Malcolm, personal communication), using the arm nomenclature of Sharma *et al.* (1969). Specific fragments of DNA, obtained by cutting DNA molecules with particular restriction enzymes, can also be hybridized to polytene chromosomes (Graziosi *et al*, 1990). Such fragments may correspond to genes or may not. The base sequence of genomic DNA is a phenotypic character which is rich in information, and the use of restriction enzymes soon revealed the existence of polymorphisms for fragment length, so-called restriction fragment length polymorphism (RFLP). Restriction fragments are now used, through *in situ hybridization*, to extend the information available from polytene chromosome maps (Ashburner, 1992).

A further advance has been the development of a technique for construction of 'low resolution maps' (Zheng *et al.*, 1991). The polytene chromosomes of *An gambiae* were microdissected into 54 segments. The segmental DNAs were digested with a restriction enzyme and the restriction fragments amplified by the polymerase chain reaction to generate sets of segmental probes, each corresponding to approximately 2% of the genome. Except for those derived from locations near the centromeres, the probes hybridized specifically with their chromosomal sites of origin. The same probes were then used to map cloned DNAs indirectly, by hybridization of segmental probe and cloned DNA in a dot blot apparatus. This proved much easier than the *in situ* hybridization of cloned DNA to polytene chromosomes; clearly, information obtained in this way will contribute to the construction of both physical and molecular maps.

(d) *DNA replication and transcription*

Nuclei that contain polytene chromosomes are in interphase and are therefore capable of DNA replication and transcription, although in polytene chromosomes only a portion of the genome is replicated.

The dynamics of DNA replication in the larval salivary glands of *Anopheles atroparvus* were investigated by exposing larvae to [^{3}H]thymidine and observing its incorporation into the polytene chromosomes. The start of the replication cycle coincided with the moult. Thus in larvae that had just ecdysed into the 3rd instar, 90% of salivary gland nuclei showed DNA synthesis. The percentage of labelled nuclei fell steadily to 10% at the end of the instar. DNA synthesis was renewed at the next moult because immediately after ecdysis to the 4th instar 70% of nuclei incorporated label. By 48 h after ecdysis DNA synthesis had ceased and the cells remained inactive for the next 48 h (Tiepolo and Laudani, 1972).

The ovarian nurse cell chromosomes of *An stephensi* were already polytene at the previtellogenic resting stage, but were small and tightly coiled. The resting stage nuclei were inactive in DNA replication but new replication cycles were initiated following a blood meal. After exposure to [^{3}H]thymidine, the nurse cell nuclei were classed as 'continuously labelled' if the polytene chromosomes incorporated [^{3}H]thymidine throughout the length of the arms, and they were considered to be in the propagative phase of the chromosome replication cycle. Nuclei were classed as 'discontinuously labelled' if some chromosome regions were labelled and others not. These were further categorized according to whether the label was predominantly over fine bands and interbands, which were considered to be in the initial phase of the replication cycle, or whether groups of bands were intensely labelled, which were considered to be in the terminal phase of the replication cycle. Early and propagative phase nuclei were first detectable at 4–6 h post-bloodmeal and they increased in frequency to peak at 10 h and declined between 12 and 16 h. Terminal phase nuclei were first detected at 8–10 h and reached peak frequency at 14 h post-bloodmeal. The proportions of the initial, continuous and terminal labelling patterns represented 0.11, 0.66 and 0.23 respectively of the total S phase of the replication cycle. The autoradiographic pattern of late-replicating regions was similar in nurse cell and salivary gland chromosomes (Redfern, 1981c).

In a further study with *An stephensi*, rates of DNA transcription in the polytene chromosomes of the nurse cells and larval salivary glands were determined from autoradiographic measurements of [^{3}H]uridine incorporation into RNA. Similar patterns of labelling were obtained from the two tissues but the nurse cell chromosomes were labelled much more intensely than those of the salivary glands. When labelling times were short the label was non-randomly distributed. Dense bands were labelled at lower intensity than other regions. Puffed regions were labelled at higher intensity than non-puffed, the labelling of puffs being proportional to puff size.

A detailed comparison was made of region 43C2 to 44C3 in the left arm of chromosome-3 in the two tissues, the region being subdivided into 21 visibly distinguishable divisions. Each division showed some level of labelling, even where puffs were not visible, therefore regions other than puffs were transcriptionally active. For every division the nurse cell chromosomes were more intensely labelled. Within the limited resolution of light-microscope autoradiography the patterns of labelling in the two tissues were found to be similar but differences were observed in individual sites. Redfern's (1981b) principal conclusions were that differences in relative rate of incorporation of [^{3}H]uridine into chromosomal RNA correlated with differences in the degree of decondensation of specific bands, and that the differences between the salivary gland chromosomes and the germ-line nurse cell chromosomes were no greater than would be expected from two different somatic tissues.

(e) Paracentric inversions

When the banding patterns of the polytene chromosomes of closely related species of *Anopheles* are compared they are rarely found to be fully homosequential. In most cases the band sequences differ due to the homozygous inversion, or reversal, of one or more lengths within a chromosome arm, so-called fixed paracentric inversions. (Inversions spanning the centromere are called pericentric, but these are unusual in mosquitoes.) The study of fixed paracentric inversions has been valuable in anopheline taxonomy, especially with sibling species which show few anatomical differences. The technique was first used in this way to corroborate the identification of some Palaearctic species of the *An maculipennis* complex, and later was used to great effect in investigations of other species complexes, notably the *An gambiae* complex, the six species of which are all characterized by at least one fixed inversion. Where sufficient data are available, the distribution of fixed inversions can be used to infer phylogenetic relationships.

In some species individuals with standard and inverted gene arrangements interbreed freely, and such paracentric inversions are described as 'floating' or 'polymorphic', contrasting with the singularity and permanence of the 'fixed' inversions that sometimes distinguish species. Populations composed of variable proportions of the three inversion karyotypes (inverted homozygote, heterozygote, standard homozygote) are said to show intraspecific inversion polymorphism. In many mosquito species, several different polymorphic inversions (i.e. affecting different parts of the chromosomes) occur together in the same population. In heterozygotes, chromosome pairing is maintained by a looping of homologous regions over the length of the inverted segment. If mating is random in a polymorphic population, the frequencies of the three karyotypes should be maintained in Hardy–Weinberg equilibrium.

Thirty-two polymorphic inversions have been described among the six sibling species of the *An gambiae* complex. No inversion polymorphism is found in *merus*, and relatively few inversions are present in *bwambae*, *melas* and *quadriannulatus*, but *gambiae* and *arabiensis* both carry multiple inversion polymorphisms. Most of these inversions occur on chromosome 2R, and some, e.g. 2Rb, 2Rc and 2La, appear to have identical break-points in the two species. The frequencies of certain chromosome arrangements are correlated with clines of climatic conditions and vegetation zones. In West Africa, for example, the standard* chromosome 2R and 2L arrangements are typical of *An gambiae* s.str. populations of rain forest and humid coastal areas. The inversion polymorphism 2La/+ is found in transition zones between forest and savanna, and the frequency of 2La rises with increasing aridity to reach 100% in drier savannas. The 2Rb/+ polymorphism is typical of humid savannas and is well represented also in drier savannas, together with 2La (Coluzzi *et al.*,

* Inversion of, for example, region 'a' on arm 2R would be annotated 2Ra or 2Ra/a for the inversion homozygote, 2Ra/+ for the heterozygote, and 2R+^{a} for the standard homozygote arrangement.

1979, 1985, 1990). In The Gambia, differences in the frequency of 2Rn inversion karyotypes of *An melas* were observed between parallel samples obtained from females resting in animal shelters and houses, and from night landing catches on man outdoors and indoors (Bryan *et al.*, 1987).

Polymorphic inversions have been reported from a number of culicine genera. For example, they occur abundantly in *Wyeomyia smithii*, in which ten different paracentric inversions were observed in one small population (Moeur and Istock, 1982).

The great body of observational data on paracentric inversions, from species of *Drosophila* and of mosquitoes, has led to hypotheses on their biological significance. The data for the *An gambiae* complex in West Africa show correlation of chromosome types with certain habitat preferences and with ecological distribution. It has been postulated that because chiasma formation is suppressed in inversion heterozygotes over the regions of inversion, thereby reducing recombination within any blocks of coadapted genes that might be located within the inversion, chromosomal inversions might stabilize physiological or behavioural changes that have an adaptive value (Coluzzi *et al.*, 1979). However, a note of caution has been sounded by Green (1982a and personal communication). It remains to be demonstrated that particular traits have an adaptive value; moreover, genetic variants may have been accidentally trapped in some inversions.

It has been postulated that paracentric inversions are an important genetic mechanism for biological differentiation and speciation in the *An gambiae* complex (Coluzzi, 1982). The finding of Hardy–Weinberg disequilibria in *An gambiae* samples in parts of West Africa, with highly significant deficiencies of heterokaryotypes, appeared to indicate the coexistence of distinct populations and incipient speciation (Bryan *et al.*, 1982; Coluzzi *et al.*, 1985). However, later studies of *An melas* in The Gambia indicated that, in some cases at least, heterozygote deficiencies could be explained by differences in resting site selection by the carriers of different karyotypes, which affected sampling accuracy (Bryan *et al.*, 1987).

1.1.4 Linkage groups

Studies of inheritance in species of *Anopheles*, *Culex* and *Aedes* have always revealed three linkage groups, corresponding to the three pairs of chromosomes. Demonstrations of linkage require the availability of two alleles of each gene, classically wild type and visible mutant, so that, with crossing, recombination of characters can occur and be observed. The construction of genetic or linkage maps, which mark the relative positions of genes on given chromosomes, is based on two premisses. First, that chiasma frequencies equal crossover rates and, second, that chiasma frequency per chromosome is proportional to chromosome length. A theoretical maximum map length is calculated by multiplying the average number of chiasmata per meiotic cell by 50. For *Aedes aegypti*, a measured value of 4.56 chiasmata per cell was used, which produced a crossover map distance of 228 map units. The relative chromosome lengths were: chromosome-1,0.27; 2,0.38; 3,0.35; so the map lengths of the three chromosomes were 62, 86 and 80 units respectively. Crossover rates obtained from three-point crosses have been used to indicate the map distances between alleles within each linkage group (Munstermann and Craig, 1979).

For most mosquito species, few visible markers have been available for use in the construction of linkage maps. Consequently, even with the addition of experimentally determinable characters such as insecticide resistance, few loci could be plotted. The situation improved with the exploitation of two polymorphisms. (a) Some enzymes occur in multiple forms, called isoenzymes or allozymes, which are coded by alleles of the same gene and which differ in electrophoretic mobility. In the small percentage of cases in which the enzymes can be assayed after electrophoresis, recombination of isoenzymes can be observed. Of the 30 markers on the linkage map of *Aedes triseriatus*, 21 are isoenzymes; of the 62 markers on the *Anopheles albimanus* map, 24 are isoenzymes (Munstermann, 1990b; Narang and Seawright, 1990). (b) The DNA restriction fragments characteristic of restriction fragment length polymorphism (RFLP) are heritable and subject to recombination. Such restriction fragments may correspond closely with particular genes, but need not. The diphenol oxidase A2 (*Dox*) gene of *An gambiae* has alleles that are distinguishable by fragment length, and therefore by electrophoretic mobility, after treatment with the restriction enzyme *Sal*I. Crossing experiments revealed the linkage association of these fragments, and so of the gene (Romans *et al.*, 1991).

Except for the visually distinguishable sex chromosomes, to associate linkage groups with chromosomes it is necessary to obtain genetic data that establish new linkages, either from induced translocations, with the consequent pseudo-linkages, or from pericentric inversions. Translocations are also needed to associate mitotic with polytene chromosomes. Such correlations have been obtained in a number of species of *Anopheles*, *Culex* and *Aedes*. No direct correlation can be expected between linkage maps and polytene chromosome maps, because the existence of achiasmate regions is ignored in linkage map construction.

In *Aedes aegypti*, linkage group I, which contains the sex locus, correlates with the shortest pair of mitotic chromosomes, which are designated chromosome-1; linkage group II correlates with the longest pair, numbered 2; and linkage group III correlates with the chromosomes of intermediate length, numbered 3 (McDonald and Rai, 1970). The linkage map of *Ae aegypti* is illustrated in Figure 1.3. Comparison of the linkage maps constructed for a number of species of *Aedes* reveals that numerous inversions (mostly paracentric) and translocations (mostly whole-arm) have arisen in the course of subgeneric divergence. Thus one whole-arm, reciprocal translocation has occurred between the chromosomes corresponding to the second and third linkage groups of *Ae aegypti* and *Ae triseriatus*. Translocations have switched the chromosomal locations of linkage groups, and inversions have rearranged the order of genes within the linkage groups, but the linkage associations of specific enzyme loci have remained intact. Linkage-group conservation of some chromosomal regions (e.g. *Gpi–Hk4–Odh*) extends across genera to *Culex* and *Anopheles* (Munstermann, 1990b; Matthews and Munstermann, 1990).

In *Culex quinquefasciatus*, linkage group I, which includes the allele for sex determination, corresponds with the shortest chromosome, number 1, which is also distinguishable by C-banding; linkage groups II and III correspond with the longest and intermediate chromosomes, 2 and 3, respectively (Bhalla *et al.*, 1974). Linkage maps have been prepared for members of the *Culex pipiens* complex (Narang and Seawright, 1982).

In *Cx tritaeniorhynchus*, linkage groups I, II and II were found to correlate with chromosomes 1 (shortest, metacentric), 2 (longer, submetacentric) and 3 (longer, metacentric) respectively. The position of the sex locus differed in strains from different geographic regions, varying between linkage groups I and III, while all other markers of those groups appeared to have remained linked. A start has been made with the linkage map (Baker *et al.*, 1971, 1977; Baker and Sakai, 1976; Narang and Seawright, 1982). In *Cx tarsalis*, linkage group I, which includes the sex locus, is associated with the longest chromosome, number 3; linkage groups II and III are associated with the intermediate and shortest chromosomes, numbered 2 and 1 respectively (McDonald *et al.*, 1978).

The investigation of linkage in anophelines has been impeded by difficulties of rearing and of obtaining single pair matings; however, the availability in some species of well-spread polytene chromosomes has been advantageous. The mitotic and polytene chromosomes have been correlated in a number of anopheline species. In a polytene complement the X chromosome is easily recognized. Distinctive banding patterns make the individual arms of the autosomes readily identifiable, and they have been designated according to chromosome number and position (right/left) as 2R, 2L, 3R and 3L. However, the occurrence of whole-arm translocations has been found, e.g. in

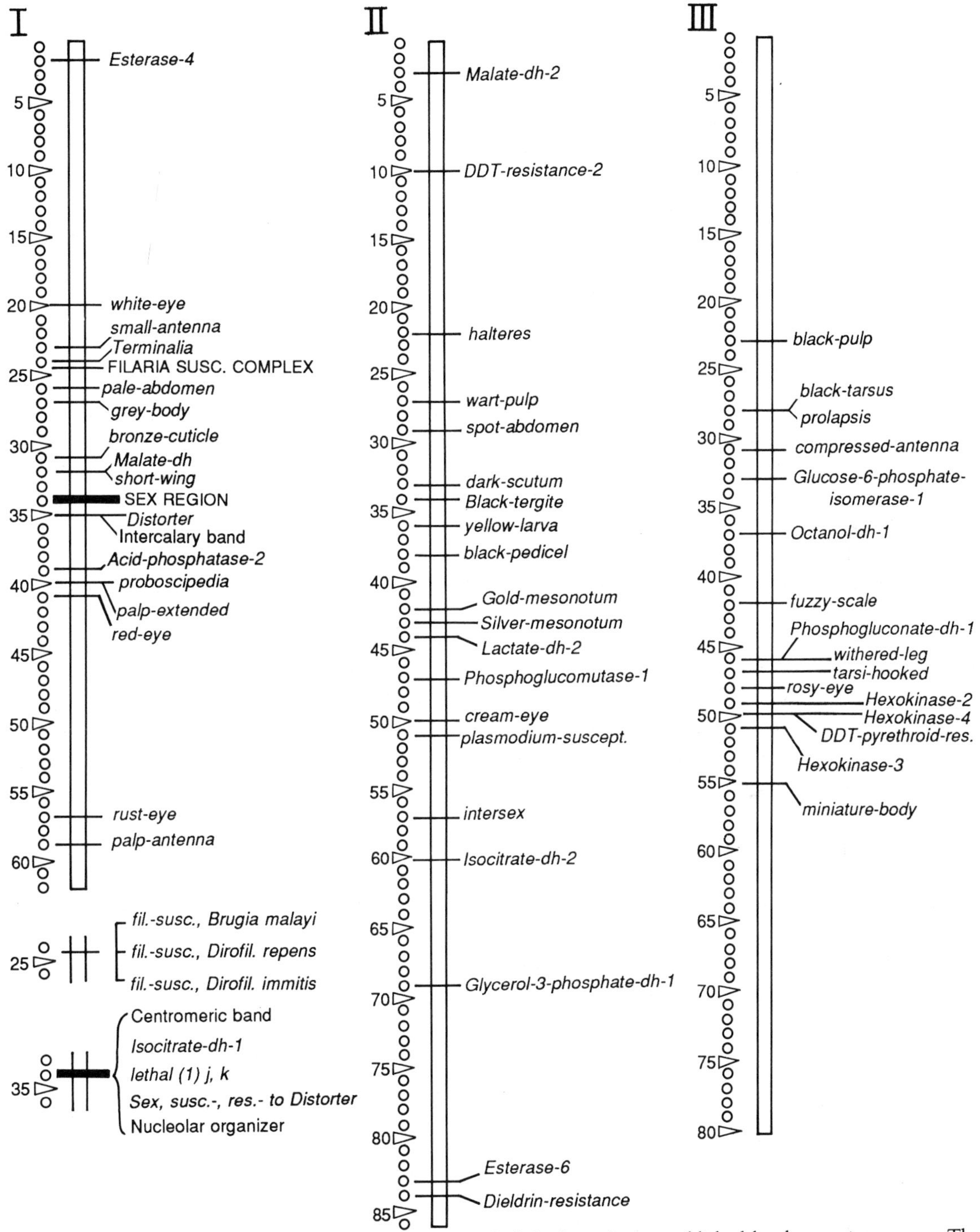

Figure 1.3 Linkage map of *Aedes aegypti*. The locations of alleles have been established by three-point crosses. The position of the nucleolus organizer was confirmed by *in situ* hybridization. Details of the filaria susceptibility complex and the sex region are inset. The sequence of *Sex*, certain other genes and the intercalary band within the sex region is not known because they occur in a stretch of chromosome within which there is no crossing over. In *Ae aegypti*, linkage groups I, II and III correspond to chromosomes 1, 2 and 3 respectively. dh, dehydrogenase; *Dirofil.*, *Dirofilaria*; fil., filaria; res., resistant; susc., susceptible/susceptibility. (After Munstermann and Craig (1979) and Munstermann (1990a), with the sex region according to Newton *et al.* (1978b).)

the subgenus *Cellia*, producing different combinations of autosomal arms in related species (Coluzzi *et al.*, 1970). One consequence of whole-arm translocations is that the arm nomenclature does not always reflect homology of arms between species; another is that the linkage group–arm name associations of one species might not hold for another. To circumvent these problems Green and Hunt (1980) introduced a new system of arm nomenclature for species of the subgenus *Cellia*. The autosomal arms that in *An gambiae* were normally designated 2R, 2L, 3R and 3L were numbered 2, 3, 4 and 5 respectively. Using paracentric inversions in the *An gambiae* complex as genetic markers, and information on linkage from other species, Hunt (1987) concluded that linkage group II was associated with arms 2+3 of *gambiae* and 2+5 of *culicifacies* and *stephensi*, and that linkage group III was associated with arms 4+5 of *gambiae* and 4+3 of *culicifacies* and *stephensi*. Linkage group I was associated with the X chromosome.

1.2 MOLECULAR CHARACTERISTICS OF THE GENOME

1.2.1 Genome size and organization

The mass of mosquito haploid genomes has been determined by Feulgen cytophotometry of germ cells in the testes and by DNA reassociation kinetics, two methods which have mostly given similar results (cf. Tables 1.1 and 1.2). The *Anopheles* haploid genome is relatively small, with a known range of 0.23–0.29 pg. Genomes of intermediate size were recorded in the single species of the Toxorhynchitinae examined and in two species of *Sabethes* and *Wyeomyia* (Culicinae, tribe Sabethini). Species of *Culex* exhibited an intra- and interspecific range from *c.* 0.5 to 1.0 pg. Species of *Culiseta* had genomes about or slightly above 1 pg. Measurements of 23 species of *Aedes*, some of which are cited in Table 1.1, produced a range of values from 0.59 to 1.90 pg. Within *Aedes* the smallest genome (0.59 pg) was found in two species of the *scutellaris*

Table 1.1 Mass of the haploid genome in a range of species, measured by Feulgen cytophotometry.

Species	*Genome mass (pg)*	*Refs*
Anopheles		
atroparvus	0.24	1
freeborni	0.29	1
gambiae	0.25	6
labranchiae	0.23	1
quadrimaculatus	0.24	3
stephensi	0.24	1
Toxorhynchites		
splendens	0.62	4
Sabethes		
cyaneus	0.79	4
Wyeomyia		
smithii	0.85	4
Culex		
pipiens	0.54, 1.02	3,1
quinquefasciatus	0.54	4
restuans	1.02	4
Culiseta		
litorea	0.92	1
melanura	1.25	4
morsitans	1.21	4
Aedes		
(*S.*) *aegypti*	0.81	2
(*S.*) *albopictus*		
(Koh Samui, Thailand)	0.62	5
(Houston, USA)	1.66	5
(*S.*) *flavopictus*	1.33	2
(*S.*) *pseudoscutellaris*	0.59	2
(*O.*) *canadensis*	0.90	2
(*O.*) *excrucians*	1.50	2
(*O.*) *stimulans*	1.44	2
(*H.*) *bahamensis*	1.37	2
(*P.*) *triseriatus*	1.52	2
(*P.*) *zoosophus*	1.90	2
Armigeres		
subalbatus	1.24	4
Haemagogus		
equinus	1.12	4

Subgenera: S, *Stegomyia*; O., *Ochlerotatus*; H., *Howardina*; P., *Protomacleaya*.

References: 1. Jost and Mameli (1972); 2. Rao and Rai (1987a); 3. Black and Rai (1988); 4. Rao and Rai (1990); 5. Kumar and Rai (1990b); 6. Besansky (1990a).

Table 1.2 Characteristics of the haploid genomes of five species. All values of genome size and complexity were determined by DNA reassociation kinetics. (Original data from Black and Rai (1988) and Warren and Crampton (1991).)

Species	*Genome mass (pg)*	*Proportion of genome*			*No. of base pairs*			*Total base pairs*
		HR	*MR*	*SC*	*HR*	*MR*	*SC*	
An quadrimaculatus	0.19	<0.01	>0.99*		3.2×10^4	1.8×10^8*		1.8×10^8
Cx pipiens	0.53	<0.01	0.06	0.94	2.4×10^5	3.4×10^7	5.0×10^8	5.3×10^8
Ae aegypti	0.83	0.20	0.20	0.60	1.6×10^8	1.6×10^8	4.8×10^8	8.0×10^8
Ae triseriatus	2.12	<0.01	<0.01	>0.99	1.7×10^5	6.1×10^5	2.0×10^8	2.0×10^8
Ae albopictus								
(Calcutta)	0.86	0.27	0.37	0.36	2.2×10^8	3.1×10^8	3.0×10^8	8.3×10^8
(Mauritius)	1.32	0.40	0.27	0.33	5.2×10^8	3.5×10^8	4.3×10^8	1.3×10^9

HR, highly repetitive; MR, moderately repetitive; SC, single copy.
* mean of MR and SC

group (*pseudoscutellaris* and *cooki*). The genome mass of most *Aedes* species fell within the range 0.9–1.3 pg. A study of 37 populations of *Ae albopictus*, also a member of the *scutellaris* group, from different geographical regions, revealed the haploid DNA content varying between 0.62 and 1.66 pg. There was no correlation of genome mass with geographic distribution, and it was postulated that the DNA content of each population was determined by the microenvironment. Within *Aedes* the largest genomes were those of *triseriatus* and *zoosophus*, both members of the subgenus *Protomacleaya*. Species in the aedine genera *Armigeres* and *Haemagogus* fell within the same range as *Aedes*.

Analysis of 28 species belonging to 11 genera of the superfamily Culicoidea demonstrated a positive correlation between genome mass and total chromosome length, a 5-fold difference in length being associated with an 8-fold difference in amount of DNA. The phylogenetic trend of evolution in the Culicoidea, as indicated by morphological and karyotypic criteria, appears to have been associated with an increase in genomic mass (Rao and Rai, 1990).

Insect genomes, like those of other eukaryotes, are composed of unique or single-copy DNA sequences (one copy per haploid genome) and of repetitive sequences. The latter are classed as moderately repeated (approx. 10^2–10^4 copies) or highly repeated (approx. 10^5–10^6 copies per haploid genome). Moderately repeated DNA consists of families of sequences which may be relatively short (e.g. 300 bp) or relatively long (e.g. 5000 bp). Each sequence is repeated many times, occurring interspersed with unique sequences and also in tandem, head to tail. The highly repeated segment of the genome includes some very short sequences, tandemly repeated in large numbers, which constitute 'satellite DNA' and which occur at centromeric and telomeric positions. It also includes more complex repetitive sequences which are widely distributed.

Genomes are usually organized in one or other of two patterns. In one pattern, short lengths (<1500 bp) of unique coding sequences are interspersed with short lengths (*c.* 300 bp) of repetitive sequences. In the other pattern, series of unique sequences totalling up to 30 000 bp are separated by stretches of repetitive sequences totalling up to 10 000 bp. The terms 'short-period interspersion' and 'long-period interspersion' have been used for these two patterns of organization. Much of the moderately repeated DNA is non-coding but some codes for rRNA and tRNAs (Section 1.2.3). All highly repetitive DNA is non-coding. Some of it is permanently hypercondensed or heterochromatinized, and some can be isolated as satellite DNA (Section 1.2.2) (Berry, 1985).

A very large proportion of the genome of *An quadrimaculatus* was found to consist of unique and moderately repeated sequences (Table 1.2). Two species of *Anopheles* examined for interspersion (*quadrimaculatus* and *freeborni*) showed the long-period pattern. About half of the repetitive sequences were conserved between *An albimanus*, *freeborni* and *quadrimaculatus*. Over 90% of the *Culex pipiens* genome consisted of unique sequences. Both *Cx pipiens* and *Cx quinquefasciatus* were unusual in showing an intermediate pattern of interspersion. Unique sequences formed almost two-thirds of the *Ae aegypti* genome, compared to about one-third in *Ae albopictus*. The three species of *Aedes* examined (*aegypti*, *albopictus* and *Ae triseriatus*) showed short-period interspersion. From 1 to 2% of repetitive sequences were conserved between *Anopheles*, *Culex* and *Aedes* (Black and Rai, 1988; Cockburn and Mitchell, 1989; Warren and Crampton, 1991).

The analyses revealed enormous variation in the amount and organization of repetitive DNA within the Culicidae. Among species of *Anopheles*, *Culex* and *Aedes* a positive correlation was obtained between genome size and percentage of repetitive DNA consisting of short repeats. This suggested that mosquito species with large genomes have much repetitive DNA, organized in a pattern of short-period interspersion, and that species with small genomes have less repetitive DNA, organized in a pattern of long-period interspersion (Black and Rai, 1988; Cockburn and Mitchell, 1989).

Among ten laboratory strains of *Ae albopictus* differing in geographical origin, the genome size ranged from 0.86 pg (Calcutta) to 1.32 pg (Mauritius). Differences in amount of highly repeated DNA accounted for most of the intraspecific differences in genome size, but even so there was 40% more single-copy DNA in the Mauritius strain than in the Calcutta strain (Table 1.2) (Rao and Rai, 1987a; Black and Rai, 1988; Warren and Crampton, 1991). Measurements of the abundance of nine highly repeated sequences in six sibling species of the *Ae scutellaris* group suggested that changes in the abundance of individual repetitive sequences can be both rapid and substantial. That may be why a disparity was found between a phylogeny constructed from similarities in abundance of highly repetitive DNA sequences and phylogenies based on other measures of systematic affinity (McLain *et al.*, 1986).

1.2.2 Heterochromatic and satellite DNA

The heterochromatic and euchromatic regions of mitotic chromosomes were described earlier (Section 1.1.1). We may usefully borrow from *Drosophila* genetics two terms that further define heterochromatin. α heterochromatin is composed of the highly repeated, simple sequences of DNA that constitute the distinctive heterochromatic regions of mitotic chromosomes, and that do not polytenize. β heterochromatin is composed of moderately repeated sequences, such as are found within the ribosomal RNA genes. It is distributed within the euchromatin, and like euchromatin can polytenize.

Chromosome regions that are composed very largely of α heterochromatin remain condensed during both interphase and mitosis. These regions are visually distinguishable in metaphase chromosomes, from their appearance under phase contrast illumination and their staining characteristics. The DNA sequences of α heterochromatin are hypercondensed, late-replicating, not transcribed, and are believed to have no functional role. Euchromatic regions are composed of single-copy sequences often interspersed with moderately repeated sequences of β heterochromatin, which may be short or long. They are condensed during mitosis but decondense during interphase, when they become transcriptionally active.

Treatment of *Aedes aegypti* metaphase chromosomes with the fluorochrome Hoechst 33258, which has affinity for AT triplets, and quinacrine HCl, which has affinity for sequences of four AT base pairs, revealed differences both between and within the pericentric and intercalary heterochromatic regions (which are described in Section 1.1.1(a)) (Marchi and Rai, 1986; Wallace and Newton, 1987).

Many of the long tandem repeats of short nucleotide sequences have a divergent GC content, which affects their buoyant density. In early investigations, in which DNA was centrifuged to isopycnic equilibrium on caesium chloride density gradients, these repetitive DNA sequences appeared as minor bands or satellites, separate from the main band of chromosomal DNA. For that reason they were called satellite DNA. Density gradient centrifugation of *Anopheles stephensi* DNA revealed at least four satellite DNAs which formed approximately 15% of the total DNA. The relative abundance of satellites I and II was used as an indicator of the extent of satellite DNA replication relative to that of main band (non-satellite) DNA. Satellites I and II were present in density gradient profiles of DNA from embryos and adult heads, were absent from profiles of DNA from larval salivary glands and adult Malpighian tubules, and were much reduced in profiles from vitellogenic ovaries. This was consistent with the known under-replication of satellite DNA in polytene chromosomes (Redfern, 1981a).

In situ hybridization experiments with mitotic chromosomes from *An stephensi* revealed that ^{3}H-labelled satellite I hybridized with the short arm of the X and Y chromosomes and with the centromeric region of chromosome-3 (Figure 1.1B). With polytene chromosomes, satellite I hybridized to a chromatin block at the centromeric end of arm 3L. There was no hybridization to the polytenized arm of the X chromosome. The 3L chromatin block replicated during the latter part of the polytene chromosome replication cycle. Redfern (1981a) concluded that the satellite I sequences fell into two classes: (a) a class at the centromere of chromosome-3 which replicates during polytenization; and (b) a class forming a major part of the heterochromatin of the sex chromosomes which does not replicate during polytenization.

Treatment of chromosome preparations from *An atroparvus* and *An labranchiae* with a number of restriction endonucleases led to complete digestion of the euchromatic arms but not the centromeres of the autosomes and of the short arm of the sex chromosomes. It also caused modification of the fluorescent banding patterns of the sex chromosomes. Structural differences, which were not otherwise apparent, were revealed in the heterochromatic arms of the sex chromosomes of the two species by the endonuclease treatment. It was concluded that chromosome divergence between the two species had involved changes, and probably rearrangements, in the heterochromatic DNA of the sex chromosomes, whereas the structural organization of the autosomes had not been dramatically modified, judging by the banding homologies of the polytene chromosomes (Marchi and Mezzanotte, 1990).

1.2.3 Ribosomal genes

In insects, as in higher eukaryotes generally, the genes for ribosomal RNA occur in multiple copies in the nucleolus organizer region of the chromosomes. The ribosomal DNA of different families of Diptera shows a similar architecture. The rRNA genes are present in a small number of chromosomal locations, often only one, and within these locations they are organized in long tandemly repeated arrays. Each of the repeat units contains (a) sequences that code for rRNA, (b) transcribed but non-coding regions, and (c) lengths of non-transcribed spacer (NTS), sometimes called intergenic spacer. The transcribed but non-coding regions include 'external transcribed spacer' (ETS), which may contain promoter sequences, and lengths of 'internal transcribed spacer' (ITS) which separate the genes. The cistron, from which the primary transcript or pre-rRNA is produced, consists of the rRNA genes plus the transcribed spacers. These spacers are degraded when the primary transcript is processed to yield mature 18S, 5.8S and 28S rRNAs. The 28S rRNA undergoes additional cleavage to yield two subunits (28Sα and 28Sβ). The 28Sβ region may contain 'insertion sequences', particularly towards its 3′ end. rRNA genes that contain insertion sequences are transcriptionally silent or produce only aborted transcripts. The constituent that separates the 28Sα and 28Sβ coding regions is not classified as an insertion sequence. Among Diptera the transcribed sequences tend to be highly conserved but the non-transcribed spacer regions are very variable. Molecular probes derived from mosquito rDNA have been used for the evaluation of gene flow and population structure.

Investigations with species of *Anopheles* and *Aedes* confirmed the similarity of the basic architecture of mosquito ribosomal DNA to that of other Diptera. Clones of *Aedes aegypti* genomic rDNA contained lengths of approx. 9.0 kb, each representing an rDNA repeat. This consisted of sequences coding for the 18S (2 kb), 5.8S (0.15 kb) and 28S (4.0 kb) rRNAs, which were essentially without introns, plus approximately 0.4 kb of external transcribed spacer and 1.0 kb of internal transcribed spacer. The non-transcribed spacer was variable in length but usually <2 kb (Figure 1.4). The rDNA units were shown to be tandemly repeated, head to tail. Restriction mapping indicated a remarkably high, but not total, similarity of the rDNA repeat between a number of strains of *Ae aegypti*, but revealed significant differences of length and constitution in the external and non-transcribed spacer (Gale and Crampton, 1989). The haploid copy number of rRNA genes in adult *Ae aegypti* was estimated at 423 ± 32 by Kumar and Rai (1990a). Gale and Crampton (1989) deduced 500 copies for larvae and adults. Park and Fallon (1990) arrived at copy numbers of 350–500 in larvae, rising to 1200 in adults, from calculations that assumed a genome mass of 1.3 pg.

The entire rDNA repeat unit of *Ae albopictus*, including the non-transcribed spacer region, contained a minimum of 15.6 kb. The 18S rRNA coding region consisted of 1.95 kb. The 28Sα and β sequences consisted of 1.75 and 2 kb

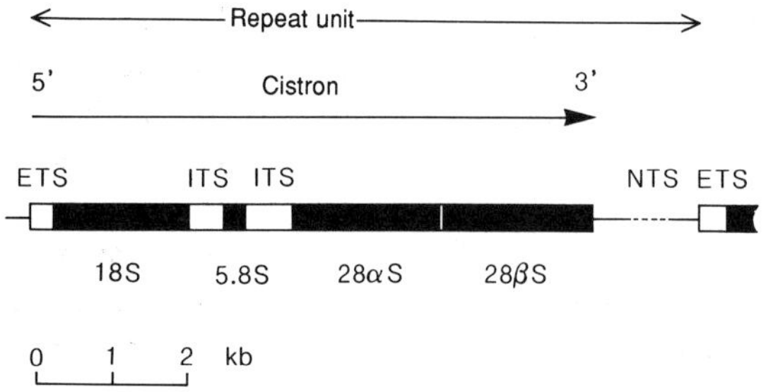

Figure 1.4 Map of a repeat unit containing rRNA genes, from *Aedes aegypti*. Coding regions are indicated by solid bars, external (ETS) and internal (ITS) transcribed spacer by hollow bars, and non-transcribed spacer (NTS) by a line. (After Gale and Crampton, 1989.)

respectively, and were separated by a 350 bp sequence which was removed from the rRNA precursor during processing. The non-transcribed spacer ranged from c. 5 to 10 kb (Park and Fallon, 1990; Baldridge and Fallon, 1991; Fallon *et al.*, 1991). A study of populations of *Ae albopictus* from different parts of the world indicated conservation of the 18S and 28S rRNA genes but extensive variation in the non-transcribed spacer. The latter consisted of two non-homologous regions. One contained multiple 190 bp *Alu*I repeats nested within larger *Xho*I repeats of various sizes. No repeats were found in the other region which gave rise to relatively fewer variants. In contrast to results reported from *Drosophila* populations, there seemed to be little conservation of the non-transcribed spacer within mosquito populations (Black *et al.*, 1989). Indeed, significant but non-directional changes in length of non-transcribed spacer were observed within populations of *Ae albopictus* during a three-year period (Kambhampati and Rai, 1991c). A haploid copy number of 820 was deduced by Rao and Rai (1990).

Comparisons between species of the *Anopheles gambiae* complex revealed a general similarity throughout the coding region of rDNA from *An gambiae*, *arabiensis* and *quadriannulatus*, but significant differences in the rDNA of *An merus* and *melas*, especially in the 18S and ITS regions. The length of the non-transcribed spacer region varied within individuals, in a manner suggesting variation in the frequency of a monomeric unit (Collins *et al.*, 1989).

The rRNA genes provide valuable diagnostic characters for identifying members of the *An gambiae* complex. The five species examined (i.e. all except *bwambae*) differ significantly in NTS and ETS architecture; consequently almost any rDNA probe that detects restriction fragments spanning these regions can be used as a species-diagnostic probe. Because of length variation in the interior of the NTS, the most useful diagnostic fragments are found at the 5′ and 3′ ends. These regions are relatively invariant within species but differ markedly between species (McLain and Collins, 1989; Collins *et al.*, 1987, 1989).

The use of synthetic oligonucleotide probes targeted at the NTS region provides a relatively simple and highly sensitive diagnostic technique to distinguish members of the *An gambiae* complex. Areas of species-specific sequence difference were found in the non-transcribed spacers. The 5′ NTS sequences did not show significant intraspecies geographic variation, and DNA fragments from this region, derived from as little as part of a leg from a living mosquito, proved adequate for species identification when amplified by polymerase chain reaction (Paskewitz and Collins, 1990).

Ribosomal DNA structure also provides information on intraspecific variability. In West African *An gambiae* s.str., no interclone variation was found in the 18S or 28S rRNA genes, except in the insertion sequences, nor in the internal transcribed spacer. However, both intragenomic and intraspecific variation were found in the non-transcribed spacer. Slight differences between clones were found in the inner region of the NTS, and these differences were substantial when clones from different geographic strains were compared (McLain and Collins, 1988). In western Kenya, hybridization of NTS sequences to genomic DNA extracted from single mosquitoes showed variation of spacer structure between populations of *An gambiae* s.str., even when they were separated by as little as 10 km, and polymorphism within certain of the populations was also found (McLain *et al.*, 1988).

The nucleolus organizer, which consists of large loops of DNA containing the rDNA multigene family, was shown by *in situ* hybridization of ^{3}H-labelled rDNA probes to mitotic chromosomes to be confined to a single chromosome per haploid genome in all species examined from eight mosquito genera, with a single exception. The ribosomal genes were situated on chromosome-1 in most species of *Aedes*, and in *Culex quinquefasciatus*, *Sabethes cyaneus* and *Wyeomyia smithii*. They were located on chromosome-2 in *Ae mediovittatus* and *Haemagogus equinus*, and on chromosome-3 in *Armigeres subalbatus* and *Tripteroides bambusa*. In *Ae triseriatus* they were located on chromosomes 1 and 3. In *An quadrimaculatus*, probes hybridized on the X and Y chromosomes, completely covering the heterochromatic arms of both (Kumar and Rai, 1990a). In preparations of *Anopheles* polytene chromosomes, the nucleolus organizer has been recognized as a puffed region at one end of the X chromosome and in close proximity to the nucleolus (Kreutzer, 1970; Chaudhry, 1986). The rDNA of *Anopheles gambiae* and *An arabiensis* appeared to be located exlusively on the X chromosome, but that of three other members of the *An gambiae* complex – *quadriannulatus*, *merus* and *melas* – did not (Collins *et al.*, 1989). Female hybrids from an *An gambiae* × *arabiensis* cross contained both parental types of the rDNA cistron, whereas male hybrids showed just the cistron structure of the female parent (Collins *et al.*, 1987; Paskewitz and Collins, 1990). It is thought that multiple copies of the rRNA genes are necessary because the RNA molecules are the final product and cannot, like mRNA molecules, be translated many times. In mosquito oocytes, as in those of some other animals, the ribosomal genes undergo massive amplification (Section 19.4).

Analysis of cytoplasmic ribosomes from *Aedes albopictus* cells showed that the small ribosomal subunit contained 28–31 proteins ranging in molecular mass from 10 to 49 kDa, and that the large subunit contained 36–39 proteins of 11–53 kDa. The largest protein in the small subunit, S1, was the predominant phosphorylated protein (Johnston and Fallon, 1985). Hybridization patterns obtained with cDNAs corresponding to ribosomal proteins L8 and L13 in *Ae albopictus* suggested that their genes occurred in low copy number (Durbin *et al.*, 1988).

1.2.4 The mitochondrial genome

Mitochondrial DNA occurs in hundreds to thousands of copies per diploid cell, compared to the essentially two copies of much of the coded nuclear DNA. Mitochondrial DNA extracted from adult *Anopheles quadrimaculatus* had the characteristic closed circular form (Cockburn and Seawright, 1988). The genetic organization of *An quadrimaculatus* mtDNA was found to be similar to that of *Drosophila yakuba*. All protein sequences that were localized occupied the same positions in both species. Considerable homology was found between protein coding regions, but none was seen between the A+T-rich origins of replication. Six tRNA genes had the same locations in the two species (Cockburn *et al.*, 1990).

One region of the mitochondrial genome of *Aedes albopictus* contains the following gene series: 12S rRNA, $tRNA^{val}$, 16S rRNA, $tRNA^{leu}$, a 9-residue spacer, and an extended reading frame (URF 1). In another region the genes for cytochrome oxidase subunit III and $tRNA^{gly}$ are separated by a short reading frame (URF 3) from a cluster of six tRNA genes (for arginine, alanine, asparagine, serine, glutamic acid and phenylalanine), which is followed by a short reading frame (URF 5). The reading frames are thought to correspond to genes encoding components of NADH dehydrogenase (HsuChen *et al.*, 1984; Dubin *et al.*, 1986).

The steady-state titre of rRNA in *Ae albopictus* mitochondria is about 50-fold that of mRNA (Dubin *et al.*, 1986). The *Ae albopictus* mitochondrial large ribosomal subunit (16S) gene consists of approximately 1335 residues, which is unusually short for such a gene. Transcripts of the gene are post-transcriptionally adenylated, producing 3′-terminal poly A tails averaging 35–36 residues. Consistent with the structure of its gene, the large ribosomal subunit has a very low G+C content (17%) and contains only two methylated residues. The 3′ half of the molecule contains many regions of primary sequence homology with *E. coli* 23S RNA and murine mt rRNA, but there is little such apparent sequence conservation in the 5′ half. Due to the low G+C content of the 3′ half and the even lower G+C content of the 5′ half, many of the *Aedes* helices are substantially richer in A,U or G,U base pairs than their bacterial or mammalial mitochondrial homologues (HsuChen *et al.*, 1984).

The 3′ ends of the mosquito large ribosomal subunit molecules were found to be remarkably homogeneous, and apparently corresponded to the end of the gene. This homogeneity contrasted with the heterogeneity of homologous mammalian rRNAs, which was thought to result from termination by transcription rather than processing. The 3′ terminus of the mosquito gene lacks sequences resembling the transcription attenuators present in the homologous mammalian genes. Dubin and HsuChen (1985) concluded that there may be differences of expression between the mosquito and mammalian mitochondrial genomes.

The small ribosomal subunit of *Aedes albopictus* was found to be smaller than such subunits in other types of organism. The 3′-terminal region of the *Ae albopictus* mt 12S rRNA has been sequenced. It showed substantial homology with homologous rRNA from other organisms, except that an entire secondary structural feature of 50–100 residues (domain G), previously thought to be universal, was absent, and the highly conserved methylated GCCCG sequence was replaced by GCCCA. The second 3′-terminal A found on one of every four 12S RNA molecules is added post-transcriptionally (Dubin and HsuChen, 1983; Dubin *et al.*, 1986).

Certain properties are characteristic of mosquito mitochondrial tRNAs; for example, an extremely low G+C content which leads to the occurrence of high A–U stems (Dubin *et al.*, 1984). A number of the tRNA genes have been sequenced. Most of the invariant or semi-invariant residues of conventional tRNA are not conserved in *Aedes* mt $tRNA^{leu}$, which also, unusually, has a UAG anticodon (HsuChen *et al.*, 1984). A striking feature of $tRNA^{val}$ is its high complement of ψ residues, numbering six per molecule, which make it the most highly modified metazoan mt tRNA known. The most unusual feature of mt $tRNA^{glu}$ is its paucity of G and C, of which it has 2 and 4 residues respectively, excluding the post-transcriptionally added 3′-terminal CCA. Analysis of anticodons indicated that U·N, U*·R wobble is operative in *Aedes* mitochondria, where U* is modified U in the first position of the anticodon, and R is G or A in the third position of codons. Unmodified C and U* in the first position of anticodons are informationally equivalent (Dubin *et al.*, 1986). *Ae albopictus* $tRNA^{ser}_{GCU}$ shows only a moderate primary sequence homology with its mammalian homologue but it resembles it in some secondary structural features, e.g. in lacking almost the entire D-arm of the standard tRNA cloverleaf (Dubin *et al.*, 1984). Restriction fragment analysis of mtDNA from members of the *Ae albopictus* and *Ae scutellaris* species groups revealed considerable divergence of mtDNA between pairs of species that did not exhibit much anatomical divergence. In contrast, among 17 widely dispersed populations of *Ae albopictus*, mtDNA polymorphism was low (Kambhampati and Rai, 1991a,b).

1.2.5 Mobile genetic elements

Mobile elements are short DNA sequences that can replicate and insert copies at random sites within chromosomes. Eukaryotes contain two main classes of mobile element: (1) transposons, which encode transposase and can replicate DNA directly; and (2) retrotransposons, which encode reverse transcriptase, and which also produce an RNA intermediate that is transcribed by the reverse transcriptase into DNA, which is then inserted at new sites. There are two subclasses of retrotransposons: (i) LTR-retrotransposons, which resemble retroviruses, and produce RNA through promotors in long terminal repeats (LTRs) at either end of the sequence; and (ii) non-LTR retrotransposons, which lack LTRs and produce RNA through promotors lying at one end of the element.

(a) Non-LTR retrotransposons in rDNA

Non-LTR retrotransposons occur in the 28S rRNA gene of *Drosophila melanogaster*. A preliminary study with *Anopheles gambiae* s.str. located insertion sequences, named 23A and 23B, at two locations in the 28S coding region. Homologous DNA appeared to be present in all individuals of *An gambiae* from all parts of Africa, although varying in genomic abundance, but it was not detected in other species of the *An gambiae* complex (Collins *et al.*, 1987, 1989). Another study revealed site-specific insertion sequences in *An gambiae* and *An arabiensis*, inserting at the same point within the 28S gene in both species, at a position 634 bp 3′ of the R1 (Type I) insertion site in *D. melanogaster*. The mosquito insertion sequences possessed characteristics of both the R1 and R2 insertion sequences of *D. melanogaster*. In both mosquito species the insertion sequences had poly A tails and a polyadenylation signal but the extreme 3′ and 5′ ends showed no other similarity to each other or to any other insertion element. In both species identical target site duplications of 17 bp were generated. The sequence TNTCCCTNNT found in this duplication was found also in the 14 bp target site duplications that flanked R1 insertion sequences in *D. melanogaster* (Paskewitz and Collins, 1989).

Analysis of rDNA from six strains of *Aedes aegypti* revealed a remarkable homogeneity, and it was deduced that there are few, if any, insertions of mobile elements in the rDNA of this species (Gale and Crampton, 1989; Crampton *et al.*, 1990).

(b) Widely dispersed non-LTR retrotransposons

An example of non-LTR retrotransposons, designated T1Ag, has been described in *Anopheles gambiae*. Approximately 100 copies of this element are interspersed throughout the genome. *In situ* hybridization of T1Ag to polytene chromosomes revealed elements on all six chromosomes and the chromocentre. Southern hybridization experiments yielded a heterogeneous pattern of bands, which was interpreted as polymorphism in the genomic location of T1Ag among individuals. No two T1Ag elements examined proved identical; most variation resulted from single-base-pair substitutions. Due to these sequence polymorphisms, identification of an unambiguous T1Ag structure was impossible, so a consensus nucleotide sequence was determined from a number of elements.

Full-length T1Ag elements are 4.6 kb in length. Analysis of a consensus sequence revealed that over 92% was occupied by two long, overlapping open reading frames; these were followed by a polyadenylation signal, AATAAA, and a tail consisting of tandem repetitions of the motif TGAAA, which defined the 3′ terminus. No direct or inverted long terminal repeats were detected. The first open reading frame, 442 amino acids in length, included a domain resembling that of nucleic acid-binding proteins. The second open reading frame, 975 amino acids long, resembled the reverse transcriptases characteristic of non-LTR retrotransposons.

The existence of approximately 100 copies of T1Ag scattered through the genome, and the variation in genomic location among field populations of *An gambiae*, pointed to a recent and ongoing mobility. No transcripts were found in *An gambiae* adult RNA, so there was no direct evidence that T1Ag elements were transcribed (Besansky, 1990a).

The T1Ag family of retrotransposable elements is interspersed and moderately repeated among at least five members of the *An gambiae* complex; *An bwambae* was not examined. The T1Ag family was clearly conserved at the sequence level among the sibling species. No T1Ag hybridizing sequences were detected by Southern blot in DNA from *An stephensi*. In four species of the *An gambiae* complex T1Ag comprised two closely related but independent subfamilies, designated α and β, defined by the presence or absence of linked sets of restrictions sites over the entire length of the element. *Anopheles merus* contained only T1Agβ sequences. The consensus maps indicated that T1Agβ had remained essentially unchanged since speciation occurred, except that in *An merus* the *Xba*I site was missing. The consensus maps of T1Agα were unique for each of four species, because of the gain or loss of one or two sites at the extreme 5′ or 3′ ends of the element. The two subfamilies occurred at different relative frequencies in the different species.

(c) *LTR-retrotransposons*

Cloning of extrachromosomal DNA from an *Aedes aegypti* cell line yielded a clone, designated pX16, with an insert size of 4.0 kb, which displayed intraspecific variation in chromosomal distribution. The refractive index of the extrachromosomal DNA showed it to be circular. The copy number of pX16 in genomes from different sources ranged from 30 in the Bangkok strain to 600 in a cell line, confirming its middle repetitive nature. Partial DNA sequence data revealed considerable homology to the consensus sequence of several retroviral and retrotransposon reverse transcriptases (Crampton *et al.*, 1990).

1.3 MEIOSIS

Meiosis takes place during gamete formation, when a single chromosomal replication is followed by two nuclear divisions, leading to the formation of four haploid gametes from a single diploid germ cell. Meiosis and fertilization are therefore complementary events.

Two processes in meiosis lead to genetic reassortment: (1) through the pairing of homologous chromosomes and crossing-over the reciprocal exchange of segments between chromatids of maternal and paternal origin is effected; and (2) through the random assortment of maternal and paternal homologues between the daughter cells from meiotic division I, each gamete receives a different mixture of maternal and paternal chromosomes.

Early in meiosis a proteinaceous structure, the **synaptonemal complex**, is formed between each pair of synapsed chromosomes. The transverse filaments and lateral elements of which the synaptonemal complex is composed give it a ladder-like appearance. Protein-containing bodies called **recombination nodules**, which occur at intervals along the synaptonemal complex, are thought to be involved in crossing-over between maternal and paternal chromatids (Alberts *et al.*, 1989).

The appearance of the chromosomes when they can first be visualized in early meiosis is thought to reflect their disposition during interphase. As described below, in early meiosis the chromosomes show somatic pairing and are in a 'bouquet arrangement', the six arms radiating from fused centromeres at one side of the nucleus and converging in a group of closely associated telomeres at the opposite side (Figure 1.5). The centromeres may not all be

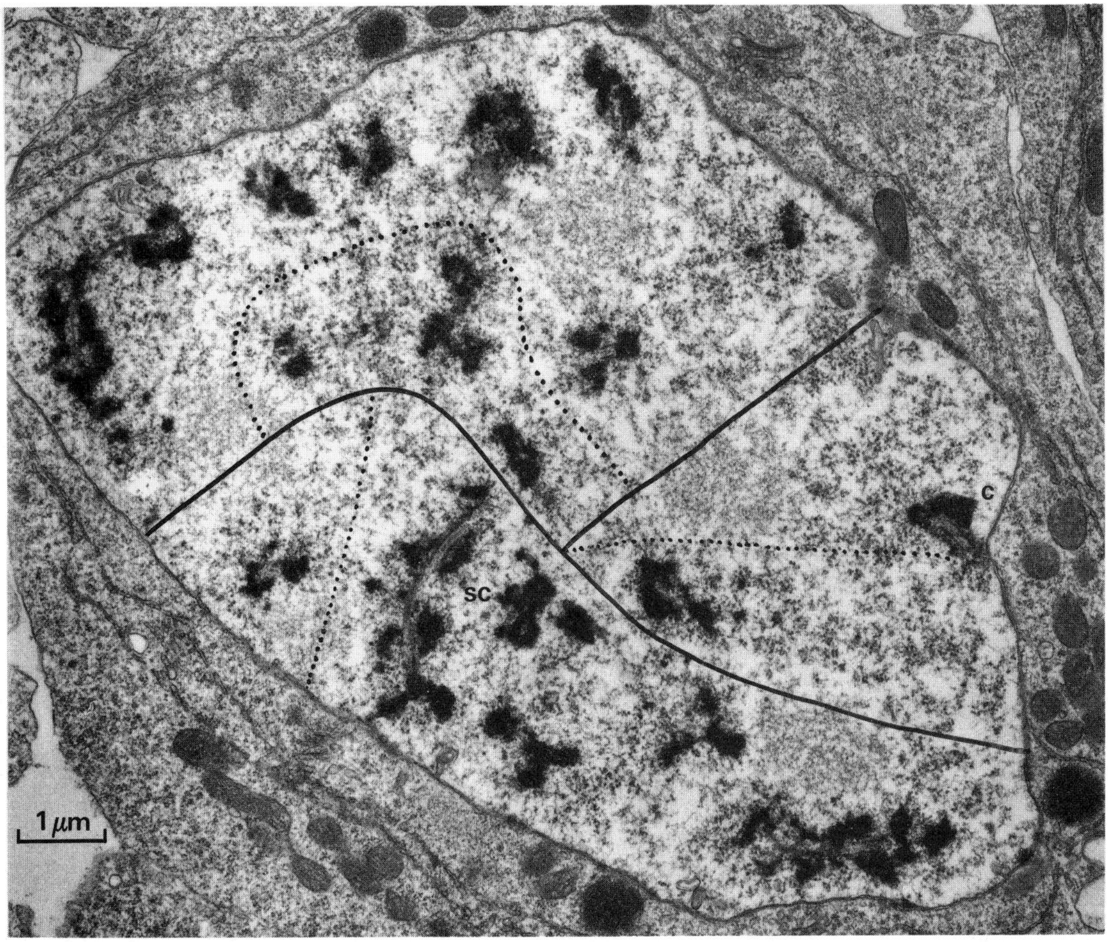

Figure 1.5 Chromosome domains within the diplotene nucleus of a spermatocyte of *Aedes aegypti*. Solid lines separate the three somatically paired chromosomes, dotted lines separate the two arms of a chromosome. c, centromeric region; sc, synaptonemal complex. (From Wandall and Svendsen (1985.) Micrograph courtesy of Dr A. Wandall.)

associated in a single mass throughout the whole of interphase, but homologous centromeres are always associated (Diaz and Lewis, 1975; Marchi *et al.*, 1981).

1.3.1 Male meiosis

Male meiosis in mosquitoes is broadly similar to that described from other animals. This Section describes those aspects of meiosis in *Aedes aegypti* spermatocytes that have contributed to knowledge of insect meiosis generally. The information is derived principally from three-dimensional ultrastructural reconstructions of spermatocyte nuclei (Wandall and Svendsen, 1985), but also from descriptions of spermatocytes that were spread, stained to reveal the synaptonemal complexes, and examined by electron microscopy (Wandall and Svendsen, 1983).

(a) Prepachytene

In mosquitoes, homologous chromosomes are already closely synapsed before the start of meiosis due to somatic pairing, and consequently the classical leptotene and zygotene stages of prophase I are omitted. Instead there is a gradual transition from somatic pairing to meiotic pairing. The term 'prepachytene' has been coined for the transitional period when the chromosomes can be identified as being in early meiotic prophase but the synaptic complexes are not fully formed.

Meiotic prophase begins in the spermatocytes of *Aedes aegypti* shortly after the last mitotic division. The chromosomes that had somatically paired at the end of mitosis were first seen in electron micrographs as solid threads, but at the start of meiotic prophase scattered spaces appeared within the chromatin, apparently between the pairing faces of the homologues. Some spaces contained fibrillar material. Subsequently synaptonemal complexes became visible in the spaces, the transverse filaments and lateral elements becoming recognizable at about the same time. Synaptonemal complex formation started at some nine initiation points per bivalent, starting later near the centromeres (Figure 1.6 A and B).

The intercalary heterochromatic region on one arm of chromosome-1 (Section 1.1.1(a)) could

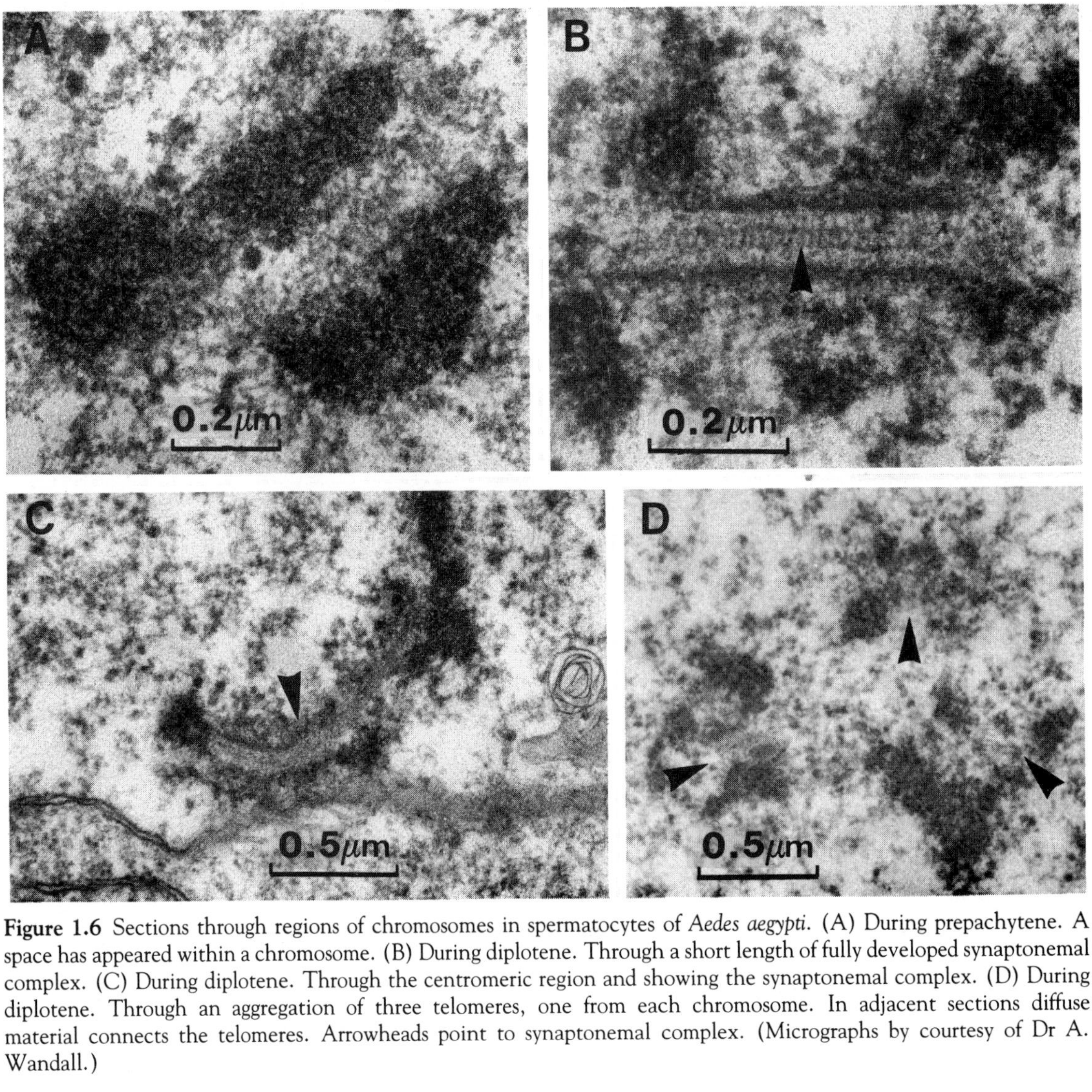

Figure 1.6 Sections through regions of chromosomes in spermatocytes of *Aedes aegypti*. (A) During prepachytene. A space has appeared within a chromosome. (B) During diplotene. Through a short length of fully developed synaptonemal complex. (C) During diplotene. Through the centromeric region and showing the synaptonemal complex. (D) During diplotene. Through an aggregation of three telomeres, one from each chromosome. In adjacent sections diffuse material connects the telomeres. Arrowheads point to synaptonemal complex. (Micrographs by courtesy of Dr A. Wandall.)

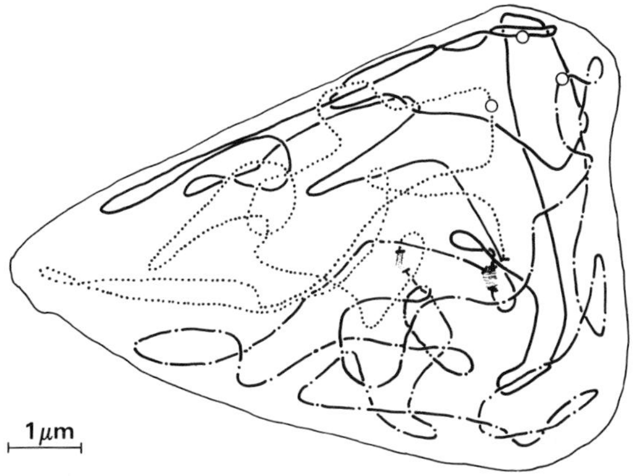

Figure 1.7 Reconstruction of the synaptonemal complexes within the pachytene nucleus of a spermatocyte of *Aedes aegypti*. Total synaptonemal complex length was 234 μm. The open circles mark the locations of the centromeres, the hatched areas represent diffuse material at the telomere ends. Bar length = 1 μm. (From Wandall and Svendsen, 1985.)

be recognized in electron micrographs before synaptonemal complex formation had started. Synaptonemal complex formed normally over the heterochromatic region, which was usually adjacent to a nucleolus. The long arm of chromosome-3 regularly showed a constant diffuse disruption to the synaptonemal complex in the region of the secondary constriction visible in the mitotic chromosome (Figure 1.1A). In mid-pachytene, according to Motara *et al.* (1985), a silver-stained nucleolus organizer was conspicuous on one bivalent in most nuclei.

Reconstruction of nuclei revealed that the chromosomes occupied separate domains within the nucleus (Figure 1.5). In all cells the centromeres were attached to the same general area of nuclear membrane. During prepachytene this area was rather large, but it became more compact during pachytene and diplotene when the centromeres were separated by only 1–2 μm. The chromatin of the centromeric regions was less electron dense than the other chromatin. In early prepachytene the telomeres were attached to the nuclear membrane, but later in prepachytene the telomeres disengaged from the nuclear membrane and were free in the nucleoplasm although remaining close together (Figure 1.6 C and D).

(b) The later stages

In surface-spread pachytene *spermatocytes* the chromosomes were in a 'bouquet configuration' with the three centromeres close together and the telomeres polarized, 180° from the centromeres, and clustered (Figure 1.7). The centromeres remained associated in the absence of nuclear membrane, suggesting an intercentromeric affinity. In most nuclei the chromosomes were bent at or near the centromere regions and were V- or U-shaped. The lengths and centromeral indices of the three meiotic chromosomes were almost identical with those of the mitotic chromosomes.

The width of the synaptonemal complexes increased from prepachytene to pachytene, reaching a maximum of 160 nm, mainly because of a thickening of the lateral elements. During diplotene there appeared to be a simultaneous shortening of the synaptonemal complexes and thinning of the lateral elements. The total length of the synaptonemal complexes decreased by more than 100 μm.

From one to four recombination nodules were found in each nucleus. Because that was only about half of the number of chiasmata seen during spermatogenesis, it was concluded that many nodules had been overlooked. Six of the seven nodules seen on chromosome-1 were located on the distal one-third of the arms, consistent with the observation that most chiasmata occur terminally (Section 1.3.3).

Kinetochores became visible on the centromeres during pachytene. They appeared first in bivalent 1, which was consistent with Bhalla's (1971) observation that the two members of bivalent 1 often separated before the other chromosomes. In early pachytene the two centriole pairs separated, and during pachytene and diplotene the centromeres faced one or other of the centriole pairs.

The desynapsis characteristic of diplotene could be recognized from the divergence of the lateral elements. There appeared to be no regular starting points for homologue repulsion. By later diplotene the centromeres could no longer be recognized, the recombination nodules had disappeared, and the synaptonemal complexes were shed from the chromosomes leading to the appearance of many small 'polycomplexes' in the nucleoplasm. By telophase they also had disappeared.

In diplotene the homologous chromosomes were held together only by relational coiling (the loose coiling of the homologous chromosomes around one another) and by the chiasmata, the points of fusion of non-sister chromatids. The chiasmata were most distinct at diakinesis and in metaphase I. In metaphase the bivalents occupied the equatorial region of a spindle and their centromeres became attached to the spindle fibres, and at anaphase the chromosomes were drawn to the two poles. The two sister chromatids that composed each chromosome showed tight somatic pairing at this stage.

The second meiotic division started after a short interphase during which the chromosomes underwent decondensation. In all species the second meiotic division proceeded to completion in the classical manner with the formation of spermatids that contained the haploid number of single-stranded chromosomes.

The following investigations of male meiosis are additional to those cited earlier. *Culex pipiens*: Lomen (1914); Whiting (1917); Moffett (1936); Grell (1946b); Jost (1971). *Aedes aegypti*: Akstein (1962); Mescher and Rai (1966); Krafsur and Jones (1967). *Ae albopictus*: Smith and Hartberg (1974). *Culex tarsalis*: Asman (1974). *Culiseta inornata*: Breland *et al.* (1964). *Anopheles stephensi*: Rishikesh (1959). Four other species of *Anopheles*: Narang *et al.* (1972).

1.3.2 Female meiosis

(a) Meiosis

Measurements of [^{3}H]thymidine incorporation showed that chromosomal replication occurred in the primary ovarian follicles of *Aedes aegypti* some two to three days before emergence (Roth *et al.*, 1973). The oocytes may be visibly distinguishable from the nurse cells before emergence or that stage may not be attained until a day or two later. When differentiated oocytes are first recognizable, the three pairs of chromosomes, which are synapsed due to somatic pairing, are in pachytene and have fully formed synaptonemal complexes. Prepachytene stages with undivided chromatin strands or incomplete synaptonemal complexes have not been seen in *Aedes* or *Culex* oocytes (Fiil, 1978b).

In pachytene the long chromosomes loop around the nucleus several times. Reconstruction of the synaptonemal complexes of *Culex quinquefasciatus* showed that the chromosomes were in a bouquet configuration with the centromeres close together and anchored to the nuclear membrane. The six telomeres were free in the nucleoplasm but associated through a common diffuse structure (Fiil, 1978b; Wandall and Svendsen, 1985; Wandall, personal communication).

Crossing over is temperature sensitive. By scoring cross over frequency after applying heat shocks to females of *Ae aegypti* which bore linked genetic markers, it was found that crossing over occurred only during a short period in early pachytene (Roth *et al.*, 1973).

By late pachytene the nuclear diameter in *Ae aegypti* has increased to 12–14 μm; the three bivalents have separated from the nuclear envelope and have shortened, forming a deeply staining mass near the nucleolus. At 40–60 h after emergence the chromosomes are in early diplotene; homologous chromosomes have expanded and parted slightly except where linked by chiasmata, and each is composed of two chromatids relationally coiled around each other. At this time the synaptonemal complexes break down into subunits which aggregate within the nucleus, either as filamentous bodies or as polycomplexes.

By the third day after emergence the oocyte's chromosomes are in the diffuse stage of diplotene, when the chromatin is a tangle of chromatids forming a sphere of 8–10 μm diameter within a nucleus of 30 μm diameter. Each chromatid is of 400–500 nm diameter and appears to have

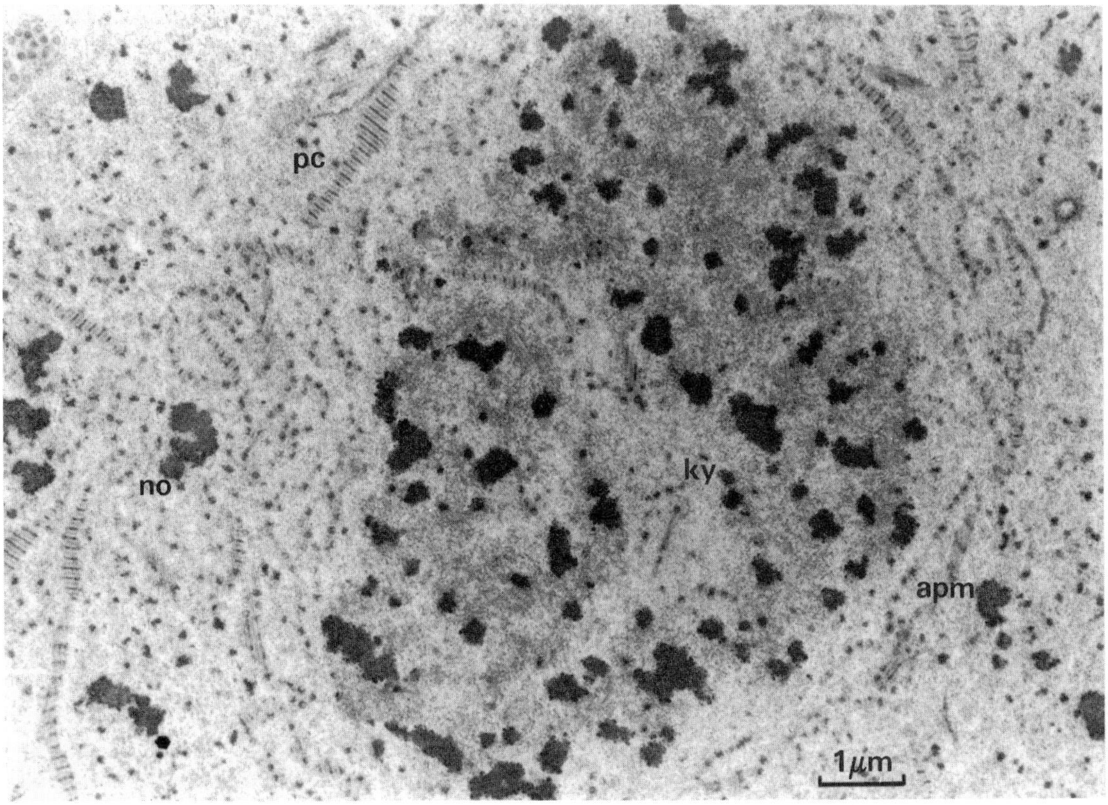

Figure 1.8 Polycomplexes and annuli in an oocyte nucleus of *Aedes aegypti*. The karyosphere (ky) is surrounded by one or more layers of polycomplexes (pc) and annuli arranged in annulated pseudomembranes (apm). The dense regions of the karyosome are mostly DNA with large RNA granules attached. In the nucleoplasm are nucleoli (no) which consist largely of RNA and a small amount of DNA. (Micrograph by courtesy of Dr A. Wandall.)

several layers of organization. Interspersed among the chromatids are 'polycomplexes' composed of up to 60 synaptonemal subunits.

Vitellogenesis commences within a few hours of the female taking a blood meal and the oocyte grows rapidly over the next 2–3 days. The nucleus also grows rapidly, first branching, then invaginating like a cup, and finally elongating into a canoe-shaped structure which, in *Ae aegypti*, is 400 μm long, 100–140 μm wide, and about 3 μm thick.

During this period the chromosomes condense and aggregate into a small ball or karyosphere, and the synaptonemal complexes disappear from the bivalents. Shortly afterwards polycomplexes and annuli, composed of the lateral elements and transverse filaments characteristic of synaptonemal complexes, polymerize and enclose the karyosphere in a multilayered capsule. Throughout a period of two to three days the chromosomes remain in this capsule, which is located in a nuclear pocket, 12–16 μm across, at the anterior pole of the oocyte and close to the nurse cells. The chromosomes are now probably in diakinesis (Roth, 1966; Fiil and Moens, 1973; Fiil, 1974). The capsule and its contents have been termed the 'germinal nucleus' and the remainder, which contains the synthetically active nucleolus, the 'vegetative nucleus' (Section 19.4).

In *Culex quinquefasciatus*, during the previtellogenic resting stage, transverse filaments of the synaptonemal complexes extend not only between the lateral elements but also between the lateral elements of different synaptonemal complexes, so that all the bivalents become interconnected by a network of pseudomembranes (Figure 1.8). As the oocyte matures, the network appears to contract so that the bivalents are drawn together. They also enclose the chromosomes within a karyosphere (Fiil and Moens, 1973). Karyosphere formation in *Anopheles gambiae* has been described by Fiil (1976a).

Shortly before the oocyte is ready for ovulation the envelope of the vegetative nucleus breaks down leaving remnants widely dispersed through the ooplasm. At this time the karyosphere or germinal nucleus has an envelope consisting of several layers of annulated lamellae (Fiil and Moens, 1973; Fiil, 1974).

The oocyte has now reached the maturation gate and it may enter a quiescent state for a period of hours or days, with its chromosomes held in metaphase I, or ovulation and oviposition may take place immediately. During oviposition the egg becomes fertilized, and shortly after oviposition the membranes surrounding the nuclei disperse and meiosis is resumed (Fiil, 1974). The first meiotic division leads to the formation of two female nuclei, and the second yields three polar bodies and a female pronucleus. This process, and the fusion of the male and female pronuclei, are described in Section 2.1.

The progress of the first meiotic division may be delayed as the ovarian follicles are held at certain developmental gates. In anautogenous mosquitoes, primary follicles containing oocytes in the diffuse stage of diplotene enter the previtellogenic resting stage. After vitellogenesis, when the oocytes are mature and awaiting ovulation, their chromosomes may be held at metaphase I.

(b) Polycomplexes

The total quantity of modified synaptonemal complex material in a maturing oocyte of *Aedes aegypti* greatly exceeds the amount of synaptonemal complex material present at the previtellogenic resting stage. The difference suggests that renewed formation of polycomplex material occurs after the blood meal. Both lateral elements and transverse filaments are synthesized. Polymerization of lateral elements leads to the formation of sheets of synaptonemal complex which become stacked one upon another and which extend far into the nucleoplasm. From the lateral elements in these stacks are formed annuli which resemble nuclear pores. The annuli polymerize further to form 'polycomplexes', and these in turn rearrange into 'annulated pseudomembranes' which come to surround the germinal nucleus (Figure 1.8). Normal capsules vary from a predominantly polycomplex-containing type to a predominately annulus-containing type. After the pachytene stage of meiotic prophase in *Culex pipiens* there is a proliferation of transverse filaments. The lateral elements do not show a corresponding increase. Pseudomembranes formed from the transverse filaments extend between the lateral elements of different synaptonemal complexes, and come to enclose the chromosomes (Figure 1.9). For a short period after the disappearance of the synaptonemal complexes from the bivalents several thousand annuli of 120–150 nm diameter are present near the chromosomes. The annuli, which do not resemble those in *Ae aegypti*, disappear and material originating mainly from transverse filaments forms a capsule around the chromosomes during diplotene (Fiil and Moens, 1973; Fiil, 1976b, 1978b).

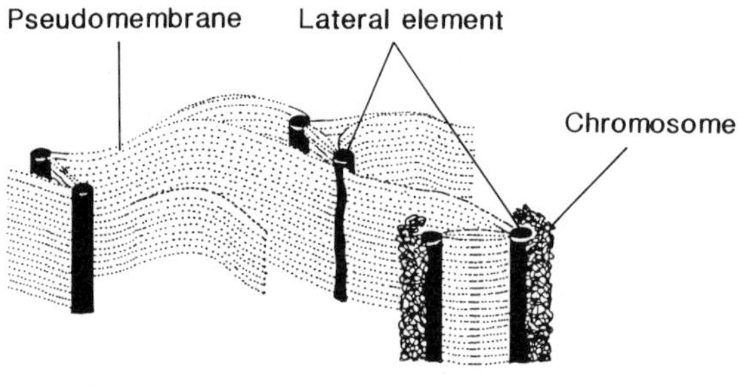

Figure 1.9 Diagram of modified synaptonemal complexes in a *Culex pipiens* oocyte after pachytene. Synaptonemal complexes are still associated with the chromosomes, but the transverse filaments, which normally lie between the lateral elements of an individual complex, become much extended and make contact with the lateral elements of other synaptonemal complexes. The bivalents thus become interconnected by pseudomembranes. (From Fiil and Moens, 1973.)

The composition of the capsule that forms around the chromosomes in *Anopheles gambiae* is not known (Fiil, 1976a).

1.3.3 Crossing over

In many families of Diptera crossing over does not occur in males during meiosis, but the Culicidae is one of the families in which chiasmata are apparent in both males and females (White, 1973). Chiasmata have been observed in both sexes in species of *Culex*, *Aedes* and *Anopheles*. In *An atroparvus* no chiasmata were observed on the short arms of the sex chromosomes, but they occurred on the long arms (Jayakar *et al.*, 1982).

Usually either no chiasmata or one chiasma is formed within an arm-pair of a bivalent. Rarely, two chiasmata form within an arm-pair; frequencies of 1.74 and 0.31% of such per total bivalents examined have been recorded for *Cx pipiens* and *An atroparvus* (Callan and Montalenti, 1947; Jayakar *et al.*, 1982). Chiasma frequencies per cell amounted to 3.1–4.1 in *Cx pipiens*, 3.0–5.2 in *Ae aegypti* and 2.4–3.8 in *An atroparvus* (Callan and Montalenti, 1947; Ved Brat and Rai, 1973; Jayakar *et al.*, 1982).

In *Ae aegypti* the rate of chiasma formation was higher in the terminal than in the interstitial region of the chromosome arms, reaching 80% in the terminal region in some strains (Ved Brat and Rai, 1973). In *An atroparvus*, comparisons of chiasma frequencies in three arm regions gave the ratios (proximal : interstitial : terminal): autosome-1, 0.06 : 0.63 : 0.38; autosome-2, 0.11 : 0.64 : 0.57; XY, 0.03 : 0.28 : 0.03 (Jayakar *et al.*, 1982). Chiasma frequency varied significantly between individuals of *An atroparvus*, suggesting genetic control (Jayakar *et al.*, 1982). Strains can differ in cross over frequency between two particular genes, and the presence of inversions has been suggested as one reason for this (Macdonald and Sheppard, 1965).

The mean chiasma frequency per bivalent was between one and two in *Cx pipiens*, *Cx tarsalis*, *Ae albopictus* and *Ae aegypti* (Callan and Montalenti, 1947; Asman, 1974; Jost, 1971; Ved Brat and Rai, 1973). In *Culiseta longiareolata* the short bivalent had a mean chiasma frequency of 1.9 and the longer bivalents had a frequency of 2.9 (Callan and Montalenti, 1947).

The presence of one chiasma is known to reduce the probability of another forming nearby. From analysis of chiasma frequencies, Callan and Montalenti (1947) concluded that in *Cx pipiens* chiasma interference extends from one arm of a chromosome across the centromere to the other arm, but that in *Cs longiareolata* it does not extend across the centromere. Owen (1949) reanalysed Callan and Montalenti's data and, using different biological premisses, concluded that their claim that interference extends across the centromere in *Cx pipiens* was not proved. From another reanalysis of Callan and Montalenti's data, Sybenga (1975) concluded that chiasma interference is much weaker in *Cs longiareolata* than in *Cx pipiens*.

Because crossing over occurs between the sex chromosomes of culicine mosquitoes their sex linkage is described as partial. *Culex tritaeniorhynchus* is exceptional in exhibiting a total or near-total absence of crossing over in all chromosomes of the female, although crossing over occurs in the male (Baker and Sakai, 1973). Among culicine mosquitoes only the females of *Cx tritaeniorhynchus* are known to have complete, or nearly complete, sex linkage.

In organisms that have sex chromosomes of the XY type, the heteromorphic chromosome has a so-called 'differential segment' that lacks chiasmata and bears the sex-linked genes, and a so-called 'pairing segment' within which the chromosomes pair and cross over but which is devoid of genes. The formation of chiasmata is thought to be necessary for regular meiosis. Crossing over is observed between the X and Y chromosomes in *Anopheles*. By following marker genes carried on autosomal segments translocated to the Y chromosome of *An culicifacies*, it was found that crossing over could occur between the homomorphic, heterochromatic long arms of the X and Y chromosomes but not between the heteromorphic, euchromatic short arms (Sakai *et al.*, 1979). In four species of *Anopheles* the autosomal bivalents always showed two chiasmata each,

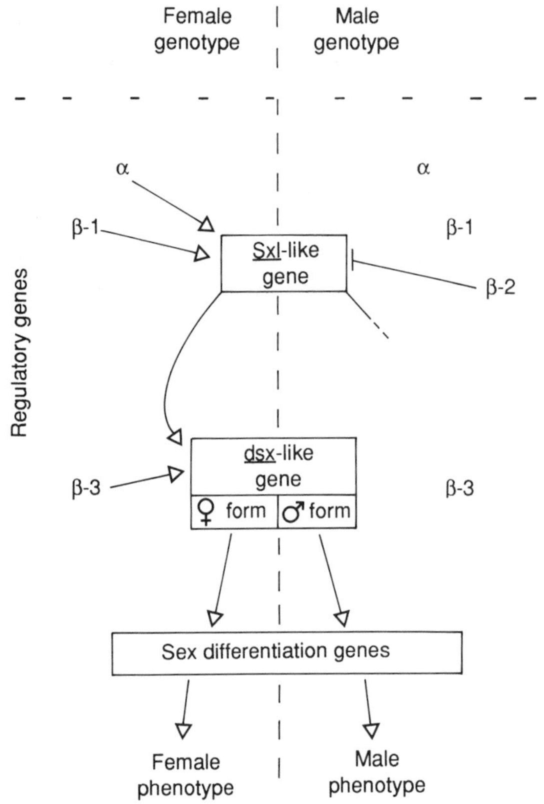

Figure 1.10 Outline of the common genetic mechanism that regulates sexual differentiation in insects. In female zygotes a genetic cascade starts with activation of an *Sxl*-like gene by maternal genes (α) and possibly also by zygotic genes (β-1.) Male zygotes produce a repressor (β-2) of the *Sxl*-like gene which is functionless in males. In females the product of *Sxl*, sometimes supplemented by the products of other zygotic genes (β-3), controls the *dsx*-like gene, promoting the splicing of its product into the female form. In males, by default, the product of the *dsx*-like gene splices into the male form. The *dsx*-like gene product, in male or female form, activates a complex of sex differentiation genes which, probably in association with homeotic genes, control the differentiation of segmentally ordered organs of male or female phenotype. (After Bownes, 1992.)

but the sex bivalent showed only one (Narang *et al.*, 1972). In *An atroparvus* a mean chiasma frequency of one per bivalent was recorded. When relative chromosome length was considered, the number of chiasmata observed in XY bivalents was not significantly different from that in the autosomes (Fraccaro *et al.*, 1976). Recombination of sex-linked genes occurs during female meiosis in *Anopheles*; the males show complete sex linkage because of the hemizygous condition of their sex chromosomes.

1.4 SEX DETERMINATION, SEXUAL DIFFERENTIATION AND SEX-RATIO REGULATION

In insects the sexual characteristics of an individual, anatomical and physiological, depend entirely upon the expression of genetic information, so immediately a zygote is formed its sex is determined by its genetic constitution. Insect hormones are not sex specific; however, they can produce sex-specific effects as a consequence of dimorphic gene programming in the responding cells. Insects show a variety of sex determination mechanisms; in *Drosophila*, for example, sex depends upon the ratio of X chromosomes to autosomes. That sex determination is cell autonomous is demonstrated by gynandromorphs, which are genotypic mosaics that develop as a combination of male and female organs.

Once embryogenesis has started, the regulation of sexual differentiation commences, terminating in the formation of an adult of male or female phenotype. The mechanism of sexual differentiation has been established in greatest detail for *Drosophila melanogaster*, owing to the many available mutants. It appears to take rather different forms in other insects, but there may be a common basis involving a set of regulatory genes, maternal and zygotic, which functions as a linear cascade (Figure 1.10). In *D. melanogaster* this contains a double switch involving the wild-type forms of the genes *Sxl* (*Sexlethal*) and *dsx* (*doublesex*), and the wild type forms of further regulatory genes including *tra* (*transformer*), *tra-2*, and *ix* (*intersex*). The product of the final gene in this cascade activates a complex of sex differentiation genes, which govern differentiation of the sexually dimorphic characteristics of the adult.

The nature of the primary determinant of sex has been established in anopheline and culicine mosquitoes, and differs strikingly between these two subfamilies. These mechanisms of sex determination would, if unmodified, lead to the production of equal numbers of males and females. However, measurements of sex ratio in wild populations sometimes show marked divergences from unity, and it is known that certain species of mosquito have genetic mechanisms for the regulation of sex ratio. Information on the genetic cascade that governs sexual differentiation in mosquitoes is sparse, but observations on culicines suggest points of similarity with *Drosophila melanogaster*.

1.4.1 Sex determination in anophelines

The nature of the primary determinant of sex is known for anophelines, but little more. The presence in *Anopheles* of distinct sex chromosomes with extensive non-pairing regions results

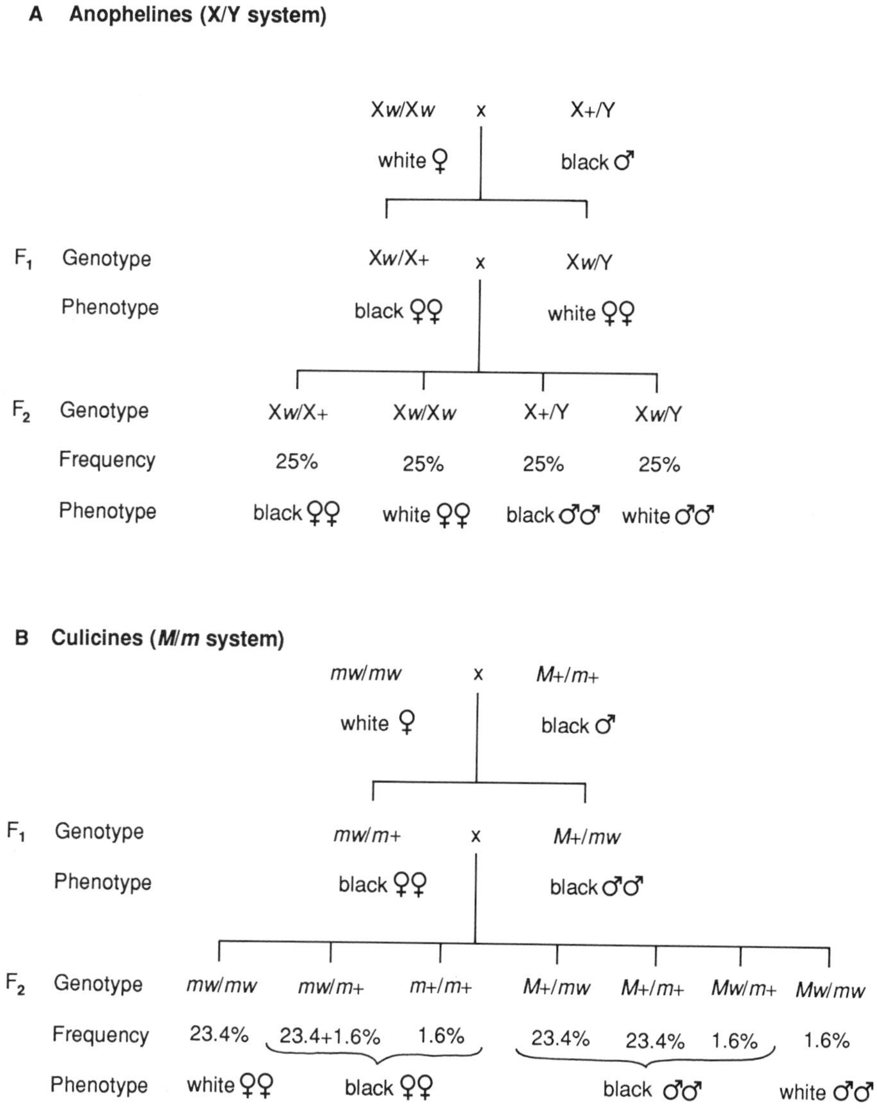

Figure 1.11 Sex linkage in anophelines and culicines as revealed by inheritance of the sex-linked recessive character *white-eye* (*w*.) (A) *Anopheles gambiae*. (B) *Culex pipiens*. The *Cx pipiens* pedigree indicates the effects of 6.4% crossing-over between the *M*/*m* and *w*/+ loci, and shows that there are two kinds of heterozygous male but only one kind of heterozygous female. (From the work of Mason (1967) and Gilchrist and Haldane (1947).)

in classical sex linkage for the eye colour genes, as has been demonstrated in *Anopheles gambiae*, *An stephensi* and *An culicifacies* (Mason, 1967; Aslamkhan, 1973; Sakai *et al.*, 1977). Figure 1.11A illustrates the inheritance of *white eye* (*w*). It can be seen that sex linkage is complete in *An gambiae*, the recessive *white-eye allele* being expressed in F_1 males in a hemizygous condition. From the F_2 genotypes it is apparent that females can carry *w* in a homozygous (X*w*/X*w*) or heterozygous (X*w*/X+) condition but that males can carry it only in the hemizygous condition (X*w*/Y). Mason (1967) concluded that all anophelines with heteromorphic sex chromosomes probably show complete sex linkage.

If a Y chromosome is present in a mammal its phenotype is recognizably male regardless of the rest of the karyotype; if there is no Y chromosome its phenotype is recognizably female. In *Drosophila* the presence or absence of the Y chromosome is not critical for sex determination since XO individuals are recognizably female. Sex is determined by the ratio of the numbers of X chromosomes to autosomes; in females X:A = 1.0, in males 0.5. The only such evidence for mosquitoes is from a triploid individual of *An culicifacies* with the complement XXY, which was male. This suggested that in *Anopheles*, unlike *Drosophila*, the number of X chromosomes is less important for sex determination than the presence or absence of the Y chromosome (Baker and Sakai, 1979).

A live gynandromorph of *Anopheles gambiae*, in which the head and thorax were female and the abdomen male, probed normally and penetrated skin but was unable to engorge. When force-mated it flexed its abdomen towards the female and grasped her terminalia with its claspers, but its aedeagus did not enter the gonotreme (Mason, 1980).

1.4.2 Sex determination and sex locus in culicines

Studies of sex determination in culicine mosquitoes were first undertaken by Gilchrist and Haldane (1947) while investigating the inheritance of the recessive, sex-linked allele *white-eye* (*w*) in *Culex pipiens* (Figure 1.11B). When white-eyed females were crossed with wild type males no white-eyed males appeared in the F_1 generation. White-eyed individuals reappeared in the F_2 generation, in normal Mendelian ratios but showing about 6.4% recombination with sex. This suggested that sex determination was controlled by a factor located on a chromosome which also carried normal homologous pairs of alleles. Analysis of crosses between the white-eyed and wild type strains suggested that sex was determined by a single pair of genes, with maleness being dominant over femaleness, so that the females were *mm* and the males *Mm*. Subsequently, a study of translocation heterozygotes in *Cx pipiens* showed that the shortest pair of chromosomes carried the sex factor and sex-linked genes (Jost and Laven, 1971). Bhalla *et al.* (1975) demonstrated crossing over between the sex locus and each of two genes located on either side of it (*eye-gap* and *maroon-eye*). This can be taken to support both the original postulate, that sex is determined by a single pair of alleles, and the alternative view that it is determined by a short segment or block of genetic material fulfilling the same function and behaving in the same way.

A similar monofactorial inheritance of sex has been reported in other culicines, i.e. in *Culex tritaeniorhynchus* (Baker, 1968), in a number of *Aedes* species, including *aegypti* (McClelland, 1966), and in *Eretmapodites quinquevittatus* (Hartberg and Johnston, 1977) and *Armigeres subalbatus* (Tadano and Mogi, 1987).

Examination of the polytene salivary-gland chromosomes of *Culex pipiens* larvae has revealed a structural difference between males and females in the shortest chromosome (chromosome-1), which correlates with linkage group I containing the sex locus. In female larvae a puff is present across the full width of the synapsed polytene chromosomes in zone 10C3 of arm 1L. In male larvae a narrow band of condensed chromatin extends across half the width of the arm in this position and a puff occupies the other half. Thus females are homozygous and males heterozygous for this condition. In individuals heterozygous for a male-linked reciprocal translocation, it was the chromosome with the distinct

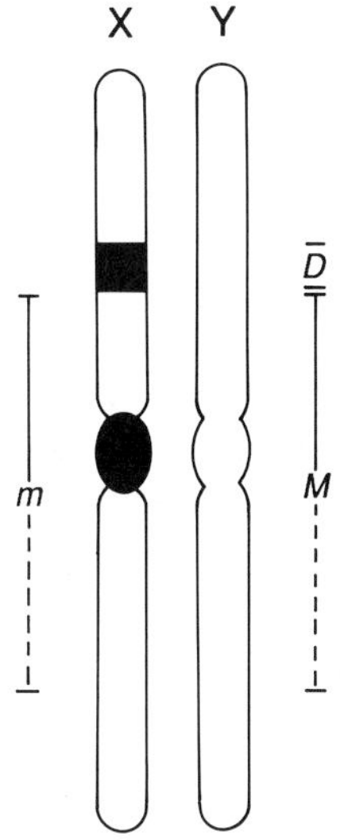

Figure 1.12 The locations of *m*, M and *D* within the sex chromosomes of *Aedes aegypti*. The region within which the sex locus occurs is bounded on one arm by the position of the intercalary band, and on the other arm by the limit of the achiasmate region. (After Wood and Newton, 1991.)

euchromatic band that was involved in the segmental exchange with the autosome. Dennhöfer (1975a) therefore concluded that the condensed state of the chromatin in zone 10C3 represents the male-determining allele (M) and that the decondensed puffed state of that zone represents the female-determining allele (*m*). The puffs were more distinct in the Hamburg strain of *Cx pipiens* than in certain other strains (Dennhöfer, 1975a). The association of the sex-determining alleles M and *m* with zone 10C3 in arm 1L was confirmed by crossover analysis of breakpoints in sex-linked translocations (Dennhöfer, 1975b).

The sex chromosomes of *Aedes aegypti* are morphologically indistinguishable from one another, but the presence of a centromeric C-band in the X but not the Y provides a reliable means of discrimination (Figure 1.1A). Most X chromosomes also possess an intercalary C-band, but the Y chromosome is polymorphic for the presence or absence of this band (Section 1.1.1(a)).

In *Ae aegypti* no crossing over occurs between the sex 'locus' and the centromere, as shown by the complete linkage between sex and centromere type (with or without C-band). That crossing over occurs at a very low frequency between the sex locus and the intercalary band was shown by the loss of *Distorter* (*D*) – which is closely associated with the intercalary band – in individuals that acquired an intercalary band on the Y chromosome (Newton *et al.*, 1978b). The sex locus is situated within a substantial central region of the sex chromosome in which chiasmata have not been seen. This region extends across the centromere. On the banded arm it extends from a position within 1–2 units of the intercalary band (on the centromeric side), and on the unbanded arm it extends to the proximal limit of chiasma formation (Figure 1.12) (Newton *et al.*, 1974; Newton *et al.*, 1978b; Wood and Newton, 1991).

1.4.3 Sexual differentiation in culicines

Information on sexual differentiation in mosquitoes can be obtained from the study of individuals in which it has been disturbed. In **sex mosaics**, or **gynandromorphs**, the individual is a genotypic mosaic, which phenotypically appears as a combination of male and female tissues. A line of demarcation between male and female organs is always apparent, thus a gynandromorph may be anterior–posterior, with for example female head and male abdomen, or it may be bilateral, or the male and female organs may have a less regular distribution. Gynandromorphs have been described from eleven culicine genera. Most specimens have the head of one sex, the abdomen of the other sex, and the thorax divided either bilaterally or obliquely (Craig and Hickey, 1967; Hall, 1987).

Crosses between strains of *Aedes aegypti* with markers in all three linkage groups confirmed that the male and female secondary sexual characters present in different parts of a gynandromorph reflect the genetic constitution of those parts. Further studies suggested that gynandromorphs are produced when two sperms fuse with products of female meiosis in a single egg. These might be the two female nuclei resulting from the first meiotic division, or the female pronucleus and one of the polar bodies resulting from the second. One sperm would be M-bearing and the other *m*-bearing, and the two diploid nuclei resulting from the fusions would each develop into half of the mosquito (Craig and Hickey, 1967). Observations on gynandromorphs in *Culex pipiens* support this explanation (Laven, 1967a).

One form of abnormality can lead to **intersexes** – individuals in which the whole body is of either male or female genotype but in which the phenotype of the sexually dimorphic organs is partly or completely transformed to that of the opposite sex. Intersexes show no clear line of demarcation between male and female organs. The affected organs are always bilaterally symmetrical. Intersexes reported from *Aedes aegypti* and *Culex pipiens*, and described below, are similar in phenotype to the *ix*, *dsx* and *tra* mutants of *Drosophila* (Bownes, 1992).

In *Aedes aegypti* a rare autosomal, recessive, temperature-sensitive gene called *intersex* (*ix*) affects the phenotype for sex. The *ix* gene is in linkage group II and appears to be sex-limited, having no known effect in females. Male larvae that were reared at 27°C developed into normal male adults. Male larvae reared at 30°C developed into intersexes, with all dimorphic parts intermediate between male and female. Male larvae reared at 35–37°C developed into pseudofemales, which looked like normal females, could be inseminated, took blood meals, and matured their ovaries. The critical period for responsiveness of *ix* individuals to high temperature was two to four days after hatching (Craig, 1965; Craig and Hickey, 1967). Bownes (1992) compared the *ix* allele of *Ae aegypti* to the *tra-2*^{ts} allele of *Drosophila*, which also switches sexual differentiation and which also is temperature sensitive during the larval stage. The *Drosophila* gene differs from *ix* of *Aedes* in that it acts not on genetic males but on genetic females – at low temperatures permitting them to differentiate normally, at intermediate temperatures producing intersexes, and at high temperatures producing pseudomales. Since *ix* of *Aedes* is epistatic over M it appears to be downstream of M in the sex determination pathway. Moreover, occurrence of the temperature-sensitive period for *ix* in the larval stage suggests that it may act further down the regulatory cascade than an *Sxl*-like gene, which would act early in development.

Abnormal development of sexual phenotype occurred in backcrosses involving F_1 males obtained by crossing *Aedes aegypti* females with males of the closely related species *Ae mascarensis*. The abnormalities included aberrant male terminalia and the feminization of genetic males. The results of backcrosses involving markers in three linkage groups suggested that the differentiation of sexual phenotype in males is regulated by a group of epistatically interacting loci distributed on all three chromosomes (Hilburn and Rai, 1982).

In several northern holarctic species in the *Aedes* subgenus *Ochlerotatus*, genotypic males develop into intersexes or pseudofemales if the male larvae are reared at abnormally high temperatures (Brust, 1968b; Horsfall, 1974). This suggests that either they possess a wild-type form of *ix* that is temperature sensitive, or that the mutant *ix* is distributed throughout the wild populations. The postembryonic development of intersex individuals of these northern *Ochlerotatus* species is described in Section 8.2.3.

Two intersex mutants have been described from *Culex pipiens*. In one, the autosomal recessive gene *zwi* (*zwitterfaktor*) transformed genotypic males into phenotypic intersexes at normal temperatures (Laven, 1957a, 1967a). In the other, the sex-linked gene *c* (*cercus*) transformed genetic females into males at normal temperatures. It was suggested that the intersexes were derived from genotypic females because phenotypic males were about as frequent as the sum of phenotypic females and intersexes. Intersexes induced by *c* were the size of normal females, and had female antennae, wings and tarsal claws; the only masculinized parts were the palps and the external and internal reproductive organs. The *c* allele was usually recessive but sometimes was manifested by heterozygotes. Homozygous mutants were infertile; sometimes heterozygotes could lay autogenous fertile eggs, presumably when the expressivity of *c* was weak (Barr, 1975; Lee *et al.*,

1975). The *c* allele of *Cx pipiens* produced similar effects on the phenotype as the mutant genes *tra* and *ix* of *Drosophila* (Bownes, 1992).

These observations indicate that culicine mosquitoes possess a number of genes, scattered through the genome, which are involved in the regulation of sexual differentiation and which are possibly organized in a cascade.

1.4.4 Regulation of sex ratio in culicines

Assuming equal segregation of the paired sex chromosomes, a cross between a homozygous recessive individual and a heterozygote will produce a 1:1 sex ratio. Such a sex ratio has often been found in strains of *Culex pipiens*, *Cx quinquefasciatus* and *Aedes aegypti* (Gilchrist and Haldane, 1947; Qutubuddin, 1953: Wood, 1961), but many populations of *Ae aegypti* and some of *Cx pipiens* had a slight to high preponderance of males (Wood, 1961; Craig and Hickey, 1967; Sweeney and Barr, 1978). Populations of *Ae punctor* in southern Britain had a significant excess of males, whereas in northern Britain they had an excess of females (Snow, 1987; Packer and Corbet, 1989b). Arctic populations of *Ae impiger* and *Ae nigripes* had a preponderance of females (Danks and Corbet, 1973). Ford (1961) pointed out that sex is a case of balanced polymorphism and that selective forces will oppose departure from the optimum proportion of males and females which is generally, but by no means always, near equality. A number of genetic mechanisms that modify sex ratio have been described in mosquitoes.

When strains of *Aedes aegypti* from different parts of the world were crossed, and the F_1 progeny inbred, the F_2 generation often had an excess of males. Experimental evidence showed that this was due to a gene called *Distorter* (*D*), which was linked to the male-determining factor M (Hickey and Craig, 1966a, b). However, sex ratio distortion occurred only when *D*, on the Y chromosome, was associated with the X haplotype m^s, a variant of *m* that was susceptible to *D*. When haplotypes of different sensitivities were isolated they were named m^{s1}, m^{s2}, m^{s3} etc. Two haplotypes giving resistance to *D* (m^{r1} and m^{r2}) were also described. It is not known whether the various m^s and m^r haplotypes represent alleles of a single gene or variants of a sequence of genes that do not recombine (Wood, 1976b; Suguna *et al.*, 1977; Newton *et al.*, 1978b).

Experimental evidence revealed a further sex-linked locus affecting sensitivity to *MD*, i.e. the *tolerance to Distorter* gene, *t*, which was linked to *m* and *red-eye*. When present, *t* largely counteracted the sensitivity to *Distorter* induced by m^s (Wood and Ouda, 1987).

Distorter is present in wild populations in many parts of Africa, in Sri Lanka, Australia and many parts of the Americas. It appears to be absent from other countries of Asia and the Pacific region and from some parts of Africa and the Americas. When *Distorter* is present in wild populations, its effects are reduced or neutralized by other genetic factors that ensure that sex ratios are not greatly distorted. These include m^r, *t* and certain autosomal factors that have a minor role (Wood and Ouda, 1987; Wood and Newton, 1991).

Examination of spermatocytes in the testes of

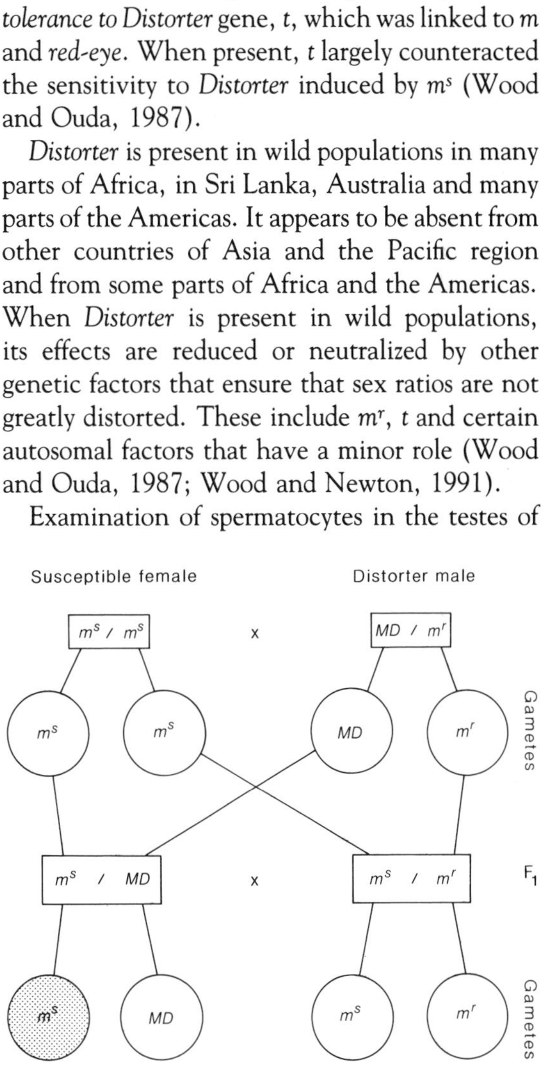

Figure 1.13 The effect of *Distorter* upon sex ratio in *Aedes aegypti*. A cross between a *Distorter* male and a susceptible female yields F_1 males of constitution m^s/*MD*. In such males a high percentage of X(m^s) chromosomes are damaged (stippled) during meiosis, causing a reduction in the number of X-bearing spermatozoa.

male progeny (MD/m^s) from $MD/m^- \times m^s/m^s$ crosses revealed fragmentation of one chromosome in the sex bivalent in a high percentage of cells undergoing meiosis. The damage was visible as early as diplotene, and showed as a preferential isochromatid breakage of the X chromosome. Breakage was restricted to four discrete sites, two in the proximal half of each arm. Restitution did not occur. Because breaks occurred more frequently in arms with chiasmata, an acentric fragment produced by breakage usually remained attached by a chiasma to the unaffected homologue. Breakage frequency was directly proportional to the sensitivity of the X chromosomes. In one cross only 8.8% of the surviving unaffected sex chromosomes were X (Newton *et al*, 1976).

Chromosome breakage leads to an absolute reduction in the number of spermatozoa produced by a *Distorter* male (Figure 1.13). Evidence that some of the acentric chromosome fragments entered spermatozoa came from scanning microdensitometric studies which revealed extra DNA in a percentage of the sperm of *Distorter* males. These sperm were probably non-functional because the observed sex ratio of sibs was more distorted than predicted from the number of chromosome breaks (Newton *et al.*, 1976; Newton *et al.*, 1978a).

Measurements of the effects of mating with *Distorter* males upon the fertility of susceptible females have produced rather variable results. When *Distorter* (MD/m^s) and normal (M/m) males were individually mated with twenty fresh, susceptible (m^s/m^s) females, at 24-h intervals, each type was capable of inseminating a mean of 8–9 females, stimulating the production of about 1500 eggs and yielding about 900 first instar larvae. However, females mated to *Distorter* males produced a greater numbers of inviable eggs (Youngson *et al.*, 1981). In another experiment egg production and egg viability were not significantly affected when susceptible females were mated with *Distorter* males, but the percentage of eggs yielding pupae was significantly reduced. That might have resulted from postzygotic mortality due to abnormal chromosomal complements in some spermatozoa. Since there was no significant correlation between pupal yield and sex ratio, postzygotic mortality could not have been the cause of the observed sex ratio distortion (Hastings and Wood, 1978; Wood and Newton, 1991).

A laboratory strain of *Culex pipiens* was found to produce a significant excess of males. Examination of pupal testes showed male meiosis to be abnormal, one of the chromosomes of the shortest bivalent being fractured in 40% of spermatocytes at anaphase I. The break occurred in both chromatids and very near the centromere. In *Cx pipiens* it is the shortest bivalent that bears the factors for sex. The proportion of broken chromosomes correlated well with the proportion of missing females, and it was concluded that the breakage had a lethal effect on female-determining germ cells (*m*) with the result that most of the surviving spermatozoa were male-determining (M). Although *m*-bearing chromosomes were affected, the character was was not expressed in the female, and the genotype of the female had no effect on the sex ratio of her immediate progeny (Sweeny and Barr, 1978).

Examination of salivary gland polytene chromosomes from *Cx pipiens* larvae of the strain showing sex ratio distortion revealed a high frequency of abnormality in zone 10C3 on chromosome-1, which is known to include the sex locus (Section 1.4.2). Sixty percent of the chromosomes were damaged or broken at that position. Unfortunately it proved impossible to sex the larvae or to distinguish between the X and Y chromosomes (Sweeney *et al.*, 1987). Sweeney and Barr (1978) proposed that sex ratio distortion in *Cx pipiens* was due to a recessive gene, *distorter*, which was expressed only when homozygous in males (M^d/m^d), but on present evidence it could equally well be ascribed to dominant (MD) and susceptible (m^s) alleles, as in *Ae aegypti*.

The phenomenon observed in *Ae aegypti* and *Cx pipiens* of a gene on a sex chromosome acting during meiosis to make that chromosome more likely to participate in fertilization is an example of **meiotic drive**. Thus males carrying the *Distorter* gene on their Y chromosomes produce an excess of sons because of X chromosome breakage during meiosis. Meiotic drive factors are attractive as agents for genetic control methods because, when coupled with a deleterious character, they can increase spontaneously in their frequency in a pest population while exerting a deleterious effect upon it. The theoretical feasibility of this was demonstrated by field-cage experiments with *Ae aegypti*. The introduction of *Distorter*, combined with a double translocation (DT_1T_3) that induced sterility, suppressed a population more rapidly than did the double translocation alone, even though at the start of the experiment the cage population carried a percentage of X chromosomes that were resistant to *D* (Curtis *et al.*, 1977; Suguna *et al.*, 1976). Sterility rates among recaptured marked females revealed that DT_1T_3 males released into the wild competed almost equally with wild males (Grover *et al.*, 1976).

Any differential mortality between males and females during development would result in an unequal sex ratio at emergence, and such a phenomenon has been described in a strain of *Aedes aegypti* from Bangkok. The strain carried two non-allelic recessive lethal genes, *l* and *k*, one close to and the other fully linked with the sex locus. In a cross between individuals heterozygous for *l* (*Ml/m+* × *ml/m+*) the lethal gene kills male larvae of constitution *Ml/ml*, giving a sex ratio of 1:2. In a cross between individuals heterozygous for *k* (M+/*mk* × *mk/m+*) the lethal gene kills female larvae of constitution *mk/mk*, giving a sex ratio of 2:1. There was evidence of heterosis preserving the frequency of the lethal genes, because the heterozygotes *ml/m+* and M+/*mk* contributed more progeny to the next generation than did the respective wild types. The two lethal genes appeared to exist in a fine balance because the overall sex ratio in the strain was close to parity (Wood, 1976a).

1.5 CYTOPLASMIC INCOMPATIBILITY

Cytoplasmic incompatibility is defined as a functional incompatibility between a maternally inherited cytoplasmic milieu and a male element. It is a widespread phenomenon among plants and animals, involving self-replicating extrachromosomal factors of a variety of types. One of the most investigated instances is that of *Culex pipiens* sensu lato, in which crosses between populations from different geographical locations may be sterile in one or both directions. In the 1950s it was shown that inheritance of the factors that governed this phenomenon were maternal and cytoplasmic. Later, the involvement of rickettsial symbionts was established.

1.5.1 Cytoplasmotypes

The sterility that occurs when incompatible strains of *Culex pipiens* cross is expressed as the production of non-hatching but mostly embryonated eggs by inseminated females. Most authors have recognized three categories of cross, although their definitions have varied. The following definitions are consistent with current usage.

1. *Compatible crosses*. Larvae hatch from >95% of rafts, with most eggs (60–100%) in these rafts hatching. Sex ratios are normal.
2. *Incompatible crosses*. Embryonic development starts in a substantial percentage of eggs (15–80%) in most rafts but larvae hatch from only 0 to <1% of eggs. Any adults that develop are female.
3. *Partly compatible (= partly incompatible) crosses*. Embryonic development starts in a substantial percentage of eggs in all rafts. In some instances all rafts have an intermediate hatch rate; in other instances some rafts have a high hatch rate but others have a zero or extremely low hatch rate. Adults that develop include males and females.

As a result of early crossing experiments involving many strains of *Cx pipiens*, a number of 'crossing types' was described, each of which showed a particular pattern of interstrain compatibility, partial compatibility and incompatibility. Few studies were made of partly compatible crosses, but incompatible crosses received detailed investigation.

Incompatible crosses may be infertile in both

directions or in one direction only. For example, the cross between populations from Oggelshausen and Hamburg was compatible in the direction Ha♀ × Og♂ but incompatible in the direction Og♂ × Ha♀. It was characteristic of incompatible crosses that insemination was normal, that the eggs were penetrated by sperm, and that embryonic development proceeded as far as segmentation and organ formation in a high percentage of the eggs, but that very few eggs hatched. In the incompatible Og♀ × Ha♂ cross no embryonic development occurred in 29.0% of eggs, embryos developed but died before hatching in 70.83%, and larvae hatched from 0.17%. The few mosquitoes obtained from incompatible crosses were always female and could be shown by marker genes to have derived all their chromosomes from the mother, suggesting that they had developed parthenogenetically (Laven, 1957b, 1967a).

When egg rafts are a few days old, any that have not been inseminated can be recognized by the degeneration of the yolk and cytoplasm. Egg rafts from incompatible crosses have a mottled appearance, some eggs being pale and others dark. The pale eggs undergo no development. Most of the dark eggs contain incompletely developed embryos showing red eye spots, and <1% of eggs hatch (Barr, 1982; Magnin and Pasteur, 1987b).

One of the first insights obtained into the inheritance of crossing type was of the importance of maternal lineage. This was illustrated in a cross between the strains from Hamburg and Oggelshausen. Female offspring from an initial Ha♀ × Og♂ cross were backcrossed to Og males. After continuous backcrossing of female progeny to Og males for over 50 generations it was shown with genetic markers that the chromosomes from the original Hamburg female parent had been replaced in the descendants with the chromosomes of the male parent strain from Oggelshausen. Even so, both male and female descendants possessed the crossing type of the original maternal strain. Laven (1957b) concluded that inheritance of crossing type was controlled by a factor in the cytoplasm. The phenomenon was one of cytoplasmic incompatibility, and its mode of transmission was called 'cytoplasmic inheritance'. In what follows we shall use the term **cytoplasmotype** (i.e. cytoplasmic crossing type) when referring to cytoplasms that determine the viability of crosses. It was discovered later that cytoplasmotypes and compatibilities were determined by a rickettsial symbiont present in the cytoplasm of female gametes.

1.5.2 Possible restorer genes

From observations on the dynamics of cytoplasmic incompatibility in wild populations of *Culex pipiens*, notably the polymorphism of cytoplasmotype within wild populations (Section 1.5.6), Rousset *et al.* (1991) postulated that maternal cytoplasm may not be the sole factor governing viability of the offspring from crosses between populations, and suggested that paternal effects in the form of nuclear restorer genes might be involved. They considered that restorer genes had not been described by earlier investigators because they had been lost from the laboratory strains, many of which had been kept for years. Consistent with this suggestion is the observation that males of a laboratory colony derived from Calicut showed full incompatibility with Paris females, whereas crosses between wild Calicut males and Paris females yielded 25% of hatching rafts (Subbarao *et al.*, 1977a).

Rousset *et al.* (1991) pointed out that if restorer genes operate only in sperm from males of appropriate cytoplasmotype, paternal nuclear effects would not have been revealed by backcrosses such as that described earlier, in which female offspring from an initial Ha♀ × Og♂ cross were backcrossed to Og males. The final offspring had Og chromosomes in Ha cytoplasm, a combination in which Og restorer genes carried by sperm would, it was postulated, be silent.

The early literature contains observations that are inconsistent with purely cytoplasmic inheritance, and which are amenable to an explanation involving restorer genes within the male nuclear genome. For example, in crosses between strains from Melbourne (Me) and Point Lonsdale (Lo), Me♀ × Lo♂ was fertile with a 98% egg hatch, whereas Lo♀ × Me♂ was sterile with zero hatch although 83% of the eggs contained embryos. Contrary to strictly cytoplasmic inheritance, the cross Lo♀ × (Me♀ × Lo♂)♂ was partly fertile with 38% hatch (Dobrotworsky, 1955). This suggests that Lo was carrying restorer genes which permitted partial compatibility between females of Lo cytoplasmotype and males carrying Me cytoplasm.

Variation of cytoplasmotype was observed among the male progeny of a small percentage of individual females from a well-characterized laboratory strain, although their female progeny all retained the maternal cytoplasmotype. Subbarao *et al.* (1977a, b) ascribed the variation among the male progeny to an unequal assortment (segregation) of the cytoplasmic determinants of incompatibility. However, polymorphism of recessive restorer genes effective in the males would be a better explanation of the phenomenon.

The existence of restorer genes in *Cx pipiens* remains to be proved or disproved. However, such genes are known to occur in other organisms, e.g. maize, and circumstantial evidence suggests that they may function in mosquitoes.

1.5.3 Rickettsiae in mosquitoes

The bacterium *Wolbachia pipientis* Hertig is the type species of a small heterogeneous genus placed in the tribe Wolbachieae and family Rickettsiaceae of the order Rickettsiales. *W. pipientis* is an obligate intracellular symbiont which appears to be universally present in wild populations of certain taxa of the *Culex pipiens* complex. A structurally similar rickettsia is present in many members of the *Aedes scutellaris* group (Figure 1.14). *Wolbachia* is found in abundance in both male and female germ-line cells, a mixture of rods and cocci (rounded cells) being present in each infected cell. The coccoid forms are up to 1.1 μm or more in diameter; the rods are variable in form, about 0.3–0.7 μm in diameter and 1.1–2.2 μm in length. They are Gram-negative and stain red with Giemsa's stain. Multiplication is by binary fission within the vacuoles of the host cells. *W. pipientis* has not been cultivated outside its natural host or in cell cultures (Hertig, 1936; Yen, 1975; Wright and Barr, 1980; Krieg and Holt, 1984).

The ultrastructure of *Wolbachia* is typical of Gram-negative bacteria. Each symbiont is surrounded by two membranes, a plasma membrane and an outer so-called cell wall. The cytoplasm contains many ribosomes and a fine reticulum of what may be strands of DNA. In the *Cx pipiens* complex each symbiont is surrounded by a membrane of host origin; in the *Ae scutellaris* group such host membranes were described as present and absent by different investigators (Yen, 1975; Wright *et al.*, 1978; Wright and Barr, 1980; Beckett *et al.*, 1978).

Wolbachia pipientis have only very rarely been found in other than germ-line cells, within which they are restricted to the cytoplasm. In the ovary they are found in both the oocytes and nurse cells. In the testis they are seen in spermatocytes. When the spermatocytes differentiate into spermatids and shed much of their cytoplasm the microorganisms are also shed, and the spermatozoa appear to contain no *Wolbachia*. *Wolbachia pipientis* have been found in all stages of the mosquito life cycle from newly-laid egg to the adult male and female. In recently-laid eggs of *Cx pipiens* they are concentrated near the micropyle. In developing embryos they become restricted to the pole cells, and in larvae that have not yet hatched from the egg they are present in moderate numbers in the germ cells. They multiply rapidly in 3rd and 4th instar larvae, and some host cells are seen to be pathological at this time. The pupae and adults are very heavily infected, particularly the females, whose oocytes and nurse cells are always infected. In pupal testes some spermatocytes are infected and others not. *W. pipientis* is transmitted transovarially from one mosquito generation to the next (Yen and Barr, 1974; Yen, 1975; Wright and Barr, 1981). Trpis *et al.* (1981) claimed that aposymbiotic adults of *Ae polynesiensis* could be obtained by rearing larvae at the near-lethal temperature of 32.5°C.

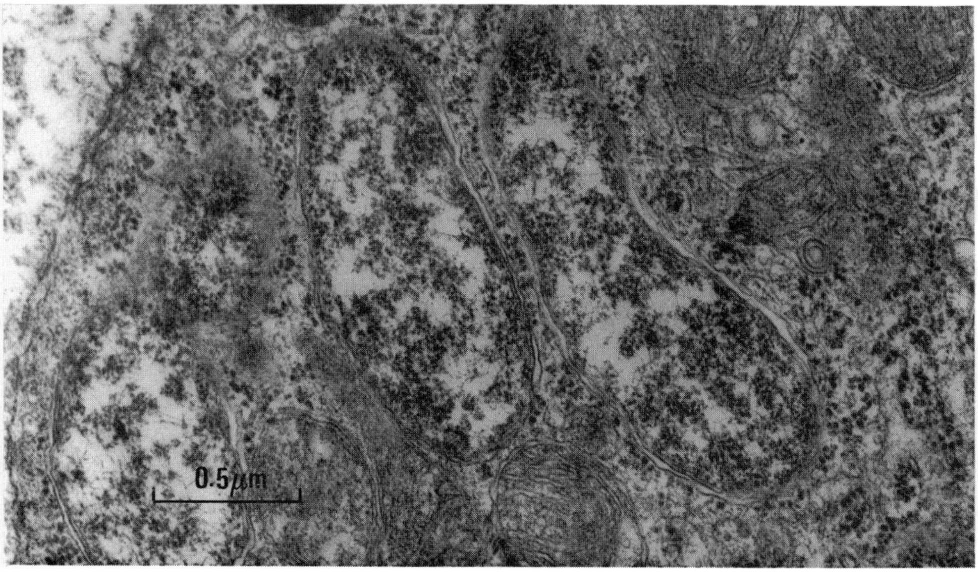

Figure 1.14 *Wolbachia* sp. infecting an oocyte of *Aedes polynesiensis*. (Micrograph by courtesy of Dr John D. Wright.)

1.5.4 Effects of rickettsiae on fertilization and embryonic development

Wolbachia-free eggs could be produced by exposing larvae of *Culex pipiens* to the antibiotic tetracycline. Eggs laid by aposymbiotic females had a reduced hatch rate but the surviving progeny were both aposymbiotic and healthy, and subsequent generations could be reared without exposure to tetracycline and remained symbiont-free. From many crossing experiments it was found that, with a single exception which is discussed later, aposymbiotic *Cx pipiens* could produce fertile progeny irrespective of parental cytoplasmotypes. Therefore, incompatibility was caused by *Wolbachia pipientis*, and cytoplasmotype was primarily a property of *W. pipientis* populations and was only secondarily a characteristic of mosquito populations (Yen and Barr, 1973, 1974; Portaro and Barr, 1975).

(a) Crosses involving aposymbiotic parents

Crosses between *Culex pipiens* and the naturally aposymbiotic species *Cx globocoxitus* (Section 1.5.5(a)) resulted in 98.6% hatch in the direction *pipiens* ♀ × *globocoxitus* ♂, but only a 1.3% hatch in the direction *globocoxitus* ♀ × *pipiens* ♂, when 95.8% of the unhatched eggs showed no embryonic development (Dobrotworsky, 1955).

Experiments with mosquitoes rendered aposymbiotic by tetracycline treatment have thrown light on the situation. Aposymbiotic males of *Cx pipiens* were compatible with all females, whatever their cytoplasmotype and whether or not the females carried symbionts (Table 1.3(c,d)). Aposymbiotic females were compatible with all aposymbiotic males, whatever their cytoplasmotype (c). However, aposymbiotic females crossed with males harbouring symbionts produced no progeny, irrespective of the compatibility relationship, and the eggs did not undergo the incomplete embryonic development that is characteristic of normal incompatible crosses (Table 1.3(e)) (Yen and Barr, 1973, 1974). Broadly similar results were obtained by Suenega (1982) and Magnin and Pasteur (1987a).

Cytoplasmic incompatibility is a property of a maternally inherited cytoplasmic (rickettsial) genome, which is able to tranform male gametes in such a way that after syngamy they test the ooplasmic genome. If they find it different the egg becomes infertile, presumably because

Table 1.3 Consequences of crossing between populations of *Culex pipiens* of compatible or incompatible cytoplasmotypes, normally and when aposymbiotic.

	Cytoplasmo-type	*Compatibility of infected cytoplasmotypes*	*Cytoplasmo-type*	*Outcome of cross*
(a)	A♀ Sym	Compatible	B♂ Sym	Fertile
(b)	B♀ Sym	Incompatible	A♂ Sym	Sterile; parthenogenetic embryos
(c)	♀ Apo	(Compat/Incompat)	♂ Apo	Fertile
(d)	♀ Sym	(Compat/Incompat)	♂ Apo	Fertile
(e)	♀ Apo	(Compat/Incompat)	♂ Sym	Sterile; no karyogamy

A and B are populations with unidirectionally incompatible cytotypes. Apo, aposymbiotic; Sym, containing symbionts; Compat/Incompat, the two cytotypes may be compatible or incompatible.

karyogamy, the fusion of the male and female pronuclei, is blocked. That it is the male gametes that test is apparent from the finding that maternal cytoplasm without rickettsiae is seen as different by sperm derived from an infected father (Table 1.3(e)).

As described earlier, the spermatozoa of *Wolbachia*-bearing males contain no visible remnant of the symbiont, but their effect on aposymbiotic eggs indicates that they are modified in some way, presumably by carrying an undetected constituent. In newly-laid eggs the *Wolbachia* are concentrated near the micropyle through which spermatozoa had entered.

(b) *Incompatible crosses*

Crosses of incompatible strains of *Culex pipiens* led to production of a substantial number of inviable embryos and a very few viable adults. Investigation of the development of inviable embryos revealed that female meiosis had been completed, and that sperm had penetrated and migrated to the centre of the egg, but that the male and female pronuclei had not fused. Despite this, the haploid female pronucleus had divided, cleavage energids had formed, and in some 75% of eggs a blastoderm had formed and development had proceeded to histological differentiation or beyond. Use of eye-colour mutations, which are visible in developing embryos, confirmed that the sperm contributed no chromosomes to the nuclei of inviable embryos (Jost, 1970a, b).

The role of sperm in activation of early developmental events was investigated by irradiating the male parents. In the compatible Hamburg × Hamburg cross, irradiation of males with 5–8 kR led to death of the eggs during the early cleavage stages, yet in the incompatible Hamburg ♀ × Paris ♂ cross, prior irradiation of the males with similar dosages did not affect the number of embryos developing to the blastoderm stage and beyond. The blastoderm nuclei of inviable embryos contained only the haploid number of chromosomes, and predominantly contained a quantity of DNA equal to that in one haploid spermatid nucleus (= C). It appeared that in incompatible crosses not only did penetration of the egg by sperm stimulate the completion of female meiosis, but some factor also stimulated development to the blastoderm stage and beyond in the absence of karyogamy (Jost 1970a, 1971, 1972). Since the non-viable embryos that developed from an incompatible cross were haploid, their failure to develop normally may have been due to recessive lethal factors which are expressed only in the hemizygous condition.

When an incompatible cross was made between wild type males of *Cx pipiens* and females heterozygous for the dominant mutation *Kuf*, the few viable adults that had developed parthenogenetically included all possible genotypes: +/+, *Kuf*/+ and *Kuf*/*Kuf*. This recombination of genes suggested to Laven (1957b) that some at least of the progeny had resulted from fusion of the products of female meiosis, i.e. fusion of the female pronucleus with a polar body or fusion of two polar bodies – a phenomenon known as automictic or meiotic parthenogenesis. Analysis of another incompatible cross involving six marker genes revealed that the four products of meiosis were not equally used in the formation of parthenogenetic females. Possibly, therefore, the few viable progeny had developed from a diploid nucleus that originated by fusion of the female pronucleus with its sister polar body (Jost, 1970b). Much remains unknown about the causal basis of incompatability.

(c) Partly compatible crosses

Little is known of the basis of partly compatible crosses. Unlike incompatible crosses, in which the few parthenogenetic individuals produced are female, the viable adults that result from partly compatible crosses include both males and females, and in some crosses at least the sex ratio is normal (Barr, 1970; Singh *et al.*, 1976).

The existence in male nuclear genomes of restorer genes, that counter the effects of cytoplasmic factors, could explain partial incompatibility. Another explanation may lie in loss of *Wolbachia* with male sexual activity. Thus, a cross between two *Cx quinquefasciatus* strains was fully incompatible when the females were mated with 1–2-day-old virgin males but was partly incompatible when sexually experienced males 6 days old or more were used. Matings with non-virgin males caused a greater reduction of incompatibility than matings with virgin males of the same age, suggesting that loss of *Wolbachia* from the testes might have contributed to the effect (Singh *et al.*, 1976).

1.5.5 Occurrence of cytoplasmotypes

Among mosquitoes, only certain species of *Culex* and *Aedes* are known to harbour *Wolbachia*, but few genera and species have been examined for the presence of the symbionts. Within *Aedes*, only species of the *scutellaris* group are known to harbour *Wolbachia* sp. *Wolbachia pipientis* is found in most members of the *Culex pipiens* complex. Rickettsial symbionts indistinguishable in ultrastructure from *W. pipientis* have been found in the ovaries of *Cx torrentium* (Larsson, 1983). *Cx torrentium* and *Cx pipiens* are anatomically very similar and they occur sympatrically. However, the male genitalia are distinct, and *torrentium* has been placed in the *trifilatus* species group of the subgenus *Culex*, whereas *pipiens* has been placed in the *quinquefasciatus* species group (Sirivanakarn, 1976).

(a) Culex pipiens *complex*

The taxonomic status of the members of the *Culex pipiens* complex is still a matter of controversy, i.e. whether individual taxa are distinct species or are subspecies or infraspecies of *pipiens* L. Fortunately, taxonomic certainty is not essential for an account of the biology of cytoplasmotypes. In this section we shall use the name *pipiens* for the taxon sometimes called *pipiens pipiens* or *pipiens* s.str., embracing both its autogenous and anautogenous forms, and the names *quinquefasciatus* Say, *globocoxitus* Dobrotworsky, and *australicus* Dobrotworsky and Drummond without commitment to taxonomic status.

From the limited microscopic evidence available, supplemented with evidence of incompatibilities, it seems probable that *Wolbachia pipientis* is present in wild populations of all but two taxa of the *pipiens* complex, and occurs throughout their ranges. The exceptions are the Australian taxa *australicus* and *globocoxitus*. The oocytes and nurse cells of these taxa lacked rickettsial symbionts but contained clumps of virus-like particles (Irving-Bell, 1974, 1977). A very few laboratory colonies of *pipiens* have been found to be aposymbiotic (Wright and Wang, 1980; Irving-Bell, 1977).

From an analysis of 350 crosses between strains of *Culex pipiens* sensu lato from Europe, North America, Asia and Africa, Laven (1967b) classified the 37 strains into 17 crossing types (cytoplasmotypes), based on patterns of compatibility and incompatibility but apparently ignoring the cases of partial compatibility. The cytoplasmotypes showed greater compatibility between populations or strains from one country or continent, and greater incompatibility between those that were geographically more remote.

Some investigators found little or no incompatibility in crosses between recently colonized parents from locations in the same geographical area. For example, strains of *quinquefasciatus* from 19 locations in southeast Asia were fully compatible in both directions (Thomas, 1971). Other investigators have reported incompatibility between populations separated by relatively short distances. When eight strains of *pipiens* collected within 100 km of Montpellier in southern France were crossed, none displayed exactly the same crossing relationships with the seven others, indicating that each was of a unique cytoplasmotype. Strains collected in sites separated by 2 km or less were bi- or unidirectionally compatible, whereas strains separated by 2–200 km were uni- or bidirectionally incompatible. Magnin *et al* (1987) tentatively concluded that the degree of incompatibility between two populations increased with distance over relatively short distances.

The existence of crossing-type polymorphism among sympatric populations was shown by the occurrence at a breeding site in Malibu, southern California, of egg rafts of *Cx quinquefasciatus* that had the appearance of rafts produced by incompatible crosses. Crossing experiments with the offspring from rafts collected at that and three other breeding sites in southern California revealed the occurrence of populations of two cytoplasmotypes, 'B' and 'C', at each of them (Barr, 1980). Of 1292 egg rafts collected at Malibu two years later, 5.6% were from incompatible crosses (Barr, 1982).

That heterogeneity of cytoplasmotype can exist among the descendants of individual females has been demonstrated in laboratory stocks. In one instance a strain was developed from a unidirectionally compatible cross between *quinquefasciatus* females reared from a single raft and males of a standard California *pipiens* strain. In the F_4 generation 10 of 60 rafts were 'incompatible', demonstrating the difference of cytoplasmotype between individuals that were derived ultimately from a single female (Barr, 1980). Other instances have been reported by Subbarao *et al.* (1977a, b) and French (1978).

(*b*) Aedes scutellaris *group*

Aedes scutellaris lends its name to a group of over 30 related species which are widely distributed on islands throughout the south Pacific ocean and in southeast Asia. *Wolbachia* were present in the gonads of most of the species examined, and absent from only two. Species in which *Wolbachia* were demonstrated by light and electron microscopy were *albopictus* (Skuse), *cooki* Belkin, *hebrideus* Edwards, *kesseli* Huang and Hitchcock, *malayensis* Colless, *polynesiensis* Marks, *pseudoscutellaris* (Theobald), *riversi* Bohart and Ingram, *scutellaris* (Walker), *tongae tabu* Ramalingam and Belkin, and *upolensis* Marks. Ultrastructural and experimental observations suggested that the symbiont had been lost from certain laboratory stocks and, in some instances, subsequently regained. *Wolbachia* could not be found in *alcasidi* Huang or *katherinensis* Woodhill (Beckett *et al.*, 1978; Wright and Barr, 1980, 1981; Wright and Wang, 1980; Trpis *et al.*, 1981; Meek, 1984; Meek and Macdonald, 1984).

Cross-mating between species of the *Ae scutellaris* group resulted in compatible, partly compatible, and uni- or bidirectionally incompatible crosses. The different species fell into a number of cytoplasmotypes. Crossing type was maternally inherited, and the experimental results were generally consistent with the hypothesis that cytoplasmotype was determined by the *Wolbachia* present. Some contradictory results could be explained by the apparent instability of the symbionts in laboratory colonies (Wade and Macdonald, 1977; Meek and Macdonald, 1984). It is important to note that all of these crosses were between species. No observations have been reported of partly compatible or incompatible crosses between populations of individual species of the *Ae scutellaris* group.

1.5.6 Dynamics of cytoplasmic inheritance

Since different cytoplasmotypes can be found within at least some populations, cytoplasmic incompatibility is an intrapopulational phenomenon; therefore, selection acts upon it directly and some genetic factor must have an advantage. As Rousset and Raymond (1991) pointed out, the real question in this connection is not 'does the maternally-inherited factor give an advantage to the individual bearing it?', but rather 'will this factor increase in frequency?'.

There is asymmetry in the transmission of cytoplasmotype, i.e. transmission occurs only through female cytoplasm. However, the fact that the sperm carry some element of their mother's cytoplasmotype, while themselves being dead ends for the maternally transmitted genes, allows this asymmetry to be used by the cytoplasmic genes to sterilize other cytoplasms and thereby increase their own frequency.

The coexistence of two cytoplasmotypes (A and B) with undirectional incompatibility has been modelled. When the cross B♀ × A♂ is sterile and the cross A♀ × B♂ is fertile, the B cytoplasmotype will normally be eliminated, regardless of its original frequency, because B females are sterilized by A males, whereas A females are fertile with all males in the population (Table 1.3). However, if the B cytoplasmotype has an advantage over A independent of the compatibility relationship, an unstable polymorphic equilibrium will exist, and one of the factors that will determine which cytoplasmotype remains is its initial frequency in the population (Caspari and Watson, 1959; Fine, 1978; Rousset and Raymond, 1991).

Rousset *et al.* (1991) investigated whether or not polymorphism equilibrium could exist where two or more cytoplasmotypes with a variety of compatibility relationships were present in an infinite, randomly interbreeding population. They found that a stable polymorphism normally could not exist between incompatible cytoplasmotypes. However, differences of fertility and viability between cytoplasmotypes could permit a stable polymorphism in certain circumstances. Considering two cytoplasmotypes c_i and c_j, when the relative numbers of offspring (a measure of compatibility) obtained in the crosses ♀c_i × ♂c_j and ♀c_j × ♂c_i are described by the parameters ϕ_{ij} and ϕ_{ji} respectively (parameters which could reflect fertility and viability differences), then a necessary condition for a stable equilibrium would be, for all i, j, that

$$\phi_{ij} + \phi_{ji} > \phi_{ii} + \phi_{jj} \qquad (1.1)$$

and hence a sufficient condition for instability would be, for some i, j, that

$$\phi_{ij} + \phi_{ji} < \phi_{ii} + \phi_{jj} \qquad (1.2).$$

Theoretical and practical studies have been undertaken into the use of cytoplasmic incompatibility as a genetic control mechanism, either as an alternative to sterile male release for population eradication, or for the replacement of susceptible populations of a disease vector with populations refractory to infection with the parasite (Laven, 1967c; Curtis, 1977; Curtis *et al.*, 1982). The polymorphism of cytoplasmotypes that has been found in many natural populations may render such ideas impractical.

2

Embryology

2.1 LARVAL EMBRYOGENESIS

Larval embryogenesis has been described from *Anopheles maculipennis* (Ivanova-Kazas, 1949), *Culex pipiens* (Christophers, 1960; Idris, 1960a; Guichard, 1971, 1973), *Cx quinquefasciatus* (Davis, 1967), *Culiseta inornata* (Harber and Mutchmor, 1970), *Aedes vexans* (Horsfall *et al.*, 1973), and *Ae aegypti* (Raminani and Cupp, 1975, 1978). These accounts are valuable for the insights they provide into the embryonic origins of the various larval organs, and also for their contribution to comparative insect embryology. Larval embryogenesis appears to be generally similar in the four mosquito genera in which it has been investigated, so for convenience the account that follows relates particularly to *Culex*; the other genera are mentioned only where their development differs or to provide additional information. Times cited are for hours elapsed after oviposition, and relate to development at 25°C.

2.1.1 The egg, maturation and fertilization

Insect oocytes are giant cells, the product of maternal ovarian development genes. Very soon after an oocyte is laid, its chromosomes complete meiosis and at that stage the oocyte is considered to have transformed into an egg. The fertilized egg is still a totipotent cell. The processes of oogenesis that culminate in the formation of mature oocytes are described in Chapters 19 and 20.

Mosquito eggs are nearly always elongate along the anterior–posterior axis. There is some variety of shape among the eggs of different genera but the internal organization of the egg appears fairly constant. At oviposition the full-grown oocyte is surrounded by the chorion which determines its shape. Internally a limited antero-posterior polarity is apparent (Figure 2.1A), and both the anterior–posterior and dorsal–ventral orientations correspond with those of the mother. The ooplasm is packed with yolk, in the form of large lipoprotein granules and smaller lipid droplets, but even so the non-yolk cytoplasm is still relatively abundant and contains many ribosomes, which indicates that the oocyte is of the 'long germ type'. Beneath the plasma membrane a layer of yolk-free cytoplasm, called the periplasm, is evident. This is 3–4 μm thick except at the poles where it is considerably thicker. The posterior periplasm contains many fine granules, of about 1 μm diameter, called polar granules or oosome. Fragments of the degenerated vegetative nucleus are still visible in the cytoplasm at the time of oviposition. The germinal nucleus or karyosphere lies in a small island of yolk-free cytoplasm, which is situated towards the anterior pole in *Anopheles maculipennis* and *Aedes aegypti* and equatorially in *Culex pipiens* and *Ae vexans*.

Female meiosis starts during oogenesis but the completion of this process is delayed until after oviposition, and at the time of oviposition the chromosomes are held in the metaphase of the 1st meiotic division (Section 1.3.2). Because the nuclei of unfertilized *Culex* eggs never progress beyond metaphase I, Jost (1971) concluded that the resumption of meiosis is stimulated by the presence of sperm.

Sperm are thought to pass through the micropyle, at the anterior end of the chorion, as the oocyte traverses the genital chamber at oviposition. Some investigators reported that polyspermy is usual in *Culex* and that from 3 to 10 sperm normally penetrate each oocyte (Idris, 1960a; Davis, 1967). However, the presence of sperm tails in the ooplasm can lead to overestimation of sperm numbers, and Jost (1971), who used Feulgen stain to distinguish the DNA of sperm heads, found that 90% of *Cx pipiens* eggs contained no more than one or two sperm. *Culiseta inornata* eggs usually contain three sperm heads (Harber and Mutchmore, 1970). Fertilization starts with the fusion of the male and female gametes (syngamy), and is completed when a male pronucleus fuses with the female pronucleus (karyogamy).

After oviposition the membranes surrounding the karyosphere disperse and meiosis is resumed (Fiil, 1974). Our knowledge of the later stages of meiosis is most complete for *Culex pipiens*, based on the work of Jost (1970a, 1971, 1972) who used quantitative staining to measure the amounts of DNA in the meiotic products and sperm heads. By convention the DNA content of a single spermatid nucleus or sperm head, in which each chromosome is believed to consist of a single chromatid, is taken as one relative unit (C). The nucleus of an unfertilized *Culex* oocyte (which is still in the course of meiosis) has a DNA content of 4C following replication of the diploid DNA content at an earlier stage.

Immediately after oviposition, in *Cx pipiens*, the oocyte nucleus is situated towards one side of the cell and at half its length, and the chromosomes are in the metaphase of the 1st meiotic division. Five to ten minutes after oviposition the chromosomes are either still in metaphase I with a DNA content of 4C, or in anaphase I having separated into two groups of 2C each. The 1st meiotic division is completed shortly thereafter with the production of two daughter nuclei. In fertilized oocytes, sperm with the DNA content C are found in the anterior half of the cell. By telophase I one sperm has migrated to the centre, level with the two nuclei that are the products of the 1st meiotic division, and a conspicuous sperm aster is formed around it. The two female nuclei enter upon the 2nd meiotic division immediately after completing the 1st, about 20 min after oviposition. One nucleus divides to yield two polar bodies. The spindle of the other nucleus is so orientated that the inner product of division is displaced into the centre of the cell; a nuclear membrane forms around this, the female pronucleus. These divisions yield three polar bodies and a female pronucleus, each with a DNA content of C. Karyogamy, the fusion of male and female pronuclei, occurs some 10–25 min after the completion of female meiosis.

In some mosquito species karyogamy occurs close to where the female pronucleus is formed, as in *Cx quinquefasciatus* where both events occur in the equatorial region. In some other species the female pronucleus first migrates; in *Ae aegypti* it is said to move anteriorly from an equatorial location before karyogamy (Raminani and Cupp, 1975), in *Ae vexans* it is said to move equatorially from an anterior location (Horsfall *et al.*, 1973). The male and female pronuclei of insects characteristically undergo DNA replication producing a 2C content in each, so that after karyogamy the zygote nucleus contains 4C (White, 1973). Presumably this occurs in mosquitoes also. Interphase nuclei in young embryos of *Cx pipiens* had a DNA content of 2C or 4C.

2.1.2 Cleavage, pole cell and blastoderm formation

Nuclear divisions occur rapidly and synchronously after karyogamy. In *Culex pipiens* a mitotic cycle is completed within 20 min at 18°C and within 15 min at 25°C. Five divisions convert the zygote nucleus to a group of 32 energids in the centre of the egg, each energid consisting of a nucleus embedded in a small volume of cytoplasm. The following two divisions yield a sphere of 64 and then 128 energids, which migrate with their cytoplasm towards the egg surface. At the 128-nucleus stage a small number of energids penetrate the posterior periplasm, binding some of the polar granules,

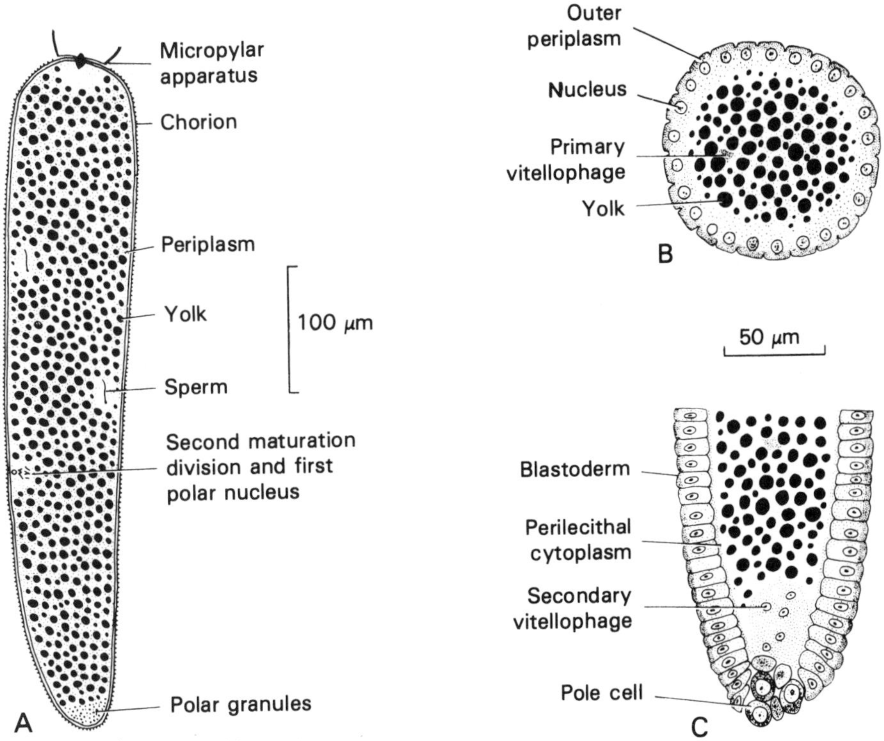

Figure 2.1 Maturation, cleavage and blastoderm formation in *Culex quinquefasciatus*. (A) Sagittal section through an oocyte undergoing the 2nd maturation division, 15 min after oviposition. (B) Transverse section through the middle region of an embryo after the 11th cleavage division, at the syncytial blastoderm stage. (C) Longitudinal section through the posterior end of an embryo at the cellular blastoderm stage. (From Davis, 1967.)

and continue their migration to the posterior pole. After the next division, the 8th, the periplasm withdraws from the posterior pole leaving a liquid-filled space into which pass those energids that have penetrated the polar plasm; by 2.5 h after oviposition they are visibly differentiated as spherical pole cells. In *Cx pipiens* 12–16 pole cells are formed (Figure 2.1C), in *Ae aegypti* 14–16. In *An maculipennis* normally four energids penetrate the posterior periplasm, and by division form a group of 20–30 pole cells. Once the energids have transformed into pole cells they do not divide again until the larval stage.

In theory the 8th cleavage division should yield 256 nuclei, but in *Cx pipiens* some 16–30 nuclei do not divide and remain in the yolk, becoming primary vitellophages. The outwardly migrating energids reach the periplasm of the anterior and posterior poles before that of the intermediate region. The nuclei then undergo five cleavage divisions at approximately 30 min intervals to form a **syncytial blastoderm** of about 3200 nuclei some 4.5 h after oviposition. During formation of the syncytial blastoderm the periplasm thickens at the expense of the internal cytoplasm, but the original periplasm remains more darkly staining than the newly added inner periplasm (Figure 2.1B). Infoldings of plasma membrane cut in between the blastoderm nuclei and by 5–5.5 h enclose them, forming a uniform layer of columnar blastoderm cells which surrounds the yolk. The blastoderm is continuous except at the posterior pole where the pole cells are situated. The egg is now at the **cellular blastoderm** stage (Figure 2.1C). A thin layer of perilecithal cytoplasm separates the blastoderm cells from the yolk. At the posterior pole about six nuclei lying in the posterior periplasm pass into the yolk as vitellophages, and several nuclei derived from the blastoderm at the anterior pole have a similar fate. They are called secondary vitellophages, although

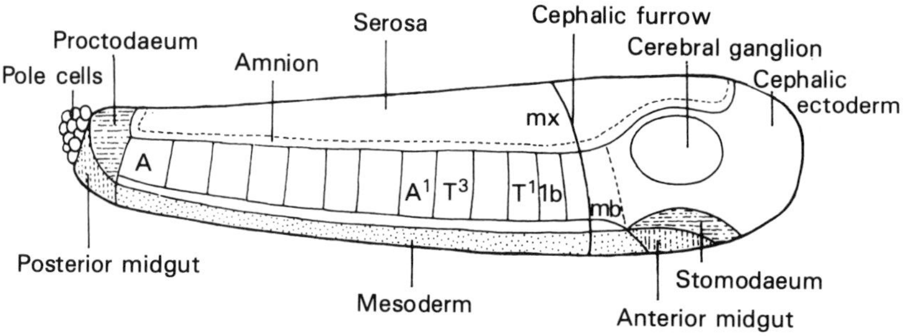

Figure 2.2 Fate map of a *Culex quinquefasciatus* embryo at the blastoderm stage, in side view, constructed on the basis of subsequent cell movement and fate. During gastrulation the cells along the ventral midline invaginate to form mesoderm, while the cells fated to form the alimentary canal invaginate near either end of the embryo. (From Anderson, 1972; after Davis, 1967.)

histologically indistinguishable from the primary vitellophages.

2.1.3 Morphogenetic movements

During embryogenesis movements of cells take place which are an integral part of development and differentiation. In insect embryos these large-scale morphogenetic movements include gastrulation, extension and retraction of the germ band, dorsal closure, and condensation of head segments. During gastrulation complex movements carry cells whose descendants will form the future internal organs from their superficial positions in the blastoderm to their definitive positions within the embryo.

In mosquitoes, as in other advanced insects, all blastoderm cells are at first columnar. The germ anlage, which comprises the presumptive embryonic areas, extends over a large part of the blastoderm. Only the dorsal region is extra-embryonic blastoderm (Figures 2.2, 2.3A, B), and not until a later stage does it become attenuated (Figure 2.3C–F)

In *Culex*, following the formation of a uniform columnar blastoderm, development proceeds directly to gastrulation, which occurs between 5.5 and 8.5 h after oviposition. The first evidence is a mid-ventral crowding of presumptive mesoderm cells, which inaugurates the **differentiated blastoderm** stage. Next a longitudinal mid-ventral groove forms as cells move to the interior (Figure 2.3B). After the groove has closed, through the ventral movement of presumptive ectoderm, the migrating cells spread out as a single layer of mesodermal cells between the embryonic ectoderm and the yolk (Figure 2.3C–E).

Aedes aegypti differs from *Culex* in that the germ anlage is formed before cellularization of the blastoderm is completed. At the time when the blastodermal cells have only peripheral and lateral boundaries, the cells that will form the future embryo can be distinguished from those of the extra-embryonic ectoderm. No gastrular groove forms; the presumptive mesodermal cells simply elongate and migrate inwards as an irregular scattered group (Raminani and Cupp, 1978).

During the early stage of gastrulation in *Culex* a number of oblique furrows appear ventrolaterally in the presumptive embryonic ectoderm (Figure 2.4). The most distinct, the cephalic furrow, lies at about 70% of the egg length from the posterior pole. By damaging the areas before and behind the cephalic furrow with UV radiation, Oelhafen (1961) proved that the cephalic furrow separates the prospective mandibular and maxillary segments.

During the invagination of mesoderm the germ anlage – or, as it is now called, the germ band – becomes greatly elongated in the first of a series of morphogenetic movements termed blastokinesis. The extending posterior region of the germ band reflects forwards over the posterior pole and subsequently extends dorsally over the yolk towards the cephalic furrow (Figure 2.4A–D). Displacement of the ectodermal component of the germ band takes place mainly by cell rearrangement with little division, whereas that of the mesoderm takes place mainly by cell division. A posterior complex of cells, consisting of the posterior midgut rudiment, proctodaeum and pole cells, is pushed ahead of the elongating germ band, and as it moves forwards it sinks into the embryo (Figures 2.3F, 2.5). The posterior end of the germ band reaches its most anterior position, about 75% of the egg length from the posterior pole, 7 h after oviposition. The various furrows disappear, from posterior to anterior successively, as the germ band extends. The position of the cephalic furrow remains constant throughout germ band extension, indicating that extension is confined to the region posterior to it.

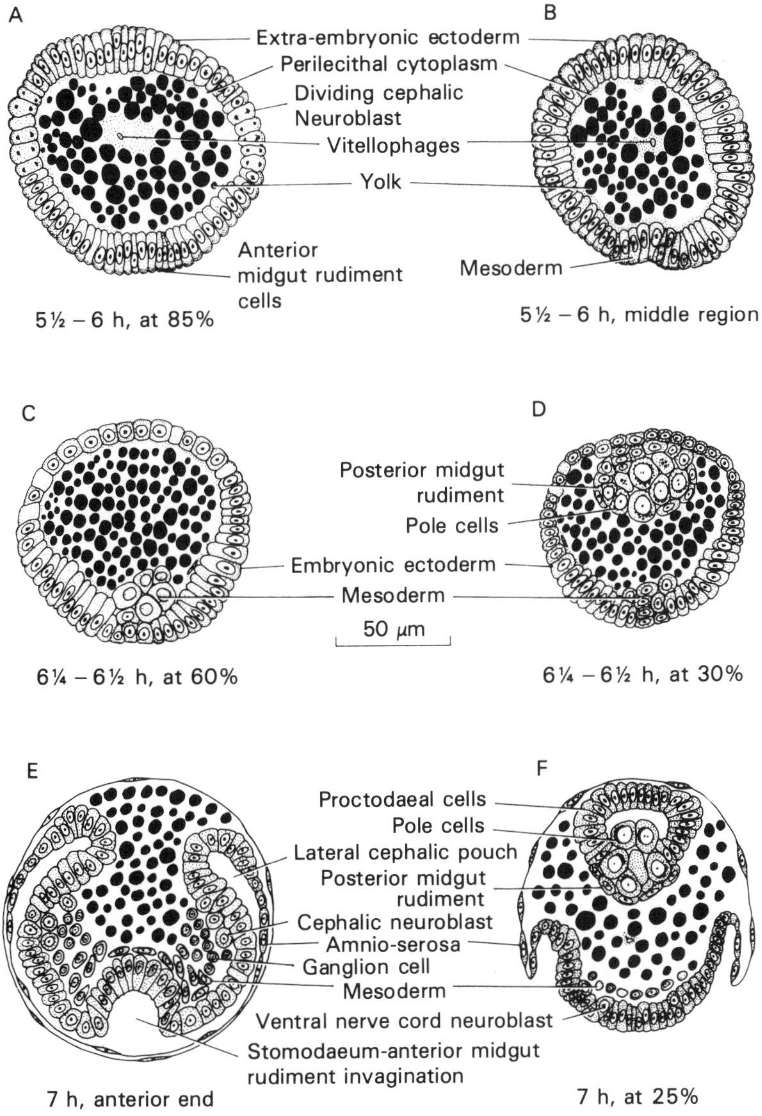

Figure 2.3 (A–F) Gastrulation and formation of amnio-serosal membrane in embryos of *Culex quinquefasciatus* incubated at 25°C. The position of each transverse section is recorded as percentage of egg length (posterior pole = 0%). (From Davis, 1967.)

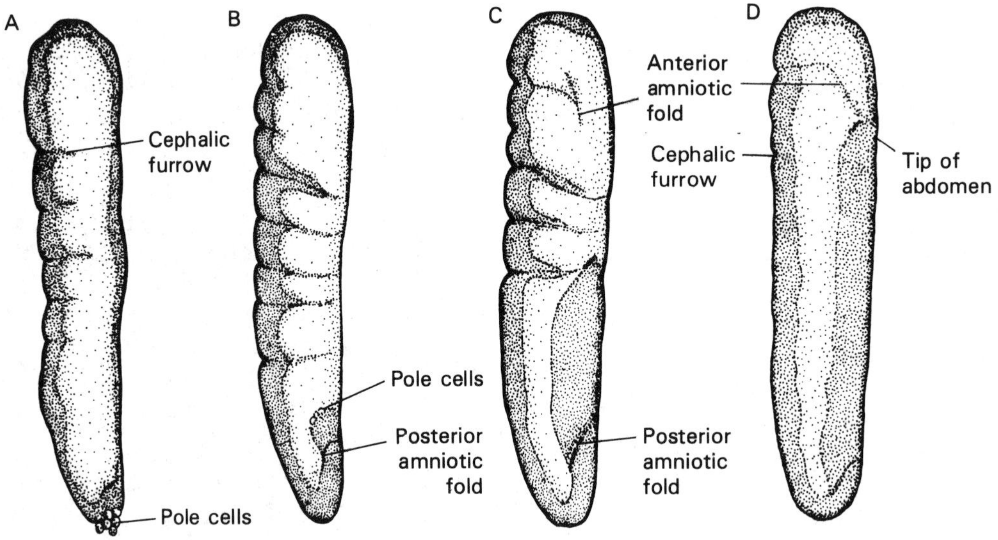

Figure 2.4 (A–D) Successive stages in the elongation of the germ band in *Culex pipiens*. All embryos are in side view. The growth of the extra-embryonic membranes and the appearance and disappearance of furrows can also be seen. (From Idris, 1960a.)

The anterior third of the embryo shows rather less change during gastrulation. Antero-ventrally, the anterior midgut rudiment and the surrounding presumptive stomodaeum, which are not histologically distinct at this time, sink into the embryo as a simple invagination. Two pairs of cephalic pouches also form (Figures 2.3E, 2.5). At an early stage of gastrulation the presumptive cephalic neuroblasts at the surface enlarge and divide. Shortly before 7 h the presumptive ventral cord neuroblasts, together with the already formed cephalic neuroblasts, pass to the interior and begin to divide yielding ganglion cells internally. During gastrulation most of the vitellophages move into the perilecithal cytoplasm to form a sparsely nucleated temporary yolk sac. The concentration of free amino acids and peptides in *Culex quinquefasciatus* eggs increases steadily during the first 16 h (Chen and Briegel, 1965), possibly due to hydrolysis of yolk proteins.

In *Culex*, segmentation occurs almost instantaneously throughout the germ band, involving both ectoderm and mesoderm. It starts about 8 h after oviposition, and segments from the mandibular to the anal appear within a period of 10 min. Cell division continues in the mesoderm until 14 h after oviposition, when the mesoderm cells lie in ventro-lateral masses above the ganglion cells of the nerve cord and in the head. Coelomic cavities have not been described by any author. By the time the embryo has finished contracting a partly segmented head lobe, three thoracic segments, and eight abdominal segments plus terminal mass can be distinguished. At an earlier stage the 8th and 9th abdominal segments could be seen to combine (Guichard, 1971), therefore the terminal segment must represent the 8th and 9th segments and the terminal mass the 10th and remnants, if any, of an 11th. Fusion of the 8th and 9th segments was observed in *Anopheles* also (Ivanova-Kazas, 1949). During the course of development six subdivisions can be distinguished, with variable clarity, within the head lobe.

Shortly after the germ band has started to extend over the dorsal surface of the egg, the edges of the presumptive extra-embryonic ectoderm grow out as the amniotic fold. Later this forms the amnio-serosal membrane which encloses the germ band completely (Figures 2.3E–F, 2.6A–C). The thin outer layer, or serosa, is continuous with the extra-embryonic ectoderm. The inner layer, or amnion, which is considered to be an extension of the germ band, is at first composed of cubical cells but these later become flattened. The posterior amniotic fold arises first and forms a hood over the caudal end of the germ band (Figure 2.4B). The direction of growth of the posterior amniotic fold is towards the posterior pole of the egg, and its growth keeps pace with the forward movement of the caudal end of the germ band, so that when the germ band has reached its full length the edge of the posterior amniotic fold is in approximately the same position as when it first arose (Figure 2.4D). The anterior amniotic fold arises later, and has only covered a small length of the head region by the time the germ band has reached its maximum length. The anterior and posterior amniotic folds eventually meet and fuse, enclosing the embryo in the amniotic cavity. The serosa later secretes a thin chitinous sac, the serosal cuticle, around the embryo (Section 3.1).

About 12 h after oviposition, in *Culex*, the elongated germ band begins to shorten, with the result that its posterior region retracts towards the posterior pole of the egg (Figure 2.7A). During this final phase of blastokinesis the whole germ band revolves through 180° on its longitudinal axis so that the ventral surface of the embryo comes to lie against the dorsal (concave) side of the chorion and the dorsal side of the embryo against the ventral (convex) side of the chorion. The anterior–posterior orientation of the embryo remains unchanged.

As a consequence of the shortening of the germ band the dorsal surface of the embryo is covered only by the extra-embryonic membranes (Figure 2.9A) which, during the later stages of germ band shortening, break ventrally and retract dorsally, eventually degenerating. Soon after 17 h the lateral edges of the embryo start to grow over the yolk, and by 23 h a thin layer of epidermal cells encloses it at the posterior end, a process called **dorsal closure**. The midgut cells do not enclose the yolk until about 25 h.

2.1.4 Organ formation

(a) Alimentary canal

Certain organs start to form before contraction of the germ band is complete, notably the gut and head appendages, but most organs develop later. The anterior region of the alimentary canal develops from two areas of blastoderm, the presumptive stomodaeum and the anterior midgut rudiment (Figure 2.2); similarly, the posterior region develops from the presumptive proctodaeum and the posterior midgut rudiment. The stomodaeum and proctodaeum grow inwards as simple epithelial tubes, carrying with them respectively cells of the anterior and posterior midgut rudiments. Beginning at 11 h the cells of the anterior midgut rudiment form a pair of lateral strands, and as the stomodaeum invaginates deeper into

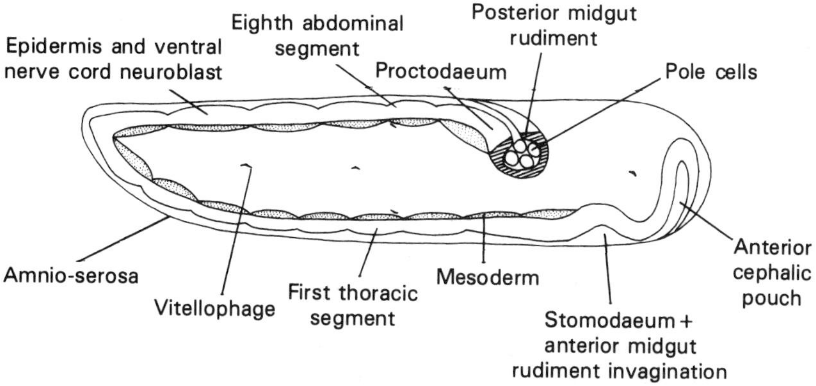

Figure 2.5 Diagram of a slightly parasagittal section through an embryo of *Culex quinquefasciatus* after 8 h incubation at 25°C. (From Davis, 1967.)

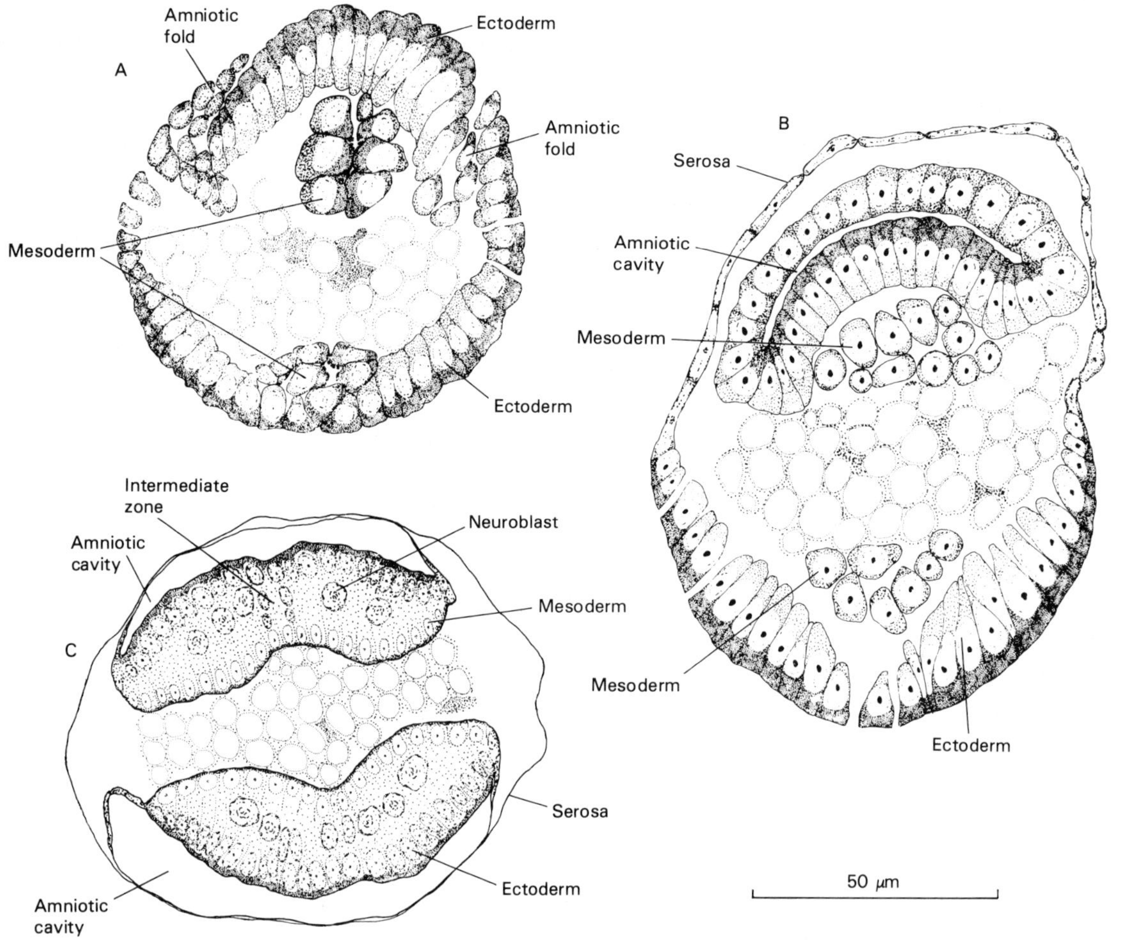

Figure 2.6 Stages in the development of the germ layers and extra-embryonic membranes in *Culex pipiens*. (A) Transverse section through the posterior region at about the stage shown in Figure 2.4A. (B) Transverse section through the posterior region at about the stage shown in Figure 2.4D. (C) Transverse section through the middle of an embryo after the completion of segmentation. (From Idris, 1960a.)

the embryo the cells of the two strands grow posteriorly between the mesoderm and the surface of the yolk (Figure 2.9B). The germ band is fully extended at this time so that the proctodaeal invagination is situated antero-dorsally. The cells of the posterior midgut rudiment, carried on the invaginating proctodaeum, also form two strands which extend posteriorly and in contact with the mesoderm. When these strands reach the posterior pole they pass ventrally and then anteriorly along the ventral surface of the yolk. By 14.5 h the strands of the anterior and posterior midgut rudiments have united, and subsequently, during dorsal closure, they spread dorsally and ventrally to form the midgut epithelium which encloses the yolk and the degenerating cells of the resorbed amnion and serosa.

Ligation experiments showed that the vitellophages play no part in midgut formation in *Culex pipiens* (Idris, 1960b). No tertiary vitellophages are formed from the midgut rudiments, but indications of yolk digestion are seen at the ends of the extending midgut rudiment strands and, later, on the inner boundaries of the midgut cells where most of the remaining vitellophages also aggregate. The stomodaeum differentiates into pharynx and oesophagus. After a little over 24 h, the end of the oesophagus, which is closed by a membrane, invaginates into the midgut and forms the oesophageal invagination (Figure 2.8). About this time the midgut remoulds itself anteriorly to form eight caeca, and the proctodaeum differentiates into the pyloric chamber, small intestine and rectum. As early as 11–12 h cell divisions near the tip of the invaginating proctodaeum presage the development of five Malpighian tubule rudiments. Also at an early stage, just inside the proctodaeal invagination, four tubercles appear which later elongate and project through the anus as sharp-pointed outgrowths, eventually becoming the anal papillae.

(b) Other organs

Differentiation of the mesodermal somites into somatic and splanchnic components occurs while the midgut strands move over the inner surface of the mesoderm (Figure 2.9), and during dorsal closure the somatic blocks of the trunk develop into muscles, fat body, and dorsal vessel. Mesoderm cells which surround the invaginating stomodaeum and proctodaeum, with others on the outer surface of the midgut rudiments, form the gut musculature. During the early stages of germ band extension the polar granules fuse and form a darkly staining mass in each pole cell. Later the pole cells, which will become primordial germ cells, separate into two groups which, at about 14 h, migrate laterally into the mesoderm of the 6th abdominal segment. As differentiation of the mesoderm occurs, each group of six to eight primordial germ cells becomes enclosed by a fine mesodermal sheath to form a gonadal rudiment (Figure 2.9C). The darkly staining masses derived from the polar granules fragment at 12 h, and some of the material becomes closely applied to the nuclear membrane. In *Anopheles maculipennis* a thick strand of mesodermal cells, the duct rudiment, extends posteriorly from each gonad rudiment, but differentiation and development of the gonads and genital ducts does not occur until postembryonic development. Exposure to

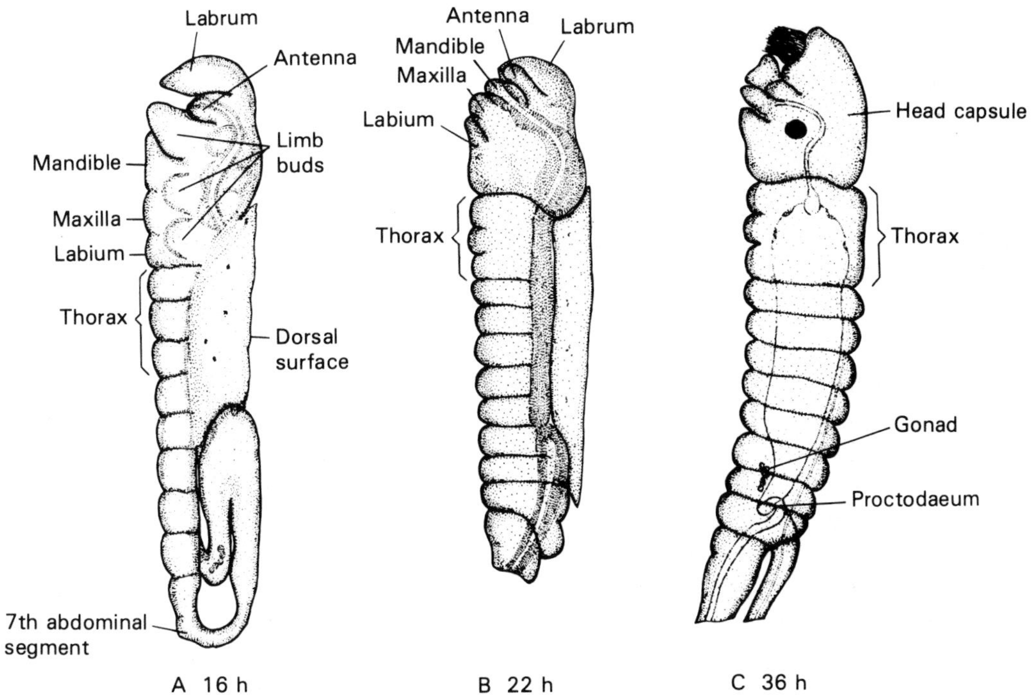

Figure 2.7 (A–C) Contraction of the germ band, segmentation, and organ formation in embyros of *Culex pipiens* incubated at 21°C. All embryos are in side view. (From Idris, 1960a.)

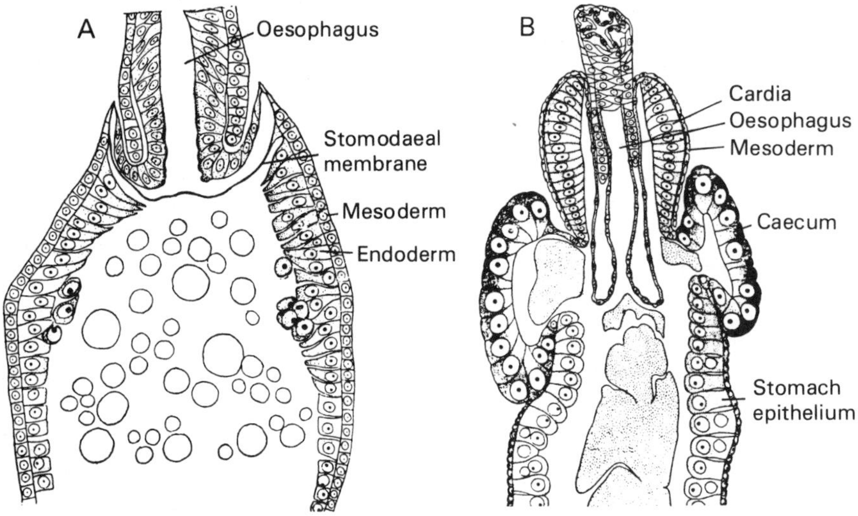

Figure 2.8 Stages in the development of the oesophageal invagination and anterior midgut in *Anopheles maculipennis*. (A) Horizontal section through the gut of an embryo after contraction of the germ band. (B) Horizontal section through the gut of a larva before hatching. (From Ivanova-Kazas, 1949.)

thermal stress during embryogenesis can lead to abnormal development of the gonads and genital ducts in aedine species and to the loss of gonads in *Culex* (Anderson and Horsfall, 1965b; Cupp and Horsfall, 1970a, b; Raminani and Cupp, 1977; Mossé and Hartman, 1980).

The ventral nerve cord, which is derived from invaginated neuroblasts, becomes segmented soon after segmentation of the germ band occurs, and its cells become arranged in two ventrolateral masses in each segment. The neuroblasts continue to divide until 18 h, and at 18 to 20 h they separate from the overlying epidermis (Figure 2.9). Neuropile appears in the ganglia and connectives from about 20 h. No fusion of ganglia occurs except anteriorly, where three ganglia fuse to form the brain and three fuse to form the suboesophageal ganglion. The tritocerebrum gives rise to the circumoesophageal connectives and the labrofrontal nerves.

The origin of the stomatogastric nervous system has been described in *Aedes vexans* and *Ae aegypti*. The frontal ganglion arises from cells which bud off from the dorsomedial surface of the pharynx. Extensions from the frontal ganglion gives rise to the recurrent nerve, hypocerebral ganglion and oesophageal nerves. The corpora allata and corpora cardiaca arise as buds from the lateral walls of the 3rd head segment which migrate posteriorly. The corpora cardiaca subsequently establish neuronal connections with the protocerebrum, while the corpora allata are situated immediately posterior to them.

In *Culex*, division of presumptive epidermal cells continues until about 18 h. During the contraction of the germ band the thickened floors of the paired anterolateral and lateral cephalic pouches evaginate and grow dorsally to form, respectively, the preoral and cephalic lobes. The preoral lobes later bear the labropalatal brushes, while the cephalic lobes fuse to form the dorsal part of the cranium and bear the antennae ventrolaterally. The mandibular and maxillary rudiments move anteriorly and laterally to lie on either side of the mouth and form the floor of the preoral cavity, while the labial rudiments fuse forming the ventral part of the head (Figure 2.7A,B). The palatum is formed (in *Ae aegypti*) from an evagination of anteromedial ectodermal cells between the labropalatal brushes. A pair of ectodermal invaginations develops in the labial segment at about 12 h, and these grow posteriorly to lie beneath the stomodaeum where they form the salivary glands. The ducts from the salivary glands fuse at the posterior end of the head, and the common salivary duct extends forward to open at the anterior end of the labium. At 18 to 20 h localized thickenings on the sides of the head form the rudiments of the larval ocelli.

At 11 h ventrolateral cells of the trunk ectoderm divide giving rise to a pair of invaginations in each segment from the first thoracic to the 8th abdominal. These are the tracheal rudiments. At 15 h the external openings of the invaginations close, and the inner cells of the tracheal rudiments move dorsally to lie among the mesoderm. Later these cells develop into the tracheal system, of which only the terminal spiracles, within the respiratory siphon, are functional. According to Guichard (1971) the respiratory siphon arises as a dorsal outgrowth from the 9th abdominal segment, and the anterior part of the 9th segment fuses with the 8th at that time. After the closure of the tracheal invaginations the outer cells of each rudiment remain attached to the epidermis and enlarge. In the thorax they become bristle cells; in the abdomen the outer cells become bristle cells but groups of about six inner cells differentiate as oenocytes. Secretion of the cuticle begins at 26 h and the hatching spine becomes distinct at 30 h.

During the 12 h before hatching histological differentiation occurs and the organs become capable of functioning, as seen in the beating of the heart. In *Culex pipiens* at this time endomitosis increases greatly, that is chromosomal replication occurs within nuclei which do not divide. A large number of cells have DNA values of 8C and 16C, especially in the intestinal tract and Malpighian tubes (Jost, 1970a).

An exponential relationship was observed between temperature and the duration of embryonic development in *Cx theileri*, a species tolerant of a wide range of temperatures. At 12°, 18°, 24° and 36°C the incubation periods were 411, 84, 41 and 27 h respectively (Van der Linde *et al.*, 1990). The incubation period of *Armigeres digitatus* was 32–34 h at 28°C (Okazawa *et al.*, 1991); that of *Cx quinquefasciatus* was 54 h at 18° and 36 h at 25°C (Christophers, 1960);

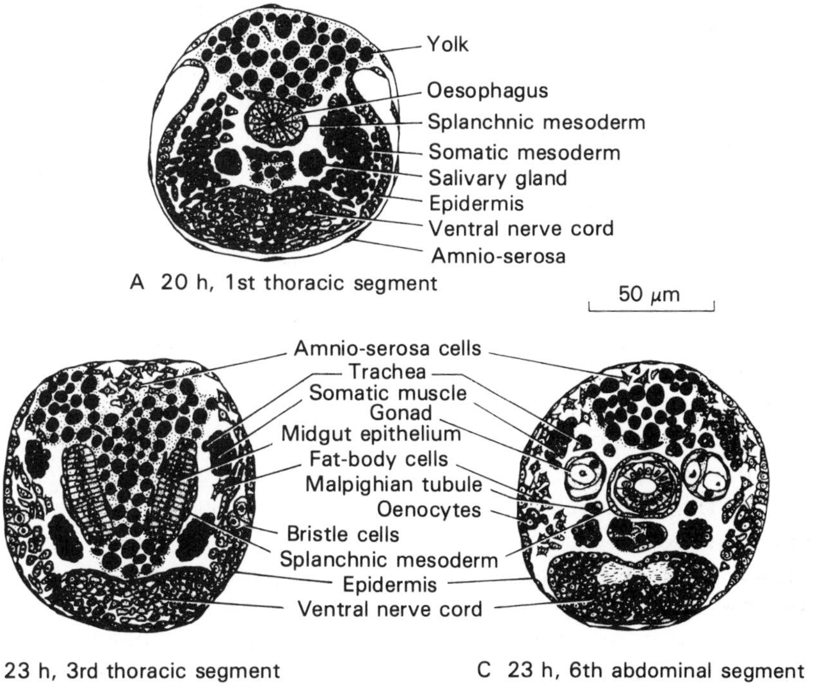

Figure 2.9 (A–C) Organ formation in embryos of *Culex quinquefasciatus* incubated at 25°C. Transverse sections. (From Davis, 1967.)

that of *Ae sticticus* was 192 h at 20° and 152 h at 25°C (Trpis *et al.*, 1973).

2.2 EXPERIMENTAL EMBRYOLOGY

At the start of larval embryogenesis the descendants of the zygote nucleus, which are believed to be equipotent, become lodged in cells which subsequently enter very different developmental pathways, culminating in the formation of a spatially coordinated organism. Two different approaches have been used by insect embryologists to investigate the processes of pattern specification.

1. Operational techniques, involving the isolation or removal of parts of the embryo, have been used to demonstrate the effects of one part of the embryo upon the development of another or to reveal capacities for independent differentiation.
2. Investigations of morphogenetic mutants of *Drosophila melanogaster* have elucidated a sequence of activities, which is in fact a cascade of gene activation, that governs pattern formation during embryogenesis.
   (a) The oocyte, already structurally polarized at oviposition, contains the products of maternal genes derived from the nurse cells. These products, which include the mRNAs of egg-polarity genes, define the spatial coordinates of the embryonic phenotype by setting up morphogen gradients in the egg.
   (b) Zygotic genes interpret the positional information provided by the morphogen gradients, progressively marking out the embryo into a series of compartments and segments. They include a number of segmentation genes which act sequentially, i.e. gap genes, pair-rule genes, and segment-polarity genes.
   (c) The products of the segmentation genes influence the expression of homeotic selector genes, first activated in the blastoderm, which maintain the differences between one segment and another and give the segments their identities.

Taken together the results from both approaches suggest that specification of body pattern is initiated and controlled by interactions between initially equipotent nuclei and cytoplasmic determinants. The latter include diffusible substances, or morphogens, coupled with regionally localized information which may have a structural or physiological nature. Cell–cell interactions constitute the final level of regulation (Boswell and Mahowald, 1985; Alberts *et al.*, 1989).

2.2.1 Experimental observations

The regulation of embryogenesis in mosquitoes has been investigated by operational but not genetic methods. Experimental techniques used on *Culex pipiens* and *Cx quinquefasciatus* have included ligation of eggs after removal of the chorion (Idris, 1960b), microbeam UV-irradiation (Oelhafen, 1961), microcauterization and centrifugation (Davis, 1970). In the following summary of these investigations positions along the anterior–posterior axis are expressed as a percentage of egg length; following convention, 0% is assigned to the posterior pole and 100% to the anterior pole. Developmental times cited are for hours elapsed after oviposition, at an ambient temperature of 25°C.

(a) Ligation during the first five cleavage divisions

Blastoderm was prevented from forming in any isolates that were devoid of energids, but it would form in front of ligatures at 30% egg length and behind ligatures at 70% egg length. Development never proceeded beyond the blastoderm stage, even when the ligatures were tied loosely.

(b) Ligation during the 6th to 8th cleavage divisions

When ligatures were tied at up to 30% egg length during this period of energid migration, tissues developed in the anterior isolate which

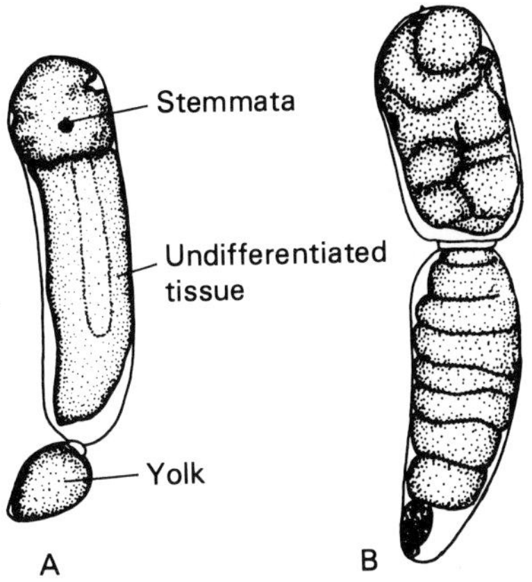

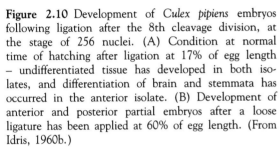

Figure 2.10 Development of *Culex pipiens* embryos following ligation after the 8th cleavage division, at the stage of 256 nuclei. (A) Condition at normal time of hatching after ligation at 17% of egg length – undifferentiated tissue has developed in both isolates, and differentiation of brain and stemmata has occurred in the anterior isolate. (B) Development of anterior and posterior partial embryos after a loose ligature has been applied at 60% of egg length. (From Idris, 1960b.)

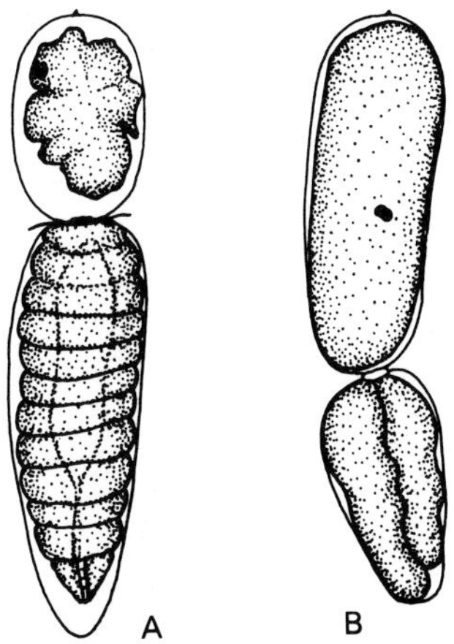

Figure 2.11 Development of *Culex pipiens* embryos after ligation during the syncytial blastoderm stage. (A) After ligation at 75% egg length, a partial embryo has developed in the posterior isolate and an unsegmented head has developed in the anterior isolate. (B) After ligation at 40% of egg length, the anterior and posterior isolates show neither segmentation nor germ-layer formation. (From Idris, 1960b.)

were unidentifiable apart from a differentiated brain and stemmata, but no development took place in the posterior isolate (Figure 2.10A). When ligatures were tied between 45 and 65% egg length a differentiated brain and stemmata developed in the anterior isolate as before and unidentifiable tissue was formed in the posterior isolate. If the ligatures were tied loosely near the middle of the egg partial embryos developed in front of and behind the ligature (Figure 2.10B).

(c) Ligation during the syncytial blastoderm stage

Ligatures that gave very large anterior or posterior isolates permitted formation of histologically differentiated partial embryos in the large isolates but only of unsegmented tissue in the small isolates. Eggs ligatured at 5% egg length yielded externally complete embryos or embryos with a very small last abdominal segment. Eggs ligatured at 10 or 18% yielded partial embryos with head, thorax, and four abdominal segments. Eggs ligatured between 25 and 55% generally did not form partial embryos, but anteriorly only an unsegmented head with differentiated brain and eyes, and posteriorly blastoderm or unsegmented tissue (Figure 2.11B). Eggs ligatured at 60 to 75% yielded posterior partial embryos which normally were complete up to the prothorax (Figure 2.11A), and which in exceptional cases were complete up to the mandibular segment.

(d) Ligation during the uniform blastoderm stage

When ligatures were tied between 5 and 65% egg length all eggs developed an anterior partial embryo, the number of segments formed depending upon the position of the ligature. When the ligature was tied below 10% egg length the eggs yielded externally complete embryos which developed to larvae. Posterior partial embryos developed only when the ligature was tied well forward, at 60% egg length or more (Figure 2.12).

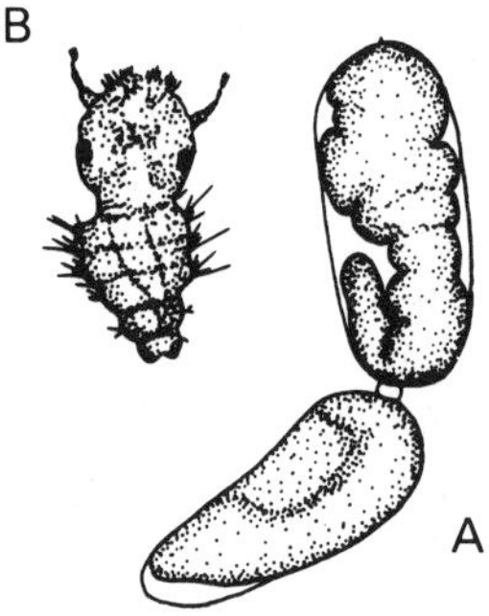

Figure 2.12 Stages in the development of *Culex pipiens* embryos after ligation during the uniform blastoderm stage at 50% of egg length. (A) At an early stage of embryogenesis – germ band has formed in both isolates but gastrulation has not occurred in the posterior isolate. (B) After hatching – an anterior isolate. (From Idris, 1960b.)

(e) Ligation during the differentiated blastoderm stage

When the ligatures were tied behind 40% egg length the posterior isolates formed only unsegmented tissue, but when they were tied between 40 and 60% the posterior isolates formed partial embryos with distinct segments. The developmental capacity of the anterior isolates was the same as that of those formed by ligation during the uniform blastoderm stage (Figure 2.13). With ligatures at 10 and 15% egg length externally complete embryos formed in front of the ligatures, and when the ligatures were tied further forward anterior partial embryos developed. Primary germ cells were missing from these embryos because the pole cells remained in the posterior isolates. If abdominal segments 6 and 7 were present gonad sheaths formed but did not contain germ cells. Midgut would develop normally from anterior midgut rudiment alone or from posterior midgut rudiment alone.

(f) UV-irradiation of posterior periplasm

If, prior to its penetration by energids, the posterior periplasm was irradiated for 10–20 s, the adult mosquitoes that eventually developed had gonads but no gametes. The testes were very small and had few nuclei; the ovaries lacked oocytes and nurse cells and apparently contained only follicle cells.

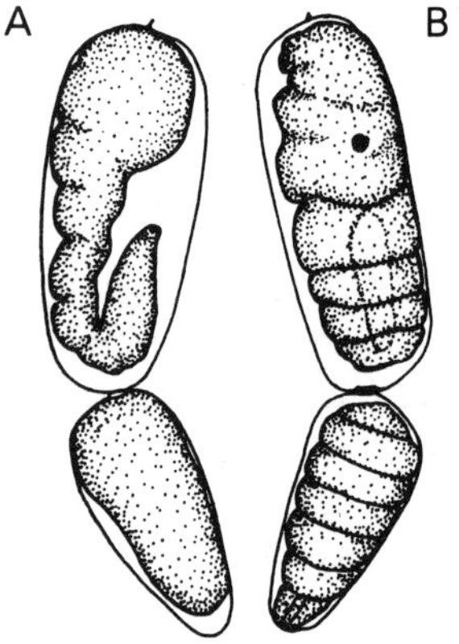

Figure 2.13 Stages in the development of *Culex pipiens* embryos after ligation during the differentiated blastoderm stage at 40% of egg length. (A) At the germ band stage. (B) After segmentation – partial embryos have formed in both anterior and posterior isolates. (From Idris, 1960b.)

(g) UV-irradiation during germ band formation

Momentary irradiation of narrow bands within the region 65–75% of egg length during the period 6–6.5 h, while the germ band was forming, caused damage which became apparent in the larva in the form of missing mouthparts, antennae or stemmata. When the posterior end of the egg was irradiated for 5 s at 6 h, the hindgut of the larvae showed considerable abnormalities or might be absent altogether. The midgut formed nevertheless, although it was sometimes very long and thin.

(h) UV-irradiation during germ band extension

Eggs were irradiated for 5 s over a band 55 μm wide at the posterior pole at 7–8 h, when the germ band was elongating. Eggs irradiated early in this period gave rise to larvae with defects on the last abdominal segments. Eggs irradiated at the end of the period gave rise to larvae with defects on the 2nd, 3rd and 4th abdominal segments. Eggs irradiated during the middle of the period gave rise to larvae with defects on the other segments of the posterior abdomen.

(i) UV-irradiation of the dorsal aspect at 18 h

Eggs which had developed for 18–18.5 h were irradiated from the dorsal aspect for 20 s over a band 125 μm wide in the region of the head and prothorax. Larvae developed which lacked developed brain, corpora allata, corpora cardiaca and prothoracic glands.

(j) Cauterization of pole cells

When the pole cells alone were cauterized during the blastoderm stage, embryos developed which lacked germ cells and gonad sheaths but were otherwise normal.

(k) Cauterization of the posterior blastoderm

Cauterization during the blastoderm stage, which caused damage from 0 to 7% of egg length, caused loss of the posterior midgut rudiment, hindgut and anal papillae. In these embryos the anterior midgut rudiment did not undergo any extensive regulation.

(l) Ligation and cauterization during germ band extension

Ligatures tied while the tip of the embryo advanced over the posterior third of the egg normally resulted in the formation of a germ band whose segments could not be identified, but in a few cases partial embryos formed. Ligatures tied at the time of maximum extension of the germ band, such that the anterior isolate contained the anterior- and posterior-most portions of the germ band while the posterior isolate contained the middle portion, led to formation in the anterior isolates of partial embryos consisting of head, anterior thoracic and terminal abdominal segments and in the posterior isolates of partial embryos consisting of posterior thoracic and anterior abdominal segments (Figure 2.14). In general the anterior isolates developed better

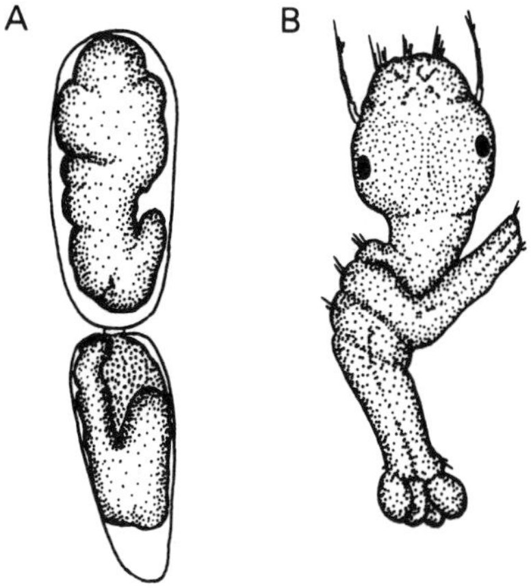

Figure 2.14 Stages in the development of *Culex pipiens* embryos after ligation at 48% of egg length at the extended germ band stage when the tip of the abdomen was at 73% of egg length. (A) At the germ band stage. (B) After hatching – an anterior partial embryo consisting of head, prothorax and 7th abdominal and following segments. (From Idris, 1960b.)

than the posterior isolates. Cauterization of the posterior pole during germ band extension caused the destruction of particular abdominal segments, depending on the timing and extent of the cauterization. Intermediate abdominal segments were destroyed and the two sections of the germ band fused. No gonad sheath formed around the germ cells unless part of the 6th abdominal segment was present.

(m) Centrifugation during cleavage

Centrifugation of eggs during the cleavage stages caused extensive redistribution of the egg contents but, even so, nearly all eggs underwent further development. Poorly differentiated muscles were distinguishable in most embryos, and ganglia in all, so gastrulation had occurred. Foregut, anterior midgut rudiments, and sometimes a differentiated oesophageal invagination were present towards the anterior end of the tissue mass.

Many embryos showed duplication of head structures; this was much more frequent after posterior (tail out) centrifugation (74% of cases) than after anterior (head out) centrifugation (8%

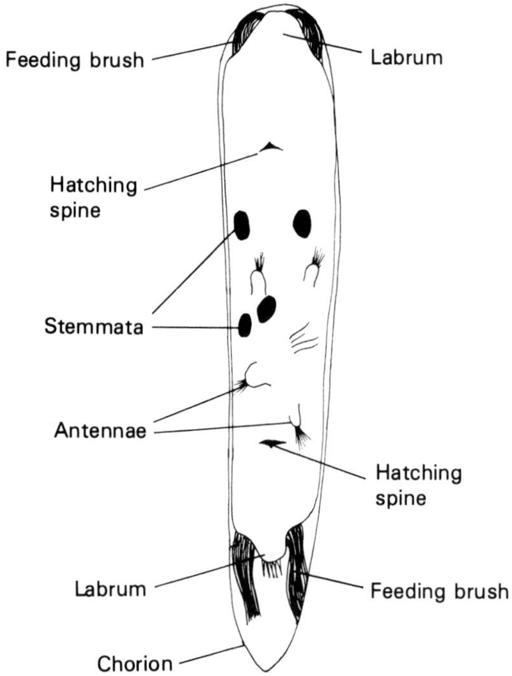

Figure 2.15 An embryo of *Culex quinquefasciatus* which showed duplication of head organs after posterior centrifugation during cleavage. (From Davis, 1970.)

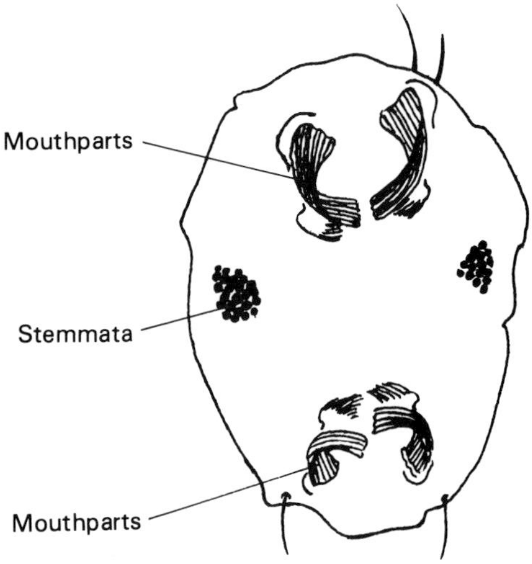

Figure 2.16 Duplication of head organs in an untreated egg of *Toxorhynchites amboinensis*. No trace of thorax or abdomen was apparent. The chorion is not shown. (From Mattingly, 1973.)

of cases). The only structures showing duplication were the lateral palatal brushes, stemmata, antennae, hatching spine and brain. The lateral palatal brushes mostly lay at the two ends of the tissue mass and the hatching spine and brain also showed marked polarity, but the antennae and stemmata showed little regularity of arrangement (Figure 2.15). A number of eggs that had been subjected to anterior centrifugation, but did not show duplication of head structures, did show a reversal of polarity.

(n) Centrifugation during blastoderm formation

No duplication of head structures or reversal of polarity was observed after centrifugation during blastoderm formation. In all embryos gastrulation had probably occurred, muscles were mostly well developed, and in over half some segmentation of the trunk had occurred. After posterior centrifugation the posterior midgut rudiment and proctodaeum were often absent, and the anterior midgut rudiment showed no evidence of regulation.

(o) Aberrant embryos with duplicated head organs

Eggs from a laboratory colony of *Toxorhynchites amboinensis* were found to contain aberrant embryos. Each embryo had paired mouthparts at either end, directed towards the middle of the egg, and some contained two groups of stemmata. No trace of thorax or abdomen was present (Figure 2.16). These embryos were described by Mattingly (1973) as consisting of two heads joined in the transverse plane.

(p) Bicaudal larvae

Examination of unhatched eggs from a colony of *Wyeomyia smithii* revealed fifteen aberrant larvae, each consisting of two abdomens fused at their 1st abdominal segments. They had no head or thorax (Figure 2.17). Externally, the abdomens were morphologically complete and normal except

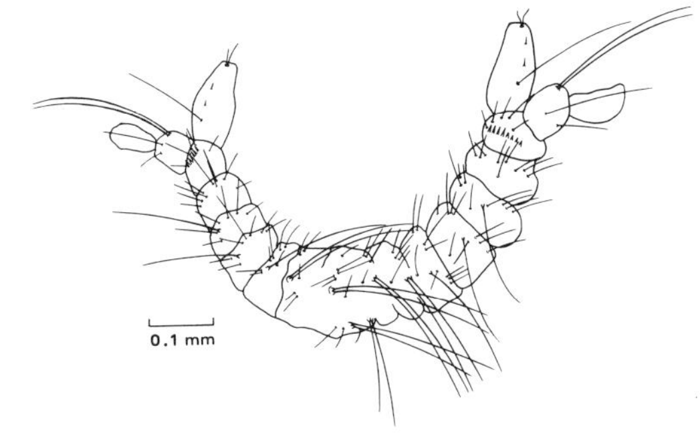

Figure 2.17 A bicaudal larva of *Wyeomyia smithii* freed from its egg shell. (From Price, 1958.)

for the displacement of bristles on the two 1st abdominal segments and the indistinctness of segmentation near the junction. The midguts of each pair were joined, and the ganglia of the two 1st abdominal segments were fused. The internal organs of each abdomen otherwise appeared to be normal, but no observations on the presence or absence of gonads were reported. The aberrant larvae were alive when removed from the eggs and made normal thrashing movements (Price, 1958).

2.2.2 Analysis of results

At the beginning of cleavage in *Culex* no regions of the egg were capable of self-differentiation, and eggs ligatured at this stage failed to develop beyond the blastoderm stage (Section 2.2.1(a)). However, after the 64-nucleus stage, ectoderm could form in large anterior or posterior isolates produced by ligation, and the brain and stemmata could self-differentiate (2.2.1(b)).

Ligation of embryos during the syncytial blastoderm stage which produced very large isolates permitted the development of histologically differentiated partial embryos whether the large isolates were anterior or posterior. When ligatures were tied between 25 and 55%, development in both isolates was minimal (2.2.1(c)), suggesting a need for morphogens from both sides of the ligature. As embryogenesis proceeded ligation was progressively less restrictive. By the cellular blastoderm stage ligation at 5–65% permitted formation of partial anterior embryos (2.2.1(d)), and by the differentiated blastoderm stage ligation at 40–60% permitted formation of partial posterior embryos (2.2.1(e)).

In *Culex* a limited amount of regulation could occur during the early stages of embryogenesis. After ligation at up to 15% of egg length during the differentiated blastoderm stage, externally complete embryos formed in front of the ligature (2.2.1(e)). After ligation at that stage midgut could develop normally from either anterior or posterior midgut rudiment alone (2.2.1(e)). Irradiation experiments showed that extension of the germ band occurred throughout the presumptive thoracic and abdominal regions, and not from a posterior growth zone alone (2.2.1(h)). Organ formation could occur during the later stages of larval embryogenesis in the absence of hormones from the brain and thoracic endocrine organs (2.2.1(i)).

Posterior centrifugation of *Culex* eggs during cleavage led to replacement of the thorax and abdomen by a mirror image duplication of the anterior head segments (2.2.1(m)), a phenomen observed naturally in *Toxorhynchites amboinensis* (2.2.1(o)). Such double cephalons are well known in chironomids, in which they are ascribed to disturbance of the relative concentrations of two anterior-region and two posterior-region cytoplasmic determinants. Inactivation of the anterior factors in chironomids can lead to the formation of double-abdomens, similar to those occurring naturally in *Wyeomyia smithii* (2.2.1(p)) (Boswell and Mahowald, 1985).

Isolation of the pole cells by ligature prevented the formation of primary germ cells (2.2.1(e)), and so did irradiation of the posterior periplasm before it had been penetrated by energids (2.2.1(f)). This is consistent with the role of polar granules as pole cell determinants, which has been proved in *Drosophila*. The presence of abdominal segment 6 or 7 was essential for the formation of the gonad sheaths (2.2.1(e) and (l)). Davis (1970) found that when the pole cells were cauterized during the blastoderm stage embryos developed which lacked both germ cells and gonad sheaths although they were otherwise normal (2.2.1(j)), and she questioned whether pole cells had been completely absent from the anterior isolates which other workers had reported to develop empty gonad sheaths in the absence of pole cells (2.2.1(e) and (f)).

The spontaneous occurrence of double cephalons and bicaudal larvae in mosquitoes (2.2.1(o) and (p)) demonstrates that at some stage during oogenesis the polarity of half of the oocyte can be reversed. By the time of germ band extension the fates of most and possibly all parts of the embryo are fully determined so that damage leads to the absence of parts from the larva (2.2.1(g). (h) and (l)).

The determination of primordia for adult structures also occurs during larval embryogenesis, and these primordia become evident during the larval stage as imaginal disks. When eggs of *Aedes sierrensis* were maintained at 33.5°C for varying periods, structural abnormalities were later observed in the larvae and the adults. Because of the different periods of exposure that were required to affect the larval and pupal structures, Cupp and Horsfall (1970a) concluded that the larval tissues were determined 48 h earlier than those of the adult.

3

The egg shell

The egg shell, which successively surrounds the oocyte, egg, embryo and pharate larva, is secreted partly by the mother and partly by the embryo. Soft and flexible when laid, it later becomes hardened and waterproof. Although the mature egg shell has rigidity and strength to provide mechanical support and protection, it permits gas exchange while minimizing water loss. A small pore, the micropyle, allows sperm to penetrate during the short period of ovulation. Eggs that are laid on the water surface are modified in a variety of ways to enable them to float.

During oviposition the longitudinal axis of the egg has the same anterior–posterior orientation as the maternal body. The head of the embryo forms at the anterior end of the egg, which is distinguishable externally by the presence of the micropyle. The side of the egg shell that is designated dorsal is, in eggs that lie horizontally, such as those of *Anopheles* and *Aedes*, the side in contact with the water or substratum. In *Anopheles* this is the curved side of the egg; in some species of *Aedes*, including *Ae aegypti*, it is the slightly more arched side. After blastokinesis, when the embryo rotates on its longitudinal axis through 180°, the ventral surface of the embryo lies against the ventral surface of the egg shell (Ivanova-Kazaas, 1949; Christophers, 1960; Horsfall *et al.*, 1973; Harbach and Knight, 1980).

3.1 STRUCTURE

The structure of the egg shell cannot be considered without first clarifying its terminology. Since the 19th century, zoologists have distinguished primary, secondary, and sometimes tertiary layers in the coverings of oocytes. These layers are classified according to their origins. *Primary* egg membranes are those formed by the oocyte itself. *Secondary* membranes are those produced by the follicle cells that surround the oocyte in the ovary. *Tertiary* membranes, produced by some animals but not all, are products of the oviduct or uterus. The primary membranes, which are extremely delicate and have physiological functions, have long been termed vitelline membranes or vitelline envelopes (Parkes, 1960). There is no evidence of primary membranes around the oocytes of insects. The secondary membranes have very different forms in different animals. In most insects they form a hard proteinaceous shell, termed the **chorion** (Snodgrass, 1935; Wigglesworth, 1950). The chorion is often composed of two layers which may not be strictly homologous in different insects. These two layers have variously been termed endochorion and exochorion (Imms, 1957) or inner and outer layers of the chorion (Hinton, 1968b).

More recently, certain investigators have termed the inner layer of the chorion of insect eggs 'vitelline membrane' or 'vitelline envelope'. This usage is contrary to zoological convention and, more seriously, introduces conceptual errors. It has been so extensively adopted by *Drosophila* geneticists that the correct terminology may be irretrievable in that field. Studies with *Aedes*, *Culex* and *Anopheles* have shown that the chorion is secreted in two phases, separated in time (Section 19.6.1). For our purposes, the region that is secreted first, and forms the inner part of the chorion, is designated endochorion, and the

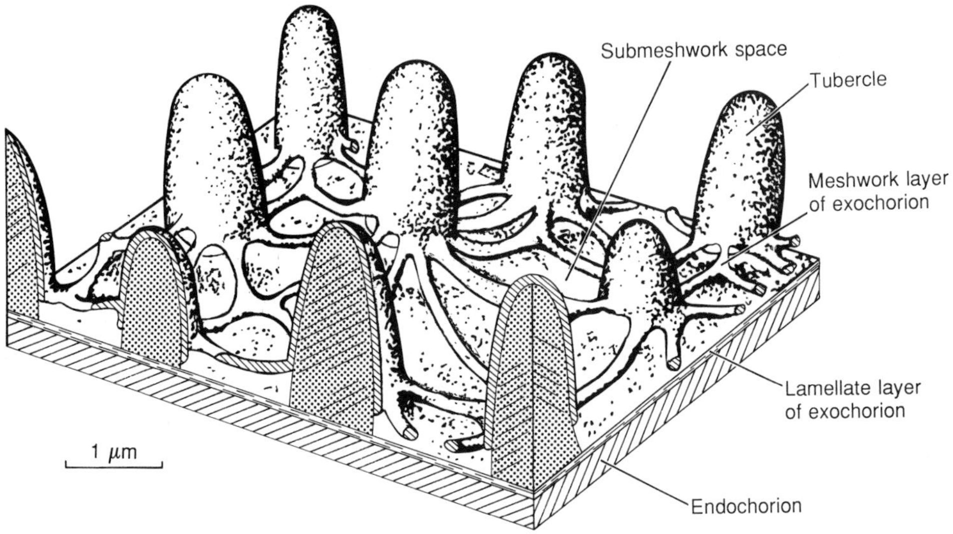

Figure 3.1 Diagram of chorion structure in a *Culex pipiens* egg half-way between the poles. (After Hinton, 1968a.)

remainder is designated exochorion. The terms inner and outer chorion are equally acceptable.

At oviposition and for some hours afterwards the mosquito egg is surrounded only by the chorion. During embryological development the serosa of the embryo secretes another layer, the **serosal cuticle**, below the chorion. The serosal cuticle, endo- and exochorion are uniform in structure over much of the egg but may exhibit substantial local modification. The exochorion has provided valuable diagnostic characters for separating closely related species since it was first used by Falleroni (1926) to distinguish *Anopheles labranchiae* and *An messeae*. Scanning electron microscopy is now commonly used to examine the exochorion surface. There have been few ultrastructural studies of the deeper layers of the egg shell.

3.1.1 *Culex*

The structure of the chorion has been investigated in species of *Culex* but the structure of the serosal cuticle of *Culex* is not known. The eggs are laid in rafts on the water surface, all standing vertically on their anterior poles and adjacent eggs locking together through chorionic protrusions. The endochorion is a continuous sheet, 0.25–0.4 μm thick. The exochorion is a compound layer, the constituents of which have been given a variety of names by different investigators. The innermost region of the exochorion, here designated the **lamellate layer**, is highly stratified when first secreted. It is about 0.05–0.1 μm thick. From the lamellate layer arise **tubercles** which are coated by the thin outermost layer, the **meshwork**. Over most of the egg the meshwork extends between tubercles as an open network of strands (Figure 3.1), but near the anterior pole it has the form of a perforated sheet about 0.11–0.15 μm thick (Hinton, 1968a; Pollard *et al.*, 1986; Sahlén, 1990). The exochorion is readily stripped away from the endochorion. Beament (1989) stated that the endo- and exochorion were attached to each other only at the two ends of the egg.

In *Cx pipiens* eggs the depth of the submeshwork space varied from 0.13 to 0.27 μm, depending on location, and tubercle height ranged from 0.90 to 1.60 μm. Tubercles with larger diameters always protruded higher above the meshwork than did those of smaller diameter. Freeze fracturing of egg rafts revealed that adjacent eggs were held together by interdigitation of tubercles over the areas of contact, the tops of the longer tubercles being forced through the meshwork of adjacent eggs. Tubercle size varied with position around the circumference of the egg. Large tubercles on the convex side of the egg locked between relatively smaller ones on the concave side. In the assembled raft larger tubercles might be compressed or tilted, while the smaller tubercles might not reach the meshwork of the adjacent egg (Sahlén, 1990).

In unlaid oocytes, at the anterior pole, the exochorion is extended to form a large closed sac, the micropyle apparatus. This contains nurse cell remnants and is traversed by a delicate membranous channel which opens over a modified region of endochorion called the micropyle plug (Figure 3.2C). The plug has a spongy texture in recently laid eggs. It is believed that during ovulation spermatozoa penetrate the egg shell via the micropyle channel and plug. During oviposition the apical region of the sac and most of the micropyle channel are lost, and subsequently the micropyle apparatus is seen as a cup-shaped corolla, about 54–64 μm wide and 24 μm deep, upon which the egg stands on the water surface (Figure 3.2D) (Christophers, 1945; Lincoln, 1965; Hinton, 1968a; Beament and Corbet, 1981).

The corolla is web-like, with prominent radial components. It is flexible when wet and can assume a range of configurations: reflexed posteriorly to surround the anterior region of the egg, or lying as a flat disk on the water surface, or protruded anteriorly like a cup. The lower or inner surface of the corolla is readily wetted by water. The upper or outer surface is hydrofuge, and its contact angle with water is >100°. The corolla has a circumference of *c.* 0.4 mm, so that a surface tension of 7.2 dyne mm^{-1} (that of clean water) would give a radial force of *c.* 3 dyne. An isolated egg weighing 0.01 mg, if floating upright,

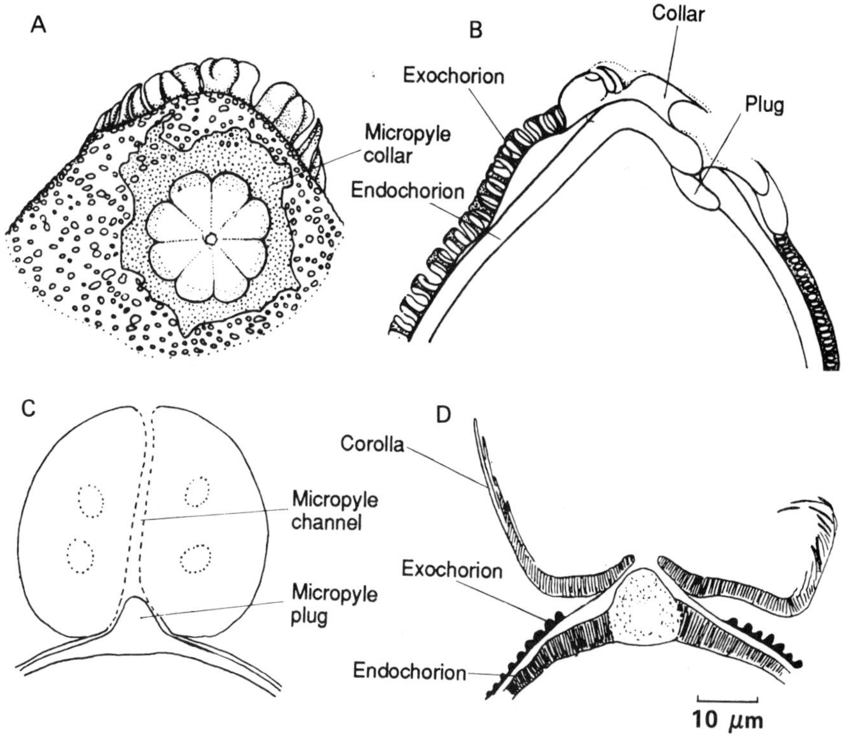

Figure 3.2 The micropyle region. (A) Surface view of the micropyle region of an *Anopheles labranchiae* egg. (B) Longitudinal section through the micropyle region of an *An labranchiae* egg before oviposition. (From Marshall, 1938, after Nicholson, 1921.) (C) Longitudinal section through the micropyle region of a *Culex pipiens* egg before oviposition. (D) Longitudinal section through the micropyle region of a *Cx pipiens* egg after oviposition. (From Christophers, 1945.)

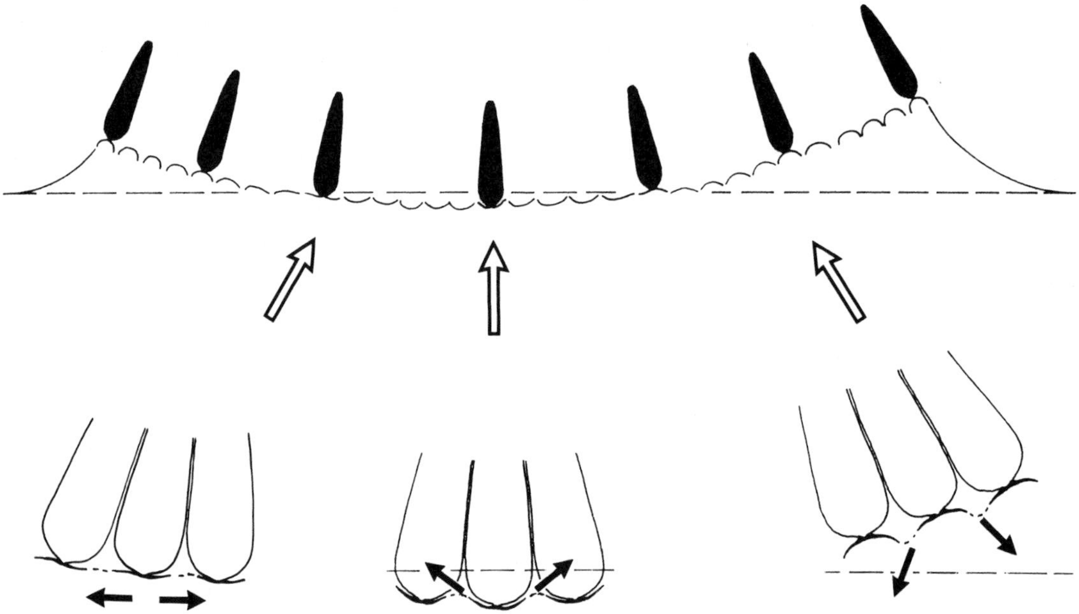

Figure 3.3 Dispositions of the corollas in a *Culex pipiens* egg raft. (Top) Diagrammatic side view of the eggs along the longitudinal midline of the raft, to show the disposition of the corollas and the water meniscus. For clarity, all of the corollas are indicated but most of the eggs are omitted. (Bottom) Diagrams illustrating the disposition of the corollas and the surface tension forces (solid arrows) at different positions in the raft (hollow arrows.) Long dashes, level of water surface away from the raft. Short dashes, local meniscuses between corollas. (After Beament and Corbet, 1981.)

would produce a downward force in the middle of the corolla of about 0.01 dyne (10^{-7} newton), a negligible force for any surface tension normally encountered on water.

Because *Culex* eggs are slightly tapered the rafts are dished, both from front to rear and from side to side. The radius of curvature of the raft is close to that of the water meniscus around emergent vegetation; consequently, when rafts drift to these boundaries they tend to remain there. The configurations of the corollas at different points along the midline of a raft surrounded by a flat water surface are shown in Figure 3.3. There are large downward forces at either end, where the corollas support columns of water, and if a raft is tilted the increased mass of supported water at the raised end provides a restoring force. In the centre, where the corollas are concave towards the eggs, corresponding upward-directed forces are produced which counteract the mass of the raft (Beament and Corbet, 1981; Beament, 1989).

Experimental evidence showed the surfaces of the submeshwork space to be wettable and the space normally to be filled with water. When rafts were floated on dye solution, dye passed through the hole in the centre of the corolla of each egg (Figure 3.2D) into the submeshwork space and became concentrated apically as water evaporated. If an egg was submerged an air film was retained between the tubercles outside the meshwork (Beament and Corbet, 1981; Beament, 1989; Sahlén, 1990). The idea that the submeshwork space is filled with air and functions as a physical gill or plastron (Lincoln, 1965; Hinton, 1968a) is erroneous. The chorion contains no visible air channels, and it must be assumed that it is permeable to respiratory gases.

The exochorion is absent from the posterior pole and at that point there is a slight round or oval depression, 8–10 μm wide, upon which an oily droplet of about 50 μm rests. Droplets from *Cx quinquefasciatus* eggs were shown to contain a volatile compound, (–)-(5*R*,6*S*)-6-acetoxy-5-hexadecanolide, to which gravid females responded as to an oviposition pheromone. The droplets appear shortly after oviposition, and reach their maximum size about 24 h later (Laurence and Pickett, 1985; Laurence *et al.*, 1985). Ether washings of *Cx tarsalis* egg rafts contained a number of unusual 1,3-diacylglycerols in which the fatty acids had acylated hydroxyl groups. It is interesting that a major hydrolytic product was *erythro*-5,6-dihydroxyhexadecanoic acid (Starratt and Osgood, 1972), from which the racemic pheromone could be synthesized directly by cyclization (Laurence and Pickett, 1982).

Beament (1989) postulated that prior to oviposition the pheromone is located between the endo- and exochorion. He observed that droplets did not appear on eggs placed on concentrated salt solution, which failed to swell, and concluded that swelling of the eggs after oviposition, due to water uptake, squeezed the oily material to the posterior pole of the egg where it formed a droplet.

3.1.2 *Aedes*

Exo- and endo*chorion* appear to be physically separate. The endochorion can be 4–5 μm thick; much thicker and tougher than that of *Culex* (Harwood, 1958). It lacks the proteinaceous crystallites present in the endochorions of many other insects. The exochorion consists of a very thin inner layer from which arises a reticulum of ridges and tubercles. In many species the reticulum forms a pattern of so-called 'exochorionic cells', which are bounded by polygonal ridges and contain tubercles highly variable in form. The ridge pattern, which is apparent on the endochorion also, possibly reflects the polygonal imprints of the ovarian follicle cells (Harwood, 1958; Powell *et al.*, 1986a; Linley, 1989). In *Ae cantans* each exochorionic cell contains a single simple tubercle. In *Ae aegypti* each contains a large, central sculptured tubercle surrounded by a number of smaller tubercles, to some of which it is connected by strands (Hinton and Service, 1969; Matsuo *et al.*, 1974). Associated with the exochorion of aedine eggs is a sticky substance, possibly a glycoprotein, which becomes hard when dry, fixing the eggs securely to the substrate (Christophers, 1960; Padmaja and Rajulu, 1981).

After the mosquito embryo has developed its extra-embryonic membranes the serosa secretes a simple sac, the serosal cuticle, between the embryo and the chorion. This occurred 16–24 h after oviposition in *Ae aegypti* eggs kept at 24°C. The serosal cuticle of *Ae aegypti* is much thinner than the endochorion, and is composed of two layers with a matrix between them. The outer layer consists of flat, overlapping, hexagonal micelles, and the inner of fibrillar micelles which run at right angles to the long axis of the egg. At the line of dehiscence of the egg cap the fibrils of the inner layer lose their parallel arrangement and become interwoven. The X-ray diffraction pattern of *Ae aegypti* serosal cuticle matched that of chitin; *Ae hexodontus* serosal cuticle gave a positive response to histochemical tests for chitin. In these two species the serosal cuticle becomes impregnated with a wax which probably reduces the rate of water loss from the egg (Beckel, 1958; Harwood, 1958; Harwood and Horsfall, 1959; Judson and Hokama, (1965).

The chorions of aedine eggs can be dissolved by brief exposure to 0.67 M sodium hypochlorite (NaOCl), leaving viable embryos within the serosal cuticles (Mortenson, 1950). The serosal cuticle is liable to disruption by sodium hypochlorite while it is being secreted but is resistant when fully formed (Judson and Hokama, 1965). The chorionic pigments of aedine eggs can be bleached by exposure to a solution of 0.037 M sodium chlorite ($NaClO_2$) and 1.45 M acetic acid at 4°C in a sealed container. This renders the egg shell transparent without affecting embryo viability (Trpis, 1970).

3.1.3 *Anopheles*

Anopheles eggs, which lie individually on the water surface, are boat-like. The flattened upper surface of each egg is not covered by water, but the remainder of the egg, which is keel-shaped, is submerged. The flat surface, which is called the deck, is surrounded by a modified region of exochorion (the frill), and is flanked on either side by two air-filled expansions of exochorion (the floats) (Figure 3.4). The blunter end of the egg is the anterior pole; the upper and lower surfaces of floating eggs correspond respectively to the dorsal and ventral aspects of the early embryo (Ivanova-Kazas, 1949). Over different regions of the eggs the exochorion is highly modified. The deck exochorion is always tuberculate, and in many species some 3–10 large lobed tubercles, of unknown function, are present at both ends of the deck. The floats are formed of hollow struts, which project outwards from the egg, and membrane which covers the struts to enclose air-filled compartments. Over the region of the egg that is below the frill, and most of which is normally submerged, the exochorion consists of perpendicular columns ending in convex caps of irregular outline, adjacent caps being connected by bridges. Where the caps are circular they resemble small mushrooms. The caps are grouped in hexagons or polygons which are thought to correspond to the boundaries of the follicle cells. The whole complex of columns and caps is filled with gas. Around the micropyle the exochorion forms a raised micropylar collar (Hinton, 1968b) (Figure 3.2A, B).

The eggs of most species of *Anopheles* bear floats which cause them to lie with the flattened dorsal surface uppermost. If these are removed the egg floats on its side (Newkirk, 1955). The eggs of several species of *Anopheles* lack floats, and in a few the frill also may be reduced or absent (Mattingly, 1969); such eggs sink readily if touched (Logan, 1953).

The gas film within the exochorionic network on the sides and lower surface of the egg resists wetting under relatively high hydrostatic pressures. In many *An atroparvus* eggs the network was not wetted after exposure to an excess pressure of 40 cm Hg (72.8 dyne cm^{-1}) in tap water for 1 h. This led Hinton (1968b) to postulate that the air-filled network served as a permanent physical gill or plastron, although only functioning in the rare event of eggs being trapped below the water surface.

Exochorion structure may change at different seasons, for example in the positions of the frill and floats (Mattingly, 1971). The diapausing winter eggs of *An walkeri* differ from the summer eggs in having larger floats and the deck exochorion reduced to two small areas at the either end of the egg (Hurlbut, 1938). In *An sergentii* eggs the presence or absence of floats depends upon the temperature prevailing during the development of the ovarian follicles, floats being formed when females are subjected to low temperatures and absent when the same females are subjected to high temperatures (Mer, 1931). Exposure of *An gambiae* s.l. to 13°C for three days immediately

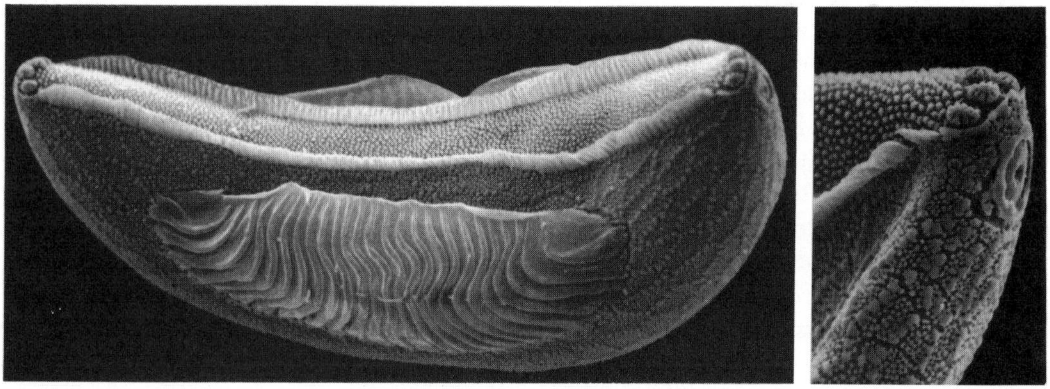

Figure 3.4 Scanning electron micrographs of eggs of *Anopheles gambiae*, in ventrolateral aspect. (Left) A frill surrounds the deck, at each end of which is a group of lobed tubercles. At the side of the egg is a float. (× 400.) (Right) Detail showing the micropyle surrounded by a micropylar collar. (× 900.) The egg shells show some preparation damage.

after gorging led to changes in the sculpturing of all eggs of the next batch to be laid. Delay of chilling until 15 h after gorging led to fewer abnormal eggs, and delay for 48 h prevented all abnormalities (Deane and Causey, 1943).

3.1.4 Other genera

Mansonia females lay their eggs in rosette-like batches on submerged vegetation, securing the posterior poles to the plant surface with cement. The egg diameter increases gradually with distance over 60% or more of the egg length from the posterior pole, this distance varying with species, and then decreases, more or less sharply so that the anterior one-third to one-sixth of the egg is tubular, some 20 μm wide, ending in a minute corolla which surrounds the micropyle. The exochorion, which is <1 μm thick, does not contain an air layer. The greater part of each egg mass is enveloped in an air bubble, but the strongly narrowed posterior regions of the eggs, where the exochorion is hydrophilic, protrude into the water (Lincoln, 1965; Boreham, 1970; Linley *et al.*, 1986).

The eggs of a number of genera, in addition to *Culex*, are laid in rafts on the water surface; viz. those of *Culiseta*, *Coquilletidia*, *Trichoprosopon*, *Uranotaenia* and *Armigeres*, although in most cases one or more subgenera employ alternative modes of oviposition (Mattingly, 1971). The boat-shaped rafts of *Culiseta* are curved like those of *Culex*, and adjacent eggs appear to link through a similar system of exochorionic tubercles. *Culiseta* eggs lack corollas, but because of the combination of a hydrophilic anterior pole with an otherwise hydrophobic egg surface, even single eggs are capable of standing upright on the water surface. *Culiseta* eggs bear terminal droplets on their posterior poles, like those of *Culex* (Van Pletzen and Van der Linde, 1981).

The sabethine mosquito *Trichoprosopon digitatum* lays rafts of some 30–80 eggs. The eggs float vertically with the anterior one-third, which has a hydrophilic exochorion, immersed. Elsewhere the exochorion is hydrophobic, but over both the hydrophilic and hydrophobic regions it is formed into numerous tubercles. At three positions equidistant around the egg, and over the middle one-third of the egg length, bands of hydrophilic exochorion about 30 μm wide extend from the hydrophilic region into the hydrophobic region of the egg shell. Over the greater part of these bands, which are termed embrasures, the exochorionic tubercles are replaced by a dense mat of long, slender, hooked filaments. Adjacent eggs are held together by a 'velcro-like' attachment of the embrasures, forming rafts which are brooded by the females (Linley *et al.*, 1990).

Chagasia eggs are boat-shaped like those of *Anopheles*, but differ in possessing a number of longitudinal floats which cover most of the surface. The egg of *Ch rozeboomi* has 6 or 8 floats and a deck of normal width (Causey *et al.*, 1945); that of *Ch bathana* has 10 floats and the deck is all but eliminated (Baerg and Boreham, 1974).

3.2 SCLEROTIZATION

Mosquito egg shells are always soft and white when laid but they harden and darken within a few hours. *Mansonia titillans* egg shells turn brown within about one hour (Linley *et al.*, 1986); those of *Anopheles punctulatus* become uniformly black after about two hours (Mackerras and Lemerle, 1949). Within two hours of oviposition the chorion of *Aedes triseriatus* eggs becomes black and almost impenetrable by a glass micropipette (McGrane *et al.*, 1988). The chorion of *Culex pipiens* eggs becomes hard and dark within one or two hours after oviposition. The free tyrosine content of *Culex* eggs falls almost to zero after oviposition, an event which Chen and Briegel (1965) associated with the sclerotization of chorionic proteins.

It has been established that it is the endochorion of *Ae aegypti* eggs that hardens and darkens, at the same time becoming less permeable to water. The exochorion remains colourless. At oviposition the endochorion has a homogenous but stratified appearance, later it appears to be formed of tightly packed flat fibrils which spiral longitudinally around the egg. Its X-ray diffraction pattern resembles that of arthropodin from the larval cuticle of *Sarcophaga* (Beckel, 1958; Harwood, 1958; Powell *et al.*, 1986b).

Chorionic proteins are secreted by the maternal follicular epithelium during embryogenesis. Details of the protein constituents of the exo- and endochorion of *Ae aegypti* are to be found in Section 19.6.1(a). The enzymes involved in sclerotization and darkening are present within the endochorion of the mature oocyte. Details of their actions are reported in Section 19.6.2.

3.3 WATER RELATIONS

3.3.1 Resistance to water loss

Mosquitoes of the tribe Aedini mostly deposit their eggs away from the water surface, and their eggs show varying degrees of tolerance to dry conditions. Eggs of *Aedes*, *Psorophora*, *Opifex* and *Haemagogus* are known to survive for several months away from water (Horsfall, 1972; Haeger and Provost, 1965; McGregor, 1965; Galindo *et al.*, 1955). However, of the many species of *Aedes*, not all have resistant eggs (Galindo *et al.*, 1951; Consoli and Williams, 1978). Eggs of *Heizmannia*, *Eretmapodites* and of certain species of *Armigeres* can survive away from water for one to three weeks (Macdonald and Traub, 1960; Barr, 1964; Lounibos, 1980). *Armigeres digitatus* eggs quickly collapse from desiccation if removed from water (Okazawa, 1991). *Psorophora* and many species of *Aedes* breed in ground pool habitats. Other species of *Aedes*, and species of *Haemagogus*, *Heizmannia*, *Eretmapodites* and *Armigeres* utilize container habitats for oviposition (Mattingly, 1971).

Floodwater mosquitoes deposit their eggs in moist soil at the margins of depressions that are exposed as the water table falls. Large numbers of *Aedes vexans* and *Ae sticticus* larvae appeared when soil in which wild females had oviposited, and which had subsequently been protected with mosquito-proof cages, was flooded after an interval of two years (Gjullin *et al.*, 1950). In a similar experiment some eggs hatched with each flooding but a percentage of the eggs of nine species of *Aedes* and *Psorophora* remained dormant and viable for four years (Breeland and Pickard, 1967). In East Africa *Ae aegypti* breeds in a variety of natural and man-made small containers, from rock holes to mollusc shells and car tyres, and the viability of the eggs varies with the type of container. In areas where there was a four-month hot dry season, egg survival was sufficient to maintain the species (Trpis, 1972a, b). In Nigeria, two granite rockpools yielded larvae of *Ae aegypti* and *Ae vittatus* 4.5 months after the last rains, following a dry season in which soil temperatures reached 40°C and air relative humidity fell as low as 5% (Irving-Bell *et al.*, 1991).

Laboratory studies suggested that survival of the pharate larvae within aedine eggs was largely determined by their ability to conserve water, although temperature was also an important factor. Ninety-eight percent of *Aedes vexans* eggs hatched after storage in a saturated atmosphere at 2°C for 5 months, but storage at lower humidities or at high temperatures reduced viability. Eggs 28 days old were much more resistant to water loss than eggs 14 days old (Brust and Costello, 1969). *Aedes vexans* eggs could survive at 20% r.h. or less for two or three weeks, and eggs that had lost water regained weight when transferred to a saturated atmosphere, particularly if the previous treatment had not been too severe (Costello and Brust, 1969). Eggs of *Psorophora ciliata* and *Ps cyanescens* kept in plastic containers at 4°C hatched after about three years (Breeland and Pickard, 1967). *Ae aegypti* eggs kept in a saturated atmosphere at 16°C for 8 months suffered little loss in weight, and a high percentage remained viable. Eggs that were held under conditions of temperature and humidity which were slightly more rigorous, lost weight more rapidly and survived less well (Meola, 1964).

Although anopheline eggs cannot resist prolonged drying, they show some tolerance of desiccation under experimental conditions, and the eggs of certain species can survive the temporary drying of a breeding site. A low percentage of eggs of *An punctimacula* hatched after 4 weeks in drying soil (Stone and Reynolds, 1939). In the laboratory, almost all *An sinensis* eggs hatched after two days of complete desiccation, and over 20% hatched after ten days exposure to 90% r.h. (Oguma and Kanda, 1976). Almost 30% of *An balabacensis* eggs survived for 21 days on moist filter paper, and 2% survived for 92 days (Wilkinson *et al.*, 1978). Eggs of *An gambiae* s.l. held in a moist environment at room temperature showed a progressive decrease in hatching, from 91% on day 2 to 1% on day 12. Samples of dry soil removed from the top 3 cm of the surfaces of temporary pools and cow hoofprints along a stream bed yielded small numbers of *An gambiae* s.l. larvae on flooding (Beier *et al.*, 1990). Tolerance of desiccation is important for *An melas* which, during the dry season, breeds only in mangrove swamps and only when these are flooded by the bi-monthly spring tides that provide water of low salinity long enough for larval growth. The females prefer damp substrates to free water for oviposition. Some 25% of eggs were viable after 12 days on a drying soil surface, but only 10% contained larvae capable of hatching (Giglioli, 1964a).

Culex pipiens eggs are surrounded by a layer of water contained in the exochorionic submeshwork space. This can be lost from the terminal eggs of a raft if their corollas lose contact with the water surface, and such eggs sometimes fail to complete development. Measurements of relative humidity at different heights above the water surface in a water butt indicated that the posterior poles of the uppermost eggs in a raft were probably exposed to an r.h. of <100% (Beament and Corbet, 1981). *Culex quinquefasciatus* and *Culiseta incidens* rafts which had dead eggs around the periphery were found after nights with strong dry winds (Guptavanij and Barr, 1985).

3.3.2 Water uptake and waterproofing

At oviposition the chorions appear to be highly permeable, and the eggs are particularly prone to water loss at that time. *Aedes aegypti* eggs that had started to darken would collapse due to water loss within 1–2 min of removal from the moist oviposition site (Morris *et al.*, 1989). Indeed, the eggs normally take up water when first laid, resulting in an increase in size and weight. *Aedes aegypti* eggs increased in weight from 5 to 12 μg after oviposition through uptake of water. Uptake was most rapid during the first two hours after oviposition but continued at an appreciable rate until the 10th hour and was complete by the 16th hour (Gander, 1951; Kliewer, 1961). Eggs of *Culex pipiens* and *Anopheles* species also took up water. It appeared that the additional water was not essential for embryonic development because *An albimanus* eggs placed on 0.2 M sodium chloride hatched although they had not increased in size (Gander, 1951; Downs, 1951).

Only a little is known about the waterproofing of mosquito egg shells, but it appears that permeability to solutes is affected as well as permeability to water, since eggs become more tolerant of salt solutions with age. The permeability of the chorion decreases during the first hours after oviposition, and this has been associated with the tanning of the endochorion. For example, in *Ae aegypti* the egg gained and lost water readily when first laid but became partly resistant to water loss and uptake as the chorion darkened. In *Aedes* the principal barrier to water loss is believed to be a wax layer associated with the serosal cuticle, because the eggs of *Ae aegypti* and *Ae hexodontus* develop maximum resistance to water loss at the time at which the serosal cuticle is formed (Beckel, 1958; Harwood and Horsfall, 1959).

Culex pipiens eggs died if exposed to HCN as solute and vapour within 11 min of oviposition, but any exposed 12 min or more after oviposition survived. Beament (1989) concluded that an impermeable wax layer was deposited inside the chorion by 12 min.

3.4 HATCHING

3.4.1 Hatching stimulus

Hatching normally follows the completion of embryonic development with little or no delay, although there is evidence from *Mansonia* that an endogenous circadian rhythm regulates the time of hatching (Section 7.4.1). If conditions are adverse, some aedine species undergo diapause in the egg stage, others enter a state of quiescence. In both cases the dormant insect is a pharate 1st instar larva within the egg shell (Volume 2). Practically all aedine eggs require a hatching stimulus, in addition to submergence in water, before the larvae they contain will burst from their shells. Early investigators found that solutions of many organic and inorganic compounds and infusions of bacteria would stimulate hatching. Later it was shown that the immediate stimulus was a fall in dissolved oxygen concentration. Changes in reduction potential independent of oxygen concentration did not stimulate hatching. Lowering the dissolved oxygen concentration by exposure to a vacuum has proved an effective method of stimulation (Gjullin *et al*, 1941; Barbosa and Peters, 1969).

A falling oxygen concentration provides a more powerful stimulus than a steady low concentration. Thus, in one experiment, exposure of *Aedes aegypti* eggs to a steady oxygen concentration of 6.2 ppm stimulated a 12% hatch, whereas a fall from 8.4 to 7.6 ppm stimulted a 90% hatch. At the soil/water interface of irrigated pasture, the oxygen concentration starts to fall immediately after flooding but several hours are required before very low levels are reached. However, in the field the eggs of floodwater species hatch shortly after flooding, presumably in response to the falling oxygen concentration (Judson, 1960). The eggs of *Ae sierrensis*, a treehole mosquito, hatched only when exposed to very low oxygen concentrations 0.25 ppm or less. A fall in concentration alone was not stimulatory (Judson *et al.*, 1966).

Male pharate larvae of *Ae aegypti* show a significant tendency to hatch before their sisters. This results in an initial distortion of the sex ratio, which declines to unity within about 3 h of immersion of the eggs (Elzinga, 1961; Mayer, 1966). Intermittent filling of treeholes after the spring thaw leads to the production of multiple broods of *Ae triseriatus* larvae before the adults first emerge in early June. Early egg hatches show a sex ratio in favour of males, which does not reverse until June. In the laboratory, males hatched significantly sooner than females in response to microbial deoxygenation. This might have been due to a difference of response threshold to dissolved oxygen concentration, or to a difference in response time (Shroyer and Craig, 1981).

Temperature can be a regulating factor for species whose eggs hatch in early spring. For nine species of so-called snow-melt *Aedes*, higher hatching rates resulted from exposure to low oxygen concentrations (0, 3, 6 ppm) as the water temperature increased between 4 and 15°C (Kardatze, 1979). Hatching of *Culiseta inornata* eggs was inhibited at 2 and 5°C but normal at 10°C (Hanec and Brust, 1967).

Anopheles melas eggs could be stimulated to hatch by suddenly increasing the temperature of the water on which they were floating, or by drastically changing the salinity of the water (Giglioli, 1964a). Agitation of the water stimulated *An squamifemur* eggs to hatch (Boreham and Baerg, 1974).

3.4.2 Hatching mechanism

Hatching has been investigated in most detail in aedine eggs, in which by first bleaching the chorion it is possible to observe the behaviour of the pharate larva. The period between first lowering of the oxygen concentration and shell rupture consists of two phases. During the first phase, which lasts 2–3 min in *Aedes aegypti* and 4–6 min in *Ae sierrensis*, there is a complete absence of visible activity, although the acetylcholine content of the larva drops. During the second phase, which lasts 2–3 min in *Ae aegypti* and 5–10 min in *Ae sierrensis*, the pharynx and oesophagus show apparent swallowing movements although no fluid is ingested until the shell ruptures. The acetylcholine content rises during the second phase and falls again after hatching (Judson *et al.*, 1965). Once the shell has ruptured the swallowing movements cause the gut to to fill with water, and as this occurs the thorax and anterior part of the abdomen are extruded from the shell. Finally the larva wriggles free (Christophers, 1960).

Aedine eggs hatch more rapidly when they have been held at high humidities, which suggests that eclosion may be affected by the extent of hydration of the egg shell. When aedine eggs are held in a saturated atmosphere, shell rupture can be induced by flushing the eggs with water-saturated nitrogen, but in the absence of free water the act of eclosion is rarely completed (Harwood and Horsfall, 1959; Judson *et al.*, 1965).

When mosquito eggs hatch, the chorion and serosal cuticle split along a particular line which, in most genera, is transverse and towards the anterior pole. This line of dehiscence lies directly over the egg burster on the head of the larva, and as a result of the split a cap comes away allowing the larva to escape. In *Aedes* eggs the fine structure of the serosal cuticle is modified at the line of dehiscence (Section 3.1.2), but no preformed line of weakness has been detected in the chorion (Judson and Hokama, 1965). Application of pressure to the anterior end of aedine eggs causes the shell to crack along the normal line of dehiscence. Caps cannot be separated from the posterior end in this manner; indeed, they cannot be separated from the anterior end during the early stages of embryonic development, when pressure causes irregular fractures (Gander, 1951; Harwood and Horsfall, 1959). When shell rupture is induced in infertile eggs, which have a normal-looking chorion but lack a serosal cuticle, the shell splits in a longitudinal direction following the orientation of the constituent fibrils (Judson and Hokama, 1965). These authors concluded that shell rupture is initiated by the egg burster penetrating the shell layers, but it remains unknown why the much thicker chorion splits along the same path as the serosal cuticle with its preformed line of weakness.

Eggs laid in rafts or batches are orientated so that hatching larvae can easily enter the water. *Mansonia* eggs are held by the posterior pole. *Culex* and *Culiseta* eggs stand with the anterior pole downwards, and the emerging larvae pass straight into the water.

After hatching, the larva of *Ae aegypti* immediately swims to the surface exposing its spiracles to the air, and within 15 to 30 min the liquid which filled the tracheal system is completely replaced by air. This appears to be an active process since it is blocked by narcotization. The hydrophobic nature of the inner surface of insect tracheae is also important for the penetration of air at hatching, for without it the water could hardly be withdrawn against the massive capillary pressure (Wigglesworth, 1938b, 1981). At hatching, the larval head is wrinkled, triangular and no wider than the thorax. Most of the cuticle is unsclerotized, and the lateral palatal brushes are tucked into the preoral cavity. The larva soon swallows water and, contracting its thorax and abdomen, drives haemolymph into the head which extends mainly in the frontal region. The cuticle of the head becomes sclerotized. The lateral palatal brushes are extended and soon start to function (Wigglesworth, 1938b; Beckel, 1958; Christophers, 1960).

4
Larval feeding

Head structures that have a generalized character in the larvae of most other nematocerous Diptera are highly elaborated in culicid larvae and permit a variety of feeding mechanisms appropriate to an aquatic medium. The feeding methods of mosquito larvae extend from particle capture, the principal method, to predation. Their food ranges from particles of <1 μm to live invertebrates of their own size. In this chapter we shall examine the food of mosquito larvae, the structure of their mouthparts, and the feeding mechanisms that they employ.

4.1 FOOD SOURCES AND FEEDING MODES

4.1.1 The food

Mosquito larvae occur in a variety of aquatic habitats, but the food these habitats provide has the same general nature, principally comprising microorganisms, particulate organic matter or detritus, and biofilm. Food resources are found at different locations within each habitat. The region of the air/water interface is enriched in both particulate organic matter and microorganisms; the term 'neuston' is used for its living components. A thin organic 'surface microlayer' covers the water surface (Section 4.3.1). Physicochemical interaction of bacteria with the surface microlayer causes bacterial enrichment of the top micrometre of the water column, producing the so-called bacterioneuston layer (Norkrans, 1980). Phytoplankton are sometimes enriched at the water surface and sometimes depleted, partly owing to vertical migration (Hermansson, 1990). Wettable particles with a density less than that of water aggregate immediately below the surface membrane, while unwettable particles rest upon it. Bacteria, Protoctista (eukaryotic microorganisms) and dendritic particles that are suspended in the water column constitute the principal food resource of some species. Some other species feed principally on the thick biofilm that covers all submerged surfaces, including those of plants (periphyton), stones (epilithon) and sediments (epipelon). These biofilms consist of a consortium of bacteria, fungi and algae embedded in a polysaccharide matrix (Lock, 1990). Larvae of a few species feed on dead invertebrates; others are predators. Many species utilize more than one food resource.

Detritus, or 'particulate organic matter', is a major food resource of aquatic insects. Most detritus originates from plant material that entered the aquatic habitat, was colonized and partly degraded by fungi and bacteria, and shredded and partly digested by the aquatic fauna. The residual plant tissue is probably refractory to digestion by mosquito larvae, although the high pH of their midguts may counter this. The particles are coated and sometimes infiltrated with microorganisms, which have a high nutritive value but which constitute only a small part of their biomass.

Particulate organic matter (POM) is classified according to size: (a) coarse (CPOM), from any size down to 1 mm; fine (FPOM), <1 mm and >50 μm; and (c) ultrafine (UPOM), <50 μm and >0.5 μm. Suspended particles ingested by mosquito larvae fall almost entirely within the ultrafine class (Section 4.4.2). By convention,

Table 4.1 Mean percentages of microorganisms and detritus (<10μm) in the alimentary canals of wild larvae. (From Walker *et al.* (1988) and Merritt *et al.* (1990).)
Within 2 h of collection from their natural habitats, 4th instar larvae were fixed in 10% formalin prior to treatment with a fluorochromatic stain (DAPI). It is probable that many protists were destroyed during ingestion and that others were digested during the 2 h holding period ($n \geqslant 5$).

Species (habitat)	*Gut contents (%)*		
	Bacteria	*Protists*	*Detritus*
Cq perturbans			
(*Typha* marsh)	95.0	0.014*	5.0
(*Typha* marsh)	97.9	0.022	2.1
Ae triseriatus			
(treehole, tyre)	79.7	<0.001	20.3
An quadrimaculatus			
(*Typha* marsh)	21.5	0.002	78.5

* Principally euglenids, diatoms and desmids.

organic particles smaller than 0.5 μm are classified as dissolved organic matter (DOM), a category which includes small particles, colloids and some bacteria in addition to solutes (Wotton, 1990a).

A guide to what mosquito larvae ingest can be obtained from examination of their gut contents, but its value is limited because some thin-walled organisms become unrecognizable soon after they have been ingested. Thus the delicate ciliates *Pleuronema* and *Homalogastra* were undetectable among the foregut contents of *Culex annulirostris* larvae five minutes after they had been ingested (Laird, 1956), and no trace of mastigotes could be found in the gut contents of *Aedes* larvae within three minutes of ingestion (Laird, 1988). Digestion by *Anopheles gambiae* and *Cx pipiens* of certain species of *Synechocccus* and *Synechocystis* (Cyanobacteria) was almost complete within thirty minutes, but digestion of other species of the same genera took longer (Thiery *et al.*, 1991). Quantitative methods for the examination of gut contents were lacking until use of a fluorochromatic stain, 4,6-diamidino-2-phenylindole (DAPI), with illumination at 365 nm, made possible the identification of organic detritus, bacteria, algae, and other protists (single celled protoctists) by their fluorescence, colour and shape (Walker *et al.*, 1988).

Examinations of the gut contents of mosquito larvae collected from natural habitats have shown that most species are not discriminatory in what they ingest, which is mainly governed by their size, method of feeding and the site. The gut contents corresponded very largely with the organisms and dendritic particles of appropriate size present at the feeding site, although sometimes with an apparent excess of the more rugged constituents (Senior-White, 1928; Pucat, 1965; Ameen and Iversen, 1978). However, within the same small container habitat, larvae of *Culex quinquefasciatus* collecting–filtering in the water column ingested free-swimming bacteria and algae while larvae of *Aedes aegypti* and *Ae polynesiensis* collecting–gathering at the substratum ingested bacterial film, algae, fungal mycelia, protists, rotifers, and the eggs of aquatic worms (Rivière, 1985). The presence of particular organisms within the larval gut does not prove that they are a food resource. Some species of algae are resistant to digestion and are discharged intact from the alimentary tracts of mosquito larvae. Indeed, it has been estimated that 75% of ingested algae may be unaffected (Howland, 1930b; Laird, 1988). A high proportion of the resistant species belong to the Chlorococcales, and their resistance has been ascribed to the presence of a carotenoid polymer, sporopollenin, in their cell walls (Marten, 1986). Non-nutritive particles found among larval gut contents included silt particles, fungal spores, insect scales, and diatom shell fragments (Senior-White, 1928; Laird, 1956).

Another approach to the investigation of larval food sources is to identify the living organisms of appropriate size present at natural feeding sites. Unfortunately fresh water biologists have tended to concentrate on the protoctists, particularly the algae, in larval habitats (Howland, 1930a; Laird, 1956; Pucat, 1965), possibly because their methods underestimated bacteria and inert particles. When the gut contents of 4th instar larvae of three species, taken from their natural habitats, were treated with a fluorochromatic stain, and the individual particles identified and counted, bacteria and detritus were numerically predominant. Very few protoctists were found, but their numbers had almost certainly been reduced by digestion (Table 4.1). However, in the case of the suspension-feeding *Coquillettidia perturbans* larvae, the biomasses of the bacterial and protist fractions were suspected to be similar despite the great difference in particle number (Walker *et al.*, 1988; Merritt *et al.*, 1990a). Bacteria may be the key food source in transient water bodies which have an initial paucity of protoctists, including algae (Laird, 1956), and they may be the most important food source in treeholes (Walker and Merritt, 1988).

The potential food source in a single biotope can be highly complex, and its composition may change dramatically with time. This is exemplified by data for a snowmelt, tundra pool in the Canadian subarctic. The pool was 2.5 m diameter, 0.5 m deep centrally immediately after the spring thaw but diminishing to 0.3 m, and was the habitat for four species of *Aedes*: *communis*, *excrucians*, *hexodontus* and *impiger*. During three weeks of daily investigation, 276 taxa (species and subspecies) of living organisms were identified among the pool biota, of which 6.5% were bacteria and 70.3% Protoctista (including 114 taxa of diatoms). The protoctists and bacteria provided all but a tiny fraction of the material found in the guts of the mosquito larvae, the remainder being chiefly fungal spores and pollen grains. A day after thawing (29 May) the pool teemed with bacteria, particularly in the vicinity of decomposing vegetation. The gut contents of newly hatched mosquito larvae were dominated by bacteria, accompanied by bactivorous protists, but particulate organic matter was also present. The initial bacterial proliferation assured a nutrient-rich environment for protoctists that had survived encasement in the ice, and the guts of the 2nd and later instar larvae were green with algae. By early June the water was greenish due to a postbacterial phytoplankton bloom, mostly diatoms. By mid-June the pool was polysaprobic (high in decomposable organic matter and low in oxygen). Diatoms and chlorophytes were plentiful in the periphyton and available to the older mosquito larvae (Laird, 1988).

The minute young of ostracods, copepods and cladocerans were a characteristic faunal component of good *Cx tarsalis* breeding waters. Analysis of these crustaceans revealed that they were rich in C_{20} polyunsaturated fatty acids, which are an essential constituent of the larval diet (Section 5.3.2). The same compounds were found in appreciable quantities in sediment and detritus from the same biotopes (Dadd *et al.*, 1988).

Larvae of the *Culex bitaeniorhynchus* species group feed selectively on filamentous algae of the Zygmentaceae, such as *Spirogyra*, which occur in masses in ponded streams and shallow permanent ponds, possibly thereby obtaining protection from predators. They also ingest organisms epiphytic on the *Spirogyra* (Laird, 1956, 1988).

Predatory mosquito larvae feed on live invertebrates, usually of their own size or smaller. Because mosquito larvae are the most common arthropods in the container habitats of many predator species, they are presumed to be their chief prey. However, predatory mosquito larvae have been reported to feed on other dipterous larvae, small dragonfly larvae and small tadpoles. They feed readily on struggling insects trapped in the surface film (Steffan and Evenhuis, 1981).

Most mosquito larvae are omnivores. Their mixed diet provides the proteins necessary for growth and also the vitamins, nucleotides and sterol that are essential for metabolism and development (Chapter 5). It seems unlikely that any class of food organism has key nutritional importance for culicids unless it is those that provide C_{20} polyunsaturated fatty acids.

4.1.2 Feeding modes

The classification of larval culicids by feeding mode into filter feeders, browsers and predators

Table 4.2 Feeding modes of mosquito larvae.

Feeding mode	*Food resource*	*Location*	*Examples*
Collecting–filtering	Microorganisms and detritus	Air/water interface	Most *Anopheles*
		Water column	Many *Culex*, *Culiseta* some *Aedes*
		Stems and roots	*Mansonia*, *Coquillettidia*
Collecting–gathering	Microorganisms and detritus	Substratum	Many *Aedes*, some *Psorophora*, *Haemagogus*, *Wyeomyia*
Scraping	Biolayers, adherent algae and protists	Surfaces of plants, stones, substratum	*Ae atropalpus*
Shredding	Leaves	Water column	*Cs inornata** *Ae triseriatus**
	Spirogyra	Water column	*Cx bitaeniorhynchus*
	Dead invertebrates	Substratum	(subgenus) *Rachisoura*
Predation	Live invertebrates	Water column	*Toxorhynchites* (subgenera) *Psorophora*, *Lutzia*, *Mucidus*

* Minor activity only.
Fuller details may be found in Sections 4.3.4 and 4.3.5.

(Surtees, 1959) provided a conceptual framework which proved useful for many years. However, it has been superseded by a more detailed classification, appropriate to all aquatic invertebrates, in which species are classified according to their functional role within the ecosystem (Cummins, 1973). In this system, 'collectors' gather particles that are in suspension or that have settled on surfaces; 'scrapers' remove food materials that are tightly adherent to plant and mineral substrata; 'shredders' gnaw plants or dead organisms; and 'predators' attack living animal prey. The particles ingested by collectors comprise both microorganisms and particles of detritus. Cummins and Klug (1979) subdivided collectors into 'filterers' and 'gatherers', to distinguish those feeding on suspended particles from those feeding on particles that had settled out on the substratum. The adaptation of these concepts for culicid larvae that is presented here is identical with that of Merritt *et al.* (1992b).

Collecting is the feeding mode most commonly used by mosquito larvae. **Collecting–filtering** (previously termed filter feeding or suspension feeding) is the removal of particles suspended in the water column or associated with the water surface. It is found characteristically among species of *Anopheles*, *Culex*, *Culiseta*, *Mansonia*, *Coquillettidia* and some other genera, and less commonly in *Aedes* (Table 4.2). **Collecting–gathering** is the resuspension and removal of particles deposited on or loosely attached to surfaces (Figure 4.1). It is found in many species of *Aedes* and in some species of *Psorophora*, *Haemagogus* and *Wyeomyia*.

Scraping is the removal and ingestion of the biofilm that adheres tightly to submerged plant and mineral surfaces, and also of the algae and stalked protists that adhere directly to such surfaces. Larvae of *Aedes atropalpus* exhibit this feeding mode in rock pools (Dyar, 1903), and it is probably more widespread.

Shredding is the gnawing and biting-off of small fragments of tissue, whether from plants

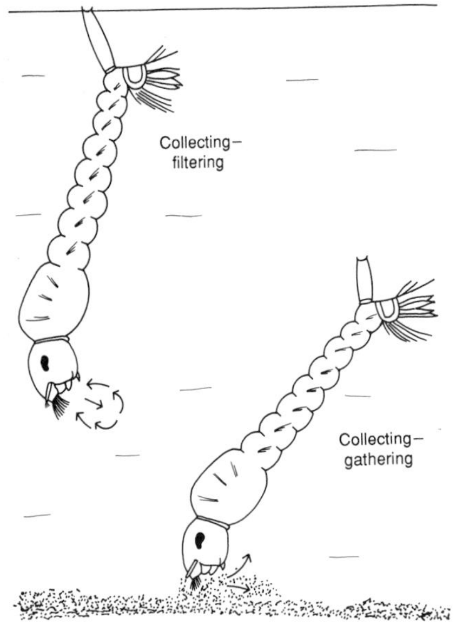

Figure 4.1 Culicine larvae exhibiting two feeding modes. (After Pucat, 1965.)

or dead invertebrates (see Section 4.3.4). The gnawing of leaves has been described in *Culiseta inornata* and *Ae triseriatus* larvae. Larvae of the *Culex bitaeniorhynchus* species group, which feed selectively on filamentous chlorophytes such as *Spirogyra*, are habitual shredders. The shredding mode is used also by larvae of *Tripteroides* (subgenus *Rachisoura*), previously classified as scavengers, which feed on dead invertebrates.

Predation is observed in larvae of *Toxorhynchites* and of the subgenera *Lutzia* (of *Culex*), *Mucidus* (of *Aedes*) and *Psorophora*, which seize and engulf other insects, often of their own size. Larvae of a number of sabethine genera are facultative predators, as are larvae of *Eretmapodites*.

Most culicines employ at least two feeding modes although they may use one predominantly. Many species, especially of *Aedes*, supplement collecting–filtering in the water column with collecting–gathering at submerged surfaces. Some also occasionally feed at the air/water interface by collecting–filtering. Larvae of *Ae triseriatus*, in simulated treeholes, spent 48% of their time collecting–filtering, 36% collecting–gathering, and about 1% shredding (Walker and Merritt, 1991). Most species of *Anopheles* are found at the air/water interface collecting–filtering, but a few, e.g. *An farauti* and *An stephensi*, spend much time at the substratum collecting–gathering (Laird, 1988; Rasnitsyn and Yasyukevitch, 1989a).

The container-dwelling larvae of some sabethine genera which have enlarged mandibles or maxillae and feed on dead invertebrates by shredding, supplement this with collecting–gathering or collecting–filtering. Among culicines, the container-dwelling larvae of *Eretmapodites* are collector–gatherers that have strongly toothed filaments in their lateral palatal brushes, and they are facultatively predatory upon competing larvae of their own and other species (Haddow, 1946; Surtees, 1959). In the laboratory, 4th instar larvae of a number of collecting–filtering species have been observed ingesting 1st

Table 4.3 Drinking rates of 4th instar larvae in tap water or sea water, in the absence of food and particulate matter (mean ± s.d.). (From the data of Aly and Dadd (1989), Asakura (1982a) and Bradley and Phillips (1977b.)

Species	*Water*	*Temp.* (°C)	*Drinking rate* *nl larva^{-1} h^{-1}*	*nl mg^{-1} h^{-1}*	*Stimulation of rate by* 0.05% *agarose* (%)	*10^{-3}M adenylic acid* (%)
Cx pipiens	Tap	22–23	48 ± 17	19	–	–
Cx quinquefasciatus	Tap	22–23	167 ± 30	73	+112	+48
An albimanus	Tap	22–23	123 ± 32	83	−20	–
Ae aegypti	Tap	22–23	309 ± 113	123	–	+95
Ae togoi	Tap	26	–	42 ± 7	–	–
	Sea	26	–	94 ± 15	–	–
Ae taeniorhynchus	Sea	27	287	82	–	–

instar larvae caught in their feeding currents (Mogi, 1978; Koenekoop and Livdahl, 1986). This phenomenon has been observed in nature also; it may have selective advantage for some container-dwelling species but does not occur in all (Seifert and Barrera, 1981; Broadie and Bradshaw, 1991).

4.1.3 Drinking

Although mosquito larvae swallow the food bolus that forms in the pharynx without any visible intake of water (Wigglesworth, 1933a; Nayar, 1966), they nevertheless imbibe substantial volumes of water when feeding and also drink extensively in the absence of food (Aly and Dadd, 1989).

Measurements of the uptake of [^{14}C]inulin or of a dye such as amaranth, added to rearing water in the absence of food, revealed that larvae drank, that the drinking rate remained constant for 1–2 h after exposure to the test agent, and that 20 h later it was still high. The drinking rates of 4th instar larvae of six species are summarized in Table 4.3. A positive correlation between larval size and drinking rate was observed in 4th instar *Culex quinquefasciatus* larvae. An increase of one day in age was accompanied by an increase of 38% in drinking rate. Older larvae that were 95% heavier drank 98% more than younger larvae (Aly and Dadd, 1989). Among saline-water species, the drinking rate of *Aedes togoi* varied with the salinity of the medium, whereas that of *Ae taeniorhynchus* did not change significantly within the range 10–200% sea water (Section 6.3.2).

Drinking rates were not significantly affected by the presence or absence of solids in the gut. *Aedes aegypti* and *Culex quinquefasciatus* larvae which were glutted with kaolin, and placed in particle-free water, drank at the same rate as larvae with empty guts. It appears that imbibed water passes down the gut between the particulate contents. Addition of kaolin particles caused water uptake by three species to increase by 70–140%. Addition of yeast extract, a phagostimulant, increased the drinking rate by 30–40%. Addition of 10^{-3}M adenylic acid stimulated increases of 48–95% (Table 4.2). The drinking rate of *Cx quinquefasciatus* larvae doubled in the presence of a colloid, 0.05% agarose, but that of *An albimanus* larvae was reduced (Table 4.3) (Aly and Dadd, 1989).

Drinking by brackish or saline-water larvae has been ascribed to the need to replace water lost osmotically across the body wall, although more water is imbibed than is needed to maintain water balance (Section 6.3.2). Drinking fresh water incurs a metabolic load and presumably therefore serves a purpose. One possible purpose would be the intake of dissolved organic matter (DOM) for its nutritive value. In some aquatic habitats the content of DOM is high relative to the particulate, in others it is relatively low. Data are more readily available for lakes and streams than for the biotopes characteristic of mosquito larvae. Such experiments as have been reported did not indicate any immediate nutritional value in DOM for mosquito larvae. *Anopheles crucians* and *An quadrimaculatus* larvae failed to reach the 2nd instar when kept in water from anopheline breeding habitats that had been filtered to remove microorganisms (Hinman, 1932). Young larvae of *Cx nigripalpus*, *Cx quinquefasciatus* and *Ae aegypti* pupated when kept in the protein-rich wastewater from dairy lagoons or fed its centrifuged sediment, but an ultrafiltrate of the wastewater which contained 5 mg protein l^{-1} was little more nutritious than tap water (Van Handel, 1986).

Aly and Dadd (1989) considered that there were orders of magnitude difference between the amount of DOM that could be imbibed by continuous drinking and the known daily weight gain of mosquito larvae, and they observed that the volumes drunk per unit time by larvae of three mosquito species were 400–3100 times less than the volumes of water cleared of particles by those species. They concluded that the nutritional intake from drinking was negligible. However, Aly and Dadd (1989) postulated that 'bottom slimes and sludges containing a few per cent of organic materials could, if drunk at high rates, contribute substantial nutriment'. Wotton (1990b) pointed out that through its uptake by bacteria and its adsorption to organic and inorganic particles, dissolved organic matter is of basic importance to the nutrition of many aquatic organisms.

4.2 STRUCTURE OF THE MOUTHPARTS AND PHARYNX

The head is well developed, with a sclerotized cranium. Due to a lengthening of the underside of the cranium the head is secondarily prognathous, consequently the cranium is turned up on the neck, and the mouthparts are directed forwards and are arranged almost perpendicular to the longitudinal axis, with the mandibles dorsal to the maxillae (Figure 4.2). The upper surface of the cranium is dominated by a large central sclerite, the dorsal apotome, formed of parts of the frons and clypeus. Anterior to it is a slender transverse labrum, and at its sides two narrow sclerites, the genae. Sutures separating the dorsal apotome and genae are lines of weakness where the cuticle splits at ecdysis (Cook, 1944; Snodgrass, 1959).

The mouthparts enclose a space, the preoral cavity, which is separated from the pharynx by the true mouth. The preoral cavity is bounded dorsally by the clypeus (part of the dorsal apotome) and labrum, laterally by the mandibles, and ventrally by the maxillae and labium. The inner surfaces of the clypeus and labrum, which together form the roof of the preoral cavity, are called the palatum (or epipharynx). Its anterior region is the labropalatum, which is separated by the tormae and intertorma from the posterior region or clypeopalatum (Figure 4.3).

The descriptions that follow in Sections 4.2.1 to 4.2.3 are appropriate to generalized particle-feeding species. A few of the named terms are not illustrated. The structural adaptations observed in the mouthparts of the different types of particle feeders and in predators are detailed in Section 4.2.4. The anatomical terms used are those recommended by Harbach and

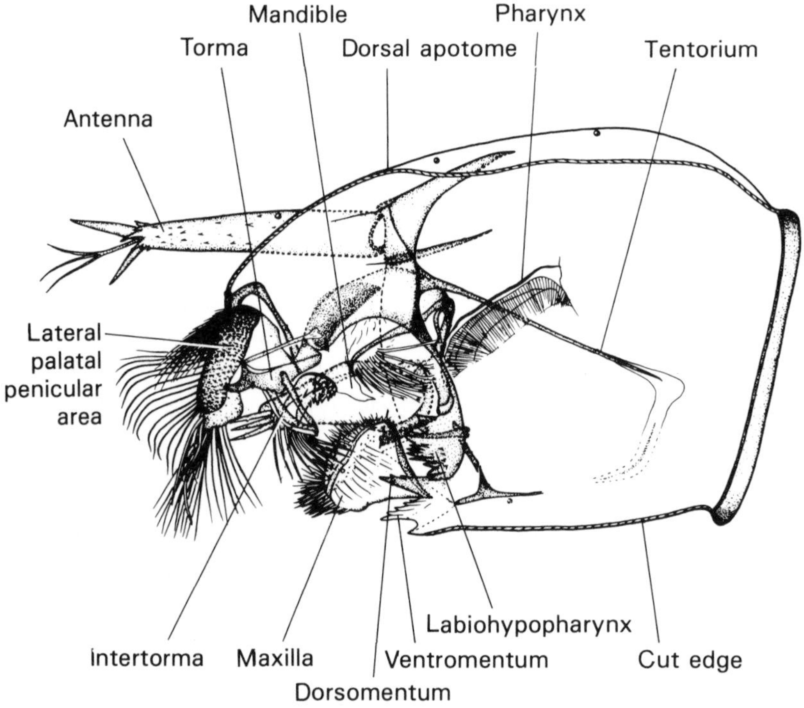

Figure 4.2 Internal view of the head of a larva of *Anopheles quadrimaculatus*, cut longitudinally to the left of the midline. (From Laffoon and Knight, 1973.)

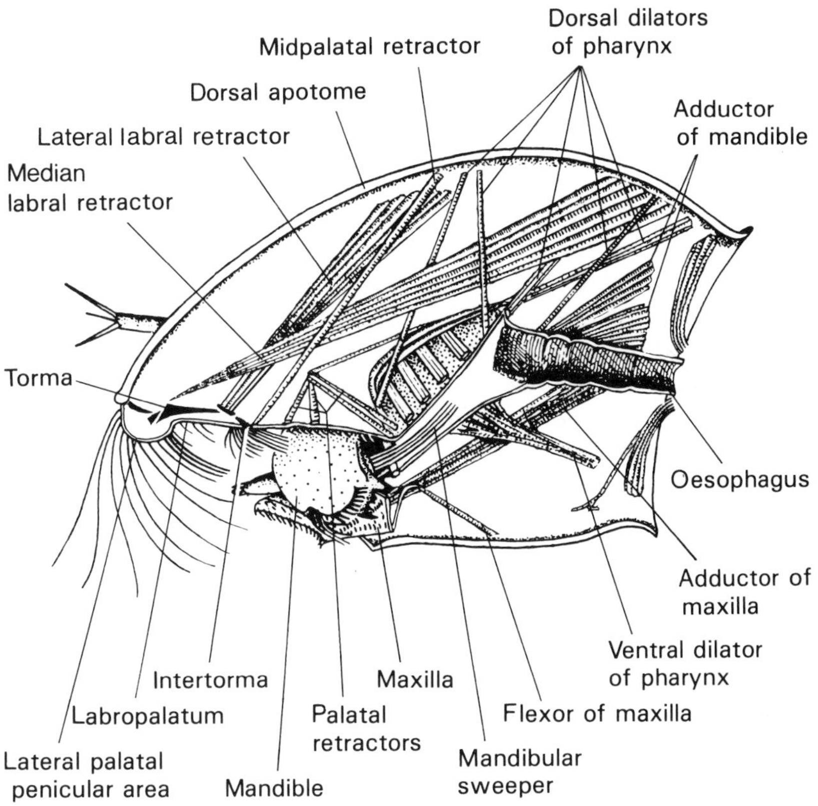

Figure 4.3 Internal view of the head of a larva of *Anopheles maculipennis*, cut longitudinally slightly to the right of the midline. (After Schremmer, 1949.)

Knight (1980), who also provided an extensive synonymy.

4.2.1 Labrum, palatum and palatal brushes

Dorsally the **labrum** is reduced to a narrow transverse sclerite, the median labral plate. Its ventral or oral surface is formed by the **labropalatum** (Figure 4.3), the anterior lip-like part of which bears three mouth brushes. A part of the lip, the anteromedial palatal lobe bears a small anteromedian palatal brush composed of short stout setae. On either side a **lateral palatal penicular area** bears a very large **lateral palatal brush** composed of long, slender filaments approx. 2–4.5 μm wide (Figures 4.2, 4.4). In *Culex tarsalis* and *Cx quinquefasciatus* the filaments in the lateral region of a lateral palatal brush are simple, being circular in cross section, whereas in the mesal region they are flattened and serrated and have a comb-like structure. In *Aedes aegypti* each brush consists of a few long, simple lateral filaments and many short mesal serrated filaments. In *Anopheles albimanus* each brush is composed of many closely packed thin filaments lacking serration. The filaments of the lateral palatal brushes pass through the cuticle of the lateral palatal penicular area and pass between a series of sclerotized rods called the lateral palatal crossbars (not illustrated). The crossbars, of about 1.2 μm diameter, are connected dorsally to the median labral plate and ventrally to the so-called **lateral palatal plate**, which is not a sclerite but a strong stretchable ligament (Figures 4.2, 4.4) (Harbach and Knight, 1977a; Manning, 1978; Dahl *et al.*, 1988; Rashed and Mulla, 1990).

Near the front of the head is a complex series of internal sclerites which is anatomically connected to the lateral palatal plates and in consequence functionally involved with the lateral palatal brushes. Anatomical and observational studies have indicated something of their interrelationships and functions but a fuller understanding awaits ultrastructural investigation. Connected to each lateral palatal plate is a sclerite of complex form, the **torma**, which is permitted only limited movement by the surrounding structures (Figures 4.2, 4.3). Between the paired tormae is a V-shaped sclerite, the **intertorma** (Figures 4.4, 4.5). Adjacent to each torma is a rod-shaped sclerite, identified by Dahl *et al.* (1988) as the **mesal intertormal apodeme** but considered to be separate from the intertorma (Figure 4.5). Two pairs of labral retractor muscles arise on the dorsal apotome. Each medial labral retractor inserts on a cuticular knob from which a ligament runs to a sclerotized ridge on the torma of its side. Each lateral labral retractor inserts broadly on the mesal intertormal apodeme of its side (Figures 4.3, 4.5).

Contraction of the lateral labral retractors displaces the mesal intertormal apodemes, freeing the posterior regions of the tormae from restraint. Simultaneous contraction of the medial labral retractors twists the tormae so that they become perpendicular to the longitudinal body axis. In consequence, force is transmitted to the lateral palatal plates, stretching them and causing the rows of filaments in the lateral palatal brushes to tilt ventrally, row by row, like the flipped pages of a book, until all the filaments are bunched together over the entrance to the preoral cavity. When the labropalatal muscles relax the tormae snap forwards and upwards, permitting elastic shortening of the stretched lateral platal plates. This abducts the filaments of the lateral palatal brushes, leaving the brushes open and the filaments fanned out and clear of the preoral cavity, while returning the tormae to their original orientations (Dahl *et al.*, 1988).

Part of the palatum is modified to form the midpalatal lobe, which carries structures used in feeding. It includes the intertorma and a number of associated small sclerites bearing arrays of fine setae and filaments which form the midpalatal brush (Figure 4.4). According to Schremmer

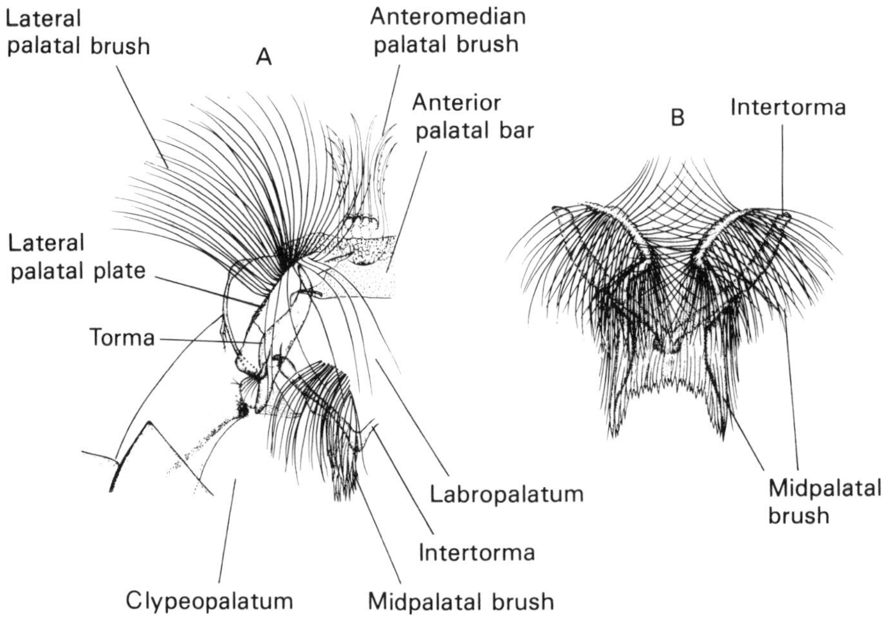

Figure 4.4 (A) Part of the lower surface of the head of a larva of *Anopheles quadrimaculatus*, after removal of mandible and maxilla. (B) Parts of the midpalatal lobe. (From Harbach and Knight, 1977b.)

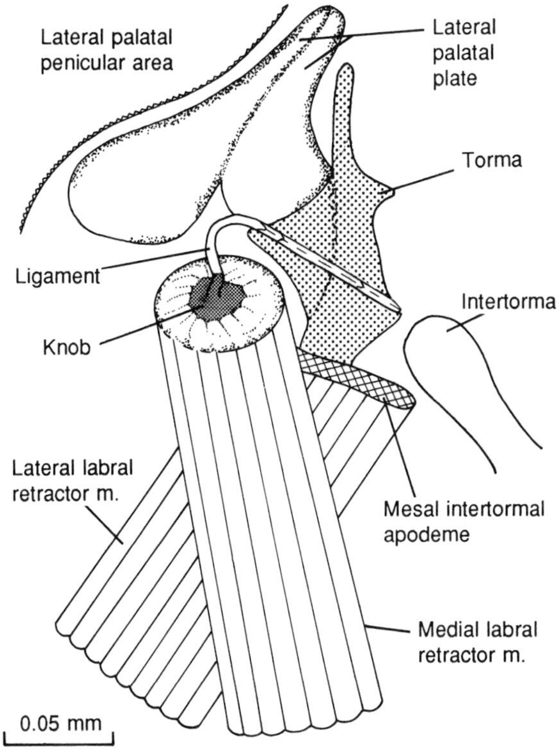

Figure 4.5 Diagrammatic dorsal view of a torma and mesal intertormal apodeme of *Culiseta morsitans*, their muscles, and associated structures. (After Dahl *et al.*, 1988.)

(1949), the midpalatal lobe is protruded into the preoral cavity by a pair of powerful midpalatal protractor muscles inserted on the end pieces of the intertorma, and withdrawn by the more delicate midpalatal retractors inserted medially on the intertorma (Figure 4.3). The actions of these muscles simultaneously arch the palatum and open the mouth.

4.2.2 The feeding appendages

The dicondylic **mandibles** are vertically hinged as a result of the inward migration of the anterior articulation. They are moved in the transverse plane by abductor and adductor muscles and do not strike one another but hammer on the labium which lies between them. The mandibles of mosquito larvae are much more elaborate than those of most insects and bear a large number of external structures. The following parts can be distinguished (Figure 4.6): (i) ventral and accessory teeth, which have incisor and molar functions respectively; (ii) a number of curved sellar setae, thought to catch large food particles and draw them into the preoral cavity; (iii) the mandibular rake, which is a series of movable flattened setae thought to clean the maxilla; (iv) a row of fine hairs, the mandibular comb, and a mandibular brush, which together are thought to clean the lateral palatal brushes; (v) a group of long hairs, the mandibular sweeper, thought to sweep food from the labropalatum into the pharynx. The sellar setae and the mandibular teeth are innervated, and a campaniform sensillum is present on the dorsal surface (Cook, 1944; Schremmer, 1949; Knight 1971; Harbach, 1977; Rashed and Mulla, 1990).

Each simplified **maxilla** consists of a so-called maxillary body, formed by fusion of the stipes, galea and lacinia, and a maxillary palpus (Figure 4.7). The dorsal surface of the maxillary body is extensively covered with short setae (maxillary pilose area), and bears series of modified setae in the maxillary brush and the laciniarastra. A number of sensilla are present at the tip of the maxillary palpus, and others occur on the maxillary body. They include tactile hairs and chemosensory pegs. The maxillae are moved by depressor and retractor muscles (Knight and Harbach, 1977; McIver, 1982).

The proximal region of the **labium** (the submentum and mentum) is largely incorporated into the ventral part of the cranium, but a small part of the mentum (the dorso- and ventromentum) projects as a sclerotized lobe which helps to enclose the preoral cavity (Figure 4.2). The distal region of the labium (mainly prementum) is situated dorsal to the mentum and below the mouth. It is united with the hypopharynx, which is a fleshy lobe, to form the composite labiohypopharynx (Figure 4.8). The salivary duct opens between the hypopharynx and the prementum. The central region of the prementum is raised and bears strongly sclerotized teeth and denticles. In non-predatory larvae the labiohypopharynx serves as an anvil upon which

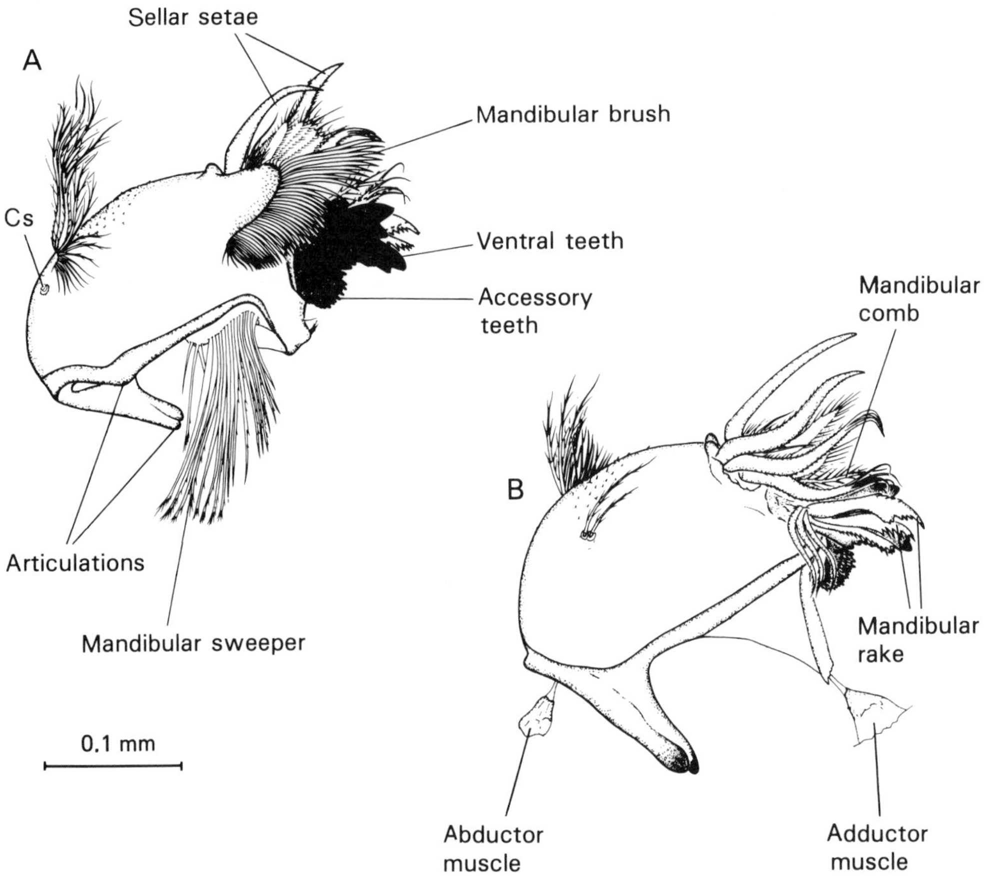

Figure 4.6 Mandible of a 4th instar larva of *Anopheles crucians*. (A) Dorsal aspect of the left mandible. (B) Ventral aspect of the right mandible. (From Harbach, 1977.)

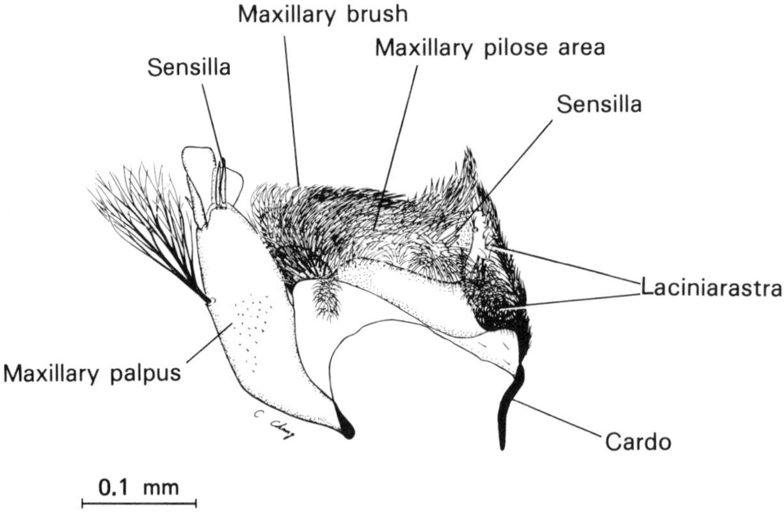

Figure 4.7 Dorsal aspect of the right maxilla of a 4th instar larva of *Anopheles crucians*. (From Knight and Harbach, 1977.)

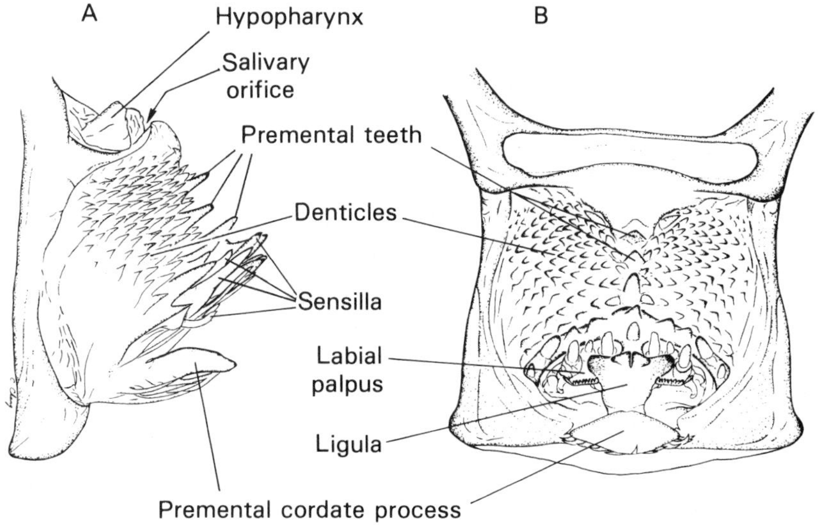

Figure 4.8 Labiohypopharynx of a larva of *Anopheles albimanus*. (A) Lateral aspect. (B) Anterior aspect. (From Harbach and Knight, 1977a.)

food particles are struck by the mandibles. The large ventral teeth of the mandible, which form the incisor area, strike against the ligula and premental teeth. The accessory teeth of the mandible, which form its molar area, strike against the denticles of the prementum. Membranous areas on either side of the prementum are thought to be the remnants of the labial palpi. Each bears four or five pegs which may be the cuticular parts of sensilla (Harbach, 1978).

4.2.3 The pharynx

The pharynx is the most anterior region of the foregut, situated between the mouth and the oesophagus, and has the nature of a thin-walled locally sclerotized sac. The mouth is surrounded by a flexible membranous area, and the mouth opening can change from a wide orifice to a small funnel-like opening. In many species a ventral oral brush is present just inside the mouth (Figure 4.9). It is thought to clean the mandibular sweepers and to retain food particles in the pharynx. Two pairs of pegs are present, one on either side of the mouth, which are possibly the cuticular parts of chemosensilla. The pharynx has an elaborate musculature (Figure 4.3). Dorsal and ventral dilator muscles enlarge the pharyngeal lumen. Intrinsic compressor muscles act antagonistically to the dilators, producing both lateral and dorsoventral compression and causing the pharynx to fold lengthways. The entrance to the oesophagus is closed by an oesophageal sphincter.

The dorsal and lateral walls of the pharynx contain four pairs of narrow, elongate pharyngeal sclerites, which run anteroposteriorly. Associated with these sclerites internally are rows of filaments which constitute the primary and secondary dorsal fringes and the primary and secondary ventral fringes (Figure 4.9). Each fringe may be composed of a number of rows of filaments. The fringes separate the central part of the pharyngeal lumen from the lateral recesses (Schremmer, 1949; Harbach and Knight, 1977b).

4.2.4 Structure and function

The mouthparts of species that are completely or substantially restricted to one feeding method exhibit characteristic specializations of structure. The mouthparts of the many species that use more than one feeding method are intermediate in structure. The following descriptions are

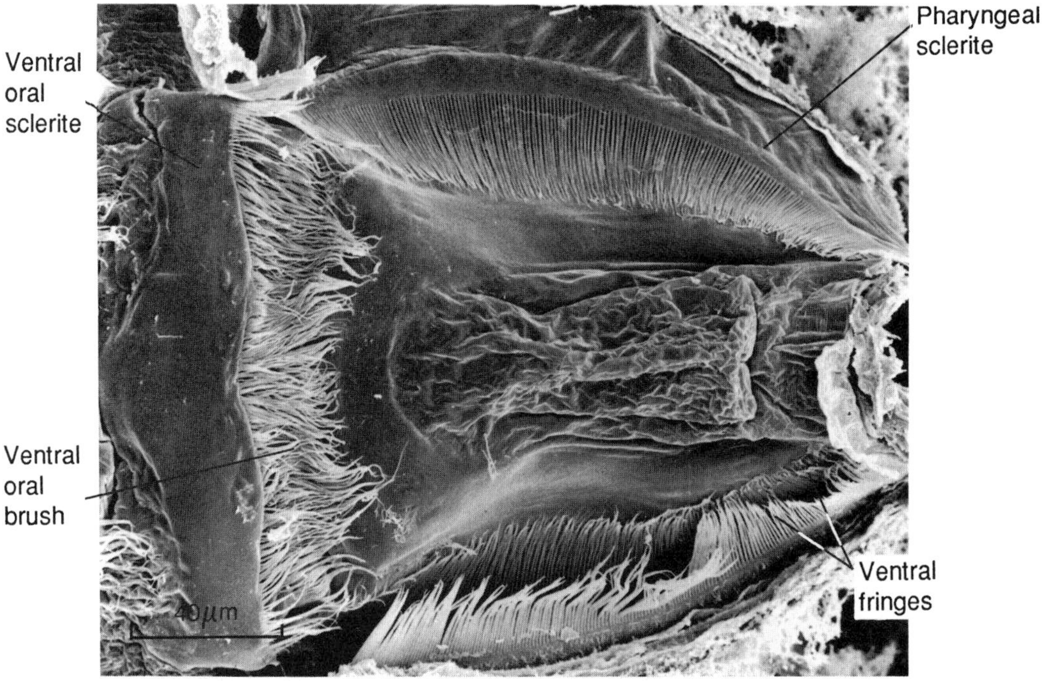

Figure 4.9 Ventral region of the pharynx of an *Anopheles quadrimaculatus* larva, seen from the dorsal aspect. (Micrograph by courtesy of Dr D. A. Craig.)

applicable to groups of species with restricted feeding habits (Cook, 1944; Surtees, 1959; Pucat, 1965; Dahl *et al.*, 1988; Widahl, 1988; Rashed and Mulla, 1990).

A. Collector–filterers that feed in the water column. (1) The head capsule is broad relative to its length. (2) The lateral palatal brush filaments are long relative to head capsule size, slender and mostly finely pectinate. (3) The mandibles are small, flat and weakly sclerotized. (4) The maxillae are large and bear many fine setae. (5) The prementum (of the labiohypopharynx) is weakly sclerotized and bears many very small teeth. (6) The antennae are long and each has a subapical tuft of setae.

B. Collector–gatherers that are adapted to resuspending settled particles. (1) The head capsule is narrow relative to its length. (2) The lateral palatal brush filaments are short relative to head capsule size, and are broader and more coarsely pectinate. (3) The mandibles are strong and bear groups of coarse setae. (4) The maxillae are smaller but bear well-developed setae. (5) The prementum is heavily sclerotized and has pointed teeth. (6) The antennae are short and stout and lack a subapical tuft of setae.

C. *Anopheles* larvae show a mixture of the characteristics of classes A and B. The lateral palatal brush filaments are slender and non-pectinate, resembling those of species that feed in the water column. However, the head capsule is narrow relative to its width, the lateral palatal brush filaments are short relative to head size, and the antennae are short and without a subapical tuft of setae, as in species that brush the substratum.

D. Predators. (1) The lateral palatal brushes are scissor-like in form and have a prehensile function; their filaments are few in number, elongate, stout and heavily sclerotized. (2) The mandibles are very large, with heavily sclerotized claws and stiff spines for grasping prey. (3) The maxillae are reduced in size but have strong setae, the palpi are small and rugged. (4) The prementum is well developed with large pointed teeth. (5) The antennae are short and have no subapical tuft, and the anal papillae are also very small.

(6) The pharynx is reduced; the pharyngeal fringes may be absent (*Toxorhynchites rutilus*) or retained (*Culex* (*Lutzia*) *halifaxii*).

4.3 FEEDING MECHANISMS

Mosquito larvae are unusual among aquatic insects in that they feed at all levels of water bodies – at the air/water interface, in the water column and at the substratum. With some exceptions, culicid larvae are gatherering and filterering collectors, and because they live in still water must generate water currents to bring food particles towards the head, in contrast to many filter feeding insects which utilize the currents of running water. The generation of currents and the capture of suspended particles by mosquito larvae cannot be understood without some knowledge of the effects upon small organisms of the fluid dynamics of water, a topic which is presented briefly in the following Section.

4.3.1 Hydrodynamics

The variable quantity that is of greatest significance to small organisms moving in water is the ratio between the forces of inertia and viscosity. Inertia is the resistance of a solid body or of a volume of fluid to change in its state of rest or motion. Inertia force is equal to the rate of change of momentum, e.g. in a fluid as it is deflected in its flow past a solid object. Viscous force is a measure of resistance to shear induced by movement of a fluid relative to a solid object. Per unit volume these terms are approximately proportional to:

inertia force: ρU^2,
viscous force: $\mu U/L$,

where ρ is density, U is the relative velocity of a fluid across a solid object, μ is the coefficient of viscosity and L is a representative linear dimension of the solid object. The relative importance of the inertia force (F_i) and the viscous force (F_v) acting on a unit volume of a fluid is given by Reynolds number (*Re*), a dimensionless number that expresses the ratio of the magnitude of these forces. Thus:

$$\frac{F_i}{F_v} \sim \frac{\rho U^2}{\mu U/L} = \frac{\rho L U}{\mu} = Re \qquad (4.1)$$

For any flow field of given form, the effect of changing ρ, L, U or μ, or of a number of these parameters together, can be described uniquely by the consequent change of *Re* alone. A 10-fold reduction in linear dimension increases relative viscous effects to exactly the same extent as a 10-fold increase in viscosity itself. High-velocity flow increases *Re*, low-velocity flow diminishes it. The characteristics of flow in any particular situation can be determined from the value of the corresponding Reynolds number. The effect of making $Re << 1$ is to make the inertia force much smaller than the viscous force, so that the viscous force is dominant in the flow field. The effect of making $Re >> 1$ is to make the inertia force dominant. The laws of motion of fluids are very different for these two cases. As a consequence of their small size and slow speed of movement, the world of mosquito larvae is dominated by the viscous force although not completely determined by it.

When velocity, viscosity and size combine to produce $Re >$ c. 10, pronounced eddies tend to form in the wake of solid objects. At $Re >$ c. 100: (a) the flow may be turbulent, i.e. it becomes unstable and the streamlines fluctuate in an irregular fashion; (b) velocity gradients are steep; and (c) viscous effects extend over short distances only. Low Reynolds number flow, at $Re << 1$, has the following characteristics: (a) flow is laminar (i.e. the streamlines are smooth) and eddies do not form in the wake of solid objects; (b) velocity gradients are gentle; and (c) viscous effects extend over long distances (e.g. movements of small solid objects generate currents over long distances). The term 'creeping motion' is used to describe fluid movement when the Reynolds number is much smaller than unity and inertia plays a negligible role in flow dynamics (Batchelor, 1967; Vogel, 1981; Tritton, 1988).

Where a solid object moves through a volume of otherwise stationary fluid, molecular interaction at the interface leads to transfer of momentum, consequently both the normal and tangential components of velocity are continuous across the boundary separating object from fluid. At the boundary itself the velocity difference between object and fluid is zero, a phenomenon known as the 'no-slip condition'. The fluid at each side of the object can be considered as moving in layers between which, at low *Re*, there is no mixing. Consequently a 'shear gradient', representing the tendency of these layers to slide over each other, exists between the surface of the object and the freestream flow. As distance to the side of the moving object increases, the difference of velocity between object and fluid increases, reaching a maximum at the position of freestream flow (where, in this case of stationary fluid, the velocity is zero). Thus, an object that moves through a stationary fluid drags some fluid into motion with it, and a stationary object in moving fluid restrains the motion of some fluid. At Reynolds numbers greater than about 100 a 'boundary layer' forms. This is a layer of fluid, adjacent to the object, which is thin compared to the diameter of the object, within which the shear gradient is very steep and within which the viscous force is important however high *Re* may be. A comparable boundary layer forms when a fluid flows past a stationary solid object.

At Reynolds numbers $<< 1$ the situation is very different because the shear gradient at the side of a moving object is much less steep. The difference of velocity between object and fluid now increases only gradually with distance from the object, and the viscous effects extend over long distances. Thus a moving solid object exerts effects many diameters to its side, and carries a large volume of fluid with it. Further, when objects have upstream-downstream symmetry, the whole flow pattern is identical upstream and downstream of the solid object. Mathematical formulations appropriate for the conditions that obtain at large Reynolds numbers, and which accurately describe the boundary layer, are inappropriate for the conditions of creeping motion.

Most early mathematical treatments of small Reynolds number flow involved a sphere as the solid object, and substantial difficulties were encountered when attempts were made to develop the principles for flow around objects of other shapes. Lamb (1911) and Batchelor (1967) established the characteristics of two-dimensional flow at small Reynolds numbers for a circular cylinder moving through a fluid in a direction perpendicular to its axis. Their equations can be developed to determine that the relative velocity of the fluid at either side of the cylinder is given by

$$u = U\left[1 + \frac{1}{\ln(7.4/Re)}\left(\frac{a^2}{2r^2} - \frac{1}{2} - \ln\frac{r}{a}\right)\right] \quad (4.2)$$

and that the relative velocity of the fluid upstream and downstream of the cylinder is given by

$$u = U\left[1 + \frac{1}{\ln(7.4/Re)}\left(\frac{1}{2} - \ln\frac{r}{a} - \frac{a^2}{2r^2}\right)\right] \quad (4.3)$$

where U is the speed of the cylinder, u is the speed of the fluid at distance r from the centre of the cylinder, a is the radius of the cylinder, and ln is logarithm to base e. Equations 4.2 and 4.3 hold for an extensive region around the cylinder but do not hold at very large distances, i.e. u/U does not tend to zero as r/a tends to infinity (Tritton (1988) and personal communication). Naturally the values of u would be altered by changes in the cross-sectional shape of the cylinder.

The value of u would also be altered by the presence of other objects. Consider an ordered array of cylinders (equivalent to an array of filaments on a setulose arthropod appendage) moving through a stationary fluid. The shear gradients around each cylinder would overlap with those around its neighbours on four sides, and so would be shallower than the gradients around a solitary moving cylinder. Consequently more fluid would be dragged along by each cylinder. Using changes in the ratio between (a) the rate of flow between two cylinders and (b) the freestream velocity, as a measure of the effects of different parameters upon the additive effects of overlapping shear gradients, Cheer and Koehl (1987) found, with cylinders of 0.1 or 1 μm diameter, spaced from 0.3 to 50 μm apart, at *Res* of 10^{-5} to 0.5, that the amount of fluid dragged between the cylinders (i) varied due to change of velocity most at *Res* of 10^{-2} to 0.5, and (ii) varied due to change of intercylinder spacing most at *Res* of the order of 10^{-2}. These are values of *Re* that we shall encounter below in descriptions of larval feeding mechanisms.

The properties of the water surface are very important for the life of aquatic insects. The phenomenon of surface tension is due to an imbalance in the normal attractive forces between neighbouring water molecules. Molecules that are near the air/water interface lack neighbours on one side and experience an unbalanced cohesive force directed away from the interface. The resultant tendency for all water molecules near the interface to move inwards is equivalent to a tendency for the interface to contract, and gives the water surface some of the characteristics of a resilient membrane. At 20°C the surface tension of pure water = 72.8 dyn cm^{-1} (or erg cm^{-2}). The surface tension diminishes with increase in temperature and in the presence of surface-active substances (Batchelor, 1967). The physical properties of the **surface membrane** and the forces affecting surface-dwelling insects have been described by Baudoin (1976). Where a hydrophobic region of insect cuticle, that has low adhesion for water, is in contact with the surface membrane, a convex meniscus forms and surface tension forces that part of the insect upwards. Where a hydrophilic region of insect cuticle, that has high adhesion for water, is in contact with the surface membrane, water will rise up the cuticle to form a concave meniscus and surface tension will operate to force that part of the insect downwards (Guthrie, 1989). The effects of these surface forces are apparent in the floating of *Culex* egg rafts (Section 3.1.1) and the suspension of culicine larvae from the surface membrane (Section 5.6.1).

All natural bodies of water are covered by an organic **surface microlayer**, which is best known from sea water but present over fresh waters also. The most stable component of the surface microlayer is a lipid film, 1–2 nm thick, largely composed of fatty acids orientated with their hydrocarbon chains extending into the air. Bonding between these chains produces considerable cohesion. Below the lipid film is a polysaccharide-protein film some 10–30 nm thick (Norkrans, 1980; Hermansson, 1990). Observations of *Anopheles* larvae feeding at the water surface (Section 4.3.2) show interaction with a surface microlayer. It remains to be clarified how the properties of the 'surface microlayer' interact with those of the 'surface membrane' in affecting surface-dwelling insects.

4.3.2 Current generation

(a) Current generation by anopheline larvae

Anopheles larvae characteristically live at the air/water interface, lying horizontally immediately below the interface, dorsal surface upwards, and held there by non-wettable setae and the spiracular lobes. In natural water bodies the water surface is covered by an organic surface microlayer, and the top micrometre of the water column is enriched with microorganisms, forming the 'bacterioneuston layer'. *Anopheles* larvae feed with their heads rotated through 180°, which places the mouthparts and preoral cavity immediately below the surface microlayer. The beating of the lateral palatal brushes produces adoral currents which carry floating particles towards the head from a distance at least equal to the length of the larva (Christophers and Puri, 1929; Renn, 1941).

The dimensions of the lateral palatal brushes of *Anopheles quadrimaculatus* were measured in a study of larval feeding mechanisms (Merritt *et al.*, 1992a). Each brush consisted of an array of about twelve rows of filaments; the number of filaments per row increased from 6 in the adoral-most row to 13 in the fifth and remaining aboral rows. The adoral filaments (0.15 mm long) were shorter than the more aboral filaments (0.184 mm long). Over most of their length, mean filament diameter was 2.0 μm. However, beyond about three-quarters of its length each filament divided into 4–6 subfilaments of 0.6 μm diameter, spaced 0.275 μm apart. Thus the filaments broadened apically and had a mean tip

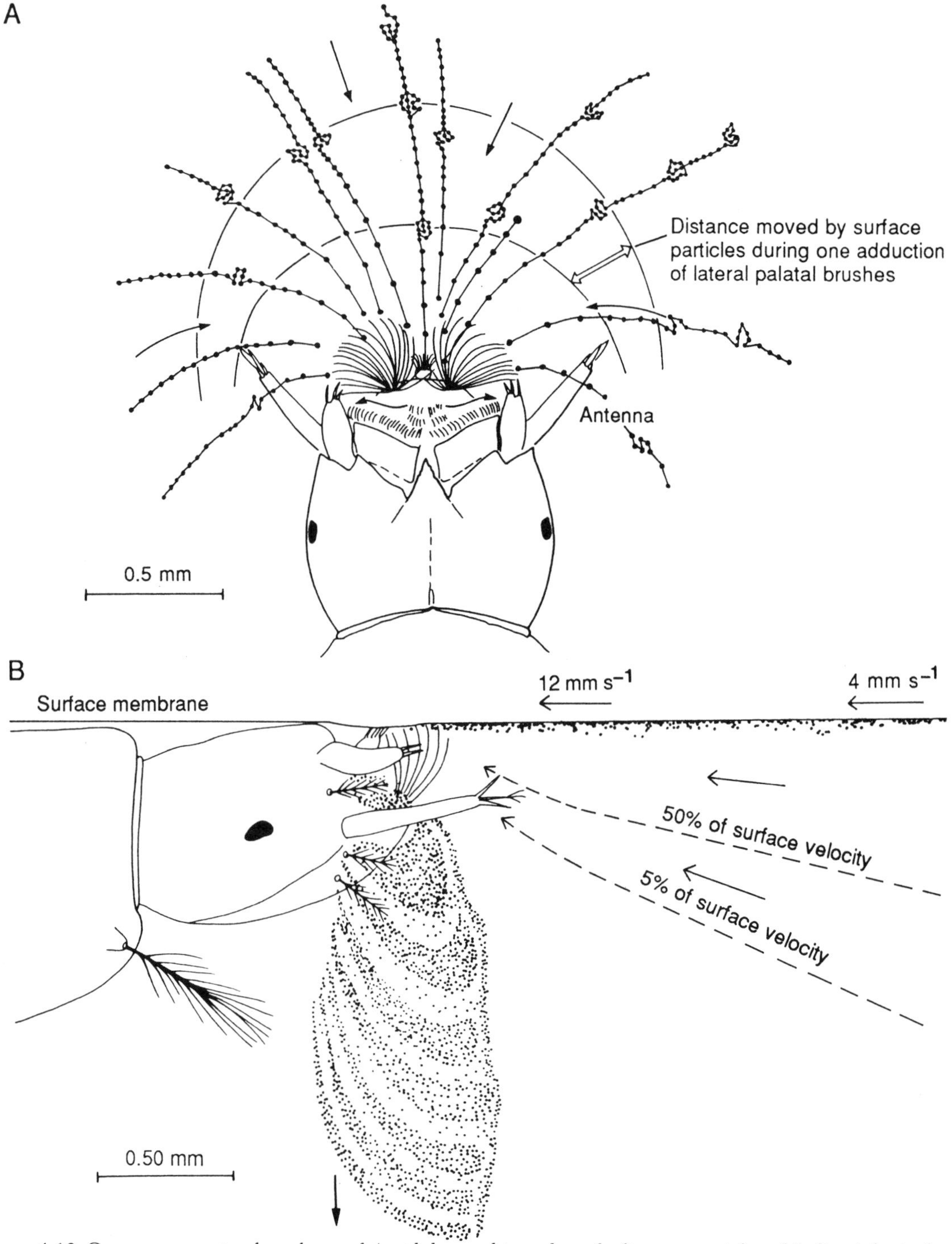

Figure 4.10 Current generation by a larva of *Anopheles quadrimaculatus* feeding on particles of Indian ink, its head rotated through 180°. (A) View from above. The lines of small dots are representative of the paths taken by individual particles. The particles move smoothly towards the head during adduction of the filament rows of the lateral palatal brushes, move only minutely backwards and sideways during their abduction. (B) Lateral view. The lateral palatal brushes are in the course of adduction. After being drawn into the preoral cavity, particle-rich water passes downwards between the cranium and the antenna (on each side), forming a lamellate outward current or plume. (After Merritt *et al.*, 1992a.)

diameter of 5 μm (range 3.6–7.1 μm). When a brush was maximally abducted (opened), not only were the filament rows splayed but the filaments within each row were splayed. In an abducted brush, the mean distance between rows at the filament tips was 37 μm, and the mean distance between filament tips within rows was 22 μm. At its greatest expansion, each brush (distally) was 0.263 mm wide and 0.356 mm thick. The distal thickness of a fully adducted brush, with compacted filaments, was 0.128 mm.

High speed cinematography was used by Merritt *et al.* (1992) to investigate current generation by *An quadrimaculatus* larvae. The following description is based principally on their account, and in some minor details on privileged access to their recordings. A single beat of the lateral palatal brushes involved adduction (retraction) and abduction (extension) phases. In its abduction movement each brush rotated slightly, moving first laterally away from the head capsule, then downwards away from the water surface, and last anteriorly. At the start of abduction the brush filaments were compacted, but during the course of the movement the filament rows progressively separated from one another – first the inner (adoral) rows and later the outer (aboral) rows. While the outer rows were still moving outwards, the inner rows started upon their adduction. As the adduction phase progressed, the filament rows moved sequentially towards the head, like the flipped pages of a book. In the brush's most extended condition only the shorter adoral filaments reached the surface membrane. As the rows of filaments closed on the head during adduction, the longer aboral filaments approached and touched the surface membrane but did not break through it. Each cycle lasted 0.2 s, giving a beat frequency of 5 Hz. During adduction the tips of the aboral filaments moved at 4.4 mm s^{-1}.

The beating of the lateral palatal brushes induced flow. This was made visible by the movement of particles which travelled towards the head from all directions, converging on the lateral palatal brushes. The movement was stepped; during filament adduction the particles moved smoothly, but during the rest of the cycle they were almost stationary, just moving minutely backwards and sideways (Figure 4.10A). Because the adoral rows of filaments started adduction while the aboral rows were still abducting, water flow commenced during the abduction phase of the cycle. The width and depth of the water currents increased with distance from the head, consequently particle velocities were negatively correlated with both distance from the head and depth (Figure 4.10B). Close to the lateral palatal brushes, particle velocities immediately below the surface membrane ranged from 10–15 mm s^{-1}. At 1.8 mm in front of the head they averaged 4.3 mm s^{-1}.

The adoral currents produced by the lateral palatal brushes did not stop at the preoral cavity but continued as two downward currents, which passed between the antenna and head capsule on each side, met below the head capsule and continued downwards without eddy formation. At their inception the downward currents probably passed over the palatum, with its midpalatal brush, and across the setulose regions of the abducted maxillae. When larvae fed on dense concentrations of particles, the outward currents had the appearance of two plumes composed of series of particle-containing lamellae, one pair of lamellae being added with each beat of the mouthparts (Figure 4.10B). Although there were many clumps of particles in the surface microlayer, these were rare in the plumes, indicating that trituration had occurred. In the relatively still water of an aquarium the plumes descended several centimetres.

The movements of the mandibles of *An quadrimaculatus* larvae were almost in phase with those of the lateral palatal brushes, but ran slightly behind. Thus, as the adducting lateral palatal brushes reached full retraction, the selar setae of the closing mandibles swept over them. The maxillae moved in opposite phase to the mandibles and lateral palatal brushes; their lateral movement was more limited than that of the mandibles. The anteromedial palatal brush moved in almost opposite phase to the lateral palatal brushes (Merritt *et al.*, 1992).

It is revealing to examine these observations in the light of the summary of fluid dynamics in Section 4.3.1. Because of the size of the mouthparts and their velocity, Reynolds number would have been low and the viscous force would have dominated. That this was so is apparent from the substantial distances over which the lateral palatal brushes generated currents. Because there were no inertial effects, particle movement started and stopped instantly when the adduction of filament rows started and stopped. The dominance of the viscous force was also apparent in the characteristics of the outward particle-carrying plume; the lamellae of the plume did not mix and its two divisions joined without eddy formation.

Under conditions giving low Reynolds numbers, the moving filaments of the lateral palatal brushes would drag significant volumes of water with them (Section 4.3.1). Consider water flow through a lateral palatal brush during an adduction movement, treating the filaments as circular cylinders having apical and midlength diameters of 5 and 2 μm respectively. From the filament dimensions, and the tip velocity of 4.4 mm s^{-1}, Merritt *et al.* (1992) estimated *Re* during adduction to be 0.002 at the subdivided filament tips and 0.009 at midlength. From equation (4.2) it transpires that water velocity at the midpoint between adjacent subfilaments or filaments, expressed as percentage of filament velocity, would be 75% at the tips and 65% at midlength. Equation (4.3) indicates that the relative water velocity at positions in front of and behind a filament, at the midpoint between filament rows, would be 80% apically and 73% at midlength. In fact, the relative velocities would be substantially higher due to the viscous effects of adjacent filaments in four directions. We can conclude that the adducting filament rows act like leaky paddles, not like sieves, and generate adoral currents. The absence of currents during abduction cannot be fully explained. The brush filaments are compacted during this phase, possibly producing a configuration large enough to induce inertia force and to raise *Re* substantially. The partial rotation that occurs during abduction may indicate a 'feathering' of the brush.

The water surface is rich in living micro-organisms and detritic particles. Very few floating particles rest on the surface membrane; most float immediately below it. Ciné film of water to which yeast, carmine or Indian ink had been added revealed a high concentration of particles immediately below the air/water interface (Merritt *et al.*, 1992a). Some early studies deserve reinvestigation. Renn (1941) distinguished two types of feeding by larvae of *An quadrimaculatus* and *An crucians*. 'Interfacial feeding ' was exhibited at surface tensions of 68 dyn cm^{-1} (68 mPa) or more, characteristic of natural water bodies, when the feeding currents approached the head in straight lines and the larvae fed on floating particles. At surface tensions of 58 dyn cm^{-1} or less, found with hay and grass infusions, the larvae exhibited so-called 'eddy feeding', generating two vortices which drove a converging current towards the head. The surface microlayer that is normally present on natural water surfaces is sometimes replaced by a thick sheet of bacteria and Protoctista which mosquito larvae are able to engulf (Christophers and Puri, 1929).

(b) Current generation by culicine larvae

The most detailed investigations have been with larvae of *Culiseta morsitans*, which feed almost exclusively by collecting–filtering within the water column, and with larvae of *Aedes communis*, which usually feed by collecting–gathering over the substratum but also use the collecting–filtering mode. Mechanisms of particle capture were investigated by studying the functional morphology of the mouthparts and head capsule (Dahl *et al.*, 1988), and by analysing video recordings of the three-dimensional flow patterns of food particles around collecting–filtering larvae (Widahl, 1992).

Contraction of the labral retractor muscles caused the tormae to turn on their axes, transmitting force to the elastic lateral palatal plates and resulting in adduction of the lateral palatal brushes, which closed on the preoral cavity with their rows of filaments packed closely together.

This was followed by a more rapid and purely elastic abduction of the lateral palatal brushes, which returned to their open condition with the rows of filaments parted. The longer adduction phase of the beat drew water towards the front of the preoral cavity from all directions, but the rapid abduction phase caused little or no water movement.

Head structures also generated a jet-like outward current, the alpha flow, which flowed approximately perpendicular to the body axis of the larva. The alpha flow terminated in a vortex ring (Figure 4.11), a type of vortex with the form of a toroid and more familiar as smoke rings. The vortex ring interacted with the adoral currents generated by the lateral platal brushes to produce a complex flow pattern around the larva. The sharp focus of the video camera gave the optical illusion of a number of separate streamlines. Thus viewing the flow field from the side gave an illusion of two separate counterflows, a beta flow below the alpha, and a gamma flow

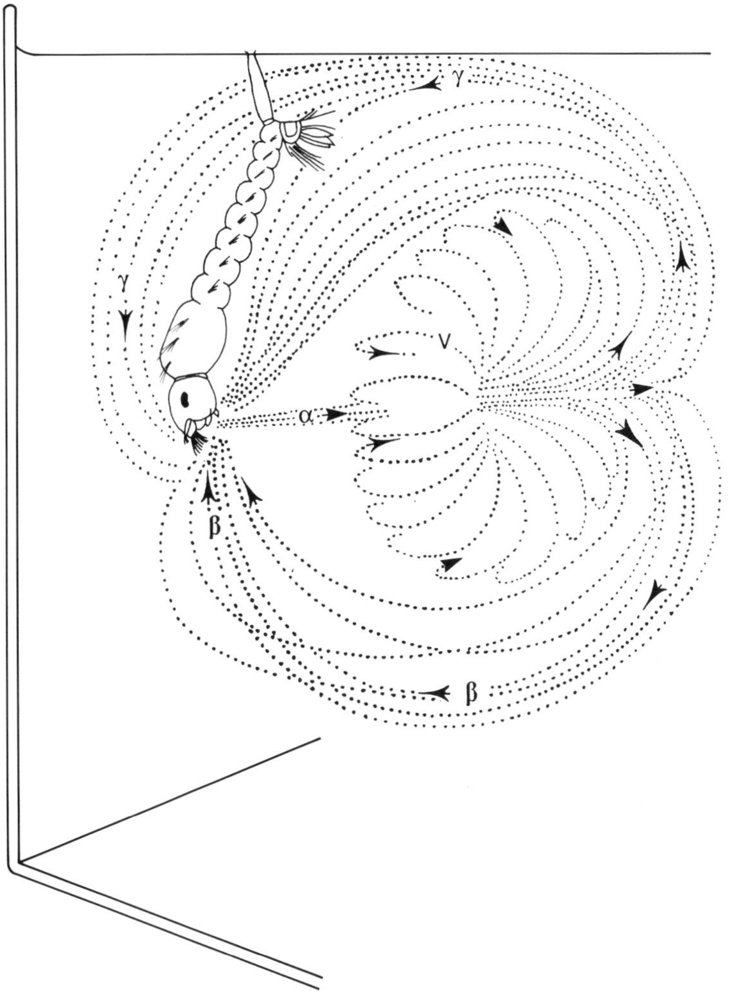

Figure 4.11 The pattern of currents generated by a culicine larva while gathering–filtering in the water column. The course of the currents was established from the flow of particles. The beating of the lateral palatal brushes causes water to flow towards the head from all directions. An outwardly directed current (α) forms a vortex ring (V) which interacts with the adoral currents forming a complex flow field around the larva. Currents which the sharply focused camera make appear as separate streamlines are labelled β and γ. (After Widahl, 1991.)

above it which parted when it reached the larval body (Figure 4.11). Viewing from above gave an illusion of two counterflows parallel to the water surface, both named delta, flowing back on either side of the alpha flow.

The part of the vortex ring nearest the larva was 3–6 mm in front of the lateral palatal brushes in *Cs morsitans* larvae ($n = 3$) and 2–4 mm in front in *Ae communis* larvae ($n = 3$). The most distant part of the vortex ring was from 10–40 mm in front of the lateral palatal brushes. Flow velocities were more rapid within the vortex ring than in the outer streamlines. Particles in suspension in the outer region of the vortex ring were entrained by the laminar counterflows and carried towards the head, but only a few reached the preoral cavity because most were captured by the alpha flow. Particles in the inner region of the vortex ring could rotate several times before either being entrained by the alpha flow or carried by counterflows into the preoral cavity.

It is possible to start on an analysis of the hydrodynamics of current generation by these culicine larvae. When the lateral palatal brushes stopped beating, particle movement close to the head capsule stopped in the same instant, and the more distant water currents stopped almost immediately, indicating that the viscous force was dominant. Estimation of Reynolds numbers near the dorsal surface of the head capsule gave values of 1–6 *Re*. Although the beat frequency of the lateral palatal brushes was lower in *Cs morsitans* than in *Ae communis* (Table 4.4), the greater filament length of the brushes of the former resulted in a much greater filament velocity. Particles in the beta flow accelerated as they approached the head capsule, and the mean velocities 0–5 mm from the head capsule were, like the alpha current, faster with *Cs morsitans* than with *Ae communis*. Formation of the vortex ring probably resulted from the discharge of the jet-like alpha current. The requirement for formation of a vortex ring is that linear momentum should be imparted to the fluid with axial symmetry (Batchelor, 1967). Increasing velocity of the alpha flow would increase the size of the vortex ring, and Widahl (1992) suggested that by creating a strong alpha flow a larva feeding in the water column extends its foraging area.

Table 4.4 Characteristics of the 4th instar larvae of two species: *Culiseta morsitans*, a collector–filterer, and *Aedes communis*, a collector–gatherer and facultative collector–filterer. (From Widahl, 1992.)

	Culiseta morsitans	*Aedes communis*
Lateral palatal brushes		
Filament length (mm)	0.90	0.26
Beat frequency (Hz)	3.4	10.0
Current velocity within 0–5 mm of the cranium		
alpha flow (mm s^{-1})	16.7	7.1
beta flow (mm s^{-1})	3.3	2.5

Expulsion of water from the pharynx was considered by Dahl *et al.* (1988) and Widahl (1992) to explain the alpha flow. However, it remains to be shown whether the pharynx can generate a current of sufficient volume and force, or whether the mouthparts are also involved.

4.3.3 Particle capture by collector–filterers

A number of mechanisms, in addition to sieving, have been postulated for the capture of water-borne food particles, e.g. capture by single filaments by direct interception, electrostatic attraction, etc. (Rubenstein and Koehl, 1977). However, for small arthropods feeding under conditions of low Reynolds number the effects of the viscous force (Section 4.3.1) must not be overlooked. Due to the 'zone of viscous effect' around moving filaments, within which there is no significant mixing, setulose appendages are functionally wide and smooth rather than comb-like. Food particles are likely to be redirected without being touched when an appendage that influences a thick layer of water around itself moves towards them (Koehl and Stickler, 1981). At *Res* $> 10^{-2}$, increases in spacing between filaments can significantly increase the leakiness of rows of filaments, so rendering them more sieve-like (Cheer and Koehl, 1987). Clearly, very precise data on dimensions, velocity and *Re* are required for functional analysis.

(a) Role of the mouthparts

Evidence discussed in Section 4.3.2 confirmed the dominance of the viscous force in the conditions under which mosquito larvae feed. The lateral palatal brushes appear to function as paddles for current generation rather than as sieves.

The lateral palatal brushes act as paddles (Section 4.3.2), and the long held view that they function as sieves in a filter-feeding role is untenable. However, although each filament row will act on the water as an almost solid object, water must enter the spaces between the filament rows as the brushes open up during abduction. Water will be squeezed out of the brushes as the adducting rows of filaments are compacted on one another at the end of adduction. Possibly, in this way, each beat of the lateral palatal brushes delivers packets of particle-containing water to the preoral cavity (Merritt *et al.* (1992). The suggestion that larvae enhance particle capture by coating the filaments of the lateral palatal brushes with mucus secreted from an epidermal gland (Merritt and Craig, 1987) was not substantiated by histochemical studies (Dahl *et al.*, 1990).

Before the principles of hydrodynamics were first applied to culicid feeding mechanisms, Schremmer (1949) proposed a mechanism of particle capture by *Anopheles* larvae based on the structure, movements and interactions of the mouthparts and on observations with a binocular microscope of larvae feeding on unicellular algae. Schremmer concluded that particles trapped on mouthbrush setae were processed along two routes: (1) lateral palatal brushes – mandibular brushes – midpalatal brush – surface of clypeoplalatum – pharynx; (2) maxillary setae – hypopharynx – pharynx. Fuller details are cited by Clements (1963). Large particles were said to be drawn into the preoral cavity by the sellar setae of the mandibles and crushed, either between the incisor regions of the mandibles and the ligula and premental teeth, or between the molar regions of the mandibles and the premental denticles. The pieces were brushed into the pharynx by the maxillae. Although Schremmer's view of the sieving capabilities of setulose structures appears erroneous, his descriptions of mouthpart interactions and particle processing are still the only ones available for *Anopheles*, and remain to be validated or disproved.

(b) Role of the pharynx

Microscopic examination of live culicine larvae indicated that the pharynx functioned like bellows. Contraction of muscles inserted on the median dorsal pharyngeal sclerite produced a rooflike expansion of the pharynx, inducing an inflow of water. After closure of the preoral cavity, the pharynx contracted, starting posteriorly, and water was expelled from a small mouth opening (Dahl *et al.*, 1988).

When freshly moulted and therefore transparent larvae of *Culex pipiens* were placed in a suspension of fine carmine particles, carmine accumulated in bow-shaped streaks at the sides of the pharynx. After a period of pumping, particles trapped on the pharyngeal fringes were deposited on the floor of the pharynx by a strong contraction, and they were then passed into the oesophagus. Schremmer (1949) concluded that, during the contraction phase, water contained in the central part of the pharynx was forced through the pharyngeal fringes, which functioned as a filter, and into the lateral recesses of the pharynx, before being discharged through the permanently open corners of the mouth.

The fate of carmine particles could not be followed in *Anopheles maculipennis* larvae because of pigmentation of the head capsule. However, Schremmer (1949) reported that the pharynx functioned like bellows, sucking water in through the open mouth and expelling it again through the permanently open corners of the same orifice, in synchrony with the beat of the mouthparts. The volume of water processed by the pharynx was much less than that drawn in by the lateral palatal brushes. The excess was discharged in laterally directed currents, one on each side of the head, which were clearly visible in experiments with Indian ink. Presumably they corresponded to the downward plume described above for *An quadrimaculatus*.

Schremmer (1949) considered that the pharyngeal fringes of *An maculipennis* larvae filtered, from water entering the pharynx, small particles that had not been ensnared by the lateral palatal brushes and appendages. Merritt *et al.* (1992a) reported that *An quadrimaculatus* larvae, unlike culicine larvae, took little or no water into the pharynx during feeding. While remaining uncommitted as to how particles were captured by the mouthparts of *An quadrimaculatus*, they concluded that groups of particles that were gathered on the palatum were pushed into the pharynx by the mandibular sweepers when the mandibles were fully adducted, and that the pharyngeal fringes functioned to comb the mandibular sweepers.

Food boluses formed rapidly in the pharynx of *An quadrimaculatus* larvae when particle density was high. One larva produced, on average, one bolus every 4.43 s. Boluses formed in the pharynx were passed into the anterior oesophagus, where three or four combined before one large bolus was passed to the midgut (Merritt *et al.*, 1992b).

4.3.4 Collecting–gathering, scraping and shredding

The collecting–gathering feeding mode has received little attention, but it has been described briefly from *Aedes apicoargenteus*, *Ae aegypti* and *Eretmapodites chrysogaster*. The lateral palatal brushes produced locomotory currents which propelled the larvae, inclined at about 45°, over the substratum. The larvae brushed the surface of the substratum with the serrated filaments of their lateral palatal brushes and the teeth and setae of their mandibles, putting particles into suspension. Currents generated by the lateral palatal brushes carried newly suspended particles towards the preoral cavity. Any large particles were manipulated by the mandibles, while the mentum served as a secondary grasping organ (Surtees, 1959).

Food acquisition by scraping presumably evolved from the collecting–gathering mode, with only relatively minor modifications to the larval mouthparts. It has been described in larvae of *Ae atropalpus*, which inhabit rock pools in northern North America and which remain submerged for long periods scraping algae off the rock face (Dyar, 1903).

Shredding is a component of the feeding behaviour of some species which are principally collector–filterers. For example, a larva of *Culiseta inornata* was described grasping the edge of a chickweed leaf with its lateral palatal brushes and striking the leaf with its mandibular teeth, chopping off small pieces of decayed material, some of which entered the preoral cavity (Pucat, 1965). *Aedes triseriatus* larvae occasionally gnaw leaves, usually the leaf veins (Walker and Merritt, 1991). In a few taxa, shredding is the characteristic feeding mode. Larvae of the *Culex bitaeniorhynchus* group feed selectively on *Spirogyra*, ingesting short lengths of these filamentous chlorophytes (Laird, 1956, 1988). Larvae of the subgenus *Rachisoura* (of *Tripteroides*) live in the liquid contents of *Nepenthes* pitchers, and feed on drowned invertebrates. They have strongly toothed maxillae which are used exclusively as clasping organs, unlike the homologous organs in predatory larvae, and triturate their food with their mandibles (Van den Assem, 1959).

4.3.5 Predation

Predatory mosquito larvae are exceptionally large and are capable of seizing and killing relatively large prey. *Toxorhynchites* larvae most commonly attack when stationary, striking at prey as it comes within range. The head is extended in the direction of prey within 45° of the longitudinal body axis, and the mandibles are opened as the head reaches its maximum extension and closed as it returns to its original position. During ingestion, the prey is always held by one mandible, and the other may be used to force the prey towards the mouth. The scissor-like lateral palatal brushes are not used to hold the prey, but occasionally are used to guide the prey into the mouth. The prey is swallowed whole (Furumizo and Rudnick, 1978;

Pichon and Rivière, 1979; Steffan and Evenhuis, 1981; Russo, 1986). The larvae of a number of sabethine genera are facultative predators. The larvae of *Runchomyia frontosa* engulf their prey (Zavortink, 1986). The larvae of some species of *Limatus*, *Wyeomyia* and *Trichoprosopon* break their prey in two, after which they are believed to suck the body fluids (Galindo *et al.*, 1951).

4.4 FEEDING RATES

The rate at which any filter feeder ingests particles will be determined by the size and concentration of particles, the animal's efficiency in separating particles from water, the rate of beat of the mouthparts, and the proportion of time that the animal spends feeding. Two parameters have been used in measurements of feeding rates: (1) rate of displacement of the gut contents, which can be seen after the ingestion of food particles of a different colour, e.g. after a change from charcoal to dried yeast, and (2) rate of disappearance of particles from the medium.

4.4.1 Phagostimulants

When *Culex pipiens* larvae were placed in distilled water, or in distilled water containing a suspension of inert particles, they filtered only intermittently, but if soluble yeast extract was added as a phagostimulant, filter feeding activity was almost continuous whether particles were present or not (Dadd, 1970a, b). The effectiveness of yeast extract as a phagostimulant increased as its concentration was increased (Rashed and Mulla, 1989).

Several nucleotides were as stimulatory to *Cx pipiens* larvae as yeast extract when added to a suspension of charcoal particles. Adenosine diphosphate and adenosine triphosphate at 10^{-4} M, adenosine monophosphate (2′, 3′, or 5′ isomer) at 2×10^{-4} M, and guanosine monophosphate at 10^{-3} M were as effective as 0.2% yeast extract. At 10^{-3} M the nucleosides adenosine and guanosine were as stimulating as their respective nucleotides at the same concentration; the same was true of 10^{-2} M uridine (Dadd and Kleinjan, 1985). *Culex pipiens* larvae will ingest a synthetic chemically defined diet. Although the omission of nucleotides from the synthetic diet appreciably reduced the rate at which it was ingested, the nucleotide-deficient diet was still strongly phagostimulatory compared to plain water. The omission of other groups of dietary ingredients, including those which on their own were moderate phagostimulants, had no effect on ingestion rates. Dadd *et al.* (1982) concluded that no single component of a complex phagostimulatory mixture may be essential to induce a high ingestion rate, but that stimulation may result from additive or synergistic effects of several moderate or weak phagostimulants. Mosquito larvae are arrested in the presence of edible materials such as cereals and meat, and of known phagostimulatory compounds, but their motility is unaffected by inert particles (Aly, 1983; Ellgaard *et al.*, 1987).

4.4.2 Particle size

The size of particles ingested by any species is a function of larval age. This was demonstrated when *Culex pipiens* larvae were placed in a series of vessels, each containing spherical latex particles of a particular uniform size. All particles had a relative density of 1.05 g cm^{-3}, and were provided at 0.5% w/v and in the presence of soluble phagostimulant. Particle sizes that permitted displacement of gut contents at a high rate were: 1st instar 0.18–1.86 μm, 2nd instar 0.36–26 μm, 3rd instar 0.36–45 μm, 4th instar 0.71–45 μm. Very long but thin objects may be ingested if appropriately presented, since nematodes averaging 240 μm long and 25 μm wide were found in the guts of 4th instar larvae, although they were normally rejected (Dadd, 1971). When provided with mineral particles plus phagostimulant, 1st and 2nd instar larvae of *Aedes triseriatus* ingested only particles of <2 μm. Third and 4th instar larvae ingested particles ranging from <2 to 25–50 μm (Merritt *et al.*, 1978; Merritt, 1987). Particles of wheat flour, fishmeal and dried blood were ingested at a similar rate by 4th instar larvae of *Culex tarsalis* and *Anopheles albimanus* although the particles were of different sizes and shapes. The wheat flour particles were oval spheres of 4–30 μm diameter, whereas the fishmeal and dried blood particles aggregated in clusters with diameters of 60–140 μm (Rashed and Mulla, 1989).

When 4th instar *Cx pipiens* larvae were given latex particles of 1.01 μm diameter suspended in a solution of yeast extract, variation of particle concentration between 0.06 and 4% w/v had no effect upon the rate at which the gut was filled. Ingestion rates fell sharply as concentrations were reduced below 0.06%, probably because of particle depletion by the feeding larvae (Dadd, 1971). Similar results were obtained with species of *Culex*, *Aedes* and *Anopheles* offered four types of inert particle at concentrations of 0.001–0.01 g cm^{-3} (Rashed and Mulla, 1989).

4.4.3 Ingestion rates

Ingestion rates have been determined from the volumes of water cleared of suspended particles over time. The rate of disappearance of *Chlamydomonas* from water containing *Anopheles subpictus* larvae indicated that the lst instar larvae cleared 0.005–0.011 cm^3 $larva^{-1}$ h^{-1}, and 4th instar larvae 0.025–0.051 cm^3 $larva^{-1}$ h^{-1} (Senior-White, 1928). Similar clearance rates have been obtained with *Anopheles* larvae in the laboratory. Fourth instar larvae of *An quadrimaculatus* and *An albimanus* placed in a suspension of 2 μm latex beads, with yeast extract as phagostimulant, cleared 0.033–0.034 cm^3 and 0.049–0.055 cm^3 $larva^{-1}$ h^{-1} respectively. It is possible that this method underestimates the feeding capabilities of anopheline larvae, which feed principally in the normally particle-rich surface microlayer. Fourth instar larvae of *Culex quinquefasciatus* and *Aedes aegypti* cleared water of latex beads at rates of 0.49–0.59 cm^3 and 0.59–0.69 cm^3 $larva^{-1}$ h^{-1} respectively. The culicine larvae cleared water of yeast cells at similar rates (Aly, 1988).

Fourth instar larvae of *An maculipennis* cleared the water surface of *Lycopodium* spores at rates of 65, 30 and 9 mm^2 min^{-1} when the spore numbers were 1–5, 21–30 and 31–40 mm^{-2} respectively. The rate of spore ingestion rose to 10 s^{-1} as the numbers increased to 21–30 mm^{-2}, but fell as the numbers increased further to 50–60 mm^{-2} (Shipitzina, 1941).

When *Cx pipiens* larvae fed on live *Chlorella* or dried yeast the gut contents were replaced in under one hour on average (Dadd, 1970a). Median times for 4th instar larvae to fill their guts with particles of wheat flour were *Ae aegypti* 42 min, *Cx tarsalis* 61 min, and *An albimanus* 100 min (Rashed and Mulla, 1989). Caution is necessary in comparing clearance or ingestion rates obtained for different species in laboratory experiments because the method of food presentation may not suit all species equally.

A positive correlation was obtained between the rate of ingestion and the proportion of time that *Cx pipiens* larvae spent particle feeding. Noting Pucat's (1965) observation that the beating rate of the mouthparts of two species of collector–gatherer varied little under constant conditions, (Dadd, 1970a, b) concluded that in his own experiments beat frequency affected the rate of ingestion much less than did the proportion of time spent feeding. Nevertheless, temperature is important. The beat rate of filter feeding larvae of *Psorophora discolor* rose from 2.8 to 10 Hz as the temperature rose from 12 to 30°C (Horsfall, 1955), and the ingestion rates of *Ae aegypti*, *Cx tarsalis* and *An albimanus* rose by 1.6–2.2-fold when the water temperature was raised from 18 to 31°C (Rashed and Mulla, 1989). Clearly, under the variable temperature conditions of natural habitats temperature is likely to affect rates of ingestion.

Aedes vexans larvae ingested inert particles of kaolin or synthetic cellulose at only one-third the rate obtained with wheat flour, fishmeal or yeast, unless a phagostimulant was added when the rates were identical (Aly, 1985a). Similarly, *Cx pipiens* larvae ingested yeast particles or algal cells more rapidly than several types of non-nutritive particle. Addition of phagostimulants such as yeast extract or yeast RNA to the suspension of inert particles produced an increase in the proportion of time the larvae spent filtering and increased their rates of ingestion (Dadd, 1970a, b; Dadd *et al.*, 1982). *Aedes vexans* larvae did not ingest food particles at a constant rate, but more slowly as they became satiated (Aly, 1985a). Starvation of *Cx tarsalis* and *Ae aegypti* larvae for 12 h increased their ingestion rates 1.6-fold and 1.8-fold respectively (Rashed and Mulla, 1989).

Almost no displacement of gut contents was apparent 4 h after the transfer of *Cx pipiens* larvae to particle-free water, and the addition of soluble phagostimulant increased the rate of displacement only slightly. Given sufficient time, it could be shown that in pure water the gut contents were displaced at about one-sixtieth the rate obtainable in a suspension of particles without phagostimulant. However, when 4th instar *Cx pipiens* larvae were transferred from charcoal suspensions to dilute *colloidal dispersions the gut contents were displaced at rates equal to those obtained in suspensions of larger particles. Upon transfer to 0.01–0.1% methyl cellulose solutions (viscosity range 1–1.5 centipoise), or to 0.01–0.6% polyacrylamide solutions (viscosity range 1–3 cP), rates of displacement of gut contents varied from just detectable to maximal. When larvae were placed in solutions of agar the displacement of gut contents varied from almost nil in 0.015% agar to near maximal in 0.025% agar. This was associated with an exceedingly small change of viscosity, from 0.97 to 1.02 cP, i.e. from about 107 to 113% of the viscosity of water (which was 0.90 cP at 25°C) (Dadd, 1975a). It is now apparent that the ingestion of colloids results from drinking (Section 4.1.2).

* Particles in the colloidal state have diameters between 1 nm and 100 nm. In dispersions of colloidal particles surface forces are an important factor in determining the properties of the phase.

5

Larval nutrition, excretion and respiration

5.1 STRUCTURE OF THE ALIMENTARY CANAL

In addition to its role as an organ of digestion and food uptake the alimentary canal of the mosquito larva has important functions of ionic and osmotic regulation and of excretion. The foregut and hindgut arise early in larval embryogenesis as two ectodermal invaginations, the stomodaeum and proctodaeum, and the midgut is formed from two groups of cells, the anterior and posterior midgut rudiments, carried on the apices of the stomodaeum and proctodaeum (Section 2.1.4). Appropriate to their ectodermal origin, the fore- and hindgut epithelia have a cuticular covering which faces the gut lumen. General accounts of the anatomy and histology of the alimentary canal (Figure 5.1) were provided by Thompson (1905), Imms (1907), Richins (1945), Schildmacher (1950), Christophers (1960) and Jones (1960); later authors are cited in the description given below. The ultrastructure of the regions of the gut that are active in ionic and osmotic regulation is described in Section 6.6.

5.1.1 The foregut

The foregut starts anteriorly at the point where the preoral cavity opens, through the mouth, into the short funnel-shaped **pharynx**. The pharynx wall is a flattened epithelium lined with cuticle and bounded by muscle fibres. Related to its functions of filtering and swallowing are two pairs of fine combs on the inner surface and a complex musculature (Section 4.2.3, Figure 4.9). The pharynx is well innervated, receiving nerves from the brain and frontal ganglion (Figure 10.6).

The **oesophagus** is a narrow muscular tube. Its flattened epithelium, which is thrown into folds, is lined with a thin cuticle and bounded by circular and occasional longitudinal muscle fibres. The posterior end telescopes into the midgut as the two-layered oesophageal invagination. Thus the oesophagus projects into the midgut and is then reflected upon itself, passing

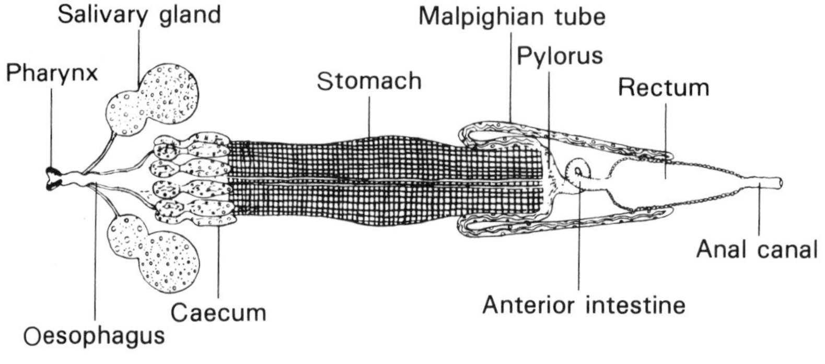

Figure 5.1 Alimentary canal and salivary glands of the larva of *Anopheles quadrimaculatus*. (After Jones, 1960.)

forwards to join the cardia, which is the most anterior part of the midgut (Figure 5.2). At the anterior end of the oesophageal invagination the circular muscle fibres are well developed, forming the annular muscle; at its posterior end the oesophageal invagination contains one or more blood sinuses. On contraction of the annular muscle the sinus becomes distended and forces back the food column in the stomach. The food column can be moved only when the larva is able to ingest particulate matter or to drink dilute colloidal suspensions (Wigglesworth, 1930; Dadd, 1975a).

The cuticular lining of the oesophageal epithelium is modified over the more posterior reflected portion of the epithelium, being thickened to form a band and extended into one or two flanges which encircle the oesophageal invagination and project posteriorly like flared skirts (Figure 5.2A). A survey of species from nine genera revealed the presence of flanges in all, although they were often absent in the first instar. Most species had two flanges, a few had one, and some species of *Anopheles* had many additional minute flanges. It has been suggested that the flanges may help propel newly formed peritrophic membrane posteriorly, or may prevent food particles from entering the pouch between the oesophagus and midgut (Wigglesworth, 1930; Richards and Richards, 1971; Richards and Seilheimer, 1977; Richards *et al.*, 1977).

A ring of undifferentiated embryonic cells situated at the junction between foregut and midgut is called the anterior imaginal ring (Figure 5.2A). Its cells, which are columnar and have characteristic staining properties, are found at the anterior end of the pouch formed by the oesophageal invagination. These cells undergo division during the larval stage, and later play an important role in the metamorphosis of the alimentary canal (Section 8.4). Two myoblastic masses situated on the surface of the alimentary canal in this region form the musculature of the adult oesophageal diverticula. Each cell mass contains two neurones, which constitute the ingluvial ganglia, and axons run from them over the gut (Romoser and Venard, 1969).

5.1.2 The midgut

In *Culex*, *Aedes* and *Anopheles* larvae the midgut has four regions: cardia, gastric caeca, anterior stomach and posterior stomach. In all regions the epithelium is composed of large microvillate columnar cells with polytene chromosomes. Regenerative (basal) cells are distributed singly or in groups at the base of the epithelium. The midgut epithelium does not have a cuticular lining, but a peritrophic membrane separates it from the food column. Longitudinal and circular muscle fibres surround the gastric caeca, and the stomach is surrounded by a lattice-like arrangement of longitudinal muscle fibres overlying circular fibres (Imms, 1907; Berger, 1938b; O'Brien, 1966a; Charles, 1987).

The **cardia** is the short length of midgut that surrounds the oesophageal invagination (Figure 5.2). The apical membrane of the cardial cells bears moderately long dense microvilli. Except in a shallow apical region, the cytoplasm is packed with rough endoplasmic reticulum (RER); it also contains a few mitochondria and many vesicles. The lateral members are connected apically by extensive septate desmosomes (Peters, 1979).

The **gastric caeca** are eight blind expansions of the midgut situated immediately behind the cardia. The shape of the caeca differs in different species, and within a species the caeca may vary in size. Four types of cells can be distinguished in the epithelium and have been named (1) resorbing/secreting cells, (2) ion-transporting cells, (3) membrane-secreting cells, and (4) imaginal cells (Volkmann and Peters, 1989a). In *Aedes aegypti* and *Culex quinquefasciatus* the resorbing/secreting cells occupy the anterior two-thirds of each caecum and the posterior part is mainly occupied by ion-transporting cells. In *Anopheles stephensi* these two cell types are intermixed. Very small regenerative cells occur between the resorbing/secreting cells and the ion-transporting cells. The ultrastructure of both the resorbing/secreting cells and the ion-transporting cells is appropriate for the transport of ions and water; it is described in Section 6.6.1. We may note here that glycogen is distributed throughout

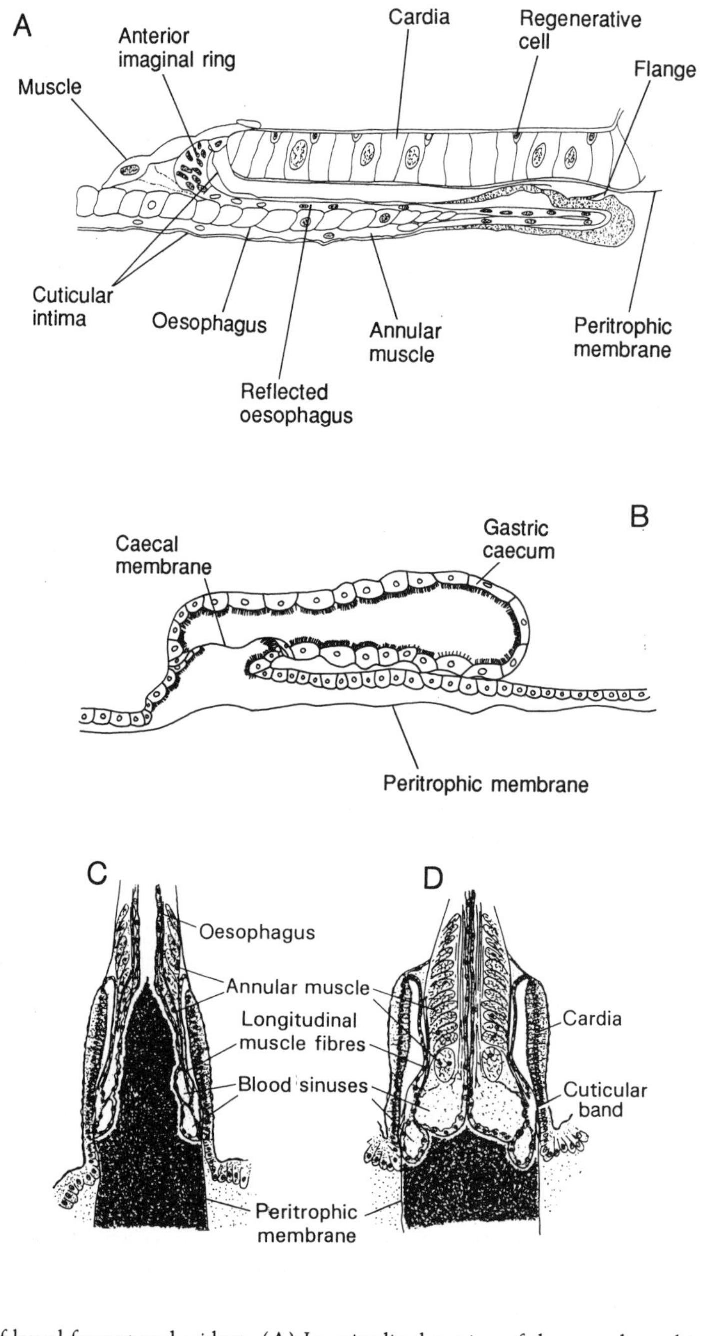

Figure 5.2 Junction of larval foregut and midgut. (A) Longitudinal section of the oesophageal invagination and cardia of *Aedes triseriatus* (after Romoser and Venard, 1967.) (B) Longitudinal section of a gastric caecum of *Aedes aegypti* (from Volkmann and Peters, 1989a.) (C, D) Oesophageal invagination of *Anopheles plumbeus* displacing food column: (C) in position of relaxation; (D) after contraction and relaxation of the longitudinal muscle, and with cardiac sphincter contracted and sinuses dilated (from Wigglesworth, 1930.)

the cytoplasm of the resorbing/secreting cells and that lipid droplets are present in the most anteriorly situated of those cells.

Two or three rows of membrane-secreting cells are arranged in a ring around the caecum. In *Ae aegypti* and *An stephensi* the ring surrounds the opening of the caecum into the midgut; in *Cx quinquefasciatus* it is situated towards the posterior end of the caecum. These cells protrude into the caecal lumen. The apical surface bears very short stumpy microvilli. A finely granular substance, thought to include chitin, is secreted by the cells and forms an amorphous membrane across the caecal lumen (Figure 5.2B). In *Cx quinquefasciatus* the membrane divides the caecal lumen into two compartments, in *Ae aegypti* and *An stephensi* it stretches across the entrance to the caecal lumen. With a thickness of 160 nm the caecal membrane is about 20% as thick as the peritrophic membrane (Volkmann and Peters, 1989a, b). These authors reported that the caecal membrane was impermeable to Evans Blue (960.8 Da) and only poorly permeable to fluorescein isothiocyanate (389.4 Da), observations which were inconsistent with those of Wigglesworth (1933a) and Ramsay (1950) unless the earlier authors had failed to identify the inner caecal lumen. The caecal membrane must nevertheless prevent the bulk exchange of fluids between the compartments of the ectoperitrophic space and the caecal lumen.

The cells of the anterior region of the **stomach** bear short tightly packed microvilli and the apical cytoplasm contains abundant mitochondria. A basal labyrinth is highly developed and the channels extend throughout almost the whole cell. Mitochondria are associated with the channels more basally (Charles and de Barjac, 1983; Charles, 1987). A short transition zone separates the anterior and posterior stomach regions. The cells of the posterior region have longer microvilli and abundant mitochondria in the apical cytoplasm. The basal labyrinth is less developed than in the anterior region. Smooth and rough ER is distributed throughout the cytoplasm (Charles and de Barjac, 1983; Charles, 1987; Cocke *et al.*, 1979). Volkmann and Peters (1989a) stated that the cells of the posterior region resemble the resorbing/secreting cells of the caecum. Small membrane-bound bodies thought to be cytolysosomes are present in cells of both regions (Davidson, 1979).

5.1.3 The hindgut

The hindgut, or proctodaeum, consists of Malpighian tubules, pylorus, anterior intestine, rectum and anal canal. Except for the Malpighian tubules, the hindgut epithelium is lined with a cuticular intima. The main functions of the hindgut are excretion and ionic and osmotic regulation through the activities of the Malpighian tubules and rectum. Five **Malpighian tubules** open into the alimentary canal behind the stomach. They run forwards from this position, usually to the 5th or 6th abdominal segment, then reverse their direction and end blindly near the posterior end of the abdomen. The tubules are composed of two cell types, primary cells and stellate cells, which differ in size and ultrastructure. The stellate cells, which are the smaller, are spaced apart and produce thin-walled regions in the tubule. In *Ae taeniorhynchus* the total number of cells in the five tubules is 62 $\pm$ 6.4 (s.d.), of which the stellate cells form 16 to 18%. The number of cells is constant from the larval to adult stage of an individual, and the number is similar in males and females. Similar ratios of primary to stellate cells were observed in the tubules of *Ae aegypti*, *Cx tarsalis* and *Cs inornata* (Satmary and Bradley, 1984). The nuclei of the primary cells contain polytene chromosomes; the nature of the chromosomes of the stellate cells is not known. Behind the openings of the Malpighian tubules is found the **posterior imaginal ring**, a band of small closely packed undifferentiated cells which become active at metamorphosis (Berger, 1938b).

The **pylorus** is a funnel-shaped length of epithelium with a cuticular lining which bears backward-pointing spines. A layer of circular muscle surrounds the posterior part forming a rather indistinct pyloric sphincter. The epithelial cells of the pylorus are small, and become increasingly small anteriorly until they resemble the undifferentiated cells of the posterior imaginal ring (Berger, 1938b; Trembley, 1951).

The **anterior intestine** (ileum) is a narrow muscular tube. Its epithelium, which is composed of squamous cells with interdigitating lateral membranes, extends into the lumen as deep folds. The folds are maintained by microtubules which arise from hemidesmosomes on the basal plasma membane and traverse the cell to insert on 'microapodemes' which project into the cell from the cuticular intima. The cytoplasm contains few mitochondria and extensive rough ER. The nuclei are endopolyploid. The outer muscle layer is composed of many small groups of muscle fibres, mostly orientated in a circular direction (Cerreta, 1976; Edwards and Harrison, 1983).

The **rectum** is a wide tube, quite distinct from the anterior intestine, and is the most densely tracheated region of the alimentary canal. It is bounded by sporadic small circular muscles. The rectal epithelium consists of large cells, with polytene nuclei. The apical cell membrane is thrown into numerous tightly packed folds or lamellae, and the basal region of the cell is modified to a basal labyrinth. In saline-water species of *Aedes*, the rectum consists of anterior and posterior regions which differ somewhat in ultrastructure (Section 6.6.3), but in all fresh-water species the epithelium is uniform throughout its length (Asakura, 1970, 1982b; Meredith and Phillips, 1973c). There are no rectal papillae in the mosquito larva.

The posterior end of the rectum tapers to join the **anal canal** which resembles the anterior intestine in ultrastructure. Its thin epithelial cells contain endopolyploid nuclei and are bounded by circular muscle. At the anterior end of the anal canal the epithelial folds extend into the lumen.

5.1.4 The peritrophic membrane

(a) Structure

The peritrophic membranes of 4th instar mosquito larvae are multilayered structures about 1 μm thick. They are bounded on the gut lumen side by a thick, finely granular layer and on the epithelium side by an exceedingly thin layer. Between these outer layers the bulk of the peritrophic membrane consists of alternate electron-dense and electron-lucent layers. The outer layer is 60 nm thick in *Aedes aegypti* and *Culex quinquefasciatus* and 100–150 nm thick in *Anopheles stephensi*. In two of these species it is continuous, but in *Cx quinquefasciatus* it is perforated with 10 nm pores spaced 10–20 nm apart. The central region of the peritrophic membrane usually contains four or five electron-dense layers, but in some species there are more. These layers have irregular round or elongate holes of 60–100 nm diameter, sometimes of 150–180 nm (Richards and Richards, 1971; Peters, 1979).

The peritrophic membrane of *Ae aegypti* is weakly birefringent. The electron-lucent layers contain microfibrils orientated in a net-like pattern with square or rhomboid apertures. Other microfibrils are orientated irregularly. Individual microfibrils are about 10 nm thick, the bundles are 130–150 nm thick, and the distance between bundles ranges from 100 to 200 nm (Zhuzhikov *et al.*, 1971; Kuznetsova, 1981). Histochemical tests showed that the two outer layers and the electron-dense layers contain carbohydrate. Lectin-binding experiments revealed the presence of *N*-acetylglucosamine and therefore of chitin in the bundles of microfibrils. Tests with asialofetuin revealed no lectins within the peritrophic membranes (Peters, 1979; Dörner and Peters, 1988).

(b) Secretion

The cylindrical peritrophic membrane is formed in the pouch between the cardia and the oesophageal invagination, and is continuously extruded in a posterior direction, extending through the stomach and into the hindgut. The substance of the peritrophic membrane is secreted by a ring of cells 8 to 10 deep, at the anterior end of the cardia, which do not differ in ultrastructure from the other cardial cells. All stages in formation of the peritrophic membrane can be seen in a single longitudinal section of the cardia, developing from traces of secretion to full thickness over a distance of about eight cells. In *Aedes aegypti* a finely granular layer is observed on top of the microvilli of the 3rd ring of cells from the anterior margin of the cardia; it grows in thickness over the 4th and 5th cells, reaching about 0.2 μm. The secretion of this material, which is seen only on the tips of the microvilli and never between them, seems to be completed a few cells later. An electron-lucent material is then secreted, followed by an electron-dense material which appears in small caps on the tips of the microvilli and more conspicuously between them. This is the outermost electron-dense layer of the peritrophic membrane. The change in secretion products is repeated two or three times. After the final secretion of electron-lucent material a very finely granular layer *c.* 10–20 nm thick is formed, which resembles the original granular layer in electron density (Peters, 1979).

The suggestion that the pouch and cuticular flanges form a press and extrusion mechanism (Wigglesworth, 1930) has been refuted on the grounds that the granular secretion never fills the space between the cardia and oesophagus, that the peritrophic membrane is already fully formed a considerable distance anterior to the cuticular ring, and that a peritrophic membrane is present in 1st instar *Ae aegypti* larvae although they lack the cuticular flanges (Richards and Richards, 1971). However, we have no alternative explanation for the backward movement of the peritrophic membrane in this region of the midgut.

(c) *Permeability*

The region within the peritrophic membrane is called the endoperitrophic space, and the narrow compartment between the membrane and the epithelium is called the ectoperitrophic space. The peritrophic membrane separates the midgut epithelium from abrasive food particles and pathogenic microorganisms, and permits the forward movement of fluid during antiperistalsis (Section 6.3.1). Peritrophic membranes show signs of degradation in the posterior parts of the stomach, and generally break up in the hindgut. Knowledge of the permeability of the peritrophic membrane is necessary for an understanding of digestion.

The characteristics of peritrophic membranes from 4th instar larvae of *Ae aegypti* were determined by Zhuzhikov (1970). When separating different solutions of inorganic ions the membranes did not generate electrical potentials, therefore the ions penetrated the membranes readily at rates determined by their diffusion coefficients. As penetration rates were similar in both directions, the membranes were not polarized. Permeability coefficients (L_p) were determined from

$$L_p = V/\Delta P \qquad (5.1)$$

where V is volume passing through the membrane in unit time and ΔP is the difference in hydrostatic pressure across the membrane. The permeability coefficient remained the same when hydrostatic pressure varied between 20 and 500 mm of water. A mean value was obtained of 3.36×10^{-8} cm^3 s^{-1} through an area of 1 cm^2 at a pressure of 1 bar, which was 2–3 times greater than the permeability coefficients of cellulose membranes.

Reflection coefficients (δ), complex ratios of reflected pressure to incident pressure which relate to the osmotic properties of compounds, were calculated from

$$\delta = \frac{L_p \cdot \Delta P - V}{L_p \cdot Pi} \qquad (5.2)$$

where Pi is osmotic pressure. To characterize permeability to non-ionic compounds, reflection coefficients were determined for di-, tri- and polysaccharides. Values of 0.0002, 0.004 and 0.078 were obtained for sucrose, raffinose and inulin respectively, compounds which have hydrated molecular radii of 0.52, 0.61 and 1.2 nm. The effective radius of the membrane pore, as determined from the reflection coefficients, was very high but differed markedly between the smallest and largest of these molecules, being 4000 nm with sucrose, 300 nm with raffinose and 25 nm with inulin. Zhuzhikov (1970) concluded that permeability to non-ionic compounds was not just a function of pore size but was regulated by interaction between the membrane and the penetrating substance, this interaction becoming greater as the size of the penetrating molecule increased. When larvae were placed in suspensions of colloidal gold, ingested particles of up to 5 nm diameter readily passed through the peritrophic membrane and accumulated in the caeca. Particles of 8 nm accumulated in smaller numbers, and 9 nm particles were rarely found in the caeca.

Permeability has also been assessed from the passage of Evans Blue and fluorescein isothiocyanate-labelled dextrans through the peritrophic membranes of different species (Peters and Wiese, 1986). Peritrophic membranes from *Ae aegypti* were permeable to Evans Blue (0.96 kDa) but not to dextrans of 2.4 kDa (Einstein–Stokes radius 1.3 nm). In contrast, peritrophic membranes from *Anopheles stephensi* larvae were permeable to dextrans of 17.2 kDa (3.2 nm) but not to those of 32 kDa (4.3 nm), and peritrophic membranes from *Culex quinquefasciatus* larvae were permeable to dextrans of 32 kDa (4.3 nm) but not to those of 37.2 kDa (4.5 nm). The authors concluded that the spacings of the microfibrillar network could not account for the observed permeabilities.

The only report of enzymes passing through peritrophic membranes is that of Detra and Romoser (1979a), who found that when peritrophic membranes were removed from *Ae aegypti* larvae and ligatured at both ends, passage of proteolytic enzyme from the residual contents to the exterior of the peritrophic membranes could be demonstrated.

5.1.5 The salivary glands

In *Anopheles albimanus* each salivary gland consists of a spherical anterior portion of 12–15 cells, with large nuclei that contain polytene chromosomes, and a larger terminal sac of 50–60 cells with smaller nuclei. The salivary ducts lack taenidia. The common duct opens on the labiohypopharynx directly in front of the mouth. A chain of nephrocytes extends between the two salivary glands. In *Culex pipiens* the anterior part of each salivary gland is elongated and composed of small cells, and the terminal sac is globular and composed of much larger cells. In *Aedes aegypti* the glands are cylindrical for most of their length but dilated towards the distal end (Jensen and Jones, 1957).

5.2 DIGESTION

In insects, digestion and the absorption of digestion products occur in the midgut. The peritrophic membrane is thought to compartmentalize digestion in some insects, permitting passage into the food bolus of enzymes such as trypsin and amylase which attack the polymeric food constituents proteins and starch, but retaining other enzymes in the ectoperitrophic space where they hydrolyse the oligomeric products of primary digestion (Terra, 1990). Whether this obtains in mosquito larvae is not known, but it is likely that the countercurrent system that operates in the ectoperitrophic space (Section 6.3.1) is important for conservation of enzymes and for absorption. An outline of the characterization of proteolytic enzymes is to be found in Section 14.1.1.

5.2.1 Gut pH

The hydrogen ion concentrations of the gut contents have been measured by placing larvae in indicator solutions, with kaolin or other adsorbent particles to assist ingestion (Table 5.1). At the anterior extremity of the midgut the contents are weakly alkaline (pH 7.5), but within a short distance the pH rises sharply. The central part of the midgut, situated in the first three abdominal segments, is the region of greatest alkalinity, and in some individuals of *Culex pipiens* the pH is >10.6. The pH falls progressively in the posterior midgut, reaching 7 or even 6 at its junction with the hindgut. In *Aedes aegypti*, the pH of the pylorus contents is 7.4–7.8 and that of the rectal contents <7.2. The pattern of pH values is similar in all instars, but in older larvae the region of high pH extends

Table 5.1 Mean pH ranges of the contents of the mid- and hindgut of mosquito larvae.

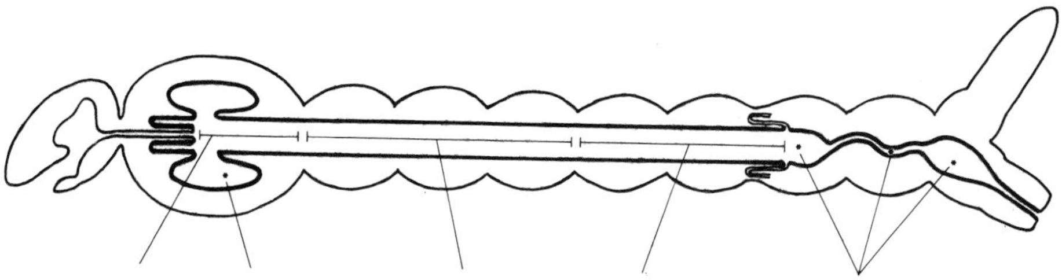

Species	*Anterior midgut*	*Gastric caeca*	*Central midgut*	*Posterior midgut*	*Hindgut*	*Ref.*
An stephensi	—	—	≃9.6	>7.0	—	3
Cx pipiens	7.5⟶10.0	8.0—8.5	>10.1⟶9.5	9.0⟶7.0	6.5—7.5	3
Ae aegypti	7.5⟶10.0	8.0—8.5	>10.1⟶9.5	9.0⟶7.0	6.5—7.5	3
Ae aegypti	?⟶ 9.7	7.6—8.0	9.9⟶9.3	8.1⟶6.2	—	2
Ae aegypti	—	7.6—8.0	—	—	7.6⟶<7.2	1
Ae atropalpus	⟶10.0	7.0	10.0⟶7.5	7.0⟶6.0	—	4
Ae scutellaris	⟶10.0	7.0	10.0⟶8.0	7.5⟶6.5	—	4
Ae epactius	⟶10.0	7.0	10.0⟶7.5	7.0⟶6.5	—	4

References: 1. Ramsay (1950); 2. Charles and de Barjac (1981); 3. Dadd (1975b); 4. Stiles and Paschke (1980).
Arrows indicate gradient of pH.

further back in the midgut (Ramsay, 1950; Dadd, 1975; Stiles and Paschke, 1980; Charles and de Barjac, 1981).

When larvae were handled or narcotized the pH of the alkaline region of the midgut started to fall within two or three minutes, suggesting that the midgut contents are only weakly buffered (Dadd, 1975b, 1976). By ligaturing larvae of *Ae aegypti* in different positions it was shown that the alkaline condition of the midgut contents is due to the activity of the gut cells in the first three abdominal segments. Presumably the region of alkalinity is normally extended from that central part of the midgut by the backward movement of the gut contents combined with the anterior flow of Malpighian tubule fluid and absorption of water in the gastric caeca. Ligation just behind the thorax caused the contents of the gastric caeca to become acid (Zhuzhikov and Dubrovin, 1969).

Few insects, apart from the larvae of Lepidoptera and of certain families of Nematocera, have strongly alkaline gut contents. Tannin–protein complexes dissociate under alkaline conditions, and it has been suggested that a very high midgut pH is therefore advantageous to herbivorous and detritus-feeding larvae (Martin *et al.*, 1980). The high pH also permits activation of *Bacillus thuringiensis* δ-endotoxin by midgut proteases and the consequent pathological disruption of the midgut epithelium.

5.2.2 Digestive enzymes

The pH activity curves for proteases from *Aedes aegypti* and *Anopheles stephensi* showed broad peaks which were maximal at pH 9 and above; the curve for *Culex pipiens* was maximal at pH 10 and above (Dadd, 1975b).

Some 12–13 electrophoretically distinct bands showing proteolytic activity were obtained from *Aedes aegypti* gut homogenates. The proteinases had molecular masses of 20–25 kDa, and all were active at alkaline pH. Their activity could be completely inhibited by DFP, showing them to be serine proteases. Hydrolysis of the specific substrates TAME and BTEE revealed that certain bands had trypsin or chymotrypsin activity and that others did not (Kunz, 1978). Because the preparations were made in the absence of proteolysis inhibitors, it is likely that a number of the bands resulted from autolysis.

Homogenates of midguts from *Culex pipiens* larvae showed trypsin, chymotrypsin, aminopeptidase and carboxypeptidase activities (Spiro-Kern and Chen, 1972). In *Ae aegypti* 85% of midgut chymotrypsin activity was found in the gut contents and 15% in the epithelium. The activity was enhanced considerably by 1 mM-Ca^{2+}, and to a lesser extent by 0.1 M-NaCl. In the absence of Ca^{2+} the pH optimum was about 8.5; in the presence of 0.03 M-Ca^{2+} it rose to at least 10 (Yang and Davies, 1971a, 1972a). In *Ae aegypti* and other species aminopeptidase is restricted to the resorbing/secreting cells of the gastric caeca and to cells in the posterior region of the stomach (Volkmann and Peters, 1989b).

In *Ae aegypti* larvae one-third of the total α-amylase, which hydrolyses 1,4-α-D-glycosidic linkages in the interior of polysaccharide chains such as those of starch and glycogen, is present in the gut (McGeachin *et al.*, 1972). Gut amylase activity was maximal betweeen pH 7 and 9 in *Ae aegypti* and *An stephensi*, and between pH 7.5 and 9.5 in *Cx pipiens* (Dadd, 1975b); its action will therefore be limited to the anterior and posterior regions of the midgut. Hinman (1933) reported invertase and lipase in the larval gut.

High activity levels were obtained for trypsin and chymotrypsin in midgut homogentates from *Ae aegypti* larvae of all instars and from pharate pupae, but by the time of larval–pupal ecdysis the activity levels were zero (Yang and Davies, 1971a, b; McGeachin *et al.*, 1972). As the larvae grew during an instar, protease activity remained proportional to body mass, and it was not reduced during larval–larval moults (Detra and Romoser, 1979b). Aminopeptidase activity increased dramatically during the first three days of larval life as the larvae increased in size, but it started to decline a day before larval–pupal ecdysis and fell during metamorphosis to the low but positive level characteristic of non-bloodfed adults (Graf and Briegel, 1982). Active trypsin, chymotrypsin and amylase were discharged in the larval faeces (Yang and Davies, 1971b).

5.2.3 Digestion, resorption and displacement of gut contents

The digestibility of microorganisms in larval food is determined by the resistant properties of their outer wall and the duration of exposure in the gut. Delicate ciliates and flagellates are rapidly digested, as may be some non-living food particles, but thick-walled organisms and much ingested detritus are relatively indigestible (Section 4.1.1). No measurements have been made of assimilation efficiency.

There is no direct evidence to show where the products of digestion are absorbed. The forward movement of fluid in the midgut and the resorption of water in the caeca (Section 6.3) may be functionally related to the uptake of digestion products. If the deposition of reserves in gut cells after feeding single foods is any indication (Section 5.4), then certain caecal cells absorb lipids, sugars and amino acids, the cells of the anterior half of the stomach absorb lipids, and the cells of the posterior half of the stomach aborb sugars and amino acids. However, this is a very questionable argument, and the observations may reflect metabolic rather than absorptive activities.

The oesophagus of *Anopheles quadrimaculatus* shows peristaltic waves every 1–30 s irrespective of food intake, but the waves are more frequent during ingestion. It seems likely that the stomach contents of mosquito larvae are forced back when contraction of the cardiac sphincter dilates the blood sinus in the oesophageal invagination. In any event, when food intake stops, the food column remains stationary in the stomach (Jones, 1960; Schildmacher, 1950; Dadd, 1970a). The hindgut contents are moved on by peristaltic movements of the anterior intestine and rectum. Every one or two minutes the stomach of *An quadrimaculatus* shows an activity cycle during which frequent small antiperistaltic waves occur for a time, followed by quiescence and then by a powerful peristaltic wave which restarts the antiperistaltic waves. This activity is apparently not related to the backward movement of the food column in the stomach but probably serves to move fluid forward for absorption in the caeca. Because the rhythmic movements of *An quadrimaculatus* gut were not stopped by cauterization of ganglia or by ligation, Jones (1960) suggested that the contractions of the alimentary canal are largely myogenic in nature and that the muscles of the fore-, mid- and hindgut act independently of each other.

The rate of passage of food can be measured by following distinctive particles through the gut. The mean times required by feeding *Cx torrentium* larvae to fill their guts were 36 min for 2nd instar larvae and 53 min for 4th instar larvae (Nilsson, 1986). Food traversed the stomachs of 4th instar *Ae aegypti* and *Cx pipiens* larvae after 20–30 min at 27°C and after 50–60 min at 20°C. Fourth instar larvae of *An maculipennis* and *An superpictus* required about 60 min at 27°C and 120–180 min at 18°C (Schildmacher, 1950). Third and 4th instar larvae of the predatory species *Cx (Lutzia) fuscanus* required 45 min for their food to traverse the stomach at 27°C (Ikeshoji, 1966). When filter feeding larvae feed in natural waters, where the particle concentrations are often low, it is likely that food will pass through the gut much more slowly.

5.3 NUTRITIONAL REQUIREMENTS

The use in the 1930s and 1940s of semisynthetic diets for mosquito larvae reared under axenic or sterile conditions paved the way for the development of fully chemically defined diets. A number of developments refined the technique and improved the quality of the diets, notably: (a) replacement of proteins with amino acids (Lea *et al.*, 1956); (b) replacement of RNA with nucleotides (Dadd and Kleinjan, 1976); (c) use of non-nutritive agarose as an aid to fluid ingestion (Dadd and Kleinjan, 1978); (d) recognition of the requirement for polyunsaturated fatty acid (Dadd and Kleinjan, 1979a); and (e) use of synthetic lecithin as carrier for arachidonic acid and cholesterol, coupled with ascorbyl palmitate as antioxidant (Dadd and Kleinjan, 1979b, 1984a). The omission of compounds, one at a time, from chemically defined diets provided a means of identifying certain compounds as essential for development to the adult stage and others as improving growth rate or viability while not being strictly essential.

The chemically defined diet developed for *Culex pipiens* allowed newly hatched larvae to be reared to adults, with 90% or more survival, at growth rates comparable to those obtained with many unrefined diets. However, the adults were weak and unable to fly. Inclusion in the diet of arachidonic acid or certain structurally related polyunsaturated fatty acids greatly increased adult viability, and the supplemented diet (Table 5.2) proved suitable for a number of species of *Culex*, *Culiseta*, *Aedes* and *Anopheles* (Dadd and Kleinjan, 1976, 1978; Dadd, 1981, 1983).

5.3.1 Amino acids and sugars

Aedes aegypti larvae failed to reach the 2nd instar when kept in media lacking protein but complete in other respects. The necessary protein could be replaced by a mixture of amino acids (Golberg and De Meillon, 1948). The following ten amino acids, as L enantiomers, are essential for *Ae aegypti*: arginine, histidine, isoleucine, leucine, lysine, methionine, phenylalanine, threonine, tryptophan and valine. The same ten amino acids are essential for growth in other insects and in many vertebrates (Singh and Brown, 1957; Lea and DeLong, 1958; Dadd, 1985). Although proline was not essential, its omission seriously retarded larval development. Proline could be detected in larvae when omitted from their diet, therefore they were capable of

Table 5.2 Composition of the synthetic dietary medium developed for *Culex pipiens* larvae by Dadd and Kleinjan (1979b) and Dadd (1980, 1981).

Composition	*g l⁻¹*
Minerals	
$K_2HPO_4.3H_2O$	0.30
$NaH_2PO_4.1H_2O$	0.30
NaCl	0.10
$MgSO_4.7H_2O$	0.20
KOH	1.50
Ca gluconate	0.05
Fe.Na EDTA	0.02
$Zn.Na_2$ EDTA	0.02
$Mn.Na_2$ EDTA	0.02
$Cu.Na_2$ EDTA	0.005
Amino acid mixture	13.90
Nucleotides	
Adenosine 2′- and 3′-phosphate	0.60
Guanosine 2′- and 3′-phosphate	0.40
Cytidine 2′- and 3′-phosphate	0.25
Uridine 2′- and 3′-phosphate	0.25
2-Deoxythymidine-5′-phosphate	0.10
Carbohydrates	
D-Glucose	2.50
Sucrose	5.00
Agarose	0.50
Vitamins	
Thiamin HCl	0.005
Riboflavin	0.005
Nicotinic acid/amide (1:1)	0.010
Pyridoxol HCl	0.005
Calcium pantothenate	0.050
Pteroylglutamic acid	0.001
Biotin	0.001
Choline chloride	0.100
Lipids	
Cholesterol	0.01
Arachidonic acid	0.0005
Ascorbyl palmitate	0.004
1,2-Dipalmitoyl-*sn*-glycero-3-phosphocholine	0.01

This diet has also proved suitable for *Ae aegypti*, *Ae sierrensis* and *Cs incidens*. For *Cs inornata* and *An stephensi* the amino acid concentration should be reduced to 0.9% w/v. For *Cx tarsalis* the amino acids should be reduced by 20%, the sugars increased by 15%, and agarose increased by 20% (Dadd, 1983; Dadd *et al.*, 1980, 1987; Dadd, personal communication).

The KOH is added to adjust the pH to 7.0.

Amino acids: 0.75 g each of L-alanine, L-arginine HCl, L-asparagine, L-aspartic acid, L-cysteine HCl, L-glutamic acid, glycine, L-histidine HCl, L-isoleucine, L-leucine, L-lysine HCl, L-methionine, L-phenylalanine, L-proline, L-serine, L-threonine, L-tryptophan and L-valine. 0.40 g of L-tyrosine.

Nucleotides: as the salts specified by Dadd (1980).

Ascorbyl palmitate is incorporated as an antioxidant.

1,2-Dipalmitoyl-*sn*-glycero-3-phosphocholine (synthetic lecithin) is used as a carrier for cholesterol and arachidonic acid.

synthesizing it but not at a rate which permitted normal growth (Singh and Brown, 1957; Singh and Micks, 1957; Lea and DeLong, 1958). The larval requirement for L-arginine could be partly met by DL-citrulline (Golberg and De Meillon, 1948).

Larvae of *Culex pipiens* failed to develop beyond the 2nd instar when provided with only the ten amino acids named as essential for *Ae aegypti*. In the absence of asparagine they did not reach the adult stage. Asparagine could not be replaced by aspartic acid, glutamic acid or glutamine. In the absence of proline larval developmental time was doubled, with heavy mortality in the later instars. It was postulated that proline was essential for the continuation of development beyond one generation, consequently both asparagine and proline were classed as essential for *Cx pipiens*. The absence of either glycine or serine retarded larval growth rates in *Cx pipiens* but satisfactory development rates could be achieved with a diet of the twelve essential amino acids supplemented with glycine or serine. Asparagine was an essential nutrient for *Cs incidens*, no larvae developing to the 2nd instar without it, but it was not required by *Ae aegypti*. High amino acid concentrations in artificial diets were deleterious. When the total amino acid concentration amounted to 1.4% w/v or more, adult vigour was reduced in *Cx pipiens* and both larval viability and adult vigour were reduced in *Cs inornata* and *Cx tarsalis*, compared with the viability and vigour observed with amino acid concentrations of 0.9 or 1% (Dadd, 1978; Dadd *et al.*, 1980).

Absence of sugar from the diet did not affect the viability of *Ae aegypti* larvae, but larval growth rates increased progressively as the concentration of D-glucose was raised from 0.1 to 1.0%. At still higher glucose concentrations the growth rate slowed. Variation of glucose and amino acid concentrations showed that at higher amino acid concentrations amino acids could replace sugars as a source of energy. With D-glucose as the only sugar, concentrations of 0.5–1.0% glucose were optimal when coupled with an amino acid concentration of 1.5%.

Sucrose could also be used (Table 5.2), probably with osmotic advantage (Sneller and Dadd, 1977, 1981; Dadd, 1980).

5.3.2 Polyunsaturated fatty acids

All insects require C_{20} polyunsaturated fatty acids (eicosanoids) in their tissues. All insects acquire C_{18} polyunsaturated fatty acids in their diet, and most can synthesize C_{20} polysaturates from them via elongation and desaturation pathways. Mosquitoes, however, are unusual in that they lack the specific desaturases needed and must acquire C_{20} polyunsaturated fatty acids through the larval diet. In mosquitoes the most readily discernible effect of lack of an essential fatty acid from the larval diet was the inability of the teneral adults to fly. In some mosquito species it also reduced survival in the immature stages. These effects could be prevented by providing

Table 5.3 The structures of polyunsaturated fatty acids that have been tested for their nutritive value for *Culex pipiens* and *Cx tarsalis*. (From Dadd, 1981, 1983; and Dadd *et al.*, 1987.)

The box drawn around the first four compounds surrounds the four *cis* double bonds that are common to the group of 'highly effective' fatty acids which, when provided in the diet, permit a majority of adults to fly and to survive for long periods. The middle box indicates the three *cis* double bonds common to the group of 'moderately effective' fatty acids with which flight is weak and few adults survive more than a day or two. The bottom box indicates the two *cis* double bonds common to fatty acids of the group considered 'slightly effective', with some emergent adults standing or hopping but unable to fly. The values of the flight capacity index are for *Culex pipiens*.

		COOH end — CH3 end	Flight capacity index
4, 7, 10, 13, 16, 19-Docosahexaenoic	22:6 n-3		71–78
5, 8, 11, 14, 17-Eicosapentaenoic	20:5 n-3		70–83
7, 10, 13, 16-Docosatetraenoic	22:4 n-6		84–93
5, 8, 11, 14,-Eicosatetraenoic*	20:4 n-6		69–100
8, 11, 14,-Eicosatrienoic	20:3 n-6		23–68
(6, 9, 12)-Linolenic	18:3 n-6		33–79
13, 16, 19-Docosatrienoic	22:3 n-3		12–26
11, 14, 17-Eicosatrienoic	20:3 n-3		0–39
(9, 12, 15) -Linolenic	18:3 n-3		6–23
13, 16-Docosadienoic	22:2 n-6		8–10
11, 14-Eicosadienoic	20:2 n-6		4–9
(9, 12)-Linoleic	18:2 n-6		0–14

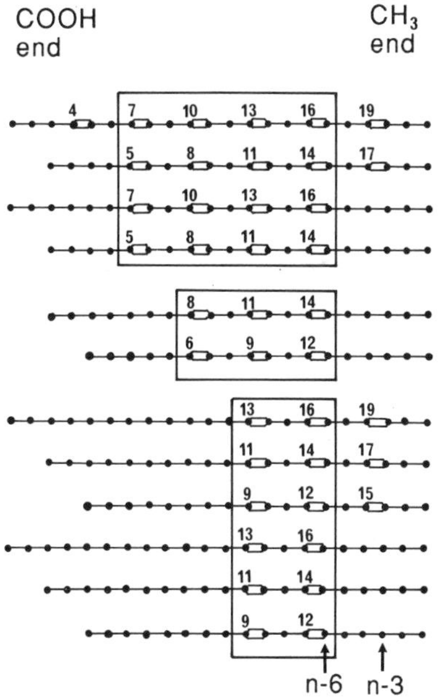

Methylene-interrupted polyunsaturated fatty acids can be represented by symbols such as 18:3 n-3 for (9,12,15)-linolenic acid and 20:4 n-6 for arachidonic acid. These symbols indicate straight chain C_{18} and C_{20} acids which, respectively, have systems of three and four methylene-interrupted *cis* double bonds starting on the 3rd and 6th carbon atoms from the methyl end of the molecule. The most important families of polyunsaturated fatty acids are the n-3, n-6 and n-9 acids. In general the acids within any single family are biosynthetically related, being interconverted by enzymatic processes of desaturation, chain elongation and chain shortening.

'Flight capacity index' is a measure of capacity for flight, with the range 0–100. It is calculated from the activities of a group of adults on emergence, with relative values of 0, 0.25, 0.5 and 1.0 for individuals that are trapped, stand, hop and fly (Dadd and Kleinjan, 1978).

* Arachidonic acid.

vitamin-level concentrations of an appropriate polyunsaturated fatty acid (Dadd and Kleinjan, 1984b; Dadd *et al.*, 1987).

(a) Chemical nature of the requirement

Arachidonic acid, $CH_3{\cdot}(CH_2)_4{\cdot}(CH{=}CH{\cdot}CH_2)_4{\cdot}(CH_2)_2{\cdot}COOH$, when present in the larval diet at 0.0005 g l^{-1}, permitted development of adults of *Culex pipiens* which were able to fly normally and to survive for long periods (Dadd and Kleinjan, 1979a). Tests with a variety of fatty acids showed that the structural characteristic that fully satisfied the requirements of mosquitoes was a series of four double bonds of *cis* configuration, terminating at the n-6 position in a C_{20} or C_{22} acid, with neighbouring double bonds separated by a saturated carbon atom. The first four compounds illustrated in Table 5.3 have this structure. Less effective were two trienoic n-6 fatty acids which permitted development of a small percentage of flying adults, most of which collapsed a day or two after emergence. Six trienoic n-3 or dienoic n-6 fatty acids enabled adults to stand or hop about on the water surface, but only rarely to fly. All monounsaturated and saturated fatty acids were completely inactive. The polyunsaturated fatty acids that were fully or partly effective in increasing adult vigour also slightly improved larval growth and survival (Dadd, 1980; Dadd *et al.*, 1987).

(b) Analyses of mosquitoes

When larvae were reared on a synthetic diet containing no polyunsaturated fatty acids, the glycerophospholipids of adult *Cx tarsalis* contained 0.1% linoleic acid (18:2) and traces of other polyunsaturates, while those of adult *Cx pipiens* contained 0.5% linoleic acid. These small amounts of polyunsaturated fatty acids are thought to have been carried over from the previous generation through the eggs. It was possibly by that means that *Cx pipiens*, *Cs incidens* and *Ae sierrensis* were able to develop to the adult stage almost equally well whether polyunsaturates were present in the synthetic diet or not; however, when they were absent the teneral adults were feeble. If *Cx tarsalis* larvae were deprived of polyunsaturated fatty acids only some 30% reached the adult stage. Few larvae of *Cs inornata* so deprived reached the pupal stage and none reached adulthood (Sneller and Dadd, 1977; Dadd *et al.*, 1980; Dadd, 1981; Dadd and Kleinjan, 1984b; Dadd *et al.*, 1987).

Individual polyunsaturated fatty acids all made up a lower percentage content of the total fatty acids of triacylglycerols than of phospholipids (Table 5.4). Usually only trace amounts of arachidonic acid and eicosapentaenoic acid (20:5 n-3) were found in the triacylglycerols of laboratory-reared *Cx tarsalis*, and never more than 0.4%, but in wild adults from 'good breeding habitats' they averaged 1.3 and 3.8% respectively. Wild adults from less satisfactory habitats tended to resemble laboratory-reared mosquitoes in their triacylglycerol and phospholipid polyunsaturated fatty acid contents (Dadd *et al.*, 1988).

Forty-five percent of glycerophospholipid fatty acids of wild adult *Culex tarsalis* obtained from 'good breeding habitats' were polyunsaturated, as were almost 40% of phospholipid fatty acids of conventionally reared laboratory adults (Table 5.4). Arachidonic acid and eicosapentaenoic acid each constituted about 6% of phospholipid fatty acids of laboratory-reared *Cx tarsalis*, whether the larvae had been fed an axenic chemically-defined diet including a polyunsaturate or a conventional laboratory diet. Wild adults obtained from 'good breeding habitats' had a similar arachidonic acid content but their eicosapentaenoic acid content was greater, constituting 18–20% of phospholipid fatty acids.

Addition of Walgreen oil (a proprietary fish oil in which eicosapentaenoic acid constituted 46% of total fatty acids) to the diet of *Cx tarsalis* larvae at 75 mg g^{-1}, tripled the adult eicosapentaenoic acid content bringing it to the level found in wild mosquitoes. When half that amount of Walgreen oil was fed to larvae, male flight capacity was enhanced by 35% but

Table 5.4 Occurrence of fatty acids in triacylglycerols and glycerophospholipids extracted from newly emerged *Culex tarsalis* which were either wild and caught in 'good breeding habitats' or derived from larvae reared in the laboratory on crude non-sterile media. Values are mean percentage content (± s.d.). (From Dadd *et al.*, 1987, 1988.)

Trivial or systematic name of acid		*In triacylglycerols*		*In glycerophospholipids*	
		Wild (N = 7)	*Reared* (N = 5)	*Wild* (N = 7)	*Reared* (N = 7)
Myristic	14:0	2.2 ± 2.2	2.5 ± 0.2	–	–
Palmitic	16:0	29.5 ± 3.1	31.7 ± 4.0	14.7 ± 1.2	13.4 ± 2.0
Palmitoleic	16:1	23.3 ± 7.6*	34.0 ± 4.9	13.7 ± 4.2	20.4 ± 6.7
Stearic	18:0	3.2 ± 1.0	1.6 ± 0.5	3.3 ± 0.6	2.8 ± 0.7
Oleic	18:1	21.4 ± 6.0	21.2 ± 2.5	18.9 ± 3.6	21.3 ± 2.8
Linoleic	18:2 (*n*–6)	3.8 ± 1.9	2.8 ± 2.8	8.4 ± 2.9*	20.8 ± 8.1
(6,9,12)-Linolenic	18:3 (*n*–6)	0.3 ± 0.3	0.3 ± 0.3	0.7 ± 0.6	1.5 ± 0.6
(9,12,15)-Linolenic	18:3 (*n*–3)	5.6 ± 4.8*	0.5 ± 0.5	9.9 ± 8.6	4.5 ± 1.8
Arachidic	20:0	0.1 ± 0.1	*t*	0.4 ± 0.2	0.4 ± 0.3
(8,11,14)-Eicosatrienoic	20:3 (*n*–6)	*n*	0.1 ± 0.1	*t*	*t*
(11,14,17)-Eicosatrienoic	20:3 (*n*–3)	*n*	*t*	*t*	*n*
Arachidonic	20:4 (*n*–6)	1.3 ± 0.5*	0.2 ± 0.2	4.7 ± 2.6	6.3 ± 2.1
(5,8,11,14,17)-Eicosapentaenoic	20:5 (*n*–3)	3.8 ± 2.4*	0.3 ± 0.2	21.3 ± 4.8*	5.8 ± 1.9
Docosanoic	22:0	*t*	0.1 ± 0.1	0.1 ± 0.2	0.2 ± 0.3
(4,7,10,13,16,19)-Docosahexaenoic	22:6 (*n*–3)	*n*	*t*	0.2 ± 0.3	0.1 ± 0.2
Unknowns (total)		3.2	0.2	1.3 ± 0.4	0.2

*$p < 0.01$ in relation to the adjacent reared value.
N = number of replicates; *t* = trace (0.1%); *n* = not tested.

remained well below that of wild males (Dadd *et al.*, 1989).

(c) Synthesis and metabolism

Experiments in which the larval dietary medium contained just one polyunsaturated fatty acid, and in which the fatty acid constituents of the glycerophosphoplipids in the emerging adults were identified, revealed that *Culex pipiens* and *Cx tarsalis* had no capability for *de novo* synthesis of polyunsaturates. However, they incorporated into tissue phospholipids polyunsaturated fatty acids of many sorts when available in the diet, and converted certain polyunsaturated fatty acids to others. Of the four highly effective fatty acids, only 20:4 n-6 (arachidonic = 5,8,11,14-eicosatetraenoic) and 20:5 n-3 (5,8,11,14,17-eicosapentaenoic) appeared unchanged in the glycerophospholipids in proportions reflecting dietary concentration. Of the other two, 22:4 n-6 appeared in the phospholipids as 20:4 n-6 (arachidonic) and to a lesser extent as 18:3 n-6 ((6,9,12)-linolenic), whereas 22:6 n-3 appeared partly unchanged but mainly as 20:5 n-3 (eicosapentaenoic) (Dadd *et al.*, 1987). Chain shortening and a saturation at the carboxyl end of the double bond sequence appeared to have occurred in both of the latter cases although animals are thought to be unable to saturate double bonds. However, other routes to the less saturated components are possible. Clearly, the essential fatty acid requirement at the tissue level was for either arachidonic or eicosapentaenoic acid.

Of the two moderately effective fatty acids, 18:3 n-6 ((6,9,12)-linolenic) appeared in the phospholipids unchanged, whereas 20:3 n-6 appeared as similar proportions of 20:3 n-6 and 18:3 n-6, indicating chain shortening. Of the slightly effective fatty acids, 22:3 n-3 and 20:3 n-3 appeared in the phospholipids as 18:3 n-3 ((9,12,15)-linolenic), while two others, 22:2 n-6 and 20:2 n-6, appeared as 18:2 n-6 (linoleic). Thus there was chain shortening within each series, n-6 or n-3, but no interconversion of n-6 and n-3 members. The C_{18} polyunsaturated fatty acids were accumulated, whether acquired from the diet or formed from other fatty acids.

Mosquitoes differ from other insects in that

they cannot satisfy their requirement for essential fatty acids with the common polyunsaturated C_{18} acids but must take in certain longer chain polyunsaturates with their food. Mosquitoes are unable to synthesize longer chain acids from the C_{18} polyunsaturates, but they are able to carry out the retroconversion of longer chain acids by chain shortening and, possibly, saturation (Dadd *et al.*, 1987).

(d) Function

In invertebrates the essential C_{20} polyunsaturated fatty acids, which are members of the eicosanoid class, serve a number of functions including a structural role as constituents of membrane glycerophospholipids and a metabolic role as precursors in prostaglandin synthesis (Stanley-Samuelson, 1991). It is likely that they serve both functions in mosquitoes. Evidence has been obtained of their involvement in prostaglandin synthesis. First, addition to a synthetic larval diet of indomethacin or phenylbutazone, inhibitors of cyclooxygenase and phospholipase A_2 respectively, retarded the development of *Culex pipiens* larvae, but to a diminishing extent as the arachidonic acid content of the diet was increased (Dadd and Kleinjan, 1984b). Second, radioactive arachidonic acid injected into adult *Aedes aegypti* was converted to $PGF_{2\alpha}$ and possibly other prostaglandins (Stanley-Samuelson *et al.*, 1989). When the diet of *Cx pipiens* larvae contained suboptimal amounts of arachidonic acid, the consequent impaired ability of teneral adults to fly could be significantly alleviated by the addition of $PGF_{2\alpha}$, although the prostaglandin could not completely satisfy the arachidonic acid requirement (Dadd and Kleinjan, 1988).

5.3.3 Sterols

Insects are unable to biosynthesize the steroid nucleus and require an exogenous source of sterols. When *Culex pipiens* larvae were provided with a synthetic diet lacking sterol they developed through no more than one instar. A number of sterols, provided singly, supported development to the adult stage, but only two of these, cholesterol and desmosterol, were effective when not formulated with lecithin as carrier or dispersant. When incorporated in the diet with a carrier (0.2 mg lecithin/100 ml medium) sitosterol, fucosterol, campesterol, 24-methylenecholesterol, stigmasterol and 7-dehydrocholesterol were as effective or almost as effective as cholesterol in promoting growth and development to the adult stage. Ergosterol, lathosterol and 22-*trans*-cholesta-5,22-dienol when formulated with lecithin were less effective than cholesterol in promoting growth but permitted a substantial percentage of individuals to reach the adult stage (Dadd and Kleinjan, 1984a).

The finding that dietary C_{28} and C_{29} phytosterols could promote growth and development in *Cx pipiens* almost as effectively as cholesterol suggested that they were converted to cholesterol. Many omniverous and phytophagous insects are able to dealkylate and convert C_{28} and C_{29} phytosterols to cholesterol, and biochemical studies have revealed separate metabolic pathways for the dealkylation and conversion of stigmasterol, sitosterol and campesterol. All of the pathways include desmosterol (24-dehydrocholesterol) as the final intermediate (Svoboda and Thompson, 1985).

Direct evidence for the dealkylation of a phytosterol had in fact been obtained earlier (Svoboda *et al.*, 1982). When larvae of *Aedes aegypti* were reared on a semisynthetic diet which included $[^{14}C]$sitosterol, about half of the radioactivity present in pupal sterols was found in cholesterol, with the balance remaining in sitosterol. When the larvae were provided with $[C^{14}]$desmosterol they formed radioactive cholesterol with high efficiency. If the enzyme inhibitor 25-azacholesterol was given with $[^{14}C]$sitosterol in the diet the incorporation of label into cholesterol was much reduced but there was substantial incorporation into desmosterol, confirming the role of Δ^{24}-sterol reductase in the pathway of cholesterol formation. Larvae fed $[C^{14}]$campesterol converted it to cholesterol with relatively low efficiency. These experiments provided the first demonstration of dipterans converting phytosterols to cholesterol.

5.3.4 Vitamins

In the absence of any one of the water-soluble B vitamins nicotinic acid, pantothenic acid, pyridoxol HCl (pyridoxine), riboflavin and thiamin HCl, *Aedes aegypti* larvae died in the 1st or 2nd instar; in the absence of biotin or pteroylglutamic acid (folic acid) they failed to pupate (Singh and Brown, 1957; Akov, 1962a). The lack of these vitamins had equally drastic effects on *Culex pipiens* except that a few adults were sometimes produced in the absence of pteroylglutamic acid or biotin, possibly because sufficient vitamin was carried in the eggs. The minimum vitamin requirements of *Cx pipiens*, which resembled those of other insects, ranged from 80 μg g^{-1} of dry dietary ingredients for nicotinic acid down to 0.3 μg g^{-1} for biotin (Kleinjan and Dadd, 1977; Dadd, 1985). The relative requirements of *Ae aegypti* larvae for five of the vitamins correlated well with the concentrations of those vitamins in the adult tissues (Micks *et al.*, 1959; Akov, 1962a). All of these compounds function as enzyme cofactors, except pteroylglutamic acid which becomes modified to act as a carrier of one-carbon units in purine synthesis (Akov and Guggenheim, 1963).

No one has reported that the lack of dietary vitamin A or its usual precursor in animals, β-carotene, has deleterious effects upon the viability of larval or adult mosquitoes. However, when *Ae aegypti* larvae were reared on a semisynthetic diet containing neither vitamin A nor β-carotene the ultrastructure of the receptor cells of the adult eye was abnormal, and the amplitude of their electrical response to light was substantially reduced. If β-carotene was included in the larval diet, at 0.077 g l^{-1}, the adult eyes were structurally and functionally normal (Brammer and White, 1969).

Choline, which is a component of some phospholipids and has other biochemical and physiological roles, is an essential nutrient for insects. In its absence, larvae of *Ae aegypti* and *Cx pipiens* failed to reach the pupal stage. The minimum requirement of *Cx pipiens* for choline was 320 μg g^{-1} of dry dietary ingredients, an amount greatly in excess of the requirements for B vitamins (Lea and DeLong, 1958; Akov, 1962b; Kleinjan and Dadd, 1977; Dadd, 1985).

5.3.5 Nucleotides

The semisynthetic diets developed for mosquitoes included RNA as an essential constituent. Dadd and Kleinjan (1977) observed that the products of alkaline hydrolysis of yeast RNA permitted *Culex pipiens* larvae to grow and develop as well as when they were provided with RNA. Although the four constituent ribonucleotides of RNA were not an effective replacement for RNA they became so when augmented with 2′-deoxythymidine 5′-phosphate, which is a principal constituent of DNA and which occurs also as a minor constituent of transfer RNA. Further, although DNA was nutritionally inadequate it became completely adequate when supplemented with the ribonucleotide uridine 5′-phosphate.

In the absence of any nucleotides, *Cx pipiens* larvae grew very slowly and only from 0 to 20% of individuals, usually no more than 5%, reached the adult stage. By providing different combinations of nucleotides in a fully synthetic diet it was shown that a mixture of three nucleotides could satisfy the need for nucleic acid. These were 2′-deoxythymidine 5′-phosphate, adenosine 5′-phosphate (which could be replaced by its usual biosynthetic precursor inosine 5′-phosphate), and either uridine 5′-phosphate or cytidine 5′-phosphate, which must be interconvertible in the larva (Table 5.5). Guanosine 5′-phosphate was not required, therefore the larvae must have been able to synthesize this nucleotide from inosine 5′-phosphate or adenosine 5′-phosphate. Orotic acid, which in other organisms is a biosynthetic precursor of pyrimidine ribonucleotides, was ineffective as a substitute for uridine 5′-phosphate and cytidine 5′-phosphate. The naturally occurring 5′-monophosphates could be replaced with commercial mixtures of the 2′- and 3′-monophosphates

Table 5.5 The effectiveness of dietary nucleotides and the homologous nucleosides and bases in meeting the nutritional requirements of *Culex pipiens* larvae fed a chemically defined diet. Compounds without a superscript number were fully effective. (From the data of Dadd, 1979.)

Nucleotide	*Nucleoside*	*Base*
2′-Deoxythymidine 5′-phosphate	2′-Deoxythymidine	Thymine[1]
Adenosine 5′-phosphate OR Inosine 5′-phosphate	Adenosine[1] Inosine	Adenine[2] (Hypoxanthine)[3]
Uridine 5′-phosphate OR Cytidine 5′-phosphate	Uridine OR Cytidine	Uracil[2] Cytosine[2]

1, Partly effective. 2, Ineffective. 3, Not tested.
2'-Deoxythymidine is also called thymidine.

produced by alkaline hydrolysis of yeast RNA, and the latter are in fact constituents of the recommended diet (Table 5.2) (Dadd and Kleinjan, 1977; Dadd, 1979).

Any or all of the three nucleotides could be replaced with the corresponding nucleosides without adverse effect, except that the use of adenosine caused a moderate reduction in the larval growth rate. Substitutions of base for nucleotide were generally unsatisfactory. When thymine was provided the larval stage was greatly prolonged, but most individuals eventually became adult. However, a single substitution with adenine, cytosine or uracil for the corresponding nucleotide allowed scarcely more development than occurred in the total absence of nucleic acid derivatives. Dietary experiments with deoxynucleotides showed that 2′-deoxyadenosine 5′-phosphate could take the place of adenosine 5′-phosphate with some loss of efficiency, but that the deoxyribose analogues of cytidine 5′-phosphate and uridine 5′-phosphate were very ineffective.

The minimum requirements of *Cx pipiens* for normal growth are a purine ribonucleotide (adenosine 5′-phosphate or inosine 5′-phosphate), a pyrimidine ribonucleoside (uridine or cytidine), and the pyrimidine nucleoside 2′-deoxythymidine (Dadd, 1979). Dadd (1985) stated, without further detail, that the needs of *Anopheles stephensi* could be met by adenosine 5′-phosphate plus either cytidine 5′-phosphate or uridine 5′-phosphate.

5.4 METABOLIC RESERVES

The fat body is the principal storage organ. In mosquitoes it is distributed as a sheet of cells applied to the body wall throughout the thorax and abdomen, as lobes which extend into the body cavity, and as sheaths around certain organs. In well-nourished larvae the fat body cells are packed with lipid droplets, glycogen deposits and vacuoles containing protein (Figure 5.3A) (Wigglesworth, 1942, 1987). Ultrastructural examination of the fat body of adults at eclosion revealed three types of proteinaceous vacuole, differing in size and in the electron density of the contents. These vacuoles disappeared within two days of eclosion (Raikhel and Lea, 1983).

Other larval tissues can also contain reserves. Use of histochemical techniques revealed the accumulation of deposits of glycogen and lipid in *Aedes aegypti* larvae which had been starved and then liberally fed with starch or olive oil. As much glycogen could be deposited in muscle sarcoplasm (Figure 5.3C) as in the fat body, and stainable deposits of glycogen could be found in the central nervous system, the cells of the posterior half of the stomach, and scattered

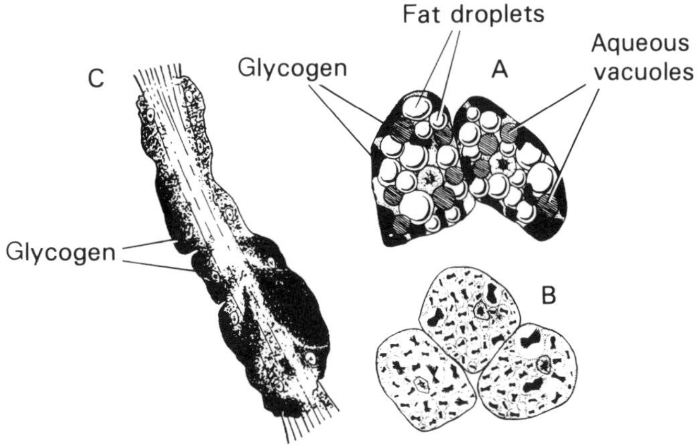

Figure 5.3 Food reserves and postulated uric acid in the tissues of *Aedes aegypti* larvae. (A) Fat body cells of a well-nourished larva. (B) Crystals, said to be of uric acid, in fat body cells of a starved larva. (C) Abdominal muscle of a larva fed on starch, stained with iodine to show deposits of glycogen. (From Wigglesworth, 1942.)

caecal cells. Lipids were stored principally in the fat body, but stainable lipid, considered to be metabolic reserve, was described in the oenocytes, the cells of the anterior half of the stomach, and scattered caecal cells (Wigglesworth, 1942).

An ultrastructural examination of the midgut of *Ae aegypti* indicated a possible storage function for the resorbing/secreting cells (Section 5.1.2), which were characterized by the presence of both glycogen and lipid. However, these cells predominated in the posterior midgut and the anterior two-thirds of the caeca (Volkmann and Peters, 1989a), giving a different pattern of glycogen and lipid deposits from that described by Wigglesworth (1942).

When well-nourished *Ae aegypti* larvae were starved, glycogen and lipid disappeared concurrently from the fat body and none remained after six days at 28°C. Over the same period the deposits in other tissues were utilized, and there was a progressive wastage of cytoplasm in all tissues as protein was utilized (Wigglesworth, 1942). At 27°C, 4th instar larvae of *Ae aegypti* survived starvation for an average of 13.3 ± 1.6 days, whereas larvae of *An sacharovi* survived for only 3.5 ± 0.4 days (Rasnitsyn and Yasyukevich, 1989b).

The carbohydrate and lipid contents of *Ae aegypti* larvae are positively correlated with larval weight (Chambers and Klowden, 1990). The total carbohydrate and lipid contents of teneral adults, detailed in Tables 15.1 and 15.2, provide a measure of the energy reserves built up during the larval stage. The massive larval abdominal muscles, which are histolysed during the first days of adult life, must be a valuable source of protein for the young adult.

5.5 EXCRETION

The Malpighian tubules and rectum are the organs principally concerned in ridding the haemolymph of unwanted substances. Water and solutes enter the Malpighian tubules from the haemolymph and pass to the rectum where certain substances are resorbed, the remainder being discharged. In saline-water species certain anions are secreted by the rectal epithelium into the lumen. The regulation of inorganic ions is described in detail in Chapter 6.

Very little is known of the nitrogenous excretory products of mosquito larvae. Uric acid is present in whole-body extracts of *Aedes aegypti* larvae, as it is in clean water in which larvae have been kept (Inwang, 1971), but it is not known whether the larvae also discharge urea and ammonium salts. Considerable quantities of a solid, postulated by Wigglesworth (1933a, b) to be uric acid, accumulate in the Malpighian tubule lumens when *Aedes aegypti* larvae are kept in salty water, but never when they are kept in fresh water when the tubules probably are better flushed.

The Malpighian tubules of *Ae aegypti* larvae are translucent until the 4th instar, when the cells in the distal part become opaque due to granular deposits in the cytoplasm. They remain opaque throughout life (Christophers, 1960). X-ray microanalysis of adult Malpighian tubules showed the concretions to contain calcium, magnesium, manganese and phosphorus (Section 13.3).

Rather similar concretions occurring in some molluscan tissues contain pyrophosphates ($P_2O_7^{4-}$) of manganese and other toxic ions, and there is evidence that they are the product of a detoxification process (Mason and Simkiss, 1982). Exposure of *Ae aegypti* larvae to 30 mM concentrations of zinc, magnesium and manganese caused high mortality in both larvae that possessed and larvae that lacked concretions (Bradley *et al.*, 1990). Clearly, exposure to a range of lower concentrations is needed to clarify this matter.

When mosquito larvae ingest soluble dyes, solid deposits of the dyes are later found in haemocytes, pericardial cells, and fat body cells. The deposits in the pericardial cells may be retained throughout life (Pal, 1944; Jones, 1954; Chevone and Peters, 1969). It has been postulated that the larval fat body also functions in storage excretion. Starvation of *Ae aegypti* larvae led to the appearance within aqueous vacuoles in the fat body cells of crystals that responded positively to certain chemical tests for uric acid (Figure 5.3B) (de Boissezon, 1930b; Wigglesworth, 1942). A mass of similar crystals formed when the fat body of well-nourished larvae was fixed, suggesting to Wigglesworth (1987) that uric acid must have been present in high concentration in the fluid contents of the vacuoles.

5.6 LARVAL AND PUPAL RESPIRATION

The open respiratory system and the requirement for access to atmospheric oxygen impose limitations on the larvae and pupae of mosquitoes as aquatic insects. In this Section we shall consider the anatomical adaptations for respiration at the water surface, and the physiological and anatomical modifications that permit some species to feed or even to live away from it.

5.6.1 Respiratory organs

(a) Larval organs

The tracheal system consists of two wide dorsal longitudinal trunks, two narrower lateral longitudinal trunks, and tracheal branches which run from them to all regions of the body. A pair of air sacs of unknown function is present near the junction of thorax and abdomen in certain species. In *Orthopodomyia pulcripalpis* the air sacs are formed by swelling of the tracheal trunks and in *Coquillettidia richiardii* by expansion of tracheal branches. *Culex* larvae show no respiratory movements, and it is thought that respiratory gases move by diffusion alone (Krogh, 1941; Keilin, 1944). The tracheoles are commonly liquid filled. Under anoxic conditions the liquid is replaced by air, a phenomenon which may not be due to osmotic forces (Wigglesworth, 1981).

Mosquito larvae have ten pairs of spiracles but only those of the 8th abdominal segment function in respiration; the larvae are therefore metapneustic. The remaining spiracles serve only for the withdrawal of tracheae when the cuticle is shed at ecdysis. The anterior of the two pairs of thoracic spiracles is mesothoracic in origin but has migrated to the prothorax (Hinton, 1947). The opening of each functional spiracle is surrounded by a spiracular apparatus which bears five flap-like spiracular lobes developed from parts of the 8th and 9th segments.

Anopheline larvae lie horizontally at the water surface with the body held to the surface membrane by a series of structures – the spiracular lobes, paired rows of stellate or palmate setae, and paired Nuttal and Shipley's organs. The spiracles are practically flush with the dorsal surface of the abdomen, where they have direct contact with the air. In all other mosquito larvae the terminal spiracles are situated within a respiratory siphon which projects dorsally from the abdomen, and in nearly all of these the larvae hang head downwards from the surface membrane, supported by the spiracular lobes. A multicellular perispiracular gland is present within the siphon in culicine larvae and adjacent to the spiracles in anophelines. The gland discharges an oily secretion which coats the spiracular lobes and atria making them hydrofuge. If the secretion is removed with diethyl ether, water enters the tracheae (Keilin *et al.*, 1935).

The two dorsal tracheal trunks extend through the respiratory siphon and each opens into an atrium lined with microtrichia. In *Aedes aegypti* the atria are completely separate and open at separate spiracles, but in *Culex pipiens* they unite just before opening at a single spiracle. Within the siphon a large forked apodeme is connected to the spiracular lobes and to the cuticular rims of the spiracles. The siphon contains five pairs of extrinsic muscles, with origins outside the siphon, and one small intrinsic muscle, and these are inserted on the apodeme, atria and spiracular lobes. Three or four pairs of chordotonal organs are situated within the siphon. When a larva leaves the water surface muscle contractions draw the atria a short distance into the tracheal trunks and also retract the five lobes, closing the spiracles. When the larva returns to the surface the lobes are pulled into their extended position by the surface tension of the water. If the lobes are removed the larva is unable to hang from the surface membrane (Wesenberg-Lund, 1918; Sautet and Audibert, 1946; Christophers, 1960; Harbach and Knight, 1980). The significance of capillarity, contact angle and surface tension for the opening of the spiracular valves and the suspension of the larva from the surface membrane were described by Brocher (1910).

(b) Pupal organs

Mosquito pupae are exceptional among insects in having only the anterior thoracic and the anterior abdominal spiracles functional. The 1st abdominal spiracles open into the ventral air space, which is enclosed by the pupal mouthparts, wings and legs and which functions in pupal buoyancy. They have no role in gas exchange (Christophers, 1960; Leung and Romoser, 1979).

Short tracheal trunks connect the dorsal longitudinal tracheal trunks with the mesothoracic spiracles within the large respiratory trumpets. The trumpets, which are formed from imaginal disks (Section 8.1.3), are double-walled. In most mosquitoes the inner and outer walls are closely

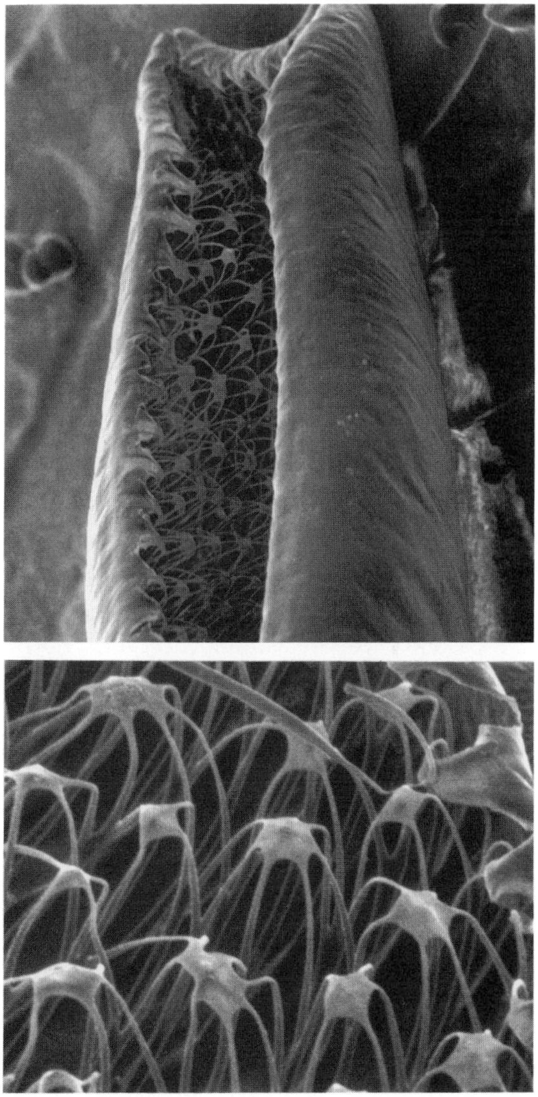

Figure 5.4 Respiratory trumpet of an *Aedes aegypti* pupa. Above. View into the pinna (× 915.) Below. Detail of the meshwork on the inner surface of the pinna. (× 1,370.) (Micrographs by courtesy of Dr P. Reiter.)

adherent, although they can be separated on dissection, but in *Harpagomyia* and some species of *Anopheles* they are well separated. The respiratory trumpets of the Culicinae are more or less cylindrical, those of the Anophelinae broader and more flap-like. In all mosquito pupae the tips of the trumpets are cut off obliquely; the trumpet region distal to the most proximal part of the spiracular opening is called the pinna, the remainder is the meatus. In many species the meatus is fully enclosed but in some a meatal cleft extends into the meatus from the spiracular opening, allowing the spiracular opening to enlarge at the surface of the water (Christophers, 1960; Reid, 1963; Harbach and Knight, 1980).

Most of the outer surface of the respiratory trumpet is hydrophilic, but the spiracular opening is surrounded by a ring of hydrophobic cuticle against which the water forms a contact angle. Over approximately the distal half of the trumpet, the inner surface is covered with an exceedingly fine and elaborate cuticular meshwork (Figure 5.4). The meshwork is also hydrophobic and it may help to prevent the entry of water when the pupa submerges. There is no evidence that mosquito pupae can survive prolonged submersion, so the meshwork is not a physical gill. Nor is it a filter, which Harbach and Knight (1980) term it.

Mosquito pupae are buoyant (Section 7.3.3.b). When at rest they float at the air/water interface, with the dorsal surface of the thorax and the first two abdominal segments against the surface membrane and the rest of the abdomen hanging downwards. A pair of multibranched setae on the 1st abdominal segment stabilize the pupa by engaging with the surface membrane, and the open ends of the respiratory trumpets project through the surface membrane (Snodgrass, 1959).

The respiratory trumpets of all species articulate freely about their insertions on the dorsal region of the thorax. In consequence, the trumpets of a pupa that is resting at the water surface lie with each aperture in the plane of the air/water interface. The trumpets are moved posteriorly just before a pupa dives and this movement, coupled with the shape of the orifice, minimizes the energy required to overcome the interfacial tension (Houlihan, 1971; Reiter, 1978b). Under normal conditions the buoyancy of a rising pupa is sufficient to enable the trumpets to break the surface membrane. As the trumpet reaches the surface, water molecules drain readily from between the two air phases. Because the contact angle of clean water is finite the whole aperture is quickly de-wetted, and lies almost in the plane of the interface. If the inner surface of the trumpets is coated with a surface active agent the air–water contact angle is zero, the interior of the trumpet becomes wetted and the pupa drowns (McMullen *et al.*, 1977).

5.6.2 Respiration of submerged larvae and pupae

(a) Cutaneous respiration

Some mosquito species can withstand prolonged submersion, making use of dissolved oxygen. The oxygen concentration can differ greatly between different natural bodies of water, and within any individual biotope it will vary both seasonally and in a daily cycle. Oxygen solubility increases with a lowering of water temperature, and the daily photosynthetic activities of aquatic plants cause marked fluctuations in oxygen production. For example, the water in irrigated ricefields in Kenya was saturated or supersaturated with oxygen during most of the daylight hours but was <20% saturated for most of the night (Reiter, 1980).

Anopheles stephensi larvae, which regularly fed by collecting–gathering at considerable depths, were much more tolerant of prolonged submersion than were *An sacharovi* and *An atroparvus* which fed only or principally at the surface (Rasnitsyn and Yasyukevich, 1989a). At 27°C, cutaneous respiration adequately compensated for the blocking of siphonal respiration by *An stephensi* larvae provided the water was >30% saturated with oxygen (Reiter, 1978a). *Aedes aegypti* larvae kept in flowing water at about 30°C and denied access to the surface lived up to 53 days and developed to the 4th instar. *Culex thalassius* larvae lived at least two weeks under such conditions but

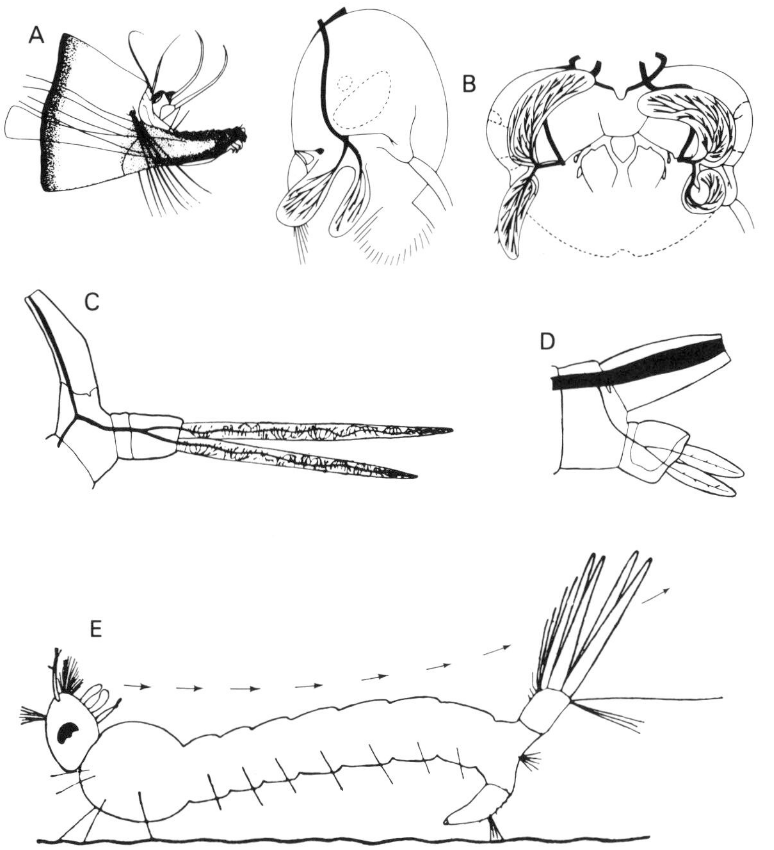

Figure 5.5 Respiratory modifications in mosquito larvae. (A) Respiratory siphon of *Coquillettidia richiardii* modified for piercing. (From Lang, 1920.) (B) Head of *Aedes argenteopunctatus* in lateral and ventral view to show cephalic gills. (C) Terminal segments of *Ae argenteopunctatus* to show modified anal papillae. (D) Terminal segments of *Ae luteocephalus* for comparison. (E) *Ae argenteopunctatus* in resting position; arrows show the path of the water current produced by the lateral palatal brushes. (From Lewis, 1949.)

Cx quinquefasciatus larvae could not survive one day (Macfie, 1917). However, larvae of both *Ae aegypti* and *Cx quinquefasciatus* were reported not only to survive after their entire tracheal systems had been filled with hexadecane but to develop to adulthood (Micks and Rougeau, 1976).

In nature the larvae of certain species remain submerged for long periods. Fourth instar larvae of *Ae flavescens* spend most of the time among vegetation at the bottom of pools (Hearle, 1929). The older larvae of *Culiseta morsitans* rarely seek the water surface but rest on the bottom with their anal papillae spread out and maintain a water current over the papillae with their palatal brushes (Wesenberg-Lund, 1921). Prolonged submersion has been described in *Ae argenteopunctatus* and *Cx sinaiticus* larvae in the Sudan (Lewis, 1949) and in *Malaya genurostris* larvae and pupae in India (Iyengar and Menon, 1948). Two pairs of large thin-walled and densely tracheated cephalic papillae extend from the ventro-lateral surfaces of the head of *Ae argenteopunctatus*, and the anal papillae are long and densely tracheated (Figure 5.5B, C). The larvae lie on the bottom, ventral side up, and water currents produced by the labral brushes pass over both the papillae on the head and the anal papillae (Figure 5.5E) (Lewis, 1949).

(b) Compressible gills

Mosquito larvae of several genera are structurally and behaviourally adapted to utilize gas bubbles trapped on the outside of aquatic plants, particularly bubbles produced by plants with extensive lacunae within their tissues. As the partial pressure of oxygen in the bubble falls, through removal of oxygen by the insect, oxygen diffuses into the bubble faster than nitrogen diffuses out of it because the exchange coefficient of oxygen between water and air is more than three times that of nitrogen (Mill, 1974). Thus the bubbles serve as compressible gills. *Culex poicilipes* larvae sometimes attach themselves to gas bubbles by the enlarged lobes of the respiratory siphon, remaining motionless for hours on end except for the action of their palatal brushes (MacGregor, 1927). *Aedeomyia* larvae cling to plants by their large antennae and hooked siphon valves; *Ad africana* larvae obtain air with equal facility from the water surface or from the film on the underside of *Pistia stratiotes* leaves. Similar habits are known in some *Mimomyia* species, e.g. *Mi splendens* (Hopkins, 1952).

(c) Penetration of plant tissues

Larvae of *Mansonia* and *Coquillettidia* (tribe Mansoniini) habitually penetrate their respiratory siphons into the submerged leaves, stems and water roots of aquatic plants that contain air-filled lacunae. The distal half of the siphon is pointed and heavily sclerotized, and contains a number of unusual cuticular structures articulated with one another (Figure 5.5A). These include six sets of projecting hooks, a flattened saw-like blade with numerous teeth, and a cuticular rod which extends for some distance into the abdomen. Powerful muscles which originate on the inner wall of the siphon are inserted on the cuticular rod (Ingram and Macfie, 1917; Wesenberg-Lund, 1918). Larvae of some species of *Mimomyia*, e.g. *hybrida* and *modesta* (tribe Ficalbiini), also have pointed siphons which are used to pierce plants (Bonne-Webster, 1932; Iyengar, 1935 a, b; Van den Assem, 1958).

Mansonia and *Coquillettidia* larvae are able to cling to plant surfaces by their bristles. Before the siphon is inserted into the root, stem or leaf it is held perpendicular to the surface and then, probably through the cutting action of the saw-blade and through pressure applied by the abdomen, the tip of the siphon is forced into the plant and is held there by its hooks and by the teeth of the saw (Wesenberg-Lund, 1918; Edwards, 1919; Laurence, 1960).

Coquillettidia perturbans larvae were significantly more abundant in sites where the oxygen concentration was below 1.5 mg l^{-1}, and it has been suggested that they exploit anaerobic environments that are inhospitable to many predators (Batzer and Sjogren, 1986; Callahan and Morris, 1987). *Mansonia uniformis* larvae have been shown to gain some protection from fish by their sessile habit (Van den Assem, 1958).

Pupae of *Mansonia* and *Coquillettidia* use their respiratory trumpets to tap the air spaces within plants. The apices of the respiratory trumpets are extended into lancet-shaped structures bearing lateral flaps which are covered with minute curved hairs. The two respiratory trumpets are kept close together, forced into plant tissues, and are held there by their curved hairs and two hooked spines (Wesenberg-Lund, 1918; Laurence, 1960). The respiratory trumpets of *Mimomyia hybrida* have a similar structure and are used in the same way (Bonne-Webster, 1932; Iyengar, 1935a).

Mansonia annulifera and *Ma uniformis* pupae must engage their respiratory trumpets in plant tissues during the brief act of larval–pupal ecdysis, while still attached to the plant by the larval respiratory siphon; free-swimming pupae are unable to engage their respiratory trumpets. During ecdysis the pupa flexes dorsally and directs its respiratory trumpets towards the point of attachment of the larval siphon. The convergent trumpet tips are suddenly jabbed into the hole occupied by the siphon, enlarging it slightly, while simultaneously the pupal abdomen flips downwards pulling out the larval siphon and leaving the pupa attached by its trumpets (Burton, 1965).

(d) Rates of oxygen uptake

Of food substances assimilated by insects, part (R) will be utilized in respiration and part (G) for growth. The very high R/G ratio of 13.8 determined for *Culex torrentium* larvae (which compares with 1.5 for simuliids) was ascribed to their active habits (Nilsson, 1986).

Mean rates of oxygen uptake by 4th instar larvae at 27°C (Q_{O_2} = μl O_2 consumed h^{-1} mg^{-1} dry weight) were: *Aedes taeniorhynchus* 7.62 ± 0.59 s.d., *Culex pipiens* 7.69 ± 1.23 and *Anopheles quadrimaculatus* 7.14 ± 0.51 (Powers and Platzer, 1984). The rate of oxygen consumption by *Ae aegypti* larvae did not change when the salinity of the medium was varied between 0 and 80% sea water (Edwards, 1982a). *Aedes aegypti* larvae exhibited a circadian rhythm of oxygen consumption which persisted for 48 h after they were transferred from LD 12:12 to continuous light. The high rate, recorded in late evening, was 35% higher than the lowest rate (Yap *et al.*, 1974).

The anaerobic capacity of wild *Cx pipiens* larvae was low; at 12°C they survived only 4–5 h of anoxia. Anoxic larvae showed a progressive loss of phosphoarginine and accumulation of large amounts of lactate (Redecker and Zebe, 1988).

Q_{O_2} values for pupae of *Ae aegypti*, of mixed age and sex, rose from 0.81 ± 0.19 at 12°C to 6.84 ± 1.33 at 45°C, with a temperature coefficient of about 3 between 12 and 25°C but decreasing at higher temperatures (Berry and Brammer, 1975). These results need to be interpreted with caution. A few hours after larval–pupal ecdysis the respiration of *An messeae* pupae falls to a very low rate but it later recovers fully (Olifan, 1949).

6

Osmotic and ionic regulation

Mosquito larvae are found in a wide range of habitats, from rain-filled pools and containers at one extreme to inland saline pools of high salinity, high alkalinity and unusual ionic composition at the other. The capacity to survive under one or other, or even both, of these extreme conditions is due to powerful transport processes in the epithelia of the alimentary canal and anal papillae which provide homeostasis of the internal environment, making mosquito larvae among the best ionic and osmotic regulators in the animal kingdom.

6.1 CHEMICAL PROPERTIES OF LARVAL HABITATS

Larval habitats can be classified in terms of gross salt content. **Fresh-water** habitats have a salinity of <0.5 or alternatively 2 parts per thousand (≡ 0.0085 M- or 0.034 M-NaCl), **brackish** habitats have a salinity intermediate between fresh water and sea water (≡ 0.55 M-NaCl), and **saline** habitats are very rich in soluble salts (Lincoln *et al.*, 1982).

Fresh-water habitats include streams, pools, marshes, treeholes, leaf axils, artificial containers, etc. The larger bodies of fresh water show marked regional variations of ion content, which may affect the nutrition of higher organisms indirectly since the algal productivity of fresh waters is correlated with their ion concentrations (Lund, 1957; Taylor, 1958). There is little evidence that the viability of fresh-water species is affected by low concentrations of particular ions and attempts to correlate the distribution of species with minor differences in ion concentration (e.g. Vrtiska and Pappas, 1984) are unconvincing. Probably the larval habitat of each species is determined by many forces. Gravid females select the larval breeding sites, when factors such as shade and water movement may be more influential than ionic concentrations.

Few fresh-water habitats of mosquito larvae have been fully analysed. In a woodland pool containing *Aedes communis* larvae all ions were at less than 0.4 mmol l^{-1} (Table 6.1) (Iversen, 1971). Water samples from treeholes containing mosquito larvae had low mean ion concentrations, with more potassium than sodium. Mean values (mM) were: Na^+, 0.08; K^+, 0.74; Ca^{2+}, 0.17; Mg^{2+}, 0.085; Cl^-, 0.04; HCO_3^- + CO_3^{2-}, 0.59 (Petersen and Chapman, 1969). The mean ion concentrations of the habitats of a number of species were reported by Hagstrum and Gunstream (1971) and Vrtiska and Pappas (1984).

Fresh-water species vary in their tolerance of salinity. *Anopheles maculipennis* larvae failed to survive in sea-water concentrations greater than 25%. In contrast, larvae of *An superpictus*, a species which normally breeds in river beds, survived well when reared in 33% sea water, showing a higher tolerance of that concentration than *An atroparvus* which is normally found in slightly brackish coastal waters (Bates, 1939). *Ae aegypti* larvae died if transferred directly from fresh water to 37% sea water, but if the concentration was raised gradually they could survive in 50% sea water (Wigglesworth, 1933b).

The larvae of most mosquito species live in fresh waters but approximately 5% of species live in brackish or saline waters. The latter belong to

Table 6.1 Concentrations (mmol/l) of inorganic ions in certain waters.

Ion	*Newcastle*[1] *tap water*	*Danish*[2] *woodland pool* (Aedes communis)	*Treeholes in* Quercus[3] muhlenbergii *(mean)* (Aedes triseriatus)	*Canadian*[4] *alkaline lake* (Aedes campestris)	*Sea water*[5] *(mean)*
Na^+	0.017	0.27	0.04	478	470.47
K^+	0.001	0.13	6.78	12	9.94
Ca^{2+}	0.025	0.25	–	0.03	10.25
Mg^{2+}	0.016	0.09	–	0.5	53.60
Cl^-	0.009	0.34	0.19	81	548.68
SO_4^{2-}	0.005	–	–	11	28.24
$HCO_3^- + CO_3^{2-}$	0.012	0	0.15	380	2.36
pH	7.2–7.4	4.4	6.9	10.2	8.1–8.3
mosmol kg^{-1} water	–	–	–	600	c. 1000

References: 1, Stobbart (1974); 2, Iversen (1971); 3, Vrtiska and Pappas (1984); 4, Meredith and Phillips (1973a); 5, Sverdrup *et al.* (1942)

the genera *Aedes, Opifex, Psorophora* (Aedini); *Aedeomyia* (Aedeomyiini); *Culex* and *Deinocerites* (Culicini); *Culiseta* (Culisetini); *Uranotaenia* (Uranotaeniini); and *Anopheles* (Anophelinae) (O'Meara, 1976). Brackish and saline habitats may be littoral, e.g. salt marshes, mangrove swamps and rock pools, or they may be inland saline pools and lakes. The littoral habitats have ion ratios similar to sea water, as do some inland water bodies. Other inland habitats contain high concentrations of ions such as Mg^{2+}, Ca^{2+}, SO_4^{2-} or $HCO_3^- + CO_3^{2-}$, and often have pH values exceeding 10; few organisms are capable of living under such severe conditions. The salinity of brackish and saline habitats is liable to fluctuate widely, decreasing as a result of rain or fresh-water runoff or increasing through evaporation. This can result in the appearance during one year of a succession of species, adapted to different salinities (Mosha and Mutero, 1982).

The following are examples of species occurring in brackish or saline habitats.

Rock pools: *Ae australis, Ae togoi, Op fuscus, Ps insularia.*
Salt marshes: *Ae detritus, Ae sollicitans, Ae taeniorhynchus, An atroparvus, An labranchiae, An sacharovi.*
Coastal lagoons: *An sundaicus, An merus, Ad catasticta, Cx sitiens, Ur lateralis.*
Mangrove swamps: *An melas.*
Crab holes: *Ae pembaensis, De cancer.*
Inland saline pools: *Ae natronius, Ae dorsalis, Ae campestris, An multicolor, Cx tenagius.*

Closely-related species may utilize very different breeding sites, e.g. fresh-water and brackish sites as in species of the *An maculipennis* complex, and fresh-water, brackish, and saline sites as in species of the *An gambiae* complex. Some species occur naturally in both fresh and brackish water, e.g. *Cs inornata, Cx tritaeniorhynchus*, and *Ae vexans*. *Cs inornata* can develop in distilled water or 70% sea water but cannot tolerate higher concentrations (Garrett and Bradley, 1984a). Other species occur in both fresh and saline waters, e.g. *Ae campestris* and *Ae dorsalis*.

Larvae of *Aedes campestris* and *Ae dorsalis* are found in fresh water pools and hypersaline lakes. On the Fraser Plateau in Canada they develop in water bodies which differ greatly in osmotic pressure and in ionic composition, for example containing either Na^+ or Mg^{2+} as major cations and either SO_4^{2-} or $HCO_3^- + CO_3^{2-}$ as major anions. One breeding site of *Ae campestris* contained 0.48 M-Na^+ and 0.38 M-$HCO_3^- + CO_3^{2-}$, and had a pH of 10.2 (Table 6.1) (Scudder, 1969; Meredith and Phillips, 1973a; Strange *et al.*, 1982). *Ae natronius* larvae live under equally severe conditions in alkaline lakes in East Africa (Beadle, 1939). Seashore species may be exposed to very severe osmotic conditions.

Larvae of the saltmarsh species *Ae detritus* have been found in 280% sea water (Beadle, 1939), and the contents of the rock pools in which *Op fuscus* breeds vary from virtually fresh water to 250% sea water (McGregor, 1965).

Aedine species generally hatch from dormant eggs in sites where ponds are regularly formed or enlarged by seasonal rain or runoff. In inland saline pools larval development often proceeds during a period when the concentration is increasing due to evaporation, and the mosquitoes may be in a race with time to complete their larval development before ionic concentrations reach toxic levels. Although the eggs can tolerate extreme conditions the lst and 2nd instar larvae cannot, but tolerance increases rapidly as larvae enter the later instars, and the pupae are highly tolerant (Kiceniuk and Phillips, 1974; Sheplay and Bradley, 1982).

In nature the larvae of many species are found in both acid and alkaline conditions. *Ae taeniorhynchus* has been recorded in habitats from pH 3.3 to 8.1, and *Psorophora confinnis* from pH 3.3–9.2 (Petersen and Chapman, 1970). The treehole species *An plumbeus* and *Ae geniculatus* have been found in waters of pH 4.4–9.3 (MacGregor, 1921; Keilin, 1932). In the laboratory *Ae flavopictus* larvae developed in media of pH 2 to 9, and *Armigeres subalbatus* larvae developed in media of pH 2–10 (Kurihara, 1959). Species found in alkaline lakes will develop normally at pH 10.5. The pH of these lakes rarely if ever rises above 10.5 because of their buffering systems, therefore it must be factors other than pH that limit the occurrence of these species in the most concentrated alkaline lakes (Strange *et al.*, 1982). Mohrig (1964) considered that pH affects the distribution of some species, but there is no direct evidence to show that pH itself is ever limiting.

Organic pollution can restrict larval breeding; few species survive in heavily polluted water but *An vagus, Ar subalbatus* and *Cx quinquefasciatus* have that ability. An autogenous strain of *Cx pipiens* developed in polluted water which rapidly killed larvae of an anautogenous strain (Roubaud, 1933). *An minimus* and *An stephensi*, both clear water species, were able to develop in polluted water in the laboratory (Muirhead-Thomson, 1941; Russell and Mohan, 1939). Ammonium ion, which is commonly present at 2 to 5 mM in sewage (Mara, 1976), is somewhat toxic to *Ae aegypti* larvae. The LC_{50} of previously unexposed larvae was 3.6 mM, but this could be increased by selection (Mitchell and Wood, 1984).

6.2 REGULATION OF THE HAEMOLYMPH

In fresh water, mosquito larvae are liable to gain water by osmosis and to lose salts; in saline waters they are liable to lose water and gain salts. These fluxes immediately bear upon the haemolymph, but the larvae can regulate the composition of their haemolymph to within narrow limits. All species appear able to regulate when in fresh water, some can regulate in fresh and brackish water, and a few in fresh and highly saline water. The composition of larval haemolymph is described in Section 9.3; its ionic content in a number of species is summarized in Table 9.2.

6.2.1 Regulation in fresh-water species

Aedes aegypti has been extensively investigated, and will be treated as representative of species restricted to fresh-water habitats. When *Ae aegypti* larvae are kept under more or less natural conditions the haemolymph contains, on average, 100 to 105 mmol/l sodium, 2 to 4 mmol/l potassium, 5 mmol/l calcium, and 50 mmol/l chloride. The haemolymph, which forms at least 60% of body weight, contains about 90% of total body sodium, about 6% of potassium, about 95% of calcium and about 75% of chloride (Wigglesworth, 1938a; Ramsay, 1953a; Stobbart, 1959, 1965, 1967; Edwards, 1982b; Barkai and Williams, 1983). The mean haemolymph osmolarity* of larvae of *Ae aegypti* and *Culex pipiens* in fresh water is about 250 mosmol l^{-1} (Wigglesworth, 1938a; Richards and Meier, 1974; Edwards, 1982b). The sum of the osmotic activities of the major inorganic ions falls short of the measured haemolymph osmotic pressure, and it has been suggested that amino acids

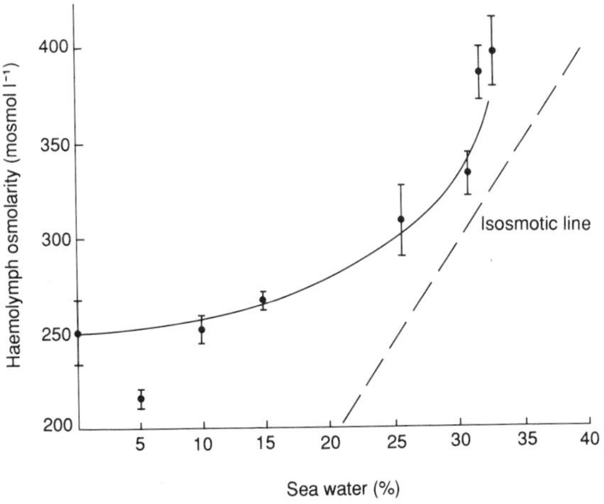

Figure 6.1 Mean haemolymph osmolarities of *Aedes aegypti* larvae reared in dilutions of artificial sea water. (After Edwards, 1982b.)

provide the balance (Edwards, 1982c). Heavy feeding leads to an increase in haemolymph osmolarity (Richards and Meier, 1974).

The capacity of larvae to regulate the osmotic pressure and ion content of their haemolymph has been measured by adapting larvae to solutions of different salinities and analysing their haemolymph. When *Ae aegypti* larvae were kept in various concentrations of artificial sea water, the haemolymph osmotic pressure varied very little as that of the medium rose through the lower range, but when the osmotic pressure of the medium approached that of the haemolymph, the haemolymph osmotic pressure increased sharply (Figure 6.1) (Edwards, 1982b). In experiments reported by Wigglesworth (1938a) and Richards and Meier (1974) the haemolymph osmotic pressure followed that of the medium precisely once it had started to rise. Thus when kept in dilute media the larvae 'osmoregulate', and when in concentrated media they 'osmoconform'. However, their capacity to survive in concentrated media is limited, with 50% sea water being the upper limit for acclimated larvae (Wigglesworth, 1933b).

Free amino acids contribute substantially to haemolymph osmotic pressure. The haemolymph amino acid concentration was considerably higher in *Ae aegypti* larvae kept in 0 than in 2% sea water, but over a higher range of sea-water concentrations the amino acid concentration was positively correlated with the salt concentration of the external medium. As the sea-water concentration was raised from 2 to 30%, the haemolymph amino acids doubled in concentration, which was sufficient to account for the observed change in haemolymph osmolarity (Edwards, 1982c).

Ion concentrations have also been measured under the different conditions. When *Ae aegypti* larvae were kept for a period in distilled water but were fed, the mean haemolymph sodium concentration fell from about 100 to 60 mM (Edwards, 1982b). Unfed larvae placed in deionized water at the high density of 10 per ml lost sodium into the water, but they achieved sodium balance when the external sodium concentration had risen to 5 μmol l^{-1}; at that stage the haemolymph contained 93 mM sodium (Stobbart, 1965). Similar experiments showed that haemolymph chloride, usually 50 mM, fell to 30 mM when unfed larvae were kept in deionized water, but that chloride balance was achieved when the external chloride concentration had risen to about 20 μmol l^{-1} (Stobbart, 1967).

Chloride was lost more rapidly than sodium when *Ae aegypti* larvae were starved in distilled water. Haemolymph sodium fell from about 100 to about 64 mM during the first 96 h after transfer to flowing distilled water, whereas chloride fell from 51 to 14 mM over a similar period with the water changed daily. After seven days in

* The osmotic properties of solutions are due to the concentration of solute particles. The unit of osmotic concentration is the osmole, an osmole being the amount of solute which if dissolved in one litre of water would exert the same osmotic pressure as one mole of an ideal non-electrolyte. Values can be expressed as osmolality (osmol kg^{-1} solvent), which is now the preferred measure, as osmolarity (osmol l^{-1} solution), or osmosity (the molar concentration of sodium chloride solution having the same freezing point or osmotic pressure). A solution of one osmole solute in one kg water has a freezing point of −1.858°C, therefore the osmolality of any solution can be determined from $\Delta t/1.858$. For example, a sample of sea water which freezes at −1.870°C has an osmolality of 1.006 osmol kg^{-1} water.

distilled water the haemolymph chloride concentration had fallen by 75% but the osmolality had fallen by only 21%, implying regulation of the non-chloride fraction (Wigglesworth, 1938a; Stobbart, 1959).

When *Ae aegypti* larvae were transferred from a medium of distilled water (but with food) to 2% artificial sea water (containing 10 mM-Na^+), haemolymph sodium increased from 60 to 80 mM. In 5–30% artificial sea water (24.5–147 mM-Na^+) the haemolymph sodium concentration was maintained at about 110 mM. If the sea-water concentration was raised above 30% the regulation of haemolymph sodium broke down and its concentration started to rise (Figure 6.2A). Haemolymph potassium varied between 2 and 4 mM in larvae reared in 0–30% sea water (0–3 mM-K^+). When larvae were kept in 85 mM-KCl the haemolymph potassium rose only to 2 mM. In larvae exposed to 0–30% sea water (0–3 mM-Ca^{2+}, 0–16 mM-Mg^{2+}), mean haemolymph calcium concentrations ranged between 8 and 19 mM, and magnesium concentrations varied between 6 and 8 mM (Ramsay, 1953a; Edwards, 1982a, b).

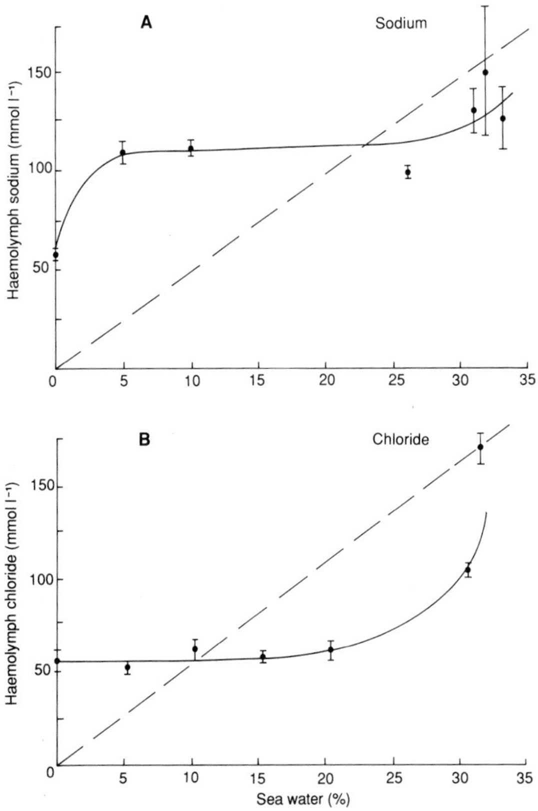

Figure 6.2 Mean concentrations of (A) sodium and (B) chloride in the haemolymph of *Aedes aegypti* larvae reared in dilutions of artificial sea water. The dashed lines indicate the sodium and chloride concentrations of the larval medium. (After Edwards, 1982b.)

When unfed *Ae aegypti* larvae were transferred from distilled water to tap water (0–1.03 mM-Cl^-), their mean haemolymph chloride concentration increased from 34 to 59 mM (Wigglesworth, 1938a). Haemolymph [Cl^-] was constant at about 50 mM in *fed* larvae kept in 0–15% artificial sea water (0–83 mM-Cl^-). However, if larvae were exposed to more than 20% sea water the haemolymph chloride concentration started to rise, and in 31% sea water it was identical with that of the medium (Figure 6.2B) (Edwards, 1982b). Clearly, unfed *Ae aegypti* larvae can survive prolonged exposure to distilled water, despite substantial loss of inorganic ions. In external media of very low to moderate ion concentrations the larvae closely regulate the ionic concentration of their haemolymph. However, if the external concentration is raised further the regulation of sodium and chloride fails.

6.2.2 Regulation in brackish-water species

The larvae of *Culiseta inornata* and *Culex tarsalis* are most commonly found in fresh-water sites but they occur also in coastal marshes of moderate salinity and can survive in up to 70% sea water. These two species resemble *Aedes aegypti* in the capacity of the larvae to osmoregulate in fresh water and to osmoconform when the salinity increases above a certain level, but they have a greater tolerance of high haemolymph osmotic pressure.

When *Cs inornata* larvae were kept in weakly brackish water their haemolymph osmotic pressure increased only slightly as the osmolarity increased between 50 and 400 mosmol kg^{-1} (≡ 40% sea water). However, in media of from 400 to 700 mosmol kg^{-1} the larvae became osmoconformers, having haemolymph isosmotic

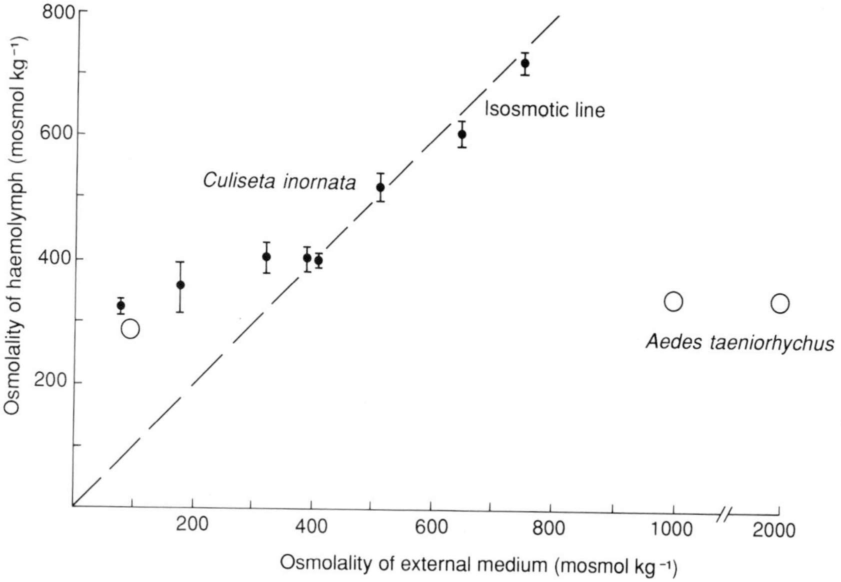

Figure 6.3 Relationships between the osmolalities of the external medium and the haemolymph of *Culiseta inornata* larvae. For comparison, three data points (O) are included for larvae of *Aedes taeniorhynchus* kept under comparable conditions. (After Garrett and Bradley, 1984a.)

with the external medium (Figure 6.3). Although the haemolymph osmotic pressure conformed at higher external osmolarities, the Na^+ and Cl^- concentrations remained closely regulated. Thus when external Na^+ and Cl^- concentrations ranged between 30 and 300 mM the haemolymph Na^+ varied only between 90 and 160 mM and the haemolymph chloride only between 60 and 100 mM. It appears that in media of higher salinity much of the increased haemolymph osmotic pressure is due to organic constituents (Garrett and Bradley, 1984a).

In dilute sea water, with osmolarities below 400 mosmol l^{-1}, larvae of *Cx tarsalis* regulated the osmotic concentration of their haemolymph, but between 400 and 700 mosmol l^{-1} they osmoconformed, like *Cs inornata*, and their haemolymph osmotic concentrations were essentially identical with those of the medium. Analyses of haemolymph samples taken from larvae reared in 5 and 60% sea water (≡ 50 and 600 mosmol l^{-1}) revealed that the inorganic ion concentrations in haemolymph from larvae in the more concentrated medium were raised only relatively slightly (Na^+ up by 25%, Cl^- up by 110%) or not at all (K^+, Mg^{2+}, Ca^{2+}). In contrast, trehalose was enriched 4.6-fold (from 8 to 37 mmol l^{-1}) and total free amino acids were enriched nearly 6-fold (from 44 to 258 mmol l^{-1}). Serine, enriched 6.4-fold, and proline, enriched 15-fold, accounted for 30% of the osmotic concentration in larvae reared in 60% sea water. In haemolymph from larvae reared at 600 mosmol l^{-1}, inorganic ions accounted for about 200 mosmol (taking ionic dissociation into account), trehalose for 37 and total free amino acids for about 240 of the total osmotic concentration of 595 mosmol l^{-1}. Unidentified osmolytes, assumed to be organic acids and proteins, accounted for about 120 mosmol l^{-1}, or 20%. Thus in a brackish environment the larvae of *Cx tarsalis* osmoconform but control the chemical composition of the haemolymph by increasing the concentrations of non-toxic osmolytes (Garrett and Bradley, 1987).

6.2.3 Regulation in saline-water species

Mosquito species that regularly breed in saline waters have a marked capacity to regulate their haemolymph. Very substantial changes in the concentration of the external medium cause only slight increases in haemolymph osmolality, and only slight changes in the concentrations of most inorganic ions.

Larvae of the rock pool mosquito *Opifex fuscus*, which in nature are subjected to widely fluctuating salinities, maintained the sodium and chloride concentrations and the osmotic pressure of their haemolymph at more or less normal levels when kept for three to five days in distilled water or in concentrations of sea water ranging from 10 to 200% (Figure 6.4). Over the range 0–36% sea water the haemolymph was hyperosmotic to the medium; at higher salinities it was hypo-osmotic. At the lower salinities at least, sodium ions accounted for most of the cation contribution to haemolymph osmotic pressure, and chloride ions accounted for about half of the anion contribution (Nicolson, 1972).

The mean osmosity of the haemolymph of *Aedes togoi* larvae in sea water was 195 mM-NaCl, about one-third that of the sea water (552 mM-NaCl). Over a wide range of external salinities the haemolymph osmosity changed only slightly, ranging from 160 mM-NaCl in fresh water to 245 mM-NaCl in 300% sea water. Among larvae adapted to different media, the haemolymph sodium concentration was 134 mM in fresh water, 151 mM in sea water, and 190 mM in 300% sea water (Asakura, 1980). Larvae kept at 23°C and transferred from fresh water to 150% sea water stabilized their haemolymph [Na^+] and [Cl^-] within 20 h, after a period of considerable overshoot. If the larvae were kept at 17°C the haemolymph ion concentrations were still rising 20 h after the transfer to concentrated sea water (Matutani *et al.*, 1983).

Larvae of the salt marsh mosquitoes *Ae detritus* and *Ae taeniorhynchus* were able to maintain the chloride concentration and osmotic pressure (Figure 6.5) of their haemolymph at nearly normal levels in both dilute and very concentrated media (Beadle, 1939; Nayar and Sauerman, 1974a). In distilled water *Ae detritus* larvae lost much of the chloride but the haemolymph osmotic pressure was rather better regulated, suggesting an increase in the non-chloride fraction (Beadle, 1939). Among *Ae taeniorhynchus* larvae reared in 10, 100 and 200% sea water the haemolymph concentrations of potassium or of magnesium were identical. However, with increases in sea water-concentration within that range the sodium concentration fell slightly and the calcium and chloride concentrations rose (Bradley and Phillips, 1977b). Upon transfer of *Ae taeniorhynchus* larvae from one medium to another, substantial

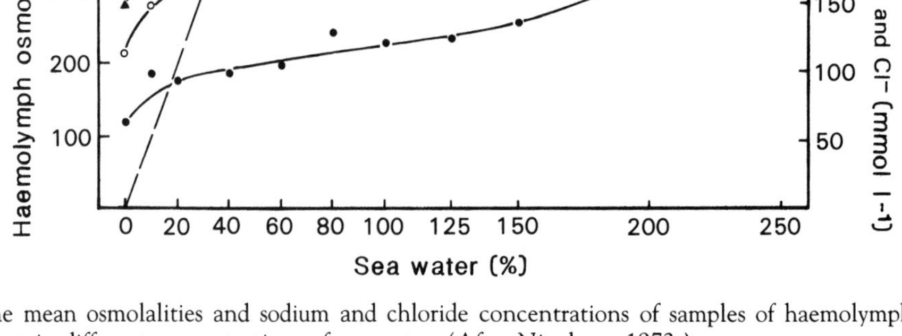

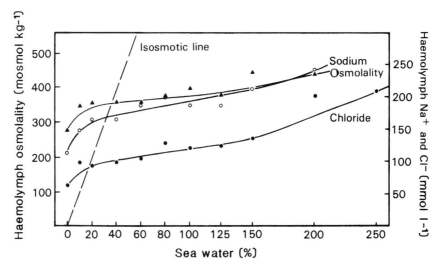

Figure 6.4 The mean osmolalities and sodium and chloride concentrations of samples of haemolymph from *Opifex fuscus* larvae kept in different concentrations of sea water. (After Nicolson, 1972.)

transient changes in haemolymph ion concentrations occurred initially, and the concentrations returned to normal only over a 12 to 24 h period (Nayar and Sauerman, 1974a; Bradley and Phillips, 1977a).

Sea water has a much higher sulphate concentration than has larval haemolymph. When *Ae taeniorhynchus* larvae were kept in sea water (31 mM-SO_4^{2-}) the haemolymph sulphate concentration was 1.8 mM, but in sulphate-enriched sea water (89 mM) haemolymph sulphate increased to 34 mM (Maddrell and Phillips, 1978). *Ae campestris* larvae may be exposed to very high concentrations of magnesium and sulphate ions in inland saline pools. After four days in media with magnesium concentrations ranging from 0.01 to 100 mM, the range of haemolymph magnesium was only 2.5–5 mM (Kiceniuk and Phillips, 1974). When external sulphate concentrations were raised from 2.5 to 73 mM, haemolymph sulphate rose from 1.5 to 6.6 mM. At extremely high sulphate concentrations regulation was less effective, and larvae held in 260 mM-sulphate contained 117 mM-sulphate in their haemolymph (Maddrell and Phillips, 1975).

The ability of *Ae campestris* larvae to regulate their haemolymph is affected by the ionic composition of the medium. Bradley and Phillips (1977a) acclimated larvae to different solutions, all of 700 mosmol l^{-1}, but with NaCl, $NaHCO_3$ or (Na + Mg)SO_4 as the major solutes. Larvae in the NaCl and $NaHCO_3$ solutions effectively regulated the osmolality and ionic composition of their haemolymph. However, the haemolymph of larvae in the (Na+Mg)SO_4 solution was almost isosmotic with the external medium, and contained high concentrations of Na^+, Mg^{2+} and SO_4^{2-}.

6.3 WATER BALANCE

Mosquito larvae differ from most insects in having a continuous water intake, due to osmotic forces and drinking. That this threatens not only water balance but body volume also was suggested by the observation that larvae with damaged ventral nerve cords increased in body weight by 8% per hour (Stobbart, 1971c). The homeostatic mechanisms employed by mosquito larvae are variants of the diuretic processes

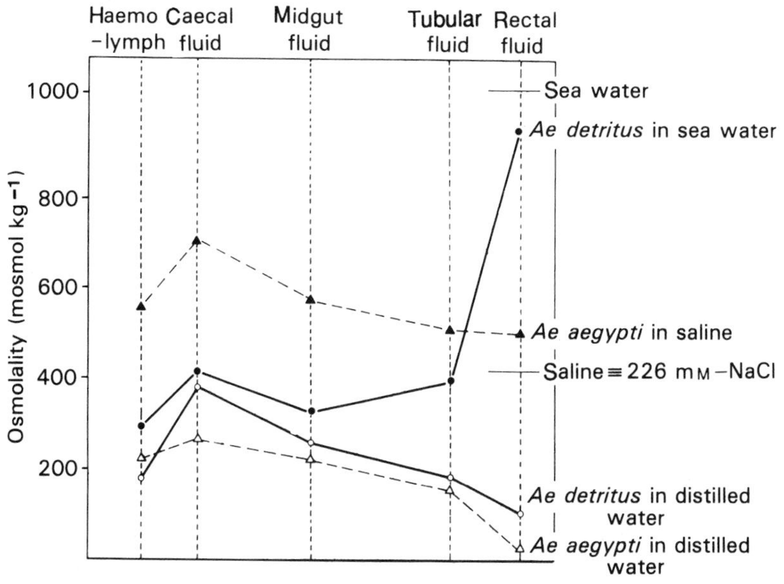

Figure 6.5 Mean osmolalities of samples of haemolymph and gut fluids from larvae of *Aedes aegypti* and *Ae detritus* kept in different media. Statistically significant differences of osmolality between haemolymph and gut fluid, mentioned in the text, are not always apparent in the figure. (After Ramsay, 1950.)

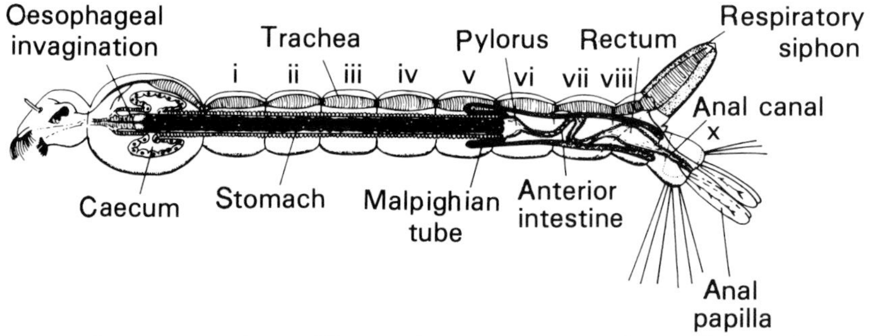

Figure 6.6 Anatomy of a mosquito larva. The numbers i, ii, etc., indicate abdominal segmentation. (From Wigglesworth, 1933a.)

evolved by insects generally and described in the opening paragraphs of Chapter 16. Key organs for diuresis are the Malpighian tubules and rectum (Figure 6.6). Three fluids produced by them are (i) tubular fluid, the unmodified secretion of the Malpighian tubules as it accumulates in the pylorus, (ii) rectal fluid, tubular fluid accumulating in and being modified by the rectum, and (iii) urine, the completed excretory fluid as discharged to the exterior (Stobbart, 1971c).

6.3.1 Water movement in fresh water species

Fresh-water mosquito larvae drink considerable volumes of water; for example, the intake by 4th instar larvae of four species ranged from 48 to 309 nl/larva/h (Table 4.3; Section 4.1.3). There is also an appreciable entry of water through the anal papillae due to osmotic forces. When *Aedes aegypti* larvae were ligatured between the 5th and 6th abdominal segments and kept in fresh water the hindmost part of the body gradually swelled, but if the larvae had previously been deprived of their anal papillae the swelling was negligible (Wigglesworth, 1933a). In larvae which were prevented from drinking, the osmotic uptake of water by individuals with normal anal papillae was estimated to equal 33% of body weight per day, whereas in larvae with hypertrophied anal papillae it equalled 190% (Shaw and Stobbart, 1963; Stobbart, 1971c).

Imbibed water moves posteriorly through the midgut, passing between the food particles (Aly and Dadd, 1989), but at some stage it appears to be absorbed into the haemolymph and then to be secreted by the Malpighian tubules. Malpighian tubules of *Ae aegypti* which had been occluded by ligature near their origins became enormously distended due to fluid secretion into their lumens. In normal larvae, fluid could be seen to accumulate in the pylorus, which is closed anteriorly by the stomach contents and posteriorly by the collapsed wall of the anterior intestine. When sufficient fluid had collected to dilate the anterior intestine also, peristaltic movements of that organ transferred the fluid to the rectum. This was repeated every 2 or 3 min, and fluid was evacuated from the rectum after longer intervals (Wigglesworth, 1933a).

The contents of the pylorus are regarded as the unmodified secretion of the Malpighian tubules, i.e. tubular fluid, because their osmolality and sodium concentration remain unaltered after the midgut has been separated by ligature. The osmolality of the tubular fluid was not significantly different from that of the haemolymph whether *Ae aegypti* larvae were kept in distilled water or in concentrated saline (Figure 6.5). When larvae were kept in distilled water the potassium concentration of the tubular fluid was almost thirty times that of the haemolymph, whereas the sodium concentration was less than one-third that of the haemolymph (Table 6.2) (Ramsay, 1951, 1953a).

The potential difference across the wall of

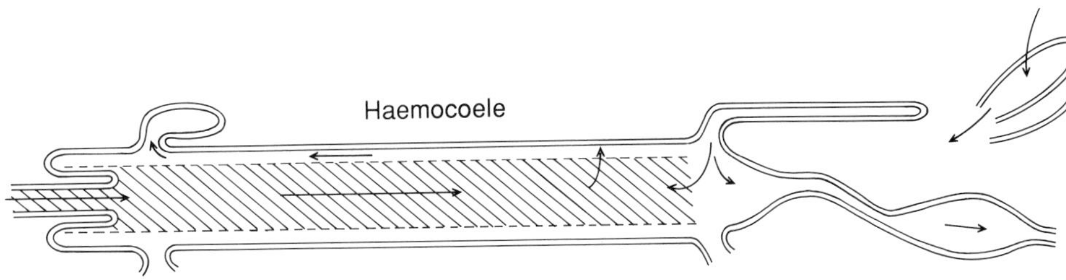

Figure 6.7 Water flow through the alimentary canal of an *Aedes aegypti* larva. The arrows indicate the intake of water through drinking and via the anal papillae, its movement within the alimentary canal, and its movement between alimentary canal and haemolymph.

Ae aegypti Malpighian tubules was 13–21 mV (lumen positive) whatever the larva's external medium. When related to the Na^+ and K^+ concentrations of the tubular fluid and haemolymph, these measurements prove that K^+ is actively transported against an electrochemical gradient but that Na^+ may enter the tubule lumen by passive diffusion (Ramsay, 1953b). These results resemble the more detailed findings obtained with Malpighian tubules from adult female *Ae aegypti* (Section 16.2.6) which revealed that potassium transport is the prime mover in the production of tubular fluid and that ions with a negative charge follow without expenditure of energy. As a result of this salt transport, and of osmotic coupling, a transport of water takes place which by solvent drag brings small molecules into the tubule lumen.

Observations on *Ae aegypti* larvae revealed that fluid from the Malpighian tubules moves both anteriorly and posteriorly (Figure 6.7). Some contents of the pylorus passed forwards due to antiperistaltic movements of the stomach, reaching the gastric caeca (Ramsay, 1951). When a fluorescent dye was imbibed it accumulated over 1–2 h in the ectoperitrophic space in the region of the 3rd to 5th abdominal segments. Antiperistaltic movements of the stomach pumped it into the anterior region of the midgut and into the basal regions of the caeca. Then the caeca contracted synchronously and forced the fluid back into the stomach. This was repeated at regular intervals (Volkmann and Peters, 1989b). The water that moved forwards was believed to return to the haemolymph via the gastric caeca. When particles of Trypan blue were ingested, dissolved dye quickly appeared in the ectoperitrophic space and was carried to the gastric caeca by antiperistaltic waves of the stomach. After two or three hours the caecal lumens contained solid masses of dye. No solid dye was present elsewhere outside the peritrophic membrane, therefore there must have been a continuous absorption of water by the caeca (Wigglesworth, 1933a; Ramsay, 1950).

It is likely that diuretic hormones are involved in homeostasis and the regulation of body volume but it is not clear how they are used in conditions of continuous diuresis. Experimental results suggested that the rate of urine production and the regulation of body volume are governed to some extent by a nervous mechanism. (1) Any operation that interrupted the nerve cord between the neck and the lst abdominal segment, which would lead to swelling, caused the antiperistaltic movements of the stomach to become more frequent, sometimes more or less continuous. From this it was deduced that the division of the flow of tubular fluid into streams of appropriate size proceeding to the midgut and rectum is achieved by nervous control of the activity of the rear end of the midgut wall. (2) Peristalsis of the anterior intestine was stimulated by increase in haemolymph volume and distention of the body wall, and also by hydrostatic pressure in the pylorus. (3) If the nervous system was damaged, resulting in excessive midgut antiperistalsis, ligation between abdominal segments iv and v caused the pylorus to dilate again and peristalsis of the anterior intestine was resumed (Ramsay, 1953b; Stobbart, 1971c).

6.3.2 Water movement in saline-water species

Aedes togoi larvae ligatured at neck and anus lost weight (0.75% body weight h^{-1}) when in sea water and gained weight (0.6% body weight h^{-1}) when in fresh water, indicating that the body cuticle is permeable to water (Asakura, 1982a). From the weight loss (1% body weight h^{-1}) of sea-water adapted larvae of *Opifex fuscus* which had been ligatured at the neck and between the 7th and 8th abdominal segments, the osmotic permeability coefficient of the body cuticle was estimated to be 2.4×10^{-2} cm h^{-1} (Nicolson and Leader, 1974).

Evidence presented in this and later sections shows that saline-water species maintain their water balance when in saline media by drinking the external medium and passing urine which is slightly more concentrated than the medium. Drinking rates have been determined by measuring the accumulation of solutes such as [^{125}I]PVP, [^{14}C]inulin, or the dye amaranth, which are swallowed with the medium but not absorbed into the haemolymph. Sea-water adapted larvae of *Ae taeniorhynchus* drank on average 82 nl mg^{-1} h^{-1} (190% of body weight per day) (Bradley and Phillips, 1977b), and of *Ae togoi* drank on average 94 nl mg^{-1} h^{-1} (225% of body weight per day) (Asakura, 1982a). In a medium resembling natural alkaline lake waters, *Ae dorsalis* drank on average 59 nl mg^{-1} h^{-1} (142% of body weight per day) (Strange *et al.*, 1982). In hypersaline Ctenocladus pond water, *Ae campestris* drank 150 nl mg^{-1} h^{-1} (300% of body weight per day). These last larvae may have been pushed to the upper limit of their osmotic tolerance, so that body volume had decreased leading to a dramatic increase in drinking rate (Kiceniuk and Phillips, 1974).

The drinking rate of *Ae togoi* varied with the salinity of the medium, thus larvae adapted to fresh water, sea water or 200% sea water drank 42, 94 and 130 nl mg^{-1} h^{-1} respectively. When transferred from fresh water to sea water *Ae togoi* larvae showed no significant increase in drinking rate until the 4th day; however, when transferred from sea water to fresh water drinking declined to the fresh-water rate within 24 h (Asakura, 1982a). In contrast, within the range 10–200% sea water, the drinking rate of *Ae taeniorhynchus* larvae was not significantly affected by the salinity of the medium, but apparently larval size determined the drinking rate. The rate of fluid intake by *Ae taeniorhynchus* larvae feeding on particulate matter was about twice that of non-feeding larvae (Bradley and Phillips, 1977b; Maddrell and Phillips, 1978).

These high rates of drinking might be thought puzzling because in weakly saline media they add to the water load and in strongly saline media they add a heavy salt load. However, in hyperosmotic media high rates of fluid ingestion reduce the extent to which urine must be concentrated to achieve osmotic homeostasis, because the greater the ratio of water ingested to water lost by osmosis, the less the urine needs be concentrated to achieve osmotic balance. The osmolality of the urine of *Ae taeniorhynchus* larvae in sea water was 11% higher than that of the sea water; its ionic composition was somewhat different (Bradley and Phillips, 1975, 1977b).

Amaranth that was imbibed in sea water by *Ae togoi* larvae entered the gastric caeca within 30 min, appeared in the stomach within one hour, and was present throughout the alimentary canal after two hours (Asakura, 1982a). Larvae of *Ae campestris* and *Ae taeniorhynchus* which were kept in saline media containing [^{125}I]PVC or [^{3}H]inulin were found to absorb from the gut almost all the water that they drank (Kiceniuk and Phillips, 1974; Maddrell and Phillips, 1978).

More water is imbibed than is lost across the body wall by osmosis, and the surplus water and its salts must be excreted. The organs responsible for this are the Malpighian tubules and the rectum. Tubules of *Ae campestris*, which were bathed in saline with an ionic composition similar to haemolymph and containing 4 mM-K^+, secreted a fluid that was almost isosmotic with the medium but contained 31 mM-K^+. Clearly, in saline- as in fresh-water species, potassium secretion is fundamental to the production of tubular fluid (Phillips and Maddrell, 1974). Tubule activity is probably stimulated by a diuretic hormone. Exposure of the head contents of *Ae taeniorhynchus* larvae to potassium-rich saline, which would depolarize neurosecretory axons, led to production of a factor that greatly enhanced the rate of fluid secretion by the Malpighian tubules (Maddrell and Phillips, 1978).

In saline-water species fluid is secreted across the rectal epithelium into the rectal lumen, adding to any tubular fluid that has passed there. Ligated recta of *Ae taeniorhynchus* showed a mean secretory rate of 19 nl h^{-1} and a highest rate of 96 nl h^{-1} (Bradley and Phillips, 1975). The mechanism of fluid secretion by the rectum is described in Section 6.5.3(b).

6.4 MECHANISMS OF ION UPTAKE

6.4.1 Ion uptake by fresh-water species

Mosquito larvae probably obtain a considerable proportion of their total salt intake with their food, but larvae that live in fresh water can take up salts directly from very dilute solutions. The lower limits of ion concentrations that permit survival have not been established, but the observation that starved *Aedes aegypti* larvae kept in small volumes of deionized water first lose ions and then come into balance with sodium at 5 μmol l^{-1} and with chloride at about 20 μmol l^{-1} (Stobbart, 1965, 1967) gives a measure of the effectiveness of two uptake systems.

The anal papillae of fresh-water species can be destroyed by placing the larvae in 5% NaCl or 20 mM-NaOH for 2 or 3 min. On returning the larvae to fresh water the papillae fall away, and within a few days the body surface heals showing only four small scars (Wigglesworth, 1933a; Ramsay, 1953a). The capacity to take up sodium, potassium, chloride and phosphate ions from the medium was markedly reduced or lost in papilla-less larvae (Ramsay, 1953a; Koch, 1938; Hassett and Jenkins, 1951). Measurements of the influx and efflux of radioactive sodium in *Ae aegypti* larvae which lacked anal papillae, or had their mouths sealed, or were unoperated, showed that 90% of the steady state exchange of sodium occurred across the anal papillae and that the remainder occurred across the gut (Treherne, 1954; Stobbart, 1959). About 90% of the steady state exchange of chloride also occurred across the anal papillae, and chloride-deficient starved larvae could achieve net uptake of chloride only if the papillae were intact (Stobbart, 1967). There is some evidence, however, of a role for the gut in ion uptake. Papilla-less sodium-deficient larvae which were fed became adapted to perform net sodium transport, presumably through the gut, 20–40 h after transfer to 2 mM-NaCl (Stobbart, 1960). The rate of uptake of radioactive phosphate was low in papilla-less larvae but the same level of radioactivity was ultimately attained as in normal larvae (Hassett and Jenkins, 1951).

Movements of dye injected into *Ae aegypti* larvae suggested to Edwards (1983) that the lumens of the anal papillae could be continuous with the general body cavity or isolated from it, possibly controlled by the ring of circular muscle at the base of each papilla. In preliminary experiments, transepithelial resistances of 2–45 kΩ were measured when haemocoele and papilla lumen were thought to be in communication, and of 60–120 kΩ when the papilla lumen was thought to be isolated. The mean value of the resistance was within the range for 'tight epithelia', indicating a capacity for maintaining high concentration gradients. When ligatured papillae were bathed in 2% sea water, mean electrical potentials relative to the external medium were −26 mV for the epithelial cell interiors and −10 mV for the papilla lumens (transepithelial potential). Under such conditions uptake of chloride ions would be against steep electrical and chemical gradients. Sodium ions could enter the epithelium down an electrical gradient but would encounter adverse electrical and chemical gradients on moving from the epithelium into the haemolymph (Edwards, 1983). Measurements of the transepithelial potential in sodium-deficient larvae indicated that both sodium and chloride were actively transported from the medium into the haemolymph but that potassium was probably taken up passively in company with chloride (Stobbart, 1974).

Sodium and chloride pumps permit Na^+ and Cl^- movement in both inward and outward directions, therefore when studying ion transport it is necessary to distinguish 'influx' and 'efflux' from 'net uptake' of ions. In *Ae aegypti* neither the influxes nor the effluxes of Na^+ and Cl^- could be explained in terms of passive diffusion, because (i) the 'calculated passive influxes' were a negligible fraction of the 'observed influxes', (ii) the 'observed effluxes' were greater than the 'calculated passive effluxes', and (iii) Na^+ efflux was an energy consuming process which was reduced to about one-seventh by KCN. The effluxes of Na^+ and Cl^- were lowest in distilled water and increased with increasing external concentrations of those ions, suggesting to Stobbart (1974) that at higher external concentrations a carrier-mediated component had become involved in efflux.

The effects of different conditions upon the rates of sodium movement in *Ae aegypti* larvae are summarized in Table 6.2. Well-nourished larvae in a sodium-rich medium exchange sodium at a fairly high rate but starved larvae have a much lower rate of exchange. The time for half exchange is about 10 h in fed larvae and about 60 h in starved larvae. In both cases 90% of the exchange occurs across the anal papillae. Sodium efflux is low in distilled water, in starved larvae at least. When sodium-deficient larvae are placed in 2 mM-NaCl a rapid net uptake of sodium occurs in fed individuals and a much slower net uptake in starved individuals, accompanied in both cases by relatively high rates of efflux. In both fed and starved sodium-deficient larvae the initial rates of influx and efflux are higher than those during steady state exchange (Stobbart, 1959, 1960).

Experimental results suggest that sodium ions enter the anal papillae of *Ae aegypti* via a saturable rate-limited system with a high affinity for Na^+, which is half saturated at 0.55 mM. The relationship between Na^+ influx and external Na^+ concentration is not accurately described by the Michaelis–Menten equation. A similar relationship exists between the external Na^+ concentration and Na^+ efflux, although efflux rates are always much lower than influx rates except at concentrations approaching zero. Chloride and hydroxyl ions in the external medium have a marked stimulatory effect upon both influx and net uptake of Na^+, indeed anions in general stimulate Na^+ influx whereas cations inhibit it (Stobbart 1965, 1967, 1971a, b).

When recently fed sodium-deficient larvae are transferred to 2 mM-NaCl the rates of influx,

Table 6.2 Sodium movements in *Aedes aegypti* larvae subjected to different conditions. Net uptake was estimated by spectrophotometric measurements of haemolymph sodium, fluxes by measurement of ^{22}Na in the haemolymph. (From the data of Stobbart, 1959, 1960.)

Type of ion movement and conditions	*Change in haemolymph Na (mmol l^{-1} h^{-1})*	
	Fed	*Starved*
1. *Steady state exchange in 2 mM-NaCl*	7.1–8.8	1.0–1.1
2. *Efflux in distilled water*	–	0.3
3. *Net uptake by Na-deficient larvae on transfer to 2 mM-NaCl (initial rate)*	39–50	7–9
4. *Efflux from Na-deficient larvae on transfer to 2 mM-NaCl (initial rate)*	49–70	3–9
5. *Total initial influx on transfer of Na-deficient larvae to 2 mM-NaCl (= 3 + 4)*	88–120	10–18

efflux and net uptake are all high at first but decline exponentially to zero over a period of 5 h, when uptake is almost complete. Uptake of Na^+ and Cl^- can occur independently but both uptake rates are faster when the ions are taken up together. When Na^+ is taken up from Na_2SO_4 solution, the Na^+ uptake is balanced electrically by loss of K^+, by exchange of H^+ and by SO_4^{2-} uptake.

The chloride uptake system in the anal papillae of *Ae aegypti* larvae has the following characteristics. When the external medium contains sodium chloride, Na^+ and Cl^- are to a large extent taken up together although the rate of net Cl^- uptake is not a fixed proportion of that for Na^+. Uptake of Na^+ and Cl^- can occur independently, but at a lower rate than when both ions are taken up together. When uptake is from potassium chloride solution the uptake of Cl^- is balanced electrically by uptake of K^+, by exchange of HCO_3^- (and possibly of OH^-), and by the loss of unknown, possibly organic, ions. In general, other anions inhibit Cl^- uptake whereas cations stimulate it. The Cl^- uptake mechanism in larvae from two populations of *Ae aegypti*, placed in sodium chloride solutions, was half saturated at 0.2 and 0.5 mM. When larvae are kept in distilled water the Cl^- efflux rate is low, but it increases progressively with increases in external [Cl^-] (Stobbart, 1967, 1971a).

Little is known about potassium uptake. Measurements of electrical potential between haemolymph and medium obtained from *Ae aegypti* larvae placed in concentrations of 2 mM-K^+ and above suggested that K^+ enters passively down an electrochemical gradient, possible via a carrier system (Stobbart, 1974). It is likely that mosquito larvae satisfy a large part of their potassium requirement from their food since potassium is present in high concentration in living microorganisms.

The anal papillae of *Ae aegypti* apparently are not involved in calcium uptake since the rate of calcium influx from dilute solutions was significantly faster in papilla-less than in normal larvae. Preliminary experiments with ligated larvae suggested that calcium was taken up across the gut and lost through the urine. The larvae had a saturable transport system for Ca^{2+} uptake which followed Michaelis–Menten kinetics. In previously fed larvae it had an affinity constant (K_m) of 0.43 mM and a maximum transport velocity (V_{max}) of 0.2 nmol h^{-1}. In starved larvae the K_m was reduced to 0.35 mM and the V_{max} to 0.12 nmol h^{-1}, possibly due to a reduction in the number of 'calcium carriers'. The average rate of Ca^{2+} uptake from 0.1 mM-$CaCl_2$ by previously fed larvae was 0.0335 nmol h^{-1} $larva^{-1}$. The transport system appeared to be energy dependent since its efficiency was decreased by ruthenium red, a selective inhibitor of Ca^{2+}-activated ATPase (Barkai and Williams, 1983).

6.4.2 Ion uptake by saline-water species

Aedes campestris larvae kept in hyperosmotic lake water were unable to achieve net uptake of $^{22}Na^+$ or $^{36}Cl^-$ via their anal papillae; uptake probably occurred via the gut after drinking. When larvae were kept for two weeks in 5 mM-NaCl they gradually developed the ability to take up ions from dilute solution, and could accumulate Na^+ and Cl^- from 0.5 mM but not from 0.05 mM-NaCl. Blocking the mouth did not abolish uptake of Na^+ or Cl^- from dilute solution but net uptake was prevented by also ligating the anal papillae. Electrical potential differences between haemolymph and various saline media suggested that uptake of both Na^+ and Cl^- was by active processes. The rates of influx and net uptake of these ions followed Michaelis–Menten kinetics, with K_m values of 2 mM and V_{max} of 1.5 nmol h^{-1} mg^{-1} body weight for both (Phillips and Meredith, 1969). The K_m value was four times higher than that reported for *Ae aegypti* by Stobbart (1965, 1967). In terms of tranport per milligram wet weight of anal papillae, V_{max} was similar for *Ae campestris* and *Ae aegypti* but the total weight of anal papillae was much greater in *Ae aegypti*, a difference that may account for the regulatory advantages which strictly fresh-water species have over saline-water species in natural waters of low salt concentrations (Phillips *et al.*, 1978) .

When *Ae campestris* larvae were transferred from a hyperosmotic medium to fresh water containing $^{22}Na^+$, no significant influx of Na^+ via the anal papillae was observed initially, but over a period of 5–15 days a large influx developed. Apparently, several days are required to induce Na^+ transport in this species. As in *Ae aegypti*, Na^+ and Cl^- transport can be turned on independently since net uptake of Cl^- occurs from sodium-free solutions and of Na^+ from chloride-free solutions (Phillips *et al.*, 1978).

Rubidium ions are actively absorbed by a saturable mechanism in the anal papillae of fresh-water adapted *Ae campestris*. Potassium ions inhibit this influx. Since competition between Rb^+ and K^+ for a common transport mechanism is known in other insects, it is likely that this uptake by anal papillae is via a K^+ pump (Phillips and Bradley, 1977).

6.5 MECHANISMS OF IONIC AND OSMOTIC REGULATION

In insects the Malpighian tubules and rectum constitute a coordinated excretory system, the tubules secreting a primary excretory fluid which is modified during its passage through the rectum. This system is exploited and further elaborated by mosquito larvae for ionic and osmotic regulation. In fresh water the active uptake of ions by the anal papillae (Section 6.4.1) is an important aspect of ionic regulation.

6.5.1 Mechanisms in fresh-water species

Fresh-water mosquito larvae have two problems in respect of ion balance. They must gain ions from the medium against a considerable concentration gradient, or from their food, and they must conserve ions while voiding the water which enters the body by osmosis and drinking. Uptake of most ions occurs via the anal papillae, while the processes of ion conservation are predominantly located in the alimentary canal. Our knowledge of ionic regulation in fresh-water species relates principally to sodium and potassium, and in considering the regulatory mechanisms it is important to remember that potassium is secreted by the Malpighian tubes to generate the flow of tubular fluid (Section 6.3.1).

When *Aedes aegypti* larvae are kept in distilled water the tubular fluid, which accumulates in the pylorus, is not significantly different in osmolarity from the haemolymph, but compared to the haemolymph it is low in sodium and extremely high in potassium (Table 6.3, Figure 6.5). That part of the tubular fluid which moves forwards from the pylorus, due to antiperistalses of the stomach, becomes significantly concentrated in the midgut caeca. That part of the tubular fluid which moves back to the rectum becomes strongly diluted, presumably through the resorption of ions (Figure 6.5). Analysis of rectal fluid indicates a low sodium and a relatively high potassium concentration, but the ability of larvae to survive in distilled water suggests that both ions must normally be almost completely resorbed before urine is discharged. It appears that fresh-water mosquito larvae rid themselves of excess water by secreting isosmotic tubular fluid, resorbing salts in the rectum, and voiding hypo-osmotic urine (Ramsay, 1950, 1953a).

When *Ae aegypti* larvae were placed in 85 mM-NaCl or 85 mM-KCl the pyloric contents had a higher than usual concentration of Na^+ or K^+, and resorption of either ion by the rectum was reduced with the consequence that it escaped from the body (Table 6.3). The K^+ concentration

Table 6.3 The effects of various external media on sodium and potassium concentrations in body fluids of *Aedes aegypti* larvae. Samples were taken after exposure to the relevant medium for 5 days except in the case of rectal fluid which was sampled after 3 weeks. (From Ramsay, 1953a.)

External medium	*Haemolymph* *(mmol l^{-1})*		*Tubular fluid* *(mmol l^{-1})*		*Rectal fluid* *(mmol l^{-1})*	
	Na	*K*	*Na*	*K*	*Na*	*K*
Distilled water	87	3	24	88	4	25
1.7 mM-NaCl	100	–	–	–	–	–
85 mM-NaCl	113	–	71	90	100	18
1.7 mM-KCL	–	4	–	–	–	–
85 mM-KCL	–	6	23	138	14	90

of the rectal contents could greatly exceed that of the haemolymph (Ramsay, 1953a).

Aedes aegypti larvae appear unable to render their rectal contents hyperosmotic to the haemolymph; consequently, regulation of haemolymph osmotic pressure fails when the larvae are exposed to a hyperosmotic medium. Under such conditions the tubular fluid, which has an osmolality not significantly different from that of the haemolymph (Figure 6.5), is slightly but significantly diluted in the rectum, presumably through resorption of salts, raising the osmolarity of the haemolymph above that of the external medium (Figures 6.1, 6.5) (Ramsay, 1950).

The loss of certain inorganic ions from larvae living in very dilute media is apparently reduced by the sequestration of ions within the haemolymph. By using ion-selective microelectrodes it is possible to measure ionic activities, i.e. the concentrations of free ions of individual ion species, as distinct from the total (free and bound) concentrations. In *Ae aegypti* larvae in 20% sea water the sodium activity of the haemolymph, about 90 mM, was close to its total sodium concentration. However, the haemolymph sodium activity of larvae reared in 0 and 2% sea water was reported to be about 20 mM, a remarkably low value in comparison with the total sodium concentrations of 60 mM in 0% sea water and 80 mM in 2% sea water. Edwards (1982b) suggested that when larvae were reared in dilute media, sodium activity was reduced by the association of sodium ions with proteins, so reducing the rate of sodium loss. In contrast, chloride activity tended to resemble chloride concentration. Even when larvae were kept in distilled water the haemolymph chloride activity (60 mM) resembled the chloride concentration, indicating that very little, if any, was bound.

The mean calcium concentration in *Ae aegypti* haemolymph was 5.2 ± 0.3 mM, representing over 95% of the calcium in the larva. Only 16% of the haemolymph calcium could be exchanged with calcium in the external medium; the remainder was considered to be bound (Barkai and Williams, 1983).

6.5.2 Mechanisms in brackish-water species

When *Culiseta inornata* larvae are kept in water of moderate salinity, i.e. above 50% sea water, their haemolymph becomes isosmotic with the external medium, but the Na^+ and Cl^- concentrations are kept below those of the medium (Figure 6.3; Section 6.2.2). The osmotic pressure of the Malpighian tube fluid closely follows that of the haemolymph. Recta which are ligatured and isolated *in vitro* accumulate fluid, showing that they are capable of secretion, but the secretion resembles the bathing fluid in osmotic pressure and Na^+ concentration. The ultrastructure of the rectal epithelium (Section 6.6.3) suggests that this is a site of ion resorption but one unlikely to be capable of hyperosmotic fluid secretion (Garrett and Bradley, 1984b). The osmoconformity of the haemolymph in hyperosmotic media is consistent with the apparent absence of a site of hyperosmotic fluid secretion, but the question remains as to how the larvae regulate haemolymph Na^+ and Cl^-. Garrett and Bradley (1984a) suggested that ions may be eliminated via the anal papillae in brackish media even though those organs are used for ion uptake in dilute media.

6.5.3 Mechanisms in saline-water species

Larvae of saline-water mosquito species can develop in water containing inorganic ions at much lower and much higher concentrations than in their haemolymph. In dilute media the larvae conserve ions, as do those of fresh-water species. In concentrated media the larvae are subject to a salt load because they imbibe the medium to maintain water balance; they must therefore excrete ions. In sea water, *Aedes taeniorhynchus* larvae have turnover times for total body sodium and chloride of 3.8 and 4.8 h respectively, indicating a very high capacity for ionic regulation (Bradley and Phillips, 1977a).

Inorganic ions that are ingested are rapidly transferred to the haemolymph. More than 95% of magnesium and sulphate ions imbibed

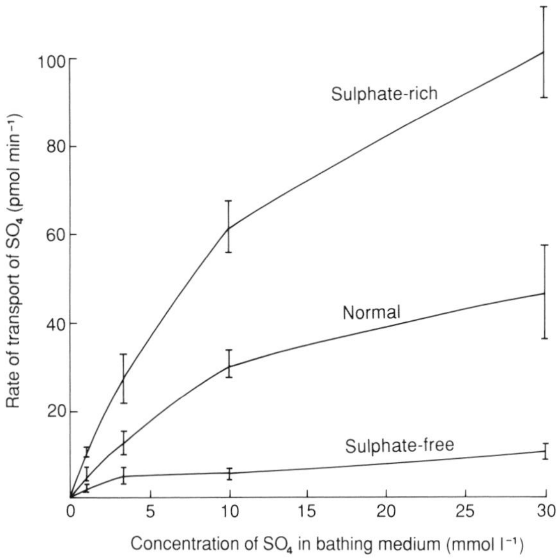

Figure 6.8 Effect of larval rearing medium on the capacity of the Malpighian tubules of *Aedes taeniorhynchus* larvae to transport sulphate.
Malpighian tubules from larvae which had been reared in sulphate-free sea water, normal sea water (31 mM-SO_4) or sulphate-enriched sea water (89 mM-SO_4) were isolated in bathing media of different sulphate concentrations, and the rates of sulphate transport were measured. Means ± s.e. (From Maddrell and Phillips, 1978.)

by *Ae campestris* in strongly saline media were rapidly absorbed from the gut (Kiceniuk and Phillips, 1974; Maddrell and Phillips, 1975). *Ae taeniorhynchus* larvae kept in sulphate-enriched sea water absorbed nearly all of the sulphate from the fluid they had imbibed. Raising the osmolality of the medium by addition of sucrose prevented the absorption of both water and sulphate from the midgut, leading Maddrell and Phillips (1978) to conclude that sulphate ions are not assimilated by an active uptake mechanism but by some form of facilitated diffusion. When *Ae detritus* larvae drank sea water (≡ 559 mM-NaCl), the mean osmosity of the caecal fluid was equivalent to 226 mM-NaCl, and that of the stomach contents to 176 mM-NaCl. Clearly, the sea water that had been imbibed lost salts in the caeca and was further modified in the stomach (Ramsay, 1950). There is histochemical evidence for the transport of chloride ions by caecal cells (Asakura, 1982a).

Studies with the saltmarsh species *Ae detritus* provided the first indication of the mechanisms of ionic regulation in saline-water mosquitoes. When kept in distilled water these larvae secreted a tubular fluid which resembled the haemolymph in osmotic pressure, but produced rectal fluid which was hypo-osmotic to the haemolymph (Figure 6.6), thus apparently regulating their ions in the same way as fresh-water species. In sea water, *Ae detritus* larvae secreted a tubular fluid which again was not significantly different in osmotic pressure from the haemolymph; this was strongly concentrated in the rectum leading to the discharge of hyperosmotic urine (Figure 6.5) (Ramsay, 1950).

(a) Malpighian tubule activity

Malpighian tubules taken from *Aedes campestris* larvae reared in pond water of relatively low salinity (170 mosmol/l), and studied *in vitro*, always secreted fluid of similar osmotic pressure to the bathing solution. The fluid produced by tubules in 4 mM-K^+ bathing solution contained 31 mM-K^+, thus these tubules resemble those of most other insects in preferentially secreting K^+ rather than Na^+. In most cases the tubule lumen was electrically negative with respect to a bathing medium which resembled haemolymph, in contrast to the tubules of *Ae aegypti* and many other insects.

Isolated tubules of *Ae campestris* could transport Mg^{2+} against a 10-fold concentration gradient and an electrical potential difference of 15 mV. Sulphate movement was not greatly affected by the transepithelial potential, and sulphate ions could be secreted against a concentration gradient. Therefore the secretion of both Mg^{2+} and SO_4^{2-} involved active transport. The tubules could transport both ions at a high rate and achieve a considerable concentration of the ions in the secreted fluid. The magnesium pump was saturated when the magnesium concentration in the bathing medium was about 5 to 6 mM; the rate of transport was half maximal at about 2.5 mM-Mg^{2+}. However, changes in magnesium concentration had no effect on the rate of fluid secretion. The sulphate pump had a high capacity but a rather low affinity, and was half saturated at about 10 mM (Phillips and Maddrell, 1974; Maddrell and Phillips, 1975).

Larvae of *Ae taeniorhynchus* also could secrete tubular fluid with a high sulphate content, but this was dependent upon prior exposure to sulphate in the external medium. Tubules from larvae reared in sulphate-free sea water had an extremely weak ability to transport sulphate ions, and much of the sulphate in their tubular fluid could have crossed the tubule walls passively. In contrast, Malpighian tubules from larvae reared in sea water secreted sulphate at a moderate rate which was positively correlated with the sulphate concentration of the bathing medium. Tubules from larvae reared in sulphate-enriched sea water secreted sulphate at twice those rates (Figure 6.8). Tubules from such larvae secreted 18.45 mM sulphate when bathed in 3.33 mM. Experiments indicated that it was not the affinity of the transport system which varied but the capacity. Evidently exposure of larvae to water containing sulphate leads to the induction of a sulphate transport mechanism. Sixteen hours exposure of larvae to sulphate-containing water induced a sulphate-transporting ability of about half the level reached on continuous exposure (Maddrell and Phillips, 1978).

(b) *Rectal activity*

The regulatory activities of the rectum have been investigated in several saline-water species by

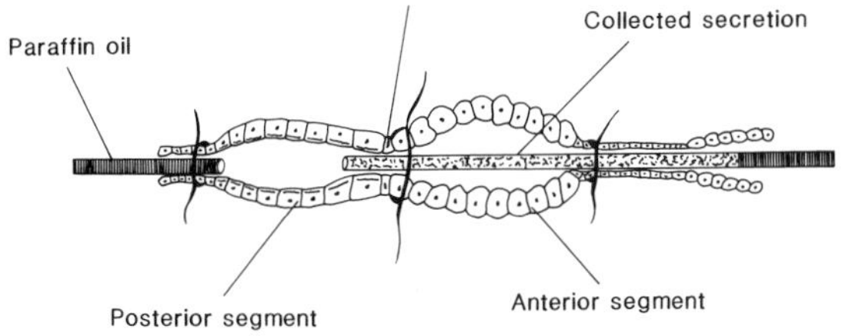

Figure 6.9 One of the arrangements of ligatures and pipettes used to remove fluid from one compartment of the rectum of an *Aedes dorsalis* larva. (After Strange *et al.*, 1984.)

dissecting out and ligating that organ, bathing it with artificial haemolymph and analysing the rectal contents (Bradley and Phillips, 1975, 1977a, b; Strange *et al.*, 1982). In more elaborate experiments, recta were bathed with artificial haemolymph, perfused with artificial tubular fluid, ligated, and cannulated with perfusion and collection pipettes so that modified perfusion fluid could be collected from the anterior or the posterior rectum, or as 'rectal secretion' after passing through the whole rectum (Figure 6.9) (Strange and Phillips, 1984; Strange *et al.*, 1984). These experiments revealed the capacity of the rectum to secrete a hyperosmotic fluid containing high concentrations of Na^+, K^+, Mg^{2+}, Cl^- and HCO_3^-, which represent most of the ions commonly found at high concentration in natural saline waters. The one ion that the rectum seems unable to transport actively is sulphate, which must be cleared from the haemolymph by the Malpighian tubules.

Rectal fluid, initially derived from the Malpighian tubes, is modified as it passes through the anterior and posterior regions of the rectum, which in saline-water species are histologically distinct (Section 6.6.3). Insertion of a cannula through the anus and into the anterior rectum of *Ae togoi* larvae, which allowed rectal fluid to bypass the posterior rectum, led to an increase in the concentration of haemolymph sodium in larvae kept in sea water but not in larvae kept in fresh water. That suggested that the posterior rectum is involved in ion excretion when larvae are in a hypertonic environment. A strong reaction to chloride ions was demonstrated histochemically in the apical lamellae of the epithelial cells of the posterior rectum of sea-water adapted larvae, but not in fresh-water adapted larvae (Asakura, 1980, 1982b).

The concentrations of particular ions in the rectal secretion were affected by, although not directly proportional to, the concentrations of those ions in the larval-rearing medium. All ions could be secreted at higher concentrations than occurred in natural or artificial haemolymph. Rectal fluid from intact whole larvae of *Ae taeniorhynchus* which had been reared in sea water, collected 1 h after the anus had been blocked with tissue adhesive, had the following composition (mM): Na^+ 435, K^+ 192, Mg^{2+} 36 and Cl^- 488 (osmolality 920 mosm). Although 19 times higher in potassium than sea water, this fluid was otherwise not dissimilar from sea water in its ionic and osmotic characteristics. Sulphate ions were not secreted in high concentration by the rectum. Because the Malpighian tubules cannot secrete a hyperosmotic fluid, the location of the sulphate-excreting system in that tissue puts a relatively low upper limit on the concentration of sulphate in the external medium which the larvae can tolerate (Bradley and Phillips 1975, 1977a).

Rectal preparations from *Ae taeniorhynchus*, in which the anterior and posterior regions were completely separated by ligatures, showed no fluid accumulation by the anterior regions, but the posterior regions filled with fluid which was strongly hyperosmotic to the bathing fluid. The rates of Na^+ and Cl^- secretion increased with increasing concentration in the bathing medium. The transport mechanisms followed allosteric rather than Michaelis–Menten kinetics. Measurement of electrical potentials between rectal lumen and haemolymph gave transepithelial potentials of 13 mV (lumen negative) for the anterior rectum and 11 mV (lumen positive) for the posterior rectum, values which were in keeping with the different functions of resorption and secretion ascribed to the two regions. Sensitivity to KCN showed that the potentials were due to active transport. Comparison of concentration gradients and electrical potentials revealed that Na^+, K^+, Mg^{2+} and Cl^- were all actively transported by the posterior segment to the lumen side (Bradley and Phillips, 1977a, c). In reviewing the experimental results with *Ae taeniorhynchus*, Bradley (1985) concluded that the posterior rectum was active only when the larvae were reared in hyperosmotic media. Capacities to actively transport different ionic species were separately inducible, and induced active transport could persist for several hours *in vitro*.

Recta of *Ae dorsalis* larvae were ligated in such a way that fluid could be removed from an anterior compartment comprising the anterior rectum and one row of cells from the posterior rectum, or from a posterior compartment comprising the posterior rectum and one row of cells from the anterior rectum (Strange *et al.*, 1984). Of the volume of fluid secreted by the whole rectum, 25% was secreted by the anterior compartment and 75% by the posterior compartment. Water movement resulted passively from the active transport of Na^+ and Cl^-. Removal of Na^+ and Cl^- from the bathing medium reduced fluid secretion by 57–92%. Calculated rates of Na^+ and Cl^- secretion were, respectively, 3 and 5 times higher in the posterior compartment than in the anterior compartment. To what extent the inclusion of a small number of posterior region cells in the anterior compartment contributed to its secretion of Na^+ and Cl^- is not known. In life the rectum of *Ae dorsalis* empties only when it becomes filled with secretion and faeces, consequently fluid secreted by the posterior rectum comes into contact with the cells of the anterior rectum.

Ligated recta from *Ae dorsalis* larvae which had been acclimated to a medium containing 250 mM bicarbonate, and which were bathed in artificial haemolymph containing 18.5 mM-HCO_3^-, secreted a fluid containing 402 mM-HCO_3^-: and 41 mM-CO_3^{2+}. The calculated total 'CO_2' secretion rate was 16.7 nmol/h/rectum. Lumen to haemocoele concentration gradients were calculated to be 21:1 for bicarbonate and 241:1 for carbonate ions, generated by the rectal epithelium against a measured transepithelial potential of 25 mV (lumen negative). To generate the observed bicarbonate and carbonate gradients by passive mechanisms, transepithelial potentials of 69 and 76 mV (lumen positive) would be required (Strange *et al.*, 1982).

The use of microcannulated rectal preparations from *Ae dorsalis* showed that the anterior compartment was the principal site of bicarbonate and carbonate secretion (Strange and Phillips, 1984; Strange *et al.*, 1984). Net Cl^- absorption was equivalent to 'total CO_2' secretion, suggesting that 'CO_2' transport was mediated by a 1:1 exchange of luminal Cl^- for haemolymph 'CO_2'. Acetazolamide, an inhibitor of carbonic anhydrase, reduced the bicarbonate secretion of rectal preparations by 80%, which suggests that it may be largely carbon dioxide that passes from the haemolymph into the rectal epithelium, to be converted to bicarbonate within the cells. Strange and Phillips (1985), who used ion- and voltage-sensitive microelectrodes to investigate the cellular mechanisms of HCO_3^- secretion and Cl^- reabsorption, concluded that the basolateral membrane of the anterior rectal cells is the site of a Cl^-/HCO_3^- exchanger which mediates transepithelial HCO_3^- and Cl^- transport.

It is clear that in hyperosmotic media saline-water mosquito larvae switch to a different mode of osmoregulation, employing the rectum as a salt gland capable of producing a hyperosmotic fluid. The larvae do not conserve water, which is readily available, but drink it and excrete the surplus ions, a strategy also employed by marine fish and marine birds. The rectum of saline-water species contains cells of two ultrastructural types which are separated into the anterior and posterior regions, the ultrastructure of the anterior region resembling that of the whole rectum of strictly fresh-water species (Section 6.6.3).

The rectum functions somewhat differently in the two species that have been most intensively investigated. In *Ae taeniorhynchus* the anterior rectum is the site of salt *resorption* for larvae acclimated to both fresh and saline waters, and the posterior rectum undertakes the hyperosmotic *secretion* of salts required for osmoregulation in saline waters (Bradley and Phillips, 1977c). In *Ae dorsalis* both regions of the rectum normally *secrete* an NaCl-rich fluid, the composition of which is modified by *ion exchange* and *reabsorptive processes* which vary with the regulatory needs of the larva (Strange *et al.*, 1984). In both species the posterior rectum secretes a strongly hyperosmotic fluid. The differences described for the anterior rectum possibly reflect the different environments of the two species, i.e. high NaCl vs Na_2CO_3 waters, which require different regulatory mechanisms. Because HCO_3^- is secreted only by Cl^-/HCO_3^- exchange, some secretion of NaCl-rich fluid may be necessary in the anterior region of *dorsalis* in high HCO_3^- environments to provide sufficient luminal Cl^- to exchange for haemolymph HCO_3^-. Essentially, Cl^- is recycled to eliminate HCO_3^-.

(c) *Anal papillae*

The possibility that anal papillae can excrete chloride ions was suggested by Phillips and Meredith (1969). The anal papillae of some *Aedes campestris* larvae were destroyed by brief exposure to 1% AgCl. When normal and papilla-less larvae were transferred from saline isosmotic with their haemolymph to 430 mM-NaCl, the larvae with no papillae developed a considerably higher haemolymph chloride concentration than the controls.

6.6 ULTRASTRUCTURE OF THE ORGANS ASSOCIATED WITH ION AND WATER TRANSPORT

Epithelial cells that function in ion and water transport have a characteristic ultrastructure. The apical plasma membrane is microvillate or is folded into numerous tightly packed lamellae. The cytoplasmic surfaces of the microvilli and lamellae may be studded with stalked particles called portasomes. The basal plasma membrane is extensively invaginated producing irregular, interconnected, pockets which form a basal labyrinth. The lateral plasma membranes are lengthened and irregularly folded, and are connected to those of adjacent cells by septate junctions which encircle the cell near the apical end. Mitochondria are abundant, and tracheation is extensive (Berridge and Oschman, 1972). The mitochondria are frequently closely associated with the apical microvilli or lamellae or with the basal labyrinth. In the basal half of the cells the association of mitochondria, singly or in pairs, with plasma membranes has the form of 'mitochondrial–scalariform junction complexes'. Scalariform junctions are characterized by a constant 15–20 nm intermembrane gap in which faint cross striations, rather than septa, are observed and by their permeability to tracers (Noirot-Timothée and Noirot, 1980; Lane, 1982).

Epithelia with these characteristics are found

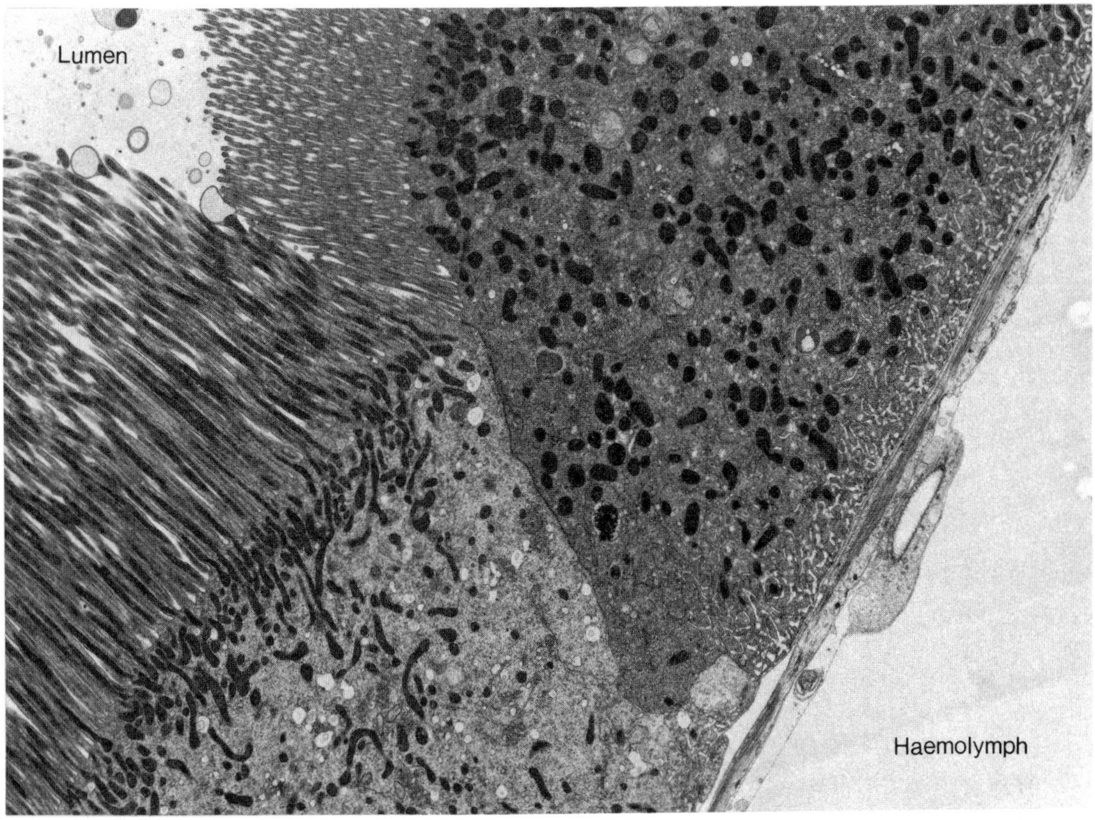

Figure 6.10 Section through a gastric caecum of a *Culex pipiens* larva, showing an ion-transporting cell (lower) and a resorbing/secreting cell (upper). × 4800. (Micrograph by courtesy of Professor W. Peters.)

in the gastric caeca, Malpighian tubules, rectum and anal papillae of mosquito larvae. The gross anatomy of all of these organs, except the anal papillae, is described in Section 5.1.

6.6.1 Gastric caeca

Two of the four cell types that are found in the gastric caeca, the resorbing/secreting cells and the ion-transporting cells, have an ultrastructure appropriate to ion and water transport (Section 5.1.2). The resorbing/secreting cells have a microvillate apical membrane, lateral membranes with gap and septate junctions, and an elaborate basal labyrinth with channels extending to the middle of the cell. Mitochondria are present immediately below the apical membrane but do not extend into the microvilli; other mitochondria are associated with the basal labyrinth. Rough and smooth ER and Golgi complexes are present in the cytoplasm (Figure 6.10). Reduction of the salt content of the larval medium, e.g. as from fresh water to distilled water, led to enlargement of the basal labyrinth and an increase in the number of mitochondria.

The ion-secreting cells have a microvillate apical membrane. Mitochondria are present in the apical cytoplasm and extend to the tips of the microvilli. The cytoplasmic side of the microvillus membrane is studded with 14 nm particles identified as portasomes, and has occasional crosslinks with the outer mitochondrial membrane. The cells have a well-developed basal labyrinth. The lateral cell membranes bear septate junctions. Smooth ER is more abundant than rough. Reduction of the salt content of the larval medium led to an increase in the length and diameter of the microvilli and an increase in the number of mitochondria within the microvilli (Volkmann and Peters, 1989a, b). The ion-transporting cells of a number of fresh-water species had a marked capacity to accumulate chloride ions in the region of the microvilli. Rearing larvae in dilute sea water led to an increase in the number of these cells (Volkmann and Peters, 1989b). In *Aedes togoi*, alkaline phosphatase activity is found at the apical border of (unidentified) epithelial cells of the gastric caeca, the activity being particularly high in larvae adapted to sea water. In sea-water adapted larvae these cells show a strongly positive reaction for chloride ions, consistent with a role in salt absorption (Asakura, 1978, 1982a).

6.6.2 Malpighian tubules

The primary cells are aligned along the tubule on opposite sides alternately, each occupying the greater part of the circumference of the tubule. Because the primary cells bulge into the tubule the tubule lumen is slit-shaped and runs a zig-zag course. The smaller stellate cells form thin-walled regions in the tubule. They are spaced apart and contribute less than 20% to the surface area of the tubule. The tubule is surrounded by a basal lamina (Christophers, 1960; Bradley *et al.*, 1982). The ultrastructure of the tubule cells is similar in larvae of the saline-water species *Aedes taeniorhynchus* (Bradley *et al.*, 1982) and the brackish-water species *Culiseta inornata* (Garrett and Bradley, 1984b), and in these species ultrastructure does not vary with change of salinity of the medium. For the fresh-water species *Aedes aegypti*, tubule ultrastructure has been described only for the adults, but resembles that of the saline- and brackish-water larvae (Mathew and Rai, 1976a; Bradley *et al.*, 1990).

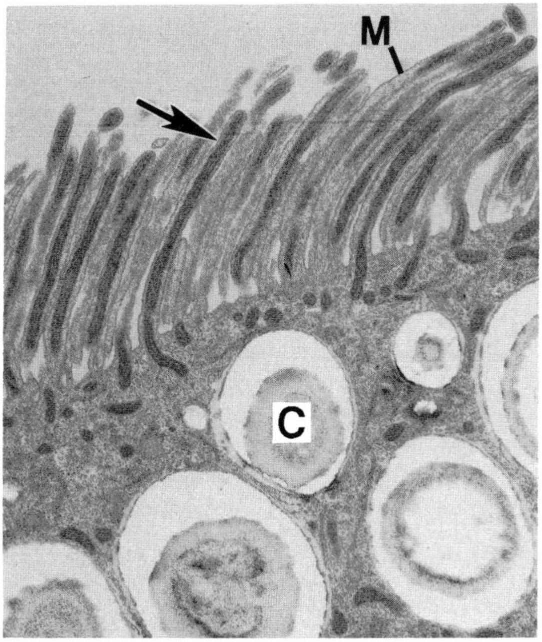

Figure 6.11 Apical region of a primary cell in a Malpighian tubule of a 4th instar larva of *Aedes taeniorhynchus* reared in sea water. C, concretion body inside a membrane-limited vacuole. M, Microvillus. Arrow, mitochondrion. × 9240. (From Bradley and Snyder, 1989.)

The primary cells have a spherical nucleus situated towards the apical end. In older larvae their cytoplasm is densely packed with mineralized membrane-bound concretion bodies (see Section 13.3). The apical membrane is extended into microvilli of two types. The most common of these are 3–4 μm long and contain mitochondria along their entire length (Figure 6.11). Microfilaments which run longitudinally between the mitochondrion and microvillus wall are believed be responsible for mitochondrial movement. Interspersed among them are shorter microvilli which lack mitochondria but which contain a core of microfilaments and an extension of the ER. The inner surface of the cell membrane of both types of microvillus is coated with knobs, or portasomes, about 25 nm long and 5 nm wide, which appear to be in contact with the mitochondria. The microvilli also contain longitudinal microfilaments from which crossbridges extend to the outer mitochondrial membrane. The basal cell membrane is extended into convoluted infoldings which penetrate about 1 μm into the cell as a basal labyrinth. Mitochondria and plasma membranes are associated in mitochondrial–scalariform junction complexes, both in the basal labyrinth and at the junctions of columnar cells. Free and bound ribosomes occur throughout the cytoplasm (Bradley *et al.*, 1982; Bradley, 1990).

The distance from the outside of the tubule to the lumen is about 30 μm at the interfaces between primary cells, and about 5 μm between primary and stellate cells. Where two primary cells abut, the basal 20 μm of lateral cell membrane is indistinguishable from the basal cell membrane, so these cells possess a basolateral membrane. The remaining apical region of the cell junction consists of a continuous or smooth septate junction, the septa of which form girdles around the cell peripheries, separating the microvillate region from that with basal infolds. Gap junctions occur just basad of the smooth septate junction. Where stellate and primary cells abut smooth septate junction extends over the apical two-thirds of the interface (Bradley *et al.*, 1982).

At low magnification the stellate cells appear as a thin band of cytoplasm separating the haemolymph and tubule lumen. The microvilli on the apical membrane are about 0.6 μm long; they have a core of microfilaments but lack the mitochondria and portasomes that characterize the microvilli of the primary cells. The basal membrane is extended into infoldings, c. 1.3 μm long, which almost reach the apical membrane. Mitochondria are positioned near the basal membrane infolds.

6.6.3 Rectum

We will first consider the rectum of *Aedes campestris* larvae which, are found in both saline- and fresh-water habitats. The rectum is a large elongate sac, with a thick cuticle-lined epithelium. It consists of ultrastructurally distinct anterior and posterior regions separated by a short

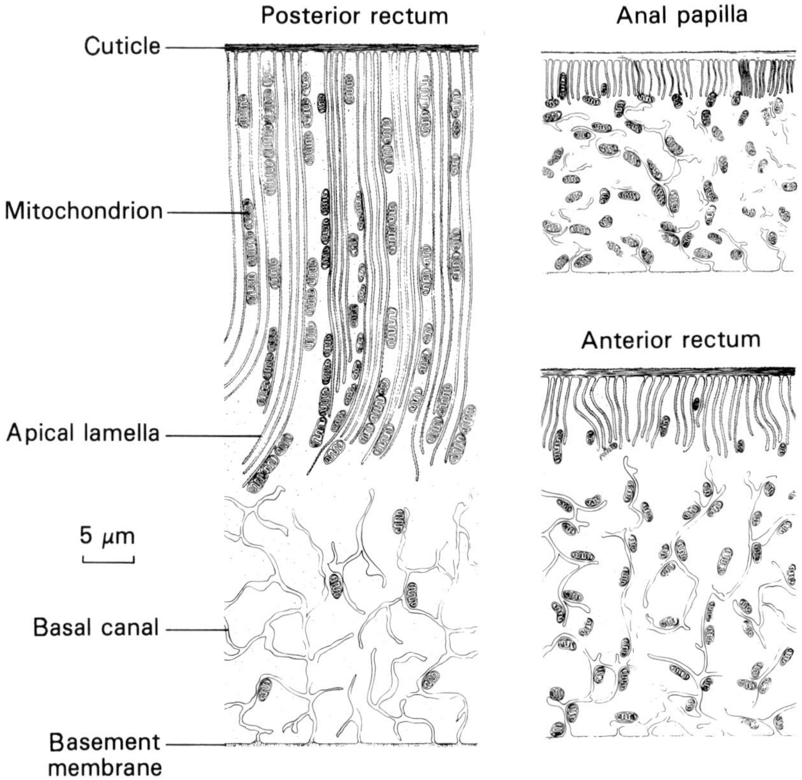

Figure 6.12 The ultrastructure of epithelial cells from the anterior rectum, posterior rectum and anal papillae of *Aedes campestris* larvae. Diagrammatic; drawn from the electron micrographs of Meredith and Phillips (1973a, c).

ring, 2–3 cells wide, of unspecialized cells. The epithelial cells of the anterior rectum are about 30 μm deep. The apical plasma membrane is thrown into a series of regular folds, or lamellae, which extend inwards for about 20% of the cell depth. The closeness of packing of the lamellae is variable. The cytoplasmic surface of each lamella is studded with portasomes of about 10 nm diam. The basal plasma membrane is highly infolded to form a basal labyrinth. The infoldings extend to the level of the apical lamellae, and the extracellular spaces which they form are dilated to a variable extent. The lateral plasma membranes are relatively unmodified but bear septate desmosomes apically and a few dilatations basally. Mitochondria are abundant and are evenly distributed throughout the cell (Figure 6.12).

The cells of the posterior rectum of *Ae campestris* are thicker, being about 60 μm deep. The apical plasma membrane is thrown into portasome-coated lamellae, which extend through about 60% of the depth of the cell. They are slightly thicker and more closely packed than in the anterior region. The basal labyrinth occupies a smaller portion of the cell than in the anterior rectum. The lateral plasma membranes are straight. Mitochondria are abundant, and about 90% are closely associated with the apical lamellae. The ultrastructure of the rectal cells is identical when *Ae campestris* larvae are reared in hyper- and hypo-osmotic media (Meredith and Phillips, 1973c).

In *Ae togoi*, a species of coastal rock pools which has been known to adapt to fresh water, the rectum is separated into anterior and posterior regions by a tubular region with an unmodified epithelium. In their ultrastructure the anterior (Figure 6.13) and posterior regions are very similar to those of *Ae campestris* (Asakura, 1982b).

In the brackish-water species *Culiseta inornata* (Garrett and Bradley, 1984b) and the fresh-water species *Aedes albopictus* (Asakura, 1970) and *Ae aegypti* (Meredith and Phillips, 1973c), the rectum is undivided and in ultrastructure resembles the anterior rectum of *Ae campestris*. In these three species the apical membrane is formed into lamellae which penetrate through about 20% of the cell. The lamellae, which are studded on the cytoplasmic surface with portasomes, are closely associated with mitochondria (Figure 6.13). The basal labyrinth extends to about 20% of the cell depth. Mitochondria are evenly distributed through the cytoplasm. In *Cs inornata* scalariform junctions are present on regions of canaliculi which are enclosed between pairs of mitochondria. Three types of intercellular junction were observed in *Cs inornata*: belt desmosomes, septate junctions, and gap junctions.

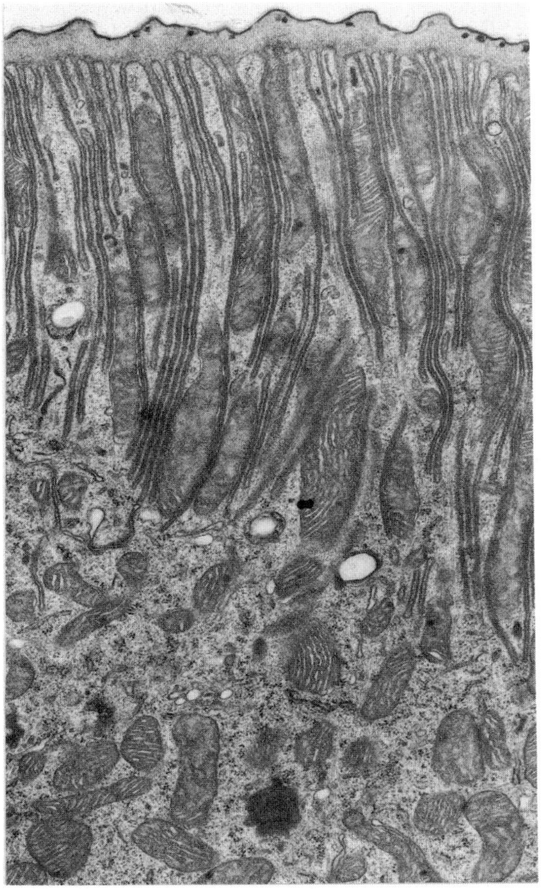

Figure 6.13 Section through the apical region of an epithelial cell in the anterior rectum of an *Aedes togoi* larva adapted to fresh water. The apical plasma membrane is deeply infolded, forming lamellae which are studded on their cytoplasmic surface with portasomes and closely associated with elongate mitochondria. × 16 000. (Micrograph by courtesy of Dr Koshu Asakura.)

6.6.4 Anal papillae

The four anal papillae are sac-like structures, consisting of epithelium covered by cuticle and situated on an extension of the terminal segment where they surround the anus (Figure 6.6). During embryonic development they arise from the proctodaeum. Muscle fibres which originate dorsally in the terminal segment are inserted inside the bases of the anal papillae. In *Ae aegypti* a narrow ring of circular muscle fibres is present at the base of each papilla, attached to the cuticle by hemidesmosomes. Experiments with injected dye indicated that these muscle fibres are able to isolate the papillar lumen from the rest of the haemocoele (Christophers, 1960; Edwards, 1983; Edwards and Harrison, 1983). Larvae of only exceedingly few species have just two anal papillae (Bradshaw and Lounibos, 1977).

In most species that breed in fresh water the anal papillae are relatively elongate, although the length can be affected by water quality. For example, mean papilla lengths for *Culex pipiens* larvae in distilled water, tap water and brackish water were 820, 360 and 200 μm respectively. Environmental effects on *Ae aegypti* were similar but less pronounced (Wigglesworth, 1938a). Fresh-water species with predatory habits have very short anal papillae. Species that are adapted for life in saline water also characteristically have short papillae; the papillae of *Ae togoi* larvae from tidal pools were 250 μm wide and 220 μm long (Meredith and Philips, 1973b).

The epithelium of the anal papillae is a syncytium, without lateral cell walls. It is thicker than the general body epithelium; in *Cx pipiens* and *Ae aegypti* the average thickness in the anal papillae was 6–7 μm, but in two saline-water species it was much thicker, amounting to 19 μm in *Ae campestris* and 60 μm in *Ae togoi*. Fine tracheoles permeate the epithelium. The apical surface of the epithelial cells is directed outwards, and the basal surface faces the papilla lumen. The cuticle that covers the apical surface consists of epi- and endocuticle. In *Ae aegypti* reared in 0 and 2% sea water the cuticle was 0.7–1.35 μm thick, in 20% sea water it was between 0.37 and 0.63 μm thick. The ultrastructure of the papilla epithelium is similar in fresh-water and saline-water species, although the thicker cells of the saline-water species show a corresponding increase in extent of the organelles. There is no evidence of neurosecretory axons within the anal papillae (Copeland, 1964; Sohal and Copeland, 1966; Mashiko and Asakura, 1968; Meredith and Phillips, 1973a; Edwards and Harrison, 1983).

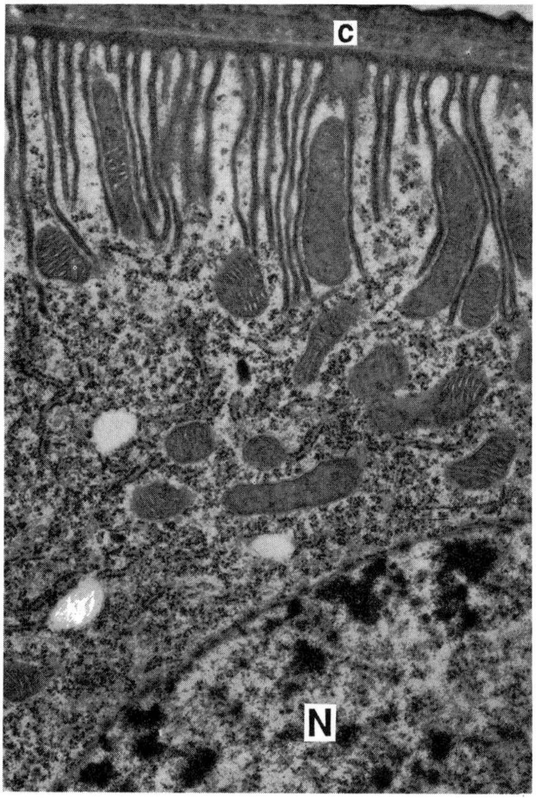

Figure 6.14 Section through an epithelial cell in an anal papilla of an *Aedes aegypti* larva reared in 0.039 M-NaCl. C, cuticle; N, nucleus. × 17 150. (From Sohal and Copeland (1966). © 1966, Pergamon Press plc. Reprinted with permission.)

The ultrastructure of the epithelial cells is similar in two fresh-water species, *Ae aegypti* (Figure 6.14) and *Cx quinquefasciatus*, a brackish-water species, *Cs inornata*, and a saline-water species, *Ae campestris*. The apical plasma membrane is folded into lamellae which extend through about 20% of the cell. Edwards and Harrison (1983), who examined *Ae aegypti*, were alone in considering that the lamellae divide apically into microvilli. Alkaline phosphatase is located in the lamellae and, in *Ae campestris* and *Cs inornata* at least, the cytoplasmic surface of each lamella is studded with portasomes. Mitochondria are much more abundant in the brackish- and saline-water species than in the fresh-water species, but in every case a population of mitochondria is closely associated with the inner tips of the lamellae. The cell surface facing the haemocoele is bounded by a basement membrane about 100 nm thick. The basal plasma membrane is extensively infolded to form interconnected canaliculi which extend about half-way across the cell. Some regions of canaliculi, usually out-pocketings, are enclosed between pairs of flattened and closely apposed mitochondria. These 'mitochondrial–scalariform junction complexes' are believed to serve in ion transport (Copeland, 1964; Sohal and Copeland, 1966; Meredith and Phillips, 1973a; Noirot-Timothée and Noirot, 1980; Edwards and Harrison, 1983; Garrett and Bradley, 1984b).

Exposure of *Ae aegypti* larvae to saline media caused ultrastructural changes in the epithelium. When larvae were reared in 0.115 M-NaCl the apical lamellae were shorter and more widely spaced. In 0.174 M-NaCl the lamellae were further reduced and the apical plasma membrane separated from the cuticle. Mitochondria occupied 30% of epithelial volume when larvae were kept in distilled water, but only 8% in larvae kept in 20% sea water. Among larvae reared in fresh water the canaliculi are relatively narrow and show few expansions; when larvae are reared in 20% sea water these spaces are enormously expanded, producing a series of large interconnected lacunae. In saline-water species the canaliculi contain many broad lacunae, and in the case of *Ae campestris* at least, these are present in larvae reared in both fresh and saline media (Sohal and Copeland, 1966; Meredith and Phillips, 1973a; Edwards, 1982a; Edwards and Harrison, 1983).

7

Growth and development

7.1 LARVAL GROWTH

Many mosquito species are *r* strategists, the larvae colonizing short-lived biotopes in which selection favours a rapid rate of population increase. Other species are *K* strategists, developing more slowly in relatively constant biotopes. In fact, all species are capable of responding to both *r* and *K* selection pressures to some extent, and a few, such as *Wyeomyia smithii* (Istock *et al.*, 1975), have high flexibility in that regard. Two reproductive parameters determine the rate of population increase: (a) the age at reproduction and (b) the extent of progeny production. Larval growth and development rates affect the former, and the conditions experienced during larval life can profoundly affect the latter. Clearly, the nature of larval growth processes is of fundamental importance in mosquito biology. The characteristics of larval growth and the effects upon it of extrinsic factors are considered in the following sections, but one important topic, larval diapause, is deferred to Volume 2.

7.1.1 Characteristics of larval growth

Mosquito larvae pass through four instars. Growth is continuous throughout each instar, as can be seen in the progressive increase in size of the body regions that are covered with thin cuticle. The heavily sclerotized parts of the body, the cranium and respiratory siphon, increase in size immediately after ecdysis but this is a manifestation of earlier growth processes. Growth of the soft parts is not always isometric; in *Anopheles sergentii*, for example, the thorax grows in length more than in breadth. The anal papillae of that species grow steadily during each instar and also increase in length abruptly after each ecdysis (Kettle, 1948).

The growth of the cranium conforms to Dyar's rule in some species, the width increasing in a regular geometric progression through the instars. Under conditions of constant temperature, growth conforms to the geometric progression

$$y = ak^n \tag{7.1}$$

where y is the final dimension, a the initial dimension, k Dyar's coefficient, and n the number of moults between the initial and final instars. The coefficient is approximately 1.5 in *Anopheles sergentii* (Kettle, 1948), *Culex territans* (De Oliveira and Durand, 1978) and *Cx quinquefasciatus* (Deslongchamps and Tourneur, 1980), and approximately 1.6 in *An quadrimaculatus* (Jones, 1953b) and *Mansonia titillans* (Nemjo and Slaff, 1984). Some species do not follow Dyar's rule. In *Aedes aegypti* (Abdel-Malek and Goulding, 1948; Christophers, 1960), *Ae atropalpus* (Bourassa, 1981) and *Cx tritaeniorhynchus* (Oka, 1955) the width of the cranium increases by a variable factor at successive moults.

Structurally, the mosquito larva is a combination of functional larval organs and incipient or slowly developing adult organs. Different larval organs grow in different ways. As described in detail in Section 7.2, some organs, such as the nervous system and the fat body, grow by cell multiplication with only slight increase in cell size, whereas other organs including the salivary glands, Malpighian tubes, anterior intestine, rectum, anal papillae and stemmata are incapable of cell division during the larval stage and grow entirely by increase in cell size. Most of the organs that grow by increase in cell size are histolysed at metamorphosis, but the Malpighian tubes, foregut and anterior intestine survive. The imaginal disks, which will form certain adult organs, develop slowly with cell division throughout the early larval instars and develop at an increased rate during the 4th instar.

The extent of larval growth is a secondary sexual character. Adult females are usually heavier than males. During the larval stage in *Ae aegypti* the mass of males increases approximately 40-fold, that of females >50-fold (Christophers, 1960). In some species the difference is not very great but in *Culiseta inornata* the mean dry weight of newly emerged females is more than twice that of males. Larvae of female genotype gain more dry weight per unit time than larvae of male genotype. In *Ae vexans*, in which the mean time from hatching to larval–pupal ecdysis at 21°C was 242 h for males and 262 h for females, the growth curves indicated that if the development time of the males were prolonged to that of the females their dry weight would still be only 89% of that of the females. In *Ae nigromaculis*, although the males and females of a brood underwent larval–pupal ecdysis at the same time the dry weight of males was only 86% of that of females (Brust, 1967).

Holometabolous insects must attain a certain mass during the larval stage if they are to commence metamorphosis and pupate. Larvae that have reached this 'critical mass' can continue to develop even if subsequently they are starved. The mechanisms by which critical mass is determined and pupation initiated are not known; weight *per se* is not necessarily the key factor. Critical masses for *Ae aegypti* larvae reared at two temperatures are detailed in Table 7.1, as are the 'potential masses', i.e. the weights at which 50% of larvae pupated when food was not limiting. At both temperatures the critical masses for male larvae were significantly lower than those for females. The results suggest that the decision to pupate occurs well before the larvae have attained their potential mass, i.e. early in the 4th instar (Chambers and Klowden, 1990).

In most species the durations of both the larval and the pupal stages are slightly shorter in males than in females, consistent with the difference in size, and males emerge before females in both wild populations and laboratory colonies. Certain investigators reported both the larval and pupal stages of *Ae aegypti* and *Cx quinquefasciatus* to be shorter in males than females (Haddow *et al.*, 1959; de Meillon *et al.*, 1967a). Others found no difference between the sexes in duration of the pupal stage in these species; however, the difference in duration of the larval stage still led to males emerging before females (Christophers, 1960; Khan and Ahmed, 1975). In *Ae nigromaculis*, *Cx nigripalpus* and *Tx rutilus septentrionalis* the males and females of a brood underwent both larval–pupal ecdysis and pupal–adult ecdysis synchronously (Brust, 1967; Nayar, 1968b; Trimble and Smith, 1978). In *Tx brevipalpis* males and females underwent larval–pupal ecdysis synchronously but females emerged shortly before males, the duration of the pupal stage being about 4% shorter in females than in males (Corbet and Ali, 1987).

Table 7.1 Critical and potential masses of *Aedes aegypti* larvae reared at two temperatures. The values for critical mass and potential mass are, respectively, the weights at which 50% of larvae pupated when food was limiting and when it was not limiting. (From Chambers and Klowden, 1990)

Temp. °C	*Critical mass* 50% (*mg*)	*(range)*	*Potential mass* (*mg*)
Females			
22°	2.63	(2.0–3.5)	4.80
32°	1.99	(1.5–3.0)	3.93
Males			
22°	1.91	(1.5–2.5)	2.94
32°	1.66	(1.0–2.5)	2.62

7.1.2 Effects of extrinsic factors on growth rates

The principal extrinsic factors that affect rates of growth and development are temperature, nutrition and larval density, but other factors such as depth of water and salinity can also be important. In nature these different factors interact in a complex way to determine developmental rates. The results of laboratory studies, in which a single factor is varied, are not always easy to interpret in ecological terms.

(a) Temperature

The relationship between temperature and rate of mosquito growth and development has been the subject of a number of investigations, and what has been discovered can be summarized in three statements.

(i) For any species, growth and development occur only within a temperature range that is defined by a lower developmental threshold and an upper lethal temperature.
(ii) Within most of this temperature range the rate of growth and development is positively correlated with temperature.
(iii) Temperature ranges vary with species.

The curve that relates the duration of the developmental period with temperature approximates to the form of a hyperbola (Figure 7.1), reflecting both the shortening of the developmental period as temperature rises and its lengthening as temperatures approach the upper lethal temperature. Over the central region of the curve the product of developmental period and temperature is constant, i.e.

$$t(T - c) = \mathrm{k} \qquad (7.2)$$

where t is the time from start of development to completion, T is the mean ambient temperature, c is the estimated developmental zero temperature, and k is the thermal constant.

Developmental velocity is defined as the reciprocal of the developmental duration (t), and is positively correlated with temperature. The curve

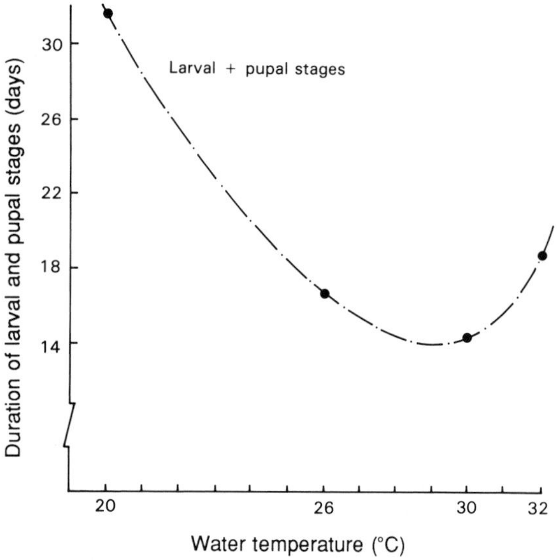

Figure 7.1 Duration of the developmental stages in *Toxorhynchites brevipalpis* as a function of temperature. The times from oviposition to emergence at four constant water temperatures. (After Trpis, 1972c. Replotted.)

relating developmental velocity with temperature is sigmoid, but over its central region it is effectively linear (Figure 7.2). Extrapolation from the linear region of the curve to the temperature axis indicates a theoretical developmental zero, which may be higher than the actual developmental threshold. Thus

$$\frac{1}{t} = \frac{T - c}{k} = \frac{1}{k}(T - c) \qquad (7.3)$$

Some early investigators developed mathematical expressions which modelled both the linear and non-linear regions of the sigmoid curve (Clements, 1963).

The times required for larval and pupal development by several *Aedes* species in forest and tundra pools in subarctic Canada were recorded by Haufe and Burgess (1956), who also determined the mean daily effective pool temperatures from measurements made at one inch depth every half hour, taking into account the movements of larvae in response to temperature gradients. The data gave a reasonable fit to the thermal summation formula (Equation (7.2)). For *Ae*

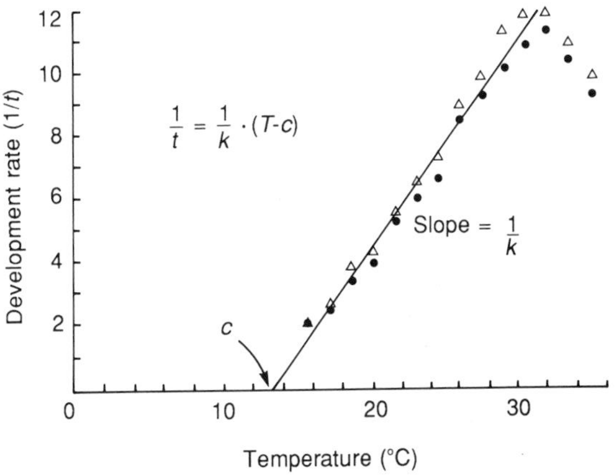

Figure 7.2 Rate of development of *Aedes aegypti* larvae as a function of temperature. The rate is calculated as the reciprocal of the time from hatching to that of the median of larval–pupal ecdysis, and is expressed as 1/development time × 10^3 h^{-1}. △, males. ●, females. (After Gilpin and McClelland, 1979.)

communis the estimated developmental zero was 38°F for males and 38.4°F for females, and the thermal constants were 271 degree-days for the males and 318 degree-days for the females (Figure 7.3). The thermal constants of other species in these forest and tundra pools ranged from 220 degree-days for *Ae hexodontus* females to 440 degree-days for *Ae excrucians* females. A close match was obtained between the number of degree-days required for development of each of the larval instars of *Ae implicatus* in the laboratory and in a natural pond when temperatures in the latter were measured at 0.5 inch depth (Stewart, 1974). An algorithm developed by Slater and Pritchard (1979) predicted with fair accuracy the development times of wild populations of *Ae vexans*, exposed to fluctuating temperatures, when mean ambient temperatures were within the range 10–14°C.

Substantial differences in rate of larval development may be found between species when reared at appropriate temperatures. In *Ae aegypti* some growth takes place at 10°C but development is not completed. Within its ecologically relevant temperature range of about 14 to 30° the rate of larval development is a function of temperature. At still higher temperatures the rate declines (Figure 7.2) (Bar-Zeev, 1958; Gilpin and McClelland, 1979). The larval stage was short in a colony of *Ae vittatus* derived from Zimbabwe, where the larvae breed in hot exposed rock pools in which water temperature cycles between 27 and 42°C. At the relatively low temperature of 31°C, the mean duration of the larval stage was 2.8 days in males and 3.0 days in females (McClelland and Green, 1970). In contrast, in the north Holarctic species *Ae flavescens*, the duration of the larval stage was 42 days at 15°C, 20 days at 20°C, the most favourable temperature, and 22 days at 25°C (Trpis and Shemanchuk, 1969). As would be expected, subarctic species are able to develop at relatively low temperatures. The lowest temperature to permit any development in *Ae impiger* is 1.1°C and in *Ae punctor* is 3.3°C; these species will develop fairly rapidly at 9°C (Haufe and Burgess, 1956). *Ae rempeli* develops during July in low arctic Canada in rock pools which have a mean water temperature of 3.7°C (Smith and Brust, 1970).

Ambient temperature can affect the various developmental stages differently. For example, the developmental rate of 4th instar larvae of *Tx brevipalpis* was reduced at 32°C, unlike that

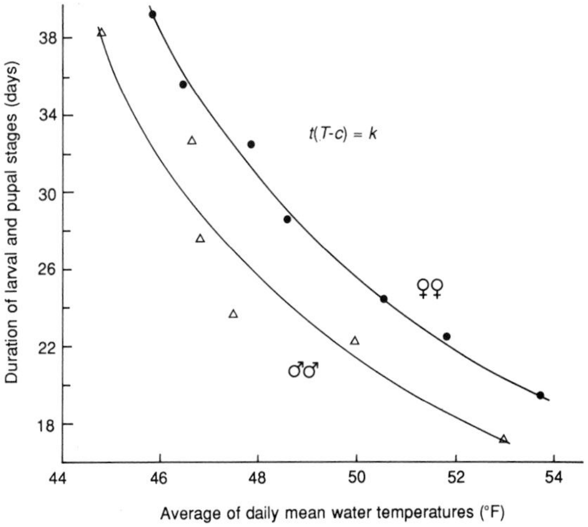

Figure 7.3 Duration of the larval and pupal stages of *Aedes communis* in nature at different temperatures, calculated as the average of the daily mean water temperatures. The curves correspond to the thermal summation formula, when k = 271 degree-days for males and 318 degree-days for females. (After Haufe and Burgess, 1956. Replotted.)

of the 1st and 2nd instar larvae (Trpis, 1972c). Similar observations have been made with *An quadrimaculatus* (Huffaker, 1944) and *Ae aegypti* (Bar-Zeev, 1958).

When the immature stages are subjected to fluctuating water temperatures, the rates of development may differ from those observed at the constant temperature which is at the mean of the fluctuating temperatures. Depending upon the duration of the alternating phases, development may be faster or slower than at the intermediate constant temperature (Headlee, 1942; Huffaker, 1944). Alternating temperature improved the survival of *Ae abserratus* in the laboratory (Brust and Kalpage, 1967) but did not affect the survival or development rates of *Cx tarsalis* in outdoor pools (Milby and Meyer, 1986).

The involvement of genetic factors becomes apparent when the growth rates of different strains are compared. For example, over a range of temperatures development was slightly but significantly more rapid in a strain of *Tx rutilus septentrionalis* from Louisiana than in a more northerly strain from Delaware (Trimble and Smith, 1978). At 16°C the larval stage of an anautogenous strain of *Cx pipiens* lasted only half as long as that of an autogenous strain, which formed slightly heavier pupae; at 25°C the difference between the strains was greatly narrowed but the autogenous strain then formed slightly lighter pupae (Lang, 1963).

(b) Nutrition

When the amount of available food is below a certain level larval growth rates are diminished. Depending on larval density, the critical factor may be the amount of food per larva or per unit volume of water.

Withholding food from 1st instar larvae of *Culex quinquefasciatus* for 12 h delayed the time of larval–pupal ecdysis by a similar period (de Meillon *et al.*, 1967a). Fasting for 12 h in every 24 had no effect on growth rate during the first three instars of *Aedes taeniorhynchus* but a minimum of 32 h feeding was necessary in the 4th instar, and fasting during the 4th instar markedly delayed pupation (Lum *et al.*, 1968). The rate of development of *Ae aegypti* larvae, cultured at low density, was positively correlated with the amount of food available (Moore and Whitacre, 1972). By under-feeding, the larval stage of *Ae aegypti* could be extended to over 4 months (Marcovitch, 1960) and that of *Toxorhynchites brevipalpis* to 5 months (Wigglesworth, 1929). Starvation prevents growth but does not necessarily prevent larval development since larvae of *Ae aegypti* and *Ae vexans*, which received no food during one instar, still underwent the next larval–larval moult (Brust, 1968a).

The capacity for rapid growth is postulated to have ecological significance. Average developmental periods for larvae of two ground-pool species, *Ae taeniorhynchus* and *Cx nigripalpus*, were 5–8 days at 27°C and under optimal conditions, whereas, at the same temperature, those for the container breeders *Wyeomyia vanduzeei* and *Wy medioalbipes* were 13–14 days, and for *Toxorhynchites rutilus* 17 days. When food was in short supply larvae of *Ae taeniorhynchus* and *Cx nigripalpus* were unable to prolong their development for much more than 21 days, but the species of *Wyeomyia* and *Toxorhynchites* could survive for many weeks or even months and without entering diapause. Frank and Curtis (1977) observed that the salt marsh habitat of *Ae taeniorhynchus* is liable to dry out. They suggested that the two ground-pool breeders can grow rapidly but, when food is scarce, have a limited ability to survive, whereas the container breeders grow more slowly and can tolerate a longer period of starvation. Under conditions of food shortage, regulation of numbers and biomass within populations of *Wy smithii* larvae was more by variation in larval growth and unequal sharing of food than by mortality; some 50–75% of larvae survived for two months under extreme nutritional stress (Istock *et al.*, 1975).

(c) Larval density

Field and laboratory observations have shown that intraspecific competion, caused by high larval density, results in longer development time, reduced pupation success and reduced pupal weight (Moore and Fisher, 1969; Siddiqui *et al.*, 1976; Reisen and Emory, 1977; Mori, 1979; Broadie and Bradshaw, 1991). A number of factors contribute to these effects. Depletion of food by resource competition affects growth rates and survival. Behavioural disturbance due to physical contact can reduce the growth rates of well-fed larvae (Dye, 1984). In some species but not all, high density leads to cannibalism among otherwise non-predatory larvae (Seifert and Barrera, 1981), and chemical inhibition by metabolites can affect growth rate and survival.

Reports of the production by mosquito larvae of so-called overcrowding or growth-retarding factors, which affect larval growth rates and cause larval mortality, have excited much interest. Their existence has been demonstrated most persuasively in *Culex quinquefasciatus*. The factors were produced by culturing 3rd and 4th instar larvae at a density of 5–7 ml^{-1} (20–27 cm^{-2} water surface), and they were bioassayed by rearing a new generation, from the 1st instar and at low density, in the filtered culture water of the earlier generation. The active constituents were ether soluble. Exposure of young larvae to one unit of active constituent, being the diethyl ether extract of the spent culture medium of one batch of 1500–2000 3rd and 4th instar larvae, prolonged the 1st instar by 4–5 days and the 2nd instar by a similar period. Analysis of the diethyl ether extract revealed growth-retardant activity, at a concentration of 2 ppm, in three hydrocarbon constituents: 3-methyloctadecane, 3-methylnonadecane and 9-methylnonadecane. These compounds could also be recovered from exuviae from the overcrowded larval cultures but not from the faeces, and it was postulated that they were derived from cuticular hydrocarbons (Ikeshoji and Mulla, 1970a, b, 1974). Whether the overcrowding factors function to regulate populations of *Cx quinquefasciatus*, as has been claimed (Ikeshoji, 1977), remains to be investigated by chemical analysis of natural breeding waters and measurements of the dynamics of wild populations.

Attempts by other investigators to demonstrate growth retarding factors by the bioassay method described above have produced variable results. Negative results were reported for *Cx quinquefasciatus* (Suleman, 1982), for an autogenous strain of *Cx pipiens* (Dadd and Kleinjan, 1974), and for *Cx tritaeniorhynchus* (Siddiqui *et al.*, 1976). Positive results were obtained with *Anopheles stephensi* (Reisen and Emory, 1977) and *Aedes sierrensis* (Broadie and Bradshaw, 1991). Weak growth-retardant activity has been reported from *Ae aegypti*. Filtered water from cultures with larval densities of 1–7 ml^{-1} lengthened the larval stage of bioassay larvae by factors of up to 1.27, although the activity levels showed no correlation with the original larval densities. At larval densities <1.2 ml^{-1} greater growth-retardant effects were produced when the food supply was not more than 0.1 mg yeast/larva/day (Moore and Whitacre, 1972). Dye (1984) also obtained a low level of growth retardation with *Ae aegypti*.

(d) Other factors

The larval stage of the fresh-water species *Cx nigripalpus* was prolonged by about one day when larvae were reared in 20% sea water (Nayar, 1968b). The rate of development of the salt-water species *Aedes taeniorhynchus* was the same in tap water and 25% sea water, but in 50 and 100% sea water the larval stage was prolonged by two days (Nayar, 1967b). *Aedes togoi* larvae developed more rapidly in water containing 10g sea salt l^{-1} than in water containing 0 or 20g sea salt l^{-1} (McGinnis and Brust, 1983). When *Ae sticticus* larvae were reared in water 80 mm deep the larval stage lasted nearly two days longer than when larvae were reared in a slanted pan with a range of depths from 0 to 80 mm. Possibly the shallow water permitted more frequent feeding at the bottom (Trpis and Horsfall, 1969).

7.1.3 Effects of extrinsic factors on size, composition and reproductive potential

The size of every mosquito species has a genetic basis, but the size attained by any individual may reflect environmental factors. Experiments with several species have shown that conditions experienced by the larvae affect the size and nutritional state of the teneral adults and may profoundly affect their reproductive potential. A number of factors are involved and their effects are additive. These factors also affect mortality rates, with implications for the net reproduction rate of the population, but only effects observable in surviving adults are considered here. Analyses of body weights (estimated from wing lengths) of females of ten species caught in light traps in New York State over a three-year period, revealed that in certain species the mean dry weight was less than half the maximum attainable. Three species that developed in temporary habitats such as treeholes, woodland pools and salt marshes, exhibited a higher dry weight variability than did a number of other species that developed in more permanent aquatic habitats such as swamps and lake margins (Fish, 1985).

(a) *Temperature*

The temperature to which mosquito larvae are exposed affects adult size, dry weight and ovariole

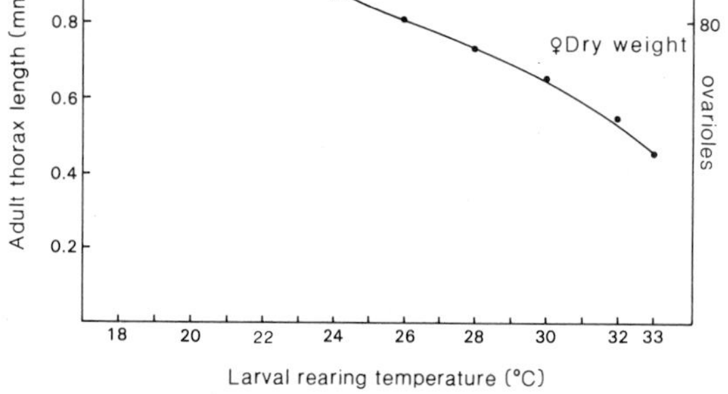

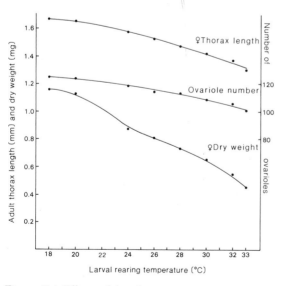

Figure 7.4 Effects of larval rearing temperature upon adult structure and dry weight in *Aedes aegypti*. Because body weight varies as the cube of the linear dimensions, the effect of rearing temperature upon dry weight appears greater than its effect upon thorax length. (From the data of van den Heuvel, 1963.)

number, all of which fall progressively as the water temperature rises. The dry weight of adult *Aedes aegypti* derived from larvae reared at 33°C was only 40% of that of adults from larvae reared at 18°C (Figure 7.4). When *Ae aegypti* larvae were held for different periods at 20°C and then transferred to 32°C, adult dry weight increased by 25% for each instar from the 2nd to the 4th for which larvae had been held at the lower temperature (van den Heuvel, 1963). The effect of rearing temperature on adult size was less marked in *Cx nigripalpus* and *Ae taeniorhynchus* which produced heaviest adults at an intermediate rearing temperature (Figure 7.5A) (Nayar, 1968c, 1969b).

The linear dimensions of adult head capsule, wing, femur, and thorax are affected in a rather complex way by larval rearing temperatures. The temperatures experienced by larvae in nature may account for the different bodily proportions of mosquitoes caught in the winter and summer, differences which can be sufficiently great to have taxonomic importance (Mer, 1937; Weyer, 1935; le Sueur and Sharp, 1991). Laboratory experiments on the effects of rearing temperature upon adult morphology have been described by van den Heuvel (1963), Nayar (1968c, 1969b), and Nayar and Sauerman (1970b). The larger females formed at lower breeding temperatures have a greater relative fitness. The relationship between maternal body size and fertility is considered in detail in Section 22.1.1, and the effects of seasonal variations in temperature upon wild populations are reported in Section 22.1.4.

Wing length has been used as an indicator of adult teneral dry weight. For *Ae triseriatus*, wing lengths could be converted to dry weight by the regression equation: dry weight = 0.009 wing length^3 − 0.017 (S.D. MCombs thesis; cited by Fish, 1985). However, a study of eight species revealed high variability in the relationship between wing length and dry weight, and indicated that separate regression formulae are required for each mosquito species and for different rearing conditions (Nasci, 1990).

Larvae of *Wyeomyia smithii* developed faster at a constant temperature of 25°C than at alternating temperatures of 18.5 and 31.5°C,

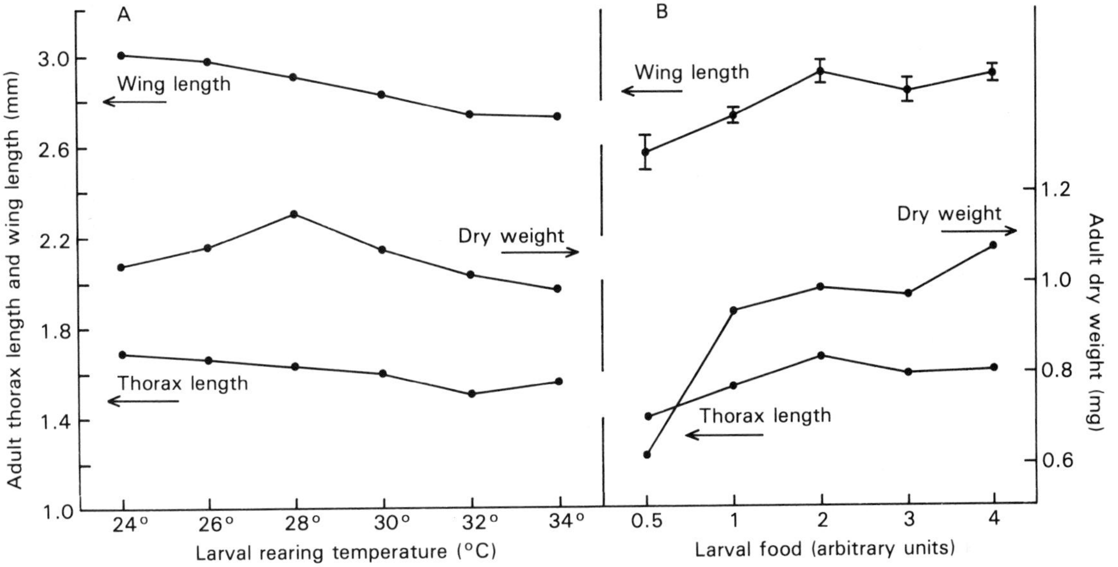

Figure 7.5 Effects of larval rearing conditions upon adult dimensions and weight in *Aedes taeniorhynchus*. (A) Effects on adult characteristics of the temperature at which well-nourished larvae were reared. (B) Effects on adult characteristics of quantity of food supplied to larvae reared at 27°C. (From the data of Nayar, 1969b.)

resulting in first reproduction at a younger age; however, fertility was far greater under the alternating conditions, and the population replacement rate was enhanced more than seven-fold (Bradshaw, 1980).

(b) Nutrition and larval density

When uncrowded batches of *Aedes taeniorhynchus* larvae were provided with five different amounts of food, increase in adult size was associated with increase in larval food up to a certain level of food supply, after which adult size stabilized (Figure 7.5B). However, adult dry weight continued to increase with the increase in larval food supply, and chemical analysis showed that lipids and glycogen accounted for all of the final increase in dry weight in the adult females and for the major part of this increase in the males (Nayar, 1969b).

Underfed *Culex nigripalpus* larvae developed into adults that were small, low in dry weight, and low in lipid and glycogen reserves. Lipid and glycogen form 20–30% of adult dry weight in well-nourished individuals of this species. Nayar (1968c) considered that most of these stores were deposited during the last part of the 4th instar. When larvae were starved for the last 20 h of the 4th instar, the adults they formed were substantially lighter than usual, contained a lower proportion of lipid, and survived for a shorter period in the absence of food. In *Cx tarsalis* the effects of nutrition and temperature can be supplementary, thus a developmental rate which is rapid due to high temperature may be further increased by higher food intake. However, poorly fed larvae kept at a high temperature may develop faster than well-nourished larvae kept at a lower temperature (Hagstrum and Workman, 1971).

When larval density is varied the effects of density and nutrition may be difficult to separate, even with food compensation. Increasing the density of *Anopheles stephensi* larvae between 0.1 and 3 cm^{-2} produced significant negative effects upon weight at emergence, size of blood meal and fertility (Reisen, 1975). Briegel (1990a, b) produced adult *Ae aegypti* of three size ranges by varying larval density and food supply. The metabolic reserves of these mosquitoes are described in Section 15.1.2 and observations on their relative fertility are reported in Section 22.2.1.

(c) Salinity

Rearing well-nourished larvae of the salt marsh mosquito *Aedes taeniorhynchus* in 50 or 100% sea water lengthened the larval stage and led to the formation of adults that were smaller, of lower weight and with a lower lipid content than adults formed from larvae reared in tap water or 25% sea water (Nayar, 1969b). The size of adult *Ae pseudoscutellaris* decreased when the larvae were reared in 30% sea water. Exposure of larvae to brackish water also modified the adult scale pattern (Marks, 1954).

7.2 CYTOGENETIC ASPECTS OF GROWTH AND DEVELOPMENT

During embryonic development the insect ovum divides into a large number of diploid cells, most of which have differentiated into specialized larval cells by the time of hatching. The postembryonic development of holometabolous insects involves growth in size of the individual and eventually metamorphosis with changes of cell type. These two processes of growth and metamorphosis may be accomplished in a variety of ways by different organs. In mosquito larvae, neurones in the central nervous system grow by cell multiplication with only a slight increase in cell size. Their nuclei show a normal interphase structure and are almost always diploid. Little histolysis of nervous tissue occurs in the pupa, and the adult nervous system develops by further cell multiplication. The fat body grows in the same way, persisting to the adult stage without change (Trager, 1937; Wooley, 1943). In contrast are tissues, mostly secretory, that grow entirely by increase in cell size. The large cells that result contain giant chromosomes, amplified by endoreduplication (see Section 1.1.2), which are able to service their synthetic requirements. However, the possession of highly specialized giant chromosomes causes problems at metamorphosis. A remarkable solution to this problem through endopolyploidy and somatic reduction has been described from some mosquito tissues and merits detailed description here. In some other tissues endopolyploidy and polyteny lead to cell death at metamorphosis, with rebuilding of the tissues by imaginal cells.

7.2.1 Endopolyploidy and somatic reduction

The epithelia of the foregut and of parts of the hindgut of mosquito larvae contain endopolyploid nuclei (Figure 7.6). During the larval stage their cells never divide but grow in size and develop giant chromosomes, although without becoming irreversibly differentiated. At metamorphosis a

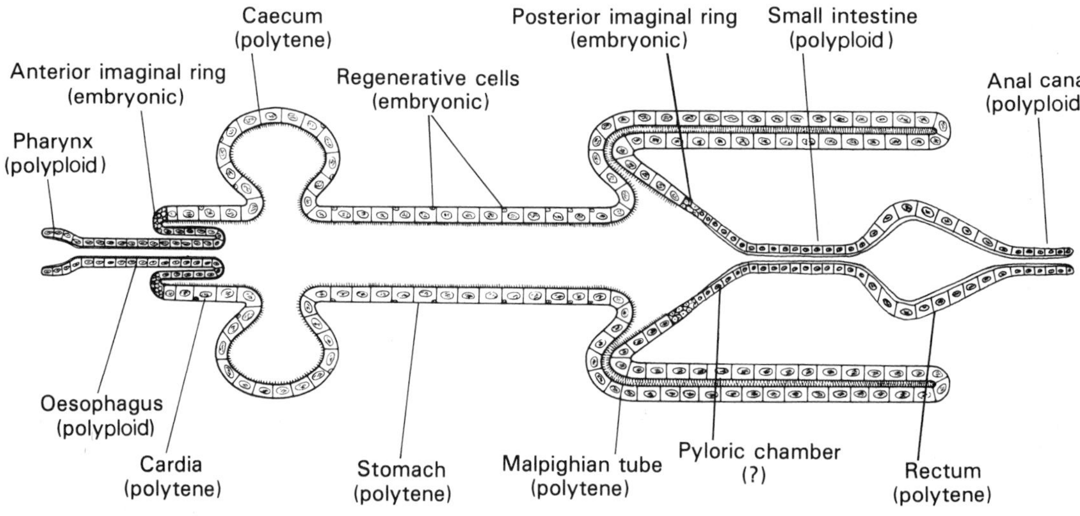

Figure 7.6 Chromosomal constitution of the epithelia of the alimentary canal of a 4th instar mosquito larva.

separation of the giant chromosomes into their constituent strands occurs and this, combined with a rapid series of cell divisions, yields a large number of small cells with normal chromosomes, a process called somatic reduction. Endopolyploidy and somatic reduction occur in the epithelia of the pharynx, oesophagus, anterior intestine and anal canal, and to a lesser extent in the epidermis, tracheae and neurilemma (Berger, 1938b; Risler, 1959, 1961).

Such endopolyploidy and somatic reduction have been described in greatest detail for epithelial cells of the foregut and hindgut of *Culex pipiens* and *Aedes aegypti* (Berger, 1938b; Grell, 1946a; Risler, 1961). During the larval stage the cells grow entirely by increase in size, and the number of cells at pupation is the same as in the 1st instar. However, the chromosomes replicate repeatedly, without formation of a spindle or any other sign of mitosis, in nuclei which have a typical reticulate interphase appearance. By the time of larval–pupal ecdysis the nuclei are all large although not of regular size. In the pupal stage these cells undergo a series of mitoses separated by interphases in which no replication of chromosomes occurs.

When the nuclei of the epithelial cells enter the prophase of the first mitotic division the chromosomes appear in the haploid, or rarely the diploid, number. During late prophase the relationally coiled homologous chromosomes fall apart, and subsequently each of the then diploid number of entities separates into many pairs of chromatids. It appears that when the chromosomes first become visible each contains a regular arrangement of sister chromatids relationally coiled in pairs, in pairs of pairs, in pairs of such pairs, and so on. In the metaphase of the first division the nuclei contain in *Cx pipiens* 48, 96, or rarely 192 chromosomes, and in *Ae aegypti* 12, 24, 48 or 96 chromosomes, the chromosomes being the same size as those in diploid nuclei ($2n = 6$). There is a direct correlation between chromosome number and nuclear size.

In early anaphase of the first divison the chromatids begin moving to the poles in small groups, and active somatic pairing is resumed as the chromatids move. Chromosomes visible in the first-division telophase nuclei are double stranded. The interphases that follow the first and subsequent mitotic divisions are all short and take place without chromosomal replication. The prophases of the later divisions differ in several ways from that of the first; for example, the chromosomes are much shorter and they do not unite to form bundles. A series of somatic reductions combined with cell divisions rapidly leads to the production of many small and presumably diploid cells, which reconstitute parts of the foregut and hindgut epithelia. This process has been observed in species of *Culex*, *Aedes*, *Orthopodomyia* and *Anopheles* (Berger, 1938a).

The cells of the epidermis are differentiated but appear to have more than one potential, because the same cells (or at least their daughters) can secrete the cuticles of larva, pupa and adult. In the epidermis of *Ae aegypti* periods of mitosis occur in each instar. During the 1st and 2nd instars the nuclei remain diploid but approximately double their volume. Such an increase in size is not found in the nuclei of the imaginal disks. During the 3rd instar some epidermal nuclei increase further in size, and these are found to be tetraploid. The number of tetraploid nuclei increases greatly at the beginning of the 4th instar. In the tergal regions of the abdomen the nuclei are almost all tetraploid, but none is tetraploid in the intersegmental regions. A long period of mitosis precedes larval–pupal ecdysis, and during this period somatic reduction leads to the formation of many small diploid cells. It is possible that de-differentiation occurs at this time. Only occasional mitoses occur in the pupal epidermis, and these are thought to be associated with the formation of hairs and scales (Risler, 1959).

The fat body cells of *Ae aegypti* are diploid during the developmental stages and in newly emerged adults, but during the first three days after eclosion the fat body cells of adult females become tetraploid or octaploid (Dittmann *et al.*, 1989).

7.2.2 Polyteny

Larval organs that contain nuclei with polytene chromosomes include the salivary glands, midgut, Malpighian tubules, rectum, abdominal muscles, trichogen cells, oenocytes and anal papillae (Figure 7.6). In every case the cells are incapable of division and, except for the Malpighian tubules, are incapable of transforming into adult organs. During larval development the cells grow very large and chromosome duplication occurs without separation of the sister chromatids, producing polytene chromosomes which may be seen in the haploid or diploid number (Trager, 1937; Berger, 1938b; Sutton, 1942; French *et al.*, 1962). At metamorphosis the Malpighian tubules alone among these organs are retained visibly unchanged in the adult, although with different physiological capabilities. All the others are autolysed and reformed by imaginal cells.

7.2.3 Imaginal cells

Most cells in mosquito larvae are differentiated and express the phenotypes of specialized larval tissues. However, a number of cells are determined, i.e. committed to a specialized course of development, but exhibit no phenotypic features. Such 'imaginal cells' play an important role in reconstruction of the insect at metamorphosis. They are characteristically small and diploid, and they may occur singly or in groups called histoblasts. Mesenchyme cells are undifferentiated cells of mesodermal origin which give rise to myoblasts, the immediate precursors of skeletal muscle.

During the 1st larval instar, division of midgut epithelial cells gives rise to columnar cells, which become polytene, and to much smaller imaginal cells, also called regenerative cells. In later instars the columnar cells do not divide but grow by increase in size, while the regenerative cells divide and by the 4th instar form an almost complete layer against the basement membrane. During the pupal stage they form a new midgut epithelium (O'Brien, 1966b).

A number of larval tissues that have polytene or endopolyploid nuclei are histolysed during the pupal stage and reformed from histoblasts. These include the epithelium of the oesophageal invagination which is reformed by the anterior imaginal ring, the pyloric epithelium which is reformed by the posterior imaginal ring, and the salivary glands. The adult abdominal muscles are formed by myoblasts (Sections 8.4, 8.5).

7.3 CUTICLE ULTRASTRUCTURE, DEPOSITION AND ECDYSIS

Moulting, with shedding and reformation of the cuticle, is essential for insect growth and development. Each moult can be considered as a sequence of events. The first is apolysis, the retraction of the epidermis from the cuticle. Then deposition of the new cuticle starts and simultaneously the inner region of the old cuticle is solubilized and resorbed. Once a substantial part of the new cuticle has been formed, ecdysis occurs – i.e. the shedding of what remains of the old cuticle. Finally the remainder of the new cuticle is secreted.

Opinions differ as to when the instars and stages of insect development begin and end. Wigglesworth (1973) considered ecdysis to be the most practical indicator of change of instar or stage, but Hinton (1973) insisted that retraction of the epidermis from the cuticle was a critical event and that instars and stages should be taken to start at the time of apolysis. He considered that during the period of a moult an insect is already in the new instar (or stage) although concealed by the cuticle of the previous instar (or stage), when it is described as pharate or hidden. Insect growth and metamorphosis are continuous rather than discontinuous processes, so definitions of the life cycle stages must be arbitrary. Hinton's (1973) definitions are used in this book.

7.3.1 Cuticle deposition and moulting

Little is known of the structure of mosquito larval cuticle other than that over the abdomen it consists of a two-layered corrugated epicuticle, a thin exocuticle and an endocuticle of some five lamellae. Pore canals are present (Cocke *et al.*, 1979). In contrast, detailed descriptions have been provided of the ultrastructure of pupal and adult abdominal cuticle and of the deposition of adult cuticle. The remainder of this section summarizes these observations.

(a) *Ultrastructure and deposition of abdominal sclerite*

This account is based on a study of *Culex pipiens* abdominal integument by Gnatzy (1970). The epidermal cells are columnar (Figure 7.7). Their apical plasma membrane bears short microvilli which vary in length at different stages of the secretory cycle and disappear between moults. The intercellular membranes are connected mainly by intermediate junctions but there are small areas of septate desmosomes and tight junctions. A basement membrane about 40 nm thick separates the basal plasma membrane from the haemocoele. The nucleus, which contains a large nucleolus, largely fills the basal to middle region of the cell. Rough ER and free ribosomes are distributed throughout the cell but are most abundant in the basal two-thirds. Golgi complexes are associated with elements of rough ER. Mitochondria are abundant in the apical third of the cell.

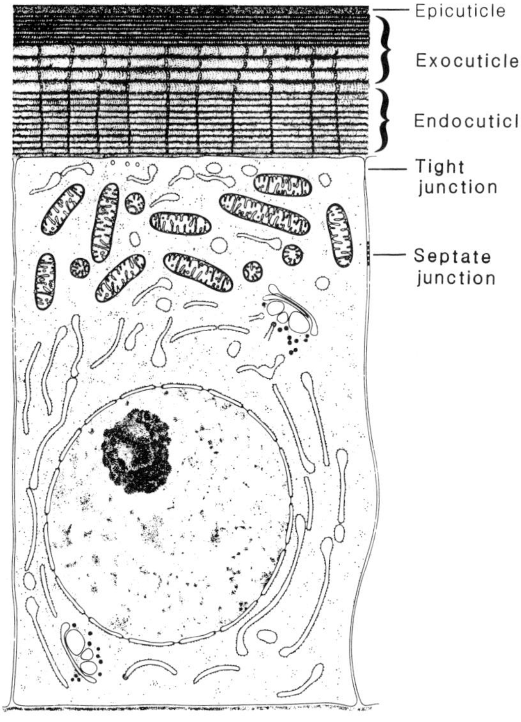

Figure 7.7 Diagram of a section through the abdominal integument of a pupa of *Culex pipiens* 12 h after larval–pupal ecdysis when the pupal cuticle is fully formed. (From Gnatzy, 1970.)

When fully formed the pupal cuticle is about 3 μm thick and consists of a thin epicuticle and a thick laminated procuticle. The epicuticle is two-layered, consisting of a distinct outer epicuticle 15–17 nm thick and an inner epicuticle 130 nm thick. It has no cement layer but there may be a very thin external extracuticular layer. The procuticle has a lamellated appearance, with thin electron-dense bands separating broader electron-lucent bands. The exocuticle, which is secreted before ecdysis, consists of visually distinct outer and inner regions. The outer region of exocuticle, which is about 600 nm thick, is formed of up to seven heavily sclerotized lamellae. The inner region, which is about 800 nm thick and formed of four lamellae, is only weakly sclerotized if at all. The endocuticle, which is deposited after ecdysis, is about 1.2 μm thick and formed of ten lamellae. Pore canals and narrower tubular filaments penetrate the procuticle.

At the time of larval–pupal ecdysis many 30 nm diameter strands are seen outside the outer epicuticle, apparently having emerged through pores in that layer. They may form a very thin extracuticular layer. The apical region of the epidermal cell contains vesicles, with fine floccular contents, which empty directly below pore canals. Secretion of endocuticular lamellae is observable immediately after larval–pupal ecdysis and, at 26°C, continues for 12 h. A space about 0.5 μm deep is present between the short microvilli of the epidermal cells and the most recently formed lamella. It contains granular material proximally and fibrous material distally, the latter becoming more or less parallel to the cuticle surface in the outermost region and assuming the typical microfibril pattern of the lamellae. Lamellae are first formed at intervals of approximately 60 min, and later at somewhat longer intervals, until ten have been laid down.

Three hours after larval–pupal ecdysis the epicuticle darkens. The outer seven lamellae of the procuticle which constitute the outer region of exocuticle darken shortly thereafter, and subsequently the epicuticle and outer exocuticle look like a single homogeneous layer. The colour of the pupal abdomen changes from white at ecdysis to grey after one hour and to dark brown after three hours. By 12 h after ecdysis the endocuticle is fully formed.

Fifteen hours after larval–pupal ecdysis apolysis starts, when the epidermis separates from the pupal cuticle forming an apolysial space in which a small amount of apolysial fluid accumulates. The inner electron-dense region of the lamella nearest the epidermis separates, as a so-called apolysial (or ecdysial) membrane, from the overlying material which exhibits a fringe of curved microfibrillar subunits. This process is repeated on the dissolution of each lamella. By 21 h after larval–pupal ecdysis only seven endocuticlar lamellae remain of the ten laid down after the previous ecdysis. By 30–36 h only five remain, and by 44 h all ten have disappeared and dissolution of the innermost lamella of the exocuticle has started. Thus more than one-third of the pupal cuticle is resorbed.

Formation of pharate adult cuticle starts during the later stages of dissolution of the pupal cuticle. Seventeen hours after larval–pupal ecdysis small 3-layered plates are deposited just above the apical membrane of the epidermal cells, which bears minute microvilli at this stage. The plates are 15–17 nm thick and consist of two electron-dense layers separated by a thicker electron-lucent layer. Addition of material at the edges of the plates leads to the formation four hours later of a continuous layer of outer epicuticle over the epidermal cells. The inner epicuticle is secreted subsequently, followed by the outer lamellae of the exocuticle. Thirty hours after larval–pupal ecdysis there is one lamella, by 36 h there are two, and by pupal–adult ecdysis at about 44 h there are four lamellae.

In adult *Aedes aegypti* the exocuticle consists of an amber outer region and an inner region which stains red with Mallory's triple stain. No blue-staining layer could be distinguished below this, indicating that there had been little or no deposition of endocuticle after emergence (Chevone and Richards, 1977).

(b) *Ultrastructure of arthrodial membrane*

Fifteen hours after larval–pupal ecdysis in *Aedes aegypti* the epidermis retracts from the pupal cuticle. Deposition of pharate adult arthrodial membrane starts within 3 h of apolysis with secretion of 3-layered epicuticular plates which later merge to form a continuous layer. By 16–18 h after apolysis secretion of procuticle has started with the appearance of loose aggregates of fibrillar material. This later forms a faintly fibrous to finely granular layer; pore canals are absent. The main region of abdominal arthrodial membrane is only 0.30 μm thick but over 'intermediate zones' next to the junctions with the tergites and sternites, where the integument is reflexed to permit the telescoping of segments, the ultrastructure is modified and the cuticle is 1.30–1.55 μm thick.

Within 1–3 h after emergence electron-translucent regions appear throughout the intersegmental cuticle, which increases in thickness from 0.3 to 3.0 μm, and loose aggregates of granular material form fibrous arrays. At this stage the cuticle starts to fold resulting in a multiple overlapping of the cuticle of the intersegmental membrane. By 12–15 h folding is essentially completed, and in the 24 h adult the intersegmental membrane is only one-sixth as long as at emergence. Immediately above the epidermis are one or two lamellae of newly secreted unfolded cuticle. The intersegmental membranes appear to remain in this condition throughout adult life. They did not unfold when the abdomen of a 24 h adult male was distended by injection of fluid or by feeding; rather the folded layers became straightened and thinned (Chevone and Richards, 1977).

(c) Oenocytes

It is generally considered that the large secretory cells found in insects, and known as oenocytes, synthesize lipids which are passed to the epidermis and then to the cuticle. In mosquito larvae large larval oenocytes, 50–60 μm diameter, occur in groups of 5 or 6 cells in pockets of parietal fat body on each side of abdominal segments 1–8. Many small imaginal oenocytes of 20 μm diameter occur on the inner surface of the ventral parietal fat body in segments 2–8 (Christophers, 1960). After larval–pupal ecdysis in *Cx pipiens* many lipid-containing vesicles appear in the large oenocytes, within tubules associated with the smooth endoplasmic reticulum. Six hours after ecdysis they occupy 5% of the cell volume and by 31 h they occupy 18%. Subsequently the vesicles disappear, simultaneously with the appearance of many cytolysosomes. No direct correlation of lipid secretion with lipid incorporation in the pharate adult cuticle could be made (Gnatzy, 1970).

7.3.2 Biochemistry of sclerotization

The biosynthetic pathways that are considered to be important in the sclerotization of insect cuticle progress from tyrosine through dopa (dihydroxyphenylalanine) and dopamine to *N*-acetyldopamine, which is believed to be the diphenolic precursor of quinones and other reactive compounds that covalently bond to proteins crosslinking them (Figure 7.8) (Andersen, 1989). Our knowledge of the mechanisms of sclerotization in mosquitoes is limited, but the information that is available on the enzymes and intermediates is consistent with the pathways outlined in Figure 7.8.

The cuticle of mosquito larvae is largely unsclerotized. Radiotracer experiments with *Aedes aegypti* showed that the tyrosyl residues present in larval cuticle could be accounted for almost entirely as polypeptide tyrosine, with a minor portion satisfying the requirements for sclerotization of mouthparts and respiratory siphon and for local melanization. However, following pupation both the new stiffened sclerites and the flexible intersegmental membranes of the pupal cuticle were sclerotized (Zomer and Lipke, 1983b).

[^{14}C]Tyrosine fed to *Ae aegypti* larvae accumulated in body fluids initially, mostly as tyrosine and dopa, but later it accumulated in cuticle. As development progressed through the late larval and pharate pupal stages, incorporation of tyrosine into pupal cuticle increased at the expense of the tyrosine content of the body fluids. There was no such change in the distribution of lysine or histidine between these two compartments, suggesting that tyrosine derivatives are utilized as intermediates for sclerotization (Zomer and Lipke, 1980).

Zomer and Lipke (1983b) fed [*ring*-^{14}C]tyrosine to *Ae aegypti* larvae, harvested the pupal cuticles, extracted soluble proteins with 8M urea, and subjected the residue to partial acid hydrolysis followed by digestion with proteolytic enzymes. Two-thirds of the radioactivity was not solubilized by 8M urea, indicating its incorporation into sclerotized protein. Further treatment isolated benzenoid moieties in which the aromatic components were covalently attached to oligopeptides. The three major classes of aromatic substituents were phenols, aryl ethers, and non-phenolic nitrogen heterocycles. Evidence that the putative crosslinks did not originate from pigment was obtained by exposing larvae to an inhibitor of melanization, following which there was no significant reduction in the formation of radioactive arylated peptides (Zomer and Lipke, 1981, 1983a, b).

In a similar investigation Sugumaran and Semensi (1987) administered [U-^{14}C]tyrosine to 4th instar larvae of *Ae aegypti* and analysed hydrolysates of the pupal and adult cuticles. Some 38% of radioactivity was recovered as tyrosine, and a further 36% in aryl-amino acid adducts. Among these adducts were found two types of crosslinking, namely covalent bonds between the nitrogen atom of the amino groups and either (a) a carbon atom of the aromatic ring (quinone tanning) or (b) the β-atom of the alkyl chain (β-sclerotization). The ratio of quinone tanning to β-sclerotization was 2:3. Other species showed different ratios.

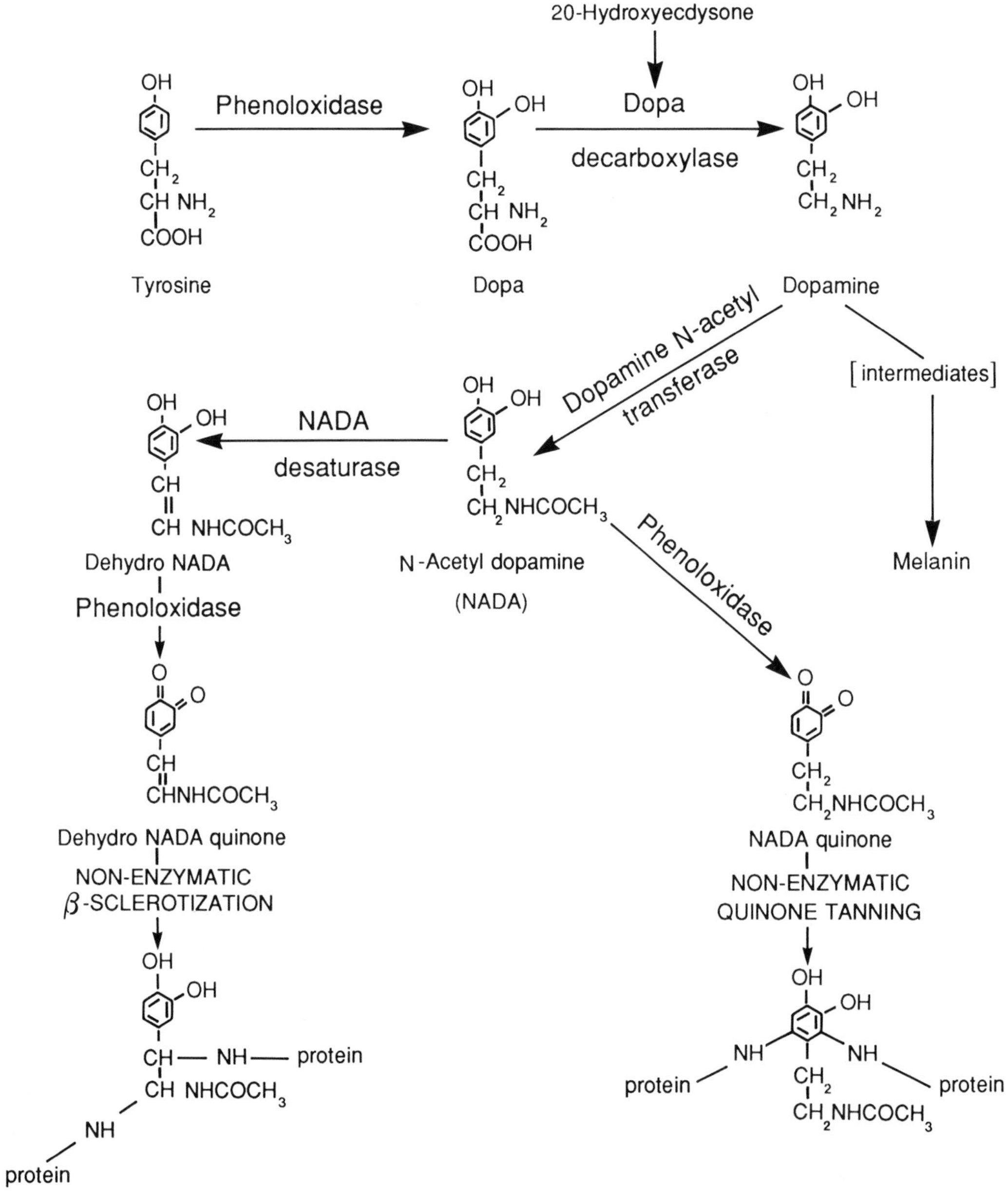

Figure 7.8 Biosynthetic pathways leading to sclerotization of cuticular proteins. Observations reported in the text provide evidence that these pathways, which have been investigated in detail in certain other insects, probably obtain in sclerotization of cuticular proteins in mosquitoes.

When tyrosine and phenylalanine were not present in the diet of *Ae aegypti* larvae in sufficient quantities, the usually sclerotized and darkened regions – cranium, saddle and respiratory siphon – were unpigmented, possibly through absence of melanin. Adults that metamorphosed from unpigmented larvae were normally pigmented (Golberg and De Meillon, 1948).

Dopa decarboxylase, more correctly called aromatic-L-amino-acid decarboxylase, has been isolated from mosquito larvae, pupae and adults and has been characterized (Schlaeger and Fuchs, 1974b; Kang *et al.*, 1980; Shampengtong *et al.*, 1988). Dopa decarboxylase activity peaked at the times of all moults. Evidence from studies with developing ovaries (Section 21.5.3) indicate that its synthesis in that organ is regulated by 20-hydroxyecdysone. Dopamine *N*-acetyltransferase, more correctly called arylamine acetyltransferase, has been found in the larvae, pupae and maturing ovaries of *Ae aegypti* (Smith *et al.*, 1989; Li and Nappi, 1992), and in the larvae, pupae and adults of *Ae togoi* (Shampengtong *et al.*; 1987a, b). It has been characterized by these authors.

7.3.3 The ecdyses

(a) Larval–larval ecdyses

At larval–larval moults the epidermis can be seen to separate from the cuticle and to secrete new cuticle and bristles beneath it while the tracheal epithelium similarly retracts from the tracheal linings and secretes new cuticular linings around them. An *Aedes aegypti* larva which is about to shed its cuticle rests at the surface and swallows water until the soft body wall becomes extremely taut. From time to time it flexes its head sharply until the cranium splits. The split follows the previously visible dorsal ecdysial line which runs from a median position on the collar around either side of the dorsum of the head to the bases of the antennae. A mid-ventral ecdysial line is also found in many species. Movements of the head free it from the old cranium and the dorsal split extends back over the thorax so that by peristaltic movements the larva is able to escape from the cast cuticle, sometimes in less than a minute. Meanwhile, the old tracheal trunks have broken across just behind their connections to the non-functional spiracles in each segment, and as the larva leaves its old cuticle the fractured pieces of the tracheal system are withdrawn, the fragments in each case passing out through the pair of spiracles immediately behind them. All spiracles except the posterior pair are then sealed until the next ecdysis. After ecdysis the tracheae fill with air and the expanded cranium hardens and slowly darkens. Air sometimes escapes into the new tracheal system as the cuticular linings are withdrawn, but in any case when the newly moulted larva hangs from the water surface, air enters through the spiracles and the fluid in the tracheal branches disappears. Feeding is interrupted for only a short period before and after each ecdysis, sometimes for less than an hour (Bekman, 1935; Wigglesworth, 1938b, 1981; Christophers, 1960).

Immediately after ecdysis the epidermal cells that secreted the lateral palatal brushes separate from the cuticle, form two sac-like invaginations, and begin deposition of the lateral palatal brushes of the following instar (Merritt and Craig, 1987).

(b) Larval–pupal ecdysis

Larval–pupal ecdysis is heralded by the appearance of the respiratory trumpets and palmate float setae of the pharate pupa as dark bodies beneath the larval cuticle. The pharate pupa comes to lie horizontally at the water surface. Shortly afterwards air appears between the larval and pharate pupal cuticles, originating from the tracheal system and leaving it via the 1st abdominal spiracles and possibly via the larval tracheal connections with the respiratory trumpets. The larval tracheae collapse at this time. Shortly afterwards the larval cuticle splits in the dorsal midline, and as the pupal thorax protrudes through the split the respiratory trumpets spring up and attach the pupa to the surface membrane. The larval cuticle and tracheal linings are shed much as at previous ecdyses, and during this period the labial and maxillary palp disks are extruded from their peripodial cavities (Wigglesworth, 1938b; Christophers, 1960; Romoser and Nasci, 1979).

After larval–pupal ecdysis the appendages of the head and thorax become cemented together, forming a shield over the front and sides of the pupa and retaining a volume of air which is outside the cuticle but enclosed by the appendages.

It is essential for the buoyancy of the pupa. The cuticle lining this ventral air space is hydrophobic, unlike the remainder of the pupal cuticle. In *Ae aegypti* the air space has a mean volume of 0.80 ± 0.034 mm^3 (Romoser and Nasci, 1979).

(c) *Eclosion*

The terms 'eclosion' and 'emergence' are equivalent, and are both used to designate the process of pupal–adult ecdysis. Some 5–6 h before pupal–adult ecdysis in *Aedes aegypti* the cibarial and pharyngeal pumps are active. Within an hour of the start of pumping the pupal cuticle starts to collapse on to the pharate adult cuticle, within 2–3 h gas starts to appear between the pupal and pharate adult cuticles, and less than an hour later gas appears in the stomach. The immediate source of the gas has not been established with certainty but it appears that fluid in the apolysial space is swallowed, drawing air from the tracheal system into the apolysial space, and that continued pumping draws air into the alimentary canal. Air withdrawn from the tracheae must be replaced by air entering through the respiratory trumpets. From the start of these activities the pharate adults become progressively more buoyant (Walker and Romoser, 1987).

At the start of eclosion the pharate adult slowly raises its abdomen to a horizontal position, and shortly afterwards the pupal cuticle splits along the mid-dorsal line of the thorax. In some species at least, a tuft of erect hairs on the thorax is the first part to emerge through the cuticle and possibly plays a part in splitting it. A minute later the thorax of the adult, which is quite dry, protrudes through the split and the body of the adult slowly rises into view. Pumping continues and the intake of air slowly distends the stomach, forcing the adult up and out of the pupal cuticle (Figure 7.9A) (Brumpt, 1941; Marshall and Staley, 1932; Walker and Romoser, 1987).

The longitudinal split in the pupal cuticle extends further back and transverse splits appear, while alternate movements of abdomen and legs disengage the body from the surrounding cuticle.

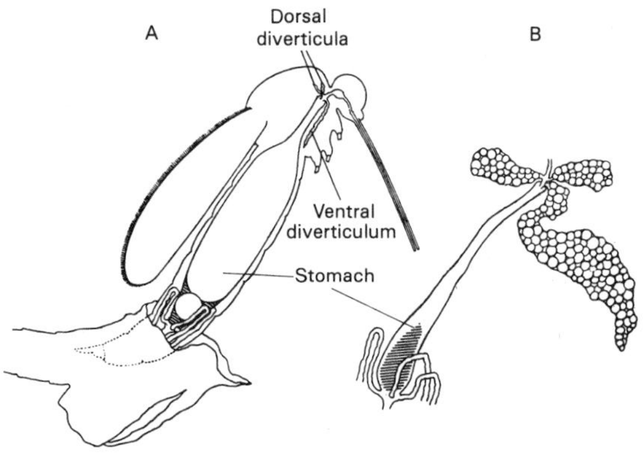

Figure 7.9 The distribution of air in the alimentary canal of a mosquito during and after emergence. (A) During emergence, the stomach is distended with air but the diverticula are empty. (B) 15 h after emergence, the diverticula are full of air and the stomach is empty. (From Marshall and Staley, 1932.)

The pupal tracheae are withdrawn through the spiracles of the emerging adult, and the linings of the fore- and hindgut are also shed. The various appendages are freed from their sheaths, in the case of the mouthparts and legs partly by muscular action, and the insect steps on to the water. The haemolymph pressure is high at this time due to the air-filled stomach; it causes the legs and wings to extend to their full size and presumably also maintains the rigidity of the legs until they harden. Adult wing length is slightly affected by the temperature prevailing during emergence, high temperature causing shorter wings and low temperature longer wings (van den Heuvel, 1963).

The process of eclosion takes about 15 minutes. When a mosquito has just emerged its abdomen is distended and the arthrodial membranes are bulging. However, diuresis and discharge of meconium start immediately (Sections 16.1.1, 16.4), and within a minute or two the abdomen starts to shrink. Ten minutes after leaving its pupal cuticle the mosquito is able to make short flights, but it usually remains for a time on the water. It can fly more or less normally after about an hour, when it flies off to another resting place. The gut diverticula, which were empty during emergence, become filled with air some time afterwards (Figure 7.9B). In *Ae aegypti* this occurs between 30 and 60 min after emergence, and results from the further swallowing of air. The stomach still contains air at that time but some four hours or more later it has disappeared, probably both by discharge through the anus and by solution in body fluids (Venard and Guptavanij, 1966; Gillett, 1983b, c).

7.4 ENDOGENOUS DEVELOPMENTAL RHYTHMS

Certain developmental events such as hatching or ecdysis may exhibit a diel periodicity, occurring during a particular phase of the 24-h day–night cycle. Some diel periodicities are believed to be controlled by physiological clocks or endogenous circadian rhythms. The test for an endogenous circadian rhythm is to show that an organism that has not previously experienced a regime with a 24-h periodicity can be induced to enter into a cycle with a period of approximately 24 h by a single change of conditions, e.g. a change from continuous light to continuous dark. The principal characteristics of circadian rhythms are (i) that they persist in the absence of any external forcing oscillation (time cue, *Zeitgeber*); (ii) that they persist with a period of approximately 24 h; and (iii) that the cycles have temperature coefficients (Q_{10}s) of about 1.0, i.e. are temperature compensated (Brady, 1974). Developmental periodicities are characteristically found in events that occur only once in the life of an individual so they are studied by observing cohorts, when the synchrony of the many individual events reveals the underlying rhythm.

7.4.1 Hatching and larval–larval ecdyses

Aedes taeniorhynchus eggs hatch within 15 min of immersion in deoxygenated water whatever light conditions they have been held under; therefore, their hatching is not controlled by any rhythm (Nayar, 1967b). In contrast, eggs of *Mansonia africana, Ma uniformis* and *Ma titillans*, when exposed to alternating light and dark, hatch only during the dark phase (Laurence, 1960; Nayar *et al.*, 1973).

At 27°C embryogenesis in *Ma titillans* was completed by 130 h after oviposition, and eggs normally hatched during a 6 h period about that time if it was then dark, otherwise they hatched early in the next dark phase. When batches of *Ma titillans* eggs were kept in continuous dark (DD), hatching was delayed and occurred between 148 and 160 h. In continuous light (LL) hatching occurred between 130 and 160 h. When eggs were held in alternating 12 h light : 12 h dark (LD 12:12) during the early part of embryogenesis and then transferred to constant dark or constant light (DD or LL), a single peak of hatching occurred around 140 h in DD and two peaks of hatching occurred around 140 and 160 h in LL. This suggested that there must be an underlying rhythm in *Ma titillans* synchronized by a change between light and dark, but whether by the change LL to DD or DD to LL was not established (Nayar *et al.*, 1973).

The timing of the larval–larval ecdyses in cohorts of *Ae taeniorhynchus* was not affected by the light regime, and these ecdyses could occur at any time of day or night (Nayar, 1967b).

7.4.2 Rhythm of larval–pupal ecdysis

If a batch of *Aedes taeniorhynchus* larvae that had hatched synchronously was fed a rich diet and reared in constant darkness (DD) and at constant temperature, the individuals eventually underwent larval–pupal ecdysis in a single rather prolonged peak. However, if the diet was restricted, larval–pupal ecdysis in DD was spread over several days and occurred in a series of peaks with a period of 21.5. h (Figure 7.10A). If undernourished larvae were given a single 4 h light pulse, whether immediately after hatching or early in the 4th instar, the rhythm became more distinct and showed a period of 22.3 h (Figure 7.10B). Thus the members of a cohort reared in darkness had an innate rhythm of larval–pupal ecdysis which could be synchronized and have its phase determined by a single light pulse. A daily pulse of 1 min duration was highly effective. Whether the stimulus was the change from L to D or from D to L was not established. Different constant temperatures did not affect the periodicity of the rhythm, its temperature coefficient (Q_{10}) over the range 22–32°C being 1.0. The rhythm of larval–pupal ecdysis could be entrained to temperature cycles. When larvae were reared in continuous darkness with 24 h temperature cycles, alternating 32°C:27°C 12:12, larval–pupal ecdysis occurred predominantly during the warm phase.

Under an LD 12:12 regime and conditions that gave optimum larval growth, the larval–pupal ecdyses occurred in two peaks with an apparent periodicity of 16 h (Figure 7.10C). However, if the larvae were stressed by a combination of low diet and high salinity the larval–pupal ecdyses were spread over several days, and the period of the rhythm became exactly 24 h (Figure 7.10D). Synchronization of larval–pupal ecdysis was observed among the members of a brood in nature, but the degree of synchrony was less than could be achieved under selected conditions in the laboratory (Nielsen and Haeger, 1954; Nayar, 1967a, b, 1968a; Provost and Lum, 1967; Lum *et al.*, 1968). The failure of the individuals reared under conditions of optimum growth to entrain to the LD 12:12 regime is unusual, but observations on *Anopheles gambiae* provide a possible explanation.

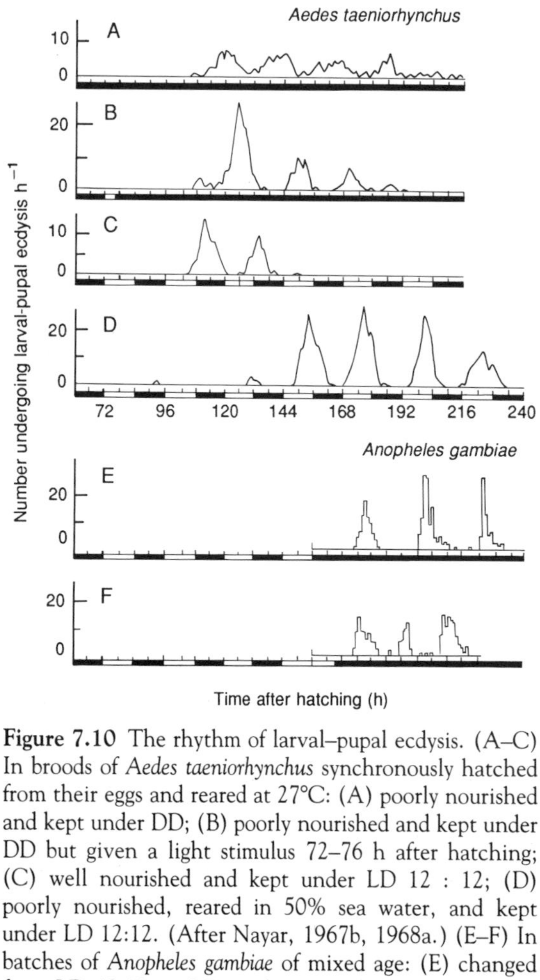

Figure 7.10 The rhythm of larval–pupal ecdysis. (A–C) In broods of *Aedes taeniorhynchus* synchronously hatched from their eggs and reared at 27°C: (A) poorly nourished and kept under DD; (B) poorly nourished and kept under DD but given a light stimulus 72–76 h after hatching; (C) well nourished and kept under LD 12 : 12; (D) poorly nourished, reared in 50% sea water, and kept under LD 12:12. (After Nayar, 1967b, 1968a.) (E–F) In batches of *Anopheles gambiae* of mixed age: (E) changed from LD 12:12 to DD at the normal time of light-off; (F) as (E) but with the final light period extended by 9 h before the change to DD. (After Jones and Reiter, 1975.)

Anopheles gambiae reared in batches under LD 12:12 showed a rhythm of larval–pupal ecdysis, ecdysis occurring in the last 6–9 h of the light phase and peaking 3–4 h before the end of the phase. Mixed batches of 3rd and 4th instar larvae reared under LD 12:12 and transferred to dim red light (DD) during the larval stage, with transfer at the normal time of light-off, exhibited a cycle of larval–pupal ecdysis with a period just under 24 h (Figure 7.10E). Prolonging the final light period by 9 h did not affect the timing of the first peak, but by about 60 h after the delayed light-off the rhythm was reset (Figure 7.10F). Between the unaffected first peak and the fully reset peak was a small peak, at about 27 h after the delayed light-off, which was postulated to consist of individuals that were sufficiently advanced to ecdyse towards the end of the first reset gate. These results suggested to Jones and Reiter (1975) that the *Zeitgeber* was the change L to D and that, if the clock was reset immediately after the period of extra light, the contribution of the clock to the timing of larval–pupal ecdysis must take place 24–30 h before ecdysis.

In *Aedes taeniorhynchus*, as in *An gambiae*, the time of ecdysis was determined not only by an endogenous gating mechanism but also by developmental processes occurring after gating had been effected. The failure of batches of synchronously hatched *Ae taeniorhychus* reared under optimal conditions and LD 12:12 to show a 24-h periodicity of larval–pupal ecdysis was probably due to the characteristics of the gating process. Well-nourished *Ae taeniorhynchus* larvae ecdysed in two peaks. The first peak may have consisted of individuals that were ready to ecdyse just before the gate closed; those in the second were possibly waiting for the gate to open. The medians of the peaks would therefore be separated by a shorter time than the period of the controlling gating process. Under stress conditions larval–pupal ecdysis was spread over a sufficient number of days, revealing the true underlying periodicity.

The time of day or night when larval–pupal ecdysis occurs varies among species in which it is believed to be gated by a circadian rhythm. When larvae were reared under LD 12:12 and at 27°C the mean time of larval–pupal ecdysis was before sunrise in *Anopheles bradleyi*, during the morning in *Aedes sollicitans, Psorophora confinnis, Culex bahamensis* and *Cx nigripalpus*, and during the afternoon in *Ae taeniorhynchus* and *An gambiae* (Nayar and Sauerman, 1970a; Jones and Reiter, 1975). Differences of photoperiod and temperature would alter these timings.

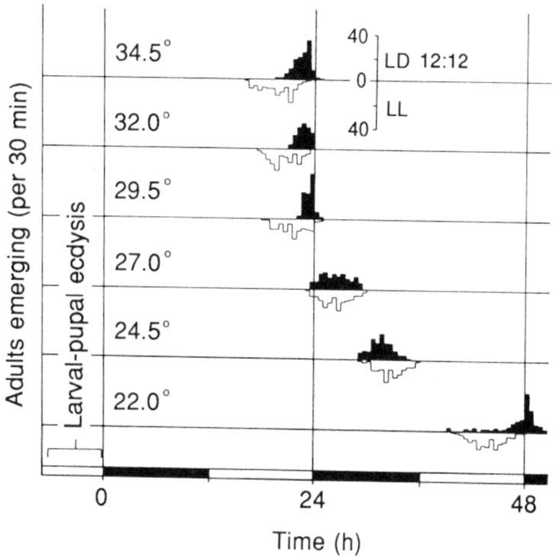

Figure 7.11 Comparison of the timing of emergence in *Anopheles gambiae*, that had undergone larval–pupal ecdysis during the last 6 h of a light phase, when either continuing in LD 12:12 (above) or transferred to LL (below.) Batches were reared at different temperatures in order to produce emergence at different times. (After Reiter and Jones, 1975.)

Gating of larval–pupal ecdysis by a circadian rhythm is not universal among mosquitoes. There is no periodicity of larval–pupal ecdysis in the *Aedes* species *aegypti, infirmatus, triseriatus* or *vexans*, nor in *Psorophora ferox, Culex salinarius, Culiseta melanura, Deinocerites cancer, Wyeomyia vanduzeei, Wy mitchelli, Toxorhynchites brevipalpis* or *Anopheles quadrimaculatus* (Haddow *et al.*, 1959; Nayar and Sauerman 1970a; Nayar *et al.*, 1978; McClelland and Green, 1970; Corbet and Ali, 1987).

7.4.3 Eclosion rhythm

Some species exhibit a diel periodicity of emergence or eclosion; others do not. At tropical latitudes *Aedes taeniorhynchus* tends to emerge during the day whereas *Culex quinquefasciatus* and *Cx nigripalpus* tend to emerge during the night (Nielsen and Haeger, 1954; de Meillon *et al.*, 1967c; Provost, 1969). *Toxorhynchites brevipalpis* emerges during the day although its larval–pupal ecdysis is arhythmic (Corbet and Ali, 1987). In *Ae aegypti* both larval–pupal ecdysis and emergence are arhythmic (Haddow *et al.*, 1959). In subarctic localities about midsummer, when nights are not fully dark, *Aedes* species such as *communis, hexodontus* and *impiger* emerge principally between 0900 and 1800 h, the time when air and water temperatures are highest (Corbet, 1966; Dahl, 1973). A causal relationship with temperature has not been established, but in southern Finland emergence of *Ae communis* was depressed when air temperatures were below 13°C (Brummer-Korvenkontio *et al.*, 1971).

A diel emergence cycle that is not determined by ambient temperature is more likely to reflect an earlier endogenous rhythm of larval–pupal ecdysis than an endogenous emergence rhythm. The developmental rate during the pupal stage is invariably temperature dependent, consequently in all species the pharate adults will be ready for emergence after a time interval determined by the ambient temperature. This is the case in *Ae taeniorhynchus* and *Cx nigripalpus* which can emerge at any hour of the day or night (Provost and Lum, 1967; Nayar, 1967b, 1968b; Nayar *et al.*, 1978). Nayar and Sauerman (1970a) considered that, in 15 other species they had investigated, the time of emergence again was determined only by the time of larval–pupal ecdysis and the temperature, and was not otherwise influenced by conditions of light and dark.

Anopheles gambiae exhibits an emergence cycle which is strongly influenced by the rhythm of larval–pupal ecdysis but which can be fine-tuned by an independent mechanism. This was demonstrated in mosquitoes caused to emerge at different times by exposure to different temperatures during the pupal stage. Mosquitoes undergoing larval–pupal ecdysis during the last 6 h of a light period were either transferred to LL or kept in LD 12:12, and were subsequently reared at temperatures between 22.0 and 34.5°C. Under both light conditions the time from larval–pupal ecdysis to emergence was temperature dependent. However, in the mosquitoes kept in LD 12:12 the time of emergence was delayed by up to 4–5 h relative to that in LL when the mosquitoes were ready for emergence during the light phase (Figure 7.11). The cue that effected the delay in emergence was shown to be the change from light to dark that had occurred about one day previously. When given to late 4th instar larvae it had no effect on the time of larval–pupal ecdysis but did affect the time of emergence (Reiter and Jones, 1975). Rather similar observations were made with *Cx nigripalpus*, which under an LD 12:12 regime emerged during the dark period. Transfer to DD or LL soon after larval–pupal ecdysis respectively advanced or retarded the time of emergence (Nayar *et al.*, 1978).

When batches of *Wyeomyia mitchelli* were reared under LD 12:12, larval–pupal ecdysis was arhythmic yet a distinct rhythm of pupal–adult ecdysis was apparent, emergence occurring mainly in the last 5–6 h of the light phase and the first 1–2 h of the dark phase. No emergence rhythm was observed when late 4th instar larvae and pharate pupae were transferred from LD 12:12 to DD or LL. However, batches that were exposed to LL or DD throughout larval life and to a single LD cycle during the pupal stage emerged rhythmically. When larvae had been reared under LD 12:12, exposure to one LD cycle shortly after larval–pupal ecdysis followed by transfer to DD was sufficient to evoke a weak emergence rhythm which persisted through the period of DD. Exposure of pupae to two, three or four LD cycles before transfer to DD increasingly consolidated the emergence rhythm. A batch that had been reared under LD 12:12 and exposed in the pupal stage to a single LD 6:12 cycle, followed by return to LD 12:12, exhibited phase shift followed by complete phase resetting (Nayar *et al.*, 1978).

There is thus among mosquitoes a variety of systems regulating emergence. There are species with no apparent periodicity, species in which the time of emergence is determined by that of larval–pupal ecdysis and the ambient temperature, species in which such entrainment can be fine-tuned by an independent mechanism during the pupal stage, and yet other species in which emergence is controlled by an endogenous rhythm.

8

Metamorphosis

As holometabolous insects mosquitoes undergo complete metamorphosis, developing through larval and pupal stages before reaching the adult state. The larvae are of simplified structure but are adapted for living and feeding in an aqueous environment. A massive increase in size occurs as the larvae develop and pass through four instars. Metamorphosis is first apparent externally after larval–pupal ecdysis but in reality is already in progress during the larval stage, notably in the early development of the imaginal disks.

8.1 THE IMAGINAL DISKS

Holometabolous larvae carry within themselves cells that are determined, i.e. developmentally committed, to form adult tissues but that remain undifferentiated while in a non-permissive hormonal milieu. Some of these are in loosely organized groups called histoblasts but many are assembled into discrete epithelial pouches known as imaginal disks. Imaginal disk cells proliferate in all larval instars, the rate increasing with age. Deposition of cuticle over the disks is an event which marks their transition from undifferentiated tissue to pupal and then adult structures. During metamorphosis the disks evert and unfold to assume the form of the adult structure (Oberlander, 1985).

The imaginal disks of the head and thorax have been described in detail from species of *Culex*, *Aedes* and *Anopheles*, in which they are very similar (Thompson, 1905; Imms, 1908; Prashad, 1918). The abdominal disks have been described in most detail from *Aedes stimulans* (Ronquillo and Horsfall, 1969; Horsfall and Ronquillo, 1970).

8.1.1 Disk development

In mosquitoes the rudiments of some imaginal disks are present in newly hatched larvae and develop slowly during the larval stage; others appear as late as the 3rd instar. Each imaginal disk is first apparent as a plate of columnar epidermal cells which grows slowly with cell division. It invaginates to form a cup, and later the base of the cup evaginates into the lumen forming a pouch (Figure 8.1). Mesenchyme cells – embryonic cells of mesodermal origin – enter the pouch which later develops into a rudimentary appendage lying in a peripodial cavity and surrounded by peripodial membrane, part of the invaginated epidermis. Much later a thin cuticle is secreted over the rudimentary appendage and over the peripodial membrane. In every case the peripodial cavity remains open to the surface. The mesenchyme cells that enter the disk give rise to the mesodermal tissues of the appendage. Thus the adult thoracic muscles develop from thoracic disks by the proliferation of cells which persist undifferentiated within the disks during the larval stage (Hulst, 1906; Imms, 1908; Prashad, 1918).

Increase in size of the imaginal disks occurs principally during the 4th instar, accompanied by numerous mitoses (Risler, 1959), and in most cases the disks are shaped only very approximately like the adult appendages at that time. The disks of the pupal paddles and labrum are formed directly by outgrowths of the epidermis and are not enclosed by peripodial membranes.

8.1.2 Developmental capacity

The developmental capacities of imaginal disks can be studied by transplanting them into larval,

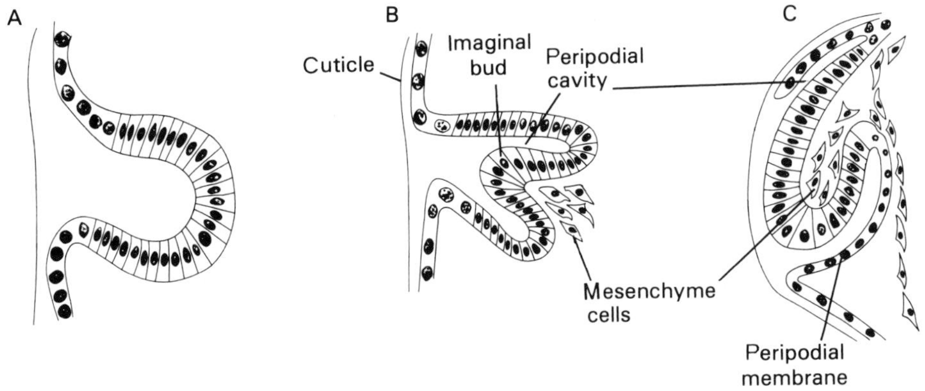

Figure 8.1 The early development of an imaginal disk as seen in transverse sections. Schematic. (A) Invagination of epidermal cells. (B) The evagination which will form an appendage. (C) Formation of the peripodial cavity and penetration of the disk by mesenchyme cells. (From Prashad, 1918.)

pupal or adult hosts and later examining them for phenotypic differentiation. For example, eye-antenna disks of *Drosophila melanogaster* transplanted from 53–64 h larvae into a larval host failed to produce adult structures when the host metamorphosed, whereas disks transplanted from older larvae did. In Diptera and Lepidoptera, disk growth in the early instars may be independent of the hormonal milieu but the increased rate of growth in the last instar is thought to be ecdysteroid dependent (Oberlander, 1985).

Foreleg disks transplanted from 4th instar *Culex pipiens* larvae into hosts of the same age differentiated into normal adult organs during the metamorphosis of the host. If the leg disk of a 4th instar larva was cut into fragments which were then transplanted separately into larval hosts, each fragment differentiated into a particular region of the adult leg, revealing that the cells of the fragment had undergone area-specific determination by the time of the transplantation. The sequence of primordia in the fragmented folded leg disk closely followed the sequence of structures in the adult leg (Figure 8.2) (Spinner, 1969).

Implantation of disks into 4th instar *Cx pipiens* larvae generally caused a prolongation of the instar, which made it possible to vary the period

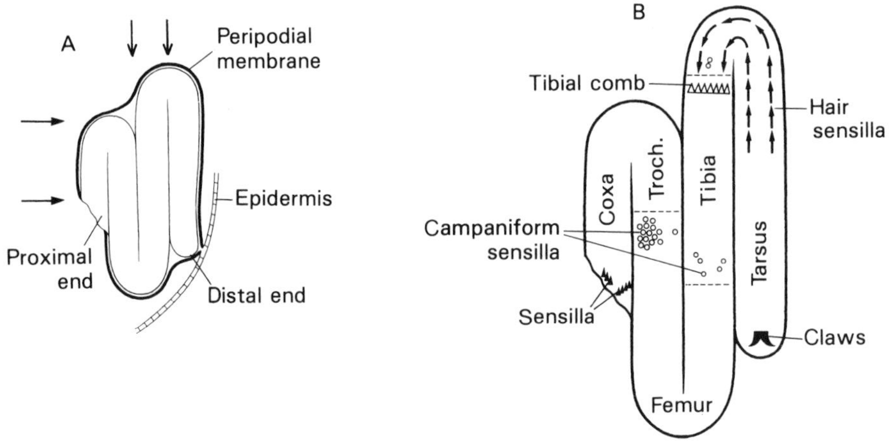

Figure 8.2 Determination of the parts of a foreleg disk in *Culex pipiens*. Schematic. (A) A foreleg disk before fragmentation. Arrows show the planes of cut. (B) Fate map of a foreleg disk. (From Spinner, 1969.)

of residence in the host. Both intact disks and disk fragments increased in size with longer implantation times, and after residence for ten days disk fragments sometimes differentiated additional structures that were appropriate to another part of the disk. Experiments showed that this represented a true regeneration of disk material. Because fragments showing regeneration always also showed an increase in size, Spinner (1969) considered that cell division had occurred and that this was a requisite for regeneration. The capacity to regenerate was restricted to the proximal part of the disk, thus fragments which contained the prospective material for coxa, trochanter and femur could, after metamorphosis, produce tibial and tarsal structures but not vice versa.

Leg disks and leg disk fragments transplanted from 4th instar *Cx pipiens* larvae into 3-day-old autogenous females, and recovered for examination after 22 days, showed no cell multiplication and no cuticle formation. However, when intact and halved disks were implanted into autogenous females within 13 h after pupal–adult ecdysis, a large percentage showed cuticle formation, and when the implants were derived from older 4th instar larvae, half of them differentiated adult structures (Spinner, 1969). Thoracic disks of late 4th instar *Ae aegypti* larvae transplanted into a larval host later showed complete differentiation. Disks transplanted into a pupal host showed only growth and cuticle formation, and disks transplanted into a sugarfed female remained almost unchanged (Bodenstein, 1945). It appears that the humoral environment in the host and the stage of development of the disk are both important factors in differentiation.

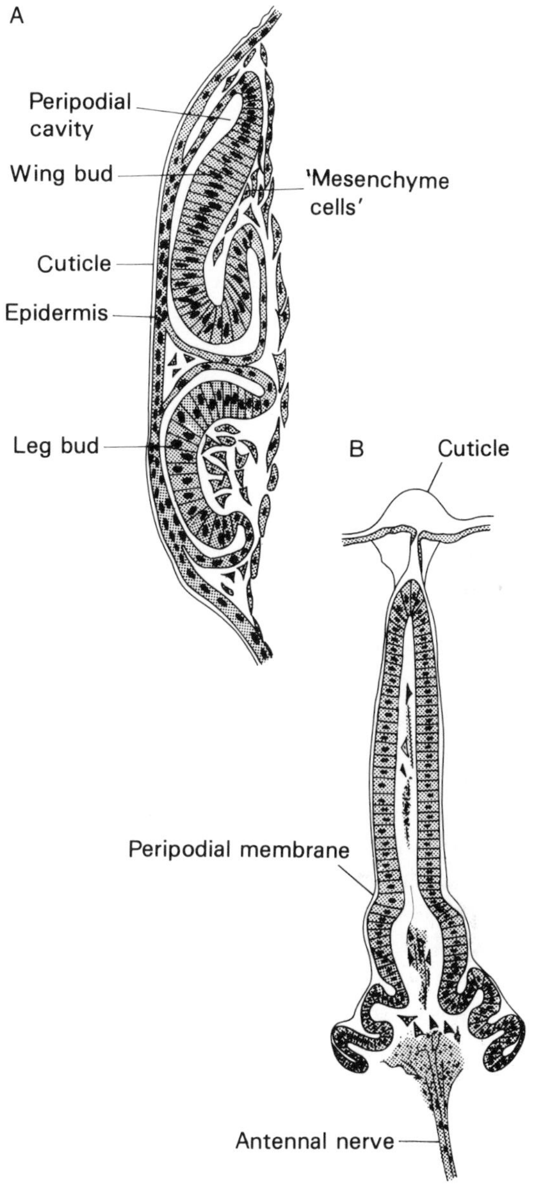

Figure 8.3 Imaginal disks. (A) Transverse section through one side of the mesothorax of a larva of *Anopheles maculipennis* showing the wing and leg disks. (B) Sagittal section through an antennal disk in a larva of *An maculipennis*. (From Imms, 1908.)

8.1.3 Characteristics of individuals disks

The head of a mosquito larva contains paired disks of the antennae, labium and maxillary palps, and an unpaired labral disk. Each thoracic segment contains a dorsal and a ventral pair of disks, the dorsal disks being variously those of the pupal respiratory trumpets, wings and halteres, and the ventral disks those of the legs (Figure 8.3A). The 8th and 10th abdominal segments contain disks which will form the pupal paddles, parts of the genital ducts and their associated glands, the male genitalia, and the cerci. The development of the disks that form the posterior reproductive tracts and genitalia is described in Section 8.2. The primordia of the compound eyes arise, like the imaginal disks, as thickenings of the epidermis. Their development, which is very specialized, is described in Section 8.3.2.

The antennal disks lie deep within the head below the larval antennae, and as each grows its peripodial membrane becomes stretched and reduced to a thin layer of tissue (Figure 8.3B). In a fully grown larva the 2nd segment of the antennal disk is much enlarged through the formation of Johnston's organ, especially in the male.

Shortly after larval–pupal apolysis the antennal and the various thoracic disks are everted from their peripodial cavities and extend along the sides of the body between the epidermis and the larval cuticle, but the labial and maxillary palp disks are not extruded until the cuticle is partly shed.

The mandibles and maxillae undergo almost no development during the larval stage. During the larval–pupal moult their epidermis develops some of the staining reactions of an imaginal disk and at ecdysis the epidermis lining these appendages is withdrawn in the form of hollow tubes (Figures 8.4, 8.5). A few hours after larval–pupal ecdysis the epidermis of the mouthparts retracts from the pupal cuticle and the mouthparts become capable of remodelling. The ventral surface of the labrum folds inwards to form the food canal of the adult. The dorsal surface of the labium rises to form a crest and cells migrating from this region enclose the salivary canal, while in the female the ridge containing the salivary canal separates as the hypopharynx.

In development of the pupal trumpets, the walls of each invaginating disk meet and fuse to form a plate consisting of a single layer of cells. This then folds, taking the form of the respiratory trumpet, and cuticle is secreted on both the inner and outer surfaces of the cells.

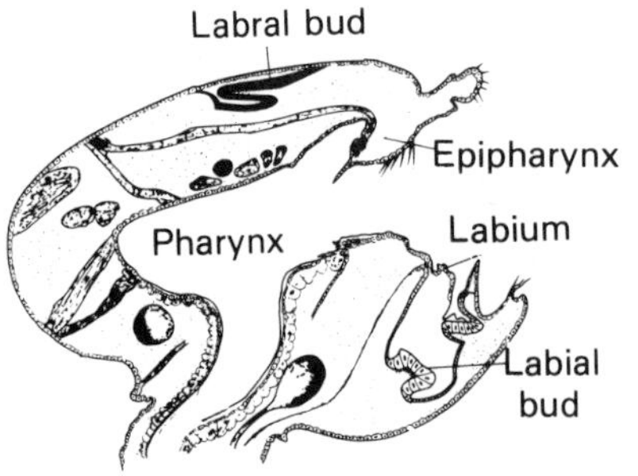

Figure 8.4 Sagittal section through the head of a fully grown *Culex* larva to show the labral and labial disks. (From Thompson, 1905.)

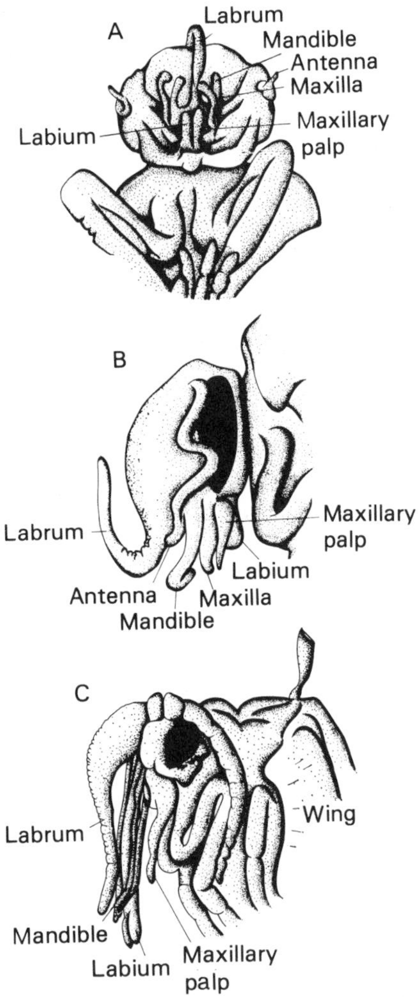

Figure 8.5 Extension of the head appendages of *Culex* during larval–pupal ecdysis as seen in individuals dissected from the larval cuticle at different stages of ecdysis. (A) Ventral aspect of pupal head at the moment when the larval cranium ruptures. (B) Side view at the moment when the respiratory trumpets are freed by rupture of the thoracic cuticle. (C) Side view just before the final release of mouthparts and legs from the larval cuticle. (From Thompson, 1905.)

The disk also forms the short tracheal branch that connects the trumpet to the tracheal trunk, secreting it around the collapsed connective from the larval mesothoracic spiracle. The trumpets become fully developed in the pharate pupa, ready to function after larval–pupal ecdysis (Prashad, 1918; Christophers, 1960).

During the larval stage the wing membranes are folded owing to lack of space but at larval–pupal ecdysis they are able to stretch. During the pupal stage the lumen between the two layers of columnar cells disappears except in particular regions which correspond to the veins of the adult wing; tracheae and blood cells appear within these remaining spaces. The wing cuticle with its setae and scales is secreted during the pupal stage, at which time the cell boundaries disappear.

The imaginal disks of mosquitoes show a relatively primitive condition compared with those of higher Diptera – most of the disks remain just below the epidermis, the peripodial cavities are never closed, the compound eyes do not develop from invaginated buds, and the large larval cranium provides space for development of the adult head. The disks that give rise to the respiratory trumpets and paddles of the pupa are adaptations to aquatic life.

8.2 SEXUAL DIFFERENTIATION

Sexual dimorphism in adult mosquitoes is expressed in a number of ways: (i) in the structure of the gonads; (ii) in the primary sexual characters which are derived from the genital imaginal disks; and (iii) in secondary sexual characters such as the adult antennae, maxillary palps, and body size. Innate behavioural differences between the sexes reflect dimorphism in the structure and function of the nervous system.

Detailed studies have been carried out on the development and differentiation of the gonads and primary sexual characters of *Aedes stimulans* by Ronquillo and Horsfall (1969), Horsfall and Ronquillo (1970), and Voorhees and Horsfall (1971). In both sexes the gonads and the anterior parts of the genital ducts develop from paired gonadal primordia which are present in the embryo. The posterior parts of the genital ducts and the external genital organs develop from imaginal disks in the 8th, 9th and 10th

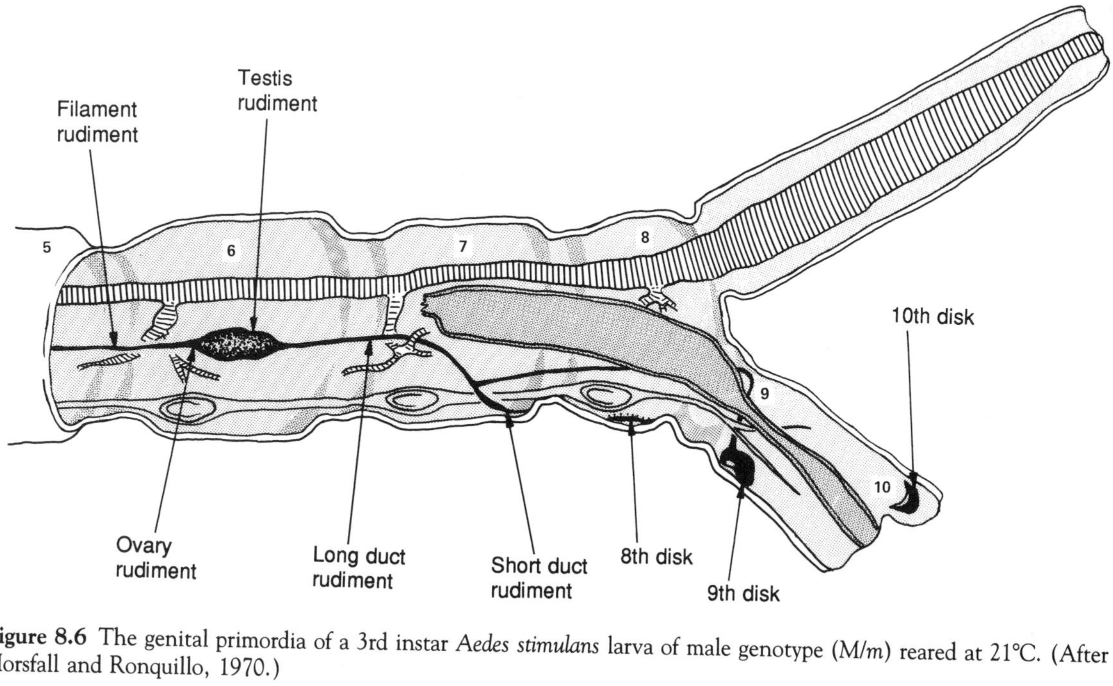

Figure 8.6 The genital primordia of a 3rd instar *Aedes stimulans* larva of male genotype (*M/m*) reared at 21°C. (After Horsfall and Ronquillo, 1970.)

abdominal segments. Some of these imaginal disks are recognizable in newly hatched larvae; others become recognizable only much later in the larval stage. The reproductive primordia found in mosquito larvae are listed in Table 8.1. As noted in Section 2.1.3, the 8th and 9th abdominal segment fuse during larval embryogenesis, therefore a degree of uncertainty must be accepted for the segmental affinities of terminal structures.

8.2.1 Development of the gonadal primordia

(a) Male genotype (M/m)

Aedes stimulans larvae of male genotype (M/m) possess primordia for both male and female organs (Table 8.1, Figure 8.6) (Horsfall and Ronquillo, 1970). In newly hatched *M/m* larvae the largest constituent of each gonadal primordium is the **testis rudiment**. Anteriorly this bears a minute cap of cells, the **ovary rudiment**, and a small **filament rudiment** containing no nuclei. Extending posteriorly from the testis rudiment is a branched strand of cells, the **long** and **short duct rudiments**. The gonadal primordium is surrounded by a fine sheath which has interstitial cells on its inner surface.

The testis rudiment grows only slightly during the first three larval instars, with an increase in the number of germ cells and interstitial cells. In the 2nd instar the interstitial cells form membranes which compartmentalize the testis rudiment, separating rows of germ cells one or two cells thick. The greatest increase in size occurs during the late 4th instar and pharate pupal

Table 8.1 Development of adult genital organs in heat-sensitive *Aedes*.

Primordia and disks	*Fate of primordia and disks in the adult mosquito*		
	Genotype m/m	*Genotype* M/m	*Stressed genotype* M/m
Gonadal primordia			
Filament rudiments (nuclei in *m/m* only)	Terminal filaments of ovaries, anterior ovarioles	Terminal filaments of testes	Terminal filaments of ovaries
Ovary rudiments	Posterior ovarioles	Disappear	Ovaries
?	–	–	Anterior lateral oviducts
Testis rudiments*	(Rudiments absent)	Testes	Disappear
Ovoid masses*	Anterior lateral oviducts	(Rudiments absent)	(Rudiments absent)
Long duct rudiments	(Rudiments absent)	Vasa efferentia	Disappear
Short duct rudiments	Posterior lateral oviducts	Disappear	Posterior lateral oviducts
8th Abdominal disks			
8th Podal buds	Disappear	Disappear	Disappear†
8th Median plate	Common oviduct, vagina, spermathecae and ducts, ventral wall of atrium	Disappears	Common oviduct, vagina, spermathecae and ducts, ventral wall of atrium
9th Abdominal disks			
9th Podal buds	Disappear	Gonocoxites, gonostyli, claspettes	Epidermal thickenings
Ampullae	(Rudiments absent)	Vasa deferentia, seminal vesicles, accessory glands	Nodules
9th Median plate	Dorsal wall of atrium, accessory gland, bursa copulatrix, postgenital lobe	Ejaculatory duct, aedeagus	Dorsal wall of atrium, accessory gland, bursa copulatrix, postgenital lobe
10th Abdominal disks			
Cercal buds	Cerci	Disappear	Cerci

* Possibly homologous in *M/m* and *m/m*.
† If heat stress is withdrawn during the 4th instar, the 8th podal buds form supernumerary male appendages.

stage when new generative tissue is added at the anterior end of the testis rudiment. The gonads reach their full size in the pupal stage. During the hour before larval–pupal ecdysis spermatocytes and spermatids become visible in the posterior compartments and by the mid-pupal stage mature spermatozoa are present throughout the length of the testis. The ovary rudiment disappears during the 4th larval instar.

As was mentioned above, in 1st instar *M/m* larvae a branched strand of cells extends posteriorly from each testis rudiment. The main component of each strand, the long duct rudiment, is a string of cells well spaced out from each other, which runs to join a small group of cells, later to become the ampulla, on the inner side of an imaginal disk in the 9th abdominal segment (Figure 8.8A). During the pupal stage the long duct rudiment is transformed into a vas efferens and the short duct rudiment disappears.

(b) *Female genotype* (m/m)

The reproductive primordia of a 3rd instar *Aedes stimulans* larva of genotype *m/m* are illustrated in Figure 8.7. The gonadal primordia, located in the 6th abdominal segment, show some complexity of structure as early as the 1st instar. At that time the largest part of each primordium is the so-called **ovoid mass**; it is not known whether this is derived from the primordial germ cells or the mesodermal cells of the embryonic gonadal primordium. Situated anteriorly on the ovoid mass is a small cap of 4 or 5 cells, the **ovary rudiment**, which itself bears a nucleated **filament rudiment** which extends forwards to attach to the alary muscles of the heart in the 2nd abdominal segment. A fine strand of cells, the **short duct rudiment**, extends backwards from the ovoid mass and is attached to the ventral body wall in the 7th segment.

The ovary rudiments grow steadily during the 2nd and 3rd instars, largely through increase in cell number. During the 3rd instar the cells of the ovary rudiments start to differentiate, and during the 4th instar they become visibly differentiated into germ cells, interstitial cells, and a ventromedial band. Cell multiplication and differentiation continue during the late 4th instar or pharate pupal stage, and the

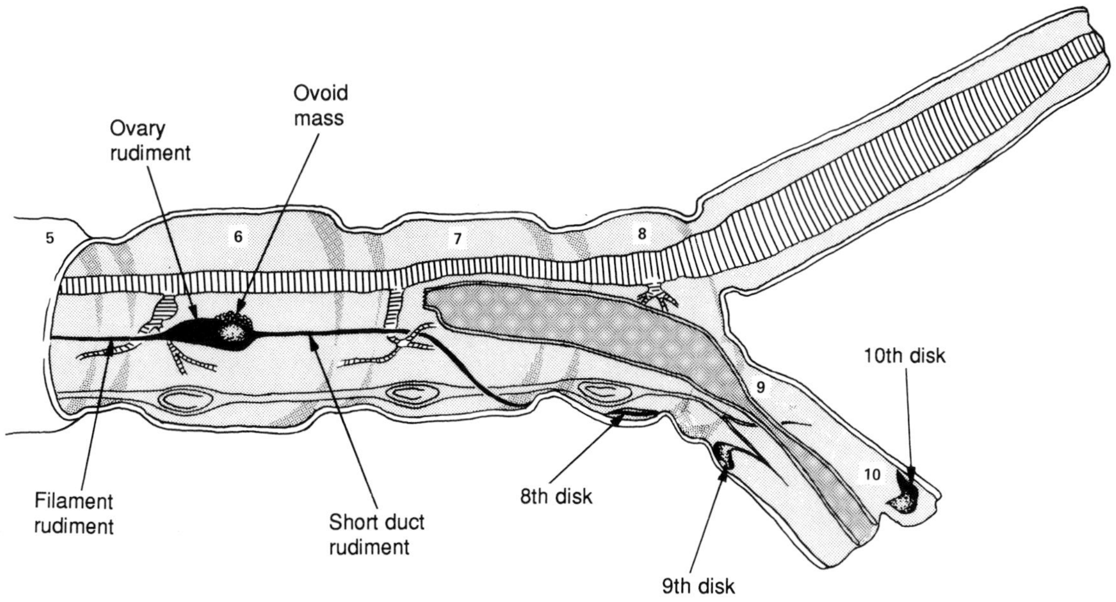

Figure 8.7 The genital primordia of a 3rd instar *Aedes stimulans* larva of female genotype (*m/m*) reared at 21°C. (After Horsfall and Ronquillo, 1970.)

ovaries attain their definitive form during the pupal stage. Within each ovary the germ cells form the germarium, the ventromedial band forms the follicular epithelium, follicular stalks, and calyx, and the interstitial cells form the ovariole sheath.

The anterior part of each filament rudiment, which lacks nuclei, forms the terminal filament of the ovary. The posterior part, which contains nuclei, is competent to form ovarian tissue, and a considerable percentage of the ovarioles are derived from this part of the filament rudiment. It seems that the nuclei within the filament rudiment are derived from pole cell nuclei. In a female of *Ae aegypti* which had been exposed to high temperature throughout embryogenesis, ovarioles formed from a filament rudiment in the 2nd and 3rd abdominal segments and from an ovary rudiment in the 5th and 6th (Cupp and Horsfall, 1970b).

In *m/m* individuals of *Ae stimulans* the anterior parts of the paired lateral oviducts form from the two ovoid masses of the gonadal primordia, while the posterior parts form from the fine strands of cells, called short duct rudiments, which extend posteriorly from them (Table 8.1). During the larval stage a certain amount of cell multiplication and differentiation occurs in the ovoid masses, and the short duct rudiments show a slight increase in cell number, but it is not until the pupal stage that any extensive development occurs. The lateral oviducts attain their imaginal form late in the pupal stage.

The mesodermal sheath that surrounds the embryonic gonadal primordium is present in the larva as a fine cellular membrane around the ovoid mass and ovary rudiment. It probably develops into the connective tissue sheath that surrounds the ovary and oviduct in the adult.

8.2.2 Development of the genital imaginal disks

In *Aedes stimulans* larvae of both genotypes (*M/m* and *m/m*) an imaginal disk appears, during the 3rd instar, as a medial pair of epidermal thickenings in the posteroventral region of the 8th abdominal segment. During the first half of the 4th instar cell division occurs in these thickenings and in the area between them, and the disk develops into two lateral **podal buds** and a **median plate** (Table 8.1). Later in the 4th instar the 8th median plate disappears from *M/m* larvae but is retained in *m/m* larvae.

In *m/m* larvae a median and two lateral invaginations form in the 8th median plate and remain as pouches until the end of the 4th instar. During the pupal stage the median pouch transforms into the median genital tract, i.e. the common oviduct and vagina. The right pouch transforms into a single spermatheca and duct, while the left forms two spermathecae and their ducts. The posterior part of the 8th median plate fuses with the 9th disk and forms the ventral wall of the atrium. Two small bands of cells which are apparent in the larva, appressed to the 8th median plate, are the primordia of the muscular sheath of the median genital tract.

In newly hatched larvae of both genotypes paired thickenings of the ventral epidermis are present in the 9th abdominal segment; these are the **9th podal buds**. In *M/m* larvae only, the 9th podal buds bear clusters of three or four cells which, in the 4th instar, will form pear-shaped vesicles called **ampullae** (Figure 8.8A, C, D). In both *M/m* and *m/m* larvae a thickening develops in the epidermis between the podal buds during the 4th instar; this is the **9th median plate** (Table 8.1).

The 9th imaginal disks show striking sexual dimorphism in their final form. We will first consider their development in *M/m* individuals (Figure 8.8A–D). During the 1st and 2nd larval instars development of the 9th podal buds is limited to a slight increase in cell number and some thickening. Early in the 3rd instar a transverse groove develops in each bud, separating a smaller posterior from a larger anterior part. During the 3rd instar the anterior part grows while the posterior part changes little, and the whole bud becomes enclosed by peripodial membrane. Subsequently the anterior part inverts, becoming cup-shaped, and during the 4th instar mesenchyme cells, which had

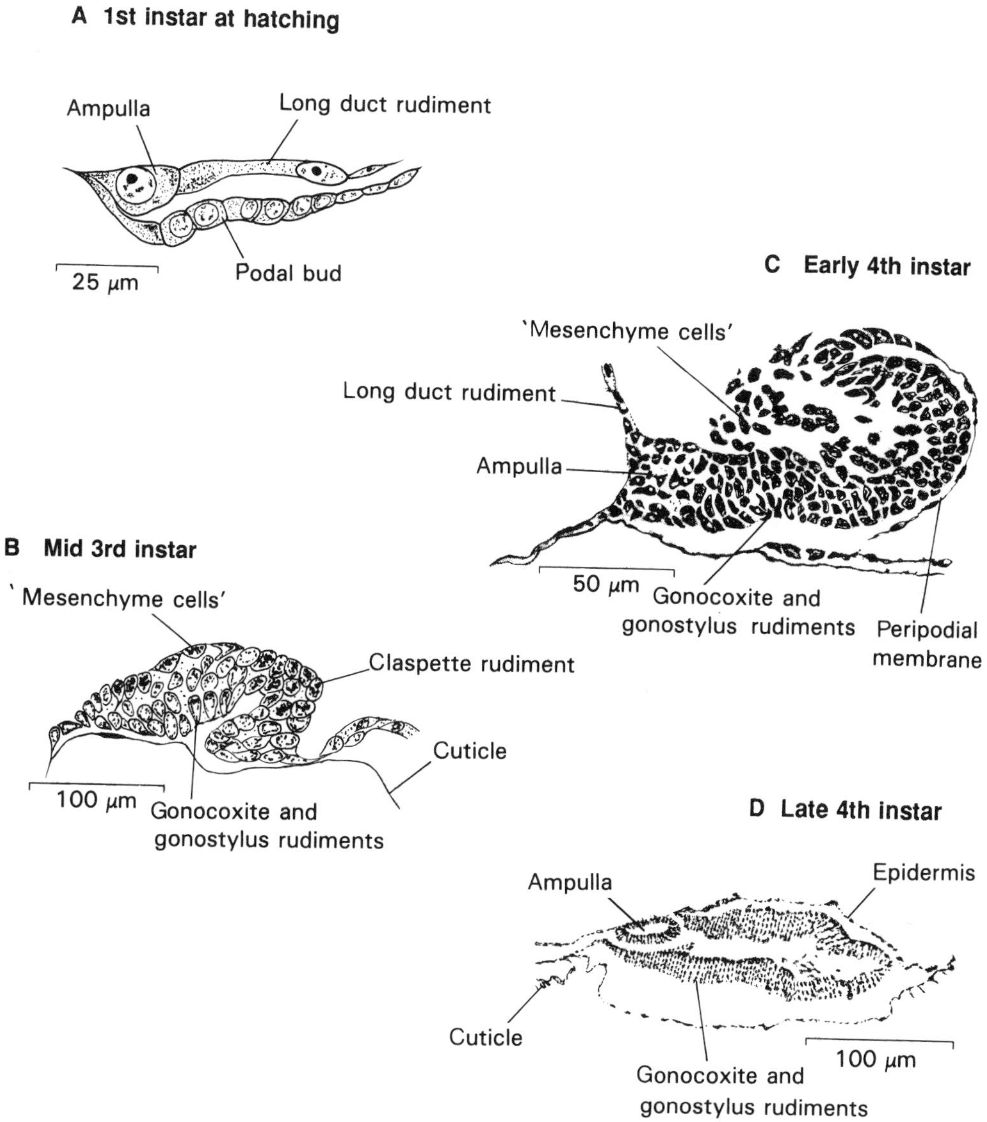

Figure 8.8 Development of an imaginal disk in the 9th abdominal segment of an *Aedes stimulans* larva of male genotye (M/m) reared at 21°C. (A) Transverse section. (B–D) Longitudinal sections. The long duct rudiment, a constituent of the gonadal primordium, later becomes a vas efferens. The ampulla later develops into a vas deferens, seminal vesicle and accessory gland. (Drawn from the photomicrographs of Horsfall and Ronquillo, 1970.)

previously been loosely adherent to the inner face of the bud, enter the cavity. Later in the 4th instar the 9th podal buds lie outside the epidermis. Differentiation into male external genitalia is completed during the pupal stage, when the small posterior part of each bud forms the claspette and the larger anterior part forms the gonocoxite and gonostylus. The 'mesenchyme cells' transform into the intrinsic muscle of the gonocoxite.

The ampullae, present in *M/m* larvae, undergo cell multiplication during the early 4th instar and then develop into vesicles which are attached to the ends of the (long duct) rudiments of the vasa efferentia. The vesicles develop outgrowths which, during the pupal stage, develop into the vasa deferentia, seminal vesicles and accessory glands. The 9th median plate first appears in the 4th instar. During the first ten hours after larval–pupal ecdysis a central invagination forms in it, giving rise to the rudimentary ejaculatory duct. Cells proliferate around this duct and grow posteriorly, forming the rudiment of the aedeagus. In time the lumen of the ejaculatory duct becomes confluent with the lumina of the vasa deferentia, seminal vesicles, and accessory glands.

In *m/m* individuals development of the 9th disks takes a different course. The ampullae never form and the 9th podal buds disappear during the pupal stage. The 9th median plate thickens at the end of the 4th instar, and at the time of larval–pupal ecdysis it has a dichotomous invagination from which the accessory gland and bursa copulatrix later develop, attaining their adult form by the middle of the pupal stage. The posterior end of the 9th median plate gives rise to the postgenital plate. The atrium, which is the tubular connection between the vagina and the external genital opening, provides entrances dorsally to the bursa copulatrix and the accessory gland. It arises from both the 8th and 9th median plates and is formed as a tube when the caudal end of the abdomen is drawn forwards. The anteroventral wall of the atrium forms from the posterior end of the 8th median plate, and the posterodorsal wall forms the 9th median plate.

The cercal buds appear in the 10th abdominal segment during the 2nd instar but little growth takes place until the 4th instar. In *m/m* pupae the flat buds form pouches and then develop into cerci. In *M/m* individuals the primordia disappear by the time of larval–pupal ecdysis.

8.2.3 Effect of heat stress on sexual dimorphism

The genetic basis of sexual differentiation is described in Section 1.4. Our knowledge of the ontogenetic development of sexual dimorphism has been extended by studies of the phenomenon of heat-induced intersexuality in species of *Aedes*.

In the *intersex* mutant of *Aedes* (*Stegomyia*) *aegypti*, an autosomal recessive gene (*ix*) causes genotypic males (*M/m*) to develop into bilaterally-symmetrical phenotypic intersexes when the larvae are reared at high temperatures. All gradations may be produced between complete maleness and almost perfect femaleness depending on the temperature and period of exposure. Individuals of female genotype (*m/m*) become definitive females regardless of temperature. The *ix* gene, which is in linkage group I, affects sexual phenotype only. In contrast, the *m* locus, also in linkage group I, is responsible for sexual genotype, i.e. the primary determination of sex (Craig, 1965; Craig and Hickey, 1967). Over 15 species of *Aedes* in the subgenus *Ochlerotatus* appear to be homozygous for a similar gene because heat-stressed *M/m* larvae develop into intersex adults (Horsfall and Anderson, 1961; Horsfall, 1974).

When *M/m* larvae were reared at high temperature, 28°C for *Ae stimulans* and 25–35° for *ix Ae aegypti*, differentiation of the female constituents of the gonadal primordia was uninhibited, and the imaginal disks developed into the female form; the purely male organs were wholly suppressed (Table 8.1). To examine this phenomenon in more detail we shall consider in turn the development of the gonadal primordia, the genital imaginal disks, and the imaginal disks of the other dimorphic organs in heat-stressed *M/m* individuals. *Ae stimulans* has been studied in most detail, but the development of *ix Ae aegypti* appears to be very similar (Horsfall and Ronquillo, 1970; Voorhees and Horsfall, 1971; Horsfall *et al.*, 1972; Olson and Horsfall, 1972; Anderson, 1967).

Each gonadal primordium of a newly-hatched *M/m* individual consists of a testis rudiment with a minute ovary rudiment and a non-nucleated filament rudiment at its tip. If *M/m* larvae are heat stressed the germ cells of the testis rudiment fail to increase in number, but the ovary rudiment starts to grow as early as the 2nd instar and during the 4th instar becomes differentiated into the parts of a normal ovary. The ovary of the intersex adult has fewer ovarioles than that of a genotypic female because it has received no contribution from the filament rudiment.

In heat-stressed *M/m* larvae, each gonadal primordium also gives rise to the anterior part of a lateral oviduct. This develops from cells, of uncertain origin, which appear between the ovarian and testis rudiments by the early 3rd instar. During the 4th instar they increase in number, forming a thick rod of cells containing small nuclei. A lumen develops through the rod during the pupal stage. The posterior part of the lateral oviduct develops from the short duct rudiment, a structure which disappears from normal males (Table 8.1) (Horsfall *et al.*, 1972).

In normal *M/m* larvae the imaginal disks of the 8th abdominal segment appear briefly but disappear during the 4th instar. In *m/m* larvae the 8th podal buds disappear as in the male, but the 8th median plate is retained and develops into the common oviduct, vagina, and spermathecae. If *M/m* individuals are heat stressed throughout larval life the 8th podal buds disappear, quite normally, but the 8th median plate develops, as in the female, to become the common oviduct, vagina and spermathecae. If heat stress is withdrawn after the end of the 3rd instar, subsequent development follows the normal male course. If heat stress is withdrawn 10–20 h after the start of the 4th instar, the 8th median plate differentiates into female organs while the 8th podal buds differentiate into supernumerary male genitalia (gonocoxites, gonostyli and claspettes), organs which are otherwise never seen (Horsfall and Anderson, 1963; Voorhees and Horsfall, 1971). Thus an elevated temperature maintains developmental plasticity for a certain period, during which a return to normal temperature permits male development. But after a certain time the transformation of *M/m* individuals to phenotypic females is fully established.

In heat-stressed *M/m* individuals the 9th imaginal disks also follow the path of female development. The 9th podal buds, which normally form the male genitalia, remain as epidermal thickenings. The ampullae become dislodged from the podal buds and are drawn forwards forming nodular masses attached to the oviduct. The 9th median plate, which normally forms ejaculatory duct and aedeagus, develops into bursa copulatrix, postgenital plate and atrium. The 10th disks develop into cerci, which are strictly female parts, instead of disappearing as in the normal male.

In short, the reproductive organs of a fully heat-stressed *M/m* individual are identical with those of a normal female except for the reduced number of ovarioles and blocked calyces (see below). Body form and organs common to males and females, such as mouthparts, antennae and claws, also become feminized. They are indistinguishable from those of normal females with the exception of the maxillary palps which are always longer than those of *m/m* individuals (Horsfall *et al.*, 1972; Horsfall, 1974).

The extent to which *M/m* and *m/m* larvae contain primordia of the genital organs of both sexes appears to be fundamental to the development of sexual dimorphism and the production of intersexes in mosquitoes. Horsfall and Ronquillo (1970) found that *M/m* larvae contain primordia of both male and female gonads and genital ducts. They stated that *m/m* larvae contain primordia of female organs only, yet they considered that the anterior lateral oviduct (which develops from the ovoid mass) is the 'analogue' of the testis. Horsfall (in correspondence) stated explicitly that the ovoid mass of *m/m* larvae is homologous with the testis rudiment of *M/m* larvae. For a fuller understanding of this problem it will be necessary to know the precise origin of the cells that form the anterior lateral oviducts in heat-stressed *M/m* individuals.

Exposure of *M/m Ae stimulans* to 28.4°C during embryogenesis, but not the larval stage, destroys the primordia of the male genital ducts in 60% of cases, producing adult males with testes but lacking vasa efferentia, vasa deferentia, seminal vesicles and accessory glands. The female primordia are not affected by high temperature since exposure to 28.4°C during both embryogenesis and the larval stage leads to the formation of complete phenotypic females (Anderson and Horsfall, 1965b).

Heat-stressed *M/m* individuals that become fully feminized can be inseminated, take a blood meal, and develop eggs with chorionic markings characteristic of the species (Craig, 1965; Anderson and Horsfall, 1963). They cannot oviposit because the anterior lateral oviducts are blocked by the remnants of the testis rudiments (Olson and Horsfall, 1972; Horsfall, personal communication).

A number of other observations have been made which extend our understanding of the development of sexual dimorphism in mosquitoes. That circulating hormones do not influence the direction of morphogenesis of the gonadal primordia was demonstrated by transplantation of gonadal primordia from *M/m Ae stimulans* into male or female larvae of *Ae vexans*, a species in which normal males are obtained even at 35°C. The transplanted primordia formed testes when the hosts were held at 18°C and formed ovaries when they were held at 27°C; the hosts' gonads developed strictly according to genotype. If maleness had been under humoral control the gonadal primordia transplanted into female larvae reared at 18°C should not have developed into males, but invariably did so. If femaleness was under humoral control the primordia placed in male hosts at 27°C should not have developed into ovaries (Anderson and Horsfall, 1965a).

When *M/m* larvae of *Ae stimulans* are heat stressed during only a part of the larval stage, certain organs will be feminized and others not, depending upon the period of exposure to heat stress and the developmental stage at which individual organs lose their plasticity and become determined. Thus, the testis rudiment is affected by high temperature in the 1st larval instar, but the genital ducts are barely affected before the 4th. The form of the antennae is determined by the 3rd instar, that of the oral stylets in the 4th. Differentiation of the testis is not required to stimulate maleness in the genital tract or genitalia since these organs can assume the male form when heat stress persists just long enough to prevent testicular growth (Horsfall and Anderson, 1964; Horsfall *et al.*, 1972).

The determination of adult form is not instantaneous throughout an individual primordium. At intermediate temperatures the primordia of the gonads, external genitalia, antennae and mouthparts may become determined at their proximal ends earlier than at their distal ends so that maleness is expressed at the base of the organ but not at the tip (Horsfall *et al.*, 1972).

When *M/m* larvae of *Ae communis* are heat stressed throughout larval life, male traits are suppressed and female traits developed to an extent that varies with temperature. Different primordia show different sensitivity thresholds, and the sequence in which dimorphic structures are wholly feminized as temperature increases between 19 and 24°C is (1) antennae, (2) oral stylets, (3) tarsal claws, (4) genital ducts, (5) external genitalia, (6) maxillary palps and (7) gonads (Brust and Horsfall, 1965).

More than 15 species in the subgenus *Ochlerotatus* have been found to produce intersexes when *M/m* larvae are subjected to heat stress, but some other species of the subgenus do not respond in this way (Anderson and Horsfall, 1963; Brust, 1968b; Horsfall, 1974). The heat-sensitive species characteristically have a subarctic and temperate zone distribution. The great majority of heat-sensitive species are univoltine; their eggs do not normally hatch the year they are laid, however often they are inundated, but hatch the following spring in pools filled with water from melting snow and ice at 5–10°C. The larvae complete development at temperatures which rarely rise above 20°C. Among the univoltine species of *Ochlerotatus*, including *communis*, *fitchii* and *punctor*, different degrees of feminization of *M/m* individuals occur when larvae are reared at temperatures between 23 and 27°C, and within a species the more northerly populations show greater temperature sensitivity (Horsfall, 1974). Heat-induced intersexes have been described in the multivoltine species *Aedes* (*Ochlerotatus*) *sierrensis* (threshold 30–31°C) (Horsfall *et al.*, 1964), and in a strain of *Aedes* (*Stegomyia*) *aegypti* originating in Kenya (threshold about 34°C) (Craig, 1965).

High temperature may not be the only extrinsic factor that can induce formation of intersexes in mosquitoes. Of 800 males of *Ae communis* collected from a natural habitat, a single individual was heavily parasitized by a fungus of the family Coelomomycetaceae, and that individual was also singular in showing intersexual characteristics in its genitalia (Aspöck, 1966). In the Chironomidae, intersexuality can be induced by mermithid parasites (Wülker, 1961).

To summarize these observations, at the beginning of postembryonic development, in certain species of *Aedes*, *M/m* individuals have the rudimentary competence to become phenotypic males or phenotypic females. Imaginal form is determined by the temperature at which the larvae develop, and all gradations of structure can be produced in dimorphic organs according to temperature and duration of exposure. In contrast, *m/m* individuals develop into females at all temperatures permitting survival. A short discussion of the genetic basis of the production of intersex individuals can be found in Section 1.4.3.

8.3 DEVELOPMENT AND METAMORPHOSIS OF THE CENTRAL NERVOUS SYSTEM AND COMPOUND EYES

8.3.1 Central nervous system

Differential growth of the supraoesophageal ganglion or brain of *Culex pipiens* during postembryonic development has been analysed using the equation

$$y = bx^k \qquad (8.1)$$

devised for allometric growth, where x and y are two dimensions, b the initial growth index, and k the ratio of the two growth rates. During the first three larval instars the brain grows less rapidly than the rest of the body (k = 0.7), but it shows positive allometric growth in the 4th instar (k = 1.8) and pupal stage (k = >1.0). The main regions of the brain are all present in the young larva of *Cx pipiens* and by differential growth they form the adult brain with its very different proportions. Different regions of the brain grow most actively at different stages of development. The mushroom bodies grow less strongly than the brain as a whole during the larval stage (k = 0.5–0.9) but much more strongly in the pupal stage (k = 1.8); the central body grows most actively in the 4th instar (k = 2.0). The antennal centres grow slowly compared to the brain as a whole during the larval stage (k = 0.8), very rapidly about the time of larval–pupal ecdysis (k = 4.0), and slowly again in the pharate and young adult (k = >1.0) (Hinke, 1961).

The brain of *Cx pipiens* is smaller in the adult female than in the male relative to total body volume. The optic lobes are smaller in the female than in the male, relative to the whole brain, but the female has relatively larger mushroom bodies, central body and antennal centres (Hinke, 1961). The nerve tracts directly associated with vision are organized in their final form during the larval stage, consistent with the early development of the compound eyes, whereas the nerve tracts associated with the adult antennae undergo extensive elaboration at metamorphosis. The suboesophageal ganglion is

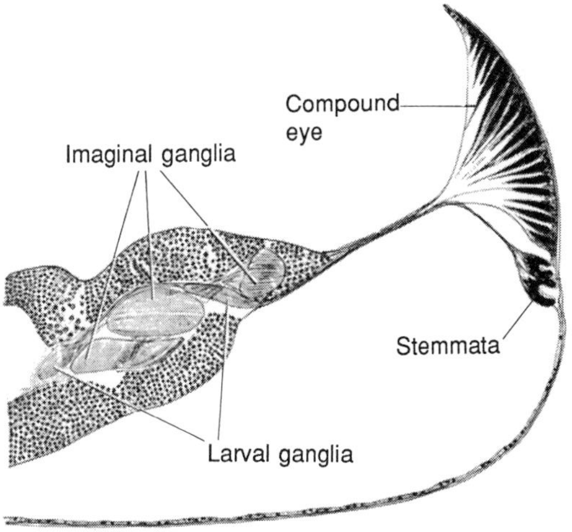

Figure 8.9 Transverse section through the optic lobe of a *Culex pipiens* larva after differentiation of the compound eye. This semi-diagrammatic figure shows the neuropile masses of the two larval and three imaginal optic ganglia, and the cortical region which contains the associated neurones. The twisted bundles of axons (chiasmata) connecting the imaginal ganglia are also shown. (From Pflugfelder, 1937.)

considerably more complex after metamorphosis, when there is also a difference in the nerves arising from it and a great increase in the size of the circumoesophageal commissures (Rogoff, 1954).

The optic lobes comprise 9% of the brain in an early 1st instar *Cx pipiens* larva, and 45% in the adult. They show strong growth in the 1st instar (k = 1.9), isometric growth during the rest of the larval stage, and further positive allometric growth in the pupa (k = 1.3) (Hinke, 1961). Development of the optic lobes in *Cx pipiens* has been described by Pflugfelder (1937). In the young larva, when stemmata can be distinguished but not ommatidia, each optic lobe contains two larval optic ganglia. The outer ganglion receives axons from the sense cells of the stemmata, and an axon bundle connects it with the inner ganglion. Later, but before the differentiation of ommatidia, three imaginal optic ganglia appear. They are first seen as distinct groups of large cells; after mitosis they form the cells of the optic ganglia. Each ganglionic cell develops an axonal process; the axons form the inner neuropile of the ganglion and the connecting strands between the optic ganglia. As the three imaginal ganglia grow in size they rotate relative to each other with a consequent twisting of the axonal connections between them, thus forming the chiasmata (Figure 8.9). When the ommatidia of the compound eye differentiate, axons from the retinular cells grow inwards and run along the stemmatal nerve before arriving at the outer imaginal ganglion. During metamorphosis the stemmata are resorbed and the larval optic ganglia are reduced in size while the imaginal ganglia increase in size.

Metamorphosis of the central nervous system in *Aedes dorsalis* and *Ae aegypti* involves only a small amount of histolysis, the principal changes resulting from growth by cell multiplication and from fusion of ganglia. The suboesophageal ganglion moves forwards to become more closely and broadly connected to the brain, the thoracic ganglia fuse with one another and with the 1st abdominal ganglion to form a single mass which still shows signs of its original segmentation, and the 8th abdominal ganglion moves forwards to fuse with the 7th (Woolley, 1943; Christophers, 1960).

8.3.2 Compound eyes

Development of the compound eyes proceeds in a sequence of events – first determination and cellular proliferation, then differentiation – which pass as waves across a region of the head epidermis, the 'prospective eye field', that will transform into the compound eyes.

(a) *Aedes aegypti*

The compound eye rudiment appears during the middle of the 1st instar as a thickening of the epidermis, called the optic placode, just anterior to the stemmata. It first forms along the posterior border of the prospective eye field and then expands anteriorly, following a wave of mitoses which advances from posterior to anterior through the prospective eye field, converting the epidermis into the densely nucleated optic placode. The entire larval stage is required for the mitotic wave to traverse the prospective eye field. The placode consists of undifferentiated cells until the end of the 3rd instar, but at that time the cells along its mid-posterior border begin to differentiate into ommatidia, and during the 4th instar the mitotic wave is followed by a wave of ommatidial differentiation. Because differentiation involves deposition of pigment, the progressive differentiation of the compound eye can be followed easily in the living insect (White, 1961).

A causal analysis of compound eye development in *Ae aegypti* was undertaken by White (1961, 1963). Microcauterization of the posterior border of the prospective eye field early in the 1st instar prevented formation of optic placode and compound eye in a large percentage of cases although microcauterization of the ventral border of the prospective eye field had no such effect (Figure 8.10A). Prospective eye epidermis whose development had been blocked in this way could be stimulated to differentiate into compound eye

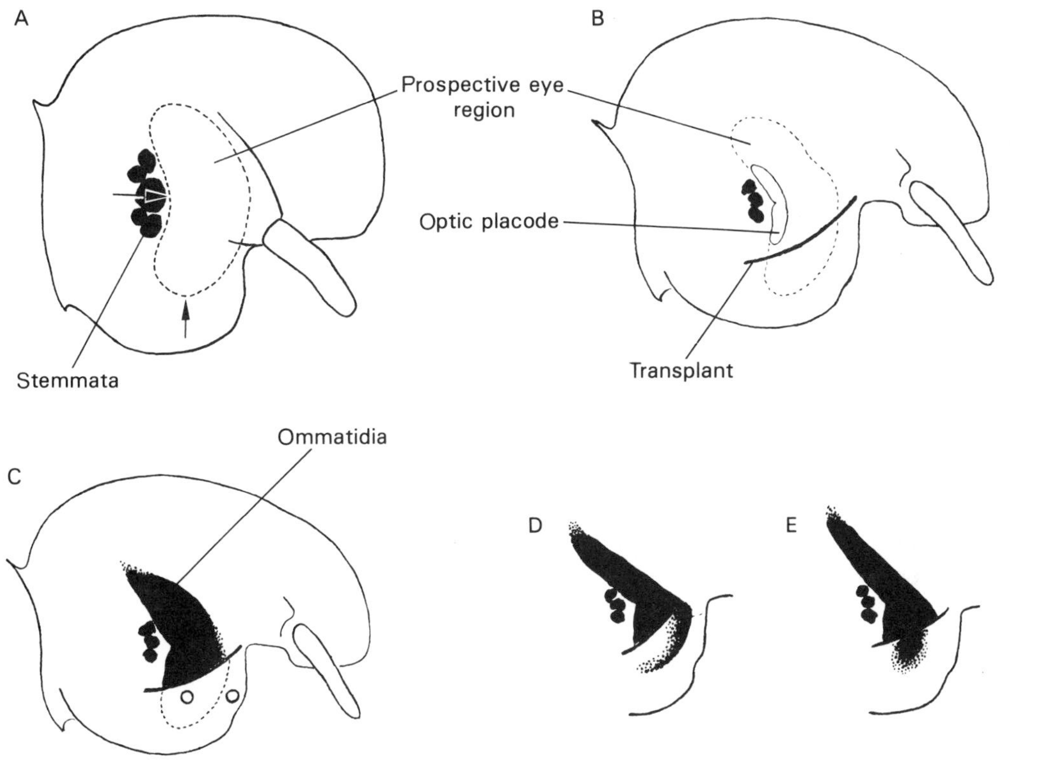

Figure 8.10 Experiments on compound eye differentiation in *Aedes aegypti*. (A) Lateral view of the head of a 1st instar larva. Cauterization at the point indicated by the horizontal arrow prevents formation of optic placode. Cauterization at the point indicated by the vertical arrow has no such effect. (B) Transplantation of epidermis from a non-optic region of the head across the prospective eye region of an early 4th instar larva. (C–E) Differentiation of ommatidia in operated larvae: (C) the transplanted tissue traversed the full width of the prospective eye region; (D) the transplanted tissue did not extend to the front of the prospective eye region; (E) the transplanted tissue had a break in the middle. (From White, 1961, 1963. Copyright © (1961, 1963) and reprinted by permission of Wiley-Liss, a Division of John Wiley and Sons, Inc.)

by implanting a fragment of optic placode from another individual. Implantation of fragments of optic placode from a *red-eye* mutant into the prospective eye fields of wild-type larvae whose optic placodes had been prevented from developing by cauterization, led to the formation of compound eyes in which a small area of red ommatidia was accompanied by an adjacent area of normal black ommatidia. In some cases the red ommatidia were surrounded by an almost full-sized black compound eye. When *red-eye* placodal fragments were implanted away from the prospective eye fields they differentiated into red ommatidia but without any black ommatidia.

It appears that a physiological differentiation centre normally forms just anterior to the stemmata, and that cauterization destroys this centre without affecting the competence of the remaining prospective eye epidermis. Optic placode, once formed, shares the properties of the differentiation centre, and removal of the optic placode will prevent all further development in the prospective eye field.

By transplanting epidermis into the prospective eye field, White (1961, 1963) showed that the expansion of the optic placode reflects the diffusion of a determination factor from the optic placode through the rest of the field, this factor stimulating the mitoses that convert the prospective eye epidermis into optic placode.

When pieces of epidermis were transplanted from various regions of the head into the path of the determination factor, only those from a limited area, the prospective eye fields, were competent to respond. Epidermis from other areas not only failed to respond but, if extended across the full width of the prospective eye field, prevented the formation of optic placode in the isolated portion. If a narrow bridge of competent tissue remained the factor could pass, demonstrating its diffusible nature (Figure 8.10B–E). When competent tissue was transplanted into the prospective eye field, correct orientation of the graft was not essential for normal eye formation, showing that the normal temporal and spatial pattern of eye development is not predetermined within the cells of the prospective eye epidermis but is superimposed on them during later development. If part of the optic placode was removed the remainder rapidly expanded to regain its normal size, and a normal compound eye subsequently formed.

The development of a compound eye therefore involves a number of separate processes. First, the prospective eye field is made competent to respond to the determining factor and a physiological differentiation centre forms anterior to the stemmata. Next, a determining factor diffuses forwards from the differentiation centre causing a rapid increase in the rate of mitosis in the competent epidermis and converting it to optic placode. Finally, a wave of ommatidial differentiation passes, in the same direction, across the optic placode.

(b) *Culex pipiens*

The optic placodes are visible in the 1st instar larva as paired thickenings of the epidermis just anterior to the stemmata. Differentiation of ommatidia starts in the 2nd instar at the posterior edge of the optic placode and proceeds anteriorly, dorsally and ventrally. Each ommatidium develops in the following manner. A group of 7 or 8 nuclei gathers below the cuticle, at first without evidence of cell boundaries. Later a cell mass moves inwards and forms a spindle-shaped body surrounded by small cells, respectively the prospective retinula cells and pigment cells. Axonal processes grow inwards from the prospective retinula cells, which are sensory neurones, and when they meet the stemmatal nerve they pass along it to the first ganglion in the optic lobe, the lamina. A mitotic figure, present distally at this time, forms the four prospective Semper's cells which look like a pointed cap over the retinula cells (Figure 8.11A). Rhabdomeres form at the inner surfaces of the retinula cells, and screening pigment starts to appear in the retinula cells and in the pigment cells. The nuclei of the retinula cells migrate below the basement membrane which bounds the optic placode. Semper's cells secrete a large liquid-filled vesicle called the cone. A densely pigmented body of unknown function, secreted by Semper's cells, now appears between the rhabdomeres. It is present only in the larval ommatidia. The ommatidia reach this level of differentiation during the 2nd instar (Figure 8.11B) when they are possibly already functional. Certainly the rhabdom is capable of movement in the 2nd instar larva, migrating inwards in the light and outwards in the dark. No further differentiation occurs during the larval stage but the number of ommatidia continues to increase until early in the pupal stage.

Transformation of ommatidia from the larval to the adult form occurs at the end of the pupal stage and during the first 12 h after emergence. Semper's cells absorb the contents of the larval cone and occupy the space. These cells then secrete the cone of the adult ommatidium and the cell bodies become much reduced, embedded within the cone. In the larva and early pupa an ommatidium is covered by an unmodified part of the cranium, but during the pupal stage the ommatidium retracts from the cuticle, a secretion fills the space, and the corneal lens starts to form in the middle of the secretion. The secretion overlying the corneal lens hardens to form an outer layer shaped like a watch glass but the secretion under the lens remains fluid throughout the pupal stage. Further changes observed in the pupa are a thickening of the pigment cells, return

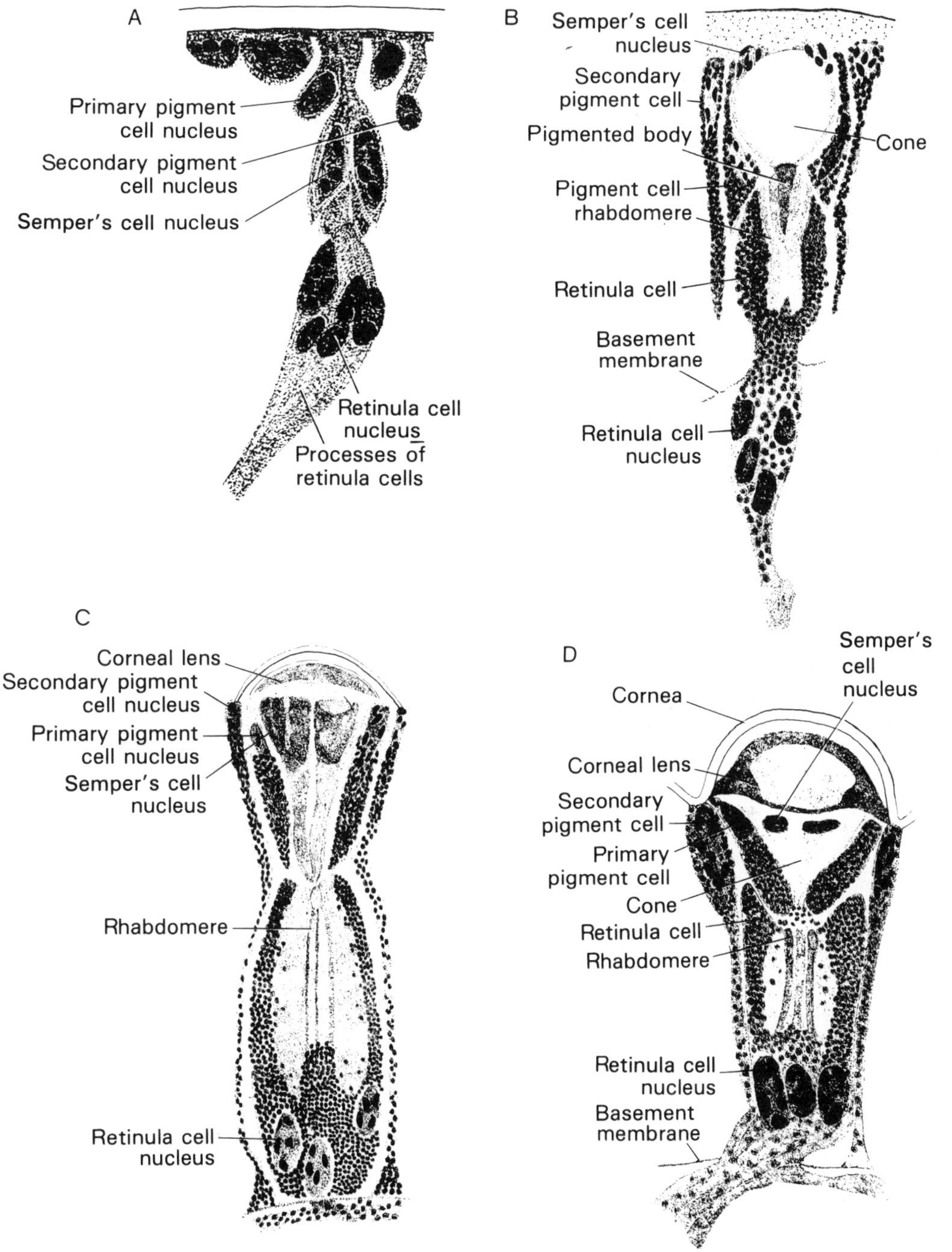

Figure 8.11 Development of an ommatidium. (A) Early differentiation of ommatidial cells in a 3rd instar larva of *Aedes aegypti*. (B) Longitudinal section through an ommatidium in a 3rd instar larva of *Culex pipiens*. (C) Longitudinal section through an ommatidium in a pupa of *Cx pipiens*. (D) Longitudinal section through an ommatidium in an adult of *Aedes aegypti*. (From Haas, 1956.)

of most of the retinula cell nuclei to a position above the basement membrane, and further differentiation of the rhabdomeres (Figure 8.11C). Shortly after emergence most of the substance of the corneal lens hardens, making the lens biconvex, but a certain amount of unhardened secretion remains below this for some time (Figure 8.11D) (Constantineaunu, 1930; Satô, 1951; Haas, 1956).

The time and manner of compound eye formation vary among mosquito species. The optic placode usually appears during the 1st instar but the rate of further development varies. Pigmented ommatidia are differentiated during the 2nd instar in *Culex pipiens* (Satô, 1951) but not before the 3rd instar in *Anopheles sinensis* (Satô, 1953) and *Ae aegypti* (Haas, 1956), the 4th instar in *Armigeres subalbatus* (Satô, 1960) and the pupal stage in *Toxorhynchites towadensis* (Satô, 1961). In larvae of *Ae aegypti* the ommatidia develop without a cone or a central pigmented body (Haas, 1956), and the cone is absent at this stage in *An sinensis* also (Satô, 1953).

8.4 METAMORPHOSIS OF THE ALIMENTARY CANAL

During the larval stage the alimentary canal grows by increase in cell size and with the formation of polytene and endopolyploid nuclei. At metamorphosis some of the tissues that have endopolyploid nuclei undergo somatic reduction and cell division to yield a large number of small, apparently de-differentiated cells which are competent to form adult tissues (see Section 7.2.1). All tissues with polytene nuclei, except those of the Malpighian tubes and some with endopolyploid nuclei, degenerate at metamorphosis and are reformed by imaginal cells. The Malpighian tubes are carried into adult life without visible change.

The metamorphosis of the alimentary canal of *Culex pipiens* and *Aedes stimulans* has been described in considerable detail (Thompson, 1905; Berger, 1938b), and brief accounts of the process in other species suggest that it is similar in these also (Samtleben, 1929; Richins, 1938, 1945; Christophers, 1960; Risler, 1961; Romoser and Venard, 1966, 1967).

8.4.1 Foregut

Metamorphosis of the foregut starts in the pharate pupa with withdrawal of the oesophageal invagination from the midgut, and at larval–pupal ecdysis the lining of the foregut is shed with the rest of the larval cuticle. The muscle coat of the oesophagus and the annular muscle of the oesophageal invagination degenerate early in the pupal stage and are replaced by imaginal muscles. The foregut epithelium is not histolysed but is remodelled into the two sucking pumps of the adult. The anterior pump develops from the epithelium of the buccal cavity and pharynx of the larva, so the name 'cibarial pump' is appropriate for this organ. The posterior pump develops from the former oesophageal epithelium, and as it corresponds in function to the pharynx of other insects it may reasonably be called the 'pharyngeal pump'. About ten hours after larval–pupal ecdysis the two dorsal diverticula of the adult gut form from the oesophageal epithelium, and they grow slowly until about the 30th hour after which they quickly reach their adult size.

Mitosis and cell division occur in the foregut epithelium of *Ae aegypti* a few hours after larval–pupal ecdysis, first in the pharynx and later in the oesophagus. By somatic-reduction divisions the big foregut cells containing endopolyploid nuclei give rise to many small cells containing diploid nuclei which form much of the pupal foregut (Risler, 1961).

At the junction of the fore- and midgut there is a histoblast in the form of a narrow band of embryonic cells, the anterior imaginal ring (Figure 8.12). This is possibly a part of the presumptive stomodaeum that was carried internally during formation of the larval foregut and that remained undifferentiated during the early larval stage. In mosquitoes the anterior imaginal ring forms only ectodermal tissues. Cell division starts in the anterior imaginal ring during the 3rd and 4th instars and continues into the early pupal stage, when it forms the oesophageal invagination

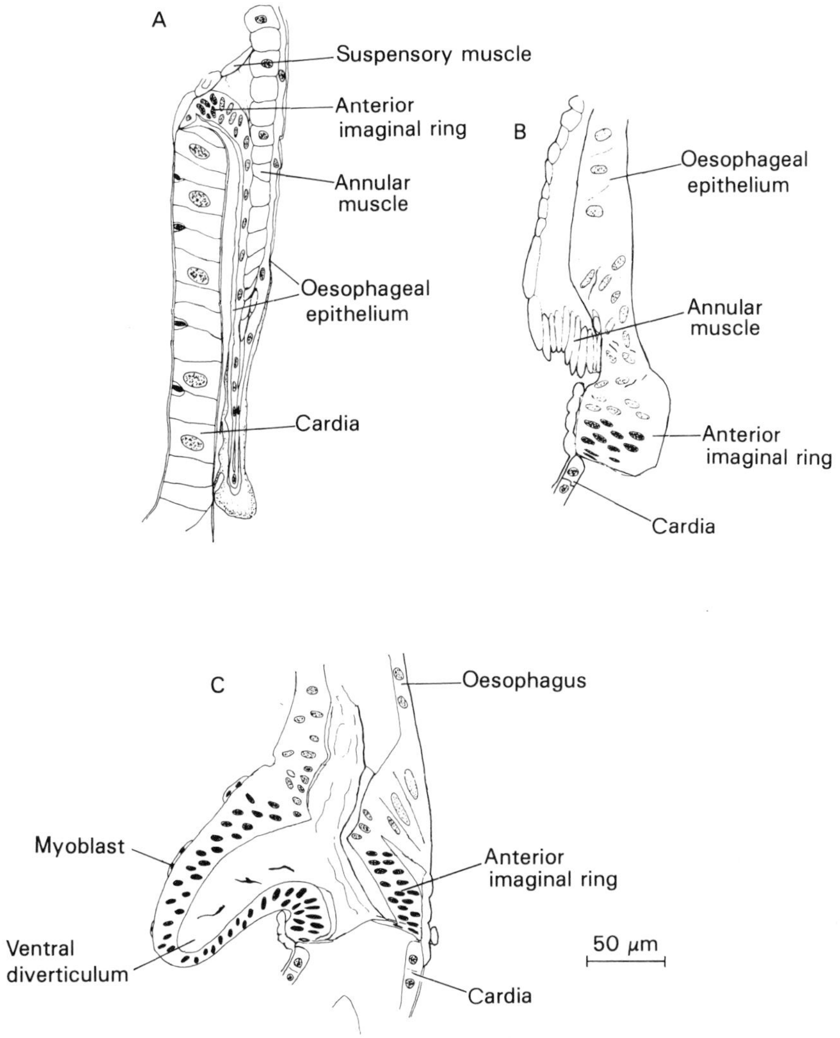

Figure 8.12A-C Metamorphosis of the alimentary canal. Longitudinal sections through the junction of the fore- and midgut in *Aedes triseriatus*, reared at 27°C. (A) 4th instar larva (one side only). (B) 3–4 h after larval–pupal ecdysis (one side only.) (C) 13–14 h after larval–pupal ecdysis. (After Romoser and Venard, 1967.)

and the ventral diverticulum of the adult gut. Myoblasts present in two small clumps on either side of the oesophagus, differentiate into fine muscle networks over the dorsal and ventral diverticula (Romoser and Venard, 1966, 1967, 1969; Clay and Venard, 1972).

8.4.2 Midgut

The midgut of early 1st instar larvae of *Ae aegypti* is composed of two distinct cell types – a layer of small epithelial cells and muscle cells. Cell division is apparent among the epithelial cells during the first hours after hatching. By 12 h after hatching some epithelial cells have increased in size, and shortly afterwards the epithelial layer is seen to be composed of large columnar cells, which have a microvillate luminal margin, and small regenerative cells, both resting on the basement membrane. The columnar cells grow only by increase in cell size and some that degenerate are replaced by growth of regenerative cells. In *Ae dorsalis* such replacement is known to occur in all larval instars. The regenerative cells are more abundant in the stomach than in the cardia and caeca. These cells undergo repeated cell divisions until at the end of the 4th instar they form a complete layer of small cells against the basement membrane in the posterior region of the stomach. After larval–pupal ecdysis the epithelial cells of the cardia, caeca and stomach separate from the basement membrane and lie free in the lumen of the midgut, later to disintegrate. The regenerative cells continue to divide and soon form the epithelium of the pupal midgut. The caeca are not replaced. The muscle network that surrounded the larval midgut persists into the adult stage (Richins, 1945; O'Brien, 1966a, b).

It is thought that some breakdown products from the larval midgut epithelium are absorbed and that the residue forms the meconium. A non-cellular membrane forms around the meconium during the pupal stage, and a similar membrane is secreted again before emergence, possibly during the pharate adult stage. These membranes appear to contain chitin and are presumed to be peritrophic membranes (Romoser and Rothman, 1973; Romoser, 1974).

8.4.3 Hindgut

The adult pylorus is formed between the 6th and 10th hours after larval–pupal ecdysis by another histoblast, the posterior imaginal ring – a zone of imaginal cells situated just posterior to the openings of the Malpighian tubes. This is possibly a component of the presumptive proctodaeum that was carried internally during formation of the hindgut and remained undifferentiated throughout the larval stage. Early in the pupal stage the cells of the posterior imaginal ring multiply by mitosis and cell division and migrate backwards forming the epithelium of the adult pylorus (Berger, 1938b). The fate of the larval pylorus has not been described.

Metamorphosis of the remainder of the hindgut begins between 4 and 8 hours after larval–pupal ecdysis in *Culex pipiens*, when by a series of somatic reduction divisions the big endopolyploid cells of the anterior intestine and anal canal produce large numbers of much smaller cells (Section 7.2.1). When this process is nearly finished, about the 18th hour, these new cells begin to invade the rectum lumen from anterior and posterior directions. The two cell migrations meet and form a new cylindrical epithelium while the polytene rectal cells are pushed out into the body cavity and destroyed. The new epithelium later differentiates into the anterior intestine, rectum and anal canal of the adult, and rectal papillae form by rearrangement of some of the cells (Berger, 1938b; Schuh, 1951). The reappearance in the cells of the anterior intestine of microtubules connecting microapodemes to hemidesmosomes is associated with refolding of the epithelium (Cerreta, 1976). The muscles of the hindgut histolyse and are replaced by adult muscles which develop from myoblasts. The cuticular linings of the foregut and hindgut are secreted just before emergence under the pupal linings.

8.4.4 Malpighian tubules

The Malpighian tubules pass from the larva to the adult without visible reorganization. Tubules removed from pupae of *Aedes taeniorhynchus* and tested *in vitro* for their responsiveness to 5-hydroxytryptamine and dibutyryl cAMP proved incapable of fluid secretion. The capacity to secrete fluid returned after pupal–adult ecdysis. The pupal tubules exhibited a shrinkage of microvilli and a retraction of mitochondria from the microvillar lumens (Bradley and Snyder, 1989).

Phosphorus-32 added to the rearing water of young *Culex pipiens* larvae accumulated in their Malpighian tubules after they had reached the 4th instar and continued to accumulate in the tubules throughout the pupal stage. The phosphorus was localized in 1–3 μm diameter cytoplasmic granules which were absent from early 4th instar larva. The phosphorus was thought to originate from histolysed organs and to be in storage form in the Malpighian tubules (Stich and Grell, 1955).

8.5 METAMORPHOSIS OF OTHER ORGANS

The number of epidermal cells increases approximately 8-fold during the 4th instar in *Aedes aegypti*, and the epidermis is carried over to the adult stage without histolysis (Risler, 1959). After larval–pupal apolysis the epidermis of the head and thorax is largely reformed into the shape of the pupa (Hurst, 1890a, b; Thompson, 1905). In pharate pupae nervous connections extend between homologous aporous articulated setae on the larval and pupal abdomens. Relocation of pupal setae, due to differential growth of the integument, is accommodated by elongation of the dendrites (Belkin, 1962).

The only functional spiracles of the larva are the terminal pair, the other nine pairs being reduced and serving only for the withdrawal of tracheae at ecdysis. At larval–pupal ecdysis the short regions of the main tracheal trunks between the terminal spiracles and the 7th abdominal segment are withdrawn and not replaced, and the epithelia surrounding them shrink to form solid cords attached posteriorly to the 9th tergite. The collapsed larval tracheae that connect the non-functional mesothoracic spiracles with the tracheal system are replaced by stout tracheae connected to the respiratory trumpets. The non-functional 1st abdominal spiracles of the larva and the collapsed tracheae running from them are replaced by functional structures, the spiracles opening into the ventral air space. The remaining pupal spiracles are non-functional. Functional spiracles develop on the meso- and metathorax and on the 2nd to 7th abdominal segments of the pharate adult (Hurst, 1890a, b; Christophers, 1960).

Great changes take place in the muscular system during metamorphosis. Most larval head muscles are histolysed but some are carried over to the adult, developing the fine structure of adult muscle and acquiring new relationships with different parts of the head. In addition, new muscles arise and the sides and floor of the head become traversed by belts of small myoblasts whose origin is uncertain. The muscles of the thorax and abdomen are carried over to the adult but the main part of the adult thoracic musculature is new, including the muscles of the appendages, largely formed within the imaginal buds, and the indirect flight muscles. The latter develop principally in the pupal stage from rudiments that are situated below the epidermis in the 4th instar larva (Thompson, 1905; Hinton, 1959; Christophers, 1960). The larval abdominal muscles degenerate early in adult life, having been used for swimming by the pupa and pharate adult. Movements of the adult abdomen are due to small adult muscles which develop from myoblasts in the pupa (Roubaud, 1932; Berger, 1938b). Histolysis of the musculature, as of other organs, takes place entirely by autolysis and without the intervention of phagocytes (Thompson, 1905; Hulst, 1906; Jones, 1954).

A ring of minute imaginal cells becomes visible at the anterior end of each salivary gland during the 1st instar. In *Anopheles albimanus* there are 12–15 cells in each ring, and the number doubles in each subsequent instar so that over 100 cells are present in each gland in the mature larva. Degeneration of the larval glands may start before larval–pupal ecdysis and continue during the following day, the posterior parts degenerating first. The imaginal cells differentiate during the first few hours after larval–pupal ecdysis when the cells increase in size and the glands become greatly elongated, but development is not completed until after emergence (Thompson, 1905; Jensen and Jones, 1957).

In *Culiseta annulata* and *Aedes aegypti* a group of five large larval oenocytes is present at hatching in the ventrolateral portion of each abdominal segment. They grow in size during the larval stage, developing polytene chromosomes, and degenerate after a period of secretory activity in the pupa. Many small imaginal oenocytes appear in the abdomen during the 4th instar; they do not grow very large but persist throughout adult life, although in reduced numbers (Hosselet, 1925; Trager, 1937; Christophers, 1960).

The dorsal vessel and the pericardial cells pass intact to the adult (Jones, 1954) although the altered structure of the head probably necessitates changes in the outlet of the aorta. The fat body passes from the mature larva to the newly emerged adult unchanged and little, if at all, diminished in size. After maturation the fat body cells no longer have protein-containing vacuoles.

8.6 HORMONES AND METAMORPHOSIS

Insects moult in response to secretion of 20-hydroxyecdysone. The complex sequence of events of a moult is set is train by 20-hydroxyecdysone but the nature of the integument that is formed is under the control of juvenile hormone (JH). If cells are to secrete larval cuticle, JH must be present in the epidermal cells during their initial exposure to 20-hydroxyecdysone. If JH is absent the cells respond to the ecdysteroid by switching to a metamorphic programme, producing first a pupal and then an adult cuticle.

Very little is known about the secretion of these two hormones by the immature stages of mosquitoes. The ecdysteroid content of *Toxorhynchites amboinensis* eggs remained steady at about 550 pg/egg during the first day after oviposition but peaked at about 1600 pg/egg during the second day, falling back to the starting level before hatching (Figure 8.13A). The ecdysteroid level fluctuated between 300 and 650 pg mg^{-1} during the early larval instars of *Tx rutilus*, possibly reflecting changes associated with moulting, and fell to 25 pg mg^{-1} for much of the 4th instar. It rose sharply to peak about two days before larval–pupal ecdysis and then declined (Figure 8.13B) (Westbrook and Russo, 1985; Russo and Westbrook, 1986).

The ecdysteroid concentration in *Aedes aegypti* was very low immediately after larval–pupal ecdysis, when it was due principally to ecdysone and 20-hydroxyecdysone, present in males at 20 and 15 pg mg^{-1} respectively and in females at 40 and 35 pg mg^{-1} (Figure 8.14). The total ecdysteroid titre started to rise within four hours of larval–pupal ecdysis and peaked at about 12 h in males and at about 16 h in females, declining slowly thereafter. Males, 12 h after larval–pupal ecdysis, contained 100 pg ecdysone and 140 pg 20-hydroxyecdysone mg^{-1}, plus

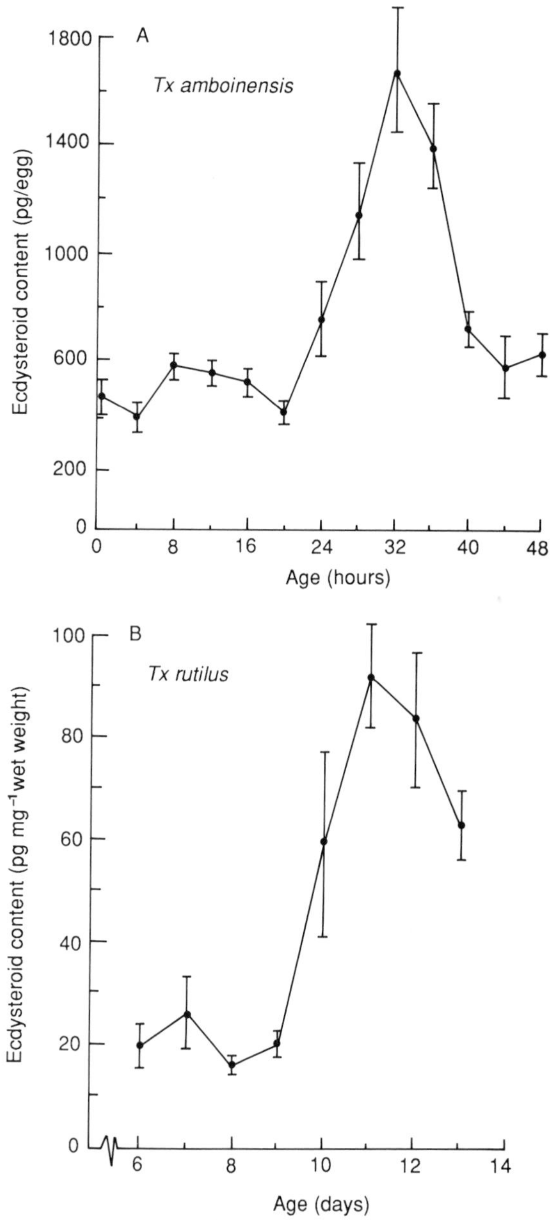

Figure 8.13 Ecdysteroid content of eggs and larvae of *Toxorhynchites*. (A) Ecdysteroid content per egg between oviposition and hatching (from Russo and Westbrook, 1986). (B) Ecdysteroid content of larvae during the 4th instar (from Westbrook and Russo, 1985).

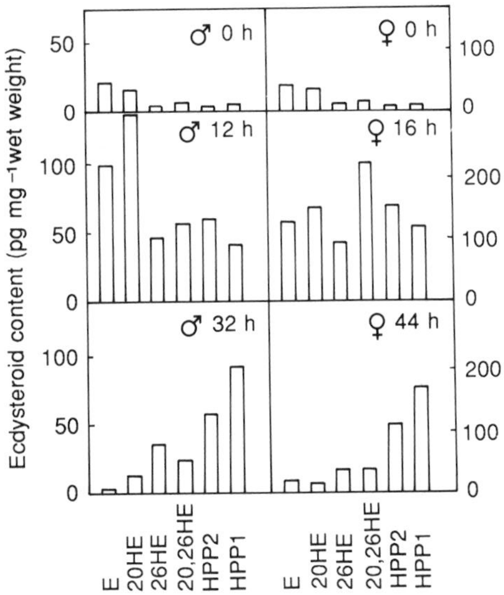

Figure 8.14 Mean concentrations of ecdysteroids present in pupae (and possibly pharate adults) of *Aedes aegypti* at three times after larval–pupal ecdysis, corrected for RIA cross reactivity and normalized to pg mg wet weight^{-1}. Mean weights of pupae were males 2.3 mg, females 4 mg. Eclosion commenced at 40 h in males and 50 h in females. E, ecdysone; 20HE, 20-hydroxyecdysone; 26HE, 26-hydroxyecdysone; 20,26HE, 20,26-dihydroxyecdysone; HPP1 and HPP2, mixtures of high polarity products and conjugates. (After Whisenton *et al.*, 1989.)

various ecdysteroid metabolites which had also increased in titre. The females, at 16 h, showed similar increases but 20,26-dihdroxyecdysone was the most abundant ecdysteroid present. Prior to pupal–adult ecdysis the concentrations of all free ecdysteroids fell, particularly those of ecdysone and 20-hydroxyecdysone, while the concentrations of the highly polar metabolic products increased concomitantly. The ecdysone: 20-hydroxyecdysone ratio was approximately 1:1 throughout the pupal stage (Whisenton *et al.*, 1989). These authors suggested that the earlier rise in ecdysone and 20-hydroxyecdysone titres seen in males led to the earlier eclosion by adult males.

8.7 MATURATION

Mosquitoes that have just emerged from the pupal cuticle are not fully formed adults, and many of their organs continue to develop for a period of hours or even days. A few examples of maturation will be described in this Section. The hormone-induced changes that occur during the gonotrophic cycle, such as development of competence in the fat body, are distinct from maturation. During the maturation phase certain larval organs that continued to function in the pupal and pharate adult stages are destroyed. The prothoracic glands, and the massive abdominal muscles that were used by the pupa and pharate adult for swimming, are histolysed at this stage.

A number of organs continue to develop after emergence. The salivary glands grow in size during the first days of adult life. The secretory content of the female glands increases during this period, as observed from the titres of apyrase, α-glucosidase and bacteriolytic factor (Section 12.2.2). The midgut epithelium of female *Aedes aegypti* is not functional at emergence but its cytodifferentiation is completed during the following two to three days with the formation of microvilli, rough ER and desmosomes, and with elaboration of the basal labyrinth. Differentiation of the midgut epithelium of males is more advanced at emergence (Hecker *et al.*, 1971a, b).

Continued development of the organs of flight after emergence is reflected in a rise of about 200 Hz in wing-beat frequency during the first days of adult life. The indirect flight muscle fibres increase in cross-sectional area at this time, and to accommodate this growth, deposition of cuticle continues over the apodemes on which the muscles are inserted. Due to daily changes in the arrangement of chitin rods the region of cuticle deposited on any day is distinguishable from that deposited the day before and the day after (Neville, 1983). In mosquitoes, the apodemes that continue to grow after emergence are (a) the bilobed phragma invaginated between the mesopostnotum and metanotum, on which the longitudinal indirect flight muscles are inserted, and (b) the fused sternal furcae 2 and 3 on which the mesothoracic dorsoventral indirect flight muscles are inserted (Figure 8.15A). A distinct line on the phragma marks its extent

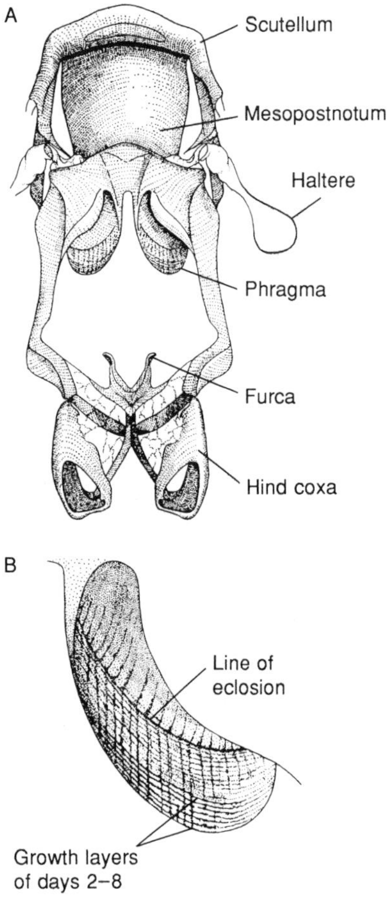

Figure 8.15 (A) Anterior view of the posterior region of the thoracic exoskeleton of *Anopheles gambiae*, showing the bilobed phragma and furca. (B) Part of the phragma of a wild specimen of *An gambiae* eight days old. (From Schlein and Gratz, 1973.)

at emergence, and growth layers added daily after emergence can be distinguished (Figure 8.15B). Such growth layers have been observed in species of *Anopheles*, *Culex* and *Aedes*. They are very distinct in stained cuticle of most wild mosquitoes but less distinct in laboratory-reared material, partly due to the constant temperature. Up to 14 daily bands have been found in the phragmata of wild and laboratory specimens. Few wild specimens show more than 8 growth layers, most between 2 and 5 (Schlein and Gratz, 1972, 1973; Schlein, 1979).

Male mosquitoes are not sexually mature at emergence. For one thing, the antennal fibrillae cling to the shafts of the flagella. During the first day after emergence the internal skeleton of the flagellomeres is developed, and males of *Ae aegypti* are able to erect their fibrillae sufficiently by 15–24 h after emergence for the antennae to function as auditory organs for the detection of females (Roth, 1948; Steward and Atwood, 1963). During the first day or two after emergence the 8th and more posterior segments of the male abdomen rotate through 180°, and until that has happened the genitalia are not orientated appropriately for copulation. Rotation is caused by the contraction of muscles that arise on one side of the abdomen and insert on the opposite side. The arthrodial membrane between the 7th and 8th segments becomes greatly stretched during rotation and the mesocuticle of adjacent sclerites shows realignment of fibres. During the following day the disrupted regions of cuticle become ultrastructurally reorganized (Chevone and Richards, 1976, 1977).

9

The circulatory system

The haemolymph has many functions. It provides the tissues that it bathes with nutrients and metabolic intermediates, and it provides a medium for the transport of their chemical wastes. It distributes hormones throughout the body. It provides defences against parasites. Certain of these activities are described in this chapter, others are touched on elsewhere.

9.1 ANATOMY

9.1.1 The dorsal vessel

The dorsal vessel is a muscular tube which extends from the head to the anterior end of the 8th abdominal segment. It consists, essentially, of a single layer of striated muscle fibres, and its structure varies little between larva, pupa and adult. The dorsal vessel has two anatomically distinct regions: the aorta, which runs through the head and thorax, and the heart, which extends through the abdomen.

The heart is formed of muscle fibres which are said to be spirally coiled. It expands slightly into eight chambers, each with a pair of ostia and associated with two groups of alary muscles (Figure 9.1). The alary muscles originate on the body wall on either side of the heart, and as they approach the heart the constituent muscle fibres fan out and anastomose, above and below the heart, with fibres from the opposite side. The alary muscles form an incomplete dorsal diaphragm which delimits a dorsal or pericardial sinus. The heart is also connected to the dorsal body wall and to other organs by fine connective tissue filaments. The chambers disappear if the alary muscles are cut, showing that they are formed by the tension of these muscles. In mosquito larvae the ostia are simple lip-like openings; in adults they have valves which are semicircular pouches bordered by muscles. The fact that the adult heart, in *Anopheles quadrimaculatus*, can beat both forwards and backwards suggests that the ostia do not determine the direction of flow (Yaguzhinskaya, 1954; Jones, 1954; Christophers, 1960).

The aorta is narrower than the heart, lacks ostia over most of its length, and has no alary muscles. The thoracic aorta of an *Aedes aegypti* adult consists of a single layer of striated muscle fibres. In any transverse section it is seen to be composed of one cell, comprising some 60 sarcomeres. The great majority of myofilaments show a circular orientation, but some strands of myofilaments run obliquely or longitudinally. On both the outer and the luminal surfaces of the circular muscle fibre the sarcolemma is invaginated at the levels of the Z bands, forming narrow clefts and producing an internal and an external pocket of sarcoplasm in each sarcomere. Mitochondria are present in the external pockets but are usually absent from the internal pockets (Clements *et al.*, 1985). Where it is associated with a neurohaemal organ, the aorta is said to have both circular and longitudinal muscle fibres (Meola and Lea, 1972a).

In the prothorax the aorta expands to form a large pulsatile chamber, and thickenings of the aorta wall at each end of the chamber have been interpreted as valves. Within and just behind the head the aorta is closely associated with parts of the neurosecretory system, forming a neurohaemal organ which contains axon terminals of the cerebral and cardiacal neurosecretory systems (Section 10.1.1). More

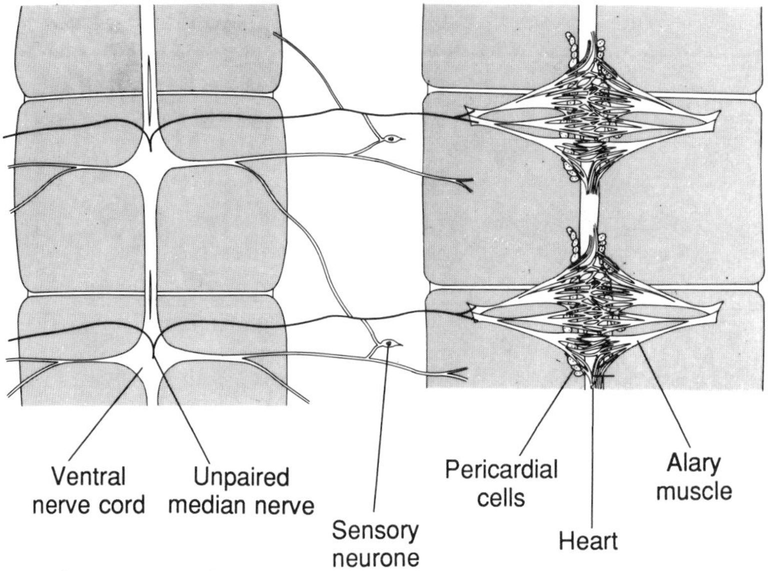

Figure 9.1 Structure and innervation of a region of the heart of an *Anopheles maculipennis* adult. The heart is drawn from its ventral aspect, and the nerve cord from its dorsal aspect. The figure also shows the ventral visceral nervous system, which is drawn in solid lines. (After Yaguzhinskaya, 1954.)

anteriorly the aorta expands into a sinus at the back of the cranium, runs between the pharynx and the brain, and terminates funnel-shaped in front of the brain. In this region the recurrent nerve forms part of the aorta wall (Jones, 1952, 1954; Christophers, 1960; Meola and Lea, 1972a; Jobling, 1987).

No motor nerves have been found supplying the dorsal vessel. The alary muscles are innervated by nerves from the ventral visceral nervous system (Figure 9.1) (Yaguzhinskaya, 1954). Branches of the segmental nerves unite with the nerves of the ventral visceral system, but it is not known whether any of their axons proceed to the alary muscles. The dorsal vessel is usually not well tracheated. However, in *Anopheles* larvae the posterior chamber of the heart receives an extremely dense mat of tracheoles arising from short outgrowths of the dorsal longitudinal tracheal trunks. These tracheoles are attached to the surface of the heart but do not penetrate it (Imms, 1907; Jones, 1954; Christophers, 1960).

9.1.2 Accessory pulsatile organs

A number of accessory pulsatile organs assist the circulation of haemolymph. Within the head of the adult mosquito a pair of antennal pulsatile organs is situated just above the antennal bases. Each consists of an ampulla, an ampulla muscle which is inserted on the aorta, and a capillary vessel which runs from the ampulla to the antenna of its own side (Figure 9.2) (Clements, 1956b). This vessel extends into the antennal flagellum (Boo and Richards, 1975). A valve in the wall of the ampulla is thought to permit

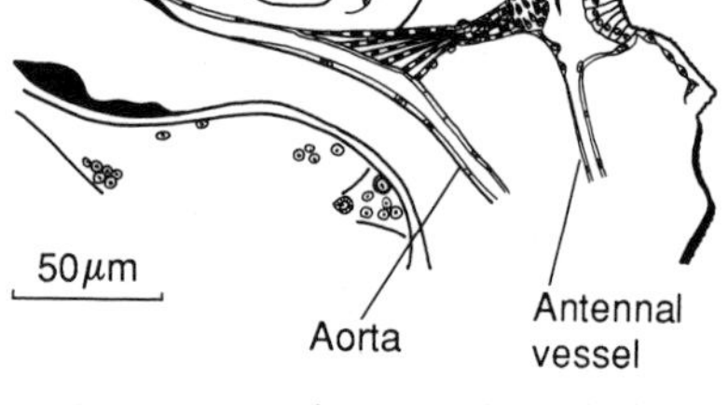

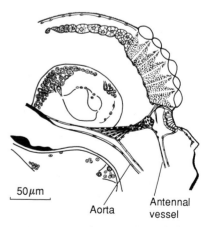

Figure 9.2 A near-sagittal section through the head of an adult *Culex pipiens*, showing one of the antennal pulsatile organs. (From Clements, 1956b.)

haemolymph to enter the ampulla when the muscle contracts. It is assumed that when the muscle relaxes elastic contraction of the ampulla pumps haemolymph into the antenna. No nerve supply to the organ has been found, and its muscle contractions are thought to be myogenic (Clements, 1956b). Haemolymph is circulated through the adult labium by a pulsatile organ which has the form of a septum containing small muscle fibres (Jobling, 1976).

A thin horizontal sheet of muscle arches across the adult scutellum. In both the intact mosquito and the isolated scutellum this membrane flutters rapidly in irregular bursts, possibly assisting the circulation of haemolymph (Jones, 1954; Christophers, 1960). No ventral diaphragm has been found in the larva, but in the pupa and adult a ventral diaphram, comprising a membrane with anastomosing muscle fibres, overlies the nerve cord in the thorax and abdomen. Its scalloped edges provide for continuity between the ventral and perivisceral sinuses (Jones, 1954). Intermittent peristaltic contractions of the lateral body wall, which were pronounced in newly emerged mosquitoes and in adults of both sexes after feeding, were thought to be caused by the ventral diaphragm (Gillett, 1982a).

9.1.3 Pericardial cells

The net-like basket formed around the heart by the alary muscles contains pericardial cells, which are a form of nephrocyte (Figure 9.1). Most of the pericardial cells are binucleate but some are syncytial, containing up to six nuclei. Typically, a series of 28 pairs of pericardial cells extends between the posterior border of the metathorax and the 8th abdominal segment. Each of the 2nd to 7th abdominal segments contains four pairs of pericardial cells, two anterior and two posterior, while the metathorax and 1st and 8th abdominal segments each contain one pair (Pal, 1944; Christophers, 1960). Jones (1954) counted 300 or more abdominal pericardial cells in larvae of four species of *Anopheles*, including *An stephensi* which Pal (1944) considered to have relatively few cells. Other thoracic pericardial cells have been reported, situated between the endocrine gland complexes and the aorta (Burgess and Rempel, 1966).

Cells of similar appearance have been called ventral nephrocytes. Some of these form a syncytial chain, of 5–8 cells, suspended between the salivary glands, or are clustered together; others overlie the ventral nerve cord (Pal, 1944; Jensen and Jones, 1957; Christophers, 1960). Evidence for the excretory function of the nephrocytes is described briefly in Section 5.5.

9.2 ORIGIN AND CHARACTERISTICS OF THE HEARTBEAT

Each heartbeat is a peristaltic contraction that passes rapidly along the whole length of the dorsal vessel. If the dorsal vessel of a mosquito is separated from the alary muscles it becomes twisted but continues to beat, showing that the alary muscles serve only to support the heart and that the peristaltic contractions are produced by the muscle fibres of the dorsal vessel. If the dorsal vessel is isolated from the nervous system it still continues to beat, therefore the cardiac rhythm originates within it, and since the dorsal vessel does not contain any neurones its contractions must be myogenic (Jones, 1954). The insensitivity of the heart of *Anopheles quadrimaculatus* to a number of drugs, including nicotine, physostigmine and adrenalin (Jones, 1956c), was consistent with cardiac automatism but not proof.

Presumably, as in other myogenic hearts, spontaneous rhythmic depolarizations generate action potentials which spread from cell to cell, activating the whole myocardium. When the dorsal vessel of an *Anopheles quadrimaculatus* larva was transected at the junction of thorax and abdomen, the aorta stopped beating but the heart continued to beat at a slow rate. When the cut was made between the 4th and 7th abdominal segments, the anterior portion of the dorsal vessel beat more slowly than usual while the posterior portion increased its beat rate. A cut made between the 7th and 8th segments prevented the posterior portion from beating. Jones (1954) concluded that a pacemaker system was present throughout the muscles of the heart, and that it was strongest at the posterior end and did not extend to the aorta.

The heart does not beat in an unhatched larva of *Aedes aegypti*; it starts to beat only when, after hatching, air first enters the tracheal system (Jones, 1977). During the larval stage of *An quadrimaculatus* the rate of beating declines progressively with age, at least from the 2nd instar (Table 9.1). The beat rate increases in the pupal stage, when beating may stop for prolonged periods, and it increases further in the adult. The heart rate varies with temperature; in 4th instar *An quadrimaculatus* it increased from 54 min^{-1} at 15°C to 139 min^{-1} at 35°C (Jones, 1954). The heart rate of feeding larvae was steady and was little affected when the larvae were disturbed. The heart rate fell when larvae were fasted, and after one hour's fasting it had fallen by about 20 beats min^{-1}. The heart rate of fasting larvae was distinctly elevated when the larvae were disturbed (Jones, 1956a). Such modification of a myogenic rhythm may be caused by neurohormones. The beat rate was about 30% of normal after larvae had been held in nitrogen for one hour (Jones, 1956b).

In mosquito larvae the peristaltic waves always run forwards over the dorsal vessel; in pupae and adults they may run forwards or backwards, but forward-moving waves predominate. In adult mosquitoes the direction of beat periodically reverses. In *An quadrimaculatus* 15–48% of the beats were posteriorly directed, and the ratio of forward to backward beats was relatively constant within an individual. When the heart of an adult *An quadrimaculatus* was transected between the 3rd and 4th abdominal segments, the anterior section beat almost exclusively in a backward direction while the posterior section beat very largely in a forward direction, both sections beating in the direction of least resistance. When ventilation movements began at the posterior end of the adult abdomen the heart beat forwards, and when they began at the anterior end the heart beat backwards. If ventilation movements were prevented by glueing the abdomen to a slide, the dorsal vessel soon stopped beating. Strong movements of the hindgut, to which the heart is tied by connective tissue fibres, could cause the heart to stop beating (Jones, 1954).

Table 9.1 Mean heart rates from representative stages of *Anopheles quadrimaculatus* at 27°C. (From Jones, 1954, 1977.)

Stage	*Instar*	*Age*	*Beats min^{-1}*
Larva	I	<24 h	131.7
	II	<24 h	134.3
	II	>24 h	107.7
	III	<24 h	118.6
	IV	24 h	96.6
	IV	48 h	72.6
	IV	72 h	70.3
Pupa	–	<24 h	109.1
Adult, male	–	Random	151.2
female	–	Random	127.5

In adults, contractions of the abdominal wall at the sterno-pleural junctions occur synchronously with the heart beat and cause slight displacement of haemolymph in the ventral abdomen. These contractions are conspicuous for a day or two after emergence and after feeding. From time to time strong peristaltic contractions of the ventral diaphragm cause substantial haemolymph flow, usually in a posterior direction (Gillett, 1982a).

9.3 HAEMOLYMPH

Haemolymph volume is most accurately assessed by measuring the dilution of injected [^{14}C]inulin solution. The mean haemolymph volume of female *Ae aegypti* was 1.3 μl at emergence, and between 0.65 and 1.0 μl in sugarfed females aged 1–7 days. Mean haemolymph volume increased to 2.4 μl immediately after blood feeding, fell to the prefeeding level within 1–2 h due to diuresis, but slowly increased again and remained at *c.* 1.5 μl between 24 and 44 h post-bloodmeal (Shapiro *et al.*, 1986). The mean haemolymph volume of female *Anopheles stephensi* was 0.34 μl on the day of emergence, when it constituted 25% of body weight. In 3-day-old sugarfed females it was similar but in 14-day-old bloodfed females it had declined to 0.19 μl (Mack *et al.*, 1979a).

Table 9.2 Composition of the haemolymph of mosquito larvae. These values are representative only, because concentrations (mmol/l) vary with age and conditions.

	Aedes aegypti	Aedes taeniorhynchus	Culex pipiens	Opifex fuscus	Anopheles quadrimaculatus	Aedes dorsalis
Medium	*Tap water*	*Tap water*	*Tap water*	*Sea water*	*Deionized water*	*Saline water 638 mosmol kg*$^{-1}$
Refs	*(1, 2, 5, 8, 11)*	*(3, 10)*	*(3, 6, 7, 10)*	*(4)*	*(3)*	*(9)*
Sodium	100	126.1	121.6	187	110.8	163
Potassium	4.2	19.3	15.2	–	14.0	9.5
Calcium	5.2	3.6	3.5	–	3.8	8.7
Magnesium	–	4.6	3.9	–	5.2	3.8
Chloride	51.3	–	47.6	121	–	56
Bicarbonate	–	–	–	–	–	8.1
Organic acids	–	–	16.8	–	–	–
Free amino acids	64.6	–	42.9	–	–	–
Glucose	8.1	–	3.5	–	–	–
Trehalose	40.0	–	24.4	–	–	–
pH	–	7.62 ± 0.14	7.51 ± 0.17	–	–	7.55 ± 0.03
mosmol kg^{-1}	270	320	320	c. 400	255	359

References: 1, Ramsay (1953a); 2, Stobbart (1967); 3, Powers *et al.* (1984); 4, Nicholson (1972); 5, Barkai and Williams (1983); 6, Wigglesworth (1938a); 7, Schmidt and Platzer (1980); 8, Bounias *et al.* (1989); 9, Strange *et al.* (1982); 10, Giblin and Platzer (1984); 11, Nizeyimana *et al.* (1986).

9.3.1 Soluble constituents

Most analyses have been of larval haemolymph, few have been undertaken on haemolymph from adults. The measurements of concentration of individual constituents reflect a balance between input and removal, but indicate nothing about rates of turnover which many be rapid. Concentrations may change with age, even within an instar, therefore the values detailed in Table 9.2 should be treated as representative only. All data that have been reported are for total haemolymph and do not distinguish the contents of the haemocytes from those of the cell-free fluid.

The concentrations of inorganic ions in the larval haemolymph of several species are detailed in Table 9.2. Sodium is the predominant cation; potassium, calcium and magnesium are present at much lower concentrations. Chloride is the predominant anion. Larvae in both freshwater and saline-water habitats continuously and actively regulate the ion composition of their haemolymph, as described in Chapter 6. Measurements with ion-selective microelectrodes showed that the activities of sodium and chloride, i.e. the concentrations of the free ions, could vary independently of their total concentrations, indicating a degree of binding to large organic molecules (Section 6.5.1).

Haemolymph protein concentrations of 50.8 and 32.75 mg ml^{-1} were recorded from late 4th instar larvae of two strains of *Culex pipiens*, the lower value coming from the strain of lower dry weight. Comparable values were 15.5 mg ml^{-1} for *An quadriaculatus*, 13.7 mg ml^{-1} for *Ae taeniorhynchus* and 53.5 mg ml^{-1} for *Ae aegypti*. In each species the protein concentrations were lower early in the 4th instar. SDS-PAGE analysis of *Cx pipiens* larval haemolymph yielded 35 discrete bands (Schmidt and Platzer, 1980; Womersley and Platzer, 1982).

The concentration of total free amino acids in the haemolymph of *Ae aegypti* larvae was c. 60 mmol l^{-1}. In 36-h-old 4th instar larvae proline (13.5 mmol l^{-1}) and threonine (8.3 mmol l^{-1}) were the most abundant amino acids. Aspartate and methionine were found only in traces. Ammonia (2.8 mmol l^{-1}) and urea (4–5 mmol l^{-1}) were also present (Edwards, 1982c; Bounias *et al.*, 1989). When larvae of a brackish-water species were reared in 60% sea water the total amino acid concentration of the haemolymph was nearly 6-fold higher than usual to enable the larvae to osmoconform (Section 6.2.2). The total free haemolymph amino acids of sugarfed adult female *Cx pipiens* fluctuated between c. 40 and 55 mmol l^{-1}, with L-proline contributing 50% of the concentration. The concentration rose to *c.* 90 mmol l^{-1}, 18 h after blood feeding (Uchida *et al.*, 1990). Haemolymph from 3–5-day-old female *Ae aegypti* contained 110 μM tyrosine. No sulphate, phosphate or glucosidic storage forms of tyrosine were found (Munkirs *et al.*, 1990).

Haemolymph from *Ae aegypti* larvae contained 2.13 g lipid per 100 ml. The most abundant lipids were tri- and monoacylglycerols, followed by free fatty acids and phospholipids. Monoacylglycerols were recovered solely as the myristate. Myristic acid and palmitic acid accounted for 79% of the free fatty acids (Gordon *et al.*, 1979). Organic acids were present in the haemolymph of *Cx pipiens* larvae at a total concentration of 16.8 mM. Malate, lactate and citrate were among eight acids identified (Powers *et al.*, 1984; Womersley and Platzer, 1984).

Glycogen, trehalose and glucose were identified among four carbohydrates in the haemolymph of 4th instar *Ae aegypti* larvae. Trehalose was the most abundant, its concentration varying with age between 17.1 and 45.5 mmol l^{-1} (Nizeyimana *et al.*, 1986). Rearing larvae of a brackish-water species in 60% sea water produced a 4.6-fold increase in haemolymph trehalose concentration (Section 6.2.2). The haemolymph of newly emerged female *An stephensi* contained 2.8 mmol l^{-1} glucose and 0.88 mmol l^{-1} trehalose, concentrations which increased to 11.1 and 1.46 mmol l^{-1} respectively four days after blood feeding. At that time the haemolymph also contained maltose, glucuronic acid and inositol (Mack *et al.*, 1979b).

Larval haemolymph is slightly alkaline. The pH values ranged from 7.37 ± 0.11 (SD) (*Toxorhynchites amboinensis*) to 7.62 ± 0.14 (*Ae taeniorhynchus*) among three species examined by Giblin and Platzer (1984). The pH of adult female *An stephensi* haemolymph was 6.50 ± 0.09 at emergence and 6.81 ± 0.08 11 days after a blood meal (Mack and Vanderberg, 1978).

9.3.2 Haemocytes

Insect haemocytes are usually classified into seven major types (Jones, 1977; Gupta, 1985). The morphological descriptions of mosquito haemocytes, summarized below, conform to a number of these established classes, but certain identifications have been questioned.

1. **Prohaemocytes** (= proleucocytes). Small, round to oval cells, usually with a smooth outline. The central nucleus fills most of the cell and is surrounded by a thin layer of cytoplasm. The cytoplasm is intensely basophilic and usually contains only few inclusions. The cells are stable. Prohaemocytes are believed to be stem cells which give rise to plasmatocytes; they may be indistinguishable from small plasmatocytes. Present in larval, pupal and adult stages of *Ae aegypti*, *Cx quinquefasciatus* (Kaaya and Ratcliffe, 1982) and *Cx hortensis* (Amouriq, 1960).
2. **Plasmatocytes** (= macronucleocytes). Polymorphic cells, variable in size, which form pseudopodia. The plasma membrane has many pinocytotic invaginations. The cytoplasm, which may appear granular, is basophilic and contains extensive RER, mitochondria, Golgi complexes and lysosomes. Plasmatocytes may be indistinguishable from prohaemocytes and granular cells. They are immune defence cells, abundant in the larval, pupal and adult stages of *Ae aegypti*, *Cx quinquefasciatus* and *Cx pipiens*. In mosquito larvae they rapidly sequestered neutral red, iron saccharate and latex particles. Some plasmatocytes in adult *Ae aegypti* contained large crystalloids (Jones and Fischman, 1970; Hall and Avery, 1978; Kaaya and Ratcliffe, 1982; Drif and Brehélin, 1983; Chen and Laurence, 1985).
3. **Granular haemocytes** (= granulocytes). Small to large spherical cells, with a relatively small nucleus. The plasma membrane is with or without processes. The cytoplasm is packed

with small or medium-sized granules. Fragile cells, believed to be involved in defence. Present in the larval, pupal and adult stages of *Cx hortensis* (Amouriq, 1960), *Ae aegypti* and *Cx quinquefasciatus* (Kaaya and Ratcliffe, 1982), in the larvae of *Ae atlanticus*, *Ps ciliata* and *Cx tarsalis* (Hall and Avery, 1978) and in adult *An stephensi* (Foley, 1978). Drif and Brehélin (1983) considered (some of) these to have been misidentifications of plasmatocytes.

4. **Oenocytoids**. Large, oval to spherical cells with completely smooth surface, homogeneous cytoplasm and eccentric nucleus; highly refringent under phase contrast illumination. The cytoplasm contains numerous microtubules, well developed rough ER, and a few mitochondria and Golgi complexes. Present but relatively uncommon in larvae of *Ae aegypti* and *Cx quinquefasciatus*. The oenocytoids rapidly sequestered neutral red *in vivo*, and exhibited phenoloxidase activity when exposed to tyrosine, dopa or dopamine (Hall and Avery, 1978; Drif and Brehélin, 1983).
5. **Spherule cells** (= spherulocytes). Round cells with vacuole-like inclusions. Reported among living cells obtained from larvae of *Ae albopictus* (Bhat and Singh, 1975) and tentatively identified among living cells from adults of *An stephensi* (Foley, 1978). Kaaya and Ratcliffe (1982) failed to find spherule cells in *Ae aegypti*, *Cx quinquefasciatus* and certain higher Diptera, and questioned their presence in Diptera.
6. **Adipohaemocytes**. Small to large, spherical, rather fragile cells, containing many possibly lipid, refringent inclusions, as well as granules which were dark under phase contrast illumination. Difficult to distinguish from fat body cells (trophocytes) which have broken off into the haemolymph during bleeding. Present in the larval, pupal and adult stages of *Ae aegypti* and *Cx quinquefasciatus* (Kaaya and Ratcliffe, 1982).
7. **Cystocytes** (= coagulocytes, thrombocytoids). Immune defence cells. Known from a number of insects as small to large, fragile, rapidly disintegrating cells, the nucleus sometimes surrounded by cisternae, the cytoplasm sometimes containing granules. Present in some Diptera but absent from larvae, pupae and adults of *Ae aegypti* and *Cx quinquefasciatus* (Kaaya and Ratcliffe, 1982).

When examining live larvae, Jones (1953a) observed no circulating haemocytes in *Ae aegypti*, extremely few in *An quadrimaculatus*, and only a small number in *Cx pipiens*. Some haemocytes were set free when pressure was applied to the larvae. He described the haemocytes of *An quadrimaculatus* as mostly attached to tissues, including tracheae, fat body and epidermis. Haemocytes were conspicuous in the anal papillae of *Ae aegypti* larvae (Jones, 1954, 1958). However, other investigators have reported much higher numbers of circulating haemocytes. Haemolymph from 4th instar *Ae aegypti* larvae contained *c.* 3000 cells mm^{-3} (Drif and Brehélin, 1983). Perfusion of 1-, 5- and 14-day-old adult females yielded from *Ae aegypti c.* 2000, 1600 and 1000 cells per individual respectively, and from *Ae trivittatus* 3200, 2200 and 2000 cells per individual (Christensen *et al.*, 1989; Li *et al.*, 1989). Perfusion of *An stephensi* yielded 9000–10 000 cells per female up to two days after emergence, but fewer than 2000 from 10-day-old females (Foley, 1978). In adult *Ae aegypti* and *Cx quinquefasciatus* most haemocytes were localized in the head and the last abdominal segment (Kaaya and Ratcliffe, 1982).

Little is known about the physiology of mosquito haemocytes. Although some mosquito haemocytes are capable of phagocytosis they are not employed in phagocytosis of degraded tissues during metamorphosis (Jones, 1954). Haemocytes of adult *Ae trivittatus* and *Ae aegypti* contain monophenol monooxygenase (Li *et al.*, 1989). Haemocytes are usually involved in wound healing in insects, but Day and Bennetts (1953) found no evidence of this in adult *Ae aegypti* in which wounds to the integument and midgut appeared to be healed through the regenerative abilities of the tissues. Total haemocyte numbers in *Ae aegypti* inoculated with *Dirofilaria immitis* microfilariae increased by *c.* 60% by the first day after inoculation and by 140% by the third, declining slowly thereafter (Christensen *et al.*, 1989). The involvement of haemocytes in the encapsulation of parasites is described in the next Section.

9.4 IMMUNE RESPONSES

Mosquitoes are subject to parasitism by fungi, protists and nematodes. The most conspicuous response to organisms that penetrate into the haemocoele, and that are too large to be phagocytized by single cells, is encapsulation of the intruders. The contribution of encapsulation to survival was demonstrated in a study in western Wyoming where larvae of virtually all snowpool *Aedes* species were parasitized by nematodes of the genus *Romanomermis* (Mermithidae), with infection rates ranging between 55 and 93%. Several species were capable of encapsulating the parasites, permitting a high percentage of infected individuals to survive to the adult stage; the others species were not and all infected individuals perished (Blackmore, 1989; Blackmore and Nielsen, 1990).

Adult female mosquitoes are intermediate hosts for filarioid nematodes of the family Onchocercidae, which can be acquired when mosquitoes feed on infected mammals. Heavy infections may be fatal. Some protection against infection is provided by the cibarial armature (Section 11.2.3). Parasites ingested by refractory mosquitoes meet immune responses, which may restrict their migration or effect their encapsulation. The susceptibility of *Aedes aegypti* to filarial infection is controlled by a complex of sex-linked recessive genes, different genes determining susceptibility to different parasites (Macdonald, 1976) (Figure 1.3). The immune response of *Anopheles* to *Plasmodium* results in encapsulation of the ookinetes (Collins *et al.*, 1986).

9.4.1 Capsule formation

In most arthropods encapsulation is initiated when granular haemocytes degranulate upon contact with foreign bodies, and release components of a phenoloxidase-activating system which adhere to them. Plasmatocytes then enclose the foreign objects within an envelope of highly flattened cells which becomes pigmented. The larvae of certain Diptera are exceptional in that encapsulation is acellular or humoral, a quickly hardening pigmented material being precipitated on to the surface of foreign objects without visible participation of haemocytes (Götz, 1986). The pigment observed in capsules is usually described as melanin, which is a polymer of small molecules with few or no crosslinks to adjacent macromolecules. It could be either catechol melanin, derived from tyrosine, or indole melanin, derived from tryptophan.

The immune response to nematodes has been investigated experimentally in adult mosquitoes, which are unusual in that both humoral and cellular elements in the haemolymph contribute to capsule formation. Microfilariae (the ensheathed 1st stage larvae) of *Brugia* spp. (Onchocercidae) that have been ingested by a susceptible host pass through the gut epithelium on their migration into the haemocoele and to the flight muscles. Most of the microfilariae (mff) shed their sheaths during this migration. The parasites complete two moults within the flight muscles before developing to the infective 3rd stage larvae.

Microfilariae of *Brugia pahangi* ingested from infected cats by *Anopheles quadrimaculatus* rapidly entered the haemocoele, most having first shed their sheaths. However, within ten minutes of ingestion both sheathed and exsheathed mff were coated with a homogeneous layer. In fixed preparations this was up to 2 μm thick, consisting in some regions of flocculent material and in others of electron-dense particles of 4–6 nm diameter. It contained no cellular constituents and was considered not to be formed by cell lysis. After two hours plasmatocytes were seen on the surface of encapsulated mff, and by 12–16 h after feeding, some acellular capsules were completely surrounded by plasmatocytes. No other cell type was involved. Completed capsules, 24–48 h after infection, consisted of an inner electron-dense, acellular layer and an outer cellular layer which showed considerable degradation with the appearance of many small vacuoles. The outer surface of the cellular layer was bounded by a double membrane. The enclosed mff were necrotic at this time (Chen and Laurence, 1985). By 48 h after ingestion an acellular layer appeared to be deposited around some of the 1st stage *B. pahangi* larvae that had penetrated the flight muscles of *An quadrimaculatus*, and this became melanized by about 72 h (Nayar *et al.*, 1989).

Within 30 min of the infection of *Armigeres subalbatus* with *Brugia malayi*, mff in the haemocoele were coated with a light brown substance, without apparent participation of haemocytes, and within 90 min they were completely enclosed in melanotic capsules. Shortly afterwards cells started to adhere to the capsules (Figure 9.3). By 24 h the cells had started to flatten and they subsequently lost their integrity. Any larvae that penetrated the flight muscles subsequently became necrotic, and within 3–5 days were enclosed in melanotic capsules (Kobayashi *et al.*, 1986a, b).

Exposure of *Brugia pahangi* mff to *Anopheles quadrimaculatus* haemolymph *in vitro* led within 5 min to the appearance of black granules on their outer surface. These increased in size, coalesced and within 20–60 min formed complete, acellular, dark brown sheaths with a mean wall thickness of 9.4 μm, of similar ultrastructure to the sheaths formed *in vivo*. If the haemolymph was deficient in haemocytes, as when it was collected by centrifugation through glass wool, acellular sheaths were not formed. Addition to the culture fluid of 10 mM diethyldithiocarbamate, an inhibitor of phenoloxidase, led to the formation of permanently transparent capsules formed of flocculent material with a mean wall thickness of 33.75 μm. The transparent material reacted weakly to the DMAB-nitrite test for tryptophan, and failed to react to Millon's test for tyrosine. It was strongly PAS positive (Chen and Laurence, 1987a, b; Chen, 1988).

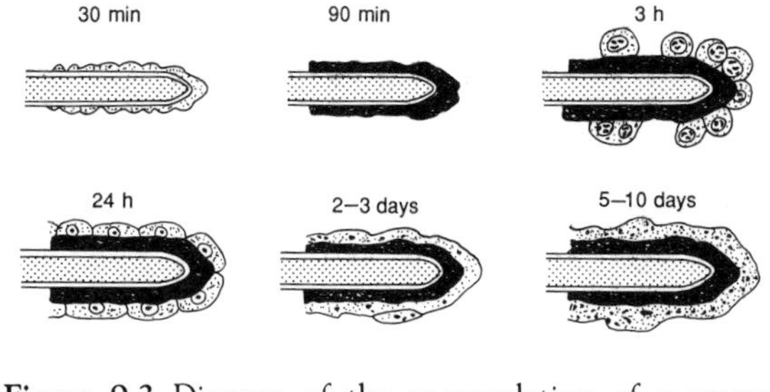

Figure 9.3 Diagram of the encapsulation of a young larva of *Brugia malayi* in the haemocoele of an adult female *Armigeres subalbatus*, seen at different times post-infection. A layer of acellular material is deposited on the larva and becomes pigmented. Haemocytes form an outer layer which loses its cellular integrity. (After Kobayashi *et al.*, 1986a.)

Microfilariae of *Dirofilaria immitis* (Onchocercidae) that are ingested in canine blood pass from the midgut to the Malpighian tubules where they invade the primary cells, occupying a vacuole which is not membrane bound, and developing to the infective stage in this intracellular location. If ingested by a refractory species, such as *Aedes sollicitans*, a percentage of ingested mff become melanized in their intracellular locations, where they have no direct contact with haemolymph (Bradley and Nayar, 1985). Transplantation of parasitized Malpighian tubules between genetically susceptible and refractory females of *Ae aegypti* showed that the refractory factors are associated with the cellular milieu of the tubules (Nayar *et al.*, 1988).

When chemically desheathed mff of *D. immitis* were inoculated into the thoraces of *Ae trivittatus*, a natural vector, haemocyte lysis and the deposition of pigmented material over the mff began within minutes, and within 24 h all inoculated parasites were fully encapsulated. Cell remnants were always observed in areas of pigment deposition. Most haemocytes that were seen were in different stages of lysis near the parasites. Within two days accumulation of cellular remnants around the capsule increased, and within four days a two-layered membrane had formed around the whole capsule isolating it from the haemolymph (Forton *et al.*, 1985; Christensen and Forton, 1986). Females of *Ae aegypti* required several days for the initiation of encapsulation reactions, and seldom were all inoculated mff destroyed. Their haemocyte populations increased 2–3-fold during the first three days after inoculation (Christensen *et al.*, 1989). Different processes may have been involved in capsule formation in these two species: capsules formed by *Ae trivittatus* were black; those formed by *Ae aegypti* were golden-brown (Munkirs *et al.*, 1990).

Females of *Anopheles gambiae*, of a strain selected for refractoriness to *Plasmodium*, encapsulated ookinetes after they had traversed the midgut wall of the mosquito but were still enclosed within the basal lamina, the normal site of oocyst formation. Granules of melanin-like substance appeared to condense on to the parasite from the fluid in the extracellular spaces of the basal labyrinth of the gut cells. Similar granules also appeared to condense from the haemolymph on to the basal lamina underlying the parasites. Haemocytes were rarely observed near completed capsules and neither haemocytes nor their remnants were constituents of the capsules. Susceptible females did not encapsulate the parasites. In both refractory and susceptible females phenoloxidase could be demonstrated within the basal labyrinth of the columnar cells and in the basal lamina surrounding the gut epithelium. However, after an infective blood meal the level of phenoloxidase activity was much reduced in susceptible but not in refractory females (Paskewitz *et al.*, 1988, 1989).

9.4.2 Mechanisms of response

Investigations into the mosquito immune system have lagged behind those with some other insects, and the experimental results that have been obtained permit only a tentative synthesis. The experimental findings are summarized below.

(a) Midgut responses

(i) Migration of *Brugia pahangi* microfilariae from the gut lumen to the haemocoele was more successful in females of a susceptible strain of *Ae aegypti* than in refractory females. The percentage of mff reaching the haemocoele was enhanced 2–3-fold when *N*-acetyl-D-glucosamine was added to an infecting blood meal imbibed by females of the refractory strain (Ham *et al*, 1991). Noting that lectin binding experiments had shown *N*-acetyl-D-glucosamine to be a constituent of both the sheath surrounding *B. pahangi* mff (Devaney, 1985) and the glycocalyx over the luminal surface of *Ae aegypti* midgut cells (Rudin and Hecker, 1989), Ham *et al.* (1991) postulated that, by binding to these sugar residues, lectins present in the gut lumen of refractory mosquitoes block the migration of mff.

(b) Induced changes and activity of haemolymph

(ii) Infection of mosquitoes with mff induced the synthesis of specific haemolymph proteins (Beerntsen and Christensen, 1990; Townson and Chaithong, 1991). Haemolymph from filaria- and bacteria-immunized donors was toxic to mff *in vitro*. Passive transfer of immunity was achieved by inoculation of mosquitoes with haemolymph from filaria-infected donors, which caused a 60% reduction in the number of larvae that developed after intrathoracic injection of mff (Townson and Chaithong, 1991). A preliminary report noted the presence of attacin-like, cecropin-like and diptericin-like sequences in the *Ae aegypti* genome, and indicated that they are expressed (Knapp and Crampton, 1990).

(c) Lectins in haemolymph

(iii) Haemolymph from *Armigeres subalbatus* pupae agglutinated erythrocytes *in vitro*, but lost this activity in the presence of certain sugars including stachyose. It therefore contained lectins. Injection of stachyose reduced the ability of mosquitoes to form melanotic deposits around mff. Lectin (haemagglutinating) activity declined with mosquito age in parallel with a fall in the capacity to melanize injected mff. Haemolymph lectin activity disappeared after incubation with mff, suggesting that the haemagglutinin binds to them (Ogura, 1986, 1987a).

(iv) Stachyose inhibited pigment deposition around mff by *Ar subalbatus* haemolymph *in vitro*, therefore haemolymph lectins may be involved in melanization. High percentages of live mff of *B. pahangi* became enclosed in melanotic coats when added to whole haemolymph from *Ar subalbatus*, but very few were encapsulated when exposed to cell-free haemolymph. *Phaseolus vulgaris* lectin showed similar binding properties to *Armigeres* haemolymph haemagglutinin. Addition of *Phaseolus vulgaris* lectin to cell-free haemolymph resulted in intense pigmentation of heat-killed mff. Therefore certain sedimentable constituents of haemolymph may be involved in the lectin release that facilitates melanization (Ogura, 1987a, b).

(v) Haemolymph taken from *Ar subalbatus* and exposed to sheathed, heat-killed mff contained many spherical structures, of <20 μm diameter, which became melanized although most of the mff were not. Melanotic spherical structures were not observed when cell-free haemolymph was used, therefore they were probably of cellular origin. Particles of melanin were densely deposited on heat-killed mff in whole or cell-free haemolymph containing *Phaseolus vulgaris* lectin, but not if *N*-acetylgalactose was also present when translucent particles became attached to the mff (Ogura, 1987a, b, 1988).

(d) Haemolymph phenoloxidase

(vi) Incubation of heat-killed mff with cell-free phenylthiourea-treated *Ar subalbatus* haemolymph did not elicit melanization, but when the mff were first exposed to supernatant of thermo-coagulated haemolymph and then incubated with PTU-treated haemolymph or L-dopa they became melanized. This suggested that during incubation with heat-treated haemolymph an enzyme, possibly prophenoloxidase, bound to the mff and that substrate became available during re-incubation with PTU-treated haemolymph (Kobayashi *et al.*, 1988).

(vii) Whole-body monophenol monooxygenase (phenoloxidase) investigated in adult *Ae aegypti*, which was probably representative of haemolymph monophenol monooxygenase, was normally in inactive proenzyme form. The proenzyme required a bivalent cation, such as Ca^{2+}, and a protease for activation (Ashida *et al.*, 1990). Melanin formation was strictly localized around parasites that invaded mosquito hosts, and extensive melanization of the haemolymph did not occur.

(viii) The haemocytes of *Ae trivittatus* showed high monophenol monooxygenase activity, and 25% were competent to bind wheat germ agglutinin. Females of *Ae trivittatus* responded within minutes to inoculation with *D. immitis* mff. The haemocytes of *Ae aegypti* showed low monophenol monooxygenase activity, and only 3% bound wheat germ agglutinin. Females of *Ae aegypti* responded slowly to inoculation with mff. Their haemocyte monophenol monooxygenase activity doubled within 24 h of the inoculation. The percentage of *Ae aegypti* haemocytes binding wheat germ agglutin increased to >40% 24–34 h after the inoculation. Only such lectin-binding haemocytes were found adhering to the parasites (Nappi and Christensen, 1986; Christensen *et al.*, 1989; Li *et al.*, 1989; Li and Christensen, 1990).

(e) Model

A tentative model of the melanotic encapsulation of mff in the haemocoele of mosquitoes was proposed by Ogura (1987b). Live mff stimulate release of lectins (e.g. haemagglutinin) from sedimentable constituents of the haemolymph. The lectins bind to the mff, to promonophenol monooxygenase, and to spherical sedimentable components of cellular origin which contain both the proenzyme and substrates for the activated enzyme. Smaller spherical components, containing the same constituents, would separate from the larger spherical components. Complexes of lectin and promonophenol monooxygenase bind to the lectins that had previously bound to the mff. Small spherical components would bind, via lectins, to the lectins already bound to the mff. The proenzyme is activated, the spherical components rupture, and many melanized granules are formed over the mff.

10

The endocrine system and hormones

In this chapter the structure and ultrastructure of the endocrine organs are described, and an outline is given of the structure, synthesis and actions of individual hormones.

10.1 THE ENDOCRINE ORGANS

10.1.1 Neuroendocrine systems

In mosquitoes, the perikarya of neurosecretory cells are situated within the central nervous system and within paired endocrine complexes at the front of the thorax, but their axons may extend to release sites far from these locations. Some neurosecretory cells release hormones into the circulatory system. The axons of others extend to specific organs or tissues for local delivery of the hormones. For descriptive purposes it is helpful to distinguish three neurosecretory systems on an anatomical basis: (a) the cerebral system, (b) the corpora cardiaca and (c) the ventral system.

During the first three larval instars of *Aedes aegypti* the discriminative stain fuchsin imparts only a faint colour to the neurosecretory cells, which are difficult to recognize. In older larvae the neurosecretory granules in the perikarya and axoplasm stain more intensely. The number of recognizable neurosecretory cells increases during the larval stage, and both their cellular and nuclear diameters increase substantially. Most neurosecretory cells persist from larva to adult with no evidence of loss or replacement during metamorphosis, but the positions of the cells change somewhat during the pupal stage (Burgess and Rempel, 1966). When stimulated, neurosecretory cells secrete both hormone and the fuchsinophilic material that characterizes them histochemically, but the extents to which hormone and fuchsinophilic material are discharged appear to be independent, at least in the case of ovarian ecdysteroidogenic hormone secretion after a blood meal (Meola *et al.*, 1970; Meola and Lea, 1971).

(a) Cerebral neurosecretory system

The neurosecretory cells of the brain have been described from *Aedes, Culex* and *Culiseta* (Burgess and Rempel, 1966; Burgess, 1971, 1973; Meola, 1970; Meola and Lea, 1972a). The perikarya of these cells are found in the protocerebrum where

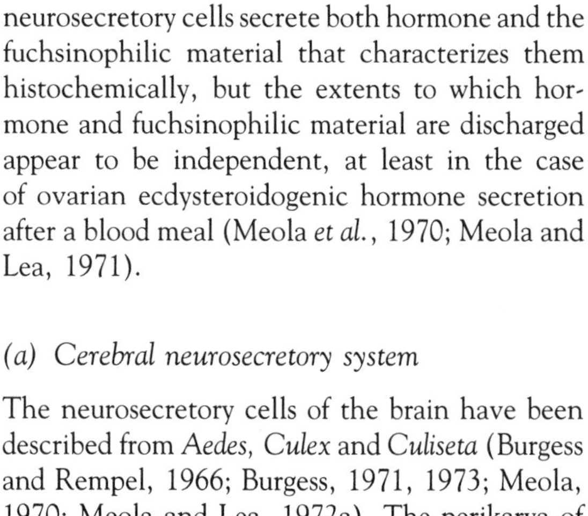

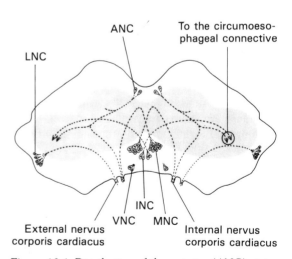

Figure 10.1 Distribution of the anterior (ANC), intermediate (INC), lateral (LNC), medial (MNC) and ventral (NVC) neurosecretory cells within the brain of an adult mosquito. Drawn from the dorsal aspect; schematic.

they are distributed in five paired groups (Figure 10.1; Table 10.1). Four paired groups occur in the pars intercerebralis, each pair flanking the deep median fissure which cleaves that part of the brain, and one paired group occurs laterally. The axons of many of the cerebral neurosecretory cells emerge from the brain in two pairs of nerves, the internal and external nervi corporis cardiaci (NCC-I and NCC-II), which run posteriorly to a neurohaemal area in the back of the head and to organs in the thorax.

The numbers of recognizable neurosecretory cells increase with each larval instar. In newly emerged adults the disposition of the cells is as follows:

1. *Medial Neurosecretory Cells* (MNC). From 8 to 12 cells per group according to species; located towards the posterior end of the brain. In *Ae sollicitans* the staining characteristics of the peripheral cells of each group differ from those of the central cells. The two axon tracts run forwards and downwards, cross each other and then run backwards to emerge from the brain in the contralateral NCC-I.
2. *Anterior Neurosecretory Cells* (ANC). Two or three large cells per group, and possibly additional smaller cells; located near the anterior edge of the brain. The axon tracts have been traced in the larva and pupa of *Culiseta inornata* in which they first run ventrolaterally, send fine branches towards the deutocerebrum, and then bifurcate. On each side, one branch then passes through the circumoesophageal connective into the ventral nerve cord where it has been traced as far as the 3rd abdominal ganglion. The other branch runs towards the posterior ventral edge of the brain, where it appears to emerge in the ipsilateral NCC-I (Burgess, 1971). The paraldehyde-fuchsin stainable component of the cells diminishes a few hours after emergence, and is not discernible after 24 h (Meola and Lea, 1972a).
3. *Intermediate Neurosecretory Cells* (INC). Two cells per group; located near the posterior edge of the brain, dorsal to the MNC. The two axon tracts cross each other in the midline, pass through the contralateral circumoesophageal connective, and run through the ventral nerve cord to the 8th abdominal ganglion. Some branching of axons occurs in the suboesophageal, thoracic, and 8th abdominal ganglia, and all branches appear to terminate within ganglia (Burgess, 1973).
4. *Ventral Neurosecretory Cells* (VNC). Two cells per group; located near the posterior edge of

Table 10.1 The cerebral neurosecretory system of mosquitoes.

Cell group	*No. of cells on each side in adult*	*Routes of axons*	*Destinations of axons*
Medial (MNC)	8 to 12	Contralateral NCC-I dorsal and ventral branches	Neurohaemal organ, thoracic endocrine organs, (?)ingluvial ganglia
Anterior (ANC)	2 or 3	(a) Circumoesophageal commissures into nerve cord (b) Ipsilateral NCC-I	(a) 3rd abdominal ganglion and beyond (b) As the MNC (?)
Intermediate (INC)	2	Contralateral circumoesophageal commissures into ventral nerve cord	8th abdominal ganglion, but branching in anterior ganglia
Ventral (VNC)	2	?	?
Lateral (LNC)	5	Ipsilateral NCC-II	As the MNC (?)

From the data of Burgess (1971, 1973) and Meola and Lea (1972a).

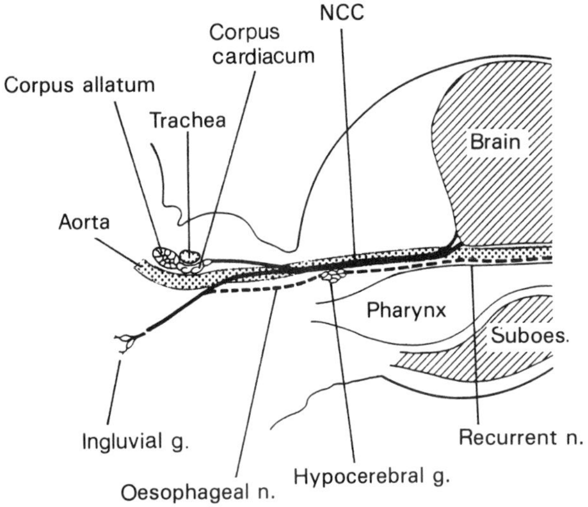

Figure 10.2 Diagram of a near-sagittal section through part of the head, the neck and the anterior tip of the thorax of an adult mosquito, showing the location of certain endocrine organs and associated nerves. (From Clements *et al.* (1985). Copyright (1985), Pergamon Press plc. Reprinted with permission.)

the brain and ventral to the MNC. Observed in adult *Aedes sollicitans*, in which they only occasionally contain fuchsinophilic material. The course of the axons is not known (Meola and Lea, 1972a).

5. *Lateral Neurosecretory Cells* (LNC). Five cells per group; located in the same transverse-vertical plane as the MNC but near the posterolateral edge of the protocerebrum. Their axons emerge from the brain in the ipsilateral NCC-II (Meola and Lea, 1972a).

In adults the paired internal and external nervi corporis cardiaci issue from the posteroventral edge of the brain and pass downwards to enter the wall of the aorta. At this point the internal and external nerves on each side fuse to form a single tract of neurosecretory axons, which runs posteriorly between the muscle fibres that constitute the aorta wall. The ventral surface of each NCC is in contact with the surface of the pharyngeal pump (Figure 10.2). More posteriorly the NCC run over the hypocerebral ganglion (of the stomatogastric nervous system), and each then divides into dorsal and ventral branches. The dorsal branch runs to a complex of endocrine organs at the front of the thorax. The ventral branch runs posteriorly and ventrally and fuses with the oesophageal nerve, continuing with it to terminate in the ingluvial (ventricular) ganglion at the junction of the foregut and midgut.

Once the NCC have entered the aorta they lose their glial sheaths and are surrounded only by a thin connective tissue matrix some 100 nm thick. Neurosecretory release sites occur throughout almost the full lengths of the NCC and their branches, forming a very elongate neurohaemal organ. Anteriorly, where the NCC run over the roof of the pharyngeal pump, release sites are present both on the ventral surface of the nerves and more dorsally where they are intimately associated with the aorta. Towards the back of the head, in the region of the hypocerebral ganglion, the number of neurosecretory release sites and preterminal swellings packed with neurosecretory granules increases greatly. In this region the muscle fibres of the aorta wall separate and allow axon terminals to penetrate, so that release sites open into the aortic lumen. Ventrally, axon terminals extend short distances from the NCC, passing around the hypocerebral ganglion, so that the neurosecretory release sites open into the haemocoele.

Release sites and preterminal swellings packed with neurosecretory granules are present, but less abdundant, throughout the branches of the NCC that run to the thoracic endocrine complex and the ingluvial ganglia (Figure 10.4A). The dorsal branch carries neurosecretory axons that terminate within the corpora allata, penetrating between the cells, and also contains anteriorly directed axons from the corpora cardiaca. Neurosecretory axons extend from the ingluvial ganglia to the salivary glands, to the dorsal and ventral diverticula, and over the surface of the midgut (Meola and Lea, 1972a; Clements *et al.*, 1985).

(b) *Corpora cardiaca*

At the very front of the thorax, where the thorax and neck meet, the paired dorsal cephalic tracheae are interconnected by a small anterior tracheal commissure. From this commissure are suspended two groups of endocrine organs, each containing neurosecretory cells, a corpus allatum and, in the immature stages, a prothoracic gland (Figures 10.2, 10.5). The neurosecretory cells are homologous with the intrinsic cells of the corpora cardiaca of other insects. In many insects the corpora cardiaca consist of intrinsic neurosecretory cells, glial cells and an aortic neurohaemal organ containing the axons of certain cerebral neurosecretory cells, with their preterminal storage sites and release sites. In a survey of the neurosecretory systems of Diptera, Normann (1983) concluded that it was appropriate that the two isolated groups of neurosecretory cells in mosquitoes be called the corpora cardiaca, although they have become separated from the aortic neurosecretory release sites. (Some earlier authors referred to the neurohaemal organ of mosquitoes as the corpus cardiacum.)

In *Aedes aegypti* each corpus cardiacum is composed of six neurosecretory cells, situated below the anterior tracheal commissure and usually at the front of the group of endocrine organs. The perikarya of the cardiacal cells are bounded only by the thin connective tissue matrix, some 50 nm thick, which surrounds the whole complex. Each perikaryon gives rise to a single process, proximally 0.9 μm in diameter, which has the typical structure of an insect axon and which runs forwards into the dorsal branch of the NCC. The perikarya are packed with neurosecretory granules, and the axons contain moderate numbers of granules as well as neurotubules (Figure 10.3).

As the axons of the cardiacal neurosecretory cells enter the allatal nerve they mix with the posteriorly directed axons of the cerebral neurosecretory cells, and consequently the distance forwards that the cardiacal neurosecretory axons run is not known. Numerous neurosecretory release sites are present on the surface of the cardiacal neurosecretory cell perikarya, and further release sites and preterminal swellings occur on the axons of these cells where they enter the allatal nerve (Figure 10.7) (Clements *et al.*, 1985).

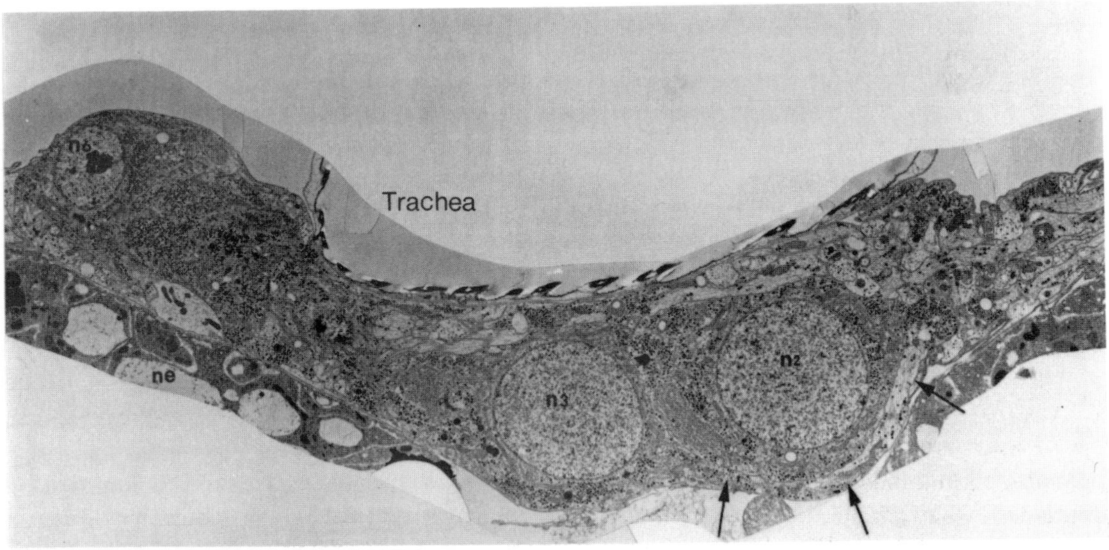

Figure 10.3 Montage of longitudinal sections through a corpus cardiacum of an adult female *Aedes aegypti*. At the right of the montage the dorsal branch of the NCC joins the corpus cardiacum. N2, N3, N6, nuclei in the sequence of cardiacal neurosecretory cells, numbered from anterior (cell 1) to posterior (cell 6); Ne, nephrocyte. Arrows, axons of the cardiacal neurosecretory cells. (From Clements *et al.* (1985). Copyright (1985), Pergamon Press plc. Reprinted with permission.)

(c) Ventral neurosecretory system

Here we place the neurosecretory cells of the suboesophageal ganglion and ventral nerve cord, which are known only poorly in mosquitoes. In insects generally, the axons that extend from neurosecretory cells in the ventral nerve cord run via the unpaired nerves of the ventral visceral nervous system to segmental neurohaemal organs in the thorax and abdomen which have been called perisympathetic organs (Raabe *et al.*, 1974).

Brief descriptions have been given of the ventral visceral nervous system of mosquitoes. In adult *Anopheles maculipennis* an unpaired median nerve arises from the dorsal surface of each abdominal ganglion, runs anteriorly for a short distance and bifurcates into left and right branches, each of which joins the posterior branch of the segmental nerve of its side in the preceding segment (Figure 9.1). The mixed nerve runs laterally, receives an axon from a neurone which is said to be sensory, and finally bifurcates before terminating on the alary muscles (Yaguzhinskaya, 1954).

In larvae of *Toxorhynchites brevipalpis* and *Eretmapodites chrysogaster* a median nerve arises from each thoracic ganglion and bifurcates into two transverse nerves. The median nerve that originates in the metathoracic ganglion runs posteriorly to beyond the 8th abdominal ganglion. In each abdominal segment it is jointed by a median nerve that arises at the anterodorsal region of the ganglion, the point of origin of the transverse nerves. These transverse nerves may be slightly swollen, and in some individuals the swellings contain droplets, said to be neurosecretory, which stain with the non-specific dyes orange G and azocarmine (Grillot, 1977).

10.1.2 Midgut endocrine system

Approximately 500 endocrine cells are distributed throughout the length of the midgut of adult female *Aedes aegypti* (Section 13.2), which is considerably more than the number of neurosecretory cells in the CNS. These midgut endocrine cells contain from a few to many hundreds of membrane-bounded granules of 60–120 nm diameter in the basal cytoplasm, which resemble those characteristic of neurosecretory cells. Two types of granule are present, solid and haloed, the latter having a clear space between the membrane and matrix. Only one type of granule is present in any cell, and the relative density of haloed granules is highest in the posterior half of the stomach. Granules clustered along the basal and basolateral cell membrane release their contents into the extracellular space by exocytosis, i.e. by fusion then opening of the granule and cell membranes (Brown *et al.*, 1985).

10.1.3 Corpora allata

The paired corpora allata are situated within the two groups of endocrine organs that are attached to the anterior tracheal commissure at the front of the thorax. They are ovoid bodies, located just posterior to the cardiacal neurosecretory cells, and up to the pupal stage are surrounded by the cells of the prothoracic gland (Figure 10.5). The cell types of the glandular complexes cannot be distinguished by light microscopy during the first two instars, but growth occurs during that period through cell division and increase in cell size. The corpora allata become distinguishable during the 3rd instar. Cell division continues in the corpora allata throughout the larval stage, and by the end of the 4th instar those of female *Aedes aegypti* contain about 48 cells, and those of males about 30. The cells contain large fuchsinophilic droplets. The corpora allata change little during the pupal stage, except for a slight increase in cytoplasmic content. Cell division continues during the first two days of adult life, and cell number increases to about 60 in females and 40 in males (Rempel and Rueffel, 1964; Burgess and Rempel, 1966).

An ultrastructural study of the corpora allata of adult *Ae aegypti* showed that they are composed of a single cell type. Some of the cells extend from the outer surface almost to the centre of the gland. The cytoplasm contains many elongate mitochondria and numerous free ribosomes. Vacuoles are abundant, and possibly represent

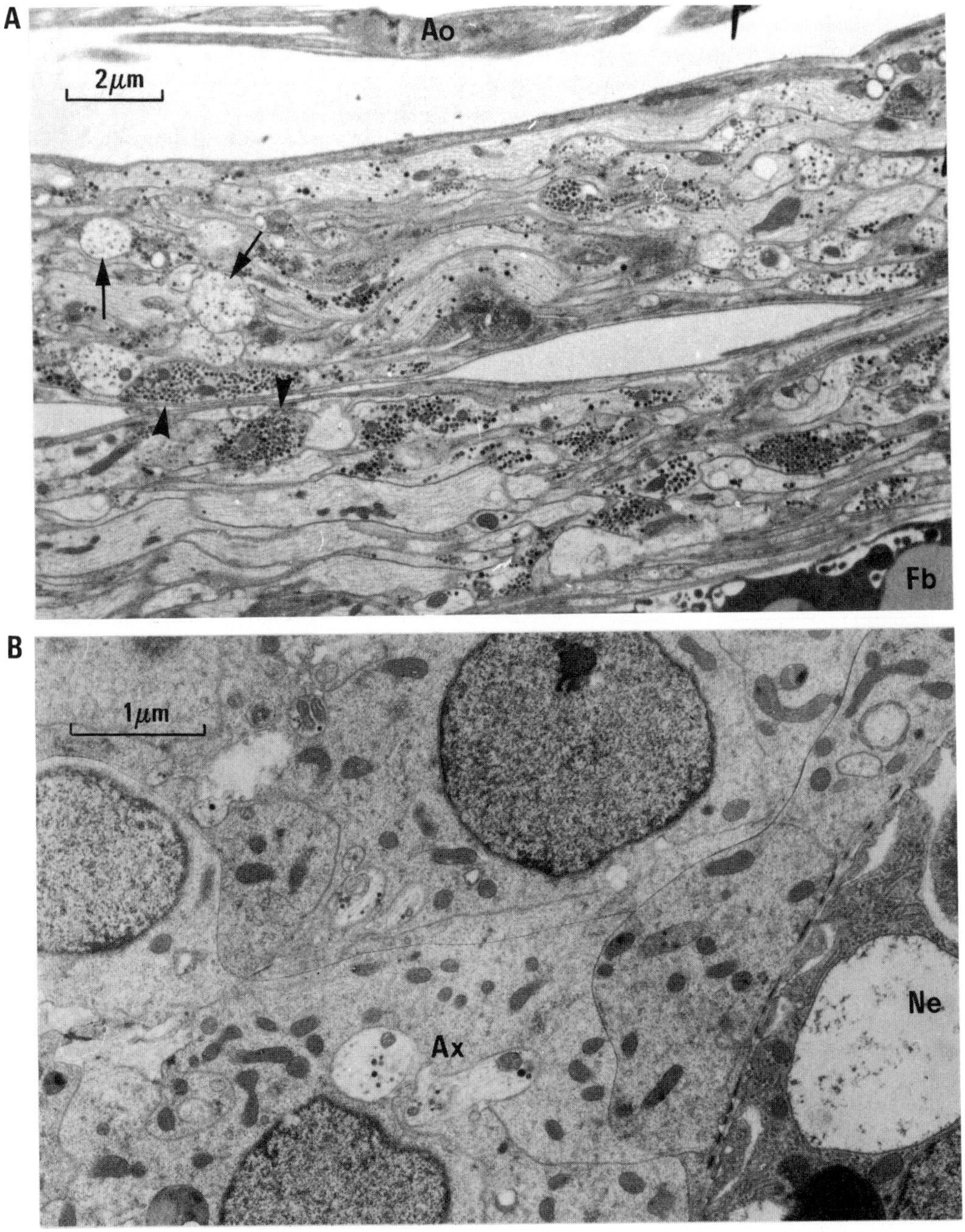

Figure 10.4 Endocrine organs of an adult female *Aedes aegypti*. (A) Longitudinal section through the dorsal and ventral branches of the fused nervi corporis cardiaci which form part of the neurohaemal organ. Arrow heads, neurosecretory release sites. Arrows, assemblages of small less electron-dense granules. (B) Montage of sections through part of a corpus allatum. Ax, neurosecretory axon; Ne, nephrocyte. (From Clements *et al.* (1985). Copyright (1985), Pergamon Press plc. Reprinted with permission.)

lipid droplets leached by solvent. Axons from the allatal nerve, which are not bounded by glia and which contain neurosecretory granules, penetrate between the cells (Figure 10.4B) (Clements *et al.*, 1985).

10.1.4 Prothoracic glands

The paired prothoracic glands are situated within the two groups of endocrine organs that are attached to the anterior tracheal commissure at the front of the thorax, where their cells surround the corpora allata (Figure 10.5). Cell division ceases in the prothoracic gland at the end of the 3rd instar, but during the 4th instar the cells increase greatly in size. By the end of the 4th instar the prothoracic glands of female larvae of *Aedes aegypti* each contain 30–35 cells.

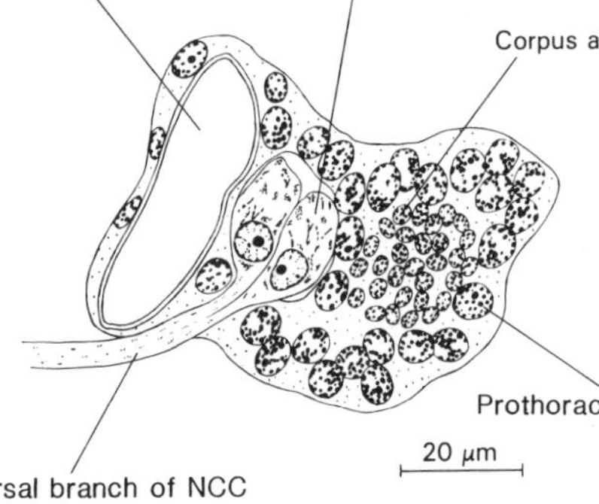

Figure 10.5 Drawing of a longitudinal section through one of the pair of thoracic endocrine complexes in a pupa of *Culex pipiens*, as seen by light microscopy.

The prothoracic glands reach their maximum size shortly after the larval–pupal ecdysis. Soon afterwards their volume starts to fall, and during the pharate-adult stage, possibly earlier, the cells start to degenerate. In *Ae aegypti* all traces of the prothoracic glands usually disappear within 24 h of emergence (Burgess and Rempel, 1966; Burgess, 1967).

10.1.5 Ovaries

Mosquito ovaries secrete at least three hormones which affect ovarian development – OEH-releasing hormone, ecdysone and a postulated oostatic hormone (see Chapter 20). The identities of the ovarian cells that secrete these hormones is not known. The ovary also secretes a decapeptide called trypsin modulating factor which inhibits the synthesis of trypsin by the midgut.

10.2 THE STOMATOGASTRIC NERVOUS SYSTEM

Since the stomatogastric nervous system is physically connected to the neuroendocrine systems, it is not inappropriate to include a description of it in this chapter. The insect stomatogastric system, which is part of the visceral nervous system, is situated predominantly in the head but extends into the thorax. Its principal constituents, the frontal, hypocerebral and ingluvial ganglia and the corpora cardiaca, arise during embryonic development as outgrowths from the dorsal wall of the stomodaeum. The system later acquires connections with the brain and the retrocerebral complex. It comprises both motor and sensory neurones and innervates the foregut, cibarial and pharyngeal musculature and the salivary glands (Penzlin, 1985).

In mosquito larvae the frontal ganglion is situated above the anterior end of the pharynx and contains about a dozen neurones surrounding a central neuropile. It is connected to the brain by frontal connectives which enter the tritocerebrum, and for some distance the connectives are associated with the labral nerves (Figure 10.6). The labral nerves are not part of the stomatogastric nervous system, but it is useful to note their distribution. Each divides into motor and sensory branches. The motor branches innervate the labropalatal muscles, the midpalatal protractors, and various pharyngeal muscles. The sensory branches arise from sensory neurones associated with certain anterior cranial setae and with two sensilla on either side of the midpalatal lobe.

A frontal nerve runs anteriorly from the frontal ganglion, sending branches to the muscles of the pharynx and clypeopalatum. A recurrent nerve runs posteriorly from the frontal ganglion, passing between the brain and foregut and sending branches to the muscles of the pharynx and oesophagus. Near the back of the head the recurrent nerve enters the hypocerebral ganglion, which contains 15–17 neurones. A pair of oesophageal nerves run posteriorly from the hypocerebral ganglion to the ingluvial (=ventricular) ganglia located at the junction of the fore- and midgut. Each ingluvial ganglion consists of two neurones embedded in a cluster of myoblasts which will form the musculature of the adult oesophageal diverticula. Very fine nerves run from the ingluvial ganglia to the oesophagus and midgut (Christophers, 1960; Burgess and Rempel, 1966; Romoser and Venard, 1969).

The great changes in the structure and musculature of the pharynx and oesophagus which occur at metamorphosis are necessarily accompanied by changes of innervation. The adult frontal ganglion contains more neurones than that of the larva and is located above the cibarial pump, near its posterior end. The labrofrontal nerves connect the frontal ganglion with the circumoesophageal connectives and suboesophageal ganglion and not, as in the larva, with the tritocerebrum. The labral nerves, branching off the labrofrontals, innervate muscles in the clypeal region and receive axons from sensory neurones located near the tip of the labrum and in the walls of the cibarial pump (Burgess and Rempel, 1966; von Gernet and Buerger, 1966).

The frontal nerve, running forwards from the frontal ganglion, innervates the clypeolabral muscle. The recurrent nerve runs posteriorly from the frontal ganglion, over the pharynx. Where it passes under the brain the recurrent nerve is intimately associated with the aorta and forms the midventral wall of that organ. The recurrent nerve ends in the hypocerebral ganglion. The hypocerebral ganglion contains the same number of neurones as in the larva, and is said to send a short nerve into the neurohaemal organ (Meola and Lea, 1972a). Two oesophageal nerves run

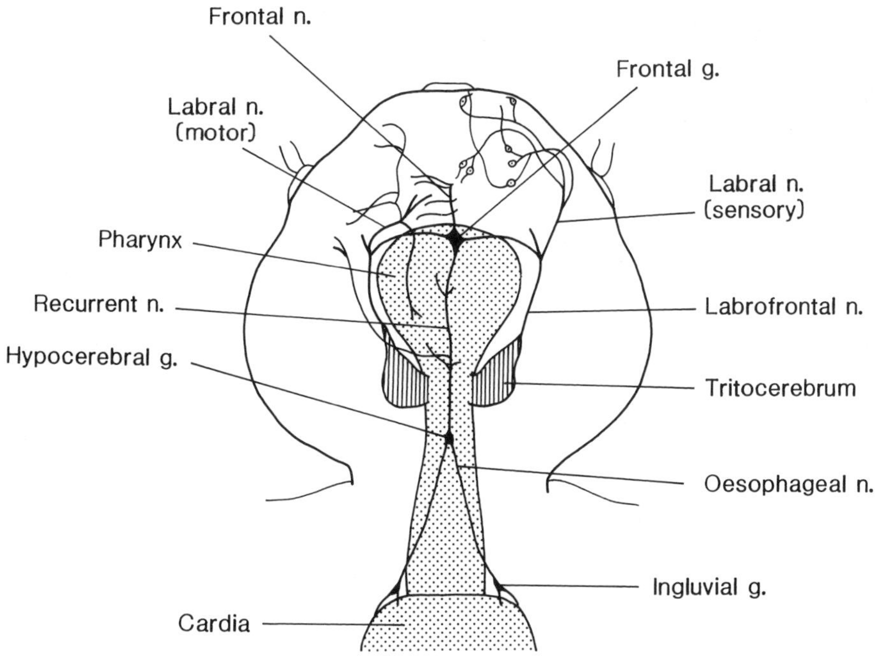

Figure 10.6 The stomatogastric nervous system and some associated nerves of a mosquito larva. Diagrammatic. g, ganglion; n, nerve. (After Burgess and Rempel, 1966.)

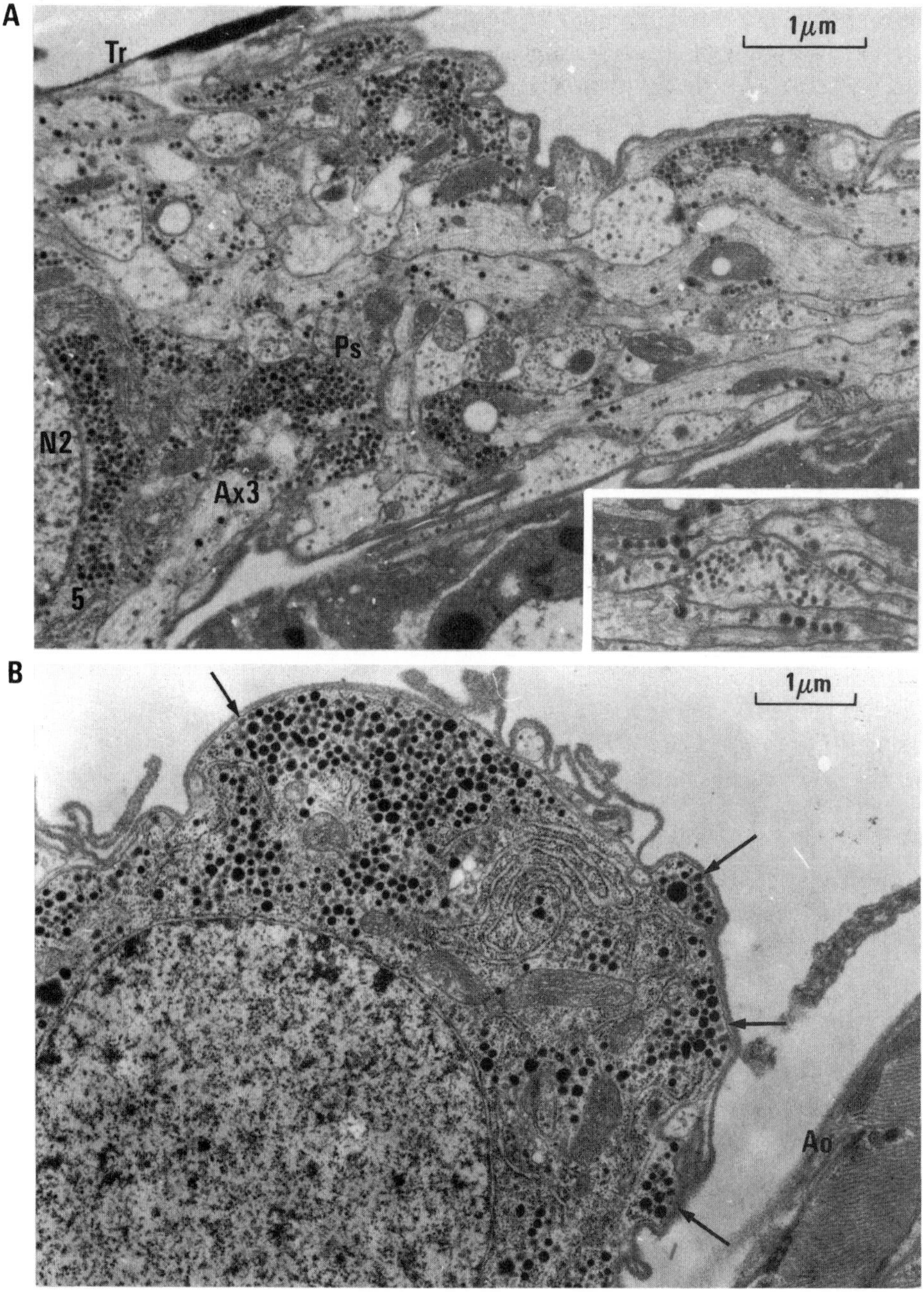

Figure 10.7 Corpus cardiacum of an adult female *Aedes aegypti.* (A) Longitudinal section through the dorsal branch of the NCC immediately anterior to a corpus cardiacum, most of which is off the figure to the left. Ax3, axon of cell 3; N2, nucleus of cell 2; Ps, preterminal storage site; Tr, anterior tracheal commissure. Insert, assemblages of small, less dense granules. (B) Longitudinal section through one side of the most anterior cardiacal neurosecretory cell (cell 1.) Arrows, possible neurosecretory release sites. (From Clements *et al.* (1985). Copyright (1985), Pergamon Press plc. Reprinted with permission.)

posteriorly from the hypocerebral ganglion and each joins the ventral branch of the fused nervi corporis cardiaci of it side before continuing to the ingluvial ganglion (Figure 10.2) (Clements *et al.*, 1985).

10.3 HORMONES

10.3.1 Peptide hormones

Of the insect peptide hormones that have been structurally characterized, some can be grouped into families of natural analogues that exhibit a common primary structure whereas the structures of others appear unique. Three representative families have the following structural or immunoreactive characteristics.

(a) *AKH/RPCH Family*. Octa- to decapeptides in which the N-terminal residue is pyroglutamic acid, the C-terminal residue is amidated, and phenylalanine and tryptophan are always present at positions positions 4 and 8.

(b) *Leucokinins*. Octapeptides with phenylalanine at position 4 and serine, tryptophan and glycine-NH_2 at positions 6, 7 and 8.

(c) *FMRFamide-related peptides* show immunoreactivity to antisera with specificity against FMRFamide (Phe–Met–Arg–Phe–NH_2).

(a) Aedes *head and midgut peptides*

Screening head extract from *Aedes aegypti* with an FMRFamide radioimmunoassay yielded three peptides that shared the carboxy-terminal sequence –Arg–Phe–$NH_2$with FMRFamide, a similarity which may have functional rather than phylogenetic significance (Matsumoto *et al.*, 1989a; Lea and Brown, 1990).

Aea-HP-I pGlu–Arg–Pro–Hyp–Ser–Leu–Lys–Thr–Arg–Phe–NH_2
Aea-HP-II Thr–Arg–Phe–NH_2
Aea-HP-III pGlu–Arg–Pro–Ser–Leu–Lys–Arg–Phe–NH_2

By comparative immunocytochemistry it was established that all cells in the CNS of adult

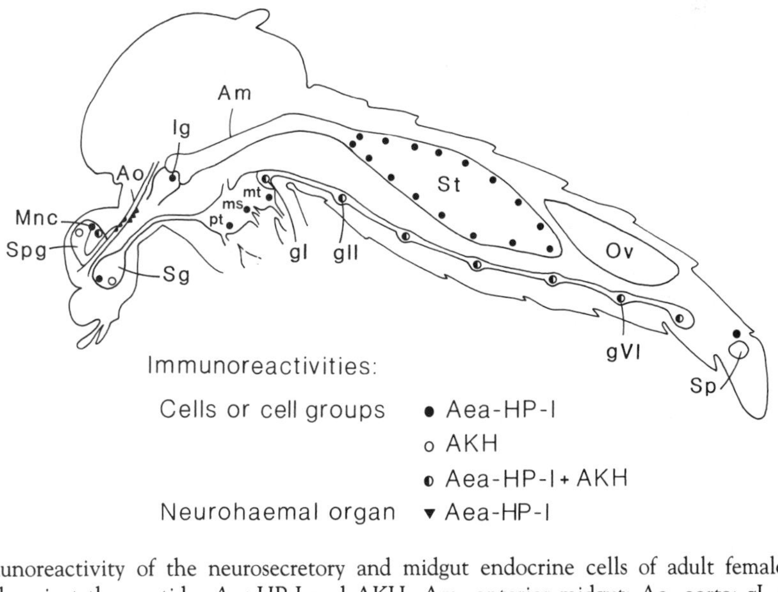

Figure 10.8 Immunoreactivity of the neurosecretory and midgut endocrine cells of adult female *Aedes aegypti* to antisera developed against the peptides *Aea*-HP-I and AKH. Am, anterior midgut; Ao, aorta; gI, gII, gVI, 1st, 2nd and 6th abdominal ganglia; Ig, ingluvial ganglion; Mnc, medial neurosecretory cells; Ov, ovary; pt, ms, mt, pro-, meso- and metathoracic ganglia; Sg, suboesophageal ganglion; Spg, supraoesophageal ganglion; Sp, spermatheca; St, stomach. (After Brown and Lea, 1989.)

female *Ae aegypti* that were immunoreactive to an *Aea*-HP-I antiserum were also immunoreactive to an FMRFamide antiserum. Approximately 120 cells in the CNS of the adult female were immunoreactive, the majority apparently being ordinary neurones in which the immunoreactive peptide may function as a neurotransmitter. Only a small percentage were neurosecretory cells (Figure 10.8). Immunoreactivity was found in paired clusters of five to eight medial neurosecretory cell perikarya, in axons in the nervi corporis cardiaci and axon terminals in the neurohaemal organ, in every case associated solely with neurosecretory granules. The neurosecretory cell perikarya of the corpora cardiaca were not immunoreactive. Immunoreactivity was present in 12–18 cell bodies posterolaterally on each side of the protocerebrum, in 4–10 cell bodies on each side at the front of the brain, in axons in the nerve tract from the tritocerebrum to the frontal ganglion, and in 3–6 cell bodies and a number of axons in each optic lobe. Immunoreactive cells were present in the suboesophageal ganglion and all ganglia of the ventral nerve cord. The cells of the ingluvial ganglia were all immunoreactive, and a single immunoreactive cell was present adjacent to the anterior-most spermatheca (Brown and Lea, 1988, 1989; Lea and Brown, 1990).

Some 15–20 cells in brain, suboesophageal ganglion and abdominal ganglia were immunoreactive to an adipokinetic hormone antiserum (Figure 10.8). Most were in the brain or the suboesophageal ganglion, and those in the brain, at least, were all normal neurones. Some of the cells that were reactive to the AKH antiserum were also reactive to FMRFamide antiserum, therefore they may secrete two peptides (Brown and Lea, 1988).

As in the CNS, all midgut cells that were immunoreactive to *Aea*-HP-I antiserum were also immunoreactive to FMRFamide antiserum. Approximately half of the 500 midgut endocrine cells responded to these antisera, all being restricted to the stomach. Blood feeding reduced the number of immunoreactive cells and the intensity of reaction (Brown *et al.*, 1986; Brown and Lea, 1989; Lea and Brown, 1990).

(b) Peptides from Culex

Fractionation of an extract of 1.2 million mosquitoes, 94% of which were *Culex salinarius*, yielded some 25 peptides which exhibited a range of pharmacological activities including myostimulation, myoinhibition, depolarization and hyperpolarization of Malpighian tubule transepithelial voltage and stimulation of fluid transport. Two peptides depolarized the transepithelial voltage of *Aedes aegypti* Malpighian tubules, and were designated 'Culekinin depolarizing peptides'. CDP-I bore a structural resemblance to the leucokinins, a family of multifunctional neuropeptides known from other insects and defined at the beginning of Section 10.3.1. CDP-II resembled the achetakinins, a family related to the leucokinins but variable in chain length and with either Ser or Pro as the third residue from the amidated C terminus.

CDP-I Asn–Pro–Phe–His–Ser–Trp–Gly–NH_2
CDP-II Asn–Asn–Ala–Asn–Val–Phe–Tyr–Pro–Trp–Gly–NH_2

(Hayes *et al.*, 1992; T.K. Hayes, personal communication).

Two other peptides present in the same extract stimulated contractions of cockroach hindgut but did not affect Malpighian tubules in a noticeable way (G.M. Holman, personal communication). These were named:

Culetachykinins:
I Ala–Pro–Ser–Gly–Phe–Met–Gly–Met–Arg–NH_2
II Ala–Pro–Tyr–Gly–Phe–Thr–Gly–Met–Arg–NH_2

(c) Ovarian ecdysteroidogenic hormone

The name ovarian ecdysteroidogenic hormone (OEH) was proposed by Lea and Brown (1990) for a neuropeptide, synthesized by some of the medial neurosecretory cells of the brain, that stimulates the ovaries to synthesize ecdysone *in vitro*, and which for many years had been called 'egg development neurosecretory hormone' (EDNH). OEH also exhibits gonadotrophic activity but whether it stimulates gonad development directly, or indirectly via ecdysone, is not known. Bioassays for OEH utilize its capacities for inducing ovaries *in vitro* to secrete ecdysone or for stimulating ovarian development when injected into decapitated autogenous females. Ecdysteroidogenic activity has been demonstrated not only in adult female mid-brain but also in whole brains from the 4th instar larvae and adult males of *Aedes aegypti* (Hanaoka and Hagedorn, 1980).

Chromatography of head extract from female *Ae aegypti* yielded a number of fractions with gonadotrophic activity and lacking FMRFamide immunoreactivity. Further purification of certain of these fractions led to the isolation of five molecular species of gonadotrophic peptide with a molecular mass of 11 kDa. One peptide was shown to possess both gonadotrophic and ecdysteroidogenic activity. These peptides were considered to be ovarian ecdysteroidogenic hormones. Although they were relatively stable the possibility remained that their heterogeneity was a result of oxidation (Matsumoto *et al.*, 1989b). Peptides with gonadotrophic or ecdysteroidogenic activity had previously been reported to have molecular masses of 6.5 kDa (Hanaoka and Hagedorn, 1980), 18.7 kDa (Borovsky and Thomas, 1985) and of 11 and 24 kDa (Whisenton *et al.*, 1987).

OEH-releasing factor is secreted by the ovaries following a blood meal and acts on the axon terminals of the medial neurosecretory cells to stimulate OEH release. Its chemical composition is not known. A factor released by the ovaries of *Culex pipiens* and *Cx nigripalpus* was active in *Aedes aegypti* but that from *Anopheles quadrimaculatus* was not (Lea and Van Handel, 1982; Borovsky, 1982).

(d) Prothoracicotrophic hormone (PTTH)

PTTH is characterized by its ability to stimulate the prothoracic glands to secrete ecdysone. Brain extract from *Toxorhynchites brevipalpis* pupae stimulated ecdysone secretion by larval *Toxorhynchites* prothoracic glands *in vitro*, as did brain extracts from *Aedes aegypti* pupae and adults (Whisenton and Bollenbacher, 1986). Extract of *Ae aegypti* heads was active in an *in vitro* assay for PTTH employing lepidopteran prothoracic glands, whereas partly purified *Ae aegypti* OEH was inactive in this assay (Kelly *et al.*, 1986b).

(e) Diuretic hormones

Three partly purified peptides, extracted from *Aedes aegypti* heads, which had molecular masses between 1.8 and 2.7 kDa, altered the transepithelial voltage of perfused Malpighian tubules *in vitro* and stimulated diuresis when injected into decapitated bloodfed females. One of these peptides, called the mosquito natriuretic factor, was postulated to be the hormone responsible for the burst of diuresis that occurs immediately after blood feeding (Section 16.2.4). Diuretic hormone activity has been demonstrated in extracts of the head and the thoracic ganglia of adult *Anopheles freeborni* (Nijhout and Carrow, 1978), and in extracts of the brain and the thoracic ganglia of *Ae taeniorhynchus* larvae (Maddrell and Phillips, 1978). Extracts of *Ae aegypti* heads possessed antidiuretic activity (Lea and Brown, 1990; Petzel and Conlon, 1991).

(f) Trypsin modulating factor

Extraction of ovaries removed from *Aedes aegypti* four days after blood feeding yielded the decapeptide Tyr–Asp–Pro–Ala–Pro–Pro–Pro–Pro–Pro–Pro–COOH, which has a molecular mass of 1047.6 Da. Injection of the decapeptide into the haemocoele of bloodfed females inhibited synthesis of trypsin by the midgut (Section 14.4.3) and consequently diminished vitellogenin incorporation into oocytes (Section 21.8). The decapeptide was named '*Aedes aegypti* trypsin modulating oostatic factor' by Borovsky *et al.* (1991). Synthetic decapeptide inhibited trypsin synthesis when injected into species of *Culex* and *Anopheles*, a sandfly, a stomoxid fly and a flea (Borovsky *et al.*, 1990). The decapeptide resembles a sequence at the putative N-terminal end of a chorionic protein encoded by the gene 15a-2 (Section 19.6.1(a)), differing only in reversal of the first two amino acid residues (Lin *et al.*,

1992). Because the postulated oostatic function is unproven, in this book the peptide will be called trypsin modulating factor (TMF).

(g) *Metabolic factor*

The medial neurosecretory cells of *Aedes taeniorhynchus* contain a factor that causes the incorporation of dietary sugar into triglyceride, the normal end-product. Removal of the medial neurosecretory cell perikarya resulted in the incorporation of dietary sugar into glycogen (Section 15.1.3).

10.3.2 Juvenile hormones

Juvenile hormone, synthesized in the corpora allata, occurs in at least four structurally-related forms in different insects – JH-0, JH-I, JH-II and JH-III. The final steps in the biosynthesis of JH-III are shown in Figure 10.9.

Extracts of larvae and adults of *Aedes aegypti* subjected to GC-MS analysis were found to contain only juvenile hormone III (Baker *et al.*, 1983: Borovsky *et al.*, 1985). Radioimmunoassay coupled with chromatographic analysis of extracts from adult *Ae detritus* showed JH-III to constitute over 95% of juvenile hormones present (Guilvard *et al.*, 1984). Corpora allata from adult female *Culex pipiens*, which were cultured *in vitro* with L-[*methyl*-^{3}H]methionine, secreted radioactive JH-III into the medium, but no JH-I or JH-II (Readio *et al.*, 1988).

In addition to its role in regulating metamorphosis, juvenile hormone affects a number of physiological processes related to mosquito ovarian development (Section 21.6), and it modifies certain adult behaviour patterns. The synthetic JH analogue Methoprene (= Altosid) is a useful experimental tool because of its resistance to metabolism.

In experiments with *Ae aegypti*, JH-III proved less active than JH-I or JH-II. In stimulating ovarian follicle growth to the previtellogenic resting stage, JH-I was 25 times, and JH-II four times more active than JH-III. In disturbing metamorphosis, JH-I was ten times, and JH-II five times more active than JH-III (Baker *et al.*, 1983). JH-I was more active than JH-III in restoring follicular growth in allatectomized *Cx pipiens*, whereas JH-III was more active in initiating adult behaviour patterns (Readio *et al.*, 1988).

[^{3}H]Juvenile hormone-III that had been topically applied to bloodfed *Ae aegypti* was metabolized to a number of products including JH-III acid, diol, and acid diol. The acid diol appeared in the urine. Haemolymph from adult females contained an esterase with some specificity for

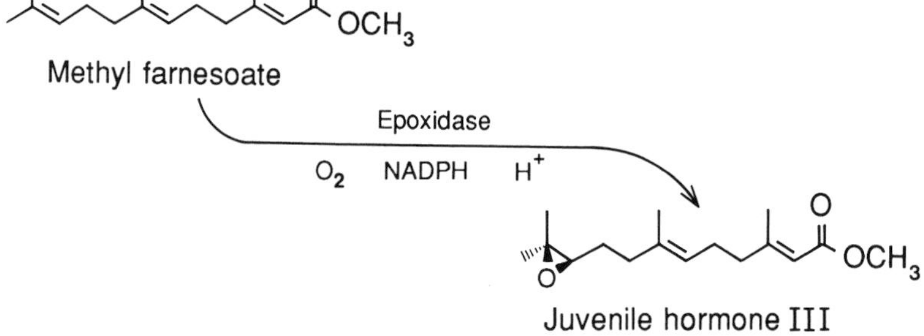

Figure 10.9 Final steps in the biosynthetic pathway of juvenile hormone III in insects.

juvenile hormone, hydrolysing JH-III to yield JH-III acid. The other haemolymph esterases did not hydrolyse juvenile hormone to a significant extent. Juvenile hormone esterase was produced by both fat body and ovary of blood-fed females (Shapiro *et al.*, 1986).

Binding protein with high affinity and specificity for juvenile hormone III was present in newly emerged adult females of *Ae atropalpus*, being found in both carcase and haemolymph. It had a molecular mass of approximately 25 kDa and a dissociation constant of 67 nM. The binding protein protected JH-III from hydrolysis by added carboxylesterases. Carcases of recently bloodfed *Ae aegypti* contained a binding protein with K_d 49 nM; vitellogenic ovaries contained one with K_d 71 nM (Thomas *et al.*, 1986). High-affinity binding proteins may be haemolymph carrier proteins or receptor proteins.

10.3.3 Ecdysteroids

The term ecdysteroid is used generically for a large group of polyhydroxylated Δ^7,6-ketosteroids that are found in plants, arthropods and other invertebrates. In insects two of these compounds may show hormonal activity, ecdysone (= α-ecdysone) and 20-hydroxyecdysone (= β-ecdysone = ecdysterone).

Radioimmunoassay (RIA) has been used to measure the total ecdysteroid content of insects. The antibodies employed react, depending on their affinities, with a range of conjugates and metabolites as well as with the natural hormones. Measurements of RIA-active materials are usually expressed as '20-hydroxyecdysone equivalents'. To identify individual compounds it is necessary to supplement the RIA with a chromatographic technique such as HPLC.

Ecdysteroids have been demonstrated in the eggs, larvae, pupae and adults of mosquitoes (Sections 8.6, 21.5). The prothoracic glands, the source of ecdysteroids in immature insects, are present in mosquito larvae and pupae but disappear from pharate adults. In adult females the ovaries are the principal site of synthesis. It is not known which cells in the mosquito ovary are the site of synthesis, but in locusts the follicular epithelial cells have been implicated (Hagedorn, 1985). 20-Hydroxyecdysone stimulates vitellogenin synthesis (Sections 20.2.3, 21.5.2), and modifies certain adult behaviour patterns.

20-Hydroxyecdysone was found by Hagedorn *et al.* (1975) to be the predominant ecdysteroid in adult females of *Aedes aegypti*, only a trace of ecdysone being found. However, both compounds were reported present in *Aedes aegypti* by Borovsky *et al.* (1986) and *Anopheles stephensi* by Redfern (1982). Ecdysone is secreted by the ovary during the vitellogenic phase of ovarian development (Hagedorn *et al.*, 1975; Hanaoka and Hagedorn, 1980), and injected [^{14}C]cholesterol can enter the pathway of ecdysone and 20-hydroxyecdysone biosynthesis (Borovsky *et al.*, 1986). Preparations of fat body + body wall and of ovaries from bloodfed *Ae aegypti* were both capable of 20-hydroxylating ecdysone *in vitro* (Hagedorn, 1985; Smith and Mitchell, 1986). The enzyme that catalyses this step, ecdysone 20-monooxygenase (EC 1.14.99.22), was present in alimentary canal, fat body and ovaries. K_m of 2–3 × 10^{-7} M were obtained with tissue homogenates. 20-Hydroxyecdysone was inhibitory, with an I_{50} of 10^{-5} M (Smith and Mitchell, 1986).

Ecdysone constituted the major portion of ecdysteroid secreted by *Ae atroparvus* ovaries *in vitro* but significant quantities of RIA-positive material that co-eluted with 20-hydroxyecdysone on HPLC were also secreted (Birnbaum *et al.*, 1984).

Ecdysteroid hormones are probably excreted as polar metabolites and conjugates. Metabolic products identified in pharate adults of *Ae aegypti* included 26-hydroxyecdysone and 20,26-dihydroxyecdysone, as well as conjugates of those hydroxylated derivatives and of ecdysonic and 20-hydroxyecdysonic acids (Whisenton *et al.*, 1989). A sulphotransferase present in larvae, pupae and adults of *Ae togoi* accepted 20-hydroxyecdysone and ecdysone as substrate (Shampengtong and Wong, 1989).

11

Adult food and feeding mechanisms

It is well known that female mosquitoes imbibe blood to obtain protein for ovary development, but less well known that mosquitoes of both sexes require plant juices as an energy source. The characteristics of culicid adults that distinguish them from the adults of related families of nematocerous Diptera stem in large measure from the blood-sucking habit, which requires modifications of female anatomy, physiology and behaviour. In this chapter we shall review the nature of the food and the manner in which mosquitoes obtain it after arriving at the source, whether plant or animal. The longer distance foraging behaviour will be considered in Volume 2.

11.1 FOOD

11.1.1 Plant juices

Plant sugars are a major food resource for mosquitoes. Floral nectar is the best-known source but mosquitoes also obtain sugars from extra-floral nectaries, damaged fruits, damaged and intact vegetative tissues and honeydew (Hocking, 1953; McCrae *et al.*, 1968, 1969; Joseph, 1970; Magnarelli, 1979a). These plant juices provide an important energy source during most of the adult life of both sexes. They are the only food resource of males.

The disaccharide sucrose and the monosaccharides D-fructose and D-glucose are the predominant constituents of nectar. The nectars of certain plant families are particularly rich in sucrose, while in the nectars of other families fructose and glucose predominate or the sugars occur in variable proportions. The disaccharides maltose and melibiose and the trisaccharides raffinose and melezitose are found less frequently (Van Handel *et al.*, 1972; Baker and Baker, 1983a, b). The records of melezitose may be dubious.

Honeydew can be an important sugar source for mosquitoes. For example, for the saltmarsh mosquito *Aedes taeniorhynchus* aphid honeydew was second in importance to nectar as a source of sugar (Haeger, 1955). Aphid and coccid honeydew can contain high concentrations of oligosaccharides that are not present in the plant sap on which the honeydew-producing insects have fed. The trisaccharide melezitose is one example, and is produced enzymatically from sucrose in the gut of certain aphids (Bacon and Dickinson, 1957). It is possible that the melezitose found in the crop contents of *Culex tarsalis* fed in the laboratory on sunflowers (Schaefer and Miura, 1972) was present on the sunflowers as a contaminant.

Amino acids are present in the nectar of most plants (Baker and Baker, 1983b) and in aphid honeydew (Auclair, 1963), but the concentrations are low and insignificant for mosquito ovary development.

Flowers of the arctic tundra that were accessible to insects sometimes contained little or no nectar, but when they had been protected from insects for 24 h the nectar usually had a sugar content of 20–50% w/v (Hocking, 1953, 1968). In three very different habitats most nectar concentrations were within that range (Baker and Baker, 1983b). The viscosities of sugar solutions increase disproportionately with increase in concentration, and they are greater at lower temperatures (Stokes and Mills, 1965). Nectar with a total sugar content of 30% w/v and which contained only fructose, glucose and sucrose, in equal proportions by weight, would have fructose and glucose concentrations both of 0.55 M, a sucrose concentration of 0.29 M, an osmolality in excess of 1.5 osmol kg^{-1}, and at 20°C a relative viscosity of approximately 3.0. It is not known whether mosquitoes feed on nectars of a restricted range of sugar concentrations, as do some other insects.

Male and female mosquitoes have been observed to ingest sucrose from dry crystalline deposits on glass surfaces. Saliva flowed from the mouthparts of the mosquito, and the sugar was ingested in a solution which appeared to move from the periphery of the labella, across its surface and into a median groove. Individuals fed for as long as 30 min without apparent diminution in the flow of saliva, and experimental evidence suggested that there was no regurgitation of gut contents. Individuals of *Culex tarsalis, Cx pipiens, Culiseta incidens* and *Cs inornata* fed readily on crystalline sugar, but *Anopheles freeborni, Aedes sierrensis, Ae aegypti* and *Ae taeniorhynchus* were not such avid feeders (Eliason, 1963). Most deposits of honeydew on leaves are dry, and chironomids and other Diptera discharge saliva when feeding on them (Downes, 1974). Saliva may also be discharged when mosquitoes feed on concentrated nectars to reduce their viscosity.

In nearly all species that have been examined, ingested plant juices are stored in the crop and not in the midgut. The mean crop capacity of wild *Ae communis* was 0.91 μl, which was equivalent to 29% of the mean weight of the unfed mosquitoes. The mean crop capacity of wild *Ae punctor* was 3.39 μl, equivalent to 62% of the mean unfed weight (Hocking, 1953). The total sugar content in distended or partly distended wild female *Culex tarsalis* varied between 0.087 and 0.45 mg per crop (Schaefer and Miura, 1972).

D-Fructose is not present in unfed or bloodfed mosquitoes, and so the presence of D-fructose in the crop or whole body, whether as the monosaccharide or as a constituent of sucrose, is taken to indicate that the mosquito has recently fed on plant juices or honeydew. D-Fructose can be identified in individual mosquitoes by its reaction at room temperature with anthrone (Van Handel, 1967a, 1972a).

Nectar feeding by mosquitoes has been observed in the high arctic as well as in temperate and tropical regions. At Zika in Uganda, out of a total of 109 known species, 60 species in 10 genera were recorded feeding on nectar. Records made over several seasons in the subarctic showed that the peaks of abundance of tundra and forest mosquitoes were synchronous with the peaks of nectar production in tundra and forest respectively (Hocking, 1953, 1968; Grimstad and DeFoliart, 1974; McCrae *et al.*, 1968, 1969).

Plant juices can be important for the survival of mosquitoes. Females of *Ae aegypti, Cx quinquefasciatus* and *Cx tritaeniorhynchus* lived much longer in the laboratory when provided with both sugar and blood meals than when given blood alone (Briegel and Kaiser, 1973; Harada *et al.*, 1976). With *Cx tarsalis* nectar feeding is particularly apparent during the autumn, at a time when there is almost no blood feeding and the females are developing the large fat bodies which provide an energy store for the winter. In California, overwintering females of *Cx tarsalis* continue to feed on nectar at a low rate throughout the winter, including the females in the diapausing component of the population (Schaefer and Miura, 1972; Reisen *et al.*, 1986). In Japan, some females of three other species of *Culex*, hibernating in caves, were found to contain nectar (Harada *et al.*, 1975).

Species of *Malaya* solicit regurgitated food from ants of the genus *Crematogaster* (Jacobson, 1909; Farquharson, 1918). The description for *Ml jacobsoni* typifies the behaviour pattern. Female mosquitoes hovered and flew vertically up and down near young bamboos on which ants were feeding on the juices of the growing tips. A mosquito would alight in front of a downward-travelling ant, insert its proboscis between the ant's mouthparts, and acquire a droplet of fluid (Macdonald and Traub, 1960).

11.1.2 Blood

The blood ingested by female mosquitoes is used principally for egg production but there is some evidence that it is also an energy source. Thus, in the laboratory bloodfed females survived longer than water-fed females; females of *Anopheles quadrimaculatus* by up to three days and of *Aedes taeniorhynchus* by up to eight days (Nayer and Sauerman, 1975b). Females of *Culex pipiens* that were provided with blood but not sugar were able to use the blood as an energy source for flight (Clements, 1955). Very few wild females of *Ae vexans* and *Cx restuans* foraged for nectar while digesting blood meals (Vargo and Foster, 1984).

Blood is essential for egg production by anautogenous females, and for this function protein is its only essential ingredient. Egg production was not increased when vitamins, nucleic acid, or sterols were added to a chemically defined diet, but the addition of sodium or potassium ions led to a doubling of egg production over that on salt-free diets (Dimond *et al.*, 1958). Females of *Ae aegypti* fed a high-protein artificial diet developed as many eggs as females fed whole bovine blood (Kogan, 1990). The nutritional properties of blood are discussed in detail in Section 22.1. It is sufficient here to state that analyses of human blood, guinea-pig blood and yolk protein from *Ae aegypti* ovaries showed a general similarity of amino acid composition. For *Ae aegypti*, L-isoleucine is the limiting factor for egg production on human, bovine or sheep blood, and the isoleucine content of haemoglobin is low in those species. For other mosquitoes, and different vertebrate hosts, other essential amino acids may be limiting.

Most species feed on mammals or birds but a few feed regularly on reptiles, amphibia, or fish (Tempelis, 1975; Heatwole and Shine, 1976). Serological methods have been used for many years to identify the sources of blood meals (Washino and Tempelis, 1983), but it is likely that these tests will be superseded by enzyme-linked immunosorbent assays (ELISA) which are specific, sensitive and easy to use in the field (Burkot *et al.*, 1981; Service *et al.*, 1986; Beier *et al.*, 1988). Individual human hosts can be identified by DNA fingerprinting after the polymerase chain reaction has been used to amplify the small amounts of human DNA that are extractable from bloodfed mosquitoes up to 10 h post-feed (Coulson *et al.*, 1990).

Methods which differ widely in convenience of use and accuracy have been used to measure the size of blood meals. The gravimetric method is simple and non-lethal but is prone to underestimate blood intake because many species commence diuresis while still feeding (Section 16.1.2). Isotopic labelling of blood with cerium-144, a radionuclide that is not excreted, solves the problem of diuresis but presents sampling problems when live hosts are used (Boorman, 1960; Redington and Hockmeyer, 1976). Haemoglobinometry, which involves the conversion of ingested haemoglobin to haemiglobincyanide, is accurate and suitable for all species that do not discharge erythrocytes during or immediately after feeding, and if total faecal haematin is assayed by this technique the females need not be killed (Briegel *et al.*, 1979). Protein assay of dissected midguts is comparable in convenience and accuracy to haemoglobinometry (Houseman and Downe, 1986). Near-infrared reflectance spectroscopy of haemoglobin requires special instrumentation but is fast, accurate and does not kill the insect (Hall *et al.*, 1990). ELISA is an accurate method when antisera are available for the host species (Konishi and Yamanishi, 1984).

Mosquitoes commonly ingest more than their own weight in blood, indeed the mean blood-meal weight can be from two to four times the mean weight of the female (Nayar and Sauerman, 1975b). The estimates of blood-meal size obtained in most early studies are suspect and will not be considered here. Service (1968a) measured gravimetrically the weight of human blood ingested by several species, and discarded the few females that he observed passing fluid while feeding. Mean blood-meal weights recorded were: *Coquillettidia richiardii* 2.51 mg (2.66 μl), *Aedes cinereus* 3.50 mg (3.71 μl), *Ae cantans* 5.73 mg (6.07 μl) and *Culiseta annulata* 5.70 mg (6.04 μl). The larger species, such as *Cs annulata* and *Ae cantans*, took larger blood meals than the smaller species such as *Ae cinereus* and *Cq richiardii*. However, the smaller species tended to ingest more blood per unit of body weight; thus *Ae cinereus* ingested on average 1.33 times its own weight, whereas *Cs annulata* ingested only 0.82 times its own weight.

A number of investigators have estimated the blood intake of *Ae aegypti*. Stobbart (1977), who used the gravimetric method but adjusted the results for fluid losses, found that the mean weight of human blood ingested was about 2.70 mg (2.55 μl). Redington and Hockmeyer (1976), who fed *Ae aegypti* on monkey blood through an artificial membrane, obtained mean volumes of 2.63 and 2.91 μl by the (unadjusted) gravimetric and the isotopic methods respectively. Using haemoglobinometry, Klowden and Lea (1978) found the replete capacities to be 5.0 ± 0.1 μl for females from well-nourished larvae, and 3.5 ± 0.1 μl for the smaller females from poorly nourished larvae.

Most anophelines, possibly all, are unable to grossly distend the midgut and abdomen with ingested blood. The maximum volume that the midgut could hold without rupturing was 2 μl in *An stephensi* and 3 μl in *An albimanus*. However, anophelines use diuresis to increase protein intake. Certain species which start to discharge clear fluid from the anus a few seconds after the start of blood ingestion appear to do so. They discharge fluid that is free of erythrocytes and has the low nitrogen concentration characteristic of a product of the Malpighian tubules. Total blood intake can be determined by comparing the nitrogen content of gut plus discharged fluid with that of the host's blood. The mean total nitrogen intake by females of *An albimanus* that had fed on a guinea pig was 56.4 μg, equivalent to 2.9 μl blood. Of this nitrogen, 97% was retained in the stomach and 3% was discharged in a large volume of clear fluid. Immediately after feeding, the females showed a weight gain equivalent to only 1.5 μl blood, therefore a *c.* 2-fold concentration had been achieved. A similar experiment with *An quadrimaculatus* indicated a concentration factor of 2.1 (Briegel and Rezzonico, 1985).

Some anophelines discharge erythrocytes during the course of blood feeding, and the discharged fluid varies from reddish-tinged to bright red according to species. Females of *An gambiae*, *An arabiensis* and *An dirus* discharged small volumes of weakly coloured fluid. Relative erythrocyte densities in host blood and midgut contents indicated concentration factors of 1.85 for *gambiae*, 1.39 for *arabiensis* and 1.23 for *dirus* (Vaughan *et al.*, 1991). Once females of *An stephensi* had imbibed about 2 μl blood, they started to discharge copious quantities of bright red fluid containing intact erythrocytes. The mean total blood consumption was 6 μl (range 2–10 μl). At the end of the meal, the midgut nitrogen concentration was much higher

Table 11.1 The volumes of blood ingested by four species of *Aedes* feeding on restrained and unrestrained rabbits. (From Klowden and Lea, 1979.)

Species	Restrained host			Unrestrained host		
	No.	Blood-meal volume (μl)		No.	Blood-meal volume (μl)	
		Mean ± s.e	Range		Mean ± s.e.	Range
Ae triseriatus	71	7.6 ± 0.3	2.0–17.5	25	3.4 ± 0.4	0.7– 7.3
Ae aegypti	117	5.0 ± 0.1	2.2– 7.0	4	1.9 ± 0.5	0.8– 3.3
Ae sollicitans	109	11.2 ± 0.3	2.0–18.6	28	10.4 ± 1.0	0.6–17.4
Ae taeniorhynchus	20	6.5 ± 0.3	2.8– 8.9	29	3.7 ± 0.5	0.2–10.3

than that of the host blood, indicating 2.3-fold concentration. The midguts contained on average 65 μg nitrogen and 500 μg haemoglobin, while the discharged fluid contained 54 μg N and 407 μg haemoglobin – indicating, for both N and haemoglobin, 55% retention of the amount ingested. Clearly, >55% of blood water was discharged. After hydrolysis, the discharged fluid showed the same molar distribution of amino acids as hydrolysed guinea-pig blood, therefore no measurable partitioning of proteins had occurred (Briegel and Rezzonico, 1985). Involvement of the bristles of the pylorus in filtration of blood by some anophelines was postulated by Vaughan *et al.* (1991).

In nature, mosquitoes are sometimes able to ingest only a fraction of the potential blood-meal volume because of the defensive behaviour of the hosts (Edman and Kale, 1971; Edman *et al.*, 1974). Klowden and Lea (1979) demonstrated this phenomenon when they used haemoglobinometry to measure the blood intake of mosquitoes feeding on caged rabbits which were either free to move or restrained by nylon stocking. The physical restraint permitted three mosquito species out of four to take significantly larger blood meals (Table 11.1). Cattle in Queensland, which were subject to heavy attack by *Cx annulirostris* and other mosquitoes, were estimated to lose approximately 166 ml blood per beast per night from December to March (Standfast and Dyce, 1968). Mosquito attack can significantly reduce the rate of weight gain in cattle (Steelman *et al.*, 1973, 1976). Indeed, there are documented cases of cattle deaths apparently due to exsanguination, possibly aggravated by hypersensitivity to salivary antigens, associated with massive attacks by *Psorophora columbiae* (Bishopp, 1933) and *Aedes sollicitans* (Abbitt and Abbitt, 1981).

11.2 STRUCTURE OF THE MOUTHPARTS AND ASSOCIATED ORGANS

Mosquito mouthparts are structurally adapted for the uptake of fluids, and those of the females are used both to probe flowers and to pierce skin. The mouthparts are extended into a proboscis (Figure 11.1A) which arises from a snout-like projection of the cranium, sometimes called the rostrum, which consists dorsally of the clypeus, laterally and ventrally of the gena and hypostoma. Most mouthpart constituents have the form of extremely long, slender stylets which are closely appressed in a bundle or fascicle. Thin, flexible areas of cuticle surround the insertions of the mouthparts on the rostrum, allowing them a limited degree of movement. Many of the muscles of the mouthparts have their origins on the endoskeleton of the head, the most important parts of which are the tentorium, the clypeal phragmata, and thickenings of the head capsule such as the subgenal ridges (Figures 11.3, 11.4, 11.9,). Studies on the anatomy of mosquito mouthparts and the feeding mechanism date back to Réaumur (1750). Important recent studies include those of Vogel (1921), Robinson (1939), Snodgrass (1944, 1959), Schiemenz (1957), Christophers (1960), Waldbauer (1962), Lee (1974) and Owen (1985). A terminology for the mouthparts and other head structures provided by Harbach and Knight (1980) is followed here largely but not entirely.

11.2.1 Female mouthparts

The following description is generally applicable to blood-sucking species. The **labium**, which is the least modified of the mouthparts, is a relatively stout organ composed of three parts: a long trough-like prementum, a pair of labella which articulate on the distal end of the prementum, and a terminal ligula (Figure 11.1E). At its extreme base the **prementum** is flattish, but elsewhere its walls curve upwards to form a trough, the premental gutter, which serves as a sheath for the stylets when they are not used in feeding (Figure 11.1C). Over its distal half the walls of the prementum overlap each other dorsally, fully enclosing the premental gutter. The basal half of the prementum is flexible, permitting it to be bent into a U-shape during blood feeding. The prementum has a large internal cavity, a part of the haemocoele, which contains muscles, nerves, tracheae and a pulsatile organ. Two pairs

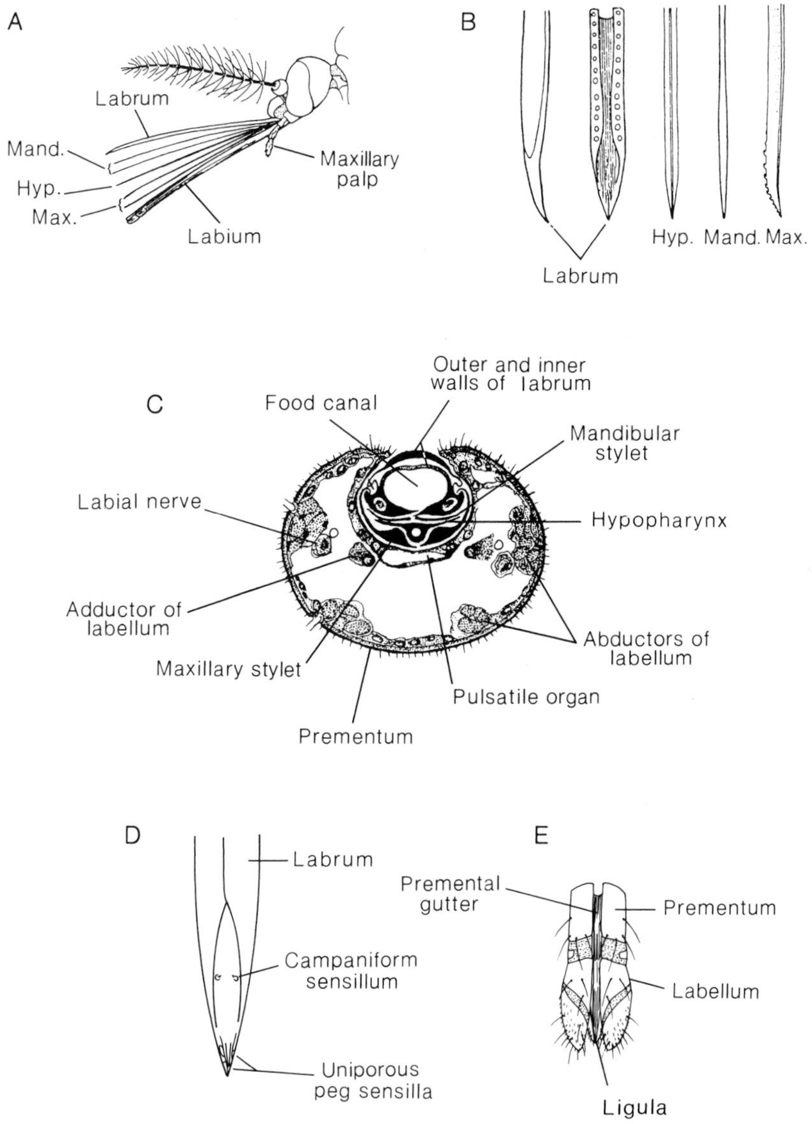

Figure 11.1 Head and mouthparts of a female *Aedes aegypti.* (A) Head and mouthparts, with the stylets removed from the labial sheath. (B) Distal parts of the stylets. (C) Transverse section of the proboscis, towards its base. (D) Tip of the labrum, from the ventral aspect, showing the sensilla. (E) Distal part of the labium, dorsal aspect. (A, B and E from Snodgrass (1944); C from Jobling (1976); and D after Lee (1974).)

of muscles are situated in the distal half of the prementum which have their insertions on the labella. One pair are abductors, which cause the labellar lobes to separate; the other pair are adductors (Vogel, 1921; Waldbauer, 1962; Jobling, 1976; Owen, 1985).

The **labella** are bead-like fleshy organs, thought to represent the labial palps of other insects. They bear a number of external sclerites, which are usually taken to indicate segmentation (Figure 11.1E), and they contain an elaborate endoskeleton. The cuticular structures, together with the labellar abductor and adductor muscles, form a system which effects the movements of the labella and ligula (Puchkova, 1977; Jobling, 1976). The cuticle on the oral surface of each labellum is hydrophilic, and along the dorsal and ventral margins of the oral surface the cuticle is thrown into a series of tight folds which are thought to function like the pseudotracheae of higher Diptera, delivering fluids to the orifice of the labral food canal. The labella bear a large number of sensilla, mostly contact chemoreceptors and mechanoreceptors but including two proprioceptors (Müller, 1968; Owen, 1971; Larsen and Owen, 1971; Owen *et al.*, 1974; Puchkova, 1976).

The **ligula**, which is a short pointed lobe between the labella (Figure 11.1E), is probably a secondarily evolved structure (Matsuda, 1965). The tip of the ligula bears a number of microtrichia but it has no sensilla and is not innervated. The tip of the stylet fascicle lies in a shallow groove which does not extend the full length of the ligula. A deep ventral groove, which extends to the top of the ligula, contains a flap of tissue. The flap can fill with haemolymph and expand out of the groove like a balloon, which it does when the ligula makes contact with water. This expansion of the ligula is thought to result from physical forces acting on the folded cuticle, supplemented by the internal pressure of the haemolymph. It is not mediated by sensory receptors (Larsen and Owen, 1971).

The six piercing stylets are the labrum, the paired mandibles, the hypopharynx, and the paired maxillae (Figure 11.1A–C). For clarity, drawings of transverse sections of the fascicle usually portray the stylets as separated. In fact, over the distal half of the proboscis at least, the stylets are in close apposition to one another (Robinson, 1939).

In more generalized insects the **labrum** forms an upper lip, below the clypeus, while the inner surfaces of these two structures constitute the **palatum** (formerly called the epipharynx), which is divided into the labropalatum and the clypeopalatum. In mosquitoes the labrum is the largest, stiffest and most dorsal stylet in the fascicle. It is a two-layered structure with a complete, sclerotized inner wall, the labropalatum, and a less extensive, sclerotized

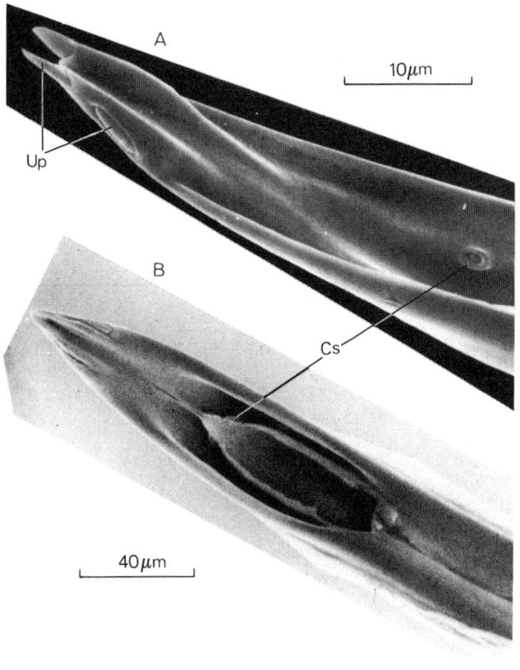

Figure 11.2 Tip of the labrum of females of (A) *Anopheles stephensi*, ventrolateral aspect. (B) *Psorophora ferox*, ventral aspect showing entrance to food canal. Cs, Campaniform sensillum. Up, uniporous peg sensilla. (From Lee and Craig, 1983b. Micrographs by courtesy of Dr R.M.K.W. Lee.)

outer wall, rather confusingly termed the labrum. Labrum and labropalatum are fused at their tips to form a solid pointed structure. Elsewhere they are connected by membranous lateral walls, but Lee (1974) stated that in *Aedes aegypti* the lateral walls are always disrupted. The sides of the labropalatum curve downwards and inwards to enclose a channel, the food canal (Figure 11.1C). The labropalatum is continuous with the clypeopalatum which forms the roof of the cibarial pump, and the food canal opens into the lumen of the pump. At the base of the proboscis the sides of the labropalatum do not meet ventrally and the food canal is closed by the underlying hypopharynx. More distally the sides of the labropalatum come together, and they have been described in different genera as meeting, interlocking, or overlapping in the midline. In this region the mandibles become inserted between the labrum and the hypopharynx (Figure

11.1C). The tip of the labrum curves downwards slightly and is sharpened-off ventrally like a quill pen (Figure 11.2). The food canal consequently opens distally as a groove, for the intake of fluids, although the groove may be closed by the mandibles. In *Psorophora ciliata* the diameter of the food canal is about 60 μm near the base of the labrum, 50 μm at a point two-thirds distant from the base, and about 40 μm near the tip (Waldbauer, 1962). In *Ae aegypti* the diameter of the food canal is about 30 μm near the middle of the labrum and about 20 μm near the tip (Lee, 1974). The dimensions reported for six other species were within the same range (Buse and Kuhlow, 1979).

The labrum, which is the only innervated stylet, bears three pairs of sensilla (Figure 11.1D, 11.2). A pair of apical sensilla, located at the tip of the stylet, and a pair of subapical sensilla are uniporous peg sensilla and are therefore contact chemoreceptors. A pair of campaniform sensilla, proprioceptors, is located near the opening of the labral food canal. All species of *Toxorhynchites* examined lacked the uniporous peg sensilla and some species lacked the campaniform sensilla also. Sensory neurones near the tip of the labrum send dendrites to the sensilla, while axons arising from the neurones join together in fine nerves which run through two canals in the labrum (Figure 11.1C) and eventually join the labrofrontal nerve (Christophers, 1960; von Gernet and Buerger, 1966; Lee, 1974; Lee and Craig, 1983b; Owen, 1985).

The **clypeus** is separated by an inflexible suture into a smaller anteclypeus and larger postclypeus. The anteclypeus extends inwards as two blade-like clypeal apodemes, which are broad at the base and curve to a point (Figures 11.3, 11.4A). Two muscles that arise on the clypeal apodemes insert on the lateral surfaces of the labropalatum. Contraction of these clypeolabropalatal muscles lifts and slightly retracts the base of the labropalatum, sliding it under the labrum. Because the labrum and labropalatum are fused distally this should result in a ventral flexion of the tip of the stylet. Schiemenz (1957) considered that this muscle also serves as a dilator of the functional mouth. An unpaired conical muscle that arises over the median surface of the postclypeus passes between the two clypeal apodemes to insert on the base of the labrum. Contraction of this clypeolabral muscle retracts the outer wall of the labrum, flexing the tip of the stylet upwards (Waldbauer, 1962; Owen, 1985).

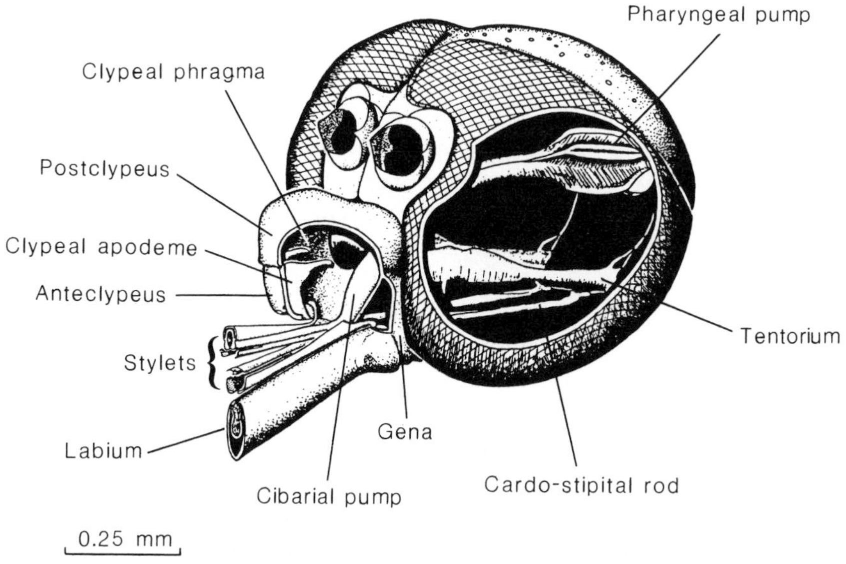

Figure 11.3 The head of a female *Culiseta annulata*, cut away to show internal sclerotized structures. (From Schiemenz, 1957.)

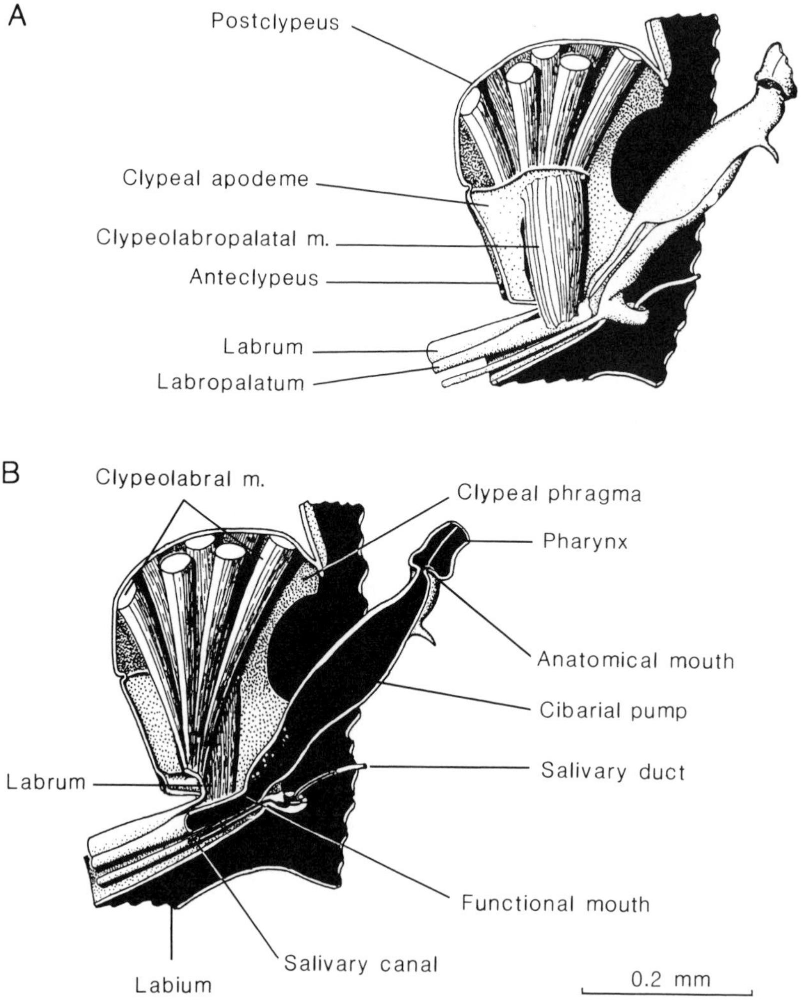

Figure 11.4 Musculature of the labrum and labropalatum in a female *Culiseta annulata.* (A) A lateral view. (B) A near-sagittal view. (From Schiemenz, 1957.)

The **mandibles** are extremely thin, delicate stylets. At the base of the proboscis the mandibles occupy a lateral position in the fascicle. More anteriorly they move medially, meet in the midline, and eventually come to lie above and below each other (Figure 11.1C). The mandibles widen and thicken slightly near their tips, which may be spatula-shaped or pointed. In most species the tips are smooth, but in species of *Anopheles* and *Armigeres* they bear a number of very fine teeth. In *Ae aegypti* the mandibles form the ventral closure for the salivary canal. A cuticular bar, the mandibular suspensorium, is connected flexibly at one end to the base of the mandible and at the other articulates with the subgenal ridge. A very long apodeme extends from this end of the bar to the mandibular protractor which arises on the postgena at the back of the cranium (Figure 11.6). In all species examined a retractor muscle which arises on the tentorium inserts on the base of the mandibular stylet. In some species another mandibular retractor, which is also inserted on the base of the mandible, arises on the postgena (Schiemenz, 1957; Wenk, 1961; Lee and Craig, 1983c; Owen, 1985).

The **hypopharynx** is a delicate flat unpaired stylet with a thickening along its midline which contains the salivary canal (Figure 11.1B,C).

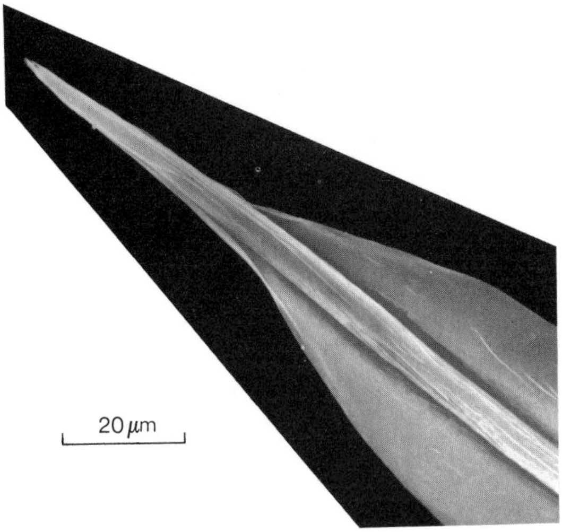

Figure 11.5 Tip of the hypopharynx of a female *Armigeres subalbatus*, ventral aspect showing the salivary canal which forms a midrib. (From Lee and Craig, 1983c. Micrograph by courtesy of Dr R.M.K.W. Lee.)

Anatomically the salivary canal is an open groove; however, over the distal part of the hypopharynx the canal is effectively closed by the overlapping of its edges, and more proximally it is enclosed by the flat edge on one side slotting into a deep groove in the other. In *Ae aegypti* the diameter of the salivary duct is from 2.5 to 3.2 μm over most of its length. Near its tip the hypopharynx is reduced to the midrib only (Figure 11.5) and the salivary canal opens at the apex. At the base of the proboscis the hypopharynx is continuous with the ventral wall of the cibarial pump, and the salivary canal opens into the lumen of the salivary pump (Figure 11.9A, C). The hypopharynx has no musculature and is incapable of independent movement (Nehman, 1968; Hudson, 1970a; Lee, 1974).

The **maxillae**, which are the principal piercing organs, are highly modified, each consisting of an internal rod, a stylet, a palp and a complex musculature (Figure 11.8). An apodeme-like rod within the head is shown by the muscles inserted on it to be the fused cardo and stipes. The cardo, which forms one end of the rod, is attached to the postgena near the posterior tentorial pit, while the other end of the rod is attached to the base of the proboscis. The lacinia has the form of a stylet. Near the base of the proboscis the paired maxillary stylets are lateral in location but over most of the length of the proboscis they lie side by side and occupy a ventrolateral position in the premental gutter. The thickened mesial edges of the two maxillary stylets flank the midrib of the hypopharynx, while the flattened outer regions curve up and around the other stylets (Figure 11.1C). The sclerotized tips of the maxillary stylets are sharply tapered. In most species their curved outer edges bear backward-pointing teeth, and in some species the inner edges bear a few forward-pointing teeth (Figure 11.7).

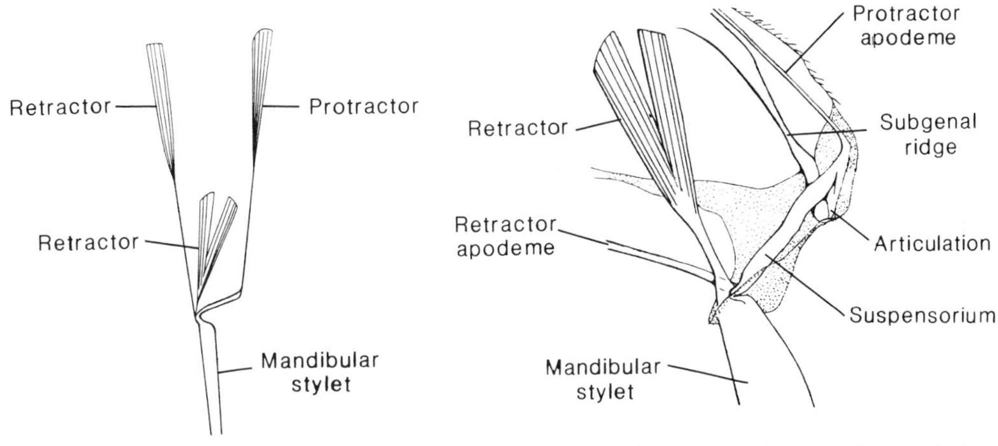

Figure 11.6 The mandible of a female *Anopheles maculipennis* with its musculature and articulation. (After Wenk, 1961.)

The outer edges of the mandibular and maxillary stylets interlock with the lateral grooves of the labrum. Because of the strength and curved form of the maxillary stylets it has been suggested that it is the interlocking of the maxillary stylets and labrum that holds the several stylets together in a fascicle (Vogel, 1921; Jobling, 1976). Robinson (1939) believed that the stylets are held together by the surface tension of a fluid that bathes them. Noting that blood-feeding *Ae aegypti* in which the salivary ducts had been cut frequently rubbed the proboscis with their forelegs, Hudson (1964) postulated that a lubricating fluid was lacking. Proof is lacking for both suggestions. It may be significant that when the stylets of *Cs inornata* were removed from the labial sheath and exposed to air, the fascicle retained its normal shape and flexibility for up to 20 days (Owen, 1963), whereas the stylets of *Cx pipiens* separated if the fascicle was immersed in oil (Hurlbut, 1966).

Four muscles are associated with each maxilla (Figure 11.8), their names indicating their points of origin and insertion. The cardinopremental muscle is inserted on the labial sclerite at the base of the prementum. Its function is to retract the prementum and maxillary stylet (lacinia), acting on the latter via the maxillary-labial articulation. The genostipital muscle is a retractor of the maxilla. The tentoriocardinal muscle is a protractor antagonistic to the genostipital muscle. The tentoriolacinial muscle is inserted on the posterior end of the lacinia; its action is to lift the base of the lacinia and with it the basal end of the prementum (Schiemenz, 1957; Christophers, 1960; Owen, 1985).

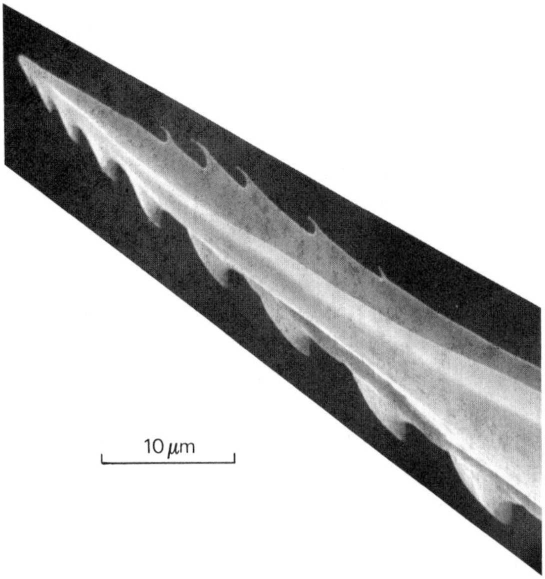

Figure 11.7 Tip of a maxilla of a female *Armigeres subalbatus*. (Micrograph by courtesy of Dr R.M.K.W. Lee.)

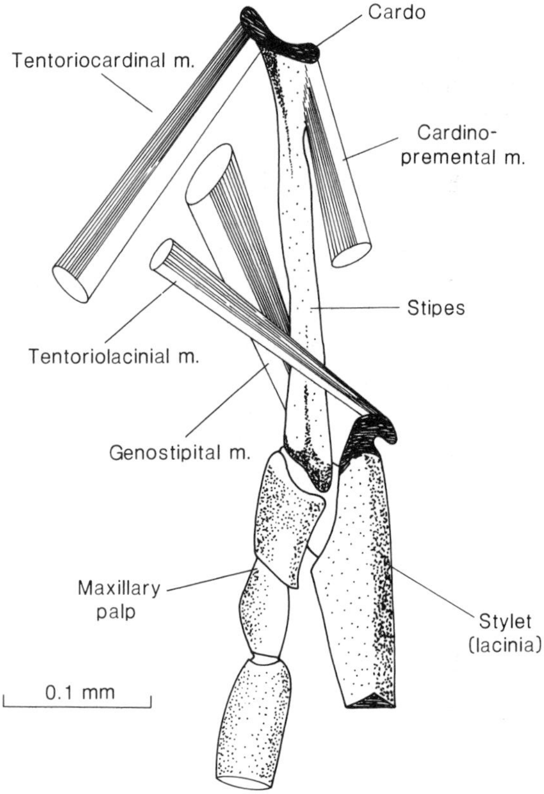

Figure 11.8 The maxilla and its musculature in a female *Culiseta inornata*. m, muscle. (After Owen, 1985.)

The maxillary palps are inserted below the clypeus and at the sides of the proboscis. They are primitively 5-segmented in mosquitoes but in some species fewer segments are apparent. The three basal segments contain intrinsic muscles. The palps of anopheline females are about as long as the proboscis, but the palps of all other female mosquitoes are short, varying from one-fifth to one-half that length. The maxillary palps bear three types of sense organ: mechanosensory setae, multiporous peg sensilla and, in *Anopheles*, a single campaniform sensillum (McIver and Charlton, 1970; McIver, 1972; McIver and Siemicki, 1975a, b).

In certain non-blood-sucking genera the female mouthparts are modified. In females of *Toxorhynchites*, which feed only on nectar, the labium is rigid, the mandibular and maxillary stylets are feeble and short but the labrum is functional (Knab, 1911; Snodgrass, 1959; J.R. Larsen, personal communication). In females of *Malaya*, which feed from ants, the labium is strongly curved, and the maxillary and mandibular stylets are absent, but the labrum and hypopharynx are present (de Meijere, 1909, 1911).

11.2.2 Male mouthparts

Male mosquitoes probe flowers and other plant organs. Their mouthparts have the same general form as those of the female, with a fascicle of stylets sheathed by the labium, but some stylets are reduced or missing. The male labium is a strong flexible organ which resembles that of the female, but the ligula is smaller and laterally compressed. The labrum, which contains the food canal, has a forked tip in most species. The mandibles are absent from some species, and in the remainder they are shortened and exceedingly slender. The hypopharynx, which contains the salivary canal, is fused over most of its length with the prementum and is fused at its tip with the ligula. Because the mandibles are greatly reduced or absent the hypopharynx forms the ventral closure of the food canal. The maxillae are present in most, possibly all, species. They are flat, very delicate stylets and are usually short or very short, but in some species they extend almost the full length of the proboscis. In nearly all male mosquitoes the maxillary palps are about the length of the proboscis and only very exceptionally, e.g. *Wyeomyia smithii*, are they as short as in the female (Vogel, 1921; Vizzi, 1953; Lee, 1974; Marshall and Staley, 1935; Lee and Craig, 1983b, c). Vizzi (1953) dismissed the idea that male mosquitoes can penetrate plant tissues to feed, pointing out that the mandibles and maxillae are too short and delicate and that the tip of the labrum is too curved and too soft for that purpose.

Labellar sensilla have not been described in male mosquitoes, but male *Culiseta inornata* give a behavioural 'labellar response' on stimulation with sugar solutions, indicating the presence of sugar receptors on the labellar lobes (Feir *et al.*, 1961). Male mosquitoes lack the labral chemosensilla of the females but have one pair of labral campaniform sensilla (Lee, 1974). The distribution of sensilla on the maxillary palps broadly resembles that on female palps (McIver, 1971; McIver and Siemicki, 1975a, b).

11.2.3 Cibarial and pharyngeal pumps and salivary valve

Two sucking organs are present in the mosquito head – the cibarial and pharyngeal pumps (Figure 11.9). The dipteran cibarial pump evolved from the bases of the mouthparts of more generalized insects, enclosing what had been the intergnathal space or preoral cavity. The embryological mouth, derived from the external opening of the stomodaeum, is located just posterior to the cibarial pump. A new functional mouth evolved between the anterior end of the cibarial pump and the remaining mouthparts. In mosquitoes a separate pharyngeal pump developed from the anterior region of the stomodaeum (Snodgrass, 1944, 1959).

The **cibarial pump** is situated below the clypeus. Its trough-shaped lower wall is continuous with the hypopharynx, and its upper wall, the clypeopalatum, is continuous with the labropalatum. The labral food canal is thus continuous with the lumen of the cibarial pump. The upper wall of the cibarial pump is largely membranous but it contains two thickened areas, the anterior and posterior hard palates (Figure 11.10A). These are separated by an extensive membranous area into which the dilator muscles are inserted. Posteriorly the upper wall ends in a

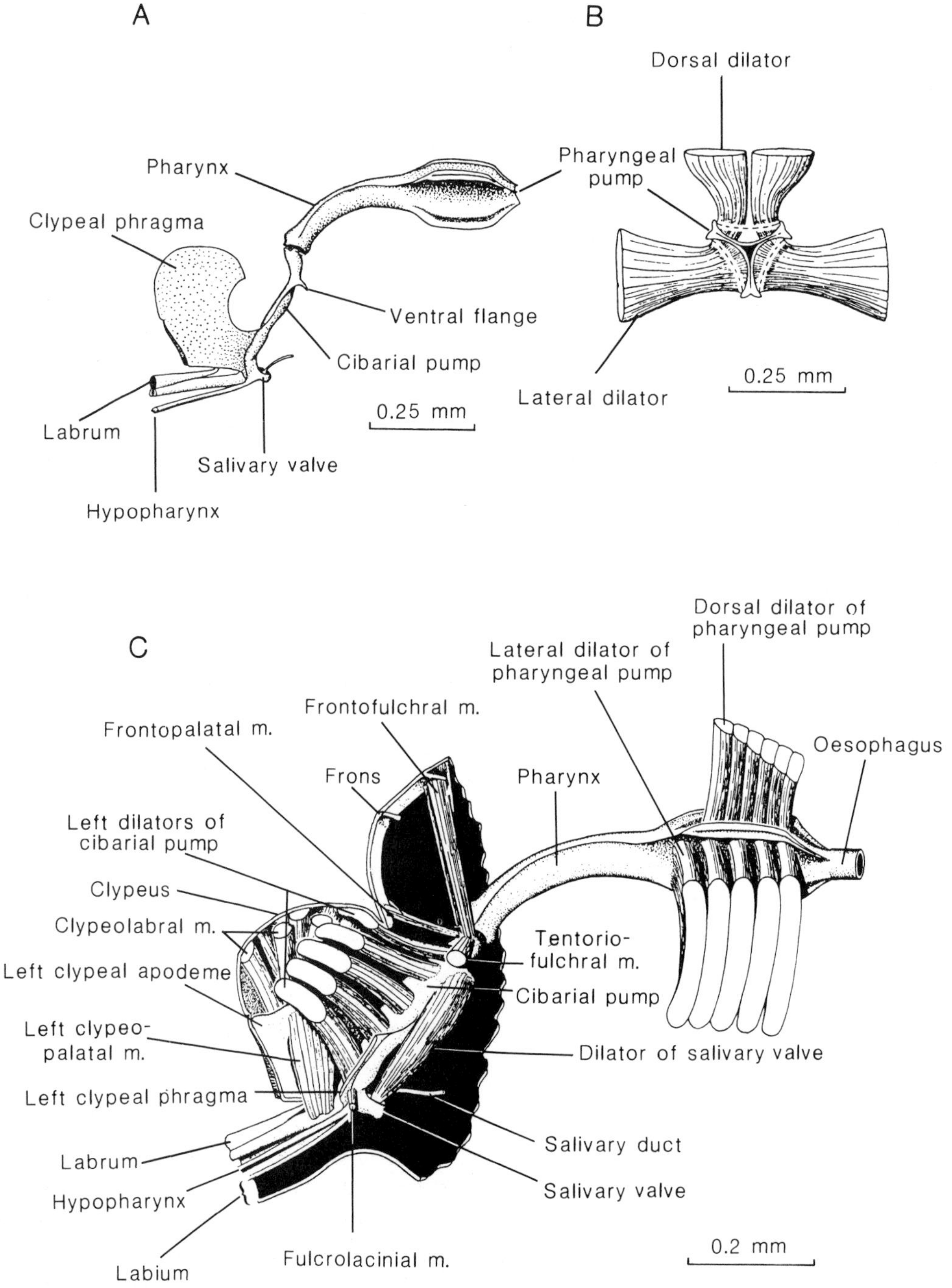

Figure 11.9 The cibarial and pharyngeal pumps of a female *Culiseta annulata*. (A) Lateral aspect of the cuticular parts of the pumps. (B) Transverse section through the pharyngeal pump. The dotted lines show the open position. (C) The musculature of the pumps. m, muscle. (From Schiemenz, 1957.)

denticulate hemispherical structure, the cibarial dome. At rest the upper wall presses into the pump through its own elasticity. Dilatation of the pump is effected by paired sets of dilator muscles which have their origins on the clypeus and their insertions on the upper wall (Figure 11.9C). The cibarial pump is braced against the pull of its dilator muscles by a pair of sclerotized sheet-like clypeal phragmata which extend from the dorsolateral walls of the postclypeus to the lateral margins of the pump (Figure 11.9A).

The floor of the cibarial pump is uniformly thickened. Posteriorly it extends as two diverging lateral processes, the lateral flanges, on which the frontofulchral and tentoriofulchral muscles are inserted. Schiemenz (1957) suggested that these muscles stabilize the cibarial pump. A ventral flange protrudes from the floor of the cibarial pump, and from this the paired dilator muscles of the salivary valve arise (Figure 11.9A, C). For a short distance behind the cibarial pump the cuticle is thin and probably forms an anterior pharyngeal valve, opened by the frontopalatal muscle and closed through the elasticity of the cuticle (Thompson, 1905; Sinton and Covell, 1927; Christophers, 1960; Waldbauer, 1962).

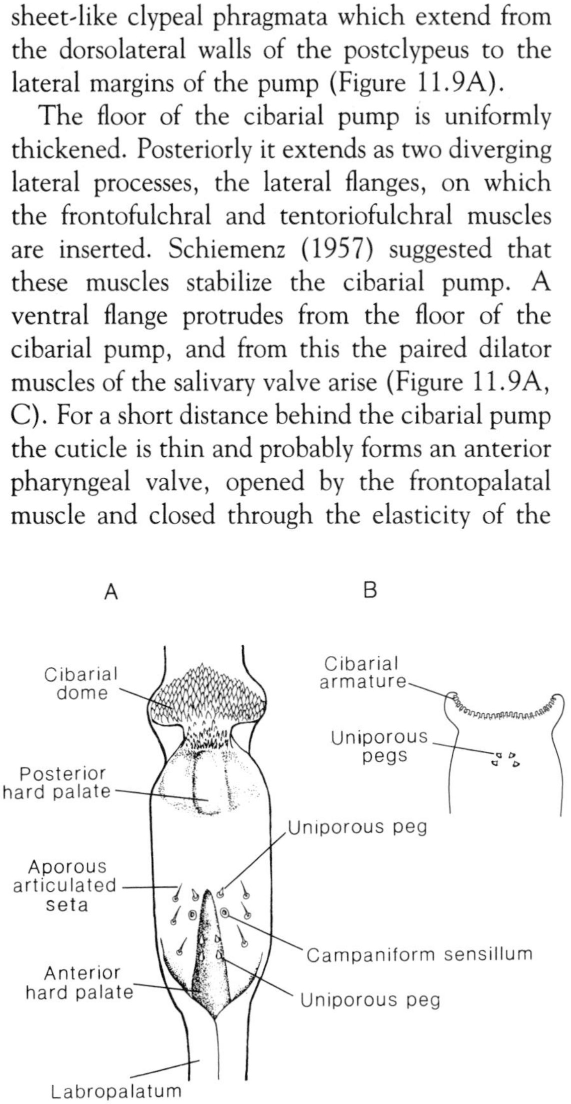

Figure 11.10 The cibarial pump of a female *Culex pipiens pallens*. (A) The dorsal inner surface. (B) The ventral inner surface. The cibarial armature of *Cx pipiens* females is weaker than that of some other species. (From Uchida (1979). Copyright (1979), Pergamon Press plc. Reprinted with permission.)

Between the lateral flanges of the cibarial pump the cuticle forms a transverse ridge which in some species bears one or two rows of teeth, the cibarial armature (Figure 11.10B, 11.11B). The cibarial dome, which is situated immediately above the cibarial armature, may have a finely denticulate surface. The structure of the cibarial dome does not vary greatly between species, but the form of the cibarial armature varies sufficiently to provide a useful taxonomic character (Sinton and Covell, 1927; Sirivanakarn, 1978). The cibarial dome and armature may be present and of similar appearance in both males and females, as in *Aedes albopictus*. In *Culex pipiens pallens* the cibarial dome is present in both sexes although smaller in the male, but the cibarial armature is present only in the female (Uchida, 1979).

The cibarial dome and armature possibly serve to protect mosquitoes from microfilariae. A high percentage of microfilariae that are ingested by a species that has a well developed cibarial armature suffer mechanical damage and are destroyed in the midgut. It appears that they are damaged by a grinding action of the cibarial dome and armature during contractions of the cibarial pump. That is not the fate of microfilariae introduced into the midgut by enema, or ingested by species that lack a well-developed cibarial armature. However, increasing complexity of the armature does not necessarily increase the lethal effect (Coluzzi and Trabucchi, 1968; McGreevy *et al.*, 1978; Buse and Kuhlow, 1979).

The more developed cibarial armatures damage erythrocytes during blood feeding. Only 2–4% of erythrocytes were haemolysed when ingested by species without a cibarial armature (e.g. *Ae aegypti*) or with a single row of cibarial teeth (e.g. *An atroparvus*), whereas 10–20% were damaged in species with two rows of teeth (e.g *An stephensi*), and 45–50% were damaged in species with one row of spoon-shaped teeth and a strongly denticulate cibarial dome (e.g. *Cx pipiens*). It was suggested that the haemolysis, which was shown not to be due to any action of salivary or

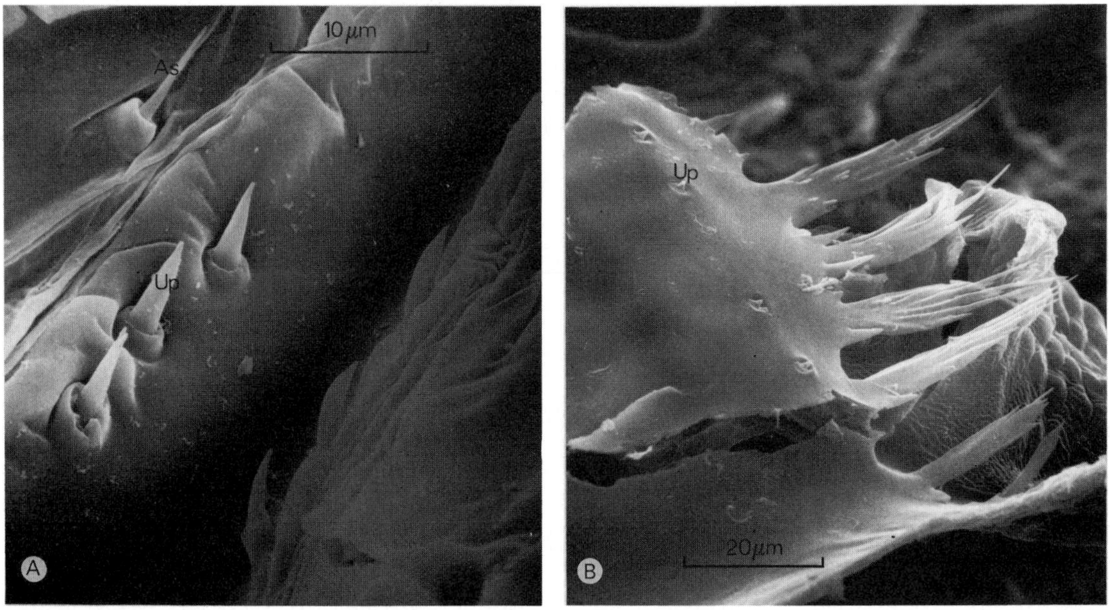

Figure 11.11 Components of the cibarial pump. (A) Part of the dorsal surface of the cibarial pump of a female *Culiseta inornata*, with aporous articulated seta (As) and uniporous peg sensilla (Up.) (B) Part of the ventral surface of the cibarial pump of a female *Anopheles farauti*, with cibarial armature and uniporous peg sensilla (Up.) (From Lee and Craig, 1983a. Micrographs by courtesy of Dr R.M.K.W. Lee.)

midgut secretions, would assist digestion (Coluzzi *et al.*, 1982).

The inner surfaces of the cibarial pump bear three types of sensillum – uniporous peg sensilla, aporous articulated setae and campaniform sensilla (Figures 11.10, 11.11) – which are respectively chemoreceptors, mechanoreceptors and proprioceptors. The numbers and distribution of the sensilla are broadly similar in male and female mosquitoes. Dendrites connect the sensilla to bipolar sensory neurones, located above and below the cibarium, and axons from these neurones join the labrofrontal nerve (von Gernet and Buerger, 1966; Dapples and Lea, 1974; Lee, 1974; Lee and Davies, 1978; McIver and Siemicki, 1981; Lee and Craig, 1983a; Uchida, 1979; Buse and Kuhlow, 1979).

The lightly sclerotized **pharynx** curves upwards and backwards from the cibarial pump, passing between the suboesophageal ganglion and the brain, and expands into the bulbous **pharyngeal pump** from which the oesophagus proceeds through the neck into the thorax (Figure 11.9A). The walls of the pump comprise three sclerotized plates, one dorsal and two ventrolateral, which are hinged to one another along their margins (Figure 11.9B). Four powerful dilator muscles can expand the lumen of the pump until it has an almost circular section, and the plates can spring inwards through their elasticity until they almost obliterate the lumen (Snodgrass, 1944; Schiemenz, 1957). A band of fine backward-pointing spines or hairs encircles the inner surface of the pharyngeal pump where it joins the oesophagus (Dapples and Lea, 1974; McGreevy *et al.*, 1978; Buse and Kuhlow, 1979). A sphincter has been described at the anterior end of the oesophagus (Nuttall and Shipley, 1903; Thompson, 1905; Christophers, 1960), but according to Schiemenz (1957) this is only the normal circular muscle of the oesophagus, and the pharyngeal pump is closed posteriorly by apposition of the ends of the three plates which are drawn together when the dilators contract, forming what may be called the posterior pharyngeal valve. The efficiency of the foregut valves is demonstrated at emergence when the stomach becomes distended with swallowed air. The male pharynx is relatively longer and thicker and the male pharyngeal pump is relatively smaller than in the female (Christophers, 1960).

The **salivary valve** is situated at the base of the hypopharynx (Figure 11.9A, C). It has a thick cup-shaped lower wall and a thin elastic upper wall which is usually collapsed into the cavity of the valve. The common salivary duct opens into the valve through the upper wall. The valve lumen narrows anteriorly and, after a sharp bend, joins the salivary canal in the hypopharynx. Two dilator muscles arise on the ventral flange of the cibarial pump and are inserted on the upper wall of the salivary valve (Robinson, 1939; Snodgrass, 1944; Schiemenz, 1957).

11.3 FEEDING MECHANISMS

11.3.1 Modes of feeding

Two modes of feeding have been distinguished in male mosquitoes and three in females:

(i) water ingestion, in which only a small volume is imbibed and passed into the midgut;
(ii) nectar feeding, in which a large volume of nectar or other plant juices is imbibed (without the fascicle being extended beyond the labella) and is passed into the crop; and
(iii) blood feeding by females, in which the fascicle is deployed beyond the labella to penetrate the skin of the host, and a large volume of blood is imbibed and passed into the midgut (Friend, 1978).

In nature mosquitoes feed on nectar or blood after locating these food sources over a distance. In this chapter we shall examine the role of the sense organs in the discrimination of different foods and the control of food intake. Mosquitoes, like some other blood-sucking Diptera, usually pass nectar and other plant juices to the crop and dorsal diverticula, but pass blood to the stomach. This separation of nectar and blood permits females to store a nectar meal in the crop, passing it slowly to the midgut for absorption, while leaving the stomach empty to receive a blood meal at any time. The feeding responses of thirsty mosquitoes (Khan and Maibach, 1971) suggest that the crop provides a reservoir for water as well as for sugar.

11.3.2 Ingestion of sugar solutions

When the tarsi of a walking mosquito touch a drop of sugar solution the mosquito stops moving or turns, and directs its proboscis towards the stimulus. The tip of the proboscis is placed in contact with the fluid, and when the chemosensory hairs on the aboral surface of the labella are stimulated the labellar lobes spread apart giving the so-called labellar response. Electrophysiological analysis of this reflex response showed that the labellar muscles are activated some 40 ms after stimulation of the aboral sensilla with sucrose. Once the labellar lobes have parted, the oral surfaces of the labella lie flat on the droplet and the external surfaces face away from it. The fluid immediately spreads over the hydrophilic oral surfaces of the labella and their chemosensilla, and over the ligula, submerging the tip of the labrum. The labrum is not extended but remains in the resting position. When touched by fluid the ligula rapidly increases in size, probably due to a combination of physical forces on the cuticle and internal pressure of haemolymph. The pumps in the head may start functioning at the time of the labellar response. When *Culiseta inornata* feeds from a drop of sugar solution the labella are held widely open for a few seconds, and are then closed while feeding is completed (Larsen and Owen, 1971; McKean, 1973; Pappas and Larsen, 1978). A similar sequence of responses was shown by water-deprived females that came into contact with droplets of water (Friend, 1978).

The instant the tarsi of female *Cs inornata* came in contact with flower heads of *Aster pilosus* the proboscis was lowered and probing into florets was initiated; at the same time the antennae were lowered to become parallel with the proboscis. The mosquitoes probed crevices on the flower head, and often probed between the corolla and stamens of florets, the insertions lasting up to 15 s. Once the proboscis was inserted the antennae and maxillary palps were raised to a 90° angle from the proboscis. The mosquitoes would probe many

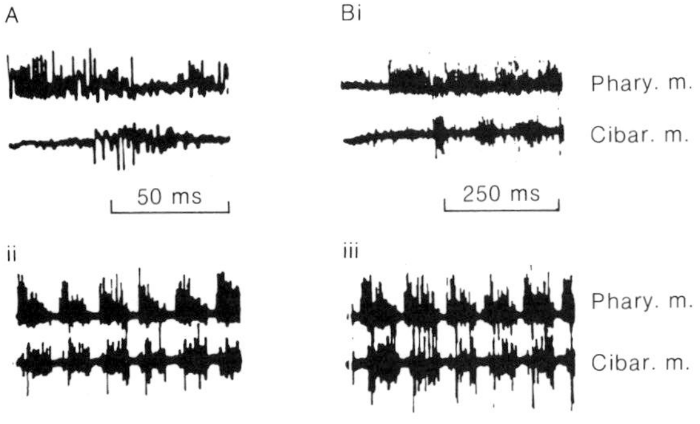

Figure 11.12 Electromyograms recorded from the cibarial and pharyngeal retractor muscles of female *Aedes vexans*. (A) The first contractions induced by stimulation of labellar sensilla with 1 M sucrose. (B)i–iii: The first and later contractions in another specimen induced by stimulation of labellar sensilla with 1 M sucrose. (From Pappas, 1988.)

florets on a single flower head, and might remain on the flower head for up to 18 min (Pappas and Larsen, 1978).

Fluid uptake is effected by the combined actions of the cibarial and pharyngeal pumps, and even the severed heads of mosquitoes are able to pump fluids (Fülleborn, 1932; Coluzzi *et al.*, 1982). In eight species of *Aedes*, *Psorophora*, *Culex* and *Anopheles* tested, pumping started when the labellar sensilla were stimulated with sucrose. Electromyograms from females of *Ae vexans* with electrodes implanted in the cibarial and pharyngeal muscles showed a response starting 60 ms after application of sucrose to the labellar sensilla, leading to bursts of muscle potentials occurring at a frequency of 11.5 Hz. Upon stimulation of the labellar sensilla with sucrose, the cibarial retractors responded about 40 ms before the pharyngeal retractors, but within 5–8 bursts of muscle potentials the two sets of muscles were approximately in phase, with contraction of the pharyngeal retractors starting between one-tenth and one-third of the way through the contraction of the cibarial retractors (Figure 11.12) (Pappas, 1988).

11.3.3 Ingestion of blood

The accessability to mosquitoes of blood vessels varies not only with host species but also with location of feeding site on the body of individual species. Below the keratinized epidermis, the dermis is highly vascularized. Small arteries of about 100 μm diameter enter the lower dermis where they join an arteriovenous meshwork, the cutaneous plexus. Arterioles pass outwards and join a second plexus, the papillary plexus, which gives rise to capillary loops that are located just below the epidermis. The capillaries are 10–20 μm in diameter. The arterioles of the mid-dermis are about 50 μm in diameter. The diameter of the venules ranges from 40–60 μm in the upper and mid-dermis to 100-400 μm in the deeper tissues (Jarrett, 1973; Wheater *et al.*, 1979). The stylet fascicle of *Aedes aegypti*, a relatively small species, is about 2 mm long and a little under 50 μm wide (Christophers, 1960; Lee, 1974). It is much too large to enter capillaries but may enter the larger arterioles and venules.

Blood feeding can be divided into four phases. (a) The *exploratory phase* – from time of alighting on the host to the time the stylets start to

Table 11.2 Mean durations of the phases of blood feeding, but with the brief withdrawal phase included in the imbibing phase. The observations were on wild females except in the case of *Aedes aegypti*.

Species	*Host*	*Mean duration of phase (s)*			*Ref.*
		Exploratory	*Probing*	*Imbibing*	
Ae aegypti	Mouse	–	42	213	1
	Guinea-pig	–	56	132	1
	Man*	3	68	220	2
	Man†	4	53	101	2
Ae africanus	Man	–	32	80	3
Ae cinereus	Man	11	25	82	4
Ae cantans	Man	8	28	150	4
Cq richiardii	Man	16	92	162	4
An plumbeus	Man	31	40	99	4

Mosquitoes were exposed to the shaven abdomen of rodents (Ref. 1) or to the human hand (3), arm (2) or arm and leg (4).
– Phase not separately recorded.
* Skin surface temperature 29°C.
† Skin surface temperature 36.2°C.

References: 1. Mellink *et al.* (1982); 2. Grossman and Pappas (1991); 3. Gillett (1967); 4. Service (1971).

penetrate the skin, the latter marked by the first tremors of the maxillary palps. (b) The *probing phase* – from first penetration of the skin to the first appearance of blood in the fascicle and cessation of palpal vibrations. (c) The *imbibing phase* – from first appearance of blood in the fascicle to the start of withdrawal. (d) The *withdrawal phase* – from first straightening of the forelegs, renewed vibration of the maxillary palps, and retraction of the stylets, to complete removal of the stylets from the skin. The mean durations of particular phases will be affected by species of host and site of biting. Some representative measurements are cited in Table 11.2. The more detailed original data show that the mean values can conceal a considerable spread of data points (Gillett, 1967; Service, 1971; Grossman and Pappas, 1991).

Our knowledge of the means by which mosquitoes pierce the skin and suck blood is derived from anatomical studies of the mouthparts, from observations on the movements of the female as she feeds, and from microscopic observations on the movements of the fascicle within the web of a frog's foot (Gordon and Lumsden, 1939; Waldbauer, 1962) or the ear of a mouse (Griffiths and Gordon, 1952; Wilson and Clements, 1965; Mellink *et al.*, 1982). Further observations on feeding were provided by Robinson (1939), Christophers (1960) and Jones and Pilitt (1973).

(a) Exploration and probing

When wild females of six species of *Anopheles*, *Aedes* or *Coquillettidia* landed on human skin, they always remained immobile for a short period (<5 s), during which any movement of the host usually caused them to fly off. This was followed by a period of exploration of the skin, an activity which might be restricted to the area where the mosquito alighted or which might extend to other areas. The mean duration of the exploratory period varied between 3 and 11 s in three species of *Aedes*, and lasted 16 s in *Coquillettidia richiardii* and 31 s in *Anopheles plumbeus* (Table 11.2).

Before penetrating the skin the mosquito repeatedly applies the tip of its labium to the skin surface. Sooner or later the tip of the proboscis is placed on the skin at a point roughly central to the points of support provided by the fore- and midlegs, and the mosquito may straighten its forelegs to bring the proboscis into a more vertical position. Shortly after the labella have been pressed against the skin the maxillary palps are raised to an angle of about 75°. The labella are kept pressed together and the fascicle penetrates the skin by a series of minute thrusts. The labium does not enter the skin but kinks back under the head (Figure 11.13). As soon as the fascicle penetrates the skin the raised maxillary palps vibrate, possibly in rhythm with other maxillary movements, and the vibrations continue until a bright red streak of blood appears in the fascicle.

Such rigidity as the fascicle possesses is provided by the labrum, to which the other stylets are appressed. It is generally agreed that it is the movements of the maxillae which are responsible for penetration of the fascicle, with the maxillary teeth acting as grappling hooks. Penetration is believed to occur by the following sequence of actions.

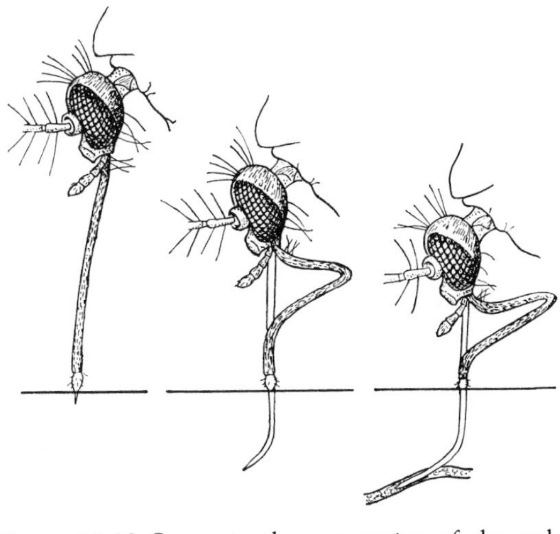

Figure 11.13 Stages in the penetration of the stylet fascicle, as observed in a female of *Aedes aegypti* feeding on the web of a frog's foot. (From Snodgrass, 1959; after Gordon and Lumsden, 1939.)

1. (a) The right maxillary protractor relaxes and the right maxillary retractors contract. Because the right maxillary stylet is anchored in the skin by its teeth, contraction of its retractors (including the powerful cardotentorial muscle) pulls the cranium towards the skin surface and thrusts the rest of the fascicle deeper into the skin.
   (b) Simultaneously with 1(a) the left maxillary retractors relax and the left maxillary protractor contracts thrusting the left maxillary stylet into the skin beyond the rest of the fascicle.
2. (a) The cycle of movements continues with contraction of the left maxillary retractors which advances the cranium and fascicle.
   (b) Simultaneously with 2(a) the right maxillary protractor contracts advancing the right maxillary stylet.

The movements of the mouthparts were too rapid for the observers to distinguish whether the left and right maxillary stylets moved alternately, as described above, or synchronously, but the fact that the maxillary palps moved alternately during penetration suggests that the same is true of the stylets. The mandibles could not be seen and it is not known whether they play an active part in piercing or simply serve to open and close the distal opening of the labral food canal.

Soon after entering the skin the tip of the fascicle was observed to bend through almost a right angle (Figure 11.13). It then tunnelled through the tissues parallel to the skin surface, probing over the whole area that could be reached. Active bending was limited to the anterior fifth of the fascicle, and was most marked in the dorsoventral plane in which the fascicle sometimes took almost a J-form. A limited sideways movement was sometimes observed but it amounted to no more than a moderate lateral curvature of the distal part of the fascicle. The fascicle was used like a probe, its direction and depth of entry being constantly altered until by chance a suitable blood supply was found. The more proximal part of the fascicle followed the course of the tunnel bored through the tissue by the cutting tip. Mellink *et al.* (1982) were alone in considering that the fascicle is uniformly stiff and inflexible, that it only bends passively and in response to host tissue resistance.

The movements of the tip of the fascicle in the dorsoventral plane are ascribed to the labrum and its muscles. Contraction of the clypeolabropalatal muscles (Figure 11.4) is thought to displace the inner wall of the labrum (the labropalatum) proximally relative to the outer wall. Because the inner and outer walls are fused at the tip this causes a ventral flexure of the tip of the labrum. Contraction of the clypeolabral muscle would have an antagonistic effect and could produce a dorsal flexure of the tip of the labrum. It has been suggested that the small amount of lateral movement shown by the tip of the fascicle might be due to the maxilla on one side protracting and retracting more rapidly than that on the other side (Waldbauer, 1962; Owen, 1985).

As the fascicle penetrated and probed the tissue, saliva was frequently discharged from its tip, appearing as a series of puffs of clear fluid which rapidly dispersed. From experiments with ^{3}H-labelled *Ae aegypti* fed on mice, it was estimated that each female injected, on average, 4.7 μg ($\equiv 4.7 \times 10^{6}$ μm^{3}) saliva into the host (Devine *et al.*, 1965). This is consistent with gland volume, recalculated from Metcalf's (1945) data as 21.6×10^{6} μm^{3} per pair of *An quadrimaculatus* salivary glands, and the observation that during blood feeding the protein content of the salivary glands falls by 25–35% in *An stephensi* and by 15–30% in *Cx pipiens* (Poehling, 1979).

Females of *Ae aegypti* probing the backs of guinea-pigs thrust their fascicles repeatedly, at 7-s intervals, into the skin, withdrawing after a mean period of 100 s if failing to locate blood. They then probed again, and if necessary again, but for not longer than mean periods of 60 s on the later occasions. Approximately 50% of females located blood during the first probe, 50% of the remainder located blood during the second probe and yet 50% again were successful during the third probe. The probability of locating blood varied at different stages of a single probe. Initially success increased with time but it later declined. If the

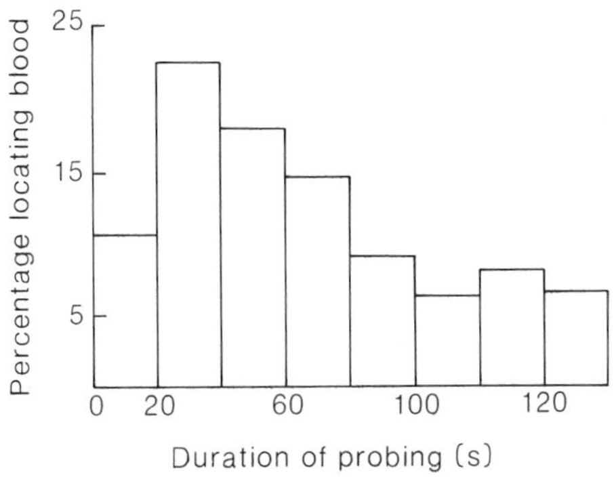

Figure 11.14 Percentage of *Aedes aegypti*, probing the back of a guinea pig, that located blood during 20 s periods after stylet insertion (n = 300.) (After Ribeiro *et al.*, 1985a.)

duration of a probe was divided into 20-s periods, the probability of finding blood was higher in the second 20-s period than the first but then declined progressively up to 100 s (Figure 11.14) (Ribeiro *et al.*, 1985a). One factor affecting the likelihood of finding blood is the density of blood vessels in the dermis. *Aedes aegypti* located blood more quickly when probing the highly vascularized ears of guinea-pigs than when probing the sparsely vascularized skin on the back (median times 44 s and 99 s) (Figure 12.4) (Ribeiro *et al.*, 1984a). Probing times did not differ significantly when the skin surface temperature of human arms varied between 29 and 36.2°C (Grossman and Pappas, 1991).

(b) Imbibing and withdrawal

Sometimes a fascicle which penetrated a blood vessel continued further into the tissue, but more commonly it either passed along the vessel, sometimes for as much as a quarter of its length, or, in the case of larger vessels, remained stationary with its tip just beyond the point of entry. Females fed in this manner from both venules and arterioles, their sucking producing a greatly accelerated blood flow within the vessels. It is not known whether saliva is discharged into the bloodstream.

When a fascicle that had lanced a blood vessel was withdrawn, an extravasation of blood into the tissue invariably occurred, producing a small pool of blood or haematoma. The insect usually detected the haematoma and immediately started sucking, drawing up blood as fast as it flowed into the tissue. This method of feeding is called 'pool feeding' in contrast to the 'vessel feeding' described above. Normally, within a few seconds of small blood vessels being lacerated the lesions are plugged by aggregations of platelets. The saliva that is discharged during blood feeding helps in blood vessel location by inhibiting this haemostasis (Section 12.2.1).

The start of imbibition is marked by the sudden appearance of a bright red streak of blood in the fascicle, and at that moment the palps stop vibrating, indicating that penetration has stopped and that imbibition has begun. During the imbibing phase the movements of the stylets cease, the palps relax and rest on the skin surface, and the mosquito becomes unresponsive to disturbance. Salivation into the bloodstream is never observed. When imbibition starts, the rate of blood flow in the blood vessel is greatly accelerated. At first the abdomen enlarges rather slowly, but later more rapidly, and with full engorgement the soft cuticle of the pleural and intersegmental regions of the abdominal integument becomes highly distended and these regions appear bright red (Jones and Pilitt, 1973).

In one investigation of feeding by *Ae aegypti*, females fed on a mouse through a fine-meshed bronze screen. Screen and mouse were part of an electric circuit which included a power source and pen recorder, and a current of 3–6 mA was recorded as long as the mosquito's mouthparts remained inserted in the skin. During engorgement the recording showed a pattern of temporary decreases in current, with a frequency of about 6 Hz, which were thought to be due to the closure of valves in the sucking pumps of the head (Figure 11.15). A pattern of briefer increases of current with a frequency of about 5 Hz also occurred, and it was postulated that these were caused by the discharge of saliva. Feeding mosquitoes showed alternating sequences of postulated suction and salivary discharge (Kashin and Wakeley, 1965; Kashin, 1966).

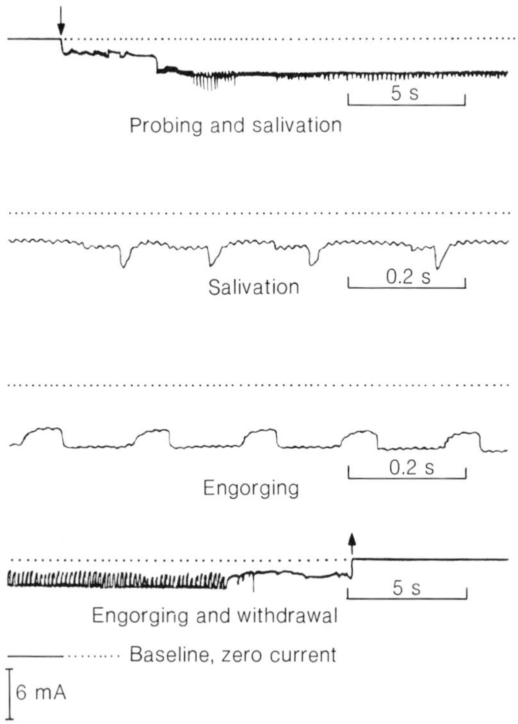

Figure 11.15 Electrical changes recorded during the feeding by individual females of *Aedes aegypti* on a mouse. Mosquito and mouse formed part of an electric circuit which was closed when the mosquito's mouthparts penetrated the skin. Downward displacements of the trace indicate increases in current, upward displacements indicate decreases in current. At the faster chart speed the traces show a slight 60 Hz a.c. ripple which should be ignored. (After Kashin, 1966.)

With a labrum length of 2 mm and mean food canal diameter of 0.03 mm, the volume of the food canal of *Aedes aegypti* is 0.0014 mm^3 (Christophers, 1960). Taking as representative values a blood meal of 3 cu. mm imbibed in 180 s, the volumetric flow rate would be 0.0166 $mm^3\ s^{-1}$, and the linear flow rate in the food canal would be 24 mm s^{-1}. The small diameter of the food canal affects the nature of fluid flow through it. This will result in a low Reynolds number, with the consequence that flow will be laminar (i.e. non-turbulent) (see Section 4.3.1). When blood flows through tubes of <0.2 mm diameter a cell-free layer forms near the wall, and the smaller the tube diameter the larger is the fraction of volume occupied by the cell-free layer. Therefore, blood that flows through the mosquito labrum will have a lower haematocrit than that in the supplying blood vessel (Kesavan and Reddy, 1985).

Variation of the skin surface temperature of human arms between 29 and 36.2°C significantly affected the imbibition rate of *Ae aegypti* (Table 11.2), but the volume of blood ingested was unaffected (Grossman and Pappas, 1991). Measurements made with a photoelectric plethysmograph placed on human skin, adjacent to a site where *Ae triseriatus* were biting, indicated that the blood volume of the skin vasculature often started to increase within 30 s of the start of probing, before reddened areas of skin were visible, and that it remained heightened throughout the period of biting. In four individuals out of ten the blood volume of bitten skin was not significantly different from that of the unbitten controls, but six individuals exhibited a mean increase in blood volume of 58.6% (range 38–155%). These six did not react when bitten by duct-transected females (Pappas *et al.*, 1986).

The fascicle may be withdrawn from the skin quickly and easily; a mean withdrawal time of three seconds was recorded by Jones and Pilitt (1973). Alternatively, the fascicle may be partly withdrawn and reinserted several times before being pulled out. The resumption of movements of the maxillary palps marks the beginning of withdrawal, and straightening of the forelegs appears to provide much of the force needed. Robinson (1939) postulated that during withdrawal repeated retractions of the maxillary stylets kept their barbed apices within the fascicle and away from the surrounding tissue. As the labium straightens the fascicle slides through the labella and returns to the premental gutter. Females fly from the host within 5 s after withdrawal of the fascicle (Service, 1971).

Vessel feeding by *Ae aegypti* was usually completed within three to four minutes, but pool feeding could take nine or ten minutes. Sometimes females utilized both methods of feeding (Gordon and Lumsden, 1939; Griffiths and Gordon, 1952; Wilson and Clements, 1965).

Mellink *et al.* (1982) observed pool feeding in regions of mouse ear that were devoid of large vessels, when it was preceded by extensive laceration of capillaries. Observations made in the field, of wild mosquitoes of several species feeding on immotile hosts, showed that in most cases feeding was completed within four minutes or less, suggesting that pool feeding was unimportant (Service, 1971; Magnarelli, 1979b).

11.4 SENSORY RECEPTION AND REGULATION OF FOOD INTAKE

In insects, feeding and the distribution of food within the gut are controlled by integrated responses of the peripheral sensilla, central nervous system and, probably, the stomatogastric system. When sucking insects start to feed, stimulation of sensilla on the mouthparts activates a central control system. This produces rhythmic patterns of motor impulses which drive the muscles of the pharyngeal and cibarial pumps. Food quality is monitored by cibarial sensilla. Feeding generally terminates in response to mechanical stimulation caused by swelling of the gut.

Mosquitoes use chemosensilla on the tarsi, labella and labrum, and inside the cibarial pump, to identify food substances. The ultrastructure and electrophysiology of these sensilla will be described in Volume 2. The sensory aspects of food selection have been investigated by electrophysiological techniques and by experiments with restrained or free-flying mosquitoes. To distinguish between the effects of phagostimulants on the sensilla of the labella and labrum, investigators have removed the fascicle from its labial sheath and presented test fluids separately to fascicle and labella, recording the percentages of mosquitoes feeding, volumes of fluid ingested, and destination of fluids. Mosquito feeding behaviour is a sequence of stimulus–response events which some experimental methods distort, so experiments that involve anaesthesia, physical restraint or the unnatural presentation of stimuli must be interpreted cautiously (Friend, 1978). Fortunately data can be obtained also from experiments with unrestrained mosquitoes, which will imbibe water or sugar solutions from exposed droplets, and pierce artifical membranes with their fascicles to drink blood or warm fluids that contain appropriate solutes.

11.4.1 Regulation of sugar intake

(a) Involvement of sensilla

The tarsi, labella, labrum and cibarial pump all bear sugar-sensitive sensilla. For the present purpose of describing the behaviour of mosquitoes that have already reached a plant food resource it is sufficient to distinguish three activities. (i) The proboscis response: searching movements made with the proboscis when the tarsi are stimulated by sugar. (ii) The labellar response: separation of the labellar lobes following stimulation by sugar of sensilla on the outer surface of the labella. (iii) Drinking: fluid intake due to activity of the cibarial and pharyngeal pumps.

In *Aedes aegypti* the male tarsi bear 60 and the female tarsi about 100 chemosensory hairs 30–40 μm long. The hairs are of three types, based on the number and type of neurones. Most chemosensory hairs are found on the fore tarsi, fewest on the hind tarsi (McIver and Siemicki, 1978). In females of *Culiseta inornata* the tarsal chemosensory hairs, which are all of one type, contain neurones sensitive to water, sucrose, and salts. Each of the labella at the tip of the labium bears three types of hair sensilla. In females of *Culiseta inornata* the outer face of each labellum bears 40–60 hairs of 30–80 μm length, which are sensitive to water, salts and sucrose. It also bears many 5 μm long hairs which are sensitive to water and salts. The oral face of each labellum bears eight hairs 6 μm long which are sensitive to water, salts and sucrose (Larsen and Owen, 1971; Owen *et al.*, 1974; Pappas and Larsen, 1976a, b). The labrum bears two pairs of chemosensilla near the tip. The sensitivity of cibarial sensilla of *Cs inornata* to sucrose was demonstrated by piercing the dorsal surface of the cibarium and, by injecting sucrose solution, inducing a labellar response (Owen, 1965).

Afferent nerve impulses were generated when the tarsal, labellar, and labral chemosensilla in *Cs inornata* were stimulated with sucrose solutions. The outer labellar hairs showed a clear increase in firing rate as the sucrose concentration increased between 0.1 and 1 M (Pappas and Larsen, 1976a). Stimulation of a sensillum on the outer surface of the labella of *Ae aegypti* with 0.1–1 M sugar solution caused a burst of action potentials with a frequency of 20–250 Hz. The rate of firing declined to a much lower constant rate within 2 to 4 s, and continued for 20 to 30 s, although in some instances pulse generation ceased within the first second. The firing rate was higher with sucrose than with glucose (Sinitsyna, 1971). Application of a drop of 1 M sucrose to a single aboral labellar hair of *Cs inornata* elicited the labellar response. If the hair underwent sensory adaptation to sucrose and the labellar response ceased, stimulation of an adjacent hair elicited a full response. Stimulation of chemoreceptors on the oral surface of the labella with 1 M sucrose also led to a labellar response (Owen, 1963; Larsen and Owen, 1971). Possibly, therefore, during nectar feeding the labellar response is maintained by the successive stimulation of different sensilla.

The threshold for the proboscis response given by *Cs inornata* upon stimulation of the tarsi with sugar was substantially higher than that for the labellar response. For both responses the thresholds for sucrose were much lower than for glucose (Table 11.3) (Feir *et al.*, 1961). The labellar threshold for sucrose declined during sugar deprivation in water-satiated *Cs inornata* (Table 11.4). The threshold increased almost 20-fold after a meal of sucrose, which was despatched to the crop, and then fell progressively over the following four days. When sucrose solution was removed from the crop by hypodermic syringe, shortly after feeding, the threshold fell from 0.65 M before removal to 0.06 M after removal (Manjra, 1971).

Implantation of electrodes into the head revealed that the muscles of the cibarial and pharyngeal pumps could be stimulated into activity by application of water or sucrose solution to certain sensilla. Application of water to both labral and labellar sensilla induced pumping in *Ae vexans*. Application of 1 M sucrose to the labral sensilla alone was ineffective, whereas application to the labellar sensilla alone stimulated pumping *Ae vexans* (Figure 11.12) and in seven other species of *Aedes*, *Anopheles*, *Culex* and *Psorophora*. However, *Culiseta inornata* required simultaneous stimulation, by sucrose, of both the labral and labellar sensilla for pumping activity to be induced (Pappas, 1988).

The sensory requirements for intake of sugar solutions by restrained females of *Cs inornata* that had had no access to sugar or water for

Table 11.3 Median threshold concentrations for evoking the proboscis response and the labellar response in *Culiseta inornata* with sugars. Test solutions were applied to the tarsi or to the outer surface of the labella of 2–3-day old restrained males and females previously provided with water but not food. (n = 250–600). (From Feir *et al.*, 1961.)

Sensilla on	*Stimulated with*	*Threshold concentrations* (M)	
		Median ± *s.d.*	*Range*
Proboscis response			
Tarsi	Sucrose	0.135 ± 0.168	0.015 – 0.5
Tarsi	Glucose	2.095 ± 0.059	0.50 – 3.0
Labellar response			
Labella	Sucrose	0.011 ± 0.223	0.002 – 0.03
Labella	Glucose	0.425 ± 0.272	0.063 – 1.0

Table 11.4 Changes in the threshold of the labellar response to sucrose in female *Culiseta inornata* experiencing sugar deprivation and sugar intake. (From Manjra, 1971.)

	Mean threshold concentration of sucrose (M)					
Treatment	*Before feeding*	*At n hours after emergence or feeding*				
		<1	24	48	72	96
Water from emergence	–	–	–	0.110	0.045	0.032
Water for 4 days then 0.5 M sucrose	0.035	0.660	0.332	0.177	0.126	0.042

24 h were investigated by Pappas and Larsen (1978). The fascicle was separated from the labial sheath, and chemosensilla on the tarsi, labrum and labella were stimulated with 0.5 M sucrose. Activity of the cibarial and pharyngeal pump muscles was recorded electrically. Stimulation of single sets of sensilla on the tarsi, labella or labrum always failed to stimulate pumping, which could be induced only by simultaneous stimulation of the labrum and labella. Simultaneous stimulation with sucrose of the labrum and either of the types of chemosensilla on the aboral surface of the labella failed to induce pumping, but pumping was induced by simultaneous application of sucrose to the labrum and to the sensilla on the oral surface of the labella (Table 11.5).

Other experiments revealed something of the role of water and sugar receptors in the regulation of drinking. Simultaneous exposure of both labral and labellar sensilla to water stimulated the drinking of only a small volume, which was not increased if the labrum was exposed to sucrose. Simultaneous exposure of the labral sensilla to water and of the labellar sensilla to sucrose stimulated the drinking of a large volume (Table 11.5) (Pappas and Larsen, 1978). Rather similar results were obtained by Hosoi (1954c) with *Culex pipiens pallens*.

Sensilla within the cibarial pump are also involved in the regulation of drinking. The tips of the fascicles of hungry or thirsty *Cs inornata* could be immersed in water or sucrose solution for as long as 5 min without inducing drinking. However, if fluid entered the cibarium, intense sucking occurred immediately. That sucking was not mediated by the labral sensilla was checked by withdrawing the stylets from the solution before the individual could accept more than a minimal amount. If the fascicle was reinserted in the solution after a short pause, as short as 5 s with some individuals, drinking was not resumed. However, momentary touching of the labellar hairs with sugar solution stimulated drinking to satiety. From such experiments Owen (1963) concluded that the cibarial sensilla were sensitive to water, glucose, sucrose and whole blood, and that simultaneous stimulation of both labral and cibarial sensilla induces prolonged sucking.

From the observations on female *Cs inornata* we may summarize the responses of mosquitoes

Table 11.5 Volumes of fluid imbibed by restrained females of *Culiseta inornata* upon simultaneous stimulation with water or 1 M sucrose of sensilla located on the labrum (in the desheathed fascicle) and on the oral surface of the labella. (Means ± s.e., n = 10). (From Pappas and Larsen, 1978.)

Fluid applied to oral face of labella	*Fluid applied to fascicle*	*Volume imbibed* (μl)
Water	Water	0.72 ± 0.28[a]*
Water	Sucrose	1.05 ± 0.40[a]
Sucrose	Water	3.45 ± 0.24[b]
Sucrose	Sucrose	3.57 ± 0.23[b]

* Values followed by the same letter are not significantly different.

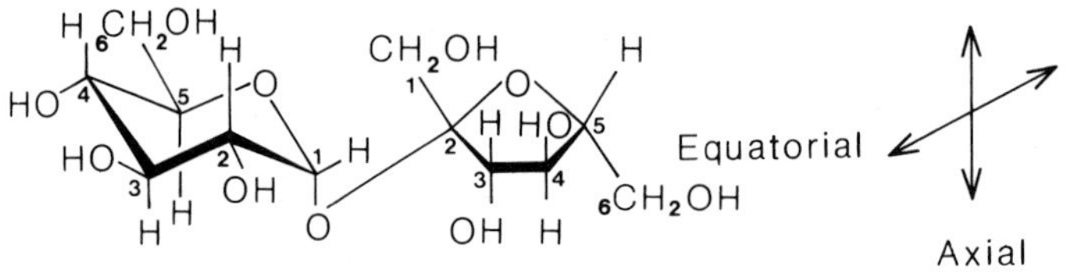

Sucrose
1-*O*-α-D-Glucopyranosyl-β-D-fructofuranoside

Figure 11.16 Nomenclature of mono- and disaccharides. Every monosaccharide has many stereoisomers, half of which are of the D series to which most natural monosaccharides conform. In aqueous solution monosaccharides undergo intramolecular reaction to yield cyclic hemiacetals, which for 6-carbon sugars may be either 5-membered or 6-membered rings. These are called pyranoses and furanoses respectively, terms which are included in the chemical names. Of the two ring systems the 6-membered pyranose form is favoured for glucose and the 5-membered furanose form is favoured for fructose. Isomers that differ only in configuration at the aldehydic carbon in a ring form are called anomers. For example, glucose exists as α- and β-anomers which, in solution, slowly come into equilibrium. When glucose is in its most favoured $^{4}C_1$ conformation the C_1 hydroxyl of the α-anomer is axial whereas that of the β-anomer is equatorial. Disaccharide linkages are classed as α- or β-linkages on the same basis. The glycoside link of sucrose, which involves the anomeric carbons of glucose and fructose, is α from glucose and β from fructose.

to sugar sources as follows. When a walking mosquito encounters sugar solution, contact chemoreceptors on the tarsi are stimulated, initiating movement of the proboscis towards the food source. Palpation of the surface brings the aboral chemosensory hairs on the labella into contact with the sugar. Stimulation of these hairs leads to spreading of the labellar lobes, which exposes the oral surface of the labella, the ligula, and the tip of the fascicle to the fluid. Stimulation of chemosensilla on both the oral surface of the labella and the labrum initiates sucking by the cibarial and pharyngeal pumps. Stimulation of chemosensilla within the cibarial pump possibly maintains the feeding response until satiety. This summary is not fully appropriate for male mosquitoes, which lack labral chemosensilla.

(b) *Structure–activity relationships*

Trioses, tetroses and heptoses did not stimulate feeding by females of *Aedes aegypti* and *Culiseta inornata* that had their probosces inserted into capillary tubes containing sugar solutions of a range of concentrations. Nearly all pentoses, hexoses, and di- and trisaccharides that were tested proved stimulating, although the median acceptance thresholds of different compounds varied considerably. Both the α- and β-anomers of pentoses and hexoses were accepted by these two species. Of the sugars commonly present in nectar, fructose and sucrose had low median acceptance thresholds for *Ae aegypti* (0.020 and 0.023 M), but that of glucose was higher (0.11 M). The most potent phagostimulant for *Cs inornata* was sucrose. The disaccharide maltose and the trisaccharides raffinose and melezitose also strongly stimulated gorging. These four carbohydrates all contain an α-glucopyranosyl moiety and (except maltose) a fructofuranosyl moiety (Figure 11.16). The monosaccharides α-glucose and fructose were less phagostimulatory, but a 1:1 mixture of α-glucose and fructose was potent. No close correlation was found between the nutritive value of sugars and the responses of *Aedes* or *Culiseta* (Salama, 1966, 1967; Schmidt and Friend, 1991).

The characteristics of the sugar receptors of *Cs inornata* were deduced from the responses of females to sugar solutions and the effects of *p*-hydroxymercuribenzoate (PHMB), which blocks pyranose receptor sites (Friend *et al.*, 1988, 1989; Schmidt and Friend, 1991). Dose–response data for intake of sugar solutions by unrestrained females feeding from open droplets are summarized in Table 11.6. That a 1:1 mixture of α-glucose and fructose was more potent than either sugar alone indicates that both sugars played a role in stimulation. The superior potency of sucrose to the glucose–fructose mixture suggests that the sugar receptor accommodated an entire sucrose molecule, presenting specific pyranose and furanose sites. Maltose was almost twice as potent as α-glucose in equimolar solutions, suggesting that the disaccharide can straddle two pyranose sites. The differential response to α- and β-glucose illustrates the specificity of the pyranose sites for the configuration of the C_1 hydroxyl group – axial in α-glucose and equatorial in β-glucose.

Both the shape and bond configuration of disaccharides appeared important for phagostimulation. Within-pair comparisons of disaccharides with 1–4 and 1–6 glycoside links showed that the α-anomers had the higher potency. Comparisons of pairs of α- or β-anomers showed that the disaccharides with a C_1–C_4 glycoside link were more potent than those with a C_1–C_6 link. The α-glycoside link produces straight molecules, whereas in the β-glycoside link the two constituents of the molecule are bent together. Molecules with a C_1–C_4 link are shorter than those with a C_1–C_6 link.

PHMB almost abolished the gorging response to α-glucose, demonstrating its effectiveness in blocking pyranose sites. It had a weak effect on the gorging response to fructose, which probably reflected the ratio of the furanose and pyranose forms of fructose in the solution. PHMB had little effect on the gorging response to 0.5M sucrose, but markedly reduced the gorging response to weaker sucrose solutions. This was taken to indicate that the furanose sites have a higher threshold than the pyranose sites but make a disproportionately large contribution to the gorging response at higher furanose concentrations.

In summary, *Cs inornata* sugar receptors have furanose- and pyranose-specific sites which can be activated independently by appropriate substrates. The pyranose sites require at least two adjacent equatorial hydroxyl groups on carbons C_2 and C_3 and an axial hydoxyl group on C_1. They are blocked by PHMB, and appear to be specific for α-glucopyranose and fructopyranose. The fructofuranose sites are not blocked by

Table 11.6 The potencies of different sugars in inducing gorging and crop filling when fed from open droplets of solution to unrestrained females of *Culiseta inornata*. Feeding started within 15 min of diet preparation ($n > 260$). (From the data of Friend *et al.*, 1988.)

Compound	*Component monosaccharides*	*Glycoside link and configuration*		*ED_{50} of gorging response (mM)*	*ED_{50} of crop response (mM)*
Monosaccharides					
β-Glucose	Glu			~500	/
α-Glucose	Glu			240[d]*	~380[i]
Fructose	Fru			200[c,d]	300[h]
α-Glu:Fru 1:1	Glu:Fru			150[b]	240[g]
Disaccharides					
Sucrose	Glu–Fru	1–2	α	62[a]	120[f]
Maltose	Glu–Glu	1–4	α	165[b,c]	260[g,h]
Isomaltose	Glu–Glu	1–6	α	~270	~620
Cellobiose	Glu–Glu	1–4	β	340[e]	/
Gentiobiose	Glu–Glu	1–6	β	~900	/
Lactose	Gal–Glu	1–4	β	~400	/

/ Response very slight or absent.
* Different letters by the ED_{50} values denote significant differences.

PHMB and have a higher threshold than the pyranose sites (Schmidt and Friend, 1991).

11.4.2 Regulation of blood intake

Female mosquitoes gorge on blood when it is applied by capillary tube to the unsheathed fascicle or offered through an artificial membrane (MacGregor, 1930; Bishop and Gilchrist, 1946). Whether stimulation of just the labral sensilla with blood is sufficient to induce drinking is not known, but certainly stimulation of both labral and cibarial sensilla is sufficient (Owen, 1963). The labellar sensilla are not required for blood intake. When the labella of *Culex pipiens* were exposed to 0.28 M glucose while the fascicle was exposed to an erythrocyte suspension, the mosquito gorged on the erythrocyte suspension despatching it to its midgut, which was the normal response to blood (Hosoi, 1954c).

When females of *Aedes aegypti* were offered warm fluids across a membrane, 10% ingested plasma but only in small quantities, whereas 38% engorged fully on an erythrocyte suspension in saline and 66% engorged fully on whole blood (Bishop and Gilchrist, 1946). In a search for phagostimulants, Hosoi (1959) fractionated ox erythrocytes and found that the active fraction contained adenylic acid. Further studies revealed that mosquitoes were stimulated to gorge by a number of adenyl nucleotides, the most active, with ED_{50} values of about 10^{-6}–10^{-5} M, being adenosine 5′-phosphate (AMP), adenosine 5′-diphosphate (ADP) and adenosine 5′-triphosphate (ATP). After these initial discoveries had been made it was found that the adenyl nucleotides are phagostimulants for nearly all groups of haematophagous insects, the compound of highest potency varying from one group to another (Friend and Smith, 1977).

Structure–activity relationship data from a number of groups of haematophagous insects suggested that binding of the adenyl nucleotide molecule (Figure 11.17) to the putative receptor proteins is probably via the phosphate chain, the amino group on C_6 of the purine moiety, and the hydroxyl groups on the 2′ and 3′ carbons of the ribose moiety (Friend and Smith, 1972). Data for culicine mosquitoes were consistent with that supposition. For *Aedes aegypti* and *Ae caspius*, potency (defined as the reciprocal of the ED_{50}) increased substantially with each increase of a phosphate group between AMP and ATP (Table 11.7). For *Culex pipiens* these three nucleotides differed little in potency, although for both *Cx pipiens* and *Cs inornata* ADP was the most active. Adenosine 5′-tetraphosphate had almost identical activity to ATP for *Ae aegypti* but was very inactive with *Cx pipiens*. Replacing the oxygen atom that links the β and γ phosphorus atoms of ATP with either NH (imido) or CH_2 (methylene) increased potency 3- to 5-fold. This change rendered the molecule non-hydrolysable, therefore the response to adenyl nucleotides does not depend upon the release of phosphate bond energy. Cyclic-AMP was substantially less potent than AMP.

Removal of hydroxyl from either the 2′ or the 3′ carbon of the ribose moiety of ATP halved its phagostimulatory potency for *Ae aegypti*, yet the absence of both hydroxyls enhanced activity. When both hydroxyl groups were present they needed to be in a *cis* relationship for activity, as they are in ribose but not in arabinose. With *Cx pipiens* the absence of hydroxyl from the 2′ carbon severely reduced potency and the absence of both hydroxyls destroyed activity. For this species, therefore, binding by the 2′ hydroxyl is critical. When the adenine of ATP was replaced with cytosine, guanine, hypoxanthine or uracil, *Ae aegypti* was totally unresponsive to 1 mM solutions and *Cx pipiens* responded only

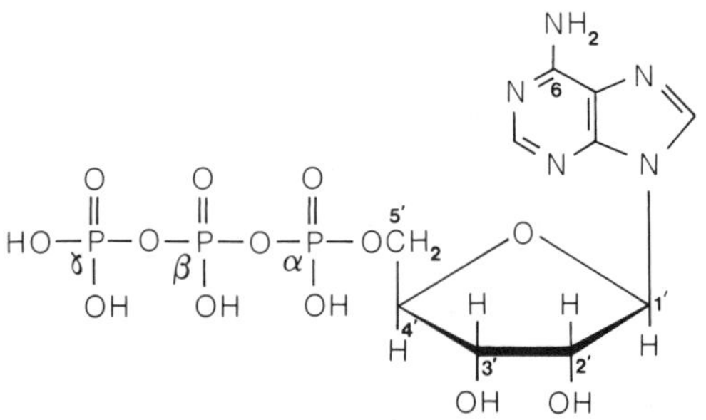

Figure 11.17 Adenosine 5′-triphosphate.

weakly to 100 mM solutions. The slightest of these changes was that from ATP to inosine 5′-triphosphate, which produced a change from the 6-aminopurine to a 6-oxopurine substituent (Galun, 1987; Galun *et al.*, 1984, 1985b, 1988). ATP does not have the highest possible affinity for the purinergic receptor of *Ae aegypti*, but it is obviously a satisfactory indicator of blood sources.

Adenyl nucleotides are known to be firmly bound inside erythrocytes, and Hosoi (1959) suggested that they may be liberated by saliva or by the surface of the sensilla. Cibarial armatures, when highly developed, damage erythrocytes, but they are downstream of the cibarial sensilla. Blood platelets, which have a high ATP content, proved very acceptable when offered in warm suspensions across a membrane, and the addition of platelets dramatically increased the otherwise negligible acceptability of human plasma and sheep erythrocytes. It seems that the apparent phagostimulation by erythrocytes may sometimes have been due to contamination with platelets. Even so, Galun and Rice (1971) estimated that the complete liberation of ATP from the platelets in human blood would barely yield a 10^{-5} M solution, and they postulated the presence of factors on the chemoreceptor surface which would cause platelet dissolution.

The responses of mosquitoes to blood, blood fractions, and solutions of phagostimulants are appreciably modified by physical and chemical factors. Few untethered females of *Culiseta inornata* ingested free blood at 20°C, but one-third engorged fully on blood at 37°C through a membrane (Friend, 1978). Culicine mosquitoes were fully responsive to adenyl nucleotides only when offered in an appropriate medium. Female *Ae aegypti* engorged maximally on ATP solutions that were isosmotic with blood and in which, as in plasma, the osmotic pressure was mainly due to NaCl and the buffering capacity to $NaHCO_3$. For this species a solution containing 0.15 M NaCl, 0.01 M $NaHCO_3$ and 5% albumin was equal to plasma as a vehicle for ATP. For *Cx pipiens* a solution of NaCl and $NaHCO_3$ alone was equal to plasma as a vehicle for ADP and ATP (Galun, 1967; Galun *et al.*, 1984, 1988).

Three *Anopheles* species – *freeborni*, *gambiae* and *stephensi* – gorged almost as actively on an ultrafiltrate of plasma, or on warm saline (0.15 M NaCl + 0.01 M $NaHCO_3$), as on whole blood. *An dirus* required the addition of 5% albumin to induce gorging. All four species were effectively

Table 11.7 Potencies of adenosine and adenyl nucleotides as phagostimulants mediating the gorging response to warm fluids offered across an artificial membrane. (From the data of Galun, 1987, and Galun *et al.*, (1985b, 1988.)

	ED₅₀ of gorging response (μM)			
Substance	Aedes aegypti	Aedes caspius	Culex pipiens	Culiseta inornata
Adenosine	–	–	>>1000	>>1000
Adenosine 5′-phosphate	463	310	27	906
Adenosine 5′-diphosphate	96	150	12	90
Adenosine 5′-triphosphate	12	9.1	24	199
Adenosine 5′-tetraphosphate	16	–	>>100	–
2′-Deoxy ATP	25	13	833	–
3′-Deoxy ATP	25	–	–	–
2′,3′-Deoxy ATP	1.2	2.1	Inactive	–
β,γ-imido ATP	2.4	0.8	20	168
β,γ-methylene ATP	3.6	4.9	48	338

Test substances were dissolved in 0.15 M-NaCl + 0.01 M-$NaHCO_3$ except for *Cs inornata* for which they were in 0.15 M-NaCl.
– Not tested.

unresponsive to ATP at concentrations up to 10^{-3} M. Thus these anophelines appear to lack purinergic receptors, and the plasma factors that enhance the responsiveness of culicines to ATP are themselves sufficient to induce gorging (Galun *et al.*, 1985a).

11.4.3 Regulation of destination and volume

When mosquitoes drink, two control processes are involved (in addition to acceptance): one regulates the destination of the fluid, the other regulates the volume that is imbibed. In general, mosquitoes exhibit the following drinking modes. When water is drunk only a small volume is imbibed, which is passed to the midgut. When mosquitoes feed on sugary plant juices a large volume is imbibed, which is passed to the dorsal diverticula and crop. When female mosquitoes feed on blood a large volume is imbibed and pumped directly to the midgut. Sphincters and valves at the entrances to the crop and midgut (Sections 13.1, 13.2) probably control the route of fluid movement. The osmolalities of nectars of high sugar concentration are several times higher than the osmolality of haemolymph, therefore storage of nectar in the impermeable crop is probably a physiological necessity.

Female *Culiseta inornata* show more control over the destination of their diets than some other mosquito species. In a series of experiments on water and sugar intake, females that were free in a cage and had been without water or food for 5–6 h were offered open drops of coloured water or sugar solution. The volumes imbibed and the destination were later assessed, volume being scored as large (>1 μl), small (0.2–1 μl) or taste (<0.2 μl). Water intake was restricted to the 'taste' and 'small' categories, and was almost always despatched to the midgut (Figure 11.8). When offered sucrose solution, the percentage of females taking large meals increased with concentration between 0.025 and 0.2 M, exceeding 90% at 0.2 M. Dilute sucrose solution was distributed, in different females, to midgut or to crop or to both, but at 0.4 M and above it was found almost entirely in the crop. Under similar conditions, and assessed over a range of concentrations, α-glucose and fructose had about half the phagostimulatory potency of sucrose. Sugars of low phagostimulatory potency, such as β-glucose, cellobiose and gentiobiose, were imbibed in only moderate volumes and were despatched to the stomach at concentrations of 0.2 M and below, being treated like water. At 0.5 M greater volumes were imbibed and some females despatched the fluid to both midgut and crop (Figure 11.18). Solutions containing both sucrose and cellobiose in different proportions, but having a total concentration of 0.5 M, were all imbibed in large volume and were increasingly despatched to the crop as the sucrose concentration increased (Friend *et al.*, 1988, 1989; Schmidt and Friend, 1991). In considering these results it is important to remember that nectar consists predominantly of sucrose, fructose and glucose in total concentrations between 20 and 50% (Section 11.1.1).

Blood feeding normally starts only after females have responded to host cues with orientated flight and have landed on the host. Heat is one of the last host stimuli to affect the female, and the combination of heat, moisture and CO_2 induces females to land and probe (Volume 2). In the laboratory, mosquitoes do not feed on diets that are at room temperature and covered by a membrane because a heat stimulus is necessary to attract them to the membrane and to induce probing. Unrestrained female *Cs inornata* imbibed 3.4 μl heparinized human blood when it was presented at 37°C and covered with an artificial membrane, but imbibed only 0.7 μl when it was presented as a free liquid at room temperature. Twenty percent of females fed on 10 mM ATP in warm Ringer under a membrane, imbibing on average 2.3 μl and despatching it to the midgut. Both warmth and a membrane were needed for mosquitoes to adopt the blood-feeding mode (Friend, 1978, 1981, 1985).

Culiseta inornata is not typical of all mosquitoes in the distribution of imbibed fluids. Unrestrained females of *Aedes aegypti*, *Ae albopictus* and *Ae vexans* almost invariably deposited blood in the midgut only, but quite high percentages of *Anopheles quadrimaculatus*, *An albimanus* and

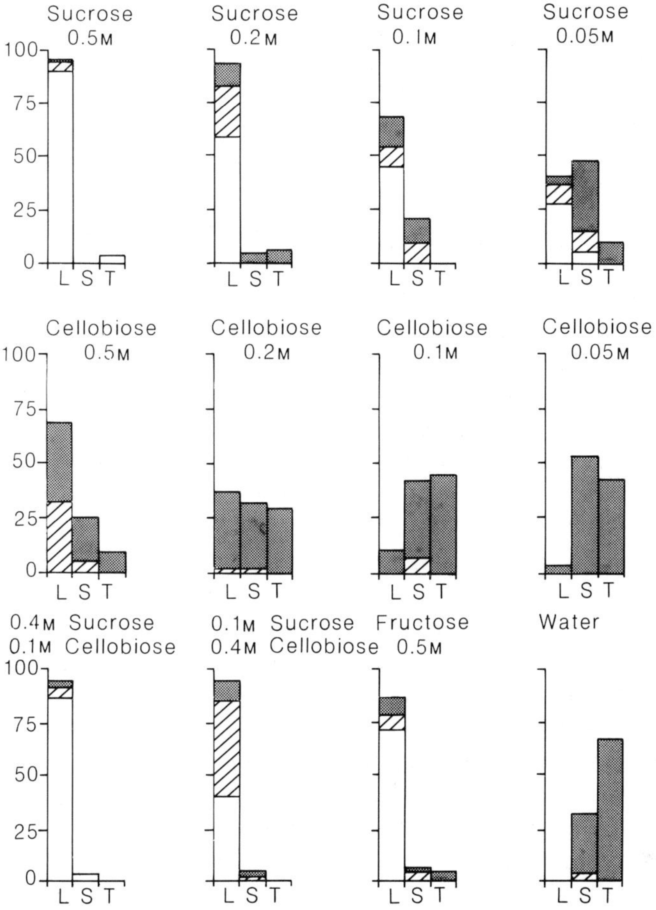

Figure 11.18 Responses of unrestrained female *Culiseta inornata* to certain sugars at various concentrations and to water. The X-axis of each histogram indicates the volume ingested: L – large (>1 μl), S – small (0.2–1 μl), and T – taste (<0.2 μl.) The Y-axis indicates the percentage in each volume class of the total number that fed. Diet destination is indicated: stippled, diet directed to the midgut; white, diet directed to the crop; cross-hatched, diet directed to both midgut and crop. (After Friend *et al.*, 1988, 1989.)

Culex quinquefasciatus deposited small or moderate quantities of blood in the crop and dorsal diverticula (Trembley, 1952). *Anopheles darlingi* is exceptional in dispatching plant juices to the stomach. In most wild female *An darlingi* the crop was inflated with air and in fewer than 1% did the crop contain glucose. In 46% the stomach contained clear fluid, and in 69% of such cases the stomach contents were positive for glucose. When females that had fed on plant juice took a blood meal, the freshly ingested blood usually pushed the plant juice to the posterior region of the stomach (Pajot *et al.*, 1975).

Mixtures of blood and glucose or of erythrocytes and glucose, ingested by unrestrained females of *Ae aegypti*, were often dispatched to both the midgut and the crop in the same insect (Day, 1954). When mixtures of haemolysed erythrocytes and sucrose were forcibly presented to the probosces of *Cx pipiens pallens*, most individuals deposited the mixture almost entirely in the midgut or almost entirely in the crop and dorsal diverticula, depending on the relative concentrations of erythrocytes and sucrose. When blood was offered to the unsheathed fascicle of *Cx pipiens pallens* and 'concentrated sugar solution' was presented

Table 11.8 Effects of cutting the ventral nerve cord in different positions upon the blood-meal weight and fertility of *Aedes aegypti*. The mean prefeeding body weight was 2.61 ± 0.14 mg. (Means ± 95% c.l.; *n* = 36–76.) (From Gwadz, 1969.)

Nerve cord cut just anterior to abdominal ganglion named	*Blood-meal weight (mg)*	*No. of eggs developed*
No treatment	2.85 ± 0.26	86.1 ± 3.3
Sham operated	2.93 ± 0.18	85.8 ± 3.0
Ganglion 6	3.61 ± 0.21	87.1 ± 4.4
Ganglion 5	4.23 ± 0.46	88.9 ± 4.2
Ganglion 4	5.14 ± 0.40	92.3 ± 2.8
Ganglion 3	7.63 ± 0.70	97.2 ± 4.0
Ganglion 2	11.99 ± 0.51	103.3 ± 4.3

to the labella, the blood was dispatched not to the midgut but to the crop (Hosoi, 1959). These results suggest a central integration of the sensory input.

The termination of feeding appears to be initiated by abdominal stretch receptors. Females of *Ae aegypti* in which the ventral nerve cord had been severed imbibed significantly larger volumes of blood from a human arm than control females. Blood meal size depended on the position at which the nerve cord had been cut; the more anterior the cut the greater the volume of blood ingested (Table 11.8). Females of *Armigeres subalbatus*, *Culex quinquefasciatus* and *Anopheles quadrimaculatus* imbibed 3–4 times the normal volume of blood when the nerve cord had been cut in front of the 2nd abdominal ganglion, and the abdomens of some females ruptured (Gwadz, 1969).

Rather similar results were obtained with sugar feeding in *Cs inornata*. When females which had a slit in the abdominal pleural membrane fed on 1 M sucrose they imbibed so much that the crop was greatly distended and forced out through the wound. When the extruded crops of three mosquitoes were pierced and the contents sucked into micropipettes, the mosquitoes imbibed from 22 to 33 μl, two feeding for 40 min. Severing the ventral nerve cord anterior to the 3rd abdominal ganglion caused hyperphagia in all sucrose-feeding females. Severance anterior to the 5th abdominal ganglion led to hyperphagia in some females but severance in front of the 6th ganglion had no effect (Owen and McClain, 1981).

When restrained females of *Cs inornata* were sequentially offered 0.25 M, 0.5 M, 1.0 M and 2.0 M sucrose during four 5-min periods, separated by intervals of 2 min, all drank to satiety at the lowest concentration. When offered the next highest concentration they responded immediately and again ceased feeding in under 5 min. This pattern of response was repeated at all concentrations, producing in 24 females a mean intake of 7.79 μl. Ten females consumed >5.5 μl and were classed as hyperphagic (Owen and McClain, 1981).

Our understanding of the regulation of mosquito feeding is incomplete but the evidence suggests the following outline. When the mouthparts contact water or sugar solution, pumping is induced by stimulation of the labellar sensilla alone or of both labellar and labral sensilla, depending on species. It seems likely that when the mouthparts of a host-seeking female contact blood, stimulation of the labral chemosensilla must be supplemented by some other host stimulus, possibly heat, to induce pumping. The sensory input is enhanced as soon as the nectar or blood contacts the cibarial chemosensilla. Input from gustatory sensilla governs the distribution of nectar to the crop and of blood and water to the midgut, possibly with involvement of the stomatogastric system. The volume of nectar or blood that is imbibed is determined by the sensory input to a central system coupled with negative feedback in the form of impulses from stretch receptors in the abdominal wall, which are believed to fire when the crop or stomach becomes distended.

12

The adult salivary glands and their secretions

Mosquitoes secrete saliva more frequently and more copiously than is easily apparent, regulating its discharge by a combination of nervous and endocrine mechanisms. The saliva has many constituents which serve a number of functions, only a few of which have been well established. The saliva is of broader interest because it contains allergens that are responsible for hypersensitive reactions to mosquito bites and because malaria parasites and many arboviruses use it as a vehicle for transmission to their vertebrate hosts.

12.1 STRUCTURE OF THE SALIVARY GLANDS

12.1.1 Gross structure

A pair of salivary glands is present in the thorax, just above the forelegs and flanking the oesophagus. In both sexes of most species each gland normally consists of three lobes, two lateral and one median. There is, however, considerable variability of form, even within species, and in an extensive study of culicines, Shishliaeva-Matova (1942) observed from one to six lobes. In *Aedes aegypti* the distal regions of the lateral lobes are often divided into two or three separate portions, each with its own duct (Janzen and Wright, 1971). In female *Anopheles quadrimaculatus* the length of the lateral lobes is about 900 μm, and that of the median lobe about 360 μm; the width of the lobes is 70–80 μm (Metcalf, 1945). Each lobe is formed of a layer of epithelial cells, disposed around a central duct and bounded externally by a basal lamina. In female mosquitoes the identical lateral lobes are formed of three distinct regions, proximal, intermediate and distal, whereas the median lobe is formed of a short neck region and a distal region (Figure 12.1A). Innervation is limited to the extreme anterior part of each gland and the adjacent part of the lateral duct (Spielman *et al.*, 1986). The ingluvial ganglia are the source of neurosecretory axons that supply the salivary glands (Meola and Lea, 1972a).

A canal or duct runs the full length of each lobe. The ducts emerging from the three lobes fuse, forming a lateral salivary duct which runs forwards to fuse with its fellow below the suboesophageal ganglion. The common salivary duct enters the salivary valve at the base of the hypopharynx (Figure 11.4). The structure of the salivary valve and the course of the salivary canal through the mouthparts were described in Section 11.2.3. The lateral salivary duct of *Culex tritaeniorhynchus* is about 15 μm wide. Transverse sections show two or three epithelial cells, bounded by a basal lamina, surrounding a cuticular duct of 3–4 μm outer diameter (Suguri *et al.*, 1972).

The salivary glands of male mosquitoes are distinctly smaller than those of the females. Each gland is tri-lobed, and by light microscopy the whole gland appears to consist of a single cell type. The secretory product collects in extracellular apical cavities and, as in the female, the apical cavities in the posterior regions of the lobes are much more dilated with secretion than those in the anterior regions. All cells in the male glands resemble those of the proximal region of the lateral lobes of the female, a similarity which extends to the electrophoretic

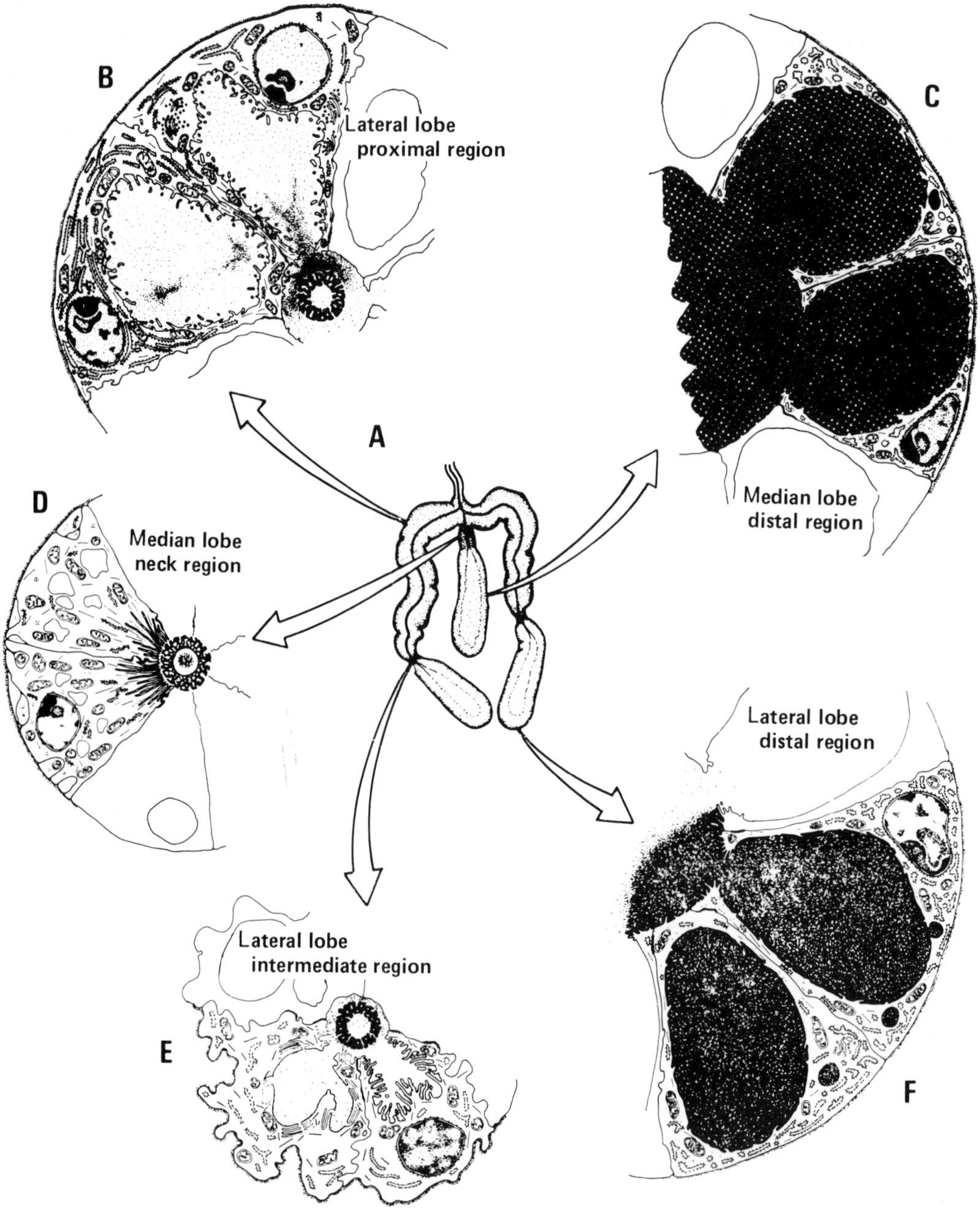

Figure 12.1 The salivary gland of an adult female *Anopheles stephensi*. (A) A whole gland. (B–F) Ultrastructure of the different regions of the gland as seen in transverse sections. (After Wright, 1969.)

mobility of their proteins (Wright, 1969; Beckett, 1988; Poehling and Meyer, 1980). In both sexes of the non-blood-sucking species *Toxorhynchites brevipalpis* and *Tx rutilus*, the whole median lobe and the intermediate region of the lateral lobes are missing (Barrow *et al.*, 1975).

Immediately after emergence the salivary glands of female *Ae aegypti* are small, thin, and unconvoluted. The main characteristics of the adult gland develop during the 36 h after emergence, when secretions appear in the extracellular cavities (Orr *et al.*, 1961). At emergence the salivary glands of *An stephensi* and *Cx pipiens* contain <25% of their eventual protein content; this increases to 50% by the 3rd day, and to 90% by the 5th day after emergence (Poehling, 1979). In adult mosquitoes the salivary gland chromosomes are not polytene. No mitotic figures were seen in the maturing salivary glands of *Ae aegypti* by Orr *et al.* (1961), who concluded that the glands grow through increase in cell volume. However, Shishliaeva-Matova (1942) frequently observed mitotic figures in the cells of the neck region and intermediate region in females of *An sacharovi*. Histochemical tests on whole salivary glands of females of *An freeborni* and *Ae aegypti* showed esterase and acid phosphatase activity, which was apparently located within the cell bodies rather than in the extracellular secretions. The greatest activity was in the intermediate region of the lateral lobes and the neck region of the median lobe (Freyvogel *et al.*, 1968).

12.1.2 Ultrastructure of the secretory cells

The ultrastructure of female salivary glands has been described in greatest detail for *Anopheles stephensi* (Wright, 1969); briefer descriptions are available for *Aedes aegypti* (Janzen and Wright, 1971), *Culex tritaeniorhynchus* (Suguri *et al.*, 1972) and *Cx pipiens* (Barrow *et al.*, 1975). Except where stated, the following account relates to *An stephensi*.

The cells of the **proximal region of a lateral lobe** are radially distributed around a narrow duct from which they are separated by a periductal space (Figure 12.1B). Each cell encloses an extracellular space, or apical cavity, which is continuous through a narrow opening with the periductal space. The apical plasma membrane, which borders the extracellular cavity, bears irregular microvilli with which mitochondria appear associated. The cytoplasm contains RER, free ribosomes, Golgi complexes and microtubules. The large spherical nucleus has a prominent nucleolus, and the outer nuclear membrane is studded with ribosomes. The cells are bound to each other apically by septate desmosomes. The duct that runs through the proximal region is about 2 μm in outer diameter and 1 μm in inner diameter. Its cuticular wall is pierced by irregularly shaped canals about 90 nm wide. In fixed tissue the contents of the apical cavity appear as moderately dense flocculent or filamentous material, which is sometimes continuous with similar material in the periductal space and within the duct lumen. In *Ae aegypti* these cells have the ultrastructural characteristics of ion- or water-transporting cells. The basal region is largely occupied by a basal labyrinth, the apical membrane bears elongate microvilli which almost fill the apical cavity, and septate desmosomes are present apically between the lateral membranes of adjacent cells.

The **intermediate region of a lateral lobe** is short and narrow, and can be recognized by the constriction it makes in the lobe (Figure 12.1A). It consists of a small number of cells, each enclosing an extracellular cavity which is continuous with the narrow periductal space (Figure 12.1E). The most striking feature of these cells is the extensive infolding of the apical plasma membrane where it faces the extracellular cavity and the periductal space. However, in the cells that are adjacent to the proximal region the apical plasma membrane is not lamellate but bears microvilli. The relatively extensive cytoplasm contains numerous mitochondria and some rough endoplasmic reticulum. The cells are linked by septate desmosomes on their lateral membranes. A perforated cuticular duct extends through the intermediate region. In fixed tissue the extracellular cavities contain a reticulum of fine filaments, but the periductal space is filled with a much denser granular or filamentous material.

In the **distal region of a lateral lobe**, the extracellular cavities are much larger than in the proximal region, and the openings that connect them to the central canal are much wider (Figure 12.1F). The apical plasma membrane bears only few microvilli. The basal cytoplasm contains some rough endoplasmic reticulum, many free ribosomes, and microtubules; mitochondria and Golgi complexes are rare. The nuclei are large and have prominent nucleoli. The extracellular cavities and the central canal contain a dense secretion. The cuticular duct widens as it extends posteriorly from the intermediate region, and its wall thins until it disappears; consequently the duct becomes replaced by a central canal which is about one-third the width of the lobe. In *Ae aegypti*, *Cx pipiens* and *Cx tritaeniorhynchus* a cuticular duct extends through the full length of all three lobes. There is no periductal space in *Cx pipiens*.

In its organization and ultrastructure the **distal region of the median lobe** (Figure 12.1C) resembles the distal region of the lateral lobes. However, its secretion appears more uniform and is more osmiophilic.

A short **neck region** connects the median lobe with the two lateral lobes. A cuticular duct, with an outer diameter of 3 μm and an inner diameter of 1.8 μm, extends through the neck region and joins the ducts of the lateral lobes. The duct wall appears porous but a continuous layer about 325 nm thick forms its inner boundary. The cells of the neck region have no extracellular cavity, and come into close contact with the duct. The apical plasma membrane is extensively folded, forming lamellae which may extend through as much as one-third of the cell (Figure 12.1D). The basal plasma membrane is irregularly folded. The cytoplasm contains many mitochondria; Golgi complexes and endoplasmic reticulum are not abundant. Radially-orientated microtubules are concentrated in the region of the infolded membrane. Large irregular membrane-bound vacuoles occur between the nucleus, which contains a prominent nucleolus, and the apical plasma membrane. Neighbouring cells are united by septate desmosomes situated close to their apical end. The neck region is the only part of the salivary gland to be well tracheated. In *Ae aegypti* the cell is extensively occupied by a basal labyrinth and associated mitochondria.

In *Aedes aegypti*, innervation is limited to the extreme anterior end of each gland and the adjacent part of the lateral duct. In the case of the lateral lobes, axons lacking glial sheaths approach cells near the point of origin of the salivary duct and terminate, widely separated from one another, in the extracellular space beneath the basal lamina. In the median lobe, unsheathed axons branch to form a plexus which surrounds the neck region before penetrating the basal lamina. All the axon terminals contain neurosecretory granules of *c.* 100 nm diameter, and vesicles of *c.* 40 nm diameter which are clustered to form synaptic plaques (Spielman *et al.*, 1986).

In *An stephensi* the cells of the distal regions of the lateral lobes and of the median lobe are clearly secretory; the cells of the proximal region of the lateral lobes were considered to be secretory by Wright (1969). The cells of the neck region of the median lobe and of the intermediate region of the lateral lobes have an ultrastructure characteristic of ion- or water-transporting cells. In *Ae aegypti* the cells of the proximal region of the lateral lobes also clearly function in ion or water transport, as do the cells surrounding the lateral salivary duct in *Cx tritaeniorhynchus*.

Examination of the salivary glands of *Cx pipiens* within four hours of blood feeding revealed that electron-dense material had been discharged from the extracellular cavities and the duct lumen of the proximal region of the lateral lobes and of the distal region of the median lobes. In contrast, the appearance of the distal and intermediate regions of the lateral lobes and of the neck region of the median lobe remained unchanged. Barrow *et al.* (1975) postulated that the product of the distal regions of the lateral lobes was continuously secreted and transported to the proximal region for storage until feeding. It appears probable that the fluid-transporting regions of the salivary glands are activated at the time of feeding, possibly by both humoral and nervous mechanisms, and that water uptake is the driving force for the discharge of saliva.

Cationic ferritin bound to the basal lamina over all three lobes but anionic ferritin failed to bind, indicating that the glands carry a net negative surface charge. Biotin- and fluorescein-labelled lectins, with a variety of sugar specificities, bound to the outside of the basal lamina but none penetrated into the underlying cells. Concanavalin A bound uniformly to the basal lamina over all lobes. In contrast, the lectins WGA, PNA and RCA bound over the median lobe and the distal region of the lateral lobes, whereas the lectin SBA bound only over the median lobe, thus various regions of the basal lamina differ in their carbohydrate constituents (Perrone *et al.*, 1986). Lectin binding patterns over the salivary glands of four species of *Anopheles* exhibited not only inter-species differences but also differences between strains of single species (Molyneux *et al.*, 1990; Mohamed *et al.*, 1991).

12.2 THE CONSTITUENTS, ACTIONS AND SECRETION OF SALIVA

12.2.1 Constituents and actions

Secretions exuded from the tip of the proboscis can be collected in warm distilled water, in oil, from sucrose wicks, and from glass slides (Allen and West, 1962; Hurlbut, 1966; Rossignol and Spielman, 1982; Mellink and van Zeben, 1976). Alternatively, mosquitoes can be forced to discharge secretion by injection with 5-hydroxytryptamine or topical application of malathion (Ribeiro *et al.*, 1984a; Boorman, 1987). The oral secretions collected by these methods are considered to be saliva.

Saliva from female *Aedes aegypti* was separated by SDS-PAGE into twelve bands which were estimated to contain as many as 20 polypeptides. The electrophoretograms appeared similar whether the secretion had been discharged into oil or into sucrose solution. Serum from bitten guinea-pigs was immunoreactive against nine of the twelve principal bands. Racioppi and Spielman (1987) considered that the major salivary proteins collected in oil are secreted into hosts during blood feeding. A 37 kDa polypeptide present in the saliva of female *Ae aegypti* was considered to be the most abundant of the secreted polypeptides. This was the final product coded for by gene D7, which was expressed only in the distal regions of the lateral lobes and in the median lobes of the female salivary glands (Table 12.1) (James *et al.*, 1991).

Mosquito saliva probably has a number of functions, being secreted during sugar feeding as well as during blood feeding. Most tests for digestive enzymes in salivary gland extracts proved negative, e.g. for lipase, amylase, trypsin and chymotrypsin (Metcalf, 1945; Nakayama *et al.*, 1985). However, the salivary glands of both male and female *Ae aegypti* contain an **α-glucosidase** (M_r 68 000) which is ejected when the mosquitoes take a sugar meal, and which accumulates in the crop. The enzyme activity was found predominantly in the proximal regions of the lateral lobes. Sucrose was hydrolysed rapidly, maltose one-fifth as fast, but the other di- and trisaccharides reported from nectar were virtually unaffected, as was trehalose. The pH optimum with sucrose as substrate was 6.0. The enzyme was identified as an α-D-glucosidase (3.2.1.20), and not a β-D-fructofuranosidase (3.2.1.26), on the grounds of its inactivity towards raffinose (Marinotti and James, 1990). Analyses of the crop contents of female *Cx tarsalis* fed 50% sucrose solution revealed that at 0 h 36% had been hydrolysed to monosaccharides, by 4 h 70% and by 24 h 100% (Schaefer and Miura, 1972). The α-glucosidase may be the final product coded for by the *Maltase-like I* gene (*Mal I*), which is transcribed throughout the male salivary glands but only in the proximal regions of the lateral lobes of the female glands (James *et al.*, 1989, 1991).

A bacteriolytic factor analogous to lysozyme was found in the proximal regions of the salivary glands of female *Ae aegypti* and in their crop contents. It was also present in male salivary glands. Rossignol and Lueders (1986) postulated that it protected mosquitoes from potentially pathogenic bacteria in sugar sources. Incubation of human erythrocytes with salivary gland extract for 18–24 h resulted in haemolysis, which was pronounced with the glands of *Cx pipiens* and *Ae punctor*, less pronounced with the glands of *Ae aegypti* and *An quadrimaculatus* (Hudson, 1964).

Evidence for a vasodilatory action by *Ae triseriatus* saliva is given in Section 11.3.3(b). Extracts of mosquitoes exhibit histamine-like pharmacological activity (Eckert *et al.*, 1951), but the absence of skin reactions in unsensitized animals bitten by mosquitoes suggested that histamine was not a constituent of saliva (McKiel, 1959; Wilson and Clements, 1965). Assays for histamine in extracts of female salivary glands produced the following results, expressed as histamine content per pair of glands: *Cx pipiens pallens* 2.4 ng (Nakayama *et al.*, 1985), *Ae aegypti* >1 ng (Wilson and Clements, 1965), *Ae albopictus* undetectable (Oka *et al.*, 1989).

When the common salivary duct was cut before females engorged, strands of fibrin were found in blood removed from the midgut of the operated females that were not present in blood from normal females. By this means the presence of an anticoagulant was demonstrated in the saliva of *Ae aegypti*, *An quadrimaculatus*, *Cx tarsalis* and (weakly) *Cx pipiens*. The anticoagulant activity was not due to inhibition of the action of thrombin on fibrinogen (Hudson, 1964). Ribeiro *et al.* (1985b) found anticoagulant activity in extracts of *An stephensi* and *An freeborni* salivary glands. Agglutination of human erythrocytes was produced by homogenates of salivary glands from females of *An quadrimaculatus*, *Cx pipiens* and four species of *Aedes* including *aegypti*. The presence of agglutinins was doubtful in *Culiseta inornata* and *Ae punctor* (Hudson, 1964).

The enzyme **apyrase** (ATP diphosphohydrolase; 3.6.1.5) is present in saliva ejected from the mouthparts of probing females. It hydrolyses ATP and ADP to AMP and inorganic phosphate. Apyrase activity is rare in animal tissues; indeed, mosquito salivary glands provide the richest known source. Apyrase from *Aedes* and *Anopheles* had similar kinetic properties. The enzyme was activated by divalent ions (Ca^{2+} being more effective than Mg^{2+}). The pH optimum was 9.0 but significant activity remained at pH 7.5. Apyrase activity was confined to the distal regions of the female salivary glands and was absent from the male glands which lack these regions (Ribeiro *et al.*, 1984a, b, 1985b; Rossignol *et al.*, 1984; Vachereau and Ribeiro, 1989).

Blood loss from lacerated small blood vessels is normally prevented by the formation, within seconds, of a plug of aggregated platelets at the site of the wound and by vasoconstriction. ADP is an important agent in this reaction (Figure 12.2) (Kroll and Schafer, 1989). Experimental and circumstantial evidence suggests that salivary apyrase functions to block ADP-induced platelet aggregation, thereby promoting haematoma formation. Saliva collected for 10 min in oil from a single mosquito was sufficient to inhibit platelet aggregation induced by ADP or collagen *in vitro* (Figure 12.3) (Ribeiro *et al.*, 1984a). Females with transected salivary ducts took longer to locate blood when probing (Figure 12.4) and took longer to imbibe blood after a source had been located (Mellink and Van der Bovenkamp, 1981; Ribeiro *et al.*, 1984a). For three species of *Anopheles* the speed of locating blood was positively correlated with titre of

Table 12.1 Protein constituents of *Aedes aegypti* saliva and their locations within the females' salivary glands.

Constituent	*Mol. mass (kDa)*	*Male*	*Female*			*Refs.*
		Whole gland	*Median lobe*	*Lateral lobe*		
				Proximal	*Distal*	
Polypeptide	37	−	+	−	+	1
Apyrase	66	−	+	−	+	2
α-Glucosidase	68	+	−	+	−	3
Bacteriolytic factor		+	−	+	−	4

+, present. −, absent.
References: 1. James *et al.* (1991); 2. Rossignol *et al.* (1984); Vachereau and Ribeiro (1989); 3. Marinotti and James (1990); 4. Rossignol and Lueders (1986).

stored apyrase in the salivary glands (Ribeiro *et al.*, 1985b). Females of *Ae aegypti* whose salivary glands were infected with sporozoites of *Plasmodium gallinaceum* probed twice as long before locating blood than uninfected females. The infected females discharged as much saliva as healthy females but had only one-third as much apyrase activity (Rossignol *et al.*, 1984). Probing *Ae aegypti* located blood significantly more rapidly when feeding on mice infected with *Plasmodium* or Rift Valley fever virus, possibly due to parasitic disruption of haemostasis, a known phenomenon (Rossignol *et al.*, 1985). No antithromboxane or antiserotonin activity could be demonstrated in *Ae aegypti* salivary glands (Ribeiro, 1987).

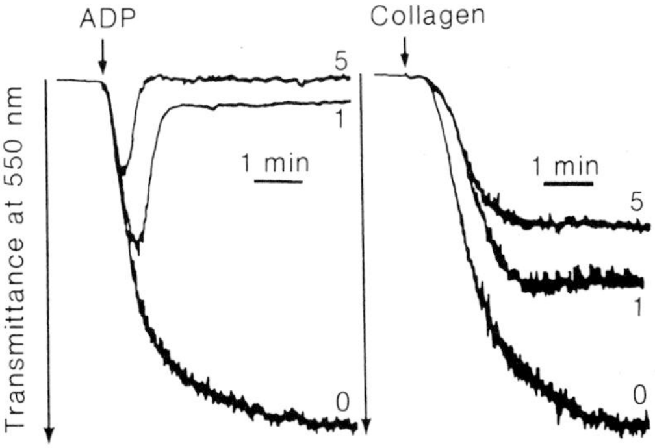

Figure 12.3 Aggregation of platelets in 100 μl citrated platelet-rich human plasma induced by ADP or collagen, in the absence or presence of saliva that had been harvested for 10 min from 1 or 5 female *Aedes aegypti*. (From Ribeiro *et al.*, 1984a.)

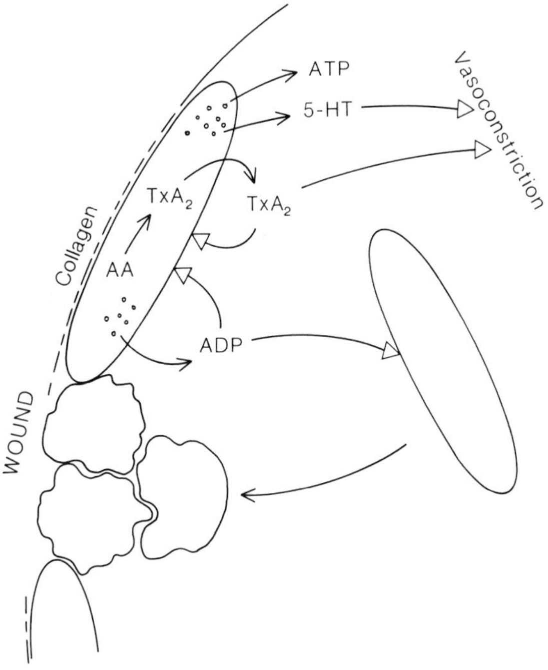

Figure 12.2 Key events in haemostasis following damage to a small blood vessel. Platelets adhere to certain subendothelial proteins that are exposed at the site of laceration, notably to collagen. The binding of collagen to surface receptors activates phospholipase A_2 which cleaves arachidonic acid (AA) from membrane phospholipids. Thromboxane A_2 (TxA_2) is formed from arachidonic acid, and is secreted from the platelets as are the contents of dense granules, ADP, ATP and 5-HT. TxA_2 and ADP bind to specific surface receptors on platelets producing the same effects as collagen and so amplifying its action. ADP induces platelets to change shape and to aggregate, producing a precisely located plug which seals the wound. TxA_2 and 5-HT stimulate vasoconstriction.

Movement of blood from the mouthparts to the midgut appeared to be normal in females with transected salivary ducts, and the fertility of the operated females was not reduced (Hudson *et al.*, 1960; Rossignol and Spielman, 1982). The biting by duct-transected females was more

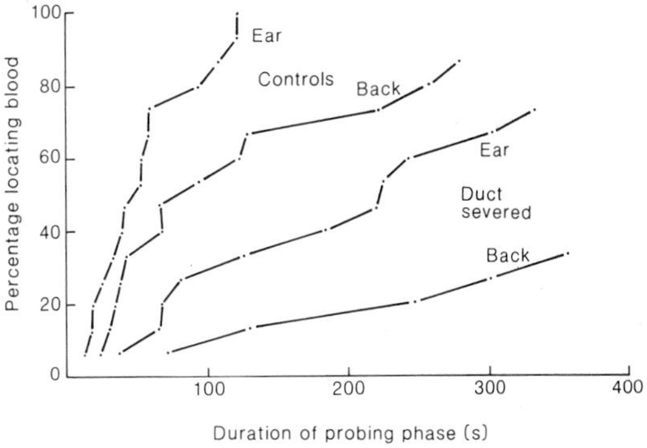

Figure 12.4 Cumulative curves of the times taken by females of *Aedes aegypti* to locate blood after starting to probe on the ear or shaved back of a guinea pig. Timing was stopped when mosquitoes withdrew their mouthparts before locating blood, and was restarted upon renewed penetration. The salivary ducts had been severed in the experimental females, whereas the necks had been pierced in the operated controls (n = 15.) (After Ribeiro *et al.*, 1984a.)

painful than that by normal mosquitoes (Hudson *et al.*, 1960).

12.2.2 Synthesis and secretion

Protein synthesis has been investigated by following the fate of [^{35}S]methionine injected into sugarfed female *Aedes aegypti*. Saliva harvested within 10 min of injection of the amino acid already contained labelled polypeptides. The salivary glands were active in protein synthesis immediately after emergence, and the rate of synthesis increased over the next three days. Over the same period the protein content of the salivary glands increased 5-fold. Both protein synthesis and protein content remained high for the first two weeks but declined slightly in the third week. Analysis of extracts from the different regions of the salivary glands, following incubation with [^{35}S]methionine, revealed that the distal regions of the lateral lobes were the most synthetically active, followed by the median lobes and the much less active anterior regions of the lateral lobes. Secretory proteins were synthesized by non-blood-feeding females at a constant rate throughout at least the first month of adult life (Racioppi and Spielman, 1987).

The secretory capabilities of the salivary glands of female *Ae aegypti* increase during the first few days after emergence. The glands exhibited no apyrase activity at emergence, and activity was still slight after one day, but reached its maximum during the second day (Ribeiro *et al.*, 1984b). α-Glucosidase activity was very low at emergence but increased steadily during the next two days. It subsequently declined somewhat, consistent with the start of sugar feeding (Marinotti and James, 1990). Bacteriolytic activity increased 6-fold during the first three days after emergence (Pimentel and Rossignol, 1990).

Flow of saliva has been observed within the stylets of non-feeding females of *Anopheles maculipennis* after removal of the fascicle from the premental gutter, the saliva being discharged from the tip of the hypopharynx and sucked up through the labral food canal. From many such observations Fülleborn (1932) concluded that mosquitoes discharge and swallow saliva when not feeding.

Secretion of saliva in response to phagostimulation was demonstrated with preparations of female *Ae aegypti* in which the mouthparts were placed in a test solution while the tip of the hypopharynx was held in the light-path leading to a photometer. Injected Na fluorescein provided a label for secreted fluid. Salivation was stimulated when the mouthparts were exposed to sucrose solution, the rate of secretion correlating positively with concentration and the optimum temperature being 22°C. ATP was also stimulatory, with 1 mM being the optimum concentration and 37°C the most suitable temperature (Spielman *et al.*, 1986).

Four hours after *Culex pipiens* had bloodfed the median lobe and the proximal region of the lateral lobes showed loss of secretory material but the distal regions of the lateral lobes remained full (Barrow *et al.*, 1975). When *Ae aegypti* bloodfed, 15–35% of the soluble proteins were discharged from the salivary glands. Proteins were discharged in greater quantites from the median lobes and the proximal regions of the lateral lobes than from the distal regions of the lateral lobes (Poehling, 1979).

To determine whether specific proteins were secreted differentially, the apyrase and α-glucosidase activities and protein content of *Ae aegypti* glands were measured immediately after sucrose and blood meals. The apyrase activity decreased significantly after a blood meal but not after a sucrose meal. In contrast, α-glucosidase activity decreased significantly after both sucrose and blood meals. Possibly α-glucosidase can be secreted alone from the proximal regions of the lateral lobes by sugar-feeding mosquitoes but blood feeding causes discharge from all regions of the glands. α-Glucosidase activity returned to the unfed control value within 1 h after a blood meal, but it required 24 h for the apyrase activity to return to normal. It was reported that almost 60% of salivary gland protein was depleted when *Ae aegypti* took a blood meal, and 40% when they fed on sucrose solution (Marinotti *et al.*, 1990). When mosquitoes were deprived of sugar sources for 48 h their α-glucosidase activity increased 5-fold, suggesting that the enzyme is normally subject to frequent depletion through sugar feeding (Marinotti *et al.*, 1990).

The observation that duct-transected mosquitoes were capable of arbovirus transmission prompted investigation of fluid secretion by duct-transected females. Normal females of *Ae aegypti* ejected saliva into oil at a rate of 3788 $\mu m^3\ s^{-1}$ (± 2397 s.d.), when measured over one to two minutes, producing droplets of 900 μm^3. When the lateral salivary ducts were cut just anterior to the glands, and externalized to open outside the neck, operated individuals ejected into oil 10% as much fluid as the operated controls. This was much more than could have been contained in the ducts. Duct-externalized females did not produce weal-and-flare reactions when fed on sensitized human subjects. Females in which the common salivary duct had been severed immediately behind the salivary valve produced no secretion. Rossignol and Spielman (1982) tentatively concluded that fluid is transported across the wall of the salivary ducts. Al-Ahdal *et al.* (1990) postulated that three proteins present in the saliva of *Culex pipiens*, but absent from salivary gland extract, were derived from the haemolymph.

The presence of synaptic and neurosecretory release sites in the axon terminals of the salivary gland innervation suggests both nervous and hormonal control of secretion. Possibly the regions that are not innervated respond to haemolymph-borne hormones.

12.3 SKIN REACTIONS OF MAMMALIAN HOSTS

The small skin eruptions that we call 'mosquito bites' are a classical example of hypersensitivity, i.e. an immune reaction that results in gross tissue changes. In this instance the allergen is mosquito saliva. The immune nature of the reaction is evident because it is shown only by previously sensitized individuals. The role of saliva was confirmed when it was demonstrated that no skin reaction was given if the mosquito's common salivary duct had been cut before feeding (Hudson *et al.*, 1960).

Four or more types of hypersensitivity have been described. Of those involved in the reactions to mosquito saliva, Types I and III depend on the interaction of antigen with humoral antibody, and are called 'immediate type reactions' (although the Type III reaction is less immediate than the Type I). Type IV hypersensitivity involves receptors bound to the T-lymphocyte surface, and because of its longer time course is called 'delayed-type hypersensitivity'.

The visible cutaneous reactions usually have the form of pruritic (itchy) weals and papules, but Arthus-type local necrotic lesions and systemic symptoms sometimes occur. The hypersensitive reactions given by mammals to mosquito bites have often been classified as (a) 'immediate responses', that are manifested instantly, peak within 30 min and disappear within a day, and (b) 'delayed responses', that are first manifested a few hours after exposure and that peak about one day after the bite. Type III hypersensitivity reactions are probably among the 'delayed responses' in this categorization.

12.3.1 Nature of the salivary antigens

People who are exposed to mosquito bites are found to be sensitized. In southern Finland, where *Aedes communis* was very abundant, 5% of infants, 35% of children and 65% of adults had IgG-class antimosquito antibodies detectable by ELISA (Ailus *et al.*, 1985). IgG and IgE antibodies with immunoreactivity to *Ae albopictus* salivary-gland antigens were found in individuals of all ages at Miki in Japan. The IgG levels were lower in individuals under two years of age (Konishi, 1990). Assays of anti-*Culex quinquefasciatus* antibodies in human sera from two regions of Orissa, India, revealed the influence of exposure to the mosquito and of age. In sera from Anugul, where *Cx quinquefasciatus* consituted only 5% of the mosquito catch, for all age groups the IgG titres were low and the IgE titres were zero or very low. In Puri, where *Cx quinquefasciatus* constituted 77% of the catch, IgG titres were moderate in children of 1–5 years and high in age groups from 10 years upwards; IgE titres were moderate up to age 16 and significantly higher in age groups from 18 years upwards (Das *et al.*, 1991).

Sera from bite-sensitized guinea-pigs reacted with nine polypeptide fractions when immunoblotted with *Ae aegypti* salivary-gland extracts (Racioppi and Spielman, 1987). In similar experiments, extracts of salivary glands from five mosquito species, that were immunoblotted with with sera from bite-sensitized human beings, were found to contain from 9 to 19 reactive polypeptides of from 14 to 127 kDa. Some antigens were shared by all mosquito species, others were more specific (Penneys *et al.*, 1989). Not all antibodies precipitate antigens, and analyses of sera from bite-sensitized animals revealed relatively few precipitating antibodies. Up to three precipitin lines were obtained in Ouchterlony immunodiffusion tests when sera from bite-sensitized rabbits were reacted with whole-body extracts of *Ae aegypti* (Wilson and Clements, 1965), and when sera from bite-sensitized guinea-pigs were reacted with extracts of *Ae aegypti* salivary glands (Racioppi and Spielman, 1987). Some at least of the sensitizing constituents of *Ae aegypti* saliva are thermostable, since guinea-pigs could be sensitized by saliva heated to 100°C for five minutes (Allen and West, 1966).

12.3.2 Types of skin reaction

Many investigations of the skin reactions of mammals to mosquito bites have relied on visual asessment of the reaction: erythema and weal in an immediate reaction, erythema and papule in a delayed reaction. Quantitative assessments have been made from measurements of the diameters of erythema, weal and papule at 30 minutes and 24 hours after a bite (Lengy and Gold, 1966; Oka, 1989).

No skin reactions are shown to the first bite that individuals receive, but once bitten they become sensitized to the salivary allergens. Human beings who are repeatedly bitten by mosquitoes pass through stages of sensitization, starting and ending with non-reactivity. Mellanby (1946) and McKiel and West (1961) reported the following sequence, the immediate response being that manifested within minutes, the delayed response that manifested after 24 h.

	Immediate response	Delayed response
Stage 1	−	−
Stage 2	−	+
Stage 3	+	+
Stage 4	+	−
Stage 5	−	−

Only young children not previously exposed to mosquito bites showed the stage 1 condition. Tests on 120 Japanese, aged 1–68 years, all of whom lived under conditions of exposure to *Aedes albopictus*, revealed individuals in stages 2–5 (Oka, 1989).

Some investigations have included skin biopsies, and a few have utilized advanced immunological techniques. Where sufficient information is available, it is possible to classify the reactions in terms of one or other of the different types of hypersensitivity.

(a) *Type I (immediate) hypersensitivity*

Type I hypersensitivity results when antigen reacts with antibody (mainly IgE) that is bound to mast cells and circulating basophils. Degranulation of mast cells causes the release or formation of a variety of chemical mediators, notably histamine, leukotrienes and prostaglandins, which increase vascular permeability and rapidly produce a visible weal and flare with pruritus at the site of exposure. The overt symptoms disappear within about one hour.

Within a few minutes of a mosquito feeding on man, rat, mouse, guinea-pig or rabbit a soft whitish weal appears around the puncture, usually surrounded by a reddish area (flare, erythema). Deep biopsies taken immediately after *Aedes aegypti* had fed at sites on the arms and back of man revealed an irregular puncture canal in the epidermis, marked oedema of the superficial dermis, occurring up to a considerable distance from the bite, marked vasodilatation, and the beginning of a perivascular infiltration of leucocytes, i.e. movement from within blood vessels to the surrounding tissue. At 30 minutes the infiltrate was composed of neutrophils, eosinophils, lymphocytes and plasma cells (Rockwell and Johnson, 1952; Goldman *et al.*, 1952; Bandmann and Bosse, 1967).

Thirty minutes after sensitized guinea-pigs had been bitten by *Ae aegypti* weals and flares had appeared. Plasma exudate was present in the epidermis, some blood vessels were dilated and others contracted, and leucocyte infiltration of the tissues had started. At 45 min plasma exudate was mainly in the dermis. Cells had migrated through the dermis and into subcutaneous tissue, and were starting to migrate laterally and towards the epidermis. At 60 min plasma exudate was not conspicuous and the blood vessels were contracted. At that time the cellular infiltration was more diffuse. Almost 90% of the cells were neutrophils, the remainder were eosinophils. By 90 min the appearance of the skin was returning to normal (French, 1972). The initial skin response of sensitized mice to the bites of *Ae aegypti* was similar to that of guinea-pigs (Mellink, 1980).

In tests with 120 Japanese naturally exposed to *Ae albopictus*, the presence in blood samples of IgE antibodies specific for *Ae albopictus* salivary gland antigens was positively correlated with the manifestation of an immediate skin reaction (whether or not a delayed reaction was shown). A high positive correlation was obtained between IgE titre and sizes of weal and erythema 30 min after the bite. Individuals who gave a delayed reaction only, or none, contained no mosquito-specific IgE (Oka, 1989).

Degranulation of mast cells at the sites of mosquito bites was observed in mice, and the exposure of the lungs of sensitized guinea-pigs to mosquito extract *in vitro* caused release of histamine and an SRS-A-like substance (Wilson and Clements, 1965). Passive transfer of anaphylactic hypersensitivity has been achieved with laboratory animals (Wilson and Clements, 1965). It was also achieved with human subjects, when the donor was a 14-year-old girl who gave a strong immediate reaction to injection with mosquito antigens but no Arthus or delayed reaction, and whose serum contained IgE antibodies specific for the mosquito antigens (Suzuki *et al.*, 1976).

(b) Type III (immune complex-mediated) hypersensitivity

Type III hypersensitivity (= Arthus reaction), which involves IgG and other antibodies, results from the persistence of antigen–antibody complexes which activate complement. It is expressed some 3–8 h after re-exposure to antigen and is characterized by erythema, oedema and infiltration of neutrophils (polymorphonuclear leucocytes).

Biopsies of human skin at the site of *Aedes aegypti* bites at six hours after biting showed an intense inflammatory reaction, with oedema and substantial perivascular infiltration of neutrophils, eosinophils and lymphocytes (Goldman *et al.*, 1952). Six hours after sensitized mice had been bitten by *Ae aegypti* the affected sites showed increased vascularization and massive cellular infiltrates, mainly of neutrophils (Mellink, 1980).

Rabbits that were exposed to the bites of *Ae communis* developed IgG antibodies which were shown, by immunoblotting, to recognize a 21.5 kDa protein, a major antigen present in the saliva of that mosquito. When exposure to the mosquitoes ceased these antimosquito antibodies gradually disappeared. Examination of sera from human beings who lived in an area populated by *Ae communis* showed an absence of antibodies against the 21.5 kDa protein in infants aged 6–8 months, whereas 45% of children aged 2–3 years contained them (Brummer-Korvenkontio *et al.*, 1990).

Among 120 Japanese naturally exposed to *Ae albopictus*, a positive correlation was obtained between exhibition of a delayed reaction to a test bite by *albopictus* and positivity in the lymphocyte transformation test (transformation of isolated lymphocytes into proliferating cells in the presence of *Ae albopictus* salivary antigens). Further, a close correlation was obtained between the size of the erythema after 24 h and the extent of lymphocyte transformation. Oka (1989) concluded that a cellular immune mechanism was involved in the delayed reaction.

A small percentage of human beings show a severe local and constitutional hypersensitivity to mosquito bites, with skin necrosis, high fever and hepatosplenomegaly (Brown *et al.*, 1938; Suzuki *et al.*, 1976). In some cases at least, the histological findings were compatible with an Arthus reaction (Hidano *et al.*, 1982). In one patient the large granular lymphocytes showed a pronounced natural killer cell activity, and natural killer cells aggregated in the lesional skin, which suggested that they were involved in the reactions of the skin and other organs (Tokura *et al.*, 1990). Very occasionally, massive and prolonged cutaneous reactions to mosquito bites have been observed in individuals who probably had an altered immune function. Such was believed to be the case with certain patients with chronic lymphocytic leukaemia, who showed severe bite reactions before the malignancy had been diagnosed (Weed, 1965). In Japan, a number of young patients who showed exceptionally severe hypersensitive reactions to mosquito bites were later found to suffer from malignant histiocytosis (Hidano *et al.*, 1982; Suenaga, 1987)

(c) Type IV (delayed) hypersensitivity

Type IV hypersensitivity is characterized by the appearance of an inflamed and oedematous papule some 24–48 h after exposure. The reactions are initiated by cells, not by antibodies. The infiltrate is mainly of basophils and contains few polymorphs.

Type IV reactions to mosquito salivary antigens, clearly separated from the type I and III reactions, are visually distinct in man and guinea-pig but not observed in rabbits and mice (Allen and West, 1966; Mellink, 1980). Biopsies of human skin taken 24 h after *Aedes aegypti* had bitten showed a reduction of oedema and an increased infiltrate of eosinophils and lymphocytes (Goldman *et al.*, 1952; Bandmann and Bosse, 1967). At 24 h and more after sensitized guinea-pigs had been bitten by *Ae aegypti*, massive wide-spread monocyte infiltrates were observed (Mellink (1980) and personal communication). Guinea-pigs that were recipients of spleen cells from animals sensitized with *Ae aegypti* saliva gave both type I and type IV reactions when bitten (Allen, 1966).

13

Structure of the adult alimentary canal

The fluid-feeding habit of adult mosquitoes is reflected in the structure and ultrastructure of the alimentary canal. In this chapter we are concerned with the alimentary canal of non-bloodfed mosquitoes. The changes that follow blood feeding are described in Chapter 14.

13.1 THE FOREGUT

During metamorphosis the anterior-most region of the culicid foregut transforms into the adult pharyngeal pump, which is described in Section 11.2.3. A short oesophagus extends from the pharyngeal pump to the oesophageal invagination in the front of the midgut (Figure 13.1). The cells of the oesophageal epithelium may be columnar, cubical or flattened depending on the degree of distension of the oesophagus. The epithelium is lined with cuticle and bounded by muscle cells, most of which are circular but some longitudinal, especially towards the posterior end. Anteriorly the oesophagus is narrow and encircled by a band of muscle cells, but posteriorly it is dilated and may be thrown into folds. The oesophageal invagination, or cardiac valve, is formed by a slight involution of the oesophagus into the midgut. Here the oesophageal wall, still lined with cuticle, is reflected forwards before joining the midgut epithelium, and what is possibly a sphincter muscle is situated between the walls of the involuted region of the oesophagus (Arnal, 1950; Christophers, 1960; Richardson and Romoser, 1972). Two ingluvial ganglia are situated at the junction of the foregut and midgut. Each is innervated by a nerve which arises through the fusion of an oesophageal nerve with a branch of the nervus corporis cardiaci (Figure 10.2).

Near its posterior end the foregut is extended into one ventral and two dorsal diverticula, the narrow openings to which are situated just anterior to the oesophageal invagination (Figure 13.1). The small dorsal diverticula occupy a space at the anterior end of the thorax between the indirect flight muscles. The very large ventral diverticulum, or crop, is situated medially, between the midgut and the ventral nerve cord, and is joined to the oesophagus by a long duct. The duct is surrounded by closely packed muscle cells which possibly serve as a sphincter, and its cuticular lining bears microchaetae 15–30 μm long. Valves were reported by Dapples and Lea (1974) between the oesophagus and each of the three diverticula. The epithelial cells of the crop are unspecialized in ultrastructure and are interconnected in places by septate and spot desmosomes. The epithelium is bounded basally by a basal lamina and apically by very thin cuticle which has an outer osmiophilic layer, believed to be epicuticle. Both the dorsal and ventral diverticula are bounded by networks of muscle cells, innervated by nerves which are possibly part of the stomatogastric nervous system (Day, 1954; Clay and Venard, 1972; Hecker and Bleiker, 1972).

The crop is highly elastic and when empty is much folded. X-rays revealed that when the crop of *Aedes aegypti* was full it occupied most of the space within the abdomen and that its duct was widely distended. When all three diverticula were full the posterior region of the oesophagus was also distended forming, with the lumens of the

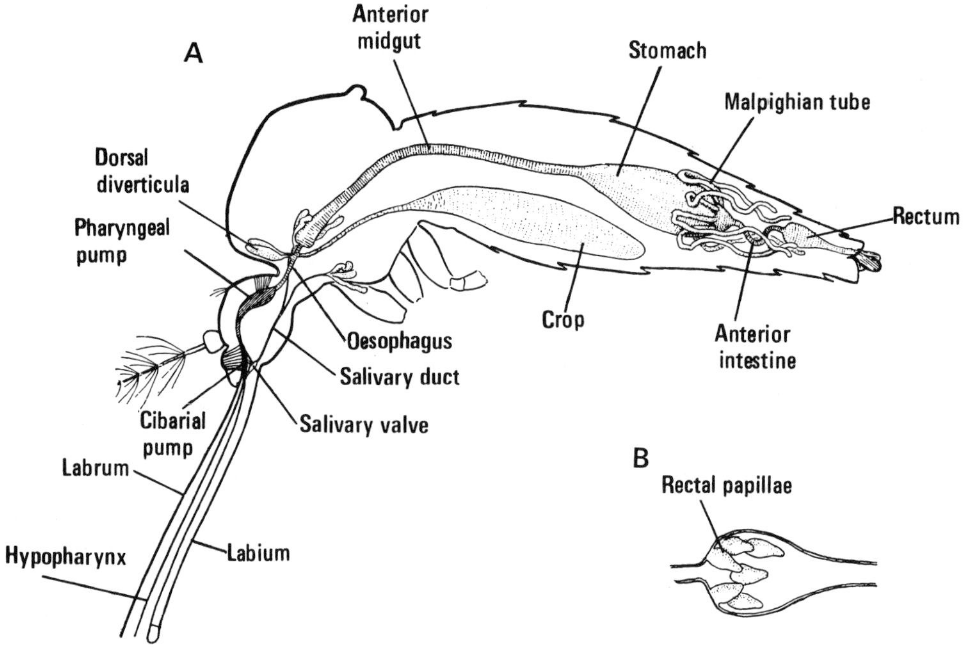

Figure 13.1 The alimentary canal of a mosquito. (A) Outline of the body of a female *Aedes aegypti* showing the alimentary canal and the left salivary gland which is displaced ventrally. (B) Rectum of a female *Ae aegypti* showing the rectal papillae. (From Snodgrass, 1944.)

diverticula, a common space which was closed anteriorly by the pharyngeal pump and posteriorly by the cardiac valve. When a female took a blood meal the posterior region of the midgut filled first, and as the rest of the midgut filled with blood it compressed the crop almost entirely within the first abdominal segment (Guptavanij and Venard, 1965).

Nuttal and Shipley (1903) concluded that the crop of *Anopheles maculipennis* is impervious to water because crops dissected out and exposed to air still contained fluid after two months. Clay *et al.* (1973) observed water loss from crops of *Eretmapodites chrysogaster* and *Aedes triseriatus* dissected out and exposed to air, but found no dilution of sucrose within the crops of *Ae triseriatus* and *An quadrimaculatus* when dissected out and placed in distilled water.

13.2 THE MIDGUT

Apart from the small bulbous cardia at its anterior end, the midgut consists of a narrow tube-like anterior region through which blood passes, and a flask-shaped posterior region or stomach which retains the blood meal, and which in culicines is capable of much distension (Figure 13.1). The gastric caeca, present in the larva, have disappeared. A basal lamina surrounds the midgut epithelium, and outside this is a delicate network of innervated circular and longitudinal muscle cells. The midgut receives axons from the ingluvial ganglia of the stomatogastric nervous system but none from the ventral nerve cord. The posterior end of the stomach is greatly constricted and is surrounded by a sphincter, forming the pyloric valve. The stomach is richly supplied with tracheae, the terminal lengths of which are tightly coiled prior to expansion of the stomach by blood feeding. The single-layered midgut epithelium is composed of three cell types. Most are 'columnar cells', which vary in shape from tall to flat depending on the degree of distension of the gut, and among these are distributed smaller numbers of undifferentiated 'regenerative cells' and approximately 500 endocrine cells (Nuttal and Shipley, 1903; Thompson, 1905; Christophers, 1960; Brown *et al.*, 1985).

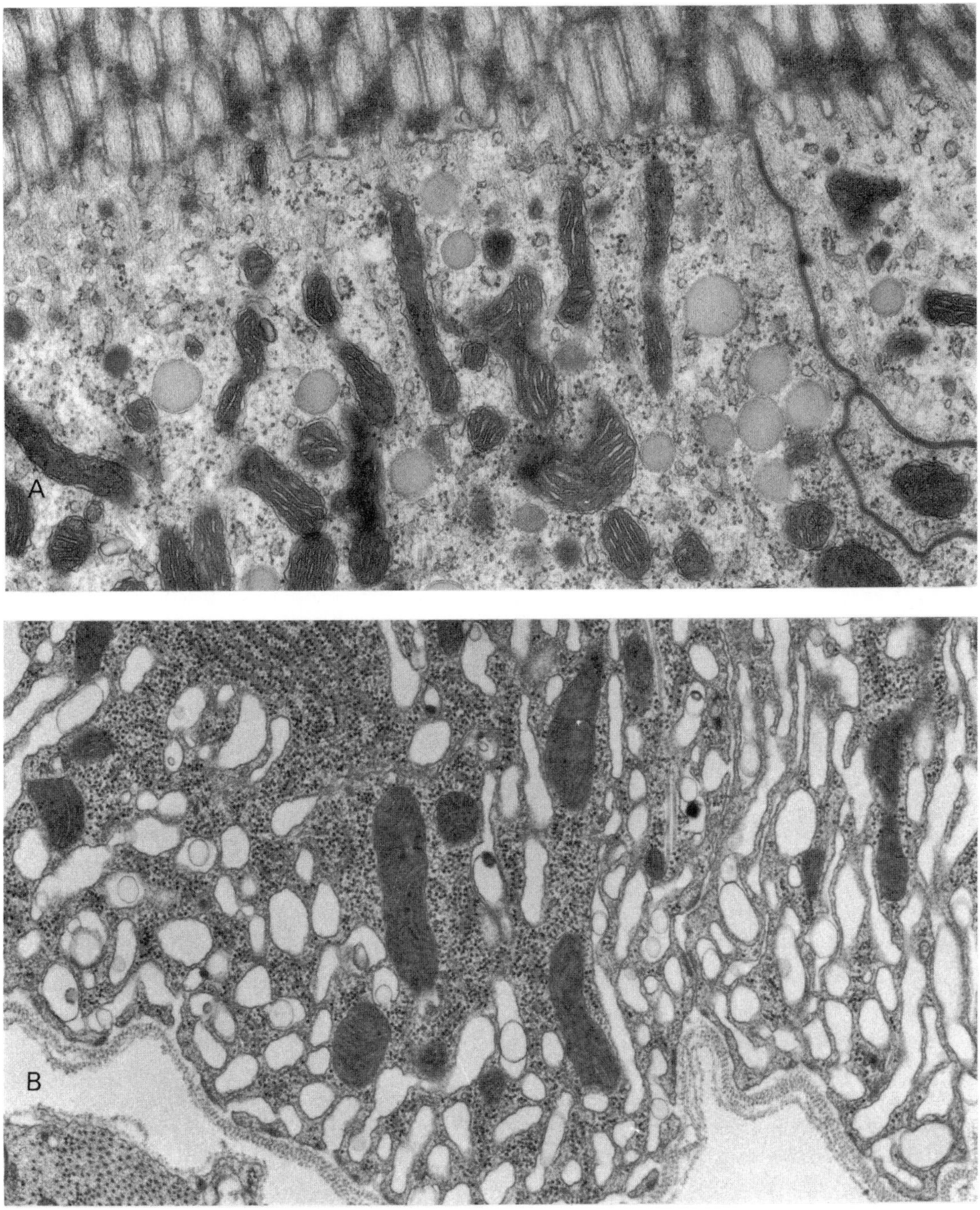

Figure 13.2 (A) Apical region of columnar cells of a female of *Anopheles stephensi* 60 h post-bloodmeal, showing dense-cored secretory vesicles of the type that contain peritrophic membrane precursors (× 25 000.) (B) Basal labyrinth of columnar cells in the posterior midgut of a non-bloodfed female of *Aedes aegypti* (× 24 400.) (Micrographs by courtesy of Dr R. Berner, Dr W. Rudin and Prof. H. Hecker.)

The midgut of female mosquitoes serves at least three functions in addition to its endocrine role: it contributes to diuresis; it is the site of synthesis and secretion of digestive enzymes and of the peritrophic membrane; and it is the site of nutrient absorption.

13.2.1 Columnar cells of females

The apical cell membrane is extended into numerous tightly packed microvilli, coated by a glycocalyx and containing microfibrils which extend into the apical cytoplasm. In *Ae aegypti*, for example, the microvilli are 4–5 μm long and of 110–140 nm diameter. Numerous mitochondria are crowded into a zone extending 2–4 μm below the apical surface. This merges with underlying cytoplasm containing arrays of rough ER, numerous free ribosomes and some smooth ER (Figure 13.2A). Lysosomes containing acid phosphatases occur sparsely in the apical cytoplasm of young mosquitoes, but in aged mosquitoes they are abundant. The basal and parts of the lateral cell membrane are extensively folded and deeply invaginated into the cell, forming a series of narrow double-walled compartments filled with mitochondria, ribosomes and microtubules, and constituting the basal labyrinth (Figure 13.2B). The ultrastructure differs in detail between genera and between species in the same genus (Bertram and Bird, 1961; Roth and Porter, 1964; Stäubli *et al.*, 1966; Hecker *et al.*, 1971a, 1974; Hecker, 1977; Houk, 1977; Rudin and Hecker, 1979).

In both anterior and posterior midgut regions the columnar cells are active in protein synthesis, evidenced by the endoplasmic reticulum and ribosomes, and undertake transport, evidenced by the microvillate apical membrane and basal labyrinth. There are, however, ultrastructural differences between the two regions. In the anterior region the columnar cells bear more microvilli, smooth ER is abundant whereas rough ER is scarce, and the basal labyrinth is more extensive. The columnar cells of the anterior midgut region do not have a storage function, and contain at most trace deposits of lipid or glycogen. The columnar cells of the posterior midgut region usually contain abundant rough ER. In *Ae aegypti* the rough ER has the form of cisternae which are tightly packed in concentric formations termed 'whorls' (Figure 13.3) (Bertram and Bird, 1961). Similar whorls are found in *An gambiae* but in *An stephensi* they are poorly developed (Stäubli *et al.*, 1966). In non-bloodfed *Cx tarsalis* the cells of the posterior midgut contain very little rough ER (Houk, 1977). Deposits are seen transiently in the cells of the posterior region during blood digestion. In *Ae aegypti* they are lipid. In *An gambiae* and *An stephensi* they contain some lipid and, in occasional specimens, substantial bodies of glycogen (Hecker, 1977). Posterior midgut cells contained large carbohydrate deposits when *An stephensi* were fed sugar for 14 days after emergence (Schneider *et al.*, 1987).

The columnar cells of non-bloodfed *Anopheles*, exemplified by *An gambiae*, *stephensi* and *maculipennis*, contain many dense-cored secretory vesicles in the apical cytoplasm, often associated with Golgi complexes (Figure 13.2A) (Hecker, 1977). The apical secretory vesicles of *An stephensi* were shown by lectin-binding

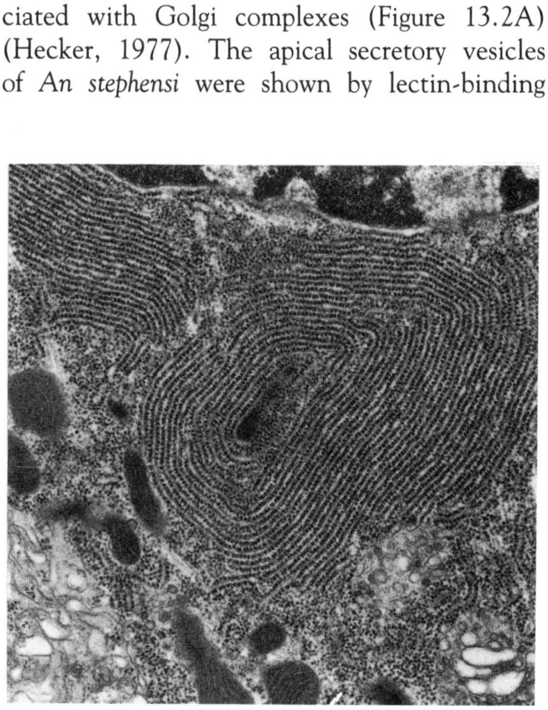

Figure 13.3 Whorl of rough endoplasmic reticulum in a columnar cell in the posterior midgut region of a non-bloodfed female of *Aedes aegypti* (× 13 000.) (Micrograph by courtesy of Prof. H. Hecker and Dr W. Rudin.)

studies to contain *N*-acetyl-D-galactosamine, a fact that supports their postulated association with the peritrophic membrane in which it is also found (Rudin and Hecker, 1989). Such secretory vesicles are absent from the midgut of non-bloodfed *Ae aegypti*. The posterior midgut cells of *Cx tarsalis* contain some dense-cored vesicles that are not apical and each of which is surrounded by a single cisterna of rough ER (Houk, 1977).

A histochemical study of *Cx tarsalis* showed the luminal surface of the midgut epithelium to be composed primarily of neutral polysaccharides and to have only a minimal anionic content. In contrast, the basolateral membranes and the basal lamina had an abundance of negative charges while the septate junctions on the lateral membranes had fixed positive charges (Houk *et al.*, 1986b, c). Lectin-binding experiments revealed that the glycocalyx of *Ae aegypti* contained *N*-acetyl-D-glucosamine, whereas that of *An stephensi* was preferentially labelled by *N*-acetyl-D-galactosamine specific lectins. There were no obvious differences between the two species in the lectins binding to Golgi membranes and other cellular organelles (Rudin and Hecker, 1989).

The posterior midgut of culicines can withstand considerable distension, its mechanical strength probably being provided by specialized regions of attachment between adjacent cells, by the basal lamina, and by the outer muscle layer. In *Ae aegypti*, over the apical half of the columnar cells the adjacent membranes of neighbouring cells run more or less straight and are largely connected by 'continuous junctions' (*zonula continua* or smooth septate junctions). A number of 'gap junctions' and 'septate junctions' are situated within the lengths of continuous junction. In the basal half of the epithelium the membranes of neighbouring cells follow an irregular course and are mostly separated by narrow intercellular clefts. At intervals the cell membranes are connected by 'spot desmosomes' (*maculae adhaerentes*) and by occasional gap junctions. 'Hemidesmosomes' occur on the basal cell membrane, facing the basal lamina (Reinhardt and Hecker, 1973). The continuous junctions of *Cx tarsalis* have a total membrane width of 24 nm and an intermembrane distance of 12 nm. Intercellular septa *c.* 4–5 wide and separated by *c.* 5 nm are sometimes apparent (Houk and Hardy, 1979).

The basal lamina surrounding the midgut epithelium is composed of a number of sheets. The thickness of the basal lamina varied according to the size of the female in *Ae triseriatus*. In large females derived from well-nourished larvae, it had a mean thickness of 0.24 μm and was composed of ten or more sheets. In small females derived from poorly nourished larvae, the mean thickness was 0.14 μm and it was composed of 4–6 sheets (Grimstad and Walker, 1991).

The basal lamina of *Ae aegypti* consisted of 4–7 sheets, and two series of electron-lucent pores could be distinguished in each sheet. One series, with pore diameters of *c.* 20 nm, gave the sheet a beaded appearance when cut in cross section. In tangential section they gave the sheet a grid-like appearance, forming two sets of grid lines which intersected at right angles. The pores of the second series, which had diameters of *c.* 7 nm, occurred within the intersections of the two sets of grid lines. After a blood meal the network was stretched and distorted (Terzakis, 1967; Reinhardt and Hecker, 1973). In *Ae dorsalis* the basal lamina consisted of 12 or more sheets. The annulae were closely aligned above one another, giving the basal lamina a distinctly banded appearance in cross section. The network was not distorted after a blood meal (Houk *et al.*, 1980). The basal lamina of engorged females of *Cx tarsalis* was permeable to horseradish peroxidase (diameter 4–6 nm) but only marginally permeable to colloidal thorium (diameter 5–8 nm) (Houk *et al.*, 1981).

The rough ER of the stomach cells forms after emergence, disappears after blood feeding and later reforms. Removal of the corpora allata from female *Ae aegypti* prevented the development of rough ER after emergence and also prevented the redevelopment of ER subsequent to its disappearance after blood feeding. Treatment with a JH analogue restored this capability, which is therefore probably regulated by juvenile hormone (Rossignol *et al.*, 1982). When females of *Ae aegypti* were given water for five days after emergence, the whorls of rough ER in the stomach columnar cells disappeared and there was a substantial reduction in the density of rough and smooth ER, but the density of ribosomes remained high. Sugar feeding for 11 days had a similar effect (Bauer *et al.*, 1977). Mono- and polyribosomes could be isolated by sucrose gradient centrifugation from the stomach cells of sugarfed females, but no polyribosomes could be obtained from the stomach cells of starved females (Gander *et al.*, 1980). However, one day after starved females had taken a blood meal the stomach cells contained large amounts of rough ER, and digestion and ovarian development were both in progress (Tadkowski and Jones, 1978).

13.2.2 Columnar cells of males

The midgut of male *Aedes aegypti* resembles that of the female; it contains columnar cells, which are half the volume of those in females, and regenerative cells. The ultrastructure of the columnar cells of the anterior midgut is broadly similar to that in the female. The stomach columnar cells of males differ from those of females: they contain much less RER, are not arranged in whorls, have fewer free ribosomes, have a less extensive basal labyrinth, and lack both spot desmosomes and the grid-like substructure of the basement membrane. At emergence the epithelium of the males seems to be more differentiated than in females (Hecker *et al.*, 1971b; Rudin and Hecker, 1976).

13.2.3 Regenerative cells

Regenerative cells occur singly between the columnar cells, on the basal side of the epithelium and in both the anterior and posterior regions of the midgut in both sexes. They are small cells containing small amounts of electron-dense cytoplasm and highly heterochromatic nuclei. The cytoplasm has a large accumulation of free ribosomes, very little endoplasmic reticulum, and a few mitochondria (Figure 13.4). Developing regenerative cells are seen occasionally, growing towards the apical surface. They contain rough ER, Golgi complex, microtubules, and developing microvilli and basal labyrinth (Hecker *et al.*, 1971a, b; Hecker, 1977; Houk, 1977).

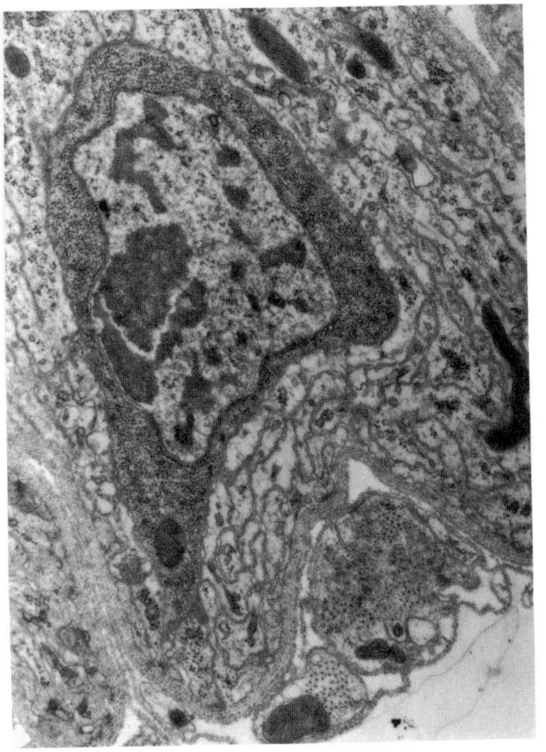

Figure 13.4 Regenerative cell and basal region of a columnar cell in the anterior midgut of a female of *Anopheles gambiae* (× 18 200.) (Micrograph by courtesy of Prof. H. Hecker.)

The rate of disappearance of columnar cells from the posterior midgut of *Culiseta melanura* was low, but columnar cells were sloughed continuously from the anterior midgut. However, no involvement of regenerative cells in tissue repair was seen, even after large numbers of cells had been sloughed through infection with virus (Weaver and Scott, 1990a). When the midgut of bloodfed *Aedes aegypti* was pierced with a needle, cell proliferation occurred for several days and the wounded epithelium healed (Day and Bennetts, 1953). Bloodfed females of *Psorophora ferox* that were parasite-free showed no proliferation of midgut cells, but females that suffered damage to the midgut from filariae showed a high level of mitotic activity and cell proliferation (Stueben, 1978).

13.2.4 Endocrine cells

Endocrine cells are distributed singly among the columnar cells and throughout the length of the midgut. In female *Aedes aegypti* they number about 500. These cells are distinguishable by being narrower (2–6 μm wide) than the columnar cells (8–10 μm wide), by the absence of a basal labyrinth and by the presence of dense-cored secretory vesicles in the basal cytoplasm (Figure 13.5). Some of the cells are electron lucent, others relatively electron dense, constituting so-called clear and dark endocrine cells. The secretory vesicles originate from Golgi complexes and their contents are released by exocytosis across the basolateral membranes. Endocrine cells are distributed at highest density in the posterior half of the stomach, at lowest density in the anterior half of the stomach, and at an intermediate density in the anterior midgut. One-quarter were found to be adjacent to a regenerative cell. Some of the larger endocrine cells extended through the thickness of the epithelium. Those that reached the midgut lumen had a microvillate apical membrane, and were designated 'open cells'. The smaller endocrine cells were positioned basally in the epithelium and were classed with the 'closed cells' that did not reach the lumen (Brown *et al.*, 1985).

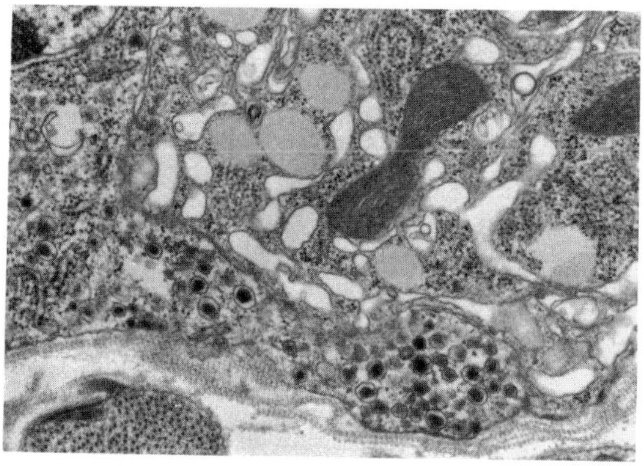

Figure 13.5 Endocrine cell and part of an adjacent columnar cell in the posterior midgut of a male of *Aedes aegypti* (× 24 000.) (Micrograph by courtesy of Prof. H. Hecker.)

Endocrine cells were found in the midguts of *Ae aegypti* larvae, and also in the midguts of autogenous adult females of *Ae atropalpus* and the non-haematophagous adult females of *Toxorhynchites* sp. (Brown *et al.*, 1985).

13.3 THE HINDGUT AND MALPIGHIAN TUBULES

The junction of mid- and hindgut is marked by the insertion, just behind the pyloric valve, of the Malpighian tubules. Both the hindgut and the Malpighian tubes are derived from the embryonic proctodaeum. The hindgut consists of pyloric chamber, anterior intestine, rectum and anal canal (Figure 13.1). In all regions the hindgut epithelium has a muscle layer on its outer surface and a cuticular lining on its luminal surface. The rectum and anal canal are innervated by the terminal abdominal ganglion, but no nerves can be seen going to the Malpighian tubules, pyloric chamber or anterior intestine (Odland and Jones, 1975; Jones and Brandt, 1981).

The pyloric chamber is a short, thin-walled, bulbous structure into which the Malpighian tubules drain. It is bounded by circular muscle fibres, concentrated posteriorly, and its cuticular lining bears numerous posteriorly directed bristles. The anterior intestine is a rather narrow looped tube. Its epithelium of cubical cells is thrown into longitudinal folds, and its muscle layer is composed mostly of circular fibres. The bladder-like rectum is bounded by a layer of circular muscle fibres and by sparser longitudinal fibres. Its irregular epithelium is extended dorsally into pear-shaped papillae – four in males and six in females – which project into the rectal lumen (Figure 13.1B). The short anal canal has a thin epithelium with a much folded cuticle and a muscle coat like that of the rectum. The anus is situated in a small area of membrane between the bases of the cerci (Nuttall and Shipley, 1903; Thompson, 1905; Trembley, 1951; Christophers, 1960; Dapples and Lea, 1974).

Ultrastructural studies of female *Aedes aegypti*

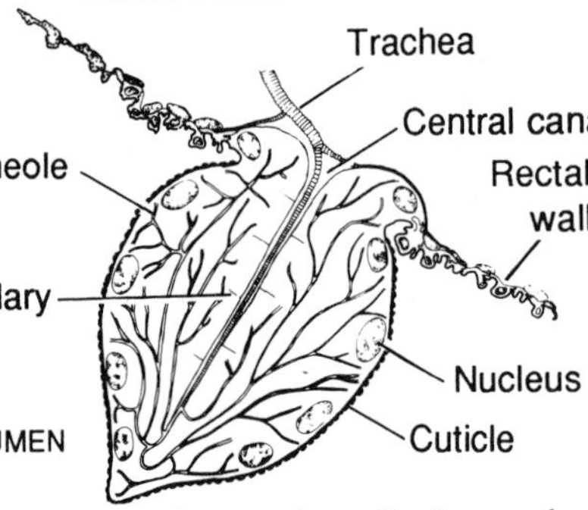

Figure 13.6 Diagram of a rectal papilla from a female *Aedes aegypti*, seen in longitudinal section. The cell boundaries are indistinct under optical microscopy. (After Hopkins, 1966.)

suggest that the epithelia of the anterior intestine and rectal papillae function in transport. The anterior intestine has a thick epithelium. The basal plasma membrane of its cells is deeply invaginated at intervals, forming compartments which contain mitochondria. The apical cell membrane is very intensively and deeply infolded, and is also associated with mitochondria (Tongu *et al.*, 1969). The general rectal epithelium of *Ae aegypti* is thin and its cells are of unspecialized ultrastructure except that the lateral cell membranes of adjacent cells are thrown into folds and connected along their length by septate desmosomes (Hopkins, 1966).

The rectal papillae are composed of a thick single-layered epithelium, which is continuous with that of the rectal wall and similarly covered with cuticle. In each papilla the epithelium surrounds a central canal which is confluent with the haemocoele (Figure 13.6). A tracheal branch runs through the canal, and near the tip of the papilla divides into further branches which turn back and subdivide, ramifying between the cells. A nerve fibre also extends through the canal. The cells of the rectal papillae are characterized by very extensive folding and invagination of the lateral cell membranes, the folds extending throughout the cell, almost to the basal and apical surfaces, and sometimes forming stacks. The two sides of each fold usually lie parallel, approximately 15 nm apart, their extracellular surfaces bearing a glycocalyx. Each cell is packed with mitochondria which are closely associated with the invaginated membranes, forming structures termed 'mitochondrial-scalariform junction complexes' by Noirot-Timothee and Noirot (1980). These are believed to function in ion transport. Adjacent cells are connected by septate desmosomes, which are most prominent where the lateral cell membranes adjoin the basal and apical surfaces. The apical plasma membrane is formed into short sparse microvilli below the cuticle (Hopkins, 1966).

The five Malpighian tubules are carried over from the larva without cell loss or apparent reorganization. The greater part of each tubule is formed of large primary cells, of which one, two or three may encompass the lumen. Smaller stellate cells are interspersed singly along the tubule. The basalolateral plasma membrane of each primary cell is extensively infolded, forming cytoplasmic compartments which extend 1.5 μm into the cell and which contain mitochondria. The apical plasma membrane is microvillate, each large microvillus being almost filled with a single mitochondrion. The central nucleus contains polytene chromosomes. The cytoplasm is densely packed with membrane-limited vesicles containing concretion bodies, spheres of concentric mineralized lamellae, which are more numerous after blood feeding. Carbonic anhydrase has been demonstrated in the vesicular membranes but not elsewhere in the principal cells. The stellate cells also have an infolded basal membrane, which forms a complex network of interconnecting channels. The apical plasma membrane of the stellate cells bears numerous short slender microvilli which are not penetrated by mitochondria. Vacuoles are lacking from the stellate cells (French *et al.*, 1962; Suguri *et al.*, 1969; Mathew and Rai, 1976a; Palatroni *et al.*, 1981).

Investigation of *Ae aegypti* Malpighian tubules showed the concretion bodies to be confined to the primary cells, roughly spherical in shape and enclosed in membrane-bound vesicles. Each had a laminar structure and was formed of concentric spheres of alternating electron-lucent and electron-opaque material. Concretion bodies of many different sizes were present throughout the cytoplasm except at the apical border and the region of basal infolds. X-ray microanalysis of tubules from two strains of *Ae aegypti* – one that possessed and one that lacked concretion bodies – revealed high concentrations of calcium, magnesium, manganese and phosphorus when concretion bodies were present (Bradley *et al.*, 1990). *Rhodnius prolixus* stores bloodmeal calcium in similar concretion bodies (Maddrell *et al.*, 1991).

The Malpighian tubules of male *Ae aegypti* are significantly shorter than and only half the diameter of those of the females, and their principal cells are without concretion bodies (Plawner *et al.*, 1991).

14

Adult digestion

14.1 ENZYMES OF THE DIGESTIVE TRACT

The midgut is the principal organ of digestion, but disaccharides ingested in nectar are at least partly hydrolysed in the crop by a salivary enzyme (Section 12.2.1). Proteins are the predominant constituents of blood, apart from water, and the mosquito midgut secretes a number of proteolytic enzymes. However, blood proteins contain carbohydrate and lipid moieties which are released during digestion, and these are hydrolysed by carbohydrases and esterases. The enzymes described in this chapter catalyse the hydrolytic cleavage of either the C–N bonds found in peptide linkages or the C–O bonds of ester and glycosidic linkages.

14.1.1 Characterization of proteolytic enzymes

Protein digestion involves two steps and two classes of enzyme. Protein molecules are cleaved into large peptides by proteinases (also called endopeptidases), and the large peptides are progressively shortened by peptidases (more specifically called exopeptidases) which remove single amino acids or dipeptides from each end. **Endopeptidases** hydrolyse peptide bonds at specific internal positions in the polypeptide chain, these positions being characterized by the nature of the amino acids adjacent to the bonds that are attacked. These enzymes are categorized according to their catalytic mechanisms. The endopeptidases trypsin and chymotrypsin are in the subclass of serine proteases, which have catalytically active serine and histidine residues at the active centre. **Exopeptidases** hydrolyse peptide bonds located at or near the ends of peptide chains, releasing an amino acid or a dipeptide. This class includes aminopeptidases which shorten peptides from the *N*-terminal end of the chain (with a free amino group), and carboxypeptidases which attack at the C-terminal end (with a free carboxyl group).

Proteolytic activity can be assayed by measuring the disappearance of proteins such as haemoglobin and casein, or spectrophotometrically with the insoluble hide powder azure. Now most assays utilize soluble chromogenic substrates containing amino acids for which the enzymes show specificity. Trypsins hydrolyse peptide bonds involving the carboxyl group of arginine or lysine, both basic amino acids, and reactivity with synthetic substrates that include such basic amino acids – e.g. *p*-tosyl-L-arginine methyl ester (TAME) and α-*N*-benzoyl-DL-arginine-*p*-nitroanilide (BAPNA) – is taken to be diagnostic of trypsin. Chymotrypsins hydrolyse peptide bonds involving the carboxyl group of a tyrosine, tryptophan, phenylalanine or leucine residue. Their synthetic substrates include benzoyl-L-tyrosine ethyl ester (BTEE). Substrates used for the assay of exopeptidases contain unprotected amino acids, e.g. L-arginine-*p*-nitroanilide for aminopeptidase. With many of the model substrates it is an ester linkage that is hydrolysed, not the amide linkage of peptide bonds. The activity of specific inhibitors can provide evidence to support the identification of enzymes, e.g. tosyl-L-lysine chloromethyl ketone (TLCK) for trypsin, and tosyl-L-phenylalanine chloromethyl ketone (TPCK) for chymotrypsin (Dixon and Webb, 1979; Applebaum, 1985; International Union of Biochemistry, 1984).

Mosquito proteolytic enzymes have also been assayed directly with [^{3}H]diisopropylfluorophosphate (DFP), which selectively and irreversibly acylates the active-site serine (Graf and Briegel, 1985; Borovsky and Schlein, 1988), and by immunoprecipitation with monoclonal antibodies of enzyme already labelled with [^{3}H]DFP (Graf *et al.*, 1988).

14.1.2 Proteolytic enzymes of the adult gut

The early investigations of mosquito proteolytic enzymes, on which our knowledge is still partly based, did not distinguish the different enzymes. Crude extract from the posterior midgut of *Aedes aegypti* showed a relatively high affinity for denatured or crystalline haemoglobin (K_m = 1.5–2 mg ml^{-1} at pH 7.9), and a lower affinity for bovine serum albumin (K_m = 19.3 mg ml^{-1}). It was almost inactive with γ-globulin (Yeates, 1980). Extracts from *Culex quinquefasciatus* midgut were broadly similar but were more reactive with bovine serum albumin and γ-globulin (Gooding, 1966a).

Midgut homogenate from bloodfed *Ae aegypti* and *Cx quinquefasciatus*, incubated with denatured haemoglobin as substrate, showed a small activity peak at about pH 5 and a broad peak of substantially higher activity between pH 7 and pH 10.5 (Gooding, 1966a). The pH of the midgut contents of bloodfed mosquitoes is known only for the period shortly after engorging and prior to enzyme secretion. Chicken blood, normally pH 7.51, had a mean pH of 7.68 (range 7.47–7.90) during the first hour after ingestion by *Cx pipiens* (Bishop and McConnachie, 1956). The pH of duck blood, normally 7.65, was little changed shortly after ingestion by four species of mosquitoes. In *Cx pipiens* it fell to a mean of pH 7.52, and in *Anopheles quadrimaculatus* it increased to a mean of pH 7.75 (Micks *et al.*, 1948). Measurements of the pH of fluid from both the endo- and ectoperitrophic spaces over the period of digestion would be informative.

(a) Trypsins

Purification of trypsins extracted from females 24 h post-bloodmeal, followed by SDS-PAGE at pH 8.4, yielded from 3 to 6 active bands in species of *Aedes* and from 1 to 3 active bands in species of *Anopheles* (Table 14.1). All had molecular masses in the range 25–36 kDa, the 30–33 kDa bands being the most prominent. Analyses of c. 12% of the amino acid sequences of *Ae aegypti* and *An quadrimaculatus* trypsins confirmed that they belonged to the subclass of serine proteases. The mosquito enzymes showed 30–40% homology with invertebrate and vertebrate trypsins. Homology between the two mosquito trypsins was only 38% for the segment sequenced. Highly conserved sequences at the

Table 14.1 Characteristics of mosquito trypsins. Extracts of midguts or whole bodies, made at 24 h post-bloodmeal, were purified by ion-exchange and affinity chromatography and analysed by SDS-PAGE. Esterolytic activity: 1 unit hydrolyzed 1 μmole TAME min^{-1}. (From Graf *et al.*, 1991).

Species	*Molecular mass (kDa)*						*Isoelectric point (pH)*	*Specific activity (Units mg^{-1})*
Ae aegypti	32.0	31.0	30.7	28.7	26.7		4.2–5.4	242
epactius	35.5	34.1	33.0	29.0			3.9–4.8	521
triseriatus	35.9	33.3	32.7	31.6	30.8	28.8	3.6–4.4	326
An albimanus	29.8	28.6					4.6–4.8	372
quadrimaculatus	29.9						4.3–4.6	1,056
stephensi	33.9	32.7	25.1				4.6–4.8	581

amino terminus and at positions 11–19 and 29–32 were separated by non-homologous sequences. Specific activity varied with species; for example, a >4-fold difference was observed between *Ae aegypti* and *An quadrimaculatus* trypsins. The mosquito trypsins had isoelectric points between pH 3.6 and pH 5.4. Polyclonal antisera against *Ae aegypti* and *An albimanus* trypsins showed complex patterns of cross-reactivity with trypsins from the six species, indicating the existence of more than one immunological class. The substantial differences between the trypsins of different mosquito species – i.e. of molecular mass, amino acid sequence, isoelectric point and specific activity – were unexpected among the members of a single insect family adapted to the digestion of vertebrate blood (Graf *et al.*, 1991).

Large numbers of apparent trypsin isoenzymes were observed in extracts of *Ae aegypti* by Graf and Briegel (1985) and Borovsky (1988). Since the analyses were undertaken without measures to prevent self-proteolysis, it appeared possible that some of the trypsin forms were artefactual. Graf *et al.* (1991) suggested that in *Aedes*, autolysis and degradation of trypsin occurred naturally during digestion, whereas in *Anopheles* the enzyme was more stable.

In *Anopheles stephensi* trypsin activity was found only in the lumen of the posterior midgut of bloodfed females (Billingsley and Hecker, 1991). About 80% of midgut proteolytic activity in bloodfed *Culex nigripalpus* was inhibited by TLCK and was ascribed to trypsin (Borovsky, 1986). Of the proteins from midguts of bloodfed *Aedes aegypti* that bound [^{3}H]DFP, and which were thereby identified as serine proteases, the binding of 77% was unaffected by the chymotrypsin inhibitor TPCK and was ascribed to trypsin. When it peaked at 24 h post-bloodmeal, the midgut trypsin-protein content was 1.4 μg (Borovsky and Schlein, 1988). Trypsins from *Culex quinquefasciatus* and *Ae aegypti* differed from bovine trypsin in that they were not stimulated by calcium ions (Spiro-Kern, 1974; Briegel and Lea, 1975; Graf and Briegel, 1985).

(b) Chymotrypsin

The midguts of larval mosquitoes have a high chymotrypsin content (Section 5.2.2), but chymotrypsin is found in low titre at most in adult midguts. Yang and Davies (1971a) found only very low chymotrypsin activity in *Aedes aegypti* midgut 24 h after blood feeding. Gooding (1969) obtained a low rate of hydrolysis of BTEE with midgut extracts from bloodfed *Ae aegypti* and *Cx quinquefasciatus*, but Spiro-Kern (1974) was unable to detect chymotrypsin in midgut homogenates from bloodfed *Cx quinquefasciatus*. Of the proteolytic activity shown by extracts of midgut from bloodfed *Cx nigripalpus*, 5–7% was inhibited by TPCK and was consequently identified as chymotrypsin (Borovsky, 1986). Of the midgut enzymes from bloodfed *Ae aegypti* that bound [^{3}H]DFP, and which were therefore identified as serine proteases, 20% remained active in the presence of the trypsin inhibitor TLCK and were therefore identified as chymotrypsin, but with an uncertainty of 8% in the assay (Borovsky and Schlein, 1988). Chymotrypsin activity was greater than trypsin activity in autogenous females of *Aedes atropalpus* and *Toxorhynchites brevipalpis*, occurring in both midgut and fat body (Masler *et al.*, 1983).

(c) Aminopeptidase

Low rates of aminopeptidase activity were measured in both the anterior and posterior midguts of unfed females of *An stephensi* (Billingsley and Hecker, 1991). In unfed females of *Culex tarsalis*, over 95% of aminopeptidase activity was membrane bound (E.J. Houk pers. comm. cited by Billingsley, 1990). The enzyme was associated with the microvillar fraction of midgut. Treatment of that fraction with papain solubilized the aminopeptidase, suggesting that it was an extrinsic enzyme on the microvilli (Houk *et al.*, 1986a).

After *An stephensi* had blood fed the aminopeptidase activity of the anterior midgut remained unchanged but that of the posterior midgut increased rapidly. Usually over 95% of the activity was within the stomach lumen (Billingsley and Hecker, 1991).

Table 14.2 Kinetic characteristics of aminopeptidases in homogenates of midgut from bloodfed *Anopheles stephensi* and *Aedes aegypti*. (From the data of Billingsley (1990) and Graf and Briegel (1982).)

Substrate		*Anopheles stephensi*		*Aedes aegypti*
Nitroanilide of	*Type*	K_m (mM)	V_{max}	*Relative affinity* (%)
Arginine	Basic	0.07	0.24	67
Lysine		0.25	0.34	49
Methionine	Non-polar	0.19	0.19	55
Leucine		1.55	0.27	100
Alanine		2.21	1.24	76
Proline		7.82	7.34	<10
Glutamic acid	Acidic	Inactive	Inactive	<10

V_{max}, nmol substrate hydrolysed min^{-1} $midgut^{-1}$.

The reactivities of aminopeptidases in homogenates of midguts from bloodfed *Anopheles stephensi* and *Aedes aegypti* were assessed with a range of amino acid *p*-nitroanilide substrates. In both cases the affinity was of a broad, non-specific nature, with high affinity shown for certain basic and non-polar substrates, but differing in detail between the two species (Table 14.2).

The aminopeptidase in homogenates of midgut from bloodfed *An stephensi* showed high activity in the pH range 7.0–8.5 with a peak at 8.0 (Figure 14.1). As with mammalian aminopeptidase, low concentrations of Mg^{2+} and Ca^{2+} slightly stimulated activity and the chelating agent 1,10-phenathroline was inhibitory. Two-thirds of the activity in homogenates of midgut from non-bloodfed females was sedimentable, therefore membrane-bound; one-third was soluble. Homogenization in the presence of detergent solubilized most of the aminopeptidase and almost doubled the activity. Column chromatography of midgut extract yielded a fraction containing aminopeptidase with native M_r 552 000, and after treatment with detergent an additional two active fractions of M_r 123 000 and 32 000 were obtained. All three showed somewhat different kinetics to the crude extracts (Billingsley, 1990).

The peak rates of aminopeptidase and trypsin activity, expressed as enzyme units per midgut, were virtually identical in *An stephensi* (1:0.9) but strikingly different in *Ae aegypti* (1:3.9), suggesting functional differences (Rudin *et al.*, 1991).

The presence of a carboxypeptidase in *Cx*

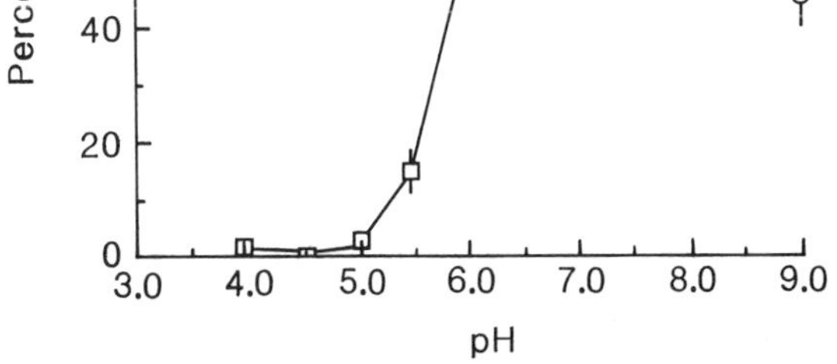

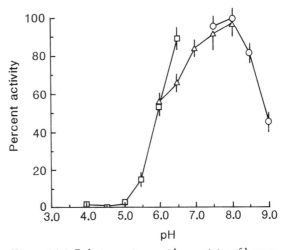

Figure 14.1 Relative aminopeptidase activity of homogenates of *Anopheles stephensi* midguts at various pH levels, with leucine *p*-nitroanilide as substrate and using three different buffers. (From Billingsley (1990). Copyright © (1991) and reprinted by permission of Wiley-Liss, a Division of John Wiley and Sons, Inc.)

pipiens midgut was reported by Spiro-Kern and Chen (1972), based on reactivities with model substrates.

14.1.3 Trypsin synthesis

Midgut from non-bloodfed *Aedes aegypti* possessed no tryptic activity and showed no evidence of trypsin precursor (Graf and Briegel, 1989). Immunocytochemical studies with a polyclonal antibody raised against *Ae aegypti* trypsin revealed no positive response in sections of female midgut fixed 15 min after blood feeding and an absence of secretory vesicles. By 12 h post-bloodmeal immunoreactivity was discernible in Golgi cisternae, in secretory vesicles associated with the Golgi complex and in free secretory vesicles. By 18 h Golgi complexes and secretory vesicles were intensely labelled but there were never more than a few secretory vesicles per profile, suggesting prompt secretion. Because trypsin activity could not be detected spectrophotometrically in isolated washed midgut epithelia from bloodfed females, Graf *et al.* (1986) concluded that the intracellular immunoreactivity was due to trypsin precursor. By 24 h synthetic activity had started to

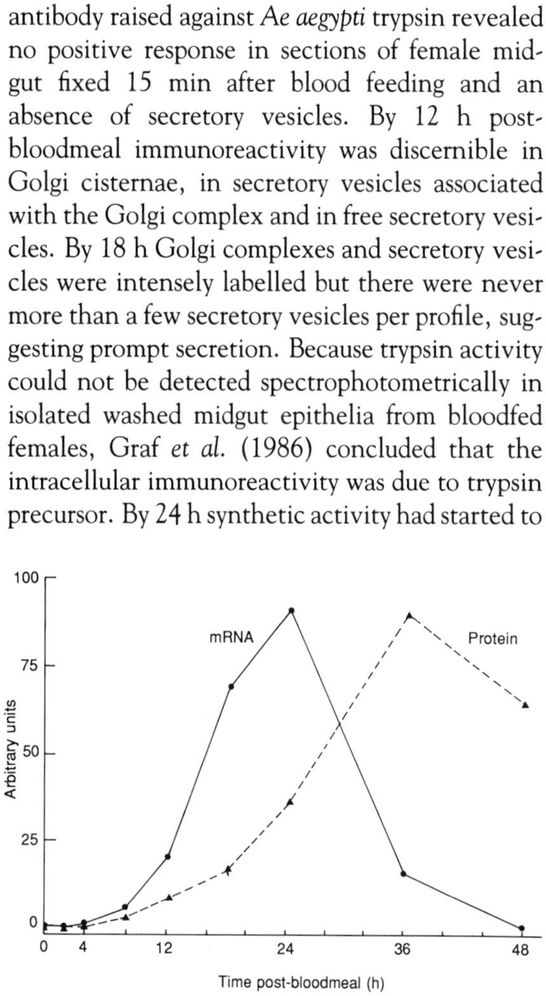

Figure 14.2 Relative amounts of trypsin-precursor mRNA and trypsin (protein) in females of *Aedes aegypti* before (time 0) and at different times after ingestion of an artificial blood meal. mRNA was measured by Northern blot analysis and trypsin by Western blotting with monoclonal antibody to 'late' trypsin. (After Barillas-Mury *et al.*, 1991.)

decline in some individuals, which had prominent lysosomes.

Trypsin synthesized at different times after blood meals, and immunoprecipitated with polyclonal antibody, was shown by SDS-PAGE to include three forms of distinct molecular mass. At five hours post-bloodmeal forms of 32 and 36 kDa were present. By six hours a 30 kDa form had started to appear, and by 12 h the 30 kDa form predominated. This work revealed that trypsin is synthesized in distinct 'early' and 'late' forms, of which the late form is the more prominent. Pulse-chase experiments revealed that the early forms were not precursors of the late form (Graf and Briegel, 1989).

In vitro translation with mRNA extracted from midguts 18 h after feeding yielded products that reacted with trypsin antiserum but had molecular masses 1–2 kDa greater than expected. Addition of microsomal membranes to the *in vitro* translation system caused a 1–2 kDa reduction in molecular mass. Apparently an early step in trypsin processing involves cleavage of 1–2 kDa peptides, possibly signal peptides (Graf and Briegel, 1989).

Screening an *Ae aegypti* midgut cDNA library with a monoclonal antibody to late trypsin led to isolation of a number of positive clones, one of which was selected for sequencing. This 856 bp sequence coded for a 255 amino acid trypsin precursor, which had a putative 15 amino acid signal peptide and a ten amino acid activation peptide. The deduced amino acid sequence of the mature enzyme, i.e. without signal and activation sequences, was homologous with that of other trypsins around the catalytic triad and also in several residues that are found only in trypsins. However, the sequence of the specificity pocket in the mosquito trypsin (-Lys-Glu-Ser-Pro-Cys-) differed from that in other trypsins (-Lys-Asp-Ser-Cys-). No mosquito trypsinogen has been isolated, but the 865 bp cDNA sequence suggests that one exists. The *Aedes* trypsin precursor differed from vertebrate trypsinogens in lacking four acidic amino acids close to the activation peptide cleavage site, which suggested that the mechanism of its activation may differ from that of vertebrate trypsinogens (Barillas-Mury *et al.*, 1991).

Quantification of the late trypsin-precursor mRNA and of late trypsin revealed that both were absent from the adult female before blood feeding. The mRNA could be detected four hours after feeding and it increased rapidly thereafter to peak at about 24 h (Figure 14.2). The trypsin was first detectable eight hours after the blood meal, maximal at 36 h, and still abundant at 48 h although the mRNA was then absent. These data indicate that transcriptional control has an important role in determining the rate of late trypsin-protein synthesis (Barillas-Mury *et al.*, 1991).

Because they were unable to detect trypsin mRNA in the midguts of non-bloodfed *Ae aegypti* but could demonstrate it at five hours post-bloodmeal, Graf and Briegel (1989) concluded that all trypsin synthesis is regulated at the transcriptional level. Experiments with inhibitors led Felix *et al.* (1991) to a slightly different conclusion. When added to a protein meal, two inhibitors of RNA polymerase only delayed trypsin synthesis. Actinomycin D did not inhibit trypsin synthesis until >10 h post-bloodmeal, and there was some evidence of superinduction prior to the inhibition. α-Amanitin did not inhibit trypsin synthesis until >8 h post-bloodmeal. In contrast, adding cycloheximide (an inhibitor of peptidyl transferase) to a protein meal totally inhibited trypsin synthesis. Felix *et al.* (1991) concluded that a first phase of trypsin synthesis starts within a few hours of feeding, when trypsin is synthesized directly from preformed mRNA, and that a second phase of trypsin synthesis starts about 7–9 h after feeding, when trypsin is synthesized from newly formed mRNA. However, the slow onset of trypsin activity (Figure 14.2) is not consistent with more than a trivial amount of synthesis being from preformed mRNA during the initial phase of synthesis.

14.1.4 Esterases and glycosidases

The presence of a number of esterases, or esterase isoenzymes, in the midgut of adult female mosquitoes has been demonstrated by PAGE. Newly emerged females of *Aedes aegypti* showed high total esterase activity in the midgut epithelium separable by electrophoresis into seven bands, three of which had the properties of acetylcholinesterase. By ten days after emergence total esterase activity had declined but several additional bands had appeared. Within three hours of blood feeding the esterase activity increased substantially, in addition to that due to esterases in the ingested blood, and the activities of the several esterase bands changed during the following three days. Most of the mosquito esterases were found in the midgut lumen, and at substantially higher concentration than in the epithelium (Geering and Freyvogel, 1974). Five esterases predominated in the midgut epithelium of non-bloodfed females of *Culex tarsalis*; of these, three were carboxylesterases, one was an acetylesterase, and one an arylesterase (Houk *et al.*, 1978, 1979). A non-specific esterase formed a constituent of the microvillar membrane (Houk *et al.*, 1986a).

A triacylglycerol lipase, capable of liberating free fatty acids from triglycerides, was present in the midgut of sugarfed females of *Ae aegypti*. It was not one of the esterases separable by PAGE. The lipolytic activity of the midgut increased substantially after feeding on mouse blood due to lipase in the blood, but it increased further during the first 15 h after feeding, indicating synthesis of lipase. Whether the lipase was in the epithelium or gut lumen was not examined. The lipolytic activity fell steadily between 15 and 50 h after feeding (Geering and Freyvogel, 1975).

The midguts of females of *Ae aegypti* and diapausing *Cx pipiens* showed negligible activity for α-amylase, which hydrolyses 1,4-α-glycosidic linkages in glycogen, starch and oligosaccharides. The whole-body α-amylase activity increased immediately after *Ae aegypti* had fed on human blood, but that was due to α-amylase in the ingested blood (Yang and Davies, 1968).

α-Glucosidase activity, assayed with *p*-nitrophenyl-α-glucopyranoside as substrate, was found in both the anterior and posterior midgut regions of *Anopheles stephensi* before and at all times after blood feeding. The enzyme activity increased in both gut regions after blood feeding, but during most of the time after blood feeding >90% of the total activity was found in the lumen of the posterior midgut (Billingsley and Hecker, 1991).

β-Fructofuranosidase (invertase), which hydrolyses sucrose to glucose and fructose, was reported to be present in the midgut and crop of starved and sugarfed females of *Ae aegypti*, but absent from the salivary glands, based on the hydrolysis of 3,5-dinitrosalicylic acid (Fisk and Shambaugh, 1954).

14.2 THE PERITROPHIC MEMBRANE

14.2.1 Secretion

Two methods of peritrophic membrane formation are known in insects; mosquito larvae use one and mosquito adults use the other. In mosquito larvae the membrane is secreted by a narrow band of cells at the anterior end of the midgut and is formed as a continuous tube (Section 5.1.4(b)). In mosquito adults all midgut columnar cells secrete the substance of the membrane which is deposited around the ingested blood mass.

Several of the earlier investigators considered that secretion of peritrophic membrane precursors started very soon after engorgement, e.g. in *Aedes aegypti* within 9, 20 or 30 min (Stäubli *et al.*, 1966; Zhuzhikov *et al.*, 1970; Bertram and Bird, 1961). In fact, the amorphous material that they observed in the space between the epithelium and the semi-compacted mass of erythrocytes was a coagulum of plasma proteins, not peritrophic-membrane precursor (Howard, 1962; Houk and Hardy, 1982; Perrone and Spielman, 1988).

The use of labelled glucosamine has helped elucidate the time scale of peritrophic membrane formation in *Ae aegypti*. When [^{3}H]glucosamine had been injected into females immediately after engorging, label became visible in the columnar cells after 30 min and was intense in the microvilli after two hours. At four hours label was present over the surface of the blood mass in the gut lumen, and had started to coalesce into a series of laminae. The peritrophic membrane was assembled most rapidly between 8 and 12 h after feeding, and was derived from precursors that were secreted continuously. By 12 h it appeared as a discrete layer completely surrounding the mass of ingested blood. Label was present on all midgut microvilli, of both the anterior and posterior regions, throughout the period of observation. The peritrophic membrane appeared to be assembled most slowly and organized most loosely where the blood mass was relatively remote from the gut wall, suggesting that it may be moulded by pressure from the gut wall (Perrone and Spielman, 1988). Other investigators stated that the peritrophic membrane of *Ae aegypti* became fully hardened more rapidly, i.e. after 4–5 h (Zhuzhikov *et al.*, 1970) or 5–8 h (Stohler, 1957; Freyvogel and Stäubli, 1965). In *Culex tarsalis* no sign of precursor material was evident until about eight hours after feeding. Secretion occurred from 8 to 12 h and consolidation of material started between 12 and 16 h, culminating in a mature peritrophic membrane by 20–24 h (Houk *et al*, 1979).

The secretory vesicles had disappeared from the apical cytoplasm 60–90 min after *Anopheles gambiae* and *An stephensi* had taken blood meals (Stäubli *et al.*, 1966). In *An gambiae* the peritrophic membrane could be visualized by electron microscopy 12 h post-bloodmeal and it was fully formed by 48 h. Membrane formation occurred concomitantly with digestion and contraction of the erythrocyte mass (Berner *et al.*, 1983). In *An stephensi* PAS-positive precursors of the peritrophic membrane appeared around the blood mass no earlier than 8 h after blood feeding (Gander, 1968). Injection of females with the RNA polymerase inhibitor α-amanitin prior to engorging prevented peritrophic membrane formation although it not stop the disappearance of apical granules (Berner *et al.*, 1983).

In both culicines and anophelines peritrophic membrane that is secreted in the anterior midgut region slowly retracts and twists, forming a plug between the anterior midgut and stomach. A similar plug forms between the stomach and hindgut (Richardson and Romoser, 1972; Romoser and Cody, 1975; Berner *et al* 1983; Perrone and Spielman, 1988).

The formation of peritrophic membranes has been reported in females that had ingested saline containing ATP or received enemas of saline or air, and this led to the conclusion that distension of the midgut epithelium provides the stimulus for secretion of precursor material by the columnar cells (Freyvogel and Jaquet, 1965; Zhuzhikov *et al.*, 1970). That idea still needs confirmation as it has been claimed that peritrophic membranes are frequently present in unfed mosquitoes, having been formed around meconium during the pupal and pharate adult stages (Romoser and Rothman, 1973; Romoser, 1974). It is clear, however, that a protein meal is not required because peritrophic membranes formed around meals of latex particles ingested with ATP (Billingsley and Rudin, 1992).

Females of *An stephensi* failed to secrete peritrophic membranes around blood meals when given blood a few hours after they had formed peritrophic membranes around meals of saline or of latex particles. Digestion of the blood meals proceeded normally despite the absence of peritrophic membranes (Berner *et al.*, 1983; Billingsley and Rudin, 1992).

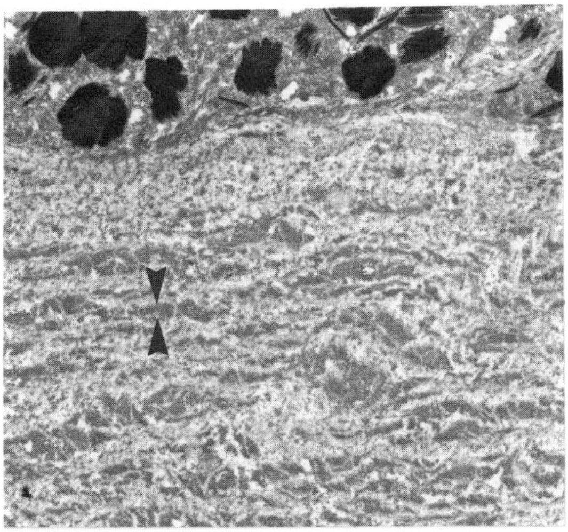

Figure 14.3 Ultrastructure of the peritrophic membrane of a female of *Anopheles stephensi* 48 h post-bloodmeal. Arrow heads indicate one of the electron-dense courtesy of Dr R. Berner, Dr W. Rudin and Dr H. Hecker.)

14.2.2 Structure and composition

Mature peritrophic membranes of females of a number of species of *Aedes*, *Culex* and *Anopheles*, examined after fixation and sectioning, were approximately 4–5 μm thick and exhibited a multilayered structure and a fibrillar substructure (Figure 14.3) (Perrone and Spielman, 1988; Houk *et al.*, 1979; Berner *et al.*, 1983). *Culiseta melanura* differed, having a peritrophic membrane which was only 1 μm thick and which was described as amorphous (Weaver and Scott, 1990a, b). *Aedes aegypti* peritrophic membrane is distinctly laminar (Sieber *et al.*, 1991), and the laminae have been said to show a parabolic arrangement of filaments, so resembling insect endocuticle (Perrone and Spielman, 1988). After treatment with acid or alkali of peritrophic membranes isolated from *Ae aegypti*, both the outer and inner surfaces showed a diffuse microfibrillar structure. The microfibrils were not regularly orientated but were interwoven forming a felt-like structure (Zhuzhikov *et al.*, 1971). Peritrophic membranes are frequently, perhaps normally, perforated or incompletely formed posteriorly, permitting defaecation (Romoser and Cody, 1975; Billingsley and Rudin, 1992). Gaps at the anterior end of the peritrophic membrane were reported in some females of *Ae aegypti* and *An stephensi* (Freyvogel and Stäubli, 1965). Some bloodfed females of *An stephensi* appeared to lack a peritrophic membrane entirely (Schneider *et al.*, 1987).

The peritrophic membranes of culicines contain chitin as a major constituent. Peritrophic membranes from *Ae aegypti*, *Ae triseriatus* and *Cx quinquefasciatus* gave positive responses to Van Wisselingh's chitosan–iodine test for chitin (Waterhouse, 1953; Freyvogel and Jaquet, 1965; Richardson and Romoser, 1972). Peritrophic membranes from *Ae aegypti* yielded *N*-acetylglucosamine, the monomeric subunit of chitin, upon acid hydrolysis (Berner *et al.*, 1983) and upon digestion with first chitinase and then β-*N*-acetylglucosaminidase. Peritrophic membranes from *Ae aegypti* which were exposed to chitinase for 36 h disintegrated (Huber *et al.*, 1991).

The presence of chitin in anopheline peritrophic membranes is doubtful. Treatment with chitinase removed material from developing peritrophic membranes of *An stephensi* (Gander, 1968), and peritrophic membranes from *Anopheles stephensi* and *An gambiae* responded positively, but not entirely typically, to the chitosan test (Freyvogel and Stäubli, 1965; Freyvogel and Jaquet, 1965). However, acid hydrolysates of peritrophic membranes from *An stephensi* yielded *N*-acetylgalactosamine and galactose (not *N*-acetylglucosamine) (Berner *et al.*, 1983), and lectin-binding experiments confirmed the presence of these compounds in the peritrophic membranes of *An stephensi* and *An gambiae* (Rudin and Hecker, 1989). Peritrophic membranes from *An stephensi* and *An gambiae* dissolved within 2–3 h of being placed in physiological saline, whereas peritrophic membranes from *Ae aegypti* were insoluble (Freyvogel and Stäubli, 1965; Freyvogel and Jaquet, 1965; Gander, 1968). An increase of calcium ions in the salt solution drastically reduced the solubility of the *Anopheles* peritrophic membrane (Berner *et al.*, 1983).

The peritrophic membrane of *Ae aegypti* starts to become fragile about 14 h after feeding, at a time when digestion is still active. The membrane is easily torn and cannot be removed intact from the blood clot (Zhuzhikov *et al.*, 1970). Females of *Ae aegypti* and *An stephensi* usually discharge their peritrophic membranes with the undigested remnants of the blood meal by about 48 h after feeding (Freyvogel and Stäubli, 1965). In *Cx nigripalpus* the peritrophic membrane is retained long after the blood residues are voided, and it appears to be attached to the gut epithelium at the junction of the anterior and posterior midguts and at the pyloric valve (Romoser and Cody, 1975). When mosquitoes take a small blood meal and engorge again a few hours later, the second blood meal partly or completely surrounds the first, and distinct peritrophic membranes can be distinguished around both (Waterhouse, 1953; Romoser *et al.*, 1989).

14.2.3 Function

The function of the peritrophic membrane that forms around the blood mass in engorged mosquitoes is not known. It is not formed soon enough to block the entry of arboviruses which infect the midgut epithelial cells within minutes of ingestion (Hardy *et al.*, 1983), nor that of microfilariae, many of which migrate into the midgut within one or two hours (Townson and Chaithong, 1991). Depending upon temperature and species, some 18–36 h after *Plasmodium* have been ingested the zygotes transform into motile ookinetes. It has been suggested that with some combinations of *Anopheles* and *Plasmodium* species the peritrophic membrane slows the exodus of ookinetes, prolonging their exposure to digestive enzymes, and that with others it may physically block ookinete exodus if fully formed in time (Ponnudurai *et al.*, 1988). Ookinetes of *P. gallinaceum* could penetrate the peritrophic membrane of *Ae aegypti* 24 h after the blood meal. At that time the peritrophic membrane was still intact and had a laminar structure, but the region of peritrophic membrane adjacent to a penetrating ookinete had lost its laminar structure and was less electron dense (Sieber *et al.*, 1991). Ookinetes that had been cultured *in vitro* secreted chitinase into the culture medium, therefore chitinase may be used to digest the peritrophic membrane (Huber *et al.*, 1991).

A number of other functions have been suggested. The peritrophic membrane may reduce contamination of the surface membrane of the gut epithelium by haematin, and may prevent mechanical damage to it by the crystals of haematin that form in the half-digested blood mass. It may help delimit the ectoperitrophic space and thereby help keep digestive enzymes away from the inhibitors in the blood mass. The plug at the anterior end of the peritrophic membrane may physically separate the anterior and posterior midgut regions, preventing movement of nectar into the region of protein digestion.

However, the normality of digestion observed in mosquitoes that lacked a peritrophic membrane argues against a major digestive function (Billingsley and Rudin, 1992). Moreover, the 'ectoperitrophic' space, thought to be important for digestion, is present around the compacted blood mass before formation of the peritrophic membrane.

14.3 DIGESTION OF THE BLOOD MEAL

14.3.1 Structural changes in the midgut cells

It is helpful to recognize different phases in the post-bloodmeal condition of the midgut. Immediately after feeding the midgut is hugely distended and for the next hour it is engaged in diuresis. A few hours later a phase of organelle synthesis occurs when additional ribosomes and endoplasmic reticulum are synthesized and organized. This leads into a phase of production of digestive enzymes and absorption of the products of digestion. Changes in the appearance and ultrastructure of the midgut epithelium during digestion of the blood meal have been described by several authors. These descriptions have been supplemented with measurements, made by H. Hecker and his colleagues, of changes in the surface areas and volumes of cellular organelles (see Hecker *et al.* (1974) and Hecker (1978) for the morphometric methods).

When a female feeds, blood passes through the anterior region of the midgut and fills the stomach which becomes very distended. Over the greater part of the stomach the epithelial cells are now squamous but at the anterior and posterior ends they are cuboidal. Examination of *Ae aegypti* revealed that before feeding the stomach cells were 5 μm wide and 25–35 μm deep, but that immediately after engorgement they became 25–35 μm wide, <5 μm deep where the nucleus produced a convexity and only 1–2 μm deep at the cell junctions (Howard, 1962). Two hours after feeding, the distances between microvilli were considerably increased, indicating that the apical cell membranes had been stretched (Bertram and Bird, 1961).

Experimental evidence indicates an increased permeability of the gut associated with blood feeding. For example, a blood constituent, possibly haemoglobin, penetrated into spaces between the septa of the continuous junctions in *Cx tarsalis* (but not in *Cx pipiens* or *Ae dorsalis*) and reached as far as the basal lamina (Houk and Hardy, 1979, 1982; Houk *et al.*, 1986c). Antibodies ingested with blood have been detected in mosquito haemolymph. Passage of rat anti-*Rickettsia* IgG antibodies into the haemolymph, within three hours of blood feeding, was demonstrated by indirect immunofluoresent assay in females of *Anopheles stephensi*, *An gambiae* and *An albimanus*, all species that fed for 10–15 min or more and that discharged blood from the anus while feeding. Only a trace of the antibody was detected in the haemolymph of *Cx pipiens* and none in the haemolymph of *An freeborni* and *Ae aegypti*, species with different feeding habits (Vaughan and Azad, 1988). However, Hatfield (1988) reported antibody transfer from ingested blood into the haemolymph of *Ae aegypti*.

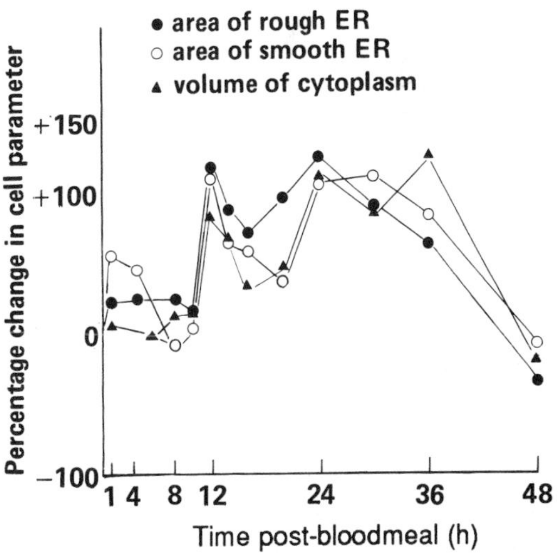

Figure 14.4 Changes in the posterior midgut cells of *Aedes aegypti* following ingestion of blood. The volume of cytoplasm and the surface areas of rough and smooth ER are expressed as percentages of the values in 3-day-old sugarfed females. (From Rudin and Hecker, 1979.)

When females of *An stephensi* fed on rats immunized against *Plasmodium falciparum*, anti-circumsporozoite IgG was present in their haemolymph immediately after feeding. The concentration peaked at 3 h post-bloodmeal, at which time the haemolymph contained 0.9–0.95 ng anti-circumsporozoite IgG per μl, which was c. 0.5% of the concentration in the host's plasma at the time of feeding. The haemolymph concentration of anti-circumsporozoite IgG was negligible by 18 h post-bloodmeal, at which time substantial quantities were still present in the blood mass although possibly trapped and not available for uptake (Vaughan *et al.*, 1990).

Blood feeding caused the folded basal lamina of *Ae aegypti* to straighten and stretch. The mesh of its grid-like substructure was expanded, and the diameters of the two series of pores were enlarged – from 7 to 20 nm and from 20 to 40 nm. Change in shape of the grid pattern indicated that the greatest expansion of the basal lamina was in the longitudinal axis of the midgut (Reinhardt and Hecker, 1973). No change in pore diameter was seen in the basal lamina of bloodfed *Ae dorsalis* (Houk *et al.*, 1980), and there was no apparent alteration in the permeability of the basal lamina of *Cx tarsalis* after blood feeding (Houk *et al.*, 1981).

The size of the stomach cells increases substantially after blood feeding. In *Ae aegypti* (at 22–25°C) the cytoplasmic volume increased suddenly about 12 h after feeding; it then fell somewhat, but 24 h after feeding rose to twice the non-bloodfed value and remained high for a further 12 h (Figure 14.4) (Rudin and Hecker, 1979). By 24 h after blood feeding the flattened epithelial cells had become more cubical, or even columnar, the surface area of the microvilli had more than doubled, and the cells contained more lysosomes (Hecker *et al.*, 1974). The volume of the basal labyrinth increased after feeding, and when measured at one hour post-bloodmeal was five-fold greater than in the unfed state. The basal labyrinth had started to shrink by four hours and had returned to normal by eight hours post-bloodmeal (Rudin and Hecker, 1979).

Between 30 minutes and two hours after blood feeding, the tightly packed whorls of rough endoplasmic reticulum present in the stomach cells of *Ae aegypti* gradually separated. Within 24 h the whorls had disappeared and rough-coated cisternae were dispersed throughout the cytoplasm, often in parallel arrays (Bertram and Bird, 1961; Hecker *et al.*, 1974). Morphometric and sucrose gradient analyses both showed about a doubling of rough ER in the stomach cells during the first day after blood feeding. Indeed, morphometric measurements showed that fluctuations in the total amounts of rough and smooth ER followed fairly closely the fluctuations in cytoplasmic volume that occurred after blood feeding (Figure 14.4). Pretreatment with α-amanitin did not greatly affect ribosome formation by midgut cells during the first day after blood feeding but it prevented the proliferation of ER membrane, consequently the normal increase in rough ER did not take place (Hecker and Rudin, 1979).

Both morphometric and sucrose-gradient analysis of *Ae aegypti* showed an increase in free and membrane-bound ribosomes in the stomach cells during the first day after feeding. Morphometric analysis showed a steady decline in the rough ER content after 24 h, but sucrose-gradient analysis indicated an increase in rough ER content at 36 h after feeding and a substantial amount remaining at 48 h (Hecker, 1978; Rudin and

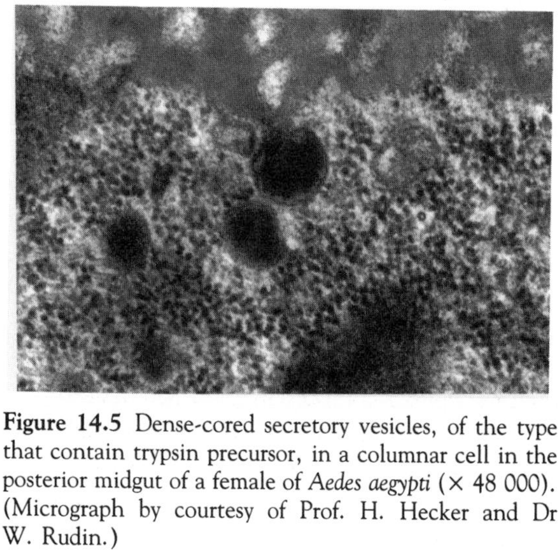

Figure 14.5 Dense-cored secretory vesicles, of the type that contain trypsin precursor, in a columnar cell in the posterior midgut of a female of *Aedes aegypti* (× 48 000). (Micrograph by courtesy of Prof. H. Hecker and Dr W. Rudin.)

Hecker, 1979; Gander *et al.*, 1980; Hecker and Rudin, 1981).

The dense-cored secretory vesicles associated with peritrophic membrane formation that are present in the apical cytoplasm of the stomach cells of *Anopheles* disappear within an hour after blood feeding (Stäubli *et al.*, 1966). In *An stephensi* secretory vesicles had started to reappear by 24 h (Hecker, 1977), but whether or not they had the same contents as the apical vesicles is not known. The sparse rough-ER-bounded secretory vesicles present in *Cx tarsalis* disappear shortly after blood feeding (Houk and Hardy, 1982).

Dense-cored secretory vesicles that exhibit trypsin immunoreactivity appear in the stomach cells of *Ae aegypti* after blood feeding, with a small brief peak at about 12 h and a more substantial peak between 24 and 36 h. They are distributed throughout the cytoplasm but are often seen in contact with the apical cell membrane, with sites of exocytosis at the bases of the microvilli (Figure 14.5) (Rudin and Hecker, 1979; Graf *et al.*, 1986). In *Cx tarsalis* similar secretory vesicles (i.e. lacking a bounding layer of rough ER) appear in the vicinity of Golgi complexes by about 12 h post-bloodmeal. Their numbers are greatest at 24 h, shortly before the peak of enzymic activity (Houk and Hardy, 1982).

Lipid inclusions start to appear in the stomach cells of *Ae aegypti* about 4 h after feeding and increase at a fairly regular rate until 20 h. They increase again substantially between 30 and 36 h, but decline by 48 h and are almost absent at 72 h. Deposits of glycogen are present 30 h after feeding; these are very much greater at 36 h but have largely disappeared by 48 h (Rudin and Hecker, 1979).

The ultrastructure of the columnar cells of the anterior and posterior midgut regions indicates that both regions are capable of synthesis and absorption. An early increase in smooth ER, more prominent in the anterior midgut, may be a sign of enhanced secretory activity. At a later stage there is a substantial increase in synthetic structures in the posterior midgut cells (Hecker, 1977; Rudin and Hecker, 1979). Some of these changes relate to secretion of peritropic membrane, others to enzyme production. The relatively greater volume density of microvilli and of basal labyrinth in the cells of the anterior midgut may reflect greater absorption and transport in that region (Hecker *et al.*, 1974).

When digestion is complete and all gut contents have been voided the ultrastructure of the stomach cells returns to its prefeeding condition. In *Ae aegypti* the cell volumes are diminished at 48 h post-bloodmeal, and rough ER and some other organelles show a concomitant reduction (Figure 14.4). The whorls of rough ER have started to reform by that time and by 72 h they are well developed (Bertram and Bird, 1961; Rudin and Hecker, 1979).

14.3.2 Times of secretion

The midguts of non-bloodfed females already possess a low level of aminopeptidase activity, and this increases rapidly after blood feeding. In *Aedes aegypti* the activity peaks at about 24 h and in *Anopheles stephensi* (Figure 14.6A) at about 30 h after blood feeding. It is possible that immediately after feeding small peptides in the blood plasma provide substrate for the exopeptidase. At three hours post-bloodmeal 13% of aminopeptidase activity in *An stephensi* was associated with the midgut epithelium and 87% was in the lumen of the posterior midgut. At six hours, 97.5% of activity was in the lumen (Graf and Briegel, 1982; Billingsley and Hecker, 1991).

Trypsin activity was first observed three hours after feeding in *Ae aegypti* (Felix *et al.*, 1991) and *An stephensi* (Billingsley and Hecker, 1991). In old females the lag phase before trypsin synthesis was greatly prolonged (Briegel, 1983). Despite a slower initial rate of secretion, trypsin peaked at about the same time as aminopeptidase, between 24 and 30 h post-bloodmeal, before falling to almost almost zero at 60 h (Figure 14.6B) (Billingsley and Hecker, 1991). Defaecation was largely responsible for the steady

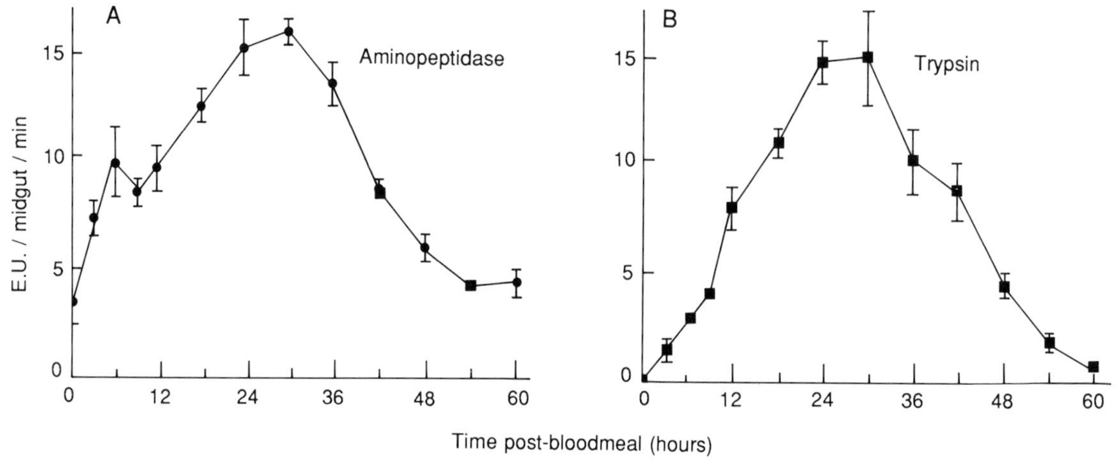

Figure 14.6 Time profiles of enzyme activities in homogenate supernatants of posterior midgut from *Anopheles stephensi*. (A) Aminopeptidase activity. (B) Trypsin activity. One enzyme unit (EU) = 1 μmol substrate hydrolysed per min per midgut. Means ± s.e. (From Billingsley and Hecker, 1991.)

decline in protease activity that followed the peak (Briegel, 1975).

There is some evidence that in *Ae aegypti* trypsin synthesis occurs in two phases, the first peaking at about ten hours post-bloodmeal and the second at about 24 h. For example, trypsins of two different molecular masses are synthesized at different times after blood feeding, and the earlier phase of trypsin synthesis is not sensitive to Actinomycin D and α-amanitin whereas the later phase is (Section 14.1.3). Meals of 5% bovine serum albumin solution stimulated a burst of trypsin secretion which peaked at 10 h post-bloodmeal, whereas meals of peptides from liver digest stimulated no secretion until after 8 h post-bloodmeal but stimulated a small peak at 18 h (Felix *et al.*, 1991). The morphometric data (Figure 14.4) indicate two phases of formation of protein-synthesis organelles with a similar time scale.

α-Glucosidase was active in both anterior and posterior midgut regions of *An stephensi* before and at all times after feeding. Activity rates in homogenates of whole midgut rose slowly during the first 18 h after feeding, and then rose more rapidly to peak at 30 h. At most times after blood feeding the α-glucosidase activity of anterior midgut homogenates was <10% of that of homogenates of whole midgut. Activity rates peaked some 6–12 h earlier in the anterior region than in the midgut as a whole (Billingsley and Hecker, 1991).

14.3.3 Rates of digestion

A variety of methods have been used to estimate the rates at which mosquitoes digest blood.

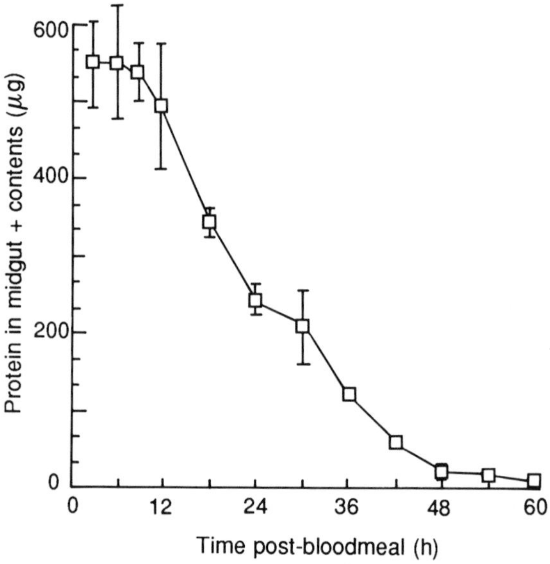

Figure 14.7 Time profile of the disappearance of protein from the midguts of bloodfed *Anopheles stephensi* kept at 27° ± 2°C. Means ± s.e. (From Billingsley and Hecker, 1991.)

Many investigators have used specific antisera to measure the rates of disappearance of plasma proteins such as albumin, α- and γ-globulins. Others have followed the fate of ingested antibodies. However, since ingested antibodies may persist for 2–3 days bound to midgut epithelium (Hatfield, 1988), data from serological methods should be interpreted cautiously and should not be used to assess the persistence of ingested protein. Different assays have been used to follow the disappearance of haemoglobin (O'Gower, 1956; Feldmann *et al.*, 1990). Probably the most satisfactory method of measuring rates of digestion is to measure the disappearance of acid-precipitable proteins (Briegel and Lea, 1975; Houseman and Downe, 1986).

When *An stephensi* which were kept at 27°C took meals of defibrinated pig's blood averaging 550 μg protein, reduction in the protein content of the midgut was first apparent nine hours after feeding. Between 12 and 24 h after feeding the protein content of the midgut fell rapidly, and by 36 h less than 20% remained (Figure 14.7) (Billingsley and Hecker, 1991). Females of *Ae aegypti* kept at 27°C completely digested meals of 4 μl rat's blood (also averaging 550 μg protein) within 36 h (Briegel and Lea, 1975). Use of polyclonal antisera with immunoblot analysis revealed that most serum proteins were undetectable 24–48 h after the ingestion of human and rodent blood by females of *Ae aegypti* kept at 26°C (Irby and Apperson, 1989).

Judging by the duration of the gonotrophic cycle under natural conditions, digestion of blood meals is commonly completed in two or three days in tropical regions, but five to eight days may be required in temperate regions. Hibernating females of *Culiseta annulata*, kept at 6–8°C, took 28–31 days to digest a blood meal (Service, 1968a, b).

Temperature affects both the rate of enzyme synthesis and the kinetics of enzyme reactions. When *Ae aegypti* were given 3 μl blood by enema and were kept at 22, 27 or 32°C, the times required to reach maximum proteolytic activity were 36, 22, and 18 h respectively. Thus this period was halved as the temperature was raised from 22 to 32°C (Briegel and Lea, 1975). Gooding (1966a) demonstrated the kinetic effects of temperature on proteinase activity, using midgut homogenates from *Ae aegypti* and *Cx quinquefasciatus*.

There is some evidence that humidity affects the rate of digestion. Observations on several Indian species of *Anopheles* showed that during the hot dry season digestion of blood took three to five days, but when the relative humidity increased the gut was cleared in one to three days (Mayne, 1928). In hibernating females of *An messeae*, in which digestion of blood proceeded normally at temperatures of 15–30°C, digestion was more rapid at higher than at lower relative humidities, but at 10°C the opposite obtained (Shlenova, 1938).

When *Ae aegypti* were given meals of 0.5–4 μl rat blood by enema, the size of blood meal had only a slight effect on the rate of proteinase synthesis during the first eight hours after feeding but the period of enzyme synthesis was substantially extended when blood-meal volumes were increased between 0.5 and 2 μl, so that a higher level of proteolytic activity was attained (Figure 14.10). The maximum level of enzyme activity increased further, but not in proportion, as the blood-meal volume increased between 2 and 4 μl. Even so, larger blood meals required a longer period for complete digestion than smaller ones (Briegel and Lea, 1975).

Proteins may fail to be digested normally if ingested with sugar. A mixture of 7.5% egg albumin in 0.5 M sucrose, ingested by female *Cx pipiens*, was treated like nectar and stored in the crop. Most females discharged the protein in their faeces within 24 h but absorbed the sugar; a few partly digested the protein and initiated ovarian development (Uchida and Suzuki, 1981). However, selection of *Ae aegypti* yielded a strain that could be maintained on meals of 10% egg albumen and 10% sucrose, producing about half as many eggs as when blood-fed (Stobbart, 1992).

The proteolytic enzymes in homogenized midguts from bloodfed *Ae aegypti* retained 80% of their activity after two days, leading Briegel

(1975) to conclude that the enzymes that are secreted during the first day after blood feeding are sufficiently stable to have substantial proteolytic potential. Proteinases were eliminated from the body throughout the period of blood digestion. A small portion was discharged during the first 24 h, and most was eliminated between 30 and 40 h after feeding. The total proteinase activity of the faeces was more than half that in the midgut at the time of maximum activity. Trypsin accounted for 75% of the proteolytic activity of the faeces; chymotrypsin was not detectable.

Digestion of the blood meal was somewhat slower in 6-day-old virgin females of *Ae aegypti* than in mated females of the same age (Edman, 1970), and the lag between feeding and trypsin synthesis was longer in virgins (Houseman and Downe, 1986). However, the injection of virgins with matrone speeded their rate of digestion, which then did not differ significantly from that of mated females (Downe, 1975).

Vertebrate blood contains trypsin inhibitors. Tests with blood from 17 species of mammals, birds and fish showed that all contained at least one inhibitor of mosquito trypsin and that most contained two. The inhibitors are protein constituents of the plasma; in bovine blood they are associated with the α_1- and α_2-globulin fractions. Inhibitory activity against partly purified mosquito trypsin was low in human sera, somewhat higher in the sera of rat, pig, cow and sheep, and very high in the sera of turkey and chicken (Huang, 1971a, b). Briegel and Lea (1975) found no difference between the levels of proteinase activity in homogenized midguts of *Ae aegypti* during the first 18 h after females had been given 3 μl of blood with higher (chicken) or lower (sheep) antitrypsin activity, and concluded that the trypsin inhibitors in vertebrate blood do not affect the digestion of blood by mosquitoes *in vivo*. However, with *Cx nigripalpus*, Van Handel and Romoser (1987) found no proteinase activity in homogenates of isolated blood masses, and found substantially lower proteinase activities in homogenates of blood-containing midguts than in fluid from the ectoperitrophic space.

Nectar from a lily, *Hemerocallis* sp., inhibited trypsin from bloodfed *Cx quinquefasciatus* when assayed *in vitro*, even after dilution to 1% (Gooding *et al.*, 1973). Interaction between trypsin inhibitors in nectar and proteolytic enzymes in bloodfed mosquitoes is likely to be prevented by the plug of peritrophic membrane at the entrance to the posterior midgut (Section 14.2.1).

Giving females of *Ae aegypti* a second blood meal can enhance or reduce trypsin secretion, depending on the timing. When meals of 1.2 mg were followed with a second meal of 1.5 mg, 5–20 h after feeding and during the period of increasing or maximum trypsin secretion, strong stimulation of trypsin secretion was observed. When the second blood meal was taken 40–50 h after the first, and after digestion of the first had been completed, a second burst of trypsin secretion was induced with peak activity only 50% of that in the earlier meal (Gass, 1977).

14.3.4 Fate of the blood mass

Within a few minutes of feeding, the erythrocytes of the blood meal are clumped into a tight mass which is surrounded by a narrow layer of cell-free serum. This concentration of erythrocytes is probably largely due to the extraction of water during post-bloodmeal diuresis. It is also possible that the salivary agglutinins discharged by some species during blood feeding are ingested with the blood and agglutinate erythrocytes within the midgut (Hudson, 1964). The blood in the midgut of recently engorged mosquitoes is bright red. A few hours after feeding, digestion begins at the periphery of the blood mass, producing a brown or black colour in that region.

The highest activities of trypsin and aminopeptidase are localized in the fluid contained in the narrow space between the midgut epithelium and the outer surface of the blood mass, or of the peritrophic membrane when formed (Figure 14.8). In *Ae aegypti* 50% of aminopeptidase activity was within the ectoperitrophic space, compared with 10% inside the peritrophic membrane and 40% in the midgut

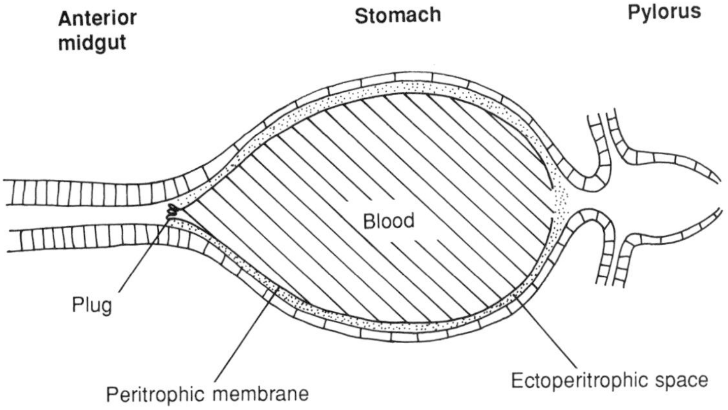

Figure 14.8 Diagram of a section through the anterior midgut, stomach and pylorus of a female mosquito during the course of digestion of a blood meal.

epithelium (Graf and Briegel, 1982). Immunoreactivity studies revealed trypsin on both sides of the peritrophic membrane, and more highly concentrated at the posterior end of the stomach (Graf *et al.*, 1986). Graf and Briegel (1982) postulated that the peritrophic membrane served to establish a compartment with concentrated enzymic activity between the protein meal and the absorbing epithelium. From his finding that most peptides inside the peritrophic membrane of *Cx nigripalpus* were of greater molecular mass than those in the ectoperitrophic space, Borovsky (1986) postulated that trypsin traverses the pores in the peritrophic membrane and attacks proteins at the periphery of the blood mass, away from most of the inhibitors in the blood and that peptides below a certain size diffuse outwards into the ectoperitrophic space where they are attacked by exopeptidases. The observations that digestion of the blood mass occurred normally in *An stephenssi* in the absence of a peritrophic membrane and in *Ae aegypti* in the presence of a greatly thickened peritrophic membrane (Berner *et al.*, 1983; Billingsley and Rudin, 1992) suggest that these hypotheses ascribe more to the peritrophic membrane than is justifiable. However, it is clear that a narrow zone forms between the epithelium and the compacted blood mass by the completion of post-bloodmeal diuresis, and that this is the region of high enzymic activity before and after formation of the peritrophic membrane. The importance of isolation

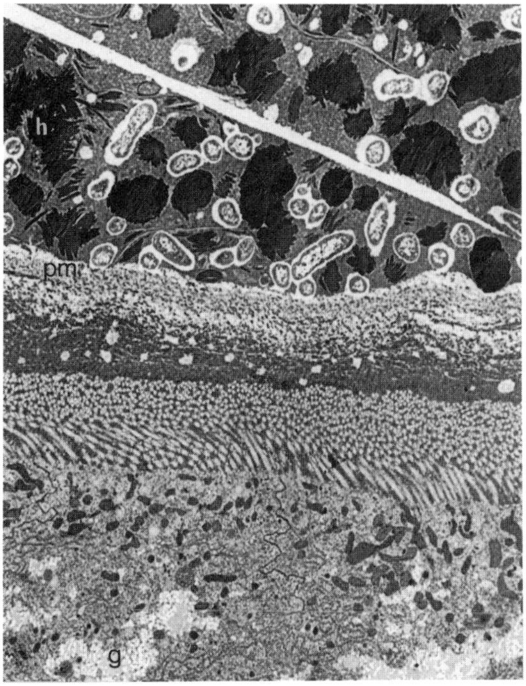

Figure 14.9 A section through the blood mass, peritrophic membrane (pm), and the apical region of a columnar cell in the posterior midgut of a female of *Anopheles stephensi* 48 h post-bloodmeal. Haematin crystals (h) and bacteria are visible within the blood mass; glycogen deposits (g) are located in the cell. (× 4800). (Micrograph by courtesy of Dr R. Berner, Dr W. Rudin and Prof. H. Hecker.)

of proteolytic enzymes from the inhibitors present in ingested blood was shown by the 45–90% inhibition of the protease activity that resulted when ectoperitrophic fluid was mixed with the residual blood mass. This observation also indicated that experiments with homogenates of blood-filled midguts may be misleading (Van Handel and Romoser, 1987).

Digestion proceeds from the periphery of the blood mass inwards, and by about one day after feeding the blood mass is brown or black except in the centre. The initially pliable blood mass usually becomes a firm clot within 24 h, by which time undigested erythrocytes are no longer found at the periphery. Digestive enzymes do not affect the inner substance of the blood mass for some time, and stained preparations show that it consists of erythrocytes bound in a meshwork of fibrin strands, like a typical blood clot. In a number of species the cibarial armature causes haemolysis of erythrocytes during feeding (Section 11.2.3), but in *Ae aegypti* trypsin appears to be the principal cause of a progressive haemolysis (Geering, 1975). During digestion the coenzyme haem is dissociated from the globin moiety of haemoglobin. In the presence of oxygen, and in the weakly alkaline medium of the midgut lumen, haem is converted to the stable derivative haematin. Sections of midgut often show clumps of crystalline haematin near the edge of the blood mass (Figure 14.9) (Hecker and Rudin, 1979; Berner *et al.*, 1983). The pigment also adheres to the peritrophic membrane and it is eventually discharged in the faeces (Briegel, 1980a). Wigglesworth (1943) found no evidence for absorption of haematin by *An maculipennis* and *Ae aegypti*, such as occurs in many other blood-sucking arthropods.

One hour after females of *Cx pipiens pallens* had taken a meal of sheep blood their haemolymph mean total amino acid concentration had fallen from 50 to 40 mmol l^{-1}, possibly due to water intake. One hour later it had returned to the prefeeding level and it subsequently continued to increase, peaking at 90 mmol l^{-1} after 18 h (at 22°C). The amino acid concentration then fell to 70 mmol l^{-1} at 48 h, and to c. 50 mmol l^{-1} after three days. Different amino acids varied in the extent to which they accumulated in the haemolymph during the period of digestion. Asparagine followed the pattern of the total amino acids, peaking at 18 h and declining thereafter; valine remained at high concentration between 18 and 48 h; isoleucine, which is low in sheep blood protein, failed to rise (Uchida *et al.*, 1990). The haemolymph amino acid concentrations obviously reflected rates of removal as much as rates of addition.

In *Ae aegypti* the midgut musculature maintains a tonic contraction while there is still blood in the stomach. Peristaltic and antiperistaltic waves occur in the hindgut every 2 or 3 s. Each sequence starts with a vigorous antiperistaltic wave which forces fluid in the pyloric chamber forwards into the posterior midgut and Malpighian tubes. The contraction then reverses direction and sucks fluid down the hindgut lumen into the rectum (Howard, 1962). At 24 h post-bloodmeal, when most of the protein has been digested, erythrocytes can still be observed in the centre of the blood mass. As the blood mass diminishes in size the peritrophic membrane appears to be pressed on to it by contractions of the midgut musculature (Stohler, 1957; Howard, 1962; Graf *et al.*, 1986). When digestion is complete the peritrophic membrane is discharged with haematin and other residues of the blood meal (Section 16.4).

14.4 REGULATION OF ENZYME SYNTHESIS

The presence of blood in the midgut induces enzyme synthesis and secretion, and attempts have been made to identify the nature of the stimuli provided by the blood meal. Details of trypsin synthesis are presented in Section 14.1.3, and the times of secretion of digestive enzymes are outlined in Section 14.3.2.

14.4.1 Effects of secretogogues

Evidence for a secretogogue control mechanism is provided by feeding experiments. Feeding *Aedes aegypti* on washed sheep erythrocytes suspended in saline induced very little proteolytic activity, which appeared many hours later. Haemoglobin solution was a better inducer than equivalent amounts of intact washed erythrocytes; proteolytic activity was greater and appeared sooner after feeding. Solutions of bovine serum albumin, sheep haemoglobin, and casein were all effective in inducing midgut proteolytic activity (Shambaugh, 1954; Akov, 1972; Samish and Akov, 1972). Protein hydrolysates had negligible stimulatory action on proteinase synthesis, as did peptone, gelatine, and histone. From the responses to the rather few proteins tested, Briegel and Lea (1975) concluded that only intact globular proteins with molecular weights greater than histone stimulate proteinase synthesis.

Meals of bovine serum albumin (BSA) induced trypsin synthesis, with a positive correlation between extent of synthesis and BSA concentration between 1 and 10%, but even 10% BSA induced less trypsin synthesis than chicken blood. BSA fragments were less active than BSA. Erythrocyte ghosts induced only exceedingly slight synthesis. Felix *et al.* (1991) concluded that it was necessary for a globular protein to be present in solution in the midgut to induce trypsin synthesis.

Enemas of 5% BSA and of major BSA fragments stimulated an early phase of trypsin synthesis whereas enemas of peptides from liver digests stimulated a late phase of synthesis. Felix *et al.* (1991) postulated that the early phase may be induced by soluble proteins of a considerable range of molecular mass, and that the late phase is induced by the presence of smaller peptides.

14.4.2 Effects of volume

When females of *Aedes aegypti* were given enemas of 0.5, 1, 2 or 4 μl blood the initial rate of protease synthesis was the same but the duration of secretion was positively correlated with blood-meal volume so that greater amounts of protease were secreted after larger meals (Figure 14.10) (Briegel and Lea, 1975). These authors attempted to find whether protease secretion is stimulated by stretching of the midgut or abdominal wall. When females of *Ae aegypti* were given different volumes of fluid by enema but the same amount of protein (1 μl of blood), both the extent and rate of proteinase secretion were nearly identical among the different females. When three groups of females were given enemas

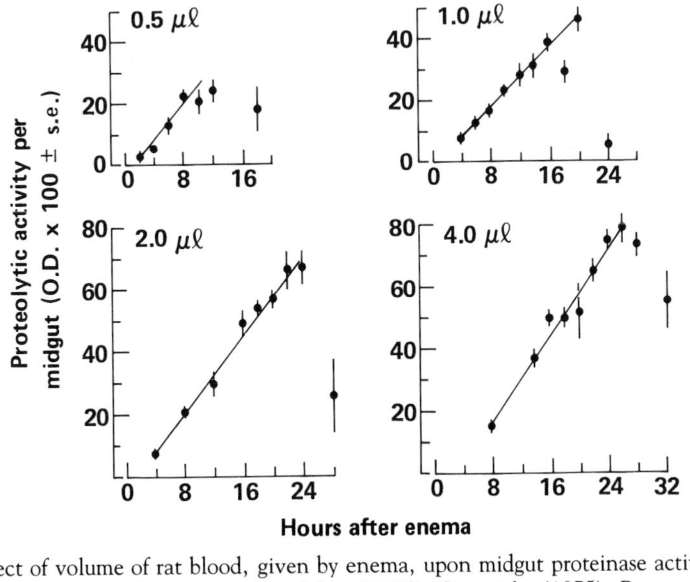

Figure 14.10 The effect of volume of rat blood, given by enema, upon midgut proteinase activity in females of *Aedes aegypti* kept at 27°C. Means ± s.e. (From Briegel and Lea (1975). Copyright (1975), Pergamon Press plc. Reprinted with permission.)

of constant volume (4 μl) but containing 1, 2 or 4 μl of blood, the rates and extents of proteinase secretion equalled those previously established for enemas of 1, 2 and 4 μl blood respectively. Briegel and Lea concluded that stretching of the midgut or abdominal wall, which might activate a neural mechanism, had no effect upon proteinase secretion.

Females of *Aedes aegypti* that engorged on a solution of 12% dextran, with ATP as phagostimulant, secreted small amounts of protease (Samish and Akov, 1972). Isolated midguts that were injected with 0.1, 1 and 2 μl sucrose solution by enema secreted trypsin in amounts that were positively correlated with volume, but smaller amounts of trypsin were synthesized than after blood stimulation. From this experiment Graf and Briegel (1989) observed that the midgut could synthesize trypsin when disconnected from all other tissues and in response to injection of a non-protein solution, and suggested that mechanical or possibly osmotic stress was the primary stimulus.

14.4.3 Hormonal regulation

Ablation of endocrine organs has provided evidence of hormonal involvement in the regulation of enzyme synthesis in *Aedes aegypti*. Prior to blood feeding, juvenile hormone induces the formation of whorls of rough endoplasmic reticulum (Section 13.2.1). Briegel and Lea (1979) removed the median neurosecretory cells (MNC), with or without their associated neurohaemal organ, from females of *Ae aegypti* and later fed the mosquitoes 3 μl blood by enema. Eight hours

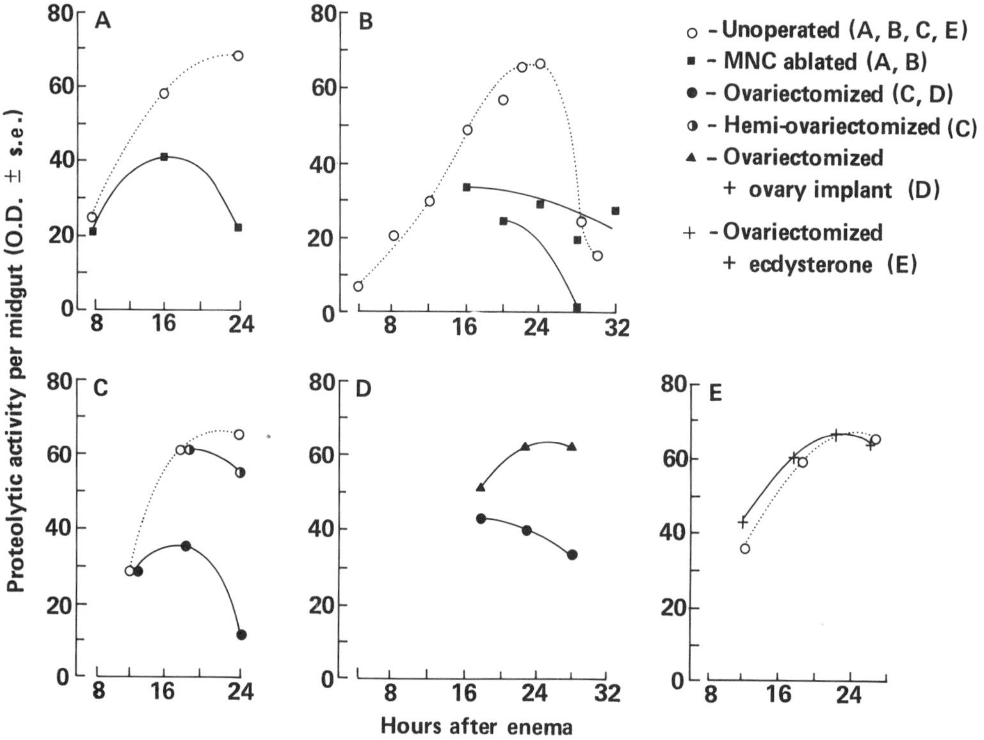

Figure 14.11 The effects of prior organ removal upon midgut proteolytic activity in females of *Aedes aegypti* given 3 μl blood by enema. (A) MNC removed at emergence; 1st blood meal. (B) MNC removed at emergence; 2nd blood meal (two sets of operated data points from two experiments). (C) Ovariectomized or hemi-ovariectomized at emergence. (D) Ovariectomized at emergence; one set with one ovary implanted. (E) Ovariectomized at emergence; 20-hydroxyecdysone injected. (From Briegel and Lea (1979). Copyright (1979), Pergamon Press plc. Reprinted with permission.)

after feeding the levels of midgut proteolytic activity were similar in the operated and control mosquitoes, but at 16 and 24 h after feeding they were much lower in the operated females than in the controls (Figure 14.11A). MNC-ablated females that completed one gonotrophic cycle and were given a second blood meal by enema, again secreted proteinases normally during the first 12 h after the enema but secreted little or no proteinase thereafter (Figure 14.11B).

In further experiments, Briegel and Lea (1979) removed the ovaries before providing a blood meal by enema, and they observed the same effect; proteinase secretion was normal up to 12 h after feeding but had declined substantially by 18 and 24 h (Figure 14.11C). The capacity of ovariectomized females to synthesize proteinases after blood feeding could be restored by implantation of an ovary from a sugarfed female or by injection of 2.5 μg 20-hydroxyecdysone (Figure 14.11D, E). The injection of 20-hydroxyecdysone into sugarfed females did not stimulate proteolytic activity. These results suggest that the early phase of proteinase synthesis may be independent of hormonal control, and that the later phase requires 20-hydroxyecdysone.

Blood protein disappeared at the same rate in MNC-ablated females that had reduced proteolytic activity as in unoperated females, leading Briegel and Lea (1979) to conclude that the endocrine-dependent proteinase activity represents excess secretion of enzymes which are not necessary for breakdown of blood proteins at the normal rate. The decapitation and ligation experiments of Gooding (1966b), Downe (1975) and Borovsky (1981a) confirmed that digestion does proceed after early removal of the brain, but these authors obtained variable results when measuring the effect on rate of digestion.

Half of the midgut endocrine cells discharge their secretion within six hours of a blood meal (Brown *et al.*, 1986), but nothing is known of their role, if any, in digestion.

The ovaries of *Aedes aegypti* synthesize a decapeptide called trypsin modulating factor (TMF) (see Section 10.3.1(f)), which inhibits trypsin synthesis. Injection into the haemocoele of 0.7 μg partly purified peptide inhibited the synthesis of certain of the DFP-reactive electrophoretic bands that were normally visible. TMF exhibited specific and saturable binding to a midgut receptor. Injection into the haemocoele immediately after a blood meal of 0.24, 1.45 or 2.86 nmol TMF caused, 24 h later, reductions in trypsin activity of 54, 66 and 86% respectively (Borovsky *et al.*, 1988, 1991a, b). Assuming a haemolymph volume of 1.5 μl, the 0.24 nmol dose that caused 54% inhibition would have had an initial concentration of 0.16 mM, and its effect would be considered pharmacological rather than physiological. Borovsky (1988) proposed that after a blood meal trypsin modulating factor is synthesized in the ovaries, and that between 24 and 55 h post-bloodmeal it is released into the haemolymph and signals the midgut to stop synthesis of trypsin.

14.4.4 Conclusions

Consideration of all the evidence suggests that secretagogues stimulate both the early and late phases of trypsin synthesis in *Aedes aegypti*, and it may be that soluble globular proteins induce the first phase and peptide products of digestion induce the second. The induction of trypsin synthesis by non-protein solutions suggests that mechanical, or possibly osmotic, stimuli can play a supplementary role but one which is normally masked. The ablation experiments suggest that hormones affect the second phase of synthesis, but because hormones such as 20-hydroxyecdysone have a number of actions it is not clear whether their effect on trypsin synthesis is direct or indirect. The position of TMF remains to be clarified.

Very little is known about the mechanisms that control synthesis of other digestive enzymes. Administering enemas of air, salt solution, or triglyceride emulsion to females of *Ae aegypti* produced a large increase in the activity of certain midgut esterases 4–24 h later, but no increase in the activity of midgut triacylglycerol lipase. Geering and Freyvogel (1975) concluded that a simple stretching of the midgut epithelium causes increased esterase production.

15

Adult energy metabolism

There have been few biochemical studies of energy metabolism in mosquitoes, although interesting work of a more physiological nature has been undertaken. It may be useful, therefore, to introduce this chapter with a brief survey of energy metabolism in insects. Glycogen and trehalose constitute the largest carbohydrate stores. Glycogen is synthesized in the fat body, and it is stored principally in that organ in a highly hydrated form. Glycogen is stored to a lesser extent within other cells, particularly muscle cells. The disaccharide trehalose is synthesized in the fat body but stored in the haemolymph in high concentration. It provides a readily available store of carbohydrate, giving up twice the amount of energy per osmotically active particle as glucose. Fat body glycogen can be converted to trehalose.

Triacylglycerols (triglycerides) are the dominant class of storage lipids, and they provide a number of advantages over glycogen as a metabolic reserve: (a) they have more than twice the energy value per unit weight, (b) they can be stored in an anhydrous form, and (c) when oxidized they liberate twice as much metabolic water as glycogen. For these reasons they tend to be the major source of metabolic energy in insects that pass through prolonged periods without feeding, e.g. in diapause. The fat body is the principal storage organ for triacylglycerols. The fatty acid complement of insect triacylglycerols is represented by eight fatty acids, three saturated (C14:0, 16:0; 18:0), three monounsaturated (C14:1, 16:1, 18:1) and two polyunsaturated (C18:2, 18:3). The monounsaturated fatty acids consist primarily of the *cis*-Δ-9-isomer and the C18 polyunsaturated fatty acids are the *cis*-Δ-9-12- and *cis*-Δ-9-12-15-isomers.

The carbohydrate and lipid reserves of insects can be derived from ingested carbohydrate, and also from ingested amino acids which constitute an important source when supplies of dietary carbohydrate are limited. Carbohydrate can be synthesized from those amino acids that, on degradation, yield pyruvate or an intermediate of the tricarboxylic acid cycle. The lipid reserves can be derived from the ketogenic amino acids, which are broken down to acetyl-CoA.

It is characteristic of adult Diptera that carbohydrate is the sole energy source for flight and that lipid is the principal energy source when at rest. Aerobic glycolysis is the major pathway by which carbohydrates are degraded in insects for energy release, glycolysis and the tricarboxylic acid cycle functioning as a single metabolic pathway. The energy contained within fatty acid molecules is made available by β-oxidation of the fatty acid and the removal of 2-carbon units in the form of acetyl-S-coenzyme A. Reviews of insect energy metabolism are provided by Candy (1985), Downer (1985) and Friedman (1985).

15.1 SYNTHESIS OF RESERVES

Mosquitoes conform with other insects in maintaining lipid and glycogen stores in the fat body. Glycogen is also stored in the flight muscles, and trehalose is stored in the haemolymph. Newly emerged adults utilize reserves carried over from the pupal stage, but soon seek out plant sugars as an important energy source. Blood meals also contribute to the energy needs of females. The observations on energy reserves in most of the following Sections refer to female mosquitoes; observations on males are reported in Section 15.1.7.

15.1.1 Composition of the reserves

The lipid, sugar and glycogen content of single mosquitoes can be estimated by extraction of total lipid with chloroform–methanol and its assay by reaction with sulphuric acid and a vanillin–phosphoric acid reagent, followed by colorimetric assay of carbohydrates in the aqueous residue with anthrone after the separation of glycogen with sodium sulphate (Van Handel, 1985a, 1985b; Van Handel and Day, 1988). The protein content of individual mosquitoes can be assayed by Kjeldahl digestion and Nesslerization, and calculated from total nitrogen with the conversion factor 6.25 (Van Handel, 1976; Briegel, 1990a). Some investigators have measured the total energy value of mosquitoes, oxidizing all organic carbon to CO_2 with hot acidified sodium bichromate and measuring spectrophotometrically the reduction of the chromium VI to the chromium III ion (Van Handel, 1972b).

The extent of the carbohydrate and lipid reserves and of the protein content of mosquitoes has sometimes been expressed as mass (mg). More often it has been expressed in terms of energy values, with calories as the unit. Because the unit 'calorie' is now obsolete all such values have been recalculated as joules (1 cal = 4.184 J). One milligram of carbohydrate or protein has an energy value of 16.74 J (4.0 cal), and 1 mg of triacylglycerol of 37.65 J (9.0 cal).

The fat body lipids of both newly emerged and sugarfed mosquitoes are predominantly triacylglycerols but 5–10% are diacylglycerols (Van Handel, 1965b; Nayar and Van Handel, 1971). Where the term triacylglycerol is used in the following pages it should be taken to include a small percentage of diacylglycerol. The principal fatty acids of triacylglycerols extracted from newly emerged *Culex pipiens* and *Cx tarsalis* were palmitic (16:0), palmitoleic (16:1) and oleic (18:1). Polyenes (polyunsaturated fatty acids) constituted only 3–4% of the total, with linoleic (18:2) at 2–3% the most abundant (Dadd *et al.*, 1987). Similar results were obtained from analyses of diapausing females of *Cx pipiens* (Van Handel, 1967b; Buffington and Zar, 1968). The fatty acids of triacylglycerols extracted from pupae and newly emerged adults of *Ae sollicitans* included about 20% of polyenes (18:2, 20:4, 20:5), but sugarfed adults synthesized only saturated and monounsaturated acids, principally palmitic (16:0), palmitoleic (16:1) and oleic (18:1), which constituted over 90% of the total. Small amounts of myristic (14.0) and stearic (18:0) acids were also synthesized (Van Handel and Lum, 1961; Van Handel, 1966). Lipids extracted from wild overwintering females of *Cx tarsalis* and *Anopheles freeborni* included approximately 75% triacylglycerols and 15% free fatty acids. In both classes the principal fatty acids were palmitic, palmitoleic and oleic, with small amounts of myristic, stearic, lauric (12:0) and myristoleic (14:1) (Schaefer and Washino, 1970; Schaefer *et al.*, 1971).

15.1.2 Reserves of teneral adults

The fat body cells, or trophocytes, of fully grown larvae of *Aedes aegypti* and *Culex pipiens* are packed with lipid droplets, glycogen granules and vesicles which possibly contain protein (Wigglesworth, 1942; Clements, 1956a), reserves which are carried over to the adult stage. The extent of the reserves depends upon the quality of larval nutrition, and there is some variability between species in the proportions of glycogen and lipid (Nayar and Sauerman, 1970b). Unlike the higher Diptera, mosquito adults do not develop an 'adult fat body' differing in structure from that of the larvae. In well-nourished adults the fat body contains deposits of glycogen and lipid but no protein-containing vesicles.

Briegel (1990a, b) analysed laboratory-reared teneral adults of *Aedes aegypti* and four species of *Anopheles* for total lipid, carbohydrate and protein. The protein must have been principally structural protein; the carbohydrate was presumed to be predominantly glycogen. Briegel particularly investigated the effect of body size on the amplitude of the reserves, producing adults

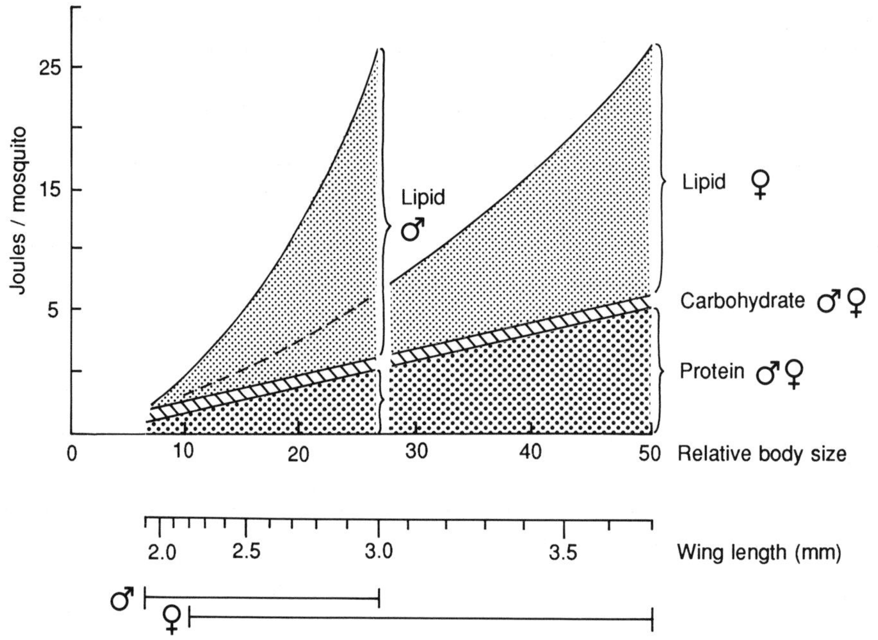

Figure 15.1 Diagrammatic representation of the protein, carbohydrate and lipid contents of teneral male and female *Aedes aegypti* measured in joules and expressed as a function of relative body size (i.e. the cube of wing length.) For both sexes, the protein and carbohydrate contents had an isometric relationship with relative body size whereas the lipid content was exponentially related. See also Table 15.1. (After Briegel, 1990a.)

of different sizes by rearing larvae at different densities. Relative body size was calculated from the cube of the wing length.

In teneral males and females of *Ae aegypti*, analysed within 3 h of emergence, the protein and carbohydrate contents had an isometric relationship with relative body size whereas the lipid content was exponentially related ($Y = aX^b$) (Figure 15.1). The lipid reserves of teneral males were always 2–4 times greater than those of females of equal size. Females that were equal in lipid content to males were at least twice their relative body size.

The sugar content of the haemolymph constitutes a significant part of the carbohydrate reserve. Newly emerged females of *Ae sollicitans*, which weighed 4 mg, contained 15 μg trehalose (0.25 J) and <0.1 μg D-glucose (Van Handel, 1969a).

The mean protein, lipid and carbohydrate contents of female *Ae aegypti* given water but starved to death (Table 15.1) were considered to be the minimum values required for survival. During starvation the females lost 14–31% of their protein and 54–84% of their lipid, depending on their size, while carbohydrates disappeared almost completely. Females that were denied water and starved died sooner, apparently of thirst, while retaining more reserves. When females were fed sucrose for 4–6 days and then starved to death their reserves fell to the same low levels as those that had never fed. During starvation structural proteins may be metabolized and used as a source of energy.

In all four species of *Anopheles* the body contents of lipid and carbohydrate were substantially smaller than in *Aedes aegypti*. The data for protein content in Tables 15.1 and 15.2 suggest that the anophelines have a different relationship between wing length and body mass than *Ae aegypti*. Further, the mean male wing length in the four species of *Anopheles* was only 0.1–0.2 mm shorter than that of females, and the composition of teneral males was not significantly different from that of teneral females. In both sexes the protein, lipid and carbohydrate contents had an

Table 15.1 Effects of larval-rearing density upon the size and composition of teneral female *Aedes aegypti*, and the effects of starvation and sugar feeding of females upon their composition. (From Briegel (1990a) with additional data and recalculation.)

Larvae were reared at 27°C. The mean composition data for the three selected size classes were derived from regression equations, relative body size being expressed as the cube of mean wing length.

A. *Effect of larval density on female dimensions*

Larval conditions	Very crowded	Crowded	Uncrowded
Larvae per pan	>>1000	400	200
Potential food input (J/larva)	<4	27.2	58.2
Wing length (mm): range	2.1–2.9	2.4–3.3	3.0–3.7
mean	2.5	3.0	3.4
Relative body size: mean	15.6	27.0	39.3

B. *Dimensions and composition of females of three size classes* (μg *and joules per female*)

Wing length (mm)	2.4		2.9		3.4	
Relative body size	13.8		24.4		39.3	
	μg	J	μg	J	μg	J
Teneral females						
Protein	112.5	1.88	230.0	3.85	392.5	6.57
Lipid	26.7	1.00	93.2	3.47	265.5	10.00
Carbohydrate	32.5	0.54	40.0	0.67	52.5	0.88
Total	171.7	3.42	362.2	7.99	710.5	17.45
Starved females						
Protein	77.5	1.30	185.0	3.10	337.5	5.65
Lipid	12.2	0.46	24.4	0.92	43.3	1.63
Carbohydrate	2.5	0.04	5.0	0.08	10.0	0.17
Total	92.2	1.80	214.4	4.10	390.8	7.45
Sugarfed females						
Protein	95.0	1.59	230.0	3.85	420.0	7.02
Lipid	125.5	4.73	295.5	11.13	533.3	20.08
Carbohydrate	40.0	0.67	92.5	1.55	62.5	2.72
Total	260.5	6.99	618.0	16.53	1115.8	29.82

isometric relationship with relative body size. Some 30–70% of teneral lipid and 40% of protein could be utilized during starvation (Table 15.2).

15.1.3 Sugar absorption and metabolism

That sugars are absorbed very rapidly was shown by the ability of exhausted *Culex pipiens* to resume flight within one minute of imbibing D-glucose (Clements, 1955). Similar observations were made on *Aedes aegypti*. Meals of 200 μg fructose, glucose or sucrose, taken by exhausted females, were barely detectable after 8 h. Sucrose was almost completely hydrolysed to fructose and glucose within 1 h of feeding, and a meal of fructose was reduced to sorbitol within 1 h (Nayar and Sauerman, 1971b).

D-Glucose absorbed from the gut is converted to trehalose, glycogen and triacylglycerols. A study of carbohydrate metabolism in *Ae sollicitans*

Table 15.2 Mean body composition of females of four species of *Anopheles* reared under uncrowded conditions, when (a) teneral, (b) starved to death and (c) fed on 20% sucrose for 2–5 days, calculated from regression equations. (Original data from Briegel (1990b), with recalculation.)

			Content/female					
Species	*Mean wing length*	*Content*	*Teneral*		*Starved*		*Sugarfed*	
			μg	*J*	*μg*	*J*	*μg*	*J*
albimanus	3.2	Protein	162.5	2.72	105.0	1.76	142.5	2.38
		Lipid	30.0	1.13	22.2	0.84	124.4	4.69
		Carbohydrate	30.0	0.50	2.5	0.04	137.5	2.30
		Total	222.5	4.35	129.7	2.64	404.4	9.37
gambiae	3.2	Protein	197.5	3.30	125.0	2.09	155.0	2.59
		Lipid	32.2	1.21	10.0	0.38	51.1	1.92
		Carbohydrate	20.0	0.32	5.0	0.08	95.0	1.59
		Total	249.7	4.84	140.0	2.55	301.1	6.10
stephensi	3.0	Protein	182.5	3.05	115.0	1.92	177.5	2.97
		Lipid	33.3	1.25	8.9	0.33	124.4	4.69
		Carbohydrate	42.5	0.71	5.0	0.08	82.5	1.38
		Total	258.3	5.01	128.9	2.34	384.4	9.04
quadrimaculatus	3.9	Protein	410.0	6.86	255.0	4.27	345.0	5.77
		Lipid	53.3	2.01	13.3	0.50	133.3	5.02
		Carbohydrate	70.0	1.17	7.5	0.13	415.0	6.94
		Total	533.3	10.04	275.8	4.90	893.3	17.73

suggested that D-fructose and D-galactose, which are present in nectar as the monosaccharides or as components of di- and trisaccharides, are absorbed from the gut and converted to the respective hexose phosphates. These are not converted to D-glucose but may be isomerized to D-glucose phosphates which are on the pathways of trehalose and glycogen synthesis. The sugar-alcohol D-sorbitol, which is found in many fruits, is readily oxidized to D-fructose in the fat body, and may then enter the same pathways (Van Handel, 1969a,b).

D-Glucose injected into *Ae sollicitans* was converted to trehalose and glycogen (Van Handel, 1969a). It seems likely that, as in many insects, monosaccharides are present only in low concentration in the haemolymph because they are rapidly converted by the fat body to trehalose and other metabolites.

Females of *Ae taeniorhynchus* from which the median neurosecretory cell perikarya (MNCp) had been removed at emergence metabolized ingested sugars abnormally. Operated females fed 17.7 J D-glucose and D-fructose synthesized glycogen and triacylglycerol normally for 6–8 h but instead of then ceasing glycogen synthesis they continued to synthesize glycogen at the same rate for 24 h; in contrast, their subsequent triacylglycerol synthesis was greatly reduced (Figure 15.2). Similar results were obtained with *Ae sollicitans* and *Ae aegypti* (Van Handel and Lea, 1965). The implantation into MNCp-ablated *Ae taeniorhynchus*, 6 h after the sugar meal, of four sets of MNCp from sugarfed females blocked further glycogen synthesis but did not stimulate further triacylglycerol synthesis (Lea and Van Handel, 1970).

When normal females of *Ae taeniorhynchus* were fed 16.7 J sucrose and 21 h later were fed a second meal of sugar, glycogen synthesis was not resumed but triacylglycerol synthesis was greatly increased. If

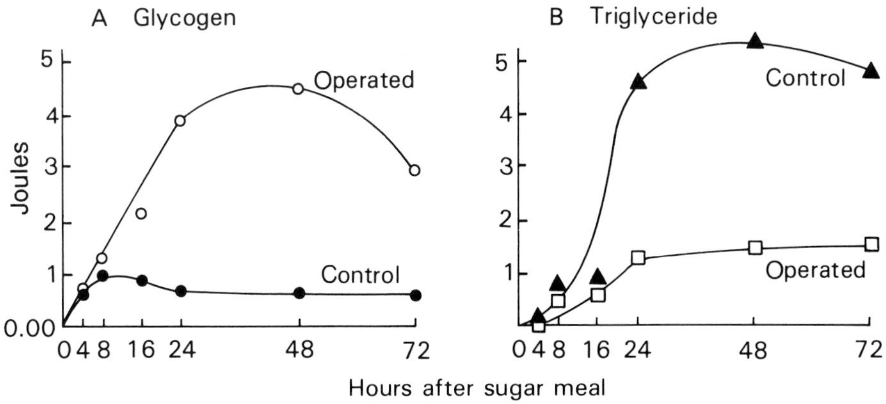

Figure 15.2 Effect of removal of median neurosecretory cells on net changes in the (A) glycogen and (B) triacylglycerol contents of female *Aedes taeniorhynchus* fed 16.7 J sugar. (From Van Handel and Lea (1965). Copyright 1965 by the AAAS.)

instead of feeding a second meal of sugar the MNCp were removed after 21 h, glycogen synthesis was resumed at the expense of further growth of the triacylglycerol pool (Van Handel and Lea 1965). Bloodfed *Ae taeniorhynchus* reacted differently to this operation. Females from which the MNCp alone or the MNCp and the cardiacal neurohaemal organ had been removed, and which fed on blood, responded like normal bloodfed females synthesizing small amounts of glycogen and large amounts of triacylglycerol (Van Handel and Lea 1970).

When the corpora allata were removed from *Ae taeniorhynchus* and the mosquitoes were later fed blood or sucrose, small amounts of glycogen and large amounts of triacylglycerol accumulated, as in normal females, therefore juvenile hormone does not affect the synthesis of these reserves (Van Handel and Lea, 1970).

It appears that in sugarfed female *Ae taeniorhynchus* synthesis of glycogen and triacylglycerols is regulated by one or more hormones from the median neurosecretory cells. A hormone inhibits glycogen synthesis about 6–8 h after ingestion of sugar, when less than 4.2 J glycogen has been formed. The MNCp must be present for normal synthesis of triacylglycerol from sugar, but their loss cannot be compensated by reimplantation. In bloodfed mosquitoes glycogen and triacylglycerol synthesis are not regulated by homones from the median neurosecretory cells.

15.1.4 Reserves from sugar feeding

Meals of sugar led to the deposition of glycogen and lipid in the fat body and of glycogen in the flight muscles (Clements, 1955, 1956a; Johnson and Rowley, 1972a). When teneral female *Aedes aegypti* were fed 10% sucrose for 3–5 days the carbohydrate and lipid reserves increased in proportion to body size, the lipids increasing most extensively (Briegel, 1990a). With four species of *Anopheles* access to sucrose for a few days led to increases in lipid deposits amounting to 160–415% of the teneral values and to similar increases in glycogen (Table 15.2) (Briegel, 1990b).

Van Handel (1965b) investigated sugar metabolism in females of *Ae sollicitans* fed glucose:fructose 1:1. Starved females which were fed 13.8 J sugar in a single meal converted about 20% to glycogen and a similar amount to triacylglycerol within a few hours. By four days after feeding the sugar content of the mosquitoes had declined to a very low level, due to energy consumption as well as to the deposition of glycogen and triacylglycerol (Figure 15.3B).

The time course of glycogen and triacylglycerol synthesis from a single meal of sugar was similar in the females of a number of other species, including *Ae aegypti* and *Ae taeniorhynchus*, but in *Psorophora confinnis* the period of synthesis of both types of reserve was more prolonged. In most species the glycogen reserves formed from a single meal did not exceed 4 J whereas the triacylglycerol reserves were greater (Van Handel and Lea, 1965; Nayar and Sauerman, 1975a).

When a female of *Ae sollicitans* was fed only 1.25 J sugar neither glycogen nor fat was synthesized; all the sugar was used to provide energy. A meal of 2.1 J led to slight glycogen deposition; a meal of 4.2 J led to increased glycogen deposition and slight triacylglycerol deposition. Increasing the sugar meal beyond 4.2 J did not greatly increase glycogen deposition but resulted in increased lipid deposition. With increasingly large single meals of sugar the maximum glycogen deposits were always formed within 8–12 h but the lipid deposits continued to increase for up to 6 days. However much sugar was fed in a single meal the glycogen deposits did not exceed 4.2 J, but sugar meals of increasing size led to ever-increasing lipid synthesis; at the extreme, starved females that received a meal of 54.5 J (3.3 mg) sugar laid down deposits of 4.2 J glycogen within a few hours and of 18 J lipid over 6 days (Van Handel, 1965b).

Females of six species of mosquitoes differing widely in size, which had continuous access to 10% sucrose solution, laid down glycogen reserves of 3–10 J per female over a period of 1–3 weeks, and lipid reserves of 12.5–23 J per female over a period of 2–4 weeks. They maintained these reserves for a large part of their lives. Free sugar remained very low, at about 0.8 J per female, in all species. However, *Ae taeniorhynchus* fed 25% sucrose solution, a sugar concentration closer to that in nectar, contained 1.65 J free sugar per female. Assessment of these results in terms of the wet weight of starved females of the same six species showed that the migratory species *Ae taeniorhynchus* developed the most substantial reserves of glycogen and lipid and that *Mansonia titillans* developed the least. *Ae sollicitans* developed substantial reserves of both types. *Ae aegypti* laid down relatively small reserves of glycogen but substantial reserves of lipid. When females of *Ae taeniorhynchus* were given continuous access to 5, 10 or 25% sucrose solution their intake was high during the first week of adult life but then declined to a much lower level (Nayar and Sauerman, 1973, 1974b, 1975a).

Magnarelli (1983) demonstrated the accumulation of energy reserves in wild mosquitoes. 'Available energy reserves' were determined by subtracting from measurements of the total energy

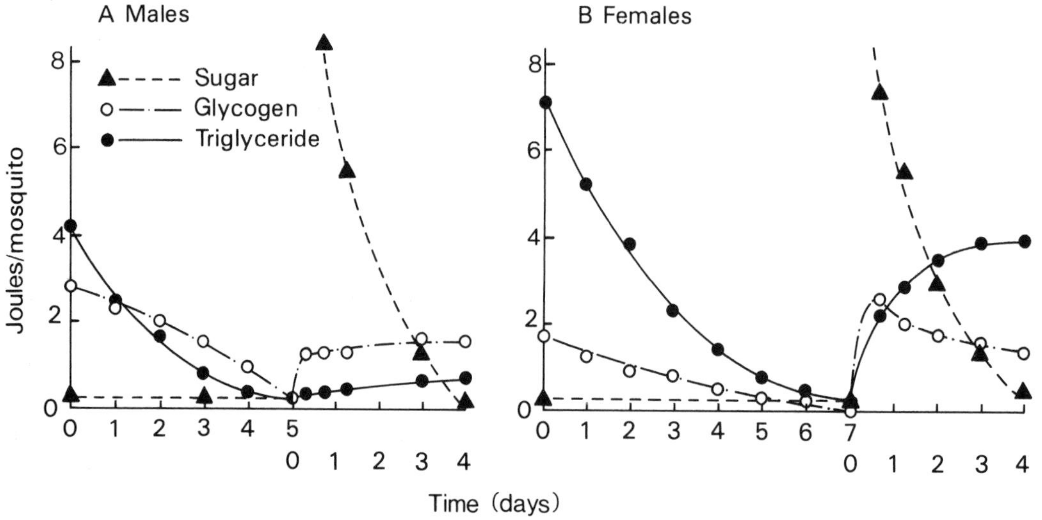

Figure 15.3 The energy values of the mean triacylglycerol, glycogen and sugar contents of (A) male and (B) female *Aedes sollicitans*, given only water for 5 and 7 days respectively after emergence, and then fed a single meal of 13.8 J sugar. Temperature 27°C. (After Van Handel, 1965b.)

value of test individuals the mean irreducible minimum values of starved mosquitoes. In two years the mean available energy reserves were 1.6–2.1 J for newly emerged and 10–15 J for blood-seeking females of *Ae canadensis*. For *Ae stimulans* the values were 3.8–11 J for newly emerged and 28–39 J for blood-seeking females. Day and Van Handel (1986) reported that the glycogen and lipid contents of individual wild females of three species were substantially and significantly lower in field-caught females than in laboratory females of the same species fed 10% sucrose.

15.1.5 Reserves from blood protein

Blood-feeding females of *Aedes sollicitans* on average ingested 13 mg rat blood, which contained 36.8 J protein. During three days this was digested and part of the digestion products converted to glycogen deposits of 2.9 J and triacylglycerol deposits of 6.7 J. Females fed 36.8 J sugar synthesized glycogen much more rapidly than the bloodfed females, but they synthesized triacylglycerols at the same slow rate (Figure 15.4A,B). Castrated females synthesized more triacylglycerol from a blood meal than normal females, presumably because they did not synthesize oocyte protein (Van Handel, 1965b). The conversion of blood-meal protein to extra-ovarian protein and yolk lipid in other species of *Anopheles* is described in Section 22.1.1(b).

A blood meal can replace nectar as an exogenous energy source. Females of *Culex pipiens* that had taken a blood meal and 24 h later been flown to exhaustion were capable of strong flight after another 24-h rest, at which time they contained deposits of glycogen in their fat bodies (Clements, 1955).

15.1.6 Diapausing females

Females that enter diapause may have hypertrophied fat bodies which contain huge lipid reserves. In some species, e.g. *Culex tarsalis* (Schaefer and Miura, 1972), females of the overwintering generation feed only on plant juices during the autumn and synthesize lipid from the ingested sugars. The glycogen content of females of *Cx pipiens* entering diapause is small compared to the lipid content (Sakurai and Makiya, 1981; Onyeka and Boreham, 1987). Diapausing females of *Cx pipiens* have a greater capacity for lipid synthesis than non-diapausing females. The two categories of female contained almost identical amounts of lipid at emergence but after 7 days on a 10% sucrose diet the diapausing females contained almost twice as much lipid as their non-diapausing counterparts (Mitchell and Briegel, 1989b).

In California females of the autumn generation of *Anopheles freeborni* feed on plant juices before entering hibernation, forming lipid reserves which reach their maximum size in October. Between September and January a small percentage of females take blood meals without developing their ovaries. Hibernating females develop extensive lipid reserves whether fed sugar alone or blood alone (Washino, 1970; Washino *et al.*, 1971). In Europe females of *An atroparvus* spend the winter in cattle sheds. They do not develop

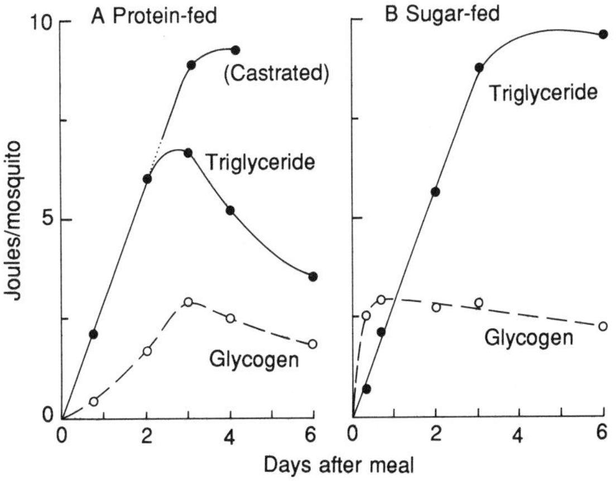

Figure 15.4 Net changes in the glycogen and triglyceride contents of previously starved females of *Aedes sollicitans* fed a single meal of (A) blood or (B) sugar, both having an energy value of 36.8 J. To demonstrate the contribution of protein, the energy values of the small amounts of lipid and carbohydrate in the blood meal were deducted from the curves in (A.) Temperature 27°C. (After Van Handel (1965b) and personal communication.)

large fat bodies through feeding on plant juices in the autumn but take occasional blood meals during the winter, developing their fat bodies to some extent but not sufficiently to withstand a long period of fasting (Swellengrebel, 1929). *Anopheles sacharovi* appears to utilize blood meals in the same way (Mer, 1931). Most diapausing females of *Cx pipiens* that were induced to feed on blood ejected the blood without digesting it (Mitchell and Briegel, 1989a).

In the southern part of the distribution range of *Culiseta inornata* the females aestivate during the hot summer months. In California, at 33–34°C, the blood- and sugar-feeding females develop increasingly hypertrophied fat bodies between April and June before ceasing activity (Barnard and Mulla, 1978). The females showed the same metabolic response when reared from egg to adult under a long-day photoperiod (LD 16:8), synthesizing very substantial lipid reserves (Barnard and Mulla, 1977).

15.1.7 Reserves of males

Van Handel and Lum (1961) claimed that males of *Aedes sollicitans* and *Ae taeniorhynchus*, unlike the females, were incapable of triacylglycerol synthesis from ingested glucose. Van Handel (1965b) reported that newly emerged male *Ae sollicitans* contained 4.2 J triacylglycerol and 2.9 J glycogen. Water-fed males, which became depleted of the glycogen and triacylglycerol carried over from the larval stage, had a poor capacity to synthesize new reserves. Starved males which were fed 13.8 J sugar in a single meal formed only 1.4 J glycogen and 0.67 J triacylglycerol (Figure 15.3A). Even prolonged feeding with concentrated sugar solution produced only small deposits of glycogen and triacylglycerol, consequently an empty crop led fairly quickly to death.

O'Meara and Van Handel (1971) stated that male *Aedes aegypti* fed sucrose solution *ad lib* were incapable of increasing their triacylglycerol deposits above the 55 μg present at emergence, but Briegel (1990a) described a massive lipid synthesis in males of *Ae aegypti* fed 10% sucrose for 2–3 days. The males of *Anopheles stephensi*, *Culex tritaeniorhynchus* and *Cx nigripalpus* could lay down substantial amounts of triacylglycerol (Shahid *et al.*, 1980; Nayar, 1982).

15.2 UTILIZATION OF RESERVES

15.2.1 Utilization at rest

Measurements have been made of the rates of disappearance of triacylglycerols and glycogen from starving mosquitoes kept in batches of 20–50 in containers of 400 cm^3, at 27°C and LD 12:12. The mosquitoes were said to remain still when not disturbed but it seems likely that they showed a circadian activity rhythm, in which case their metabolism would not have been completely representative of resting individuals. In males and females of most species investigated, the rate of depletion of both triacylglycerols and glycogen was exponential, i.e. utilization was proportional to the amount remaining at any time. In newly emerged females of *Ae sollicitans* both triacylglycerols and glycogen were depleted exponentially, but in the males an exponential rate was observed only with triacylglycerols (Figure 15.3). Starving male *Ae sollicitans* had completely depleted both triacylglycerols and glycogen after five days, when death from starvation started to occur. Starving females started to die on the 7th day (Van Handel, 1965b; Nayar and Pierce, 1977). The rate at which reserves are lost is temperature dependent. In *Ae sollicitans* starving from emergence at temperatures between 10 and 30°C, the logarithms of the times in which 50% had died and in which 50% of the energy content had disappeared varied with the reciprocal of the absolute temperature, with Q_{10} values between 2.1 and 2.3 (Van Handel, 1973).

Newly emerged males of *Ae atropalpus* contained 3.0 J triacylglycerol, 0.92 J glycogen and 0.54 J trehalose. At 20°C, over a period of eight days, unfed males utilized the carbohydrate reserves faster than the lipid. Their mean energy consumption was 3.76 J day^{-1}, or approximately 10.5 J h^{-1} g^{-1}. The autogenous females, during the period of egg maturation, utilized 1.0 J day^{-1}, or approximately 16.7 J h^{-1} g^{-1} (Van Handel, 1976).

The metabolic reserves of non-feeding, hibernating females of *Culex pipiens* from southern England were analysed by Onyeka and Boreham (1987) at two-weekly intervals during two winters. The females showed a certain amount of flight activity during the winter. They entered hibernation with a lipid content of some 1.3–1.6 mg, representing about 60% of their dry weight, and between September and March lost about 85% of it. The mean rate of loss was approximately 12% per month, and the form of the loss curve appeared exponential, suggesting that the rate of loss was proportional to the size of the lipid store. The glycogen content of the mosquitoes was much lower than the lipid content, amounting to 0.074 mg per female in early November and 0.054 mg in late January. That indicated a mean rate of glycogen loss of approximately 10% per month. A similar rate of lipid utilization by hibernating *Cx pipiens* has been reported, as well as a reduction in the water content of the mosquitoes (Buxton, 1935; Mitchell and Briegel, 1989a). In the shorter winter of northern California the rates of lipid utilization by a number of overwintering species were similar to or greater than those reported above for *Cx pipiens* (Schaefer and Washino, 1969, 1974: Schaefer *et al.*, 1971).

The nutritive value of a variety of carbohydrates to female *Aedes taeniorhynchus* and *Ae aegypti* was assessed by measuring the survival times of mosquitoes given continuous access to 10% solutions of different compounds. Compounds that gave good survival were:

monosaccharides	D-fructose, D-galactose, D-glucose, D-mannose
disaccharides	maltose, melibiose, sucrose, α-α-trehalose
trisaccharides	melezitose, raffinose
sugar alcohols	D-mannitol, D-sorbitol.

The list includes all sugars previously cited as occurring in nectar (Section 11.1) and also D-sorbitol which occurs in many fruits. *Ae aegypti* rejected certain of these carbohydrates unless they were offered in a mixture with L-sorbose, a phagostimulant which itself had no effect on survival. A number of other carbohydrates, when ingested, failed to extend survival beyond that on water alone (Nayar and Sauerman, 1971a, b), but these are not common in plant juices available to mosquitoes. Broadly similar results were obtained by other investigators who fed 5% sugar solutions to *Aedes aegypti* (Weathersby and Noblet, 1973; Stewart *et al.*, 1979) and *Culiseta inornata* (Salama, 1967), but with poorer survival on D-galactose and melibiose. Females of a number of species which were fed 10% sucrose solution *ad lib.* survived for 5 to 10 weeks, indicating that an appropriate sugar can satisfy all energy requirements and that metabolism continues without undue loss of tissue nitrogen (Nayar and Sauerman, 1975a).

The median neurosecretory cells and corporata allata did not affect the utilization of glycogen or triacylglycerol in 'resting' non-gravid females of *Ae taeniorhynchus*; the reserves disappeared at the same rate from normal, MNCp-ablated, and allatectomized females (Van Handel and Lea, 1965, 1970).

Kurtti *et al.* (1979) used microcalorimetry to measure heat generation and heat uptake by resting *Ae aegypti*. At an ambient temperature of 25°C, the basal metabolic rate (BMR) of newly emerged females was 2.1 × 10^{-6} J s^{-1} mg^{-1} live weight, and that of bloodfed females was 4.6 × 10^{-6} J s^{-1} mg^{-1} (fed) live weight. Thermogram scans of individual females, one to two days old, showed a low but steady BMR and about twelve positive and negative transients, i.e. of cooling and heating character, during the 22 h scan. Over the scan period (+)18.4 mJ of total heat were expended as positive transients, and (−)31.4 mJ as negative or exothermic transients. These total heats would be equivalent to the evaporation of 8 μg water and the metabolism of *c.* 5 μg hexose respectively, although they could not definitely be assigned to those origins. During the first six hours post-bloodmeal the females produced large but infrequent cooling transients, totalling on average (+)18.4 mJ, while the heating transients, due to metabolism, amounted to (−)130 mJ per individual.

15.2.2 Utilization during flight

Unfed mosquitoes, one or two days old, have sufficient energy reserves to fly for several hours on a flight mill. To identify the fuel used in flight, Nayar and Van Handel (1971) measured the amounts of triacylglycerol, glycogen and trehalose in males and females of *Aedes sollicitans* and *Ae taeniorhynchus* that had been flown to exhaustion. Two sets of unflown controls were also analysed, one at the outset and the other at the termination of the flight period; consequently the amount of fuel used for flight could be distinguished from that used during the same period at rest. The consumption of lipid by the flown mosquitoes was no greater than that used by unflown controls, therefore lipid did not provide any energy for flight. The energy for flight came from glycogen. At exhaustion the glycogen reserves had usually been largely consumed, although sometimes the mosquitoes became exhausted before the reserves had been depleted that far. Before flight the mosquitoes contained 10 ± 1 μg (0.17 J) trehalose, and at exhaustion 7–8 μg remained; the trehalose pool therefore made little net contribution to flight energy.

Females of *Ae taeniorhynchus* that were flown not to exhaustion but for a standard period of 4.5 h, and then analysed, showed a substantial depletion of glycogen but no depletion of triacylglycerol and virtually no net loss of sugars as a result of the flight. The mean distance flown during the 4.5 h rose from 4.1 km during the 1st week of life to 6.5–7.0 km in the 3rd to 5th weeks of life. Glycogen depletion during sustained vigorous flight for 4.5 h during the 2nd to 5th weeks amounted to 0.25–0.37 J km^{-1} or 121–167 J h^{-1} g^{-1}. There was a linear relationship between mean flight speed during the 4.5 h flight and metabolic rate as measured by glycogen consumption (Nayar and Sauerman, 1972). Comparative studies on six species which differed in size also showed an approximately linear relationship between the mean flight speed of the different species and glycogen utilization expressed as J h^{-1} g^{-1} (Nayar and Sauerman, 1973).

Male and female *Culex tarsalis*, which were kept under LD 8:16 or LD 16:8 with continuous access to sucrose, showed daily fluctuations in glycogen content. They tended to deplete their glycogen reserves during the early part of the dark phase, the period of active flight, and to build up their glycogen reserves during the later part of the dark phase or during the light phase, the period of rest (Takahashi and Harwood, 1964). Females of *Ae taeniorhynchus*, which were fed 6.3 J D-glucose immediately before a flight of 6 h duration, had increased stores of glycogen and lipid at the end of the flight, showing an ability to synthesize those substances while flying. When females were fed radioactive glucose immediately before flying, the $^{14}CO_2$ collected during flight accounted for a large percentage of the energy produced, indicating that in recently fed mosquitoes crop sugars are the major source of flight energy (Nayar and Van Handel, 1971).

Unfed females of *Cx pipiens* that were flown to exhaustion were able to fly an appreciable distance on the following day (Clements, 1955). Quantitative analyses of *Ae taeniorhynchus* showed that there was no synthesis of glycogen during rest after flight to exhaustion, so the capacity for further flight, which that species also showed, was possibly due to the transfer of residual glycogen to the flight muscles during rest (Nayar and Van Handel, 1971). Blood feeding leads to the synthesis of glycogen, and blood meals have been shown to fuel flight over substantial distances (Clements, 1955; Nayar and Van Handel, 1971).

Birtwisle (1971) measured the concentrations of glycolytic intermediates in the thorax of *Ae aegypti* at rest, but no kinetic studies have been undertaken to show the changes that occur at the start of flight, which would indicate the metabolic pathways that lead to ATP synthesis during flight. However, the presence of glycerol-3-phosphate dehydrogenase (NAD^+) in adult thoraces (Pryor and Ferrell, 1981) is consistent with the postulate that the glycolytic pathway of mosquito flight muscle includes a glycerol phosphate shuttle to oxidize cytosolic NADH.

The indirect flight muscles of *Cx tarsalis* flown to exhaustion exhibited substantial ultrastructural differences from those of unflown controls. After flight to exhaustion the intra-fibrillar deposits of glycogen were depleted and the sarcoplasm contained only sparse granules of glycogen. Moreover, the mitochondria appeared to have broken down; although cristae were evident no outer mitochondrial membranes could be discerned (Johnson and Rowley, 1972b).

16

Adult diuresis, excretion and defaecation

This chapter is concerned with the processes that rid the adult body of unwanted substances and ensure homeostasis of the haemolymph under conditions of water and salt loading. The diuretic and excretory processes of adult mosquitoes, like those of the larvae, are variants of the mechanisms that have evolved in insects in general and which we shall briefly review. Insects have a two-part excretory system in which an initial secretion of the Malpighian tubules (tubular fluid), which is isosmotic with the haemolymph, is modified in the rectum before being expelled as urine. Fluid secretion by the Malpighian tubules is driven by the active transport of potassium ions across the epithelium into the tubule lumen. Chloride ions follow passively, and potassium chloride is normally the predominant solute in the tubular fluid. The concentrations of other solutes, including sodium, sugars and amino acids, are low relative to their concentrations in the haemolymph. The tubular fluid enters the pylorus from where some may be moved forwards for resorption in the midgut. The remainder is moved posteriorly to the rectum where it is modified by selective absorption or secretion of water, ions and metabolites, producing urine which may be hyper- or hypo-osmotic to the haemolymph. The volume of fluid excreted is determined by endocrine control of both Malpighian tubule secretion and rectal absorption, different hormones inducing diuretic and antidiuretic activity (Spring, 1990).

Discharge of excretory products may give the impression of two distinct activities, the rapid production of urine containing principally water and salts (diuresis) and the voiding of semi-solid nitrogenous wastes, but these two activities involve related or interactive physiological processes. Through the capacity of the hindgut to resorb water, excretion can be accomplished with a minimum of water loss, permitting water conservation when required. Equally, the Malpighian tubule–hindgut system permits diuresis on occasions of water load and regulates the volume and composition of the extracellular fluid at all times. Defaecation of undigested food residues and other discarded products of the midgut provides a third eliminatory process.

The disposal of waste or surplus substances by adult mosquitoes occurs most conspicuously after emergence and after each blood meal, and nectar feeding also is followed by the disposal of water. After emergence the teneral mosquito voids waste substances that accumulated during metamorphosis and discharges water and salts to reduce its haemolymph volume, which is still extensive like that of the developmental stages, to the lower volume characteristic of the adult stage. Once this fluid has been discharged the remaining body water is carefully conserved.

The large volumes of blood ingested by female mosquitoes not only make flight difficult but also produce water- and salt-loads which threaten the homeostasis of the haemolymph. The excretory system has the capacity to discharge over 40% of the water and sodium contained in the ingested blood plasma within one hour of feeding. During the first day after feeding deamination of amino acids produces excess nitrogen which is voided in the form of low molecular weight nitrogenous metabolites. Defaecation occurs later.

Authors have not been mutually consistent in the use of terms associated with excretion and diuresis, so it is necessary to define these terms as they are used here.

1. Tubular fluid – the unmodified secretion of the Malpighian tubules.
2. Urine – the completed excretory fluid as discharged to the exterior.
3. Excreta – nitrogenous waste including uric acid, urea and ammonium ion, usually discharged in a semi-solid state but sometimes in the urine.
4. Faeces – undigested food residues and other discarded products of the midgut.
5. Meconium – nitrogenous and other waste products remaining in the gut from the period of metamorphosis.

The term 'excretion' is used here in two senses: broadly, for all eliminatory processes apart from defaecation, and more narrowly for the elimination of nitrogenous compounds.

16.1 DIURESIS AFTER EMERGENCE AND AFTER FEEDING

Insect epithelia that are adapted for the passage of water and ions characteristically have a microvillate apical cell membrane, a deeply infolded basolateral cell membrane, and numerous mitochondria which are associated with both the microvilli and the basal labyrinth. Such an ultrastructure is found in the midgut, the Malpighian tubules, the anterior intestine and the rectal papillae of adult mosquitoes (Sections 13.2, 13.3). At metamorphosis the rectum of the larval mosquito is anatomically transformed to that of the adult while the Malpighian tubules pass to the adult without cell loss or visible change. However, the secretory abilities of the adult tubules differ from those of the larval tubules.

The high osmotic permeability of the Malpighian tubule cell membranes possibly allows solute and water flows to be coupled by osmosis. Thus the movement of water from the haemolymph and across the tubules may be a passive response to osmotic gradients produced in the cells and in the lumen by the active transport of potassium ions which is known to occur. The tubules secrete a potassium-rich fluid at a rate that is sensitive to the potassium concentration of the medium. This system is undoubtedly supplemented by other mechanisms of active transport. There is evidence also of a slow paracellular movement of larger molecules and charged molecules through the intercellular junctions (O'Donnell and Maddrell, 1983; O'Donnell *et al.*, 1984).

Bloodfed mosquitoes need to rid themselves of the high sodium load gained from the ingested plasma and the potassium gained in the blood cells. Adult mosquitoes differ from the majority of insects and from their own larvae in that, in the absence of secretagogues, their Malpighian tubules actively transport both sodium and potassium at similar rates. Under conditions of rapid diuresis the excretory mechanism changes to one based on sodium secretion.

Diuretic hormone activity has been demonstrated in a number of insects, and there is evidence that diuretic hormones are peptides originating in the central nervous system. Mosquitoes release a diuretic hormone during blood feeding. This is thought to stimulate the Malpighian tubule cells to produce cAMP which acts as a second messenger, inducing further changes which produce fluid flow across the tubular epithelium.

The phenomenon of diuresis in bloodfed mosquitoes resembles that in another bloodsucking insect, *Rhodnius prolixus*. The fluid secretion rates of *Rhodnius* Malpighian tubules are said to be the highest reported for any animal (Florey, 1982), but the peak rates of urine flow exhibited by mosquitoes imply similar tubular secretion rates. Detectable amounts of diuretic hormone are found in the haemolymph of individuals of *Rhodnius* that have been imbibing blood for as little as 15 s (Maddrell and Gardiner, 1976). The speedy diuretic response of mosquitoes to blood feeding suggests that hormone release may be equally rapid in them.

Most studies of diuresis in adult mosquitoes have been on *Aedes aegypti*; in the following

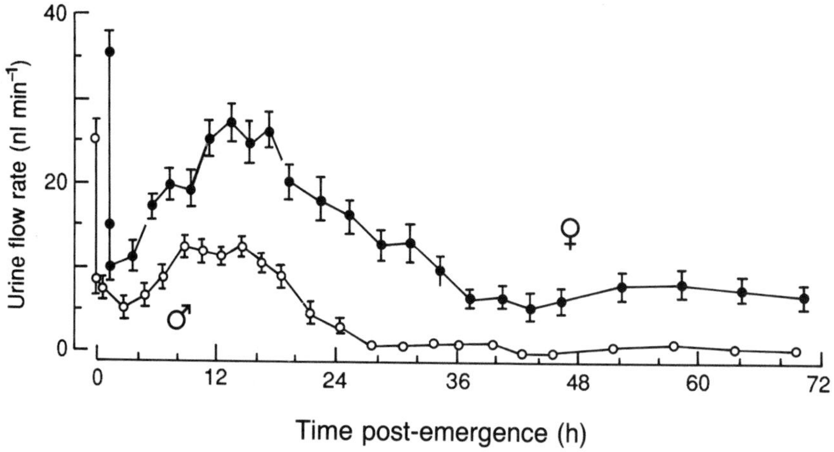

Figure 16.1 Mean rates of urine flow in unfed male and female *Aedes aegypti* during the first three days of adult life. The data points for males and females are shifted slightly to the left and right respectively. (From Gillett, 1983a.)

sections, where no species is named, *Ae aegypti* is the species concerned.

16.1.1 Diuresis after emergence

The teneral mosquito, newly emerged from its pupal skin, is disadvantaged in flight by the weight of haemolymph carried over from its juvenile stages, and discharge of droplets of clear fluid from the anus is the first observable action it performs. The droplets vary in diameter, and an individual may expel a few large drops or many small drops during equal time periods. By counting numbers of droplets and classifying each as large, medium or small (mean measured volumes 34, 25 and 13 nl respectively), it was possible to determine rates of urine production (DeGuire and Fraenkel, 1973; Jones and Brandt, 1981; Gillett, 1982b).

Most *Aedes aegypti* ejected a droplet just before or at the moment the abdomen was freed from the pupal cuticle, and immediately afterwards they discharged urine at a high rate, females at >40 nl min^{-1} and males >25 nl min^{-1}. However, this post-emergence burst of urine flow was short-lived and declined sharply within 20 min (Figure 16.1). From about 3–4 h post-emergence the rate of urine flow started to increase again. In females it increased slowly to a maximum >25 nl min^{-1} by 12–14 h post-emergence, and declined from 15 to 16 h onwards. In males urine flow restarted at the same time but ended earlier; further, the flow rates were lower. After completion of the 'day-1-peak' the rate of urine flow stabilized at 6.9 ± 0.5 nl min^{-1} in unfed females and at 0.6 ± 0.1 nl min^{-1} in unfed males (Gillett, 1983a).

Male and female *Anopheles gambiae* exhibited a post-eclosion burst of diuresis, which peaked between 1 and 5 min after the adult left the pupal skin. In most individuals the burst ended within 12 min, by which time the abdomen had lost its distended appearance (Goma, 1964).

Decapitation within 1 min of leaving the pupal cuticle reduced or eliminated the post-emergence burst of diuresis in *Ae aegypti* (Jones and Brandt, 1981). Decapitation at any time from 1 to 14 h post-emergence greatly reduced urine flow in the day-1-peak, the flow rate varying slightly with time of decapitation (Gillett, 1983a).

16.1.2 Diuresis after feeding

Meals of nectar and other plant juices are stored in the crop and their water content is discharged slowly. Sugar solution imbibed by *Ae aegypti* was first seen in the midgut about 30 minutes after the crop had filled, and females that drank 10% sugar solution discharged no fluid or at most a few droplets during the first 30 min after feeding (Jones and Madhukar, 1976; Jones and Brandt, 1981). Sugarfed females of *Ae taeniorhynchus* lost weight at a slow constant rate which averaged 2.2–2.7 μg min^{-1} during the first 50 min and 1.7 μg min^{-1} between 50 and 90 min after feeding (Nayar and Bradley, 1987). This was probably greater than their transpiration rate since the transpiration rate of female *Ae aegypti* with sealed anus was 0.55 μg min^{-1} (Stobbart, 1977).

Like many mosquitoes, females of *Aedes aegypti* that feed to repletion ingest more than their own weight in blood. Diuresis often commences some 50–75 s after the start of feeding, and up to ten or more droplets of urine may be passed before feeding has ended. Extrusion of a droplet of urine, which takes a fraction of a second, involves a tilting upwards and telescoping of the terminal abdominal segments, rectal peristalsis, splaying of the cerci, and opening of the anus. The anus closes suddenly upon completion of the rectal peristalsis, segments 7 and 8 move slightly, and the droplet is forcefully ejected, often over 5–10 mm. Strains of *Aedes aegypti* differ in the rate of production and uniformity of size of the excretory droplets. The fluid discharged by bloodfed *Ae aegypti* must have passed through the Malpighian tubules since trypan blue or cerium–144, if present in the ingested blood, do not appear in the urine. The first one or two drops of urine may be cloudy, and these contain uric acid; the remainder are clear and free of uric acid but contain ninhydrin-positive substances (Boorman, 1960; Redington and Hockmeyer, 1976; Stobbart, 1977; Jones and Brandt, 1981; Mellink *et al.*, 1982).

Rates of urine production by bloodfed mosquitoes have been measured by counting the number of droplets discharged. Experiments with one strain of *Ae aegypti* showed droplet volume to be consistently within the range 10–12 nl, so urine production could be calculated directly from droplet numbers. Urine production passed through three phases which differed in rate of flow: (i) a **peak phase** of very rapid flow in the first few minutes after the blood meal, (ii) a **post-peak phase** of declining flow rate, and (iii) a **late phase** when the flow rate was slightly above the control level. In one female the peak phase of diuresis lasted a little over ten minutes, with the flow rate reaching a maximum of 54 nl min^{-1} six minutes after the start of blood feeding. The post-peak phase lasted about 60 min, and by the end of it the urine flow rate had fallen to about 2–6% of that at the peak (Figure 16.2, Table 16.1B) (Williams *et al.*, 1983; Petzel *et al.*, 1987).

Urine production has also been recorded by weighing females at short intervals after the blood meal. One gravimetric measurement of urine production by *Ae aegypti* revealed the same three phases of urine production, but with the peak phase lasting 25–35 min. By some 60–120 min post-bloodmeal the weight losses could be accounted for by transpiration (Stobbart, 1977). Whenever the droplet method has been used it has indicated a short initial peak (Boorman, 1960; Jones and Brandt, 1981; Williams *et al.*, 1983), but strain differences cannot be ruled out as the cause of this difference.

Comparisons of the osmolality and the sodium, potassium and chloride concentrations of *Ae aegypti* haemolymph with those of human plasma (Table 16.1A) provide a measure of the immediate osmotic and ionic loads imposed on the

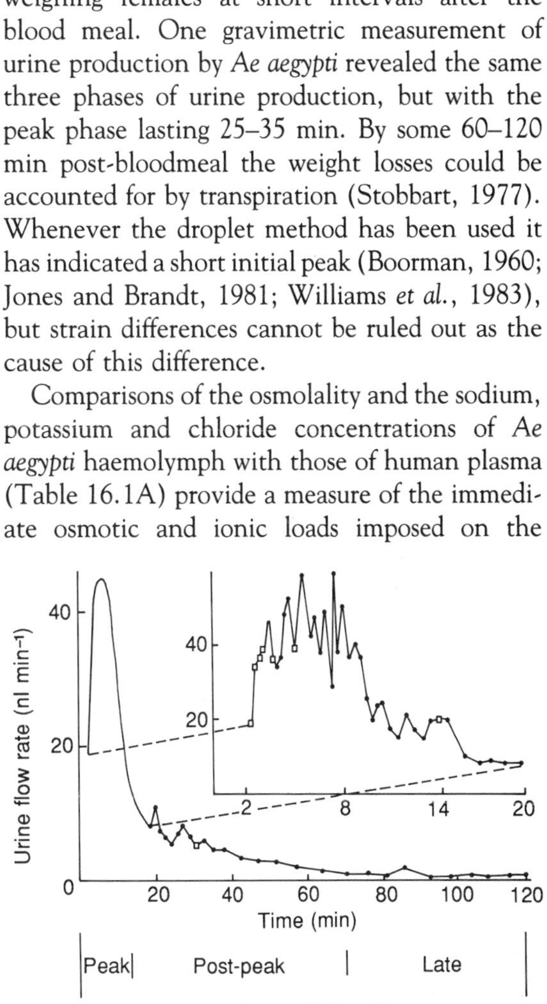

Figure 16.2 Urine flow rate in a single female *Aedes aegypti* during the first two hours after feeding to repletion on human blood, the blood meal commencing at time 0. Inset, an expansion of the time scale for the peak phase of the diuresis. (From Williams *et al.*, 1983.)

Table 16.1 Composition of the haemolymph and urine of *Aedes aegypti* and of human blood plasma. (From Williams *et al.*, 1983.)

(A) Composition of haemolymph from non-bloodfed mosquitoes (means ± s.e.) and of clinical ranges of human plasma.

Parameter	*Haemolymph*	*Human plasma*
Osmolality (mosmol kg^{-1})	354 ± 14	280–295
[Na^+] (mM)	96 ± 7	136–146
[K^+] (mM)	6.5 ± 1.0	3.5–5.5
[Cl^-] (mM)	135 ± 19	96–106

(B) Urine flow rate and composition at selected times during diuresis following a meal of human blood by *Aedes aegypti* (means ± s.e.).

Parameter	*Phase of diuresis*		
	Peak	*Early post-peak*	*Late post-peak*
Time post-bloodmeal (min)	6	26	61
Flow rate (nl min^{-1})	54.4 ± 5.2	11.1 ± 4.3	2.6 ± 0.5
Osmolality (mosmol kg^{-1})	309 ± 7	217 ± 16	298 ± 41
[Na^+] (mM)	175 ± 8	132 ± 16	106 ± 23
[K^+] (mM)	4.2 ± 0.4	16 ± 7	59 ± 16
[Cl^-] (mM)	132 ± 11	88 ± 9	177 ± 22

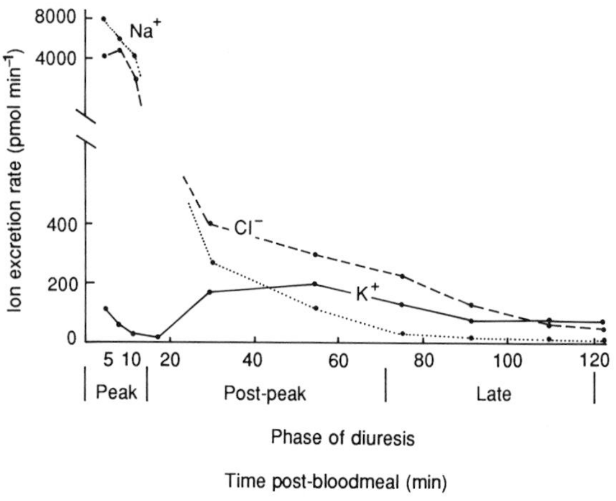

Figure 16.3 Ion excretion rates as a function of time after the start of a blood meal taken by a female *Aedes aegypti* on a human host. Assayed in urine samples collected from the individual used in Figure 16.2. The first Cl^- datum point is probably too low due to an artefact in the assay (K.W. Beyenbach, personal communication.) (After Williams *et al.*, 1983.)

bloodfed females. The osmolality of human plasma was significantly lower than that of the haemolymph, its sodium concentration was significantly higher. Williams *et al.* (1983) measured the rates of water and ion excretion of bloodfed females of *Ae aegypti*, and their observations provide an insight into the mosquito's homeostatic capabilities. Energy dispersive spectra showed Na, K and Cl to be the only elements in urine with atomic weights >20. The osmolality of the urine could be accounted for largely by the salts of sodium and potassium. These were mostly chlorides but in many urine samples the chloride concentration was less than the sum of the sodium and potassium concentrations; in one experiment the mean anion concentration deficit amounted to 47 mM during the peak phase and 60 mM during the early post-peak phase (Table 16.1B).

The three phases of diuresis were characterized not only by flow rate but also by the ionic composition and osmolality of the urine. The peak phase was characterized by rapid excretion of a urine high in sodium, low in potassium, and with an osmolality slightly below that of haemolymph (Figure 16.3, Table 16.1). Williams *et al.* (1983) postulated that the high rate of urine flow probably prevented the hindgut from modifying the composition of its contents to any great extent, so that the urine would not differ much from the tubular fluid. Potassium excretion increased during the post-peak phase, when its concentration in urine exceeded that in human plasma (Figure 16.3, Table 16.1). This suggests that the erythrocyte cell membranes become leaky within 30 min of ingestion.

The late phase of diuresis was characterized by low rates of urine flow, a relatively high potassium excretion rate, and urine of variable osmolality. The very low urine osmolalities (<100 mosmol kg^{-1}) that were sometimes observed, and the low flow rate, suggest that fluid remained in the hindgut for relatively long periods of time. By the time that experiments were terminated, 120 min post-bloodmeal, the urine had usually become isosmotic with the haemolymph again.

Urine sodium concentration and flow rate were positively correlated; urine potassium concentration and flow rate were negatively correlated but with a strongly non-linear relationship. Potassium concentrations were low when flow rates exceeded 20 nl min^{-1} and were variable at lower flow rates, sometimes exceeding 100 mM. The urine chloride concentration showed no significant correlation with flow rate. Urine flow rate and osmolality were positively correlated. During the peak phase urine osmolality was marginally higher than that of human plasma, and slightly below the mean haemolymph osmolality but within its range (300–400 mosmol kg^{-1}). As flow rates declined the urine became increasingly hypo-osmotic to the haemolymph, falling below 100 mosmol kg^{-1} in some cases. Excretion of 'free water' occurred during the post-peak phase, when the urine was hypo-osmotic to the ingested plasma as well as to the haemolymph. When the flow rates became very low the urine became isosmotic with the haemolymph again.

In Table 16.2 the volume and electrolyte content of the blood meal are broken down into the loads contained in plasma and erythrocytes, which are dealt with by the mosquito at different times after ingestion of the blood meal, and compared with the volume and electrolyte loads excreted in urine. The excreted volume represented only 23% of the total ingested load but 42% of the ingested plasma load. Some 80% of volume reduction occurred during the first 20 min after feeding. The amount of potassium excreted exceeded the plasma potassium load. The experimental mosquitoes were estimated to contain 0.6 μl haemolymph. More than 0.6 μl of urine was excreted during the first 20 min after feeding, therefore a volume at least as great must have been absorbed from the blood meal during this period. The ultrastructure of the midgut epithelium is consonant with a transport function. During the first hour after blood feeding changes occur in the

Table 16.2 Mean volumes and electrolyte loads in the meal of human blood taken by females of *Aedes aegypti* and in the urine collected during 102 ± 10 min from the beginning of the meal. (From Williams *et al.*, 1983.)

Parameter	*Ingested load*			*Excreted load*	
	Total	*Erythrocytes*	*Plasma*	*Urine*	*% of ingested plasma load*
Volume (μl)	3.5	1.6	1.9	0.8	42
[Na$^+$] (nmol)	284	14	270	120	44
[K$^+$] (nmol)	149	140	9	13	144
[Cl$^-$] (nmol)	300	110	190	111	58

ultrastructure of the rectal papillae, notably a swelling of the terminal regions of the basal labyrinth. Ultrastructural changes continue up to 12 h post-bloodmeal and the altered state persists for another 24 h at least (Hopkins, 1967).

The capacity for fluid secretion *in vitro* by tubules of *Ae taeniorhynchus*, in response to 5-HT and cAMP, was lower in tubules taken from adults than in those from larvae. Tubules dissected from 3-day-old females immediately after blood feeding showed a significantly greater transport capacity than tubules from sugarfed females of the same age. This effect of blood feeding persisted in 4-day-old females. Unlike the situation in some other insects, the increased capacity for fluid secretion was not associated with ultrastructural reorganization of the microvillate cell membrane (Bradley and Snyder, 1989).

Anopheles freeborni showed a striking capacity for post-bloodmeal diuresis. Diuresis started immediately after cessation of feeding and continued at a constant rate, between 35 and 45 nl min^{-1} depending on the female, for almost 30 min. Diuresis ended by 45–50 min after the cessation of feeding. Females that took small blood meals discharged urine over the same time period but at a lower rate, suggesting that the rate of diuresis reflects the rate at which water and ions pass into the haemolymph from the gut. Fluid transfer from midgut to haemolymph continued for some time after diuresis had ended; the midgut continued to shrink while the volume of haemolymph around it increased proportionately (Nijhout and Carrow, 1978).

16.2 MECHANISMS OF TUBULAR FLUID PRODUCTION

The excretory system of mosquitoes must meet complex requirements. The blood-feeding habit necessitates the rapid excretion of water and sodium ions without loss of potassium immediately after feeding. During the course of digestion, potassium ions must be eliminated at a regular slow rate. In the absence of fluid intake, salts and water must be circulated internally without loss.

16.2.1 Methods

Two types of experimental preparation have been used to investigate the properties of adult female *Aedes aegypti* Malpighian tubules – non-perfused and perfused. In the non-perfused Ramsay preparation, a tubule was severed from the alimentary canal and placed in a 50 μl drop of saline which was under oil. The open end of the tubule was pulled into the surrounding oil so that secreted fluid that flowed from the open end accumulated in the oil as an aqueous droplet (Figure 16.4A). The dimensions of the secreted droplets were measured optically; their contents could be analysed by electron-probe microdroplet techniques and wavelength dispersive spectroscopy. Test substances could be added to the bathing saline and their effects on fluid and ion secretion rates measured (Williams and Beyenbach, 1983).

The electrical properties of the Ramsay preparation could be investigated by use of a microelectrode inserted into a principal cell or inserted through cells into the tubule lumen, and with a reference electrode in the drop of bathing saline. Penetration to the tubule lumen permitted measurement of the transepithelial electrical potential difference or voltage (V_t). Impaling a primary cell permitted measurement of the basolateral membrane potential (V_{bl}). The apical membrane potential (V_a) could be estimated from the difference between V_t and V_{bl}. Membrane conductances could be evaluated from the changes of V_{bl} following changes in the bathing saline (Sawyer and Beyenbach, 1985: Aneshansley *et al.*, 1988).

In a modification of the non-perfused preparation, isolated tubules were fixed to the wax bottom of a saline flow-bath by means of their tracheal attachments. The primary cells were impaled with a microelectrode to measure membrane potentials and the microelectrode was advanced into the tubule lumen to measure V_t (Sawyer and Beyenbach, 1985).

Perfused preparations were set up by first cutting a segment 0.5–2 mm long from a Malpighian tubule, drawing its ends into the mouths of glass holding pipettes, and submerging the whole in a saline bath of 0.3 ml capacity (Figure 16.4B). One end of the tubule segment was cannulated with a perfusion pipette (tip diameter 10–15 μm) which contained a silver wire electrode. The other holding pipette, into which perfused fluid emerged, served as a collection pipette. Alternatively the tubule segment could be cannulated with a double-barrel pipette; the tubule lumen would be perfused with saline from one barrel which also served to measure voltage, and current could be injected into the tubule lumen through the other barrel. The salines that bathed and perfused the tubule were usually identical, but test substances could be introduced into the bath saline or perfusate.

The transepithelial voltage (V_t) of perfused tubule preparations was measured directly between electrodes in contact with the tubule contents and the bath saline. The transepithelial electrical resistance (R_t) was measured by cable analysis after the injection of short pulses of current. Basolateral membrane voltage could be measured with conventional microelectrodes inserted into the principal cells. The fractional resistance of the basolateral membrane of tubule primary cells ($f_{R_{bl}}$), *defined as the ratio of the basolateral membrane resistance over the sum of the basolateral* (R_{bl}) and apical membrane resistances (R_a), was determined from the voltage deflection across the basolateral membrane, measured with an intracellular microelectrode after a transepithelial current pulse. Thus,

$$f_{R_{bl}} = \frac{\Delta V_{bl}}{\Delta V_t} = \frac{\Delta V_{bl}}{\Delta V_{bl} + \Delta V_a} = \frac{R_{bl}}{R_{bl} + R_a} \qquad (16.1)$$

The electrical characteristics of perfused tubule preparations could be measured when the tubules were perfused with bathing saline, i.e. in the absence of transepithelial ion gradients when all transepithelial voltages arose from the transport activities of the epithelial cells (Williams and Beyenbach, 1984: Petzel *et al.*, 1985: Aneshansley *et al.*, 1988; Hegarty *et al.*, 1991).

16.2.2 Characteristics of spontaneously secreting tubules

Malpighian tubules of *Aedes aegypti* secreted fluid spontaneously as soon as they had been transferred from the mosquito into a saline droplet

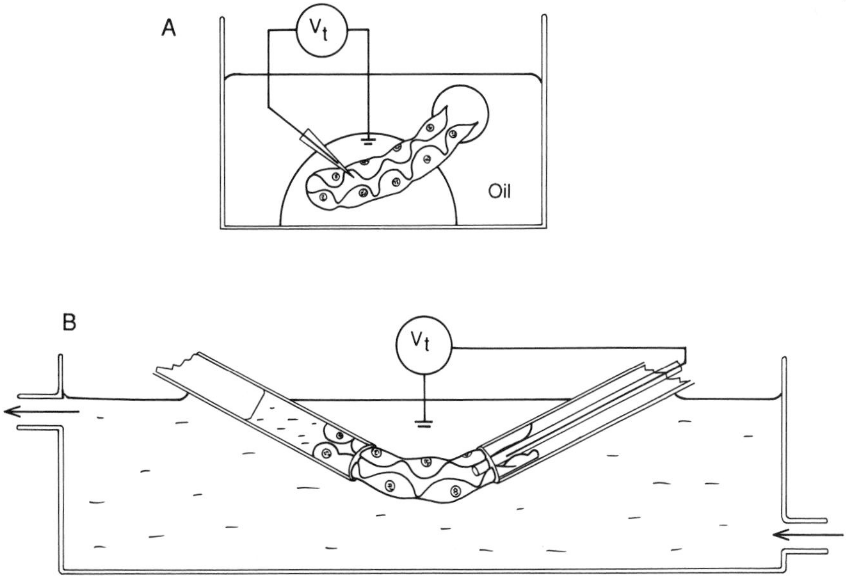

Figure 16.4 Two Malpighian tubule preparations. (A) Diagram of a Ramsay preparation, showing also penetration of the tubule with a microelectrode for measurement of the potential difference between the tubule lumen and the bathing saline. (B) Diagram of a perfused preparation. (After Aneshansley *et al.*, 1988.)

under oil. The rate of fluid secretion varied between tubules, but for any tubule the rate of secretion was constant for 5 h or more. This variation apart, no differences in electrical or secretory properties have been found between the five Malpighian tubules of individual *Ae aegypti*, consistent with their anatomical similarity. Under control conditions with Ramsay preparations, the mean rate of spontaneous fluid secretion by single tubules from adult females varied from 0.65 to 0.79 nl min^{-1}. The secreted fluid was nearly isosmotic with the bathing saline and consisted of rather similar concentrations of NaCl and KCl. Its $[Na^+]$ represented 50–60% of that of the bathing saline. The $[K^+]$ was some 30–40 times greater than, and the $[Cl^-]$ about the same as, that of the saline (Table 16.3). Na^+, K^+ and Cl^- accounted for over 90% of the measured osmolality (Williams and Beyenbach, 1983; Petzel *et al.*, 1985; Hegarty *et al.*, 1991). Such an isosmotic high-potassium fluid is secreted by most isolated insect Malpighian tubules.

Measurements of transepithelial and transmembrane voltages and resistances have been made using perfused *Ae aegypti* tubule preparations exposed to symmetrical saline solutions. Tubules consistently generated lumen-positive transepithelial voltages, ranging from *c.* 25 to 90 mV. Values of the apical membrane voltage were consistently higher than those of the basolateral membrane. Transepithelial resistance ranged between *c.* 4 and 24 kΩ.cm. Under control conditions, the values of the fractional resistance of the basolateral membrane ($f_{R_{bl}}$) and the apical membrane ($1\text{-}f_{R_{bl}}$) were usually similar (Table 16.4) (Hegarty *et al.*, 1991). In one series of experiments a mean transepithelial voltage (V_t) of 30 mV (lumen positive) was recorded. The mean basolateral membrane voltage was −77 mV and the mean apical membrane voltage −107 mV (both cell negative). The mean transepithelial resistance (R_t) was 14.9 kΩ.cm (Petzel *et al.*, 1987). Measurement of V_t in non-perfused preparations, in which the tubule lumens contained secretory fluid, revealed a positive voltage similar to that in symmetrically-perfused tubules (Sawyer and Beyenbach, 1985). The transepithelial voltage was rapidly inhibited by dinitrophenol, consistent with its being due to active transport. Williams and Beyenbach (1984) concluded, from the value of R_t and the ability of the tubules to generate transepithelial concentration gradients (Table 16.3), that the *Ae aegypti* Malpighian

Table 16.3 Electrochemical characteristics of Malpighian tubule preparations in the absence and presence of 1 mM dibutyryl cAMP, derived from fluid composition data from Ramsay preparations and electrical measurements on perfused preparations. Transepithelial potential difference (V_t). Transepithelial chemical potential difference ($\Delta\mu$). Transepithelial electrochemical potential difference ($\Delta\tilde{\mu}$) calculated as the sum of V_t and $\Delta\mu$. (From Williams and Beyenbach, 1984.)

Treatment	*Ion*	*Concentration (mM)*		V_t	$\Delta\mu$	$\Delta\tilde{\mu}$
		Bath	*Lumen*	*(mV)*	*(mV)*	*(mV)*
Control	Na^+	159	94	+53*	−13†	+ 40*
	K^+	3	91		+87*	+140*
	Cl^-	157	161		− 1*	+ 52†
cAMP	Na^+	159	178	+99*	+ 3*	+102*
	K^+	3	17		+44*	+143*
	Cl^-	157	185		− 4*	+ 95†

* Potential difference opposes secretion of the ion.
† Potential difference favours secretion of the ion.

Table 16.4 Electrophysiological characteristics of perfused *Aedes aegypti* Malpighian tubule preparations bathed and perfused with identical salines, in the absence of secretagogues (controls) and when bathed in 10^{-4} M dibutyryl cAMP in the absence or presence of 10^{-4} M bumetanide. (From Hegarty *et al.*, 1991.)

Parameter	*Control*	*cAMP*	*p*	*Control*	*cAMP + bumetanide*	*p*
Basolateral membrane						
Fractional resistance	0.56	0.32	<0.001	0.53	0.42	<0.05
Voltage (mV)*	–62.9	–33.9	<0.001	–55.2	–23.0	<0.001
Apical membrane						
Voltage (mV)*	110.5	105.6	n.s.	92.7	90.3	n.s.
Transepithelial						
Resistance (kΩ.cm)	14.3	9.8	<0.01	9.8	7.9	<0.01
Voltage (mV)†	47.6	71.8	<0.05	37.5	67.2	<0.001

*Cell negative.
†Lumen positive.
n.s., not significant

tubule should be classified as a moderately tight epithelium.

The passive permeabilities of tubules were investigated by measurements of transepithelial diffusion potentials, observed following changes of ion concentration. Lowering the bath sodium concentration from 159 to 9 mM caused a drop of 23 ± 6 mV in V_t, whereas lowering the luminal sodium concentration had no significant effect. Equally substantial changes to bath or lumen chloride concentrations had no effect on transepithelial voltage. Williams and Beyenbach (1984) concluded that bath [Na^+] was important for the maintenance of V_t.

Conductances of the basolateral membrane were investigated by measuring the effects on basolateral membrane voltage (V_{bl}) of changes in the ion concentrations of the bathing saline. A 5-fold decrease in bath [Na^+] hyperpolarized the membrane (cell negative) by 10 mV whereas a 4.4-fold increase in bath [K^+] depolarized it by 8 mV. The study revealed that the basolateral membrane was permeable to Na^+ and K^+ with both conductances of similar magnitude, consistent with the similar Na^+ and K^+ secretion rates measured under control conditions. The intracellular electrical potential appeared to be dominated by transmembrane diffusion potentials for K^+, judging by the negative V_{bl} and the depolarization induced by raised bath [K^+] (Sawyer and Beyenbach, 1985).

Only little is known about the apical membrane. Experimental evidence consistent with the presence of an electrogenic pump, presumably a H^+-ATPase, in the apical membrane of the principal cells of *Ae aegypti* tubules was reported by Beyenbach and Pannabecker (1991) and Pannabecker *et al.* (1992). Evidence consistent with the presence of chloride channels in the apical membrane was obtained with patch-clamp methods. The putative chloride channels had 25 pS conductance in symmetrical salines (160 mM-Cl^-). With asymmetrical salines the reversal potentials of the single channel current suggested that the channels were highly selective for Cl^- (Wright and Beyenbach, 1987).

The fractional resistance of the basolateral membrane ($f_{R_{bl}}$), i.e. the ratio of the basolateral membrane resistance to the total transcellular resistance, was 0.63 ± 0.11. This value of 63% indicated that under control conditions the conductance of the apical membrane was greater than that of the basolateral membrane. Because the fractional resistance was a function of membrane specific resistance and membrane surface area this was not a surprising result given the microvillate nature of the apical membrane (Petzel *et al.*, 1987).

Williams and Beyenbach (1984) calculated the transepithelial electrochemical potential differences ($\Delta\tilde{\mu}$) for Na^+, K^+ and Cl^- as the sum of the transepithelial voltage (V_t) and the transepithelial chemical potential difference ($\Delta\mu$) (Table 16.3). From the values of $\Delta\tilde{\mu}$ they concluded that in the absence of secretagogues both sodium and potassium were secreted against their respective electrochemical gradients, whereas chloride movement could be explained by passive secretion. It was not possible to estimate the rate of chloride secretion that could be driven by V_t because the magnitude of the chloride conductance was not known.

Generally, epithelial tissues that function in isosmotic water and ion transport have the ultrastructure and electrochemical properties characteristic of 'leaky epithelia', i.e. a microvillate apical cell membrane and infolded basal cell membrane, low transepithelial voltage and resistance, small transepithelial ionic and osmotic gradients, and high permeability to solutes and water. The Malpighian tubules of *Aedes aegypti* are remarkable in combining the ultrastructural characteristics and capacity for high rates of salt and water transport of a 'leaky epithelium' with the electrochemical characteristics of a moderately 'tight epithelium', i.e. high transepithelial voltage, moderate transepithelial electrical resistance, and significant transepithelial sodium and potassium gradients. On balance the mosquito Malpighian tubule is foremost a transporting epithelium; it secretes Na^+ and K^+ by active transport across an epithelium which is nearly impermeable to the electrodiffusion of Na^+, K^+ and Cl^- but highly permeable to water (Maddrell, 1980; Aneshansley *et al.*, 1988).

Studies with *Aedes taeniorhynchus* showed that the secretory characteristics of the adult Malpighian tubules differ from those of the larvae. In Ramsay preparations the sodium concentration of the tubular fluid was positively correlated with that of the bathing medium (Figure 16.5). However, the tubular fluid secreted by adult tubules was markedly richer in sodium than that secreted by larval tubules, although the larvae have a striking capacity for sodium excretion, being adapted to the saline waters of salt marshes. Adult female tubules secreted five times faster than the smaller male tubules, and female tubular fluid was significantly richer in sodium than that of males. Larvae of *Ae taeniorhynchus* reared in sulphate-enriched water had a striking ability to secrete sulphate at very high concentration (Section 6.5.3(a)), but this ability was absent from male and female adults derived from such larvae (Maddrell, 1977; Maddrell and Phillips, 1978).

The rate of fluid secretion by tubules of adult *Ae taeniorhynchus* was sensitive to potassium concentration. When the bathing saline [K^+] was 5–10 mM (said to be the concentration in haemolymph) adult female tubules secreted 1–2 nl fluid per minute, but in 44 mM-K^+ the secretion rate was 5 nl min^{-1}. Maddrell (1977) thought this inappropriate since, under the conditions obtaining in haemolymph, fluid secretion would be considerably below the maximum rate; however, the measurements were made in the absence of excretagogues.

16.2.3 Responses to cyclic AMP

The functional importance of cAMP in *Aedes aegypti* Malpighian tubules was shown by the striking effects of dibutyryl cyclic AMP (db-cAMP), a

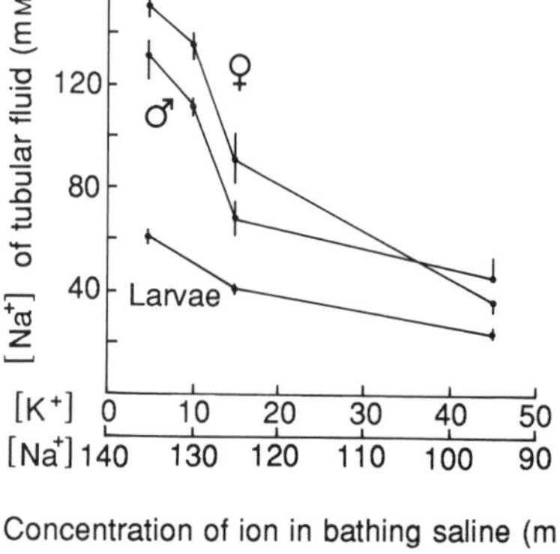

Figure 16.5 The sodium concentration of tubular fluid secreted by Malpighian tubules from larvae and male and female adults of *Aedes taeniorhynchus* when subjected to bathing salines of different sodium and potassium concentrations. Means ± s.e. (From Maddrell, 1977.)

membrane-permeable derivative. Addition of 0.1 or 1 mM db-cAMP to the bath rapidly changed the resistance and voltage profiles of the tubular epithelium (Table 16.4). The principal effect was a reduction in the fractional resistance of the basolateral membrane and a consequent depolarization of the basolateral membrane voltage. Because the apical membrane fractional resistance and voltage were unaffected, the transepithelial resistance decreased and the transepithelial voltage hyperpolarized ($V_t = V_{bl} + V_a$). The cAMP-induced hyperpolarization of V_t was reduced by a lowering of bath [Na^+], but was not affected by a lowering of bath [Cl^-], showing that cAMP increased the basolateral sodium conductance (Williams and Beyenbach, 1984; Sawyer and Beyenbach, 1985; Petzel *et al.*, 1987; Hegarty *et al.*, 1991).

When 0.1 or 1 mM db-cAMP was included in the saline that bathed Ramsay preparations, the rate of fluid secretion promptly increased. Electron probe analysis of the secreted fluid revealed that the exogenous cAMP had significantly increased the luminal [Na^+] and significantly decreased the luminal [K^+] without significantly affecting the luminal [Cl^-] (Table 16.5A). Because the fluid secretion rate increased in the presence of exogenous cAMP, the effects of cAMP on net ion secretion rates, calculated as the product of fluid secretion rates and ion concentrations, differed from its effects on luminal ion concentrations: sodium and chloride secretion rates were significantly enhanced but the rate of potassium secretion was little changed (Table 16.5B) (Williams and Beyenbach, 1983; Hegarty *et al.*, 1991). The primary cells were estimated to secrete the equivalent of their sodium content more than ten times per minute when stimulated with cAMP (Aneshansley *et al.*, 1988).

In the presence of db-cAMP, as under control

Table 16.5 Ion concentrations (A) and fluid and ion secretion rates (B) of tubular fluid secreted by isolated Malpighian tubules of female *Aedes aegypti* prepared by Ramsay's method. (From Williams and Beyenbach, 1983; and Petzel *et al.*, 1985.)

(A) Mean ion composition of tubular fluid from tubules bathed in saline without secretagogue (control) or containing dibutyryl cAMP (1 mM) or natriuretic factor.

Ion	*Concn. in saline (mM)*	*Concentration in tubular fluid (mM)*			
		Control	*cAMP*	*Control*	*Natriuretic factor*
Na^+	159	94	178*	76	148*
K^+	3	91	17*	114	31*
Cl^-	157	161	185	181	181

(B) Fluid and ion secretion rates of tubules bathed in saline without secretagogue (control) or in saline containing dibutyryl cAMP (1 mM) or natriuretic factor.

	Secretion rate			
Fluid/Ion	*Control*	*cAMP*	*Control*	*Natriuretic factor*
Tubular fluid (nl min^{-1})	0.79	2.9*	0.65	2.3*
Na^+ (pmol min^{-1})	69	454*	51	344*
K^+ (pmol min^{-1})	68	43	75	68
Cl^- (pmol min^{-1})	118	451*	119	416*

* Significantly different from control.

conditions, both sodium and potassium were secreted against their respective electrochemical gradients; only chloride secretion could be explained by passive movement (Table 16.3). No transepithelial chloride diffusion potentials were observed when transepithelial chloride concentration differences were imposed, confirming that there was no significant transepithelial electrodiffusion of Cl^- (Williams and Beyenbach, 1984).

That intracellular cyclic AMP ($cAMP_i$), generated within the tubule cells, functioned in the regulation of fluid secretion was demonstrated by the actions on Ramsay preparations of forskolin (a non-specific stimulant of the adenylate cyclase system) and theophylline (an inhibitor of phosphodiesterase). These secretagogues significantly increased both $[cAMP]_i$ and the rate of fluid secretion. In the absence of secretagogues $[cAMP]_i$ was 340 pmol mg protein^{-1} and the fluid secretion rate 0.56 nl min^{-1}. Fluid secretion rates reached their mean maximum of about 2 nl min^{-1} when $[cAMP]_i$ was driven to 900 pmol mg protein^{-1}; driving $[cAMP]_i$ higher with additional secretagogue did not stimulate the secretion rate further (Petzel *et al.*, 1987).

Bumetanide is an inhibitor of electrically neutral Na^+–K^+–$2Cl^-$ co-transport. When applied alone to *Ae aegypti* tubules, 0.1 mM bumetanide significantly increased the $[Na^+]$ and decreased the $[K^+]$ in secreted fluid, but because the increase in $[Na^+]$ was of similar magnitude to the decrease in $[K^+]$ there was no net change in cation concentration, and hence none in anion concentration or in rate of fluid secretion. However, when applied with db-cAMP bumetanide was a potent inhibitor of cAMP-stimulated transepithelial electrolyte and fluid secretion.

Exposure of tubules to db-cAMP plus bumetanide caused greater depolarization of the basolateral membrane than exposure to cAMP alone, but it produced no increased effect on the basolateral membrane fractional resistance, and the transepithelial resistance fell by 19% rather than 31% (Table 16.4). These results suggest that the actions of bumetanide are electrically neutral, and indicate that the electrophysiological effects of db-cAMP, like the secretagogue effects, depend on the presence of an intact bumetanide-sensitive transport system.

16.2.4 Peptides, eicosanoids and pharmacological agents

The addition of a crude saline extract of mosquito heads to the bathing saline of an isolated Malpighian tubule altered its electrophysiological characteristics, stimulated fluid secretion, and significantly increased the rates of sodium and potassium secretion (Williams and Beyenbach, 1983, 1984). Interpretation of these results was impossible because of the wide range of pharmacologically active substances in the extract, consequently attempts were made to isolate and purify peptide neurohormones that were believed to be present. HPLC of saline extracts from the heads of 3–10-day-old male and female *Aedes aegypti* yielded three fractions that affected the transepithelial voltage of perfused tubules. Their active constituents were shown to be peptides with molecular weights between 1.8 and 2.7 kDa (Petzel *et al.*, 1985).

The actions of fractions I, II and III on tubule preparations, which were rapidly reversible on wash-out from the bath, are summarized in Table 16.6. Fraction I depolarized V_t but had no effect on $[cAMP]_i$ nor upon the rates of fluid or ion secretion *in vitro*. Fraction II depolarized V_t, caused a slight but significant increase in $[cAMP]_i$, stimulated the rate of fluid secretion 3- to 4-fold, and increased the sodium and chloride secretion rates. It did not alter the potassium secretion rate but it significantly reduced $[K^+]$ in the tubular fluid. Fraction III hyperpolarized V_t after an initial transient depolarization, caused a massive increase in $[cAMP]_i$, and had the same effects on fluid and ion secretion as fraction II. As far as they were examined, the actions of fraction III on the tubule primary cells were almost identical with those of dibutyryl cAMP, and it was suggested that cAMP had the role of second messenger to fraction III. Because fraction III stimulated secretion of NaCl-rich as opposed to KCl-rich fluid (Table 16.5), the name **mosquito natriuretic factor** was proposed for it (Petzel *et al.*, 1985, 1986, 1987).

To examine their actions *in vivo*, fractions I, II and III were injected into females that had been bloodfed and immediately decapitated to prevent (further) release of endogenous diuretic hormone. All three fractions significantly increased the volume of urine discharged compared with the saline-injected controls. Fraction I showed the lowest activity, fraction III the greatest (Wheelock *et al.*, 1988).

These results suggest that the active peptide in fraction III is a natriuretic hormone which activates the adenylate cyclase system in Malpighian tubules and thereby raises the internal cyclic AMP concentration, increases the sodium conductance of the basolateral membranes, and so stimulates secretion of an NaCl-rich fluid (Petzel *et al.*, 1987).

Aedes aegypti Malpighian tubules are responsive to leucokinins, a family of octapeptides present in the head of the cockroach *Leucophaea maderae* and which increase the motility of cockroach hindgut. Exposure of perfused *Ae aegypti* tubules to seven of the leucokinins at 3.6×10^{-7} M caused a fall in transepithelial voltage in each case; leucokinin-4 was active at 3.6×10^{-9} M. Depolarization was dependent on extracellular chloride concentration. In tests with five of the peptides, lowering bath $[Cl^-]$ to 10 mM greatly reduced the leucokinin-induced depolarization. At concentrations between 10^{-11} and 10^{-9} M, leucokinin-8 significantly reduced the rate of fluid secretion but had no effect on transepithelial voltage. At concentrations of 10^{-8} M or more it depolarized the tubules, and at 10^{-7} M or more it increased the rate of fluid secretion. Hayes *et al.* (1989) pointed to similarities between the actions of fraction II and the leucokinins on mosquito tubules (Table 16.6) and postulated that fraction II and the leucokinins might be structurally related.

Subsequently, peptides that depolarized the Malpighian tubule transepithelial voltage were isolated from mosquitoes. They bore structural similarities to the leucokinins (Section 10.3.1 (b)), and were designated culekinin depolarizing peptides (CDP). The ED_{50} of CDP-I for tubule depolarization was 2×10^{-10} M, and this action was chloride dependent (Hayes *et al.*, 1992; T.K. Hayes, personal communication).

A peptide in *Ae aegypti* head extract exhibited antidiuretic activity. It decreased fluid secretion by tubules in Ramsay preparations by 76%, reduced forskolin-stimulated fluid secretion by 67%, and caused a 21 mV hyperpolarization of transepithelial voltage (Petzel and Conlon, 1991).

Eicosanoids (oxygenated metabolites of certain C_{20} polyunsaturated fatty acids) regulate many aspects of mammalian renal function. Exposure of *Ae aegypti* Malpighian tubule preparations to 0.1 mM concentrations of specific inhibitors of eicosanoid biosynthesis reduced basal fluid secretion rates. ETYA (5,8,11,14-eicosatetraynoic acid), which

Table 16.6 Actions upon perfused Malpighian tubule preparations of three HPLC-fractions of *Aedes aegypti* head extract, dibutyryl cAMP and two peptides. (From the data of Williams and Beyenbach, 1983, 1984; Petzel *et al.*, 1985, 1987; Wheelock *et al.*, 1988; Hayes *et al.*, 1989, 1992.)

Test substance	*Transepithelial voltage*	*Is effect on voltage Cl⁻ dependent*	*[cAMP]$_i$*	*Rate of transepithelial fluid secretion*
Fraction I	Depolarized	nt	Unchanged	Unchanged
Fraction II	Depolarized	nt	Raised	Stimulated
Fraction III*	Hyperpolarized	nt	Raised	Stimulated
cAMP (1 mM)	Hyperpolarized	No		Stimulated
Leucokinin-8	Depolarized	Yes	nt	Inhibited/Stimulated†
CDP-I	Depolarized	Yes	nt	Stimulated

nt, not tested
* Mosquito natriuretic factor
† Dose Dependent action

inhibits all eicosanoid biosynthesis, reduced fluid secretion by more than 50%. Esculetin, a lipoxygenase inhibitor, had little or no effect on fluid secretion, but the epoxygenase inhibitor SKF525A produced a small but significant reduction in fluid secretion. Two competitive inhibitors of cyclooxygenase, indomethacin and naproxin, yielded similar results to ETYA. The effects of all eicosanoid biosynthesis inhibitors were reversed after washing and upon addition of cAMP. Petzel and Stanley-Samuelson (1992) concluded that products of the cyclooxygenase pathway, in particular prostaglandins, modulate basal rates of fluid secretion by Malpighian tubules.

Addition of a pulse of 250 pmol 5-hydroxytryptamine to the irrigating saline of a perfused tubule preparation caused a transient depolarization of the transepithelial potential. With larger doses the depolarization was followed by a brief hyperpolarization. Irrigation of a Ramsay preparation with 10^{-6} M 5-HT depolarized the transepithelial voltage almost to zero and weakly stimulated fluid secretion. Irrigation with 10^{-5} M 5-HT caused a slight but significant increase in $[Na^+]$ in the tubular fluid and a similar decrease in $[K^+]$, while $[Cl^-]$ remained unchanged (Veenstra, 1988).

Exposure of Ramsay preparations to 10^{-4} M ouabain for 30 min significantly inhibited the rate of fluid secretion, indicating the presence on the basolateral membrane of an Na^+–K^+ pump (Hegarty *et al.*, 1991).

16.2.5 Characteristics of male tubules

In Ramsay preparations, Malpighian tubules from male *Aedes aegypti* had a basal secretory rate one-sixth that of female tubules. The concentrations of Na^+, K^+ and Cl^- in the tubular fluid were broadly similar in the two sexes. Extracts of male and female heads generally had the same effects upon Malpighian tubules in terms of sodium, chloride and water secretion, and of transepithelial voltage and resistance, whether the tubules were male or female. Differences were observed between male and female tubules in the secretion of potassium in response to partly purified head extracts but might have been artefactual. Cyclic AMP stimulated sodium, chloride and water secretion by male tubules as it did with female tubules. Plawner *et al.* (1991) concluded that the Malpighian tubules of males and females had similar ion-transport mechanisms for NaCl and fluid secretion and that the diuretic hormones and receptors were also similar. Quantitative differences in secretion rates reflected the larger size of the female tubules.

16.2.6 Models of tubular fluid secretion

To satisfy the separate requirements of Na^+ discharge during post-bloodmeal diuresis and K^+ production from erythrocytes during digestion, any model of tubular fluid secretion must permit independent processing of Na^+ and K^+. For our purpose, we shall adopt the theoretically valid but unproven concept that the driving force in fluid secretion is an active transfer of cations across the apical (luminal) face of the cell, and that the presence in the tubule lumen of high concentrations of Na^+, K^+ and Cl^- draws water out of the cell by osmosis (Maddrell, 1991). Cation transport generates lumen-positive voltages, and anion transport generates negative voltages, hence the transepithelial voltage reflects the relative rates of cation and anion transport. If the dominant driving force for NaCl and KCl transport is active transport of cations, the transepithelial voltage will be lumen-positive, and increasing the rate of cation transport will induce hyperpolarization of the transepithelial voltage. Thermodynamic data indicate that isolated, spontaneously-secreting mosquito Malpighian tubules secrete Na^+ and K^+ from the bath to the tubule lumen by active transport and Cl^- by passive transport. However, different peptides extracted from mosquitoes can hyperpolarize or depolarize the tubule transepithelial voltage, therefore it is likely that more than one model of tubular fluid secretion is needed.

A tentative model of ion movements in the Malpighian tubules of *Aedes aegypti*, under conditions of spontaneous secretion and of stimulation by natriuretic factor, is illustrated in Figure 16.6. Experimental evidence has provided preliminary

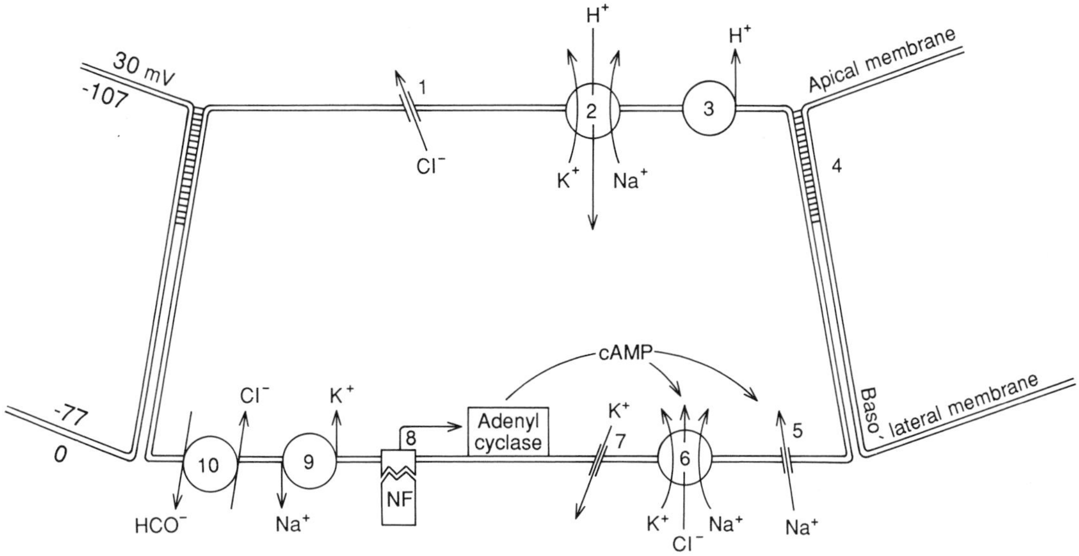

Figure 16.6 Tentative model of mechanisms controlling movement of ions through the principal cells of the Malpighian tubules of adult female *Aedes aegypti*. Representative voltages are indicated. 1. Postulated chloride channel. 2. Postulated antiport ion exchanger. 3. Postulated H^+-ATPase. 4. Smooth septate junction. 5. Sodium channel. 6. Postulated bumetanide-sensitive Na^+–K^+–$2Cl^-$ carrier. 7. Potassium channel. 8. Receptor for natriuretic factor. 9. Na^+/K^+ pump, ouabain sensitive. 10. Postulated SITS-sensitive Cl^-/HCO_3^- exchange.

evidence for a variety of transport systems in the basolateral membrane. The existence of sodium and potassium channels was shown by the dominant K^+ conductance and by the smaller Na^+ conductance which increased in the presence of cAMP. The inhibition of fluid secretion by oubain revealed the presence of a Na^+/K^+ carrier. The fact that db-cAMP lowered the fractional resistance of the basolateral membrane and increased the basolateral sodium conductance suggests that cAMP selectively activated certain ion channels. The antagonistic effects of bumetanide indicate that the actions of cAMP were in part mediated by bumetanide-sensitive transport, possibly a Na^+–K^+–$2Cl^-$ co-transport system. There is circumstantial evidence for the presence of SITS-sensitive sites of chloride entry, which possibly act via Cl^-/HCO_3^- exchange.

Stimulation of the principal cells by natriuretic factor caused cAMP synthesis and a consequent increase in sodium, chloride and fluid secretion rates. A number of lines of evidence suggest that under stimulated conditions the entry of sodium, potassium and chloride ions is linked – for example, the relationship of sodium and potassium secretion rates to bath chloride concentration, and the action of bumetanide on cAMP stimulated cells. It is thought that cAMP enhances both the rate of Na^+ entry via sodium channels and the entry of Na^+, K^+ and Cl^- via the electrically neutral Na^+–K^+–$2Cl^-$ carrier. Efflux of K^+ through potassium channels in the basolateral membrane possibly limits potassium secretion by stimulated tubules.

Because cAMP increased the rate at which sodium was secreted against its transepithelial electrochemical gradient, and caused an increase in V_t that was dependent upon bath $[Na^+]$, it appears that cAMP-stimulated sodium secretion, coupled with the passive movement of chloride, is the prime factor responsible for increases in the rate of fluid secretion in post-bloodmeal diuresis. We may conclude that the natriuretic factor present in fraction III is a genuine diuretic agent, selectively stimulating the secretion of Na^+, Cl^- and water.

The leucokinins and culekinins are much weaker diuretic agents and their effects are nonspecific, increasing the secretion of Na^+, K^+ and

Cl^-. These peptides depolarize the transepithelial voltage, apparently by increasing transepithelial Cl^- conductance. Any increase in permeability of the tubule wall to Cl^- would increase the availability of Cl^- for secretion with Na^+ and K^+ and would increase the rate of fluid secretion. The mechanisms by which leucokinins and culekinins modulate Cl^- conductance, and the site of the Cl^- conductance in the epithelium, remain to be elucidated.

The inhibition of potassium secretion by bumetanide suggests that the Na^+–K^+–$2Cl^-$ carrier is involved in potassium entry into the cell. In unstimulated tubules, the Na^+–K^+–$2Cl^-$ carrier has a significant role in potassium secretion but only a minor role in sodium and chloride secretion. The ouabain-sensitive Na^+/K^+ pump is also likely to have a role in selective potassium secretion. The increased fluid secretion rate observed on treatment with ouabain is consistent with the Na^+ that entered the cell via the Na^+–K^+–$2Cl^-$ carrier not being returned to the haemolymph through the ouabain-sensitive pump. The resulting increased internal ion concentration would raise the rate of ion transport across the luminal membrane, inducing faster water flow.

For insects generally, little is known about transport systems in the apical (luminal) membrane of Malpighian tubules. The idea that sodium and potassium are actively extruded into the tubule lumen by a cation pump situated in the apical membrane has been superseded by the concept of the extrusion of hydrogen ions by a H^+-ATPase. Associated with the ATPase are antiport systems, ion-exchangers that allow H^+ to re-enter the cell in exchange for Na^+ and/or K^+. The potential difference across the luminal membrane caused by the extrusion of H^+ induces the passive outward movement of Cl^-, and the high osmotic concentration in the tubule lumen draws water out of the cell (Maddrell, 1991). For mosquito tubules, it can be stated that there is preliminary evidence of the presence of H^+-ATPase and of chloride channels in the apical membrane (Section 16.2.2). The characteristics of mosquito-tubule transepithelial electrochemical potentials indicate that Na^+ and K^+ secretion involves active processes whereas Cl^- moves passively across the apical membrane, serving electroneutrality as the counter ion of Na^+ and K^+ transport.

16.3 REGULATION OF DIURESIS

Pharmacological evidence suggests that mosquitoes, like other insects, have diuretic and antidiuretic hormones, and a number of questions naturally follow from this. Where are these hormones synthesized, and where released? How is their release regulated, and what are their actions on the target organs? Experimental investigations have taken us a short way towards the answers to these questions and suggest that, as in the larva, water movement is regulated by the combined actions of hormones and the mechanical activities of parts of the alimentary canal.

Every 3–5 s weak antiperistaltic waves passed along the posterior midgut of unfed females of *Aedes aegypti*. Every 2 or 3 s an antiperistaltic wave in the pylorus forced fluid into the posterior midgut. Then a peristaltic wave which travelled from the pylorus to the rectum drove fluid posteriorly through the hindgut lumen. The pylorus thus functioned as a two-way pump. After engorgement the antiperistaltic waves of the posterior midgut ceased and did not resume until the midgut was empty (Howard, 1962). The contractions of the hindgut of *Ae aegypti* were described by Odland and Jones (1975) as strong, myogenic and rhythmic with a frequency of 10–15 min^{-1}. Contractions started at the anterior end of the pyloric chamber and developed into peristaltic waves which swept over the anterior intestine, stopping at the junction with the rectum. Each peristaltic wave was normally succeeded by a smaller antiperistaltic wave. During diuresis, peristaltic waves passed along the rectum and anal canal. When they reached the end of the anal canal the anus opened and a drop of urine was discharged on to the postgenital lobe (Stobbart, 1977).

Abdominal distension could be the initial stimulus that leads to the induction of diuresis. When females of *An freeborni* were fed saline they discharged urine over the same time-course as bloodfed females, but injection of 2–4 μl of saline into the haemocoele failed to induce the rapid phase of diuresis. Instead fluid was discharged at a slow rate for several hours (Nijhout and Carrow, 1978). This points to the gut rather than the body wall as the site of putative stretch receptors.

Extracts of the central nervous system have been tested for diuretic hormone. Heat-treated homogenates of heads and of thoracic ganglia of *Anopheles freeborni* proved to be equally effective in stimulating diuresis in Ramsay preparations but were inactive when injected into intact females (Nijhout and Carrow, 1978). Heat-treated head extract from *Aedes aegypti* induced diuresis in Ramsay preparations and also when injected into bloodfed decapitated females; three chromatographic fractions of head extract were active *in vitro* and *in vivo* (Table 16.6) (Wheelock *et al.*, 1988).

Bloodfed females were decapitated in investigations into the importance of the head for the control of diuresis. When females of *Ae aegypti* were decapitated soon after feeding, the rates of urine production first measured for the operated females were substantially lower than the control and pre-decapitation rates, and total urine production amounted to only 15–20% of the control quantity (Stobbart, 1977; Wheelock *et al.*, 1988). When females of *An freeborni* were decapitated 1 min after blood feeding the first measurements of urine production by the decapitated females were substantially below the control values in most cases. Urine was subsequently produced at the constant low rate of 6–7.5 nl min^{-1} for 5–7 h, by which time the operated females had discharged as much as the controls. When females were decapitated 10 or 20 min after feeding diuresis continued at the control rate for an additional 1 min and then declined to a low rate (Nijhout and Carrow, 1978). These results can be interpreted in terms of disappearance of a diuretic hormone. Clearly the head is necessary to maintain high rates of diuresis.

To find whether diuretic hormone was continuously present in the haemolymph or was released after blood feeding, haemolymph from sugarfed and bloodfed *Ae aegypti* was tested for its effect on the transepithelial voltage of perfused tubules and on the urine flow rate of Ramsay preparations. Haemolymph from bloodfed females consistently and significantly stimulated fluid secretion whereas haemolymph from sugarfed females was inactive. However, both categories of haemolymph affected the transepithelial potential (V_t). The identity of the depolarizing agent was unknown but Wheelock *et al.* (1988) concluded that a diuretic hormone was released into the haemolymph only after blood feeding, acting on the Malpighian tubules to stimulate fluid secretion.

Pinching the nerve cord of *Ae aegypti* in the 1st abdominal segment 2–4 min after blood feeding caused an immediate fall in the rate of urine flow, and pinching the anterior midgut had a similar effect. On the grounds that the site of midgut injury was posterior to the release sites of diuretic hormones, Stobbart (1977) concluded that damage to the stomatogastric nervous system must have been the cause of the failure of diuresis. He postulated that, as in the larva (Section 6.3.1), active diuresis required a change from antiperistalsis to peristalsis in the midgut musculature.

In further experiments Stobbart (1977) pinched the ventral nerve cord in the 1st abdominal segment before blood feeding, measured the diuretic rate of operated bloodfed females for a few minutes and then blocked the alimentary canal between the midgut and pyloric chamber by pressure from a blunt needle. In 12 operated but dummy-clamped females the low diuretic rate induced by pinching the nerve cord was unaffected by clamping, but in 7 out of 29 operated and clamped females the diuretic rate increased significantly, on average 4-fold, after clamping. Stobbart concluded that in unfed females tubular fluid normally moved forwards into the midgut for recycling to the haemolymph, and that in bloodfed females a change, under nervous control, from predominantly antiperistaltic to peristaltic contractions was an important element in post-bloodmeal diuresis.

Cyclic AMP almost certainly plays a role in the induction of the short post-bloodmeal burst of diuresis, which is characterized by urine with a high sodium concentration. The intracellular cAMP concentration of the Malpighian tubules increases soon after the start of blood feeding, rising from 340 pmol to peak at 550 pmol mg protein^{-1} after about 5 min, and preceding the peak rate of NaCl diuresis by 1 min. It then declines to the control level at 9 min, before rising again to a second apparent peak at 25 min (Petzel *et al.*, 1987). Treatment of Malpighian tubule preparations with the natriuretic factor stimulated an increase in $[cAMP]_i$ and production of fluid with a high sodium concentration, and injection of this factor into bloodfed decapitated females stimulated diuresis (Table 16.6). The natriuretic factor could be the hormone responsible for the post-bloodmeal burst of diuretic activity.

Whether the phase of increased potassium excretion that occurs from 30–70 min after feeding is stimulated by a different hormone or is simply a stage in the return to the normal secretion of tubular fluid is quite unclear. It has been suggested that 'fraction I', which stimulated fluid secretion in mosquitoes decapitated after a blood meal but did not stimulate secretion in isolated Malpighian tubules, might act by inhibiting fluid resorption by the rectum (Wheelock *et al.*, 1988).

16.4 EXCRETION AND DEFAECATION AFTER EMERGENCE

Newly emerged mosquitoes contain meconium in the posterior half of the stomach in the form of fluid and semi-solid matter surrounded by peritrophic membrane. The meconium includes residues from the histolysis of the larval alimentary canal and nitrogenous wastes (Romoser, 1974) and retains chymotrypsin and trypsin activity (Briegel, 1983). The fluid discharged by *Aedes aegypti* immediately after emergence is clear but contains some uric acid. The mean weight of uric acid discharged during the first 45 min after emergence was 0.07 μg, and it did not differ significantly between males and females (DeGuire and Fraenkel, 1973; Jones and Brandt, 1981). Various species void the semi-solid meconium during the first 1–2 days after emergence, the time of discharge varying with temperature, but remnants of peritrophic membrane can persist long after the meconium has disappeared (Rosay, 1961; Venard and Guptavanij, 1966; Romoser and Cody, 1975).

During the first 48–72 h after the emergence of *Ae aegypti* uric acid was the major nitrogenous substance discharged. The excreta of sugarfed adults of *Aedes*, *Culex* and *Anopheles* contained uric acid, urea, ammonia, protein and amino acids. Older sugarfed mosquitoes excreted little nitrogenous matter; over a 5-week period the nitrogen content of sugarfed female *Ae aegypti* remained more or less constant (Terzian *et al.*, 1957; Irreverre and Terzian, 1959; Thayer and Terzian, 1971).

Most uric acid is voided via the Malpighian tubules and hindgut but some accumulates within fat body cells. Little uric acid was found in the fat body of young adults of *Culex pipiens* but the uric acid deposits increased with age and were particularly extensive in hibernating females (de Boissezon, 1930b).

16.5 COMPOSITION OF EXCRETA AND FAECES AFTER BLOOD FEEDING

Mosquitoes ingest considerably more protein in a blood meal than is subsequently found in their mature ovaries. This is because the amino acid balance of blood protein limits the amount that can be converted to yolk protein and because some blood protein must be utilized for lipid synthesis. In consequence much of the ingested protein is deaminated and the surplus nitrogen excreted in metabolites with a high nitrogen content and low molecular weight, *viz.* uric acid ($C_5H_4N_4O_3$), urea (CH_4N_2O) and ammonium ion (NH_4^+). In *Aedes* and *Culex* all surplus nitrogen from the blood meal is voided, but in *Anopheles* some appears to be retained as extra-ovarian protein (Section 22.1.1(b)).

Little is known about the transport of nitrogenous metabolites in mosquitoes, only that after blood feeding minute uric acid granules appear in the lumens of the Malpighian tubules and that the rectum progressively fills with spherical masses and granules of uric acid (Wigglesworth, 1932).

Early investigators identified the nitrogenous metabolites present in the excreta and faeces but they were unable to account for all ingested nitrogen (Briegel, 1969; Thayer and Terzian, 1971; Thayer *et al.*, 1971; France and Judson, 1979). However, when gaseous ammonia was trapped all nitrogen in the blood meal could be accounted for from the amounts recovered in the excreta, faeces and mature ovaries (Table 16.7; Figure 16.7). Because excretion and defaecation occur at different times after blood feeding the excreta and faeces can be analysed separately.

When females of *Ae aegypti* fed rat blood were denied access to water during the period of digestion the major constituent by weight of the excreta was uric acid, with ammonium ion second and urea third. The ratios differed when computed as mol per female, when ammonium ion became the major constituent (Table 16.7). If the females were given access to water during the period of digestion their metabolism changed somewhat, with the result that more urea was produced at the expense of uric acid and ammonium ion. Uric acid, urea and ammonium ion were voided during the same time period, therefore the uricotelic, ureotelic and ammonotelic pathways operated simultaneously. The variability in the extent of ureotely revealed a degree of plasticity in excretory metabolism. The molar ratios of the various nitrogenous metabolites varied slightly with host identity (Figure 16.8). Despite its nucleated erythrocytes chicken blood gave rise to less uric acid than did rat blood (Briegel, 1986a).

After feeding on chicken blood females of *Culex quinquefasciatus* discharged 0.048, 0.096 and 0.036 μmol total amino acid on the 1st, 2nd and 3rd days post-bloodmeal, which constituted 16% of the recovered nitrogen. The ratio of free to conjugated amino acids was approximately 1:1, and most was discharged over the same period as uric acid. During the 1st and 2nd days histidine constituted 70% of the free amino acids and glycine constituted >60% of the conjugated amino acids. Arginine was present only in traces (Briegel, 1969). In *Aedes aegypti* 8–10% of discharged nitrogen was in the form of amino acids, the ratio of free to conjugated amino acids being 3:2. Histidine constituted 50–66% of the total molarity voided and about 90% of all amino acid nitrogen voided. Arginine constituted 5–6% and glycine ≈15% of the total molarity voided (Briegel, 1986a).

Histidine contains three nitrogen atoms per molecule and therefore is a significant vehicle for nitrogen excretion. Since mosquitoes are incapable of histidine synthesis the excretory histidine cannot be a metabolic product. When fed 4 μl human or guinea-pig blood females of *Ae aegypti* incorporated into their ovaries 14% of the histidine in the human blood and 20% of that in the guinea-pig blood (Briegel, 1985). It seems probable that most or all of the bloodmeal histidine that was not used for oogenesis was not deaminated but eliminated directly.

Small amounts of arginine ($C_6H_{14}N_4O_2$) are also excreted unchanged. The lower concentration of arginine than of histidine in the excreta reflects its lower concentration in mammalian blood and greater incorporation into the ovaries (France and Judson, 1979; Briegel, 1985, 1986a). The conjugation of glycine with toxic compounds is well known, so it may be that the conjugated

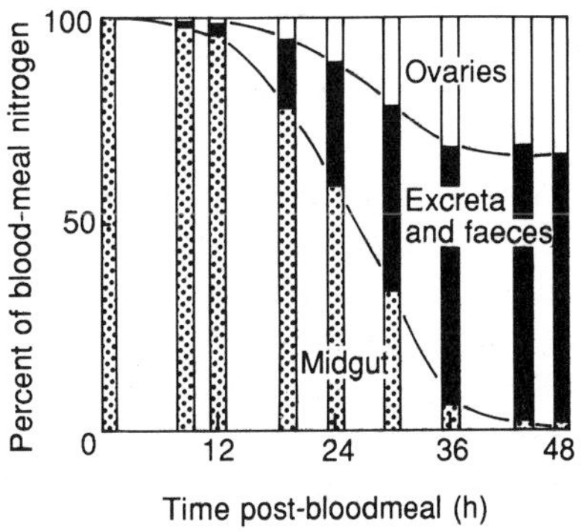

Figure 16.7 Temporal distribution of blood-meal nitrogen during an ovarian cycle between the midgut, excreta + faeces, and ovaries of females of *Aedes aegypti* given 4 μl rat blood by enema. (From Briegel (1986a). Copyright (1986), Pergamon Press plc. Reprinted with permission.)

Table 16.7 Distribution of blood-meal nitrogen between the ovaries and various excretory and faecal constituents 70 h after *Aedes aegypti*, which had either been denied water or given access to water, were fed 3 μl rat blood by enema. When water was provided some ammonia was lost and the recovery of nitrogen was incomplete. Faecal proteins were hydrolysed before assay and are recorded as amino acids. (From Briegel, 1986a.)

	Denied water				*Provided with water*			
	μg N/♀	%N		nMol/♀	μg N/♀	%N		nMol/♀
Blood meal	85.1	100			85.1	100		
Ovaries	36.3	43			36.7	43		
Excreta + faeces	50.5	59			38.6	44		
Uric acid	22.4	26	60%	400	9.1	11	43%	160
Urea	7.1	8		250	13.5	16		480
Ammonium ion	9.7	11		690	6.6	8		468
Amino acids	10.3*	12		340	5.4†	6		220
Haematin	2.1	3		38	2.2	3		38

* 68% of amino N was in histidine (166 nMol/female).
† 43% of amino N was in histidine (54 nMol/female).

glycine found in the excreta was part of a detoxification product.

The faeces contained haematin, digestive enzymes and the remnants of the peritrophic membrane. Upon the digestion of haemoglobin haem is oxidized to haematin and is not absorbed, consequently this source of nitrogen is eliminated in the faeces. The output of faecal haematin showed an exact stoichiometric relationship with the haemoglobin input, four moles of haematin being produced for each mole of dietary haemoglobin, irrespective of blood volume or host species. Small amounts of protease were discharged throughout the period of digestion, but most was voided with the haematin. The digestive enzymes that were discharged included trypsin, aminopeptidase and carboxypeptidase, still in their active forms. The total of this proteolytic activity was over half that in the midgut at the time of maximum activity (Briegel, 1975, 1986a; Graf and Briegel, 1982).

Quantitative analyses made during an ovarian cycle of *Ae aegypti* revealed that the total nitrogen content of the pooled excreta, faeces and mature ovaries always equalled the blood-meal nitrogen intake; consequently the mean nitrogen content of the female body before the blood meal and after oviposition were identical. In one experiment, with females fed 3 µl rat blood, the total nitrogen content per female was 60.7 ± 3.8 µg before feeding and 60.8 ± 3.33 µg after oviposition. Of the 82.8 ± 3.3 µg nitrogen ingested, 29.7 ± 6.7 µg was utilized for yolk protein synthesis and 48.2 ± 4.1 µg was excreted. Clearly the ingested nitrogen was either utilized for oogenesis or excreted.

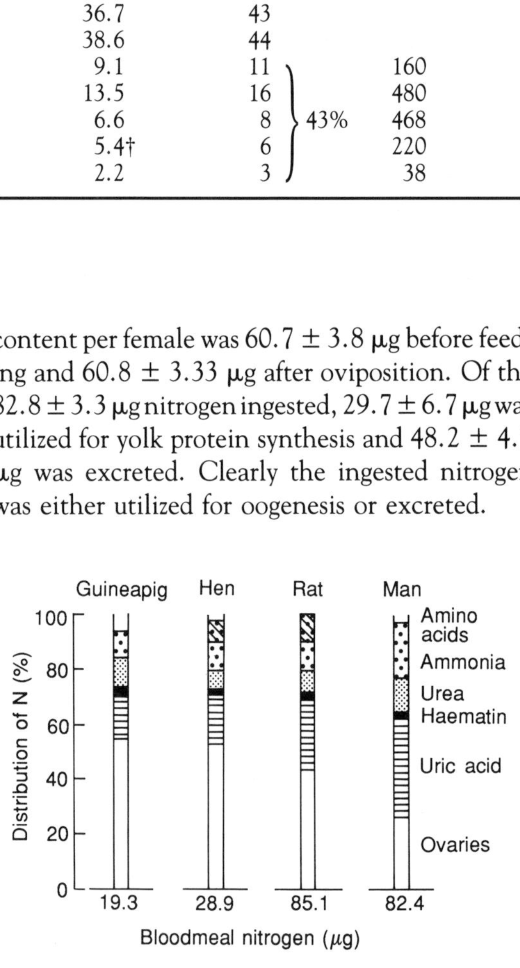

Figure 16.8 Budget of dietary nitrogen in *Aedes aegypti* at the end of an ovarian cycle. Females were given 1 µl guinea-pig or hen blood or 3 µl rat or human blood, and the percentage incorporation of nitrogen into the mature ovaries and the excretory and faecal constituents was determined. The examples were chosen on the basis of maximum nitrogen recovery (⩽ 93%). Except with rat blood, excreted amino acids were not accounted for. (From Briegel (1986a). Copyright (1986), Pergamon Press plc. Reprinted with permission.)

The extent of nitrogenous excreta produced by *Ae aegypti* from the blood of four different hosts reflected the dynamics of yolk synthesis. At one extreme only 26% of the nitrogen in human blood, which is characterized by its low isoleucine content, was incorporated into the ovaries while 70% appeared in nitrogenous metabolites. At the other extreme, 55% of the nitrogen in guinea-pig blood was incorporated into the ovaries and 35% appeared as nitrogenous metabolites (Figure 16.8) (Briegel, 1986a). Complete nitrogen budgets for *Anopheles albimanus* and *An stephensi* revealed that approximately 80% of blood-meal protein was catabolized and excreted through the three major pathways that terminated in uric acid, urea and ammonia (Briegel, 1990b).

16.6 REGULATION OF EXCRETION AND DEFAECATION AFTER BLOOD FEEDING

Processing of the blood meal includes three phases of elimination of unwanted substances. Immediately after feeding excess water and ions are discharged. Later, as protein digestion starts, nitrogen excretion also starts and excretion peaks when the rate of proteolysis reaches its maximum. Finally, after digestion has been completed residual substances in the midgut are voided. Excretion and defaecation can be distinguished visually because the excreta, coloured white or pale yellow by the uric acid, appear first and the dark brown or black faeces, coloured by haematin, appear later; however, there is some overlap.

Females of *Aedes aegypti* kept at 27 or 28°C discharged excreta gradually from about 12 to 36 h post-bloodmeal, and discharged faeces during a relatively short period between 24 and 48 h post-bloodmeal (Figure 16.9). Occasionally up to 90% of the haematin was voided within a period of 3–4 h, but in most instances the process lasted 8–12 h. Small amounts of proteinase were discharged between 12 and 24 h post-bloodmeal but most was voided during the period of haematin discharge. The time of defaecation varied slightly with size of blood meal. By the time haematin had been voided the alimentary canal was empty. It was extremely rare for oviposition to start before all haematin had been voided from the mid- and hindgut (Briegel, 1975, 1986a; Gillett *et al.*, 1975).

Females of *Ae aegypti* that were decapitated immediately after blood feeding did not undergo vitellogenesis but nevertheless showed a similar temporal pattern of uric acid excretion to control females, but with a lag of four hours or more. The decapitated females excreted additional uric acid which corresponded precisely with the amount of nitrogen usually utilized in the production of eggs (Briegel, 1980b).

Females decapitated 32 h after feeding defaecated normally at 50–60 h whereas females decapitated 28 h after feeding retained most of the haematin in the gut. Briegel (1980b) concluded that possibly, between 28 and 32 h after feeding, nervous stimulation of the alimentary canal controlled subsequent defaecation. The time of defaecation is an inherited character. Strains of *Ae aegypti* could be classified as early, intermediate or late according to whether the time by which 50% of females had discharged haematin (the DT_{50}), was 28–32, 33–39 or >40 h post-bloodmeal. When strains were crossed the linear regression of filial on maternal DT_{50} values confirmed the heritability of the trait. Strains that discharged haematin early also discharged uric acid early (Briegel, 1986b).

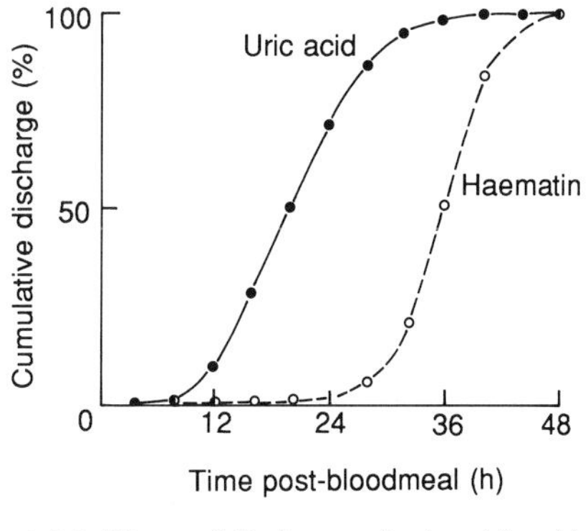

Figure 16.9 Times of discharge of uric acid and haematin by females of *Aedes aegypti* fed 3 μl rat blood and kept at 27°C. Excreta and faeces were collected at 4 h intervals, and the data are plotted as cumulative percentage of total compound discharged. (After Briegel (1986a). Copyright (1986), Pergamon Press plc. Reprinted with permission.)

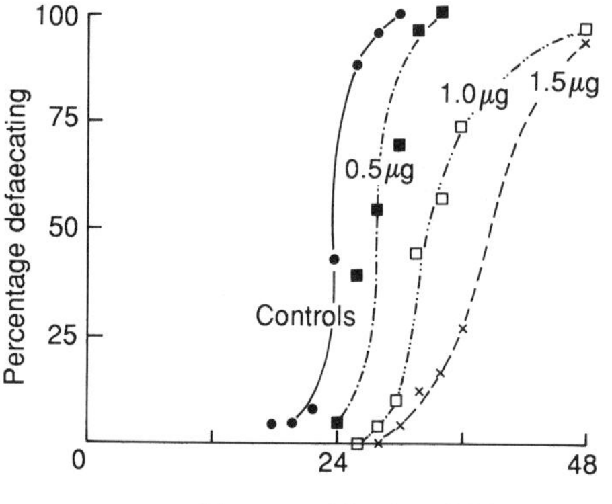

Figure 16.10 Effects of injection with different amounts of 20-hydroxyecdysone upon the time of onset of defaecation in ovariectomized females of *Aedes aegypti*. Control females were injected with saline. The mosquitoes had fed to repletion on human blood and were kept at 28°C. (From Cole and Gillett, 1979.)

Strains of *Ae aegypti* differed by up to six hours in the time by which 50% of females had released ovarian ecdysteroidogenic hormone (OEH). A positive correlation was obtained between that parameter and the time post-bloodmeal by which 50% of females had discharged haematin (Briegel, 1986b), so it seems that the time of OEH release affects the timing of physiological processes that occur up to 36–48 h later.

The involvement of 20-hydroxyecdysone in the control of defaecation has been postulated by Gillett and collaborators, who recorded the time of onset of haematin discharge under different experimental conditions. Females of *Ae aegypti* that took very small blood meals and failed to mature their ovaries defaecated 12 h or more before the controls. Early discharge of the gut contents was also caused by removal of the MNC perikarya or of the ovaries prior to the ingestion of a full blood meal, and by decapitation immediately after blood feeding. All of these operations prevented vitellogenesis. The gut contents of small feeders were not discharged prematurely if 0.1 μg 20-hydroxyecdysone was injected. Similarly the effects of decapitation and ovariectomy could be prevented by injecting 20-hydroxyecdysone 4 h post-bloodmeal. A positive correlation was obtained between the dose of 20-hydroxyecdysone injected into ovariectomized females, within the range 0.5–1.5 μg, and the delay in onset of defaecation (Figure 16.10). It was concluded that the 20-hydroxyecdysone secreted after blood feeding not only stimulates vitellogenesis but also prolongs the period of retention of the blood meal (Gillett *et al.*, 1975; Cole and Gillett, 1978, 1979). Similar results were obtained with *Anopheles freeborni* (Rosenberg, 1980).

17

Structure of the gonads and gonoducts

17.1 STRUCTURE OF THE TESTES AND MALE GENITAL DUCTS

In the more generalized insects each testis is composed of a number of tubular follicles enclosed in an epithelial sheath which is often pigmented. Each follicle is bounded by an epithelial layer and filled with cysts, each of which initially contains a single germ cell. Mitotic and eventually meiotic divisions occur synchronously in all the germ cells of any cyst with formation of a clone. In the Diptera each testis consists of a single follicle surrounded by a sheath. Mosquito testes are pear-shaped or elongate bodies, situated dorsolaterally in the 5th and 6th abdominal segments and connected by terminal filaments to the heart and alary muscles (Snodgrass, 1959; Christophers, 1960; Phillips, 1970a).

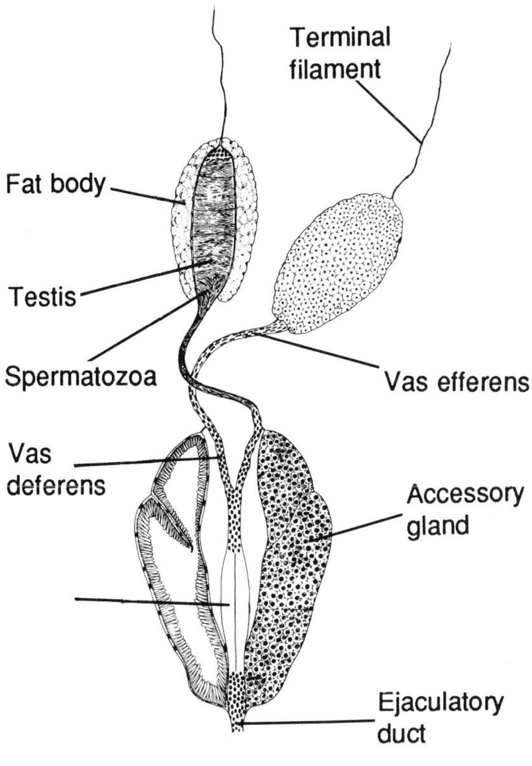

Figure 17.1 The male internal reproductive organs of *Aedes stimulans*. (From Anderson and Horsfall (1963). Copyright © (1963) and reprinted with permission of Wiley-Liss, a Division of John Wiley and Sons, Inc.)

In *Anopheles stephensi* the testis wall is formed of a tough outer coat and an inner squamous epithelial layer. The outer coat is syncytial and its cytoplasm contains densely aggregated yellow and black pigment granules. The testis lumen is divided into cysts by oblique septa which are formed by infoldings of the epithelium, each two-layered and composed of a few very thin cells. The cysts contain spermatogonia and spermatocytes, and all germ cells within a single cyst are more or less at the same stage of development. Older spermatocytes and differentiating spermatids lie free within the posterior part of the testis lumen where there are no septa (Rishikesh, 1959).

Testicular cysts have also been described in *Culex pipiens* (Lomen, 1914; Whiting, 1917), *Cx tarsalis* (Asman, 1974) and *Aedes aegypti* (Jones, 1967). Testicular cysts are not formed in *Anopheles atroparvus*, in which the anterior portion of the testis forms a germarium, containing spermatogonia, which opens through a narrow orifice into a single large lumen within the testis. In the lumen, cells at the same stage of maturation are aggregated in clusters. The outer coat of the testis contains xanthommatin and an ommin (Cadeddu and Laudani, 1974).

The anterior portions of the genital ducts, the vasa efferentia, are derived from the testis rudiments (Section 8.2.1(a)). The posterior regions, comprising the vasa deferentia, seminal vesicles and ejaculatory duct, with the accessory glands, are derived from imaginal disks of the 9th abdominal segment (Section 8.2.2). Despite their different origins the vasa efferentia and vasa deferentia may appear similar. The vasa efferentia are always narrow thin-walled tubes. The vasa deferentia may be slightly thicker walled and in some species they are muscular, e.g. *An quadrimaculatus*, *Ae aegypti*, *Ae stimulans*, and *Psorophora howardii*. The vasa deferentia may open independently into the seminal vesicles of their own side, as in *Ae stimulans* (Figure 17.1), or they may first fuse to form a short common duct which supplies both vesicles, as in *Ae aegypti*. In *Ae aegypti* nerves from the terminal ganglionic mass supply the seminal vesicles, accessory glands and ejaculatory duct (Christophers, 1960; Hodapp and Jones, 1961; Lum, 1961a; Anderson and Horsfall, 1963).

The seminal vesicles are distensible regions of the genital ducts, in many species fused externally but retaining separate lumens. In *Ae aegypti*, during the first 24 h after emergence, the epithelial cells of the seminal vesicles appear to be secretory (Hodapp and Jones, 1961). In newly emerged males of various species the vasa deferentia and seminal vesicles are filled with a fluid which provides a medium for the spermatozoa. Once spermatozoa become active in the testes it is only a matter of 15–30 min before they are found in the seminal vesicles. As spermatozoa enter the seminal vesicles those organs become increasingly distended and their once thick walls transform to thin membranes (Lum, 1961b). Spermatozoa are stored not only in the seminal vesicles but also in the anterior ducts and the posterior parts of the testes.

Adjacent to the seminal vesicles are two large accessory glands which unite at their bases and open into the anterior end of the ejaculatory duct (Figure 17.1). Information on the structure and secretions of the accessory glands will be given in Volume 2.

17.2 STRUCTURE OF THE OVARIES AND OVIDUCTS

Two ovaries are situated dorsolaterally in the posterior region of the abdomen, each composed of functional units called ovarioles. The ovaries are connected by lateral oviducts to the common oviduct. Each ovary is penetrated by a central chamber lined with an epithelium continuous with that of the lateral oviduct. This is the calyx, from which the ovarioles radiate (Figure 17.2). Two tracheae enter each ovary and branch profusely before ending as tracheoles within the ovarioles. Nerves run from the terminal ganglionic mass to the lateral oviduct and the base of the ovary and to the common oviduct. The descriptions of ovaries and oviducts that follow, principally relate to species of *Anopheles* and *Culex* (Nicholson, 1921; Christophers, 1960; Curtin and Jones, 1961; Detinova, 1962; Giglioli, 1963, 1964b). The structure of the more distal reproductive organs – which function in copulation, insemination and oviposition – will be given in Volume 2.

17.2.1 The ovariole

The anterior part of each ovariole is occupied by a germarium, within which mitotic divisions of germ cells take place, while the remainder, the vitellarium, contains one or two follicles or egg chambers. Mosquito ovarioles are classed as meroistic because they contain nurse cells as well as oocytes, and are further categorized as polytrophic because a group of nurse cells is enclosed with an oocyte in each ovarian follicle. Follicular stalks connect adjacent follicles to one another, and connect the first follicle to the calyx. The ovariole is surrounded by an ovariolar sheath which extends beyond the germarium as a terminal filament. It is sometimes necessary to specify the series of follicles, and Feinsod and Spielman (1980b) have introduced two terms for this purpose. The more mature follicles, which are adjacent to the calyx, they called 'proximate follicles', and the younger follicles 'penproximate follicles'. After each ovarian cycle there is a new

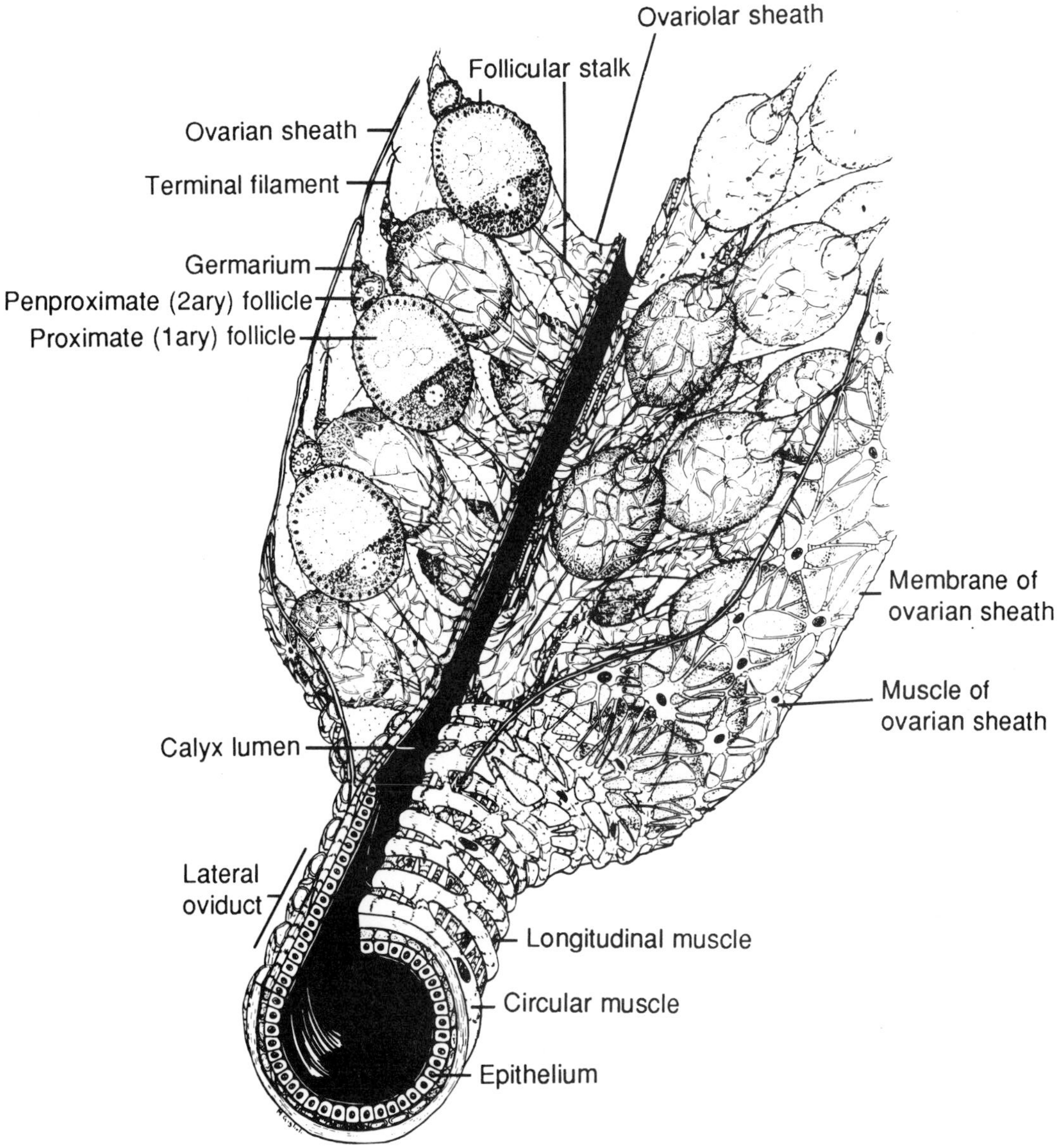

Figure 17.2 Part of an ovary and lateral oviduct of a nulliparous female of *Anopheles melas*, opened to show the calyx. (From Giglioli, 1964b.)

series of proximate follicles, which were formerly the penproximate follicles. The proximate and penproximate follicles of nulliparous females are often called primary (1ary) and secondary (2ary) follicles respectively.

The germarium, follicles and follicular stalks which compose each ovariole are invested by a basal lamina, which is sometimes called the tunica propria (Bertram and Bird, 1961; Anderson and Spielman, 1971). The idea that the most distal follicular stalk (or pedicel) is a hollow tube which opens into the calyx and provides a channel for the passage of the mature oocyte at ovulation (Giglioli, 1964b, 1965) appears to have been disproved. Lehane and Laurence (1978) found that the follicular stalk is not tubular but rod-like, and that, except in the later stages of the ovarian cycles, the calyx wall is a solid structure without perforations.

The ovariolar sheath is a thin membrane with an outer muscle layer composed of isolated cell bodies interconnected by muscle fibres. The muscle cells are continuous with those of the calyx (Figure 17.2). Proximally the ovariolar sheath extends as a terminal filament, connected to the ovarian sheath. The ovariolar sheath is attached to the germarium, is closely apposed to but not attached to the follicles, and between the ultimate follicle and the calyx has the form of a wide sleeve. The whole ovary is surrounded by an ovarian sheath which is also a network of muscle cells over a membrane, but in this case the muscle cells are continuous with those of the circular muscle layer of the lateral oviduct. Proximally the ovarian sheath forms the long suspensory ligament which inserts on the 4th abdominal tergite.

The number of ovarioles ranges from fewer than 50 to 500, varying with species but affected also by the size of the individual female (see Section 22.1).

17.2.2 The ovarian follicle

The formation of ovarian follicles from primordial germ cells and mesoderm is described in Section 19.1. Each is composed of one oocyte, seven nurse cells and an outer epithelium (Figure 17.3). As in other insects, the function of the nurse cells is to synthesize maternal ribosomes and mRNAs for transfer to the oocyte.

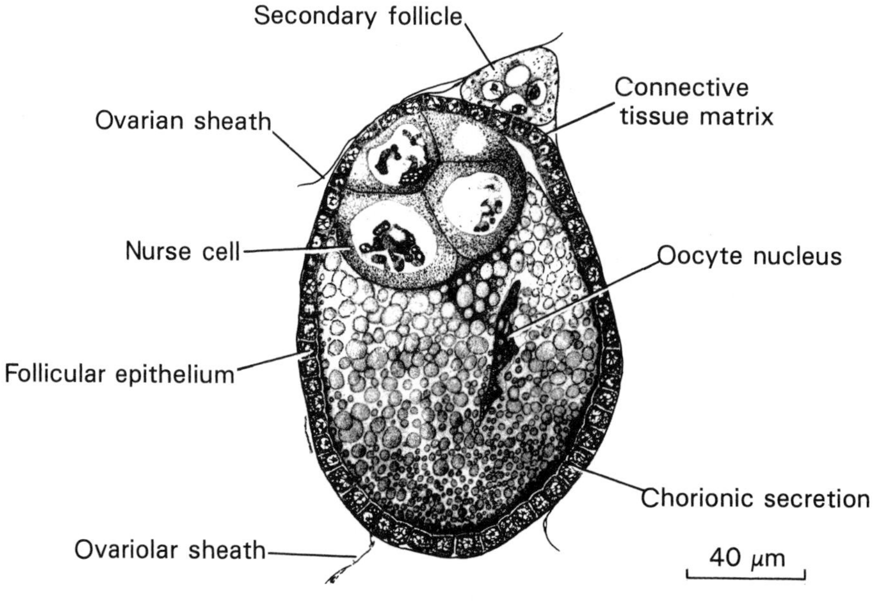

Figure 17.3 Longitudinal section of an ovarian follicle of *Anopheles atroparvus* during vitellogenesis. (From Nicholson, 1921.)

In the previtellogenic follicles of *Aedes aegypti*, the epithelial cells are cuboidal, and have large nuclei but rather sparse cytoplasm, which is rich in ribosomes but relatively poor in RER and Golgi complexes. Their basal cell membranes rest on a basal lamina, and they are joined neighbour to neighbour by desmosomes at their apical margins. Channels of about 20 nm diameter pass between adjacent epithelial cells, connecting the haemocoele with the small perioocytic space.

The epithelial cells are connected to one another by intercellular bridges, and a single cell may be joined to more than one other cell in this way. The bridges are barrel shaped, 0.25–0.5 μm long, and project into both of the interconnected cells. The cytoplasm within the bridges may contain ribosomes, microtubules and endoplasmic reticulum (Roth and Porter, 1964; Anderson and Spielman, 1971). In *Culex quinquefasciatus* the number of bridges between the epithelial cells increased greatly during the period of active cell division in the first day after blood feeding, and the follicular epithelium came to contain clusters of over 30 interconnected cells. Most of these cells had two bridges but up to five bridges per cell were seen (Fiil, 1978a).

The nurse cells of *Ae aegypti* are large (27 μm diameter), and their nuclei alone have the width of three epithelial cells. The nurse cell cytoplasm is rich in ribosomes and also contains many mitochondria, occasional lipid droplets and a few elements of rough ER. The oocyte and the seven nurse cells are a syncytium, being connected by intercellular bridges (Section 19.1). The measured diameters of intercellular bridges ranged from 0.9 to 2.6 μm. Each bridge has an electron-dense rim confluent with the plasma membrane, and ribosomes, mitochondria and other organelles can be seen within them. The oocyte is directly connected to three nurse cells by intercellular bridges, and a part of the oocyte may protrude finger-like between the nurse cells, with all three bridges situated at the tip of the protrusion (Fiil, 1974; Mathew and Rai, 1976b).

In *Ae aegypti*, the basal lamina surrounding the epithelial layer is formed of regularly orientated protein particles, each consisting of four subunits. Each protein particle is connected to four others, and the whole structure forms into pores which, in fixed tissue, are approximately square and about 20 nm in their largest dimension (Raikhel and Dhadialla, 1992).

Conspicuous changes take place in the oocytes of the ultimate follicles during a gonotrophic cycle. The ooplasm becomes packed with yolk and ribosomes, the oocyte increases steadily in size, and the chorion is deposited.

17.2.3 The calyces and oviducts

The calyces, lateral oviducts and common oviduct are all tubular structures consisting of an inner

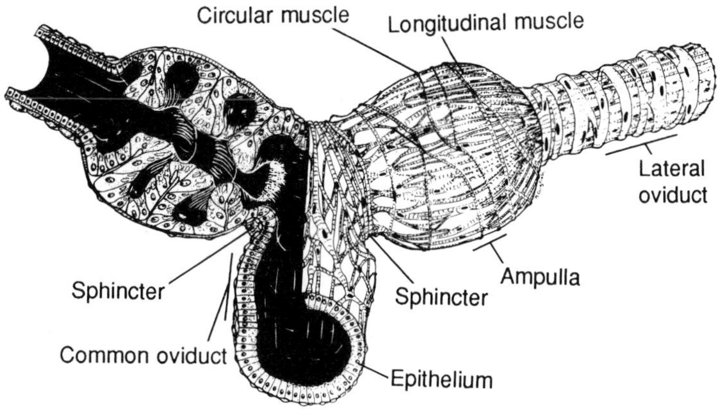

Figure 17.4 The lateral oviducts, common oviduct and ampullae of a multiparous female of *Anopheles melas*, partly cut away on the left side. (From Giglioli, 1963.)

epithelial layer and an outer muscular layer. The calyx runs through the axis of the ovary and is then continuous with the lateral oviduct. In unfed nulliparous females it has the form of a flattened tube which does not have a well-developed lumen. Giglioli (1964b), using light microscopy, could not detect an epithelium in the calyx of *Anopheles melas* but Lehane and Laurence (1978), with electron microscopy, detected a thin epithelium in *Aedes aegypti*. The muscle layer of the calyx contains anastomosing longitudinal muscle fibres which have cellular connections with the muscle cells of the ovariolar sheaths (Figure 17.4).

In non-bloodfed females of *Ae aegypti* the lateral oviduct is a flattened tube with an incomplete lumen. Its epithelial cells are squamous, contain large numbers of microtubules, and are connected to one another apically by extensive septate desmosomes (Lehane and Laurence, 1978). The lateral oviducts of *Anopheles melas* consist of an epithelium of cuboidal cells surrounded by a layer of slender longitudinal muscle and a stronger outer layer of circular muscles. At the base of the ovary the circular muscle layer divides into two concentric sheaths or collars. The inner collar extends by some 3–4 circular bundles of muscle fibres over the base of the calyx and then terminates. The outer collar gives rise to the muscular network of the ovarian sheath (Figure 17.2). The ovarian and ovariolar sheaths are syncytia of striated anastomosing muscle fibres and stellate cell bodies on acellular membranes (Figure 17.4) (Giglioli, 1963, 1964b).

The common oviduct of *Anopheles* is a large median tube which opens into the atrium through the gonopore, and in this region the circular muscles of the atrium thicken to form a strong sphincter. The wall of the oviduct has three layers: an inner secretory columnar epithelium, an intermediate layer of slender anastomosing longitudinal muscle fibres, and an outer layer of circular muscle fibres. At the posterior end of the common oviduct the circular muscle fibres are gathered into closely spaced rings, but anteriorly they lose their regularity and form a network (Figure 17.4). The common oviduct divides anteriorly and dilates to form paired ampullae from which lateral oviducts run to the ovaries. The outer circular muscles of the ampullae are few and are attenuated but the longitudinal muscles are more extensively developed. The secretory epithelium lining the ampullae becomes increasingly enlarged and folded during successive ovarian cycles (Detinova, 1962; Giglioli, 1963).

18

Spermatogenesis and the structure of spermatozoa

Differences of structure are often found between the spermatozoa of the higher taxa, e.g. at the family level. Mosquito sperm are very exceptional in having a 9 + 9 + '1' arrangement of flagellar tubules. They are also unusual in their mode of swimming, the sperm-tail propagating two waves that differ in frequency and amplitude.

18.1 SPERMATOGENESIS

During the successive stages of spermatogenesis a small number of primordial germ cells multiply and become differentiated to form many thousands of spermatozoa (Table 18.1). The earliest phase of spermatogenesis, the formation of primary spermatogonia from primordial germ cells, is condensed to a single step in the table because little is known of this phase in mosquitoes. With the formation of spermatogonia the germ cells enter upon a multiplication phase that leads to the development of spermatocytes which undergo meiosis, with the halving of chromosome numbers which is essential for gamete formation. Male meiosis is described in detail in Section 1.3.1. Finally, the haploid products, the spermatids, undergo cellular differentiation to form spermatozoa.

The testis rudiments form during embryogenesis when groups of pole cells move anteriorly and become enclosed by mesodermal sheaths (Section 2.1.4(b)). The pole cells are now called primordial germ cells, and in *Aedes vexans* there is a small increase in the number of these cells during embryogenesis (Horsfall *et al.*, 1973). Spermatogenesis is cystic in insects. Spermatogonia become enclosed individually within sacs formed of a single layer of mesodermal cells, and subsequently divide and multiply.

Table 18.1 An outline of the stages of spermatogenesis in mosquitoes. Little is known about the earliest phase, the development of primordial germ cells to primary spermatogonia.

Diploid	Primordial germ cells	some differentiation and mitosis
Diploid	Primary spermatogonia	mitoses
Diploid	Secondary spermatogonia	
Diploid	Primary spermatocytes	meiosis I
	Secondary spermatocytes	meiosis II
Haploid	Spermatids	spermiogenesis
Haploid	Spermatozoa	

Spermatogenesis occurs mainly during the larval and pupal stages of mosquitoes but species differ in the timing of that process. *Aedes stimulans*, which shows three main periods of testis growth, is probably typical of many species. Late in the 1st instar the testes increase slightly in length and the germ cells become globular. During the 2nd instar and the first half of the 3rd instar new germ cells are formed anteriorly, accompanied by a doubling in length of the testis, and the germ cells become enclosed in cysts. From the middle of the 4th instar to the middle of the pupal stage new generative tissues are added anteriorly, new cysts are formed and uninterrupted mitotic activity is observed among the germ cells. Spermatozoa may appear in the most posterior cysts as early as five hours after larval–pupal ecdysis, and by 95 h after the ecdysis (at 21°C) spermatozoa are present through most of the length of the testis. Some cyst walls at the posterior end of the testis become histolysed, and sperm lie free within the testis cavity and in the most proximal part of the vas efferens (Horsfall and Ronquillo, 1970).

Spermatogenesis proceeds more rapidly in *Anopheles atroparvus*. At the beginning of the 4th instar the most advanced germ cells are primary and secondary spermatogonia, but after 72 h spermatocytes are present, and after 96 h both spermatids and spermatozoa are found. The germ cells are not enclosed in cysts but cells at the same stage of development are aggregated in clusters (Cadeddu and Laudani, 1974). Spermatogenesis occurs equally early in *Culiseta inornata* (Warren and Breland, 1963).

A single mosquito testis can contain germ cells at all stages of differentiation. The germ cells within any one cyst are all at the same stage of development, and the more posterior cysts contain the more developed germ cells. The developing spermatids are aligned side by side in nearly perfect register so that, in transverse sections, all are cut at the same level (Phillips, 1969). In *Ae aegypti* the mean numbers of cysts per testis during each of the first four days of larval life were 4, 7, 14 and 20, and the mean number remained at about 20 until the 1st day of adult life, after which it declined. Spermatids were sometimes found in pharate pupae but in most individuals they did not appear until half-way through the pupal stage. Spermatids sometimes appeared in one testis before they appeared in the other. About half of the pharate adult males contained spermatozoa in their most distal cysts. After the 1st day of adult life the spermatozoa actively moved about within the terminal cyst, often in dense spinning whorls (Jones, 1967). Mature cysts contained, on average, about 700 spermatozoa (Jones and Wheeler, 1965). This is not a number that would be expected from the repeated doubling in number of germ cells, but it may be a consequence of cell deaths. Lomen (1914) commented on the frequency with which degenerating germ cells could be seen during spermatogenesis in *Culex pipiens*. The seminal vesicles of three unmated males of *Ae aegypti* contained from 3700 to 6300 sperm, mean 5130, and more sperm were present within the genital ducts (Jones and Wheeler, 1965).

Spermatogenesis may continue during the adult stage but at a reduced rate. In *Ae aegypti* new cells are formed after the male has become sexually mature, and some spermatogenesis continues for more than 10 days, but the rate at which new sperm are formed is negligible compared with that in the immature adult. Spermatogenesis continues in sexually mature males of *Cx quinquefasciatus* at a slightly higher rate than in *Ae aegypti*, but in *Armigeres subalbatus* spermatogenesis has stopped completely by the time the male is capable of copulation (Hausermann and Nijhout, 1975).

The progress of spermatogenesis in *Cx pipiens*, at 25°C, was followed by autoradiography after injection of [^{3}H]thymidine into males whose testes contained germ cells at all stages of development. Radioactivity was found in the spermatogonia and primary spermatocytes four hours after the injections, and the testes were monitored thereafter at 24-h intervals. From 28 to 76 h post-injection, radioactivity was found only in the primary and secondary spermatocytes. From 100 h until the 8th day after injection, radioactivity was found in spermatids, and it was first found in spermatozoa on the 9th day. Cells that were labelled when

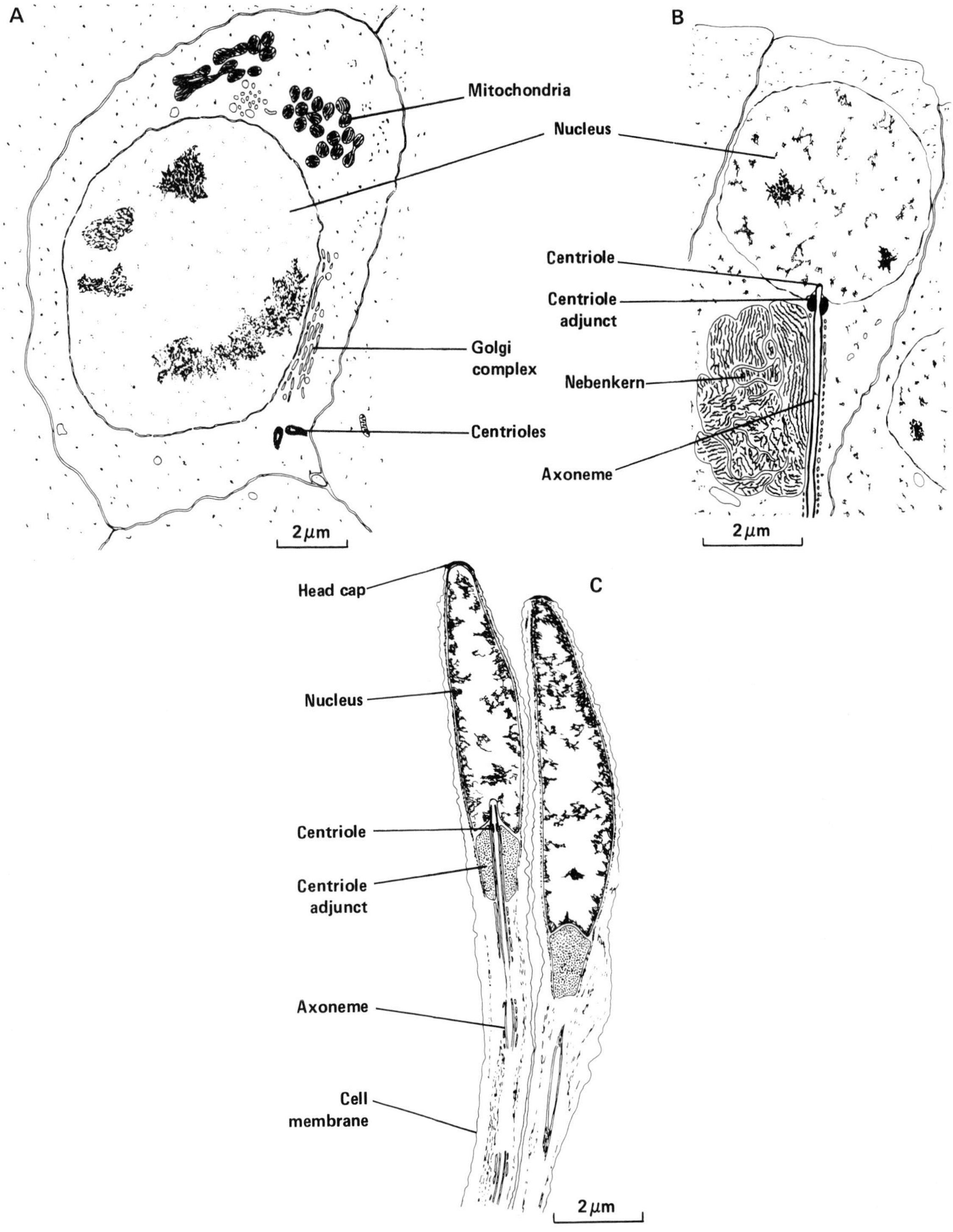

Figure 18.1 Spermiogenesis. Mosquito spermatids at three stages of cytodifferentiation. (A) Early, with paired centrioles (a rare condition) and grouped mitochondria. (B) Later, with developing centriolar adjunct and nebenkern. (C) Advanced. Drawn from the electron micrographs of Breland *et al.* (1966, 1968) and Tongu *et al.* (1968).

they were primary spermatocytes transformed to spermatozoa on the 8th day after injection, and those labelled as spermatogonia reached maturity on the 9th day (Sharma *et al.*, 1970).

The cytodifferentiation of spermatids to spermatozoa is called **spermiogenesis**. Krafsur and Jones (1967) used phase contrast microscopy to investigate this process in *Ae aegypti*, and aspects of it have been described by other authors who used electron microscopy (Breland *et al.*, 1966, 1968; Tongu *et al.*, 1968; Phillips, 1970b; Swan, 1981). Young spermatids are spherical if isolated, but when packed in a cyst their shape is irregular. They are often, perhaps always, joined in pairs or in larger groups by intercellular bridges which result from incomplete cytokinesis. Early in spermiogenesis the spermatid has a large nucleus, and its cytoplasm contains groups of mitochondria, Golgi complexes, a centriole and other organelles (Figure 18.1A). The centriole consists of nine groups of three microtubules fused into triplets. Very rarely, two centrioles are present, close to one another. The mitochondria soon fuse to form a single body, the nebenkern, which retains their crystalline structure. A small finger-shaped invagination develops in the nucleus, and the centriole becomes located at the entrance to this. In transverse section the centriole is seen to consist of nine triplet microtubules connected by radial links to a single central microtubule or rod (Phillips, 1970b). The centriole is thought to form a basal body from which the axoneme develops, but the centriole as such probably disappears before the end of spermiogenesis. The nine doublets are the first parts of the axoneme to appear. The accessory tubules then develop as curved outgrowths from subfibre B of each doublet, and the single central fibre appears last (Swan, 1981).

A small electron-dense body, the centriole adjunct, appears next to the centriole. The nebenkern is situated near the base of the axoneme (Figure 18.1B). The cell elongates away from the nucleus, giving the spermatid an anterior–posterior axis. The nebenkern divides into two portions, and as the cell grows the nebenkern and the microtubule system both increase in length, the extending nebenkern leaving two ribbons of so-called mitochondrial derivative between itself and the nucleus. The nucleus becomes ovoid, and the cytoplasm around it contains microtubules. The mitochondrial derivatives become circular in cross section, and rod-like.

Next the nucleus becomes longer and narrower, assuming a spindle shape, the amount of cytoplasm around it is reduced and the microtubules disappear (Figure 18.1C). The centriole adjunct becomes increasingly conspicuous and develops into a sheath which surrounds the mitochondrial derivatives and the flagellar tubules in the region behind the nucleus. When the spermatids of *Ae aegypti* reach their final length they are somewhat over 200 μm long (Krafsur and Jones, 1967).

18.2 SPERMATOZOA

18.2.1 Ultrastructure

Insect spermatozoa characteristically consist of head and tail regions. The greatest part of the head is composed of a highly condensed haploid nucleus, but it also bears an apical acrosomal vesicle which contains hydrolytic enzymes to assist penetration of the oocyte. Extending through most of the length of the tail are the flagellum, which provides motility, and mitochondrial derivatives, filled with crystalline proteinaceous material, to power it. The anterior-most part of the flagellum is juxtaposed to the nucleus and is embedded in a centriolar adjunct. No centrioles persist in mature insect spermatozoa (Phillips, 1970a).

Measurements of spermatozoan length in ten mosquito species indicated that *Culex quinquefasciatus* sperm were the shortest at 220 μm and *Culiseta inornata* sperm the longest at 570 μm (Breland *et al.*, 1968). Estimates of head lengths (possibly including the centriole adjunct) made from micrographs ranged from 9 μm in *Cx quinquefasciatus* to c. 26.5 μm in *Psorophora cyanescens* (Breland *et al.*, 1966, 1968). Maximum tail widths approximated 0.55 μm in *Ae*

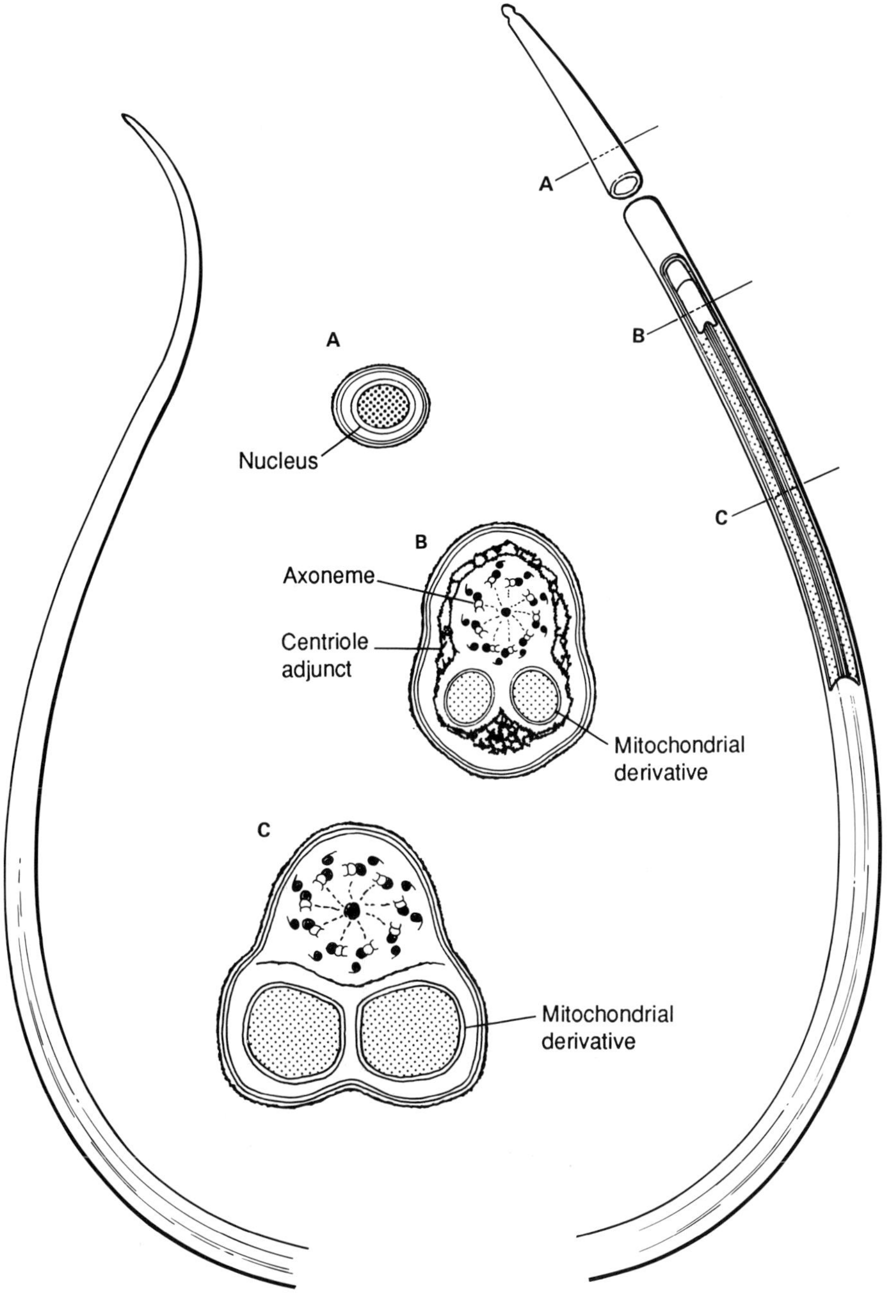

Figure 18.2A-C Structure of mosquito spermatozoa.

aegypti and 0.6 μm in *Cx pipiens* (Clements and Potter, 1967; Tongu *et al.*, 1968). The characteristics of mosquito sperm ultrastructure have been described for species of *Culex*, *Aedes* and *Anopheles* (Núñez, 1963; Clements and Potter, 1967; Tongu *et al.*, 1968; Phillips, 1969; Dallai and Afzelius, 1990).

The acrosome of *Aedes mariae* sperm is an approximately spherical vesicle, about 0.2 μm wide, with moderately electron-dense contents.

The tip of the nucleus fits into a shallow concavity in the acrosome, while a posterior extension of the acrosome occupies a recess in the nucleus. There is no subacrosomal material (Dallai *et al.*, 1984).

The limiting membrane of the mosquito spermatozoon is about 30 nm thick and is composed of a unit membrane surrounded by a surface coat. The head region consists almost entirely of a homogeneously electron-dense nucleus within a nuclear envelope and bounded by the limiting membrane of the spermatozoon (Figure 18.2A). A deep invagination in the posterior region of the nucleus receives the accessory tubules of the flagellum (Figure 18.2).

Over a short length of the spermatozoon, just behind the head region, a sac-like sheath, the centriole adjunct, surrounds the flagellar tubules and the mitochondrial derivatives (Figure 18.2B). In transverse sections it appears to consist of two thick layers separated by a narrow space. At its posterior end the centriole adjunct is incomplete and does not enclose the whole of the microtubule complex.

The flagella and cilia of animals and plants almost invariably contain a system of microtubules, or axoneme, composed of an array of nine doublet microtubules arranged cylindrically around a pair of single microtubules. Radial spokes extend from each doublet to the inner pair. The microtubule walls are formed of longitudinal series of protofilaments, themselves assembled from tubulin molecules. The walls of the two central microtubules are composed of 13 protofilaments. Each doublet consists of one complete and one partial microtubule, designated A and B tubules respectively. Each A tubule is formed of a ring of 13 protofilaments, while the B tubule is formed of 11.

In most insects the sperm tail conforms to this 9 + 2 pattern but with the addition to the axoneme of an outer ring of 9 singlet microtubules, called accessory microtubules, giving a 9 + 9 + 2 pattern. The number of protofilaments in the accessory tubules varies in different insects, but in most it is 16 (Phillips, 1974; Dallai and Afzelius, 1990). Mosquito sperm-tails have a highly unusual pattern, 9 + 9 + '1', the central pair of microtubules being replaced by a single rod with an electron-opaque core (Figures 18.2B, 18.3). Moreover, each accessory tubule is composed of 15 protofilaments. At present, mosquitoes are known to share these two characteristics with *Bibio* (Diptera, Bibionide) alone. Transverse sections of mosquito sperm show tubule B bearing two short (dynein) arms, the accessory tubule bearing a single arm, and a block of electron-dense material attached to tubule A and stopping just short of the accessory microtubule. As the tail region narrows posteriorly the

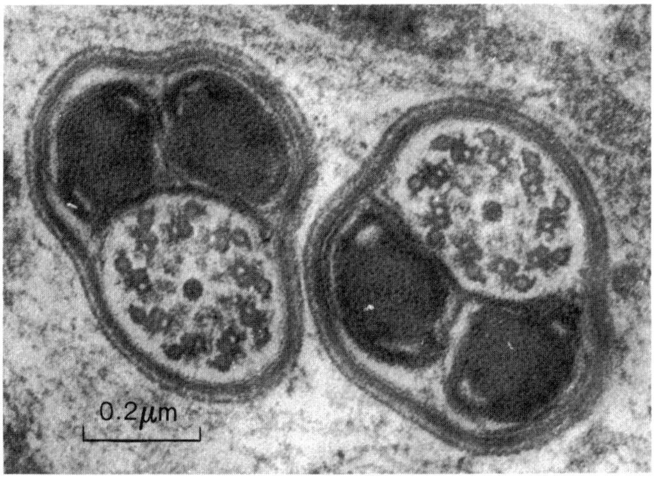

Figure 18.3 Transverse sections through spermatozoa of *Aedes aegyti* within a spermatheca.

mitochondrial derivatives disappear, and further back the axoneme becomes reduced, losing first the accessory microtubules and subsequently the central rod and some of the doublets (Núñez, 1963; Clements and Potter, 1967; Phillips, 1969; Dallai and Afzelius, 1990).

The two mitochondrial derivatives, which are of equal size, are largely composed of dense paracrystalline material. In transverse sections of *Cx pipiens* sperm this had a lattice structure with a periodicity of 8 nm in three preferred directions, appearing to form angles of 60° with one another (Núñez, 1963). An electron-dense sheet separates the mitochondrial derivatives from the microtubules. The mitochondrial derivatives terminate before the end of the tail leaving a short end piece which contains only microtubules.

18.2.2 Motility

The spermatozoa of most animals propagate quasi-sinusoidal tail waves of a single waveform. Mosquito-sperm tails propagate two waves differing in frequency and amplitude, a waveform that has been observed in the sperm of some other insects. In *Aedes notoscriptus*, small-amplitude fast waves of 34 Hz were superimposed upon large-amplitude slow waves of 3.4 Hz beat frequency (Swan, 1981). In *Ae aegypti*, the smaller component had a wavelength of 12 μm and an amplitude of vibration of 1.3 μm, while the larger component had a wavelength of 45 μm, an amplitude of 12 μm, and a propagation rate of 66 $\mu m\ s^{-1}$ (1.46 Hz). The most active sperm swam at *c.* 25 $\mu m\ s^{-1}$ (Linley and Simmons, 1981).

Mosquito sperm swim in an approximately straight line and rotate as they move. In *Ae aegypti* the longer wavelength constituent of the wave-form was described as helical (Linley and Simmons, 1981). In *Culex* sperm, the plane of displacement was the same over short lengths of the tail but not identical over the full length (Phillips, 1974).

Aedes notoscriptus sperm moved actively in buffered saline for only a few minutes. Treatment of sperm with detergent resulted in removal of the plasma membrane and cessation of movement. Demembranated sperm that were treated with a reactivation solution containing ATP regained their motility, remained active for over an hour, appeared to move normally and propagated a double wave. This showed that the plasma membrane was not involved in generation of the double waves. Sonication in the reactivation medium damaged the tails of some sperm. These sperm propagated a double wave up to the point of damage but propagated only the small-amplitude fast wave beyond the damaged region (Figure 18.4). Because the large-amplitude slow wave was not propagated beyond regions of damage, it was presumed not to be generated autonomously but to be generated by an anteriorly located structure. In contrast, the low-amplitude fast wave was generated autonomously beyond regions of damage, and was presumed to be generated and propagated by the axoneme (Swan, 1981).

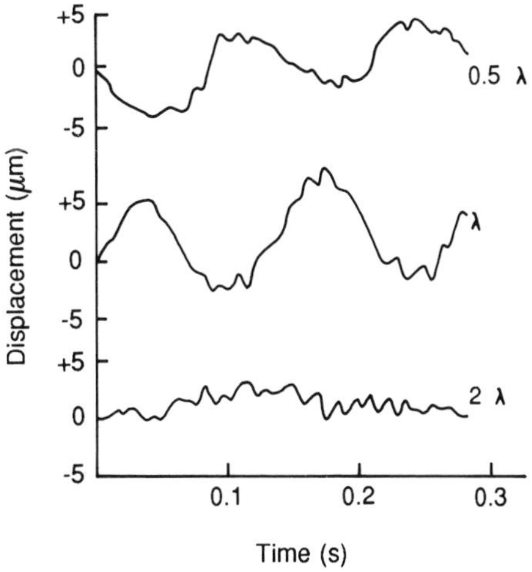

Figure 18.4 Wave propagation in an *Aedes notoscriptus* sperm-tail. Movements of the tail of a sperm that had been demembranated, reactivated and sonicated were measured at three points behind the head equal to 0.5, 1 and 2 wavelengths (λ) of the large-amplitude wave. The plots at 0.5λ and λ contain a large-amplitude low-frequency wave of 3.4 Hz on which is superimposed a low-amplitude high-frequency component of approx. 34 Hz. The sperm-tail was damaged mid-way between points λ and 2λ from the end of the sperm-head. Only the low-amplitude high-frequency wave propagated beyond the point of damage. (From Swan (1981). Copyright © (1981) and reprinted by permission of Wiley-Liss, a Division of John Wiley and Sons, Inc.)

19

Oogenesis

Ovarian follicles are the productive units of insect ovaries. In the meroistic ovaries of higher insects each follicle consists of an oocyte and a number of nurse cells, all of oogonial origin, and a bounding epithelium which is derived from the mesoderm. Formation of a follicle starts within the germarium when a stem line oogonium gives rise to a cystoblast, a cell whose progeny will cease dividing after a specific number of mitoses. The cell divisions of the cystoblast and its daughter cells are incomplete, so that a clone of cystocytes connected by intercellular bridges is produced. Once this cluster of cells has become surrounded by a layer of somatic cells an ovarian follicle has formed. One of the enclosed cystocytes becomes differentiated as an oocyte, the remainder as nurse cells.

The oocyte starts upon meiosis, which progresses to a limited extent during oogenesis. A more conspicuous change in the oocyte is its many hundred-fold increase in volume, which results from the incorporation of yolk derived from the fat body. The nurse cells do not undergo meiosis; in the Diptera their chromosomes either undergo numerous endomitotic cycles within the nuclear envelope, producing many dispersed chromosomes, or undergo endoreduplication cycles with the formation of polytene chromosomes. Both conditions provide an immensely enhanced capacity for RNA synthesis which is activated during oogenesis, when the intercellular bridges allow a stream of cytoplasm rich in ribosomes to flow from the nurse cells into the oocyte. In most insects with polytrophic meroistic ovaries there is, consequently, no need for amplification of the oocyte's rDNA, and the oocyte's chromosomes are condensed into a small karyosphere during oogenesis. As will be described, mosquitoes are exceptional in condensing their oocyte chromosomes into a karyosphere while also developing a nucleolus which extends throughout the enlarged oocyte and which is active in RNA synthesis.

During the later phases of follicle maturation structural proteins that will assemble to form the chorion are deposited over the surface of the oocyte by the follicle cells. The chorion remains soft until after oviposition, and a pore, the micropyle, permits the entry of sperm as the egg is laid.

19.1 THE FORMATION OF OVARIAN FOLLICLES

During mosquito embryogenesis pole cells form after cleavage energids have entered the posterior periplasm with its polar granules. The pole cells migrate in two groups and settle in the 6th abdominal segment, where they become surrounded by mesodermal sheaths and form the gonad rudiments (Section 2.1.4.(b)). The pole cells are now called primordial germ cells. By the 4th larval instar the ovaries are already distinct and contain cell clusters within a matrix of smaller cells, the interstitial tissue. By the early pupal stage each ovary has reformed into a group of ovarioles. Later in the pupal stage each ovariole differentiates into a germarium and a vitellarium comprising one, two, or even three ovarian follicles.

The mosquito germarium consists of a central mass containing comparatively large nuclei, surrounded by a layer in which smaller nuclei are irregularly scattered. The larger nuclei are those of the primordial germ cells and of oogonia derived from them and termed cystoblasts. The smaller, peripheral nuclei give rise to follicular epithelial cells. At intervals clusters of eight cells, partly surrounded by epithelial cells, appear at the posterior end of the germarium. The clustered cells are cystocytes, and each cluster is derived from a cystoblast through three successive mitotic divisions. Eventually the whole body will separate from the germarium as an ovarian follicle, when it will be surrounded completely by the epithelium and a basement membrane.

Cytokinesis is incomplete in each of the three cell divisions that give rise to the cystocytes; consequently the eight siblings form a syncytium interconnected by cytoplasmic bridges or ring canals. In *Aedes aegypti* the planes of the three divisions are such that the eight interconnected cystocytes form not a linear but a branched grouping; thus two cells each have three intercellular bridges, two cells each have two, and four cells each have one. One of the cells with three intercellular bridges differentiates as the oocyte (Fiil, 1974). Following differentiation the chromosomes of the oocyte start upon meiosis, but the nurse cell chromosomes undergo endoreduplication and become polytene. Structures identified as polycomplexes, which are concerned with the meiotic process (Section 1.3.2.(b)), have been observed in a small percentage of *Ae aegypti* nurse cell nuclei (Roth, 1966), which suggests that certain nurse cell nuclei start upon meiosis. Once the oocyte and nurse cells have differentiated, the synchrony that typified earlier cell division is lost and the nurse cells within a single follicle fall out of phase with one another in their polytene cell cycles. Although endoreduplication of nurse cell chromosomes may start during the pupal stage, it is not completed until some days after emergence. In *Anopheles atroparvus* the nurse cell nearest to the developing oocyte is larger than the others and its nucleolus-organizer locus is the most active, as demonstrated by the large puff in region 5 of the X chromosome. Among the remaining nurse cells both the size of the nucleus and the degree of polyteny diminish as the cells are located further from the oocyte (Kreutzer, 1970). In histological sections, the oocytes of the primary follicles can be distinguished from the nurse cells before emergence (*Aedes aegypti*) or at the time of emergence (*Anopheles gambiae, Culex quinquefasciatus*), and they are therefore already differentiated (Christophers, 1960; Roth, 1966; Fiil, 1976a; 1978a).

At a certain stage of development the first follicle becomes distinguishable owing to a constriction which starts to separate it from the germarium. Follicles separate synchronously from all germaria, the first batch to appear forming the series of primary follicles. Each follicle remains connected to its germarium by a follicular stalk consisting of a row of cells, formed by a local proliferation of epithelial cells within a strand of basement membrane. In *Anopheles*, at least, the follicular stalk connects with the follicle asymmetrically, and the micropyle apparatus of the egg forms immediately below it (Imms, 1908; Nicholson, 1921; Christophers, 1960; Parks and Larsen, 1965, Shalaby, 1971).

Species differ in the extent to which the primary follicles have separated from the germaria at the time of emergence. In *Cx pipiens pallens* and *Cx tritaeniorhynchus* the primary follicles are still largely enclosed within the germaria at emergence (Oda, 1968; Yajima, 1973); in *An sinensis* and *Ae aegypti* they are visibly distinct but still closely attached to the germaria (Taketomi, 1967; Gwadz and Spielman, 1973). Large females of *Ae togoi* emerge with their primary follicles separated from the germaria. The primary follicles of *Mansonia uniformis* are even more advanced at emergence, with the follicular epithelium actively in cell division (Laurence and Simpson, 1974). There is evidence that, within a species, the quality of larval nutrition determines the stage of development of the ovarian follicles at emergence (Mer, 1936).

With each ovarian cycle a batch of oocytes matures and a new set of follicles forms within the germaria, separates and starts upon development. In *Ae aegypti* a second set of follicles appears in the germaria at the time that the primary follicles enter the previtellogenic resting stage. About one day after a blood meal these secondary follicles separate from the germaria. During the 3rd and 4th days they grow and develop to the previtellogenic resting stage while the primary follicles produce mature oocytes. A set of tertiary follicles appears within the germaria during this period (Gwadz and Spielman, 1973; Feinsod and Spielman, 1980b).

Other species were found to differ in the times of formation and development of the later sets of follicles. In *Cx pipiens* the secondary follicles made only slight growth after the blood meal when the primary follicles developed to maturity, and failed to complete their previtellogenic growth as long as primary oocytes were retained. The secondary follicles resumed development after oviposition and reached the previtellogenic resting stage within 48 h (Readio and Meola, 1985). In *An quadrimaculatus* growth of the secondary follicles to the previtellogenic resting stage also occurred in two steps, one before and one after oviposition (Meola and Readio, 1988). In contrast, the secondary follicles of *An stephensi* had already separated from the germaria by the time the primary follicles were at the previtellogenic resting stage, and as the primary follicles developed to maturity after blood feeding the secondary follicles developed to the previtellogenic resting stage (Redfern, 1982).

19.2 THE OVARIAN CYCLE

In mosquitoes, egg production is a cyclic process. Batches of sister follicles separate successively from the germaria of the many ovarioles, and the follicles of each batch develop synchronously through a sequence of steps until the oocytes they contain are fully formed. The steps from the formation of a batch of sister follicles to the formation of a batch of mature oocytes constitute an ovarian cycle. In anautogenous mosquitoes all ovarian cycles except the first extend over more than one gonotrophic cycle (see below). Species differ in the details of their reproductive physiology, but characteristically after a blood meal the proximate series of follicles, i.e. those nearest the oviducts (Feinsod and Spielman, 1980a), mature and are laid, the penproximate follicles develop to the previtellogenic (stage II) resting stage, and the follicles of the next series separate from their germaria and rest at stage I. Thus two ovarian cycles will normally be in progress in one reproductively active female.

The most important developmental events during an ovarian cycle are: (i) differentiation of the oocytes; (ii) reductional division of the oocytes' chromosomes; (iii) synthesis of RNA by oocytes and nurse cells and its deposition in the ooplasm; (iv) synthesis of yolk by the fat body and its deposition in the ooplasm; and (v) formation of the chorion.

Beklemishev (1940) introduced the different concept of the 'gonotrophic cycle', which starts with the search for a host and the taking of a blood meal, involves the maturation of a batch of oocytes, and ends with oviposition. Field workers often record the duration of the 'oviposition cycle', i.e. the period between one oviposition and the next.

19.2.1 Phases and stages of ovarian development

Many investigators use changes in the linear dimensions of ovarian follicles as a measure of growth, usually recording the mean length of the follicles or of the contained yolk mass (Figure 19.1), but this provides little information about the developmental state of the follicles. For descriptive purposes, the ovarian cycle can be divided into sequential phases and stages using physiological and anatomical criteria (Table 19.1). Modifying the suggestions of Troy *et al.* (1975), we can define four developmental phases:

1. *Previtellogenic phase.* The period during which follicles separate from the germaria, cell types become visibly distinct, the oocytes become competent to incorporate vitellogenin, and a little growth occurs prior to entry into a resting stage.
2. *Initiation phase.* The few hours immediately after a blood meal when follicle growth recommences, with formation of additional

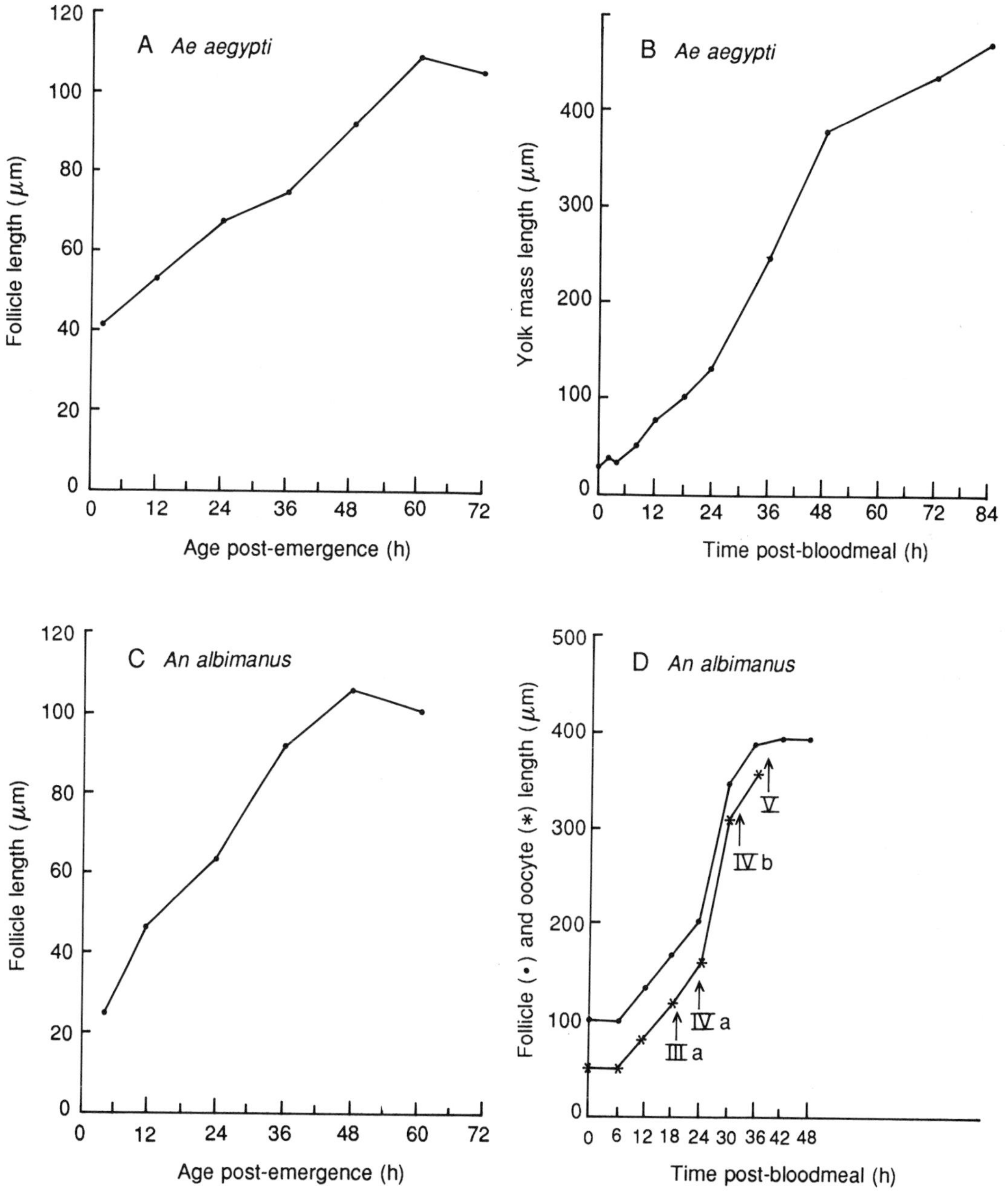

Figure 19.1 Growth of primary ovarian follicles recorded as changes in mean follicle length. (A) In sugarfed *Aedes aegypti* after emergence. (B) In *Ae aegypti* after a blood meal. (C) In sugarfed *Anopheles albimanus* after emergence. (D) In *An albimanus* after a blood meal. All mosquitoes were kept at 27°C. (After Hagedorn *et al.*, 1977; Yonge and Hagedorn, 1977; Lu and Hagedorn, 1986.)

microvilli and coated pits, and vitellogenin synthesis starts.

3. *Trophic phase*. The main period of vitellogenin synthesis and incorporation with rapid oocyte growth, ending with the cessation of vitellogenin synthesis and formation of an intact endochorionic layer between oocyte and epithelial cells.
4. *Post-trophic phase*. When the oocyte assumes its mature form and becomes surrounded by an unhardened chorion.

Christophers (1911) divided the course of ovarian development of *Anopheles* into five stages, based on the appearance of follicles under a microscope, and it is now possible to relate Christophers' stages to some of the physiological events of oogenesis. Later authors elaborated Christophers' original scheme for other mosquitoes (Kawai, 1969; Watts and Smith, 1978; Clements and Boocock, 1984), and a modern version is given below.

1. *Stage G*. The follicle has partly separated from the germarium but the oocyte is not entirely surrounded by epithelium.
2. *Stage Ia*. The follicle has separated from the germarium and the oocyte is entirely surrounded by follicular epithelium. The oocyte is not visibly distinct from the nurse cells.
3. *Stage Ib*. The oocyte is distinguishable from the nurse cells by its smaller nucleus and prominent nucleolus. At a magnification of 200× no lipid droplets are visible in the ooplasm.
4. *Stage IIa*. Some very small refractile lipid droplets are visible in the ooplasm at 200× magnification but not at 50×.
5. *Stage IIb*. The ooplasm contains fine lipid droplets, clearly visible *en masse* at 20× to 50× magnification but not distinct at 10×; these cause the oocyte to appear slightly dark under transmitted light but white by reflected light. No large droplets are visible in the ooplasm, and no ooplasmic inclusions stain with neutral red. The oocyte's nuclear membrane and nucleolus can still be distinguished.
6. *Stage IIIa*. The ooplasm is clouded with inclusions distinct, *en masse*, at 10× magnification. At first the inclusions are small. Subsequently, large yolk spheres appear

Table 19.1 Relationships of the phases and stages of ovarian development and of the developmental gates.

Phase of development	*Christophers' stage*	*Developmental gate*	*Stimulus for further development*
Previtellogenic phase	G	Germarial gate	20-Hydroxyecdysone
	Ia		
	Ib	Stage I gate	Juvenile hormone
	IIa, IIb	Previtellogenic gate	Uncertain
Initiation phase	IIIa	Stage III gate	OEH + 20-hydroxyecdysone
Trophic phase	IIIb		
	IVa		
	IVb		
Post-trophic phase	V	Maturation gate	Syngamy

which stain yellowish-brown with neutral red. These yolk spheres are of similar diameter to the epithelial cell nucleoli, and are about 0.2 to 0.3 times the diameter of the oocyte's nucleolus. The oocyte's nucleus becomes obscured by the yolk. The oocyte progressively occupies up to 50% of the length of the follicle.

7. *Stage IIIb.* The follicle has increased in size considerably but without changing shape. The oocyte occupies >50% to 75% of the length of the follicle.
8. *Stage IVa.* The follicle has grown further and has become slightly narrower, starting to assume the shape of a mature oocyte. The oocyte occupies 90% of the follicle length. The nurse cells still appear intact and occupy about 10% of the follicle length.
9. *Stage IVb.* The follicle assumes the shape of the mature oocyte, or almost so, but has not reached full length. The nurse cells have degenerated and at first stain bright crimson with neutral red; later in this stage they disappear. Chorionic structures start to appear.
10. *Stage V.* The oocyte reaches its full length. The follicular epithelium degenerates. Chorionic structures such as the floats (*Anopheles*), micropyle cup (*Culex*) or surface sculpturing (*Aedes*) become fully formed.

This description of ovarian stages is based on the development of primary ovarian follicles in *Ae aegypti* and *Cx pipiens*. *Anopheles gambiae* shows the following differences: there is no distinct stage Ib characterized by differences between oocyte and nurse cells, and by stage IIa the oocyte half fills the follicle.

19.2.2 Developmental gates

In an ovarian cycle periods of growth and development alternate with periods of arrest. Arrested follicles resume development only under specific physiological conditions, which allows development of batches of follicles to be coordinated with the cycles of feeding and oviposition. In anautogenous females, developing follicles normally undergo at least three periods of arrest, and other points of potential arrest can be demonstrated experimentally. The concept of developmental gates is helpful in this context. These gates are points of potential or actual arrest at which follicular development will be halted, without detriment to the follicles, unless or until specific physiological conditions occur (Table 19.1; Figures 21.2, 21.4). These might entail exposure to specific stimuli or, conceptually at least, the absence of specific inhibitors. Follicles held at a particular developmental gate are considered to have entered the corresponding resting stage.

(a) Germarial gate

In the earliest stage at which ovarian follicles are visibly distinct they are still closely attached to the germaria (Christophers' stage G). Normally the primary follicles separate from the germaria shortly after emergence, and each subsequent batch of follicles leaves the germarial resting stage when a blood meal restarts the ovarian cycle. The stimulus for development beyond the germarial gate is thought to be provided by 20-hydroxyecdysone (Section 21.9).

(b) Stage I gate

After follicles have separated from the germaria they grow slightly but may soon enter a state of arrest. Follicles in the stage I resting stage are about 40–50 μm long, their epithelia consist of relatively few cells, and their oocytes are visibly distinct but do not contain lipid or other droplets in the ooplasm. In many species the primary follicles of well-nourished females normally pass the stage I gate without a stop, and the gate can be demonstrated only by experimental procedures such as ligation of the abdomen at emergence. The stimulus that promotes development beyond the stage I gate is provided by juvenile hormone (Section 21.1).

The primary follicles of undersized females, derived from unfed or overcrowded larvae, may be arrested at the stage I gate until a meal of sugar or blood is taken. Treatment of small unfed females with juvenile hormone will stimulate the primary follicles to develop beyond the stage I gate (Mer, 1936; Feinsod and Spielman, 1980a, b). The proximate follicles of females entering hibernation are arrested at the stage I gate or possibly at an even earlier stage (Oda, 1968; Sanburg and Larsen, 1973; Spielman and Wong, 1973). In most anautogenous females the penproximate follicles are probably held at the stage I gate when the proximate follicles are held at the more advanced previtellogenic gate. *An sinensis* and *Cx tritaeniorhynchus summorosus* are unusual in that most females are reported to have their primary follicles in the stage I resting stage when they first engorge, yet they develop these follicles to maturity on the one blood meal (Taketomi, 1967; Yajima, 1973).

(c) Previtellogenic gate

The primary follicles may develop directly to Christophers' stage IIa or IIb after emergence, or may develop to this stage only after the females have fed. At this stage the follicles are about 100 μm long and the ooplasm contains fine lipid droplets. The primary follicles of anautogenous females then enter the previtellogenic resting stage, remaining in it until the female has taken a blood meal. The primary follicles of autogenous females pass the previtellogenic gate without arrest, demonstrating the conditional nature of the gates.

(d) Stage III gate

In most anautogenous species that have been examined, follicles that pass the previtellogenic gate and start to deposit yolk in the oocytes either develop to maturity or are resorbed. There have been a few reports of follicles entering arrest during the course of vitellogenesis, and they indicate the existence of a stage III gate.

Evidence of one or more gates operative during the trophic phase was found in two cave-dwelling West African mosquitoes. Females of *An caroni* emerged with their primary follicles at stage I; they engorged after two days, when the follicles developed to stage IIb or IIIa. A second blood meal was taken three days after the first, and the primary follicles developed to stage IIIb. A third blood meal taken three days after the second permitted maturation of the ovaries, some eight days after the first blood meal (Pajot, 1964). Females of *An hamoni* required four blood meals to mature their first batch of eggs. They emerged with their primary follicles at stage I, and the first blood meal permitted development to stage IIb, the previtellogenic resting stage. The second blood meal permitted development to stage IIIb, the third to stage IV, and the fourth to stage V. At the constant temperature of 24.5°C in the cave, ovarian development was completed 11 days after the first blood meal (Adam *et al.*, 1964; Adam and Vattier, 1964). Krafsur (1977) reported wild females of *An funestus* with ovaries resting at stages IIIa and IIIb.

In some females of *Aedes aegypti* that take a small blood meal, the ultimate follicles enter arrest in stage IIIa but resume development after a further blood meal (Lea *et al.*, 1978; Clements and Boocock, 1984). Experimental evidence is consistent with the idea of a stage III gate in *Ae aegypti*. A critical period for oocyte maturation occurs some 3–10 h after blood feeding, prior to which decapitation, which blocks OEH release, causes arrest at stage III. There is evidence that 20-hydroxyecdysone stimulates development from stage III to the mature follicle (Sections 21.4, 21.5).

(e) Maturation gate

Oocytes that have reached stage V and are ready for ovulation are often described as mature, yet their chromosomes are still in the metaphase of the first meiotic division. Meiosis is resumed only after the oocytes have been penetrated by sperm at oviposition (Chapter 2). Thus the oocytes are held in a state of nuclear stasis from the time they are ready for ovulation until syngamy stimulates completion of the maturation divisions. The maturation resting stage normally lasts only a few hours, until the appropriate activity rhythm leads to ovulation and oviposition. However, if females are unable to oviposit, the stage V oocytes can remain viable within their ovaries for 15 days or more (Gillett, 1955; Woke, 1955; Shalaby, 1959).

19.3 EFFECTS OF EXTERNAL FACTORS

Nutrition, photoperiod and temperature are factors that determine whether or not a female will develop her ovaries and at what rate.

The quality of larval nutrition affects previtellogenic development of the primary follicles in newly emerged adults. For example, large females of *Anopheles sacharovi*, *An atroparvus* and *Aedes aegypti*, presumably derived from well-nourished larvae, were able to develop their primary follicles to the previtellogenic resting stage without feeding, therefore at the expense of reserves accumulated during larval life. Small females had to feed on sugar or blood before their primary follicles could develop to the privitellogenic resting stage (Mer, 1936; Detinova, 1944; Feinsod and Spielman, 1980a). In nature the females of some species commonly take two blood meals during a gonotrophic cycle. Most wild females of *An gambiae* needed two blood meals to develop their first batch of eggs, although, in the laboratory, females from well-nourished larvae would develop their first batch of eggs after one large meal (Gillies, 1954; Fiil, 1976a). The same appeared to be true of *Ae aegypti* (Macdonald, 1956; Feinsod and Spielman, 1980a). Additional blood meals may be taken by semi-gravid or gravid females, which in *Ae aegypti* may constitute about 40% of wild females caught biting (Macdonald, 1956; Yasuno and Tonn, 1970; Hervy, 1977).

Temperature markedly affects the rate at which ovarian follicles develop from the previtellogenic resting stage to maturity. In the northerly Moscow Oblast females of *An maculipennis* required 79 days to complete 12 gonotrophic cycles, whereas in southern USSR, in Stalingrad Oblast, only 42 days were required (Detinova, 1962). The ambient temperature at Taveta, in Tanzania, affected two species differently. During the cool–dry season, when mean temperatures ranged between 23.3 and 24.7°C, *Anopheles funestus* had a 3-day gonotrophic cycle; for the rest of the year, when temperatures exceeded 25.5°C, it had a 2-day cycle. In contrast, the gonotrophic cycle of *An gambiae* lasted two days throughout the year (Gillies, 1953). In Rangoon, *Cx quinquefasciatus* oviposited at sunrise and sunset. Females ovipositing at sunset had fed after midnight three nights before; those ovipositing at sunrise had fed before midnight two nights earlier (De Meillon *et al.*, 1967b). The gonotrophic cycle of *Chagasia bonneae*, in Brazil, may last only two days (Wilkes and Charlwood, 1979); in Assam, females of *An vagus* were found ready to lay their eggs 24 h after taking a blood meal (Muirhead-Thomson, 1951). However, ovarian development is not rapid in all tropical species; in *Coquillettidia fuscopennata* it took seven days at 23°C (Gillett, 1961). The temperate region species *Culiseta annulata*, which feeds throughout the year, has a gonotrophic cycle of 5–6 days in midsummer but in winter this lasts 23–26 days (Service, 1968a). Some species can complete ovarian development within a short period even at low temperatures. On Ellesmere Island in the high arctic, during July when the mean temperature was 6.7°C, *Ae impiger* required 6 days or less after emergence to become gravid and *Ae nigripes* required 8 days or less (Corbet, 1965).

Exposure to certain photoperiods causes the females of some species to hibernate, when their primary follicles do not undergo normal previtellogenic development. When females of *Cx pipiens* were exposed to a photophase of 9–12 h per day during the pupal and adult stages, their primary follicles did not develop to the previtellogenic resting stage (Sanburg and Larsen, 1973). It appears that it is the failure of the corpora allata to secrete juvenile hormone that is the immediate cause of developmental arrest in the diapausing females (Spielman, 1974).

19.4 rRNA SYNTHESIS IN OOCYTE AND NURSE CELLS

In many organisms ribosomes that are assembled during oogenesis provide the machinery for protein synthesis during early embryogenesis; in these

Table 19.2 Key events in the oocyte nuclei of the primary follicles of *Aedes aegypti* during oogenesis. The times are approximate and relate to mosquitoes kept at 27°C. (From the data of Roth (1966), Roth *et al.* (1973), Anderson and Spielman (1973), Fiil and Moens (1973) and Fiil (1974).)

Time	*Events*
Before emergence	Replication of chromosomal DNA; formation of synaptonemal complexes; start of meiosis
At emergence	Chromosomes in pachytene
40–60 h post-emergence	Chromosomes in early diplotene; nucleolus synthesizes rRNA
By 72 h post-emergence	Follicles enter previtellogenic resting stage; meiosis arrested at diffuse stage of diplotene
4 h post-bloodmeal	Nucleolus resumes rRNA synthesis
By 8 h post-bloodmeal	Meiosis is resumed; polycomplexes start to extend into nucleoplasm
13–20 h post-bloodmeal	Nucleolus fragments; chromosomes enclosed in karyosphere then in germinal nucleus
24–60 h post-bloodmeal	Oocyte and nucleus grow rapidly; nucleolus extends throughout nucleoplasm
About 65 h post-bloodmeal	Oocyte attains maximum length; nuclear envelope disintegrates
By 68 h post-bloodmeal	Oocyte assumes final form and enters maturation resting stage; meiosis arrested in metaphase I
Immediately after oviposition, upon sperm entry	Capsule surrounding chromosomes disperses; meiosis resumes
Within 30 min of oviposition	Meiosis completed; fusion of male and female pronuclei

organisms large amounts of rRNA are synthesized and accumulated during oogenesis. In insects with panoistic ovaries the oocyte chromosomes and nucleoli are modified for rRNA synthesis during oogenesis. In contrast, in most insects with meroistic ovaries the oocyte chromosomes are condensed in a karyosphere and the role of rRNA synthesis is taken over by the nurse cells, the rRNA being transported via intercellular canals to the cytoplasm of the oocyte (Cave, 1982). Mosquitoes are one of the few groups of insects with meroistic ovaries in which rRNA is synthesized by both the oocytes and the nurse cells.

In eukaryotes, ribosomal RNA is produced by the continuous transcription of multiple copies of rDNA (Section 1.2.3) and is immediately packaged with ribosomal proteins to form ribosomes. Transcription occurs in the nucleolus organizer, a part of one or more chromosomes which contains the repeated arrays of rDNA. The packaging takes place in the nucleolus. Nucleolus and nucleolus organizer are absent at times when chromosomal DNA is condensed. They are visible during the early stages of meiotic prophase I, when in some organisms the nucleolus organizer appears as an extra-chromosomal DNA body within the nucleolus.

Mosquito oocytes grow in volume several hundred-fold within a short period after the female has ingested blood, so exceptional measures are necessary to produce the huge number of ribosomes required. Polyteny enables the nurse cell chromosomes to amplify their transcription rates, but that mechanism is not available to the oocytes, which have recourse to another system of enhanced rRNA synthesis.

19.4.1 The oocyte nucleus

Before females of *Aedes aegypti* have emerged, chromosome replication has occurred and meiosis has started in the oocytes of the primary follicles. When the primary follicles enter the previtellogenic resting stage their oocyte chromosomes are in diplotene. A few hours after blood feeding the oocyte chromosomes start to condense and they become enclosed within a small ball or karyosphere of about 10 μm diameter. As the oocyte grows rapidly, over a period of two or three days, the nucleus also expands and it eventually becomes boat-shaped with a length of 400 μm, width of 100–140 μm and thickness of about 3 μm. Throughout this period the chromosomes remain in the karyosphere which is located in a nuclear pocket, 12–16 μm across, at the anterior pole. The karyosphere is enclosed by a thin capsule composed of remnants of synaptonemal complex (see Section 1.3.2), and is now termed the 'germinal nucleus', distinct from the large 'vegetative nucleus'.

At emergence the nucleolus is a centrally located sphere of 3 μm diameter, composed of dense granules arranged in spheroidal sheets. After three days it has grown to 7 μm diameter and strands extend from it far into the nucleoplasm. Increased nucleolar activity is one of the earliest consequences of blood feeding. By four hours after the female has bloodfed a cavity has appeared within the nucleolus, and from about this time [^{3}H]uridine is incorporated into its dense periphery, indicating that synthesis of rRNA is taking place. The nucleolus increases massively in size and between 13 and 20 h after engorgement it breaks up into many fragments. As the oocyte and its nucleus grow rapidly between 24 and 60 h nucleolar fragments become distributed throughout the nucleoplasm (Table 19.2).

A large extra-chromosomal DNA body has not been observed in mosquitoes (Fiil, 1976a), but it is clear that the nucleolus organizer becomes physically separated from the chromosomes which are contained in the ensheathed karyosphere. It appears that after separating from the chromosomes the DNA of the nucleolus organizer is further replicated giving rise to widely dispersed rDNA-containing fragments. The incorporation of [^{3}H]uridine into this material throughout the whole period of vitellogenesis indicates continuous rRNA production. Injection of Actinomycin D into females 7 or 30 h post-bloodmeal blocked uridine incorporation completely. After the oocyte has reached its maximum length the nuclear envelope breaks down leaving remnants dispersed throughout the ooplasm (Roth, 1966; Anderson and Spielman, 1973; Fiil and Moens, 1973; Fiil, 1974, 1976b).

After females of *Anopheles gambiae* have taken a blood meal the oocyte nucleolus increases greatly in size and fragments, with the result that huge numbers of small pieces, often spherical, are found throughout the canoe-shaped nucleus in the final stages of oogenesis. The structure of the fragments resembles that of small regions of the main body of the nucleolus, i.e. of an electron-lucent thinly particulate centre surrounded by an electron-dense fibrillar zone which has a granular periphery (Fiil, 1976a). It is likely that these three zones consist of rDNA, of rDNA being transcribed, and of maturing ribosomal precursor particles (Alberts *et al.*, 1989).

The separation of the oocyte nucleus into a small 'germinal nucleus' and a massive boat-shaped 'vegetative nucleus' has been described from species of *Anopheles, Culex, Culiseta* and *Aedes*, and is thought to be unique to mosquitoes (Nicholson, 1921; Nath, 1924; Bauer, 1933; Fiil and Moens, 1973; Fiil, 1974, 1976a).

19.4.2 The nurse cells

Measurements of [^{3}H]uridine incorporation into *Culex quinquefasciatus* ovaries revealed a period of RNA synthesis during the previtellogenic phase. All three cell types in the primary follicles were synthesizing RNA but synthesis was most active in the nurse cells. The rate of synthesis started to increase at about 6 h after emergence, reached a maximum at about 12 h, and returned to a low level by 20 h. Most was 28S, 18S and 4–5S RNA, and was therefore predominantly ribosomal, and indeed the nurse cell cytoplasm became filled with ribosomes. Most of the rRNA that was synthesized during the first 48 h after emergence was conserved for at least 5 days and could be detected in the fertilized egg, but rRNA synthesized during the remainder of the previtellogenic phase was turned over (Frelinger and Roth, 1971).

In *Anopheles stephensi* the nurse cell chromosomes were already polytene at the previtellogenic resting stage. Judged by rates of [^{3}H]thymidine incorporation they were inactive in DNA replication at that stage but new replication cycles were initiated following blood feeding. Thymidine incorporation had started by 6 h post-bloodmeal, and reached a high rate by 8 h which was maintained until 34 h post-bloodmeal. The polytene chromosomes increased in size during this period (Redfern, 1981c). In *An gambiae* the nurse cell polytene chromosomes became more distinct during stages IIIb to IVa and degenerated during stages IV to V (Coluzzi, 1968; Fiil, 1976a).

The nurse cells within a follicle may develop differently. In *An gambiae* the chromosomes of some nurse cells became polytene before the female had bloodfed, those of the other cells after blood feeding (Fiil, 1976a). In vitellogenic females of *An superpictus* only the three nurse cells closest to the oocyte possessed well-developed polytene chromosomes (Coluzzi *et al.*, 1970). In females of *An atroparvus* examined 24 h after engorging, the nurse cells closest to the oocyte exhibited more nucleolar activity than those further away. The nurse cell closest to the oocyte was the largest of all, and its nucleolus organizer was the most active (Kreutzer, 1970).

In *Aedes aegypti* the nurse cells enlarged during the second day after blood feeding, and their nuclei incorporated uridine more actively at that time than did any other ovarian organelles. They were believed to synthesize most of the RNA produced within the ovarian follicles (Anderson and Spielman, 1973), and this presumably passed to the oocyte through the intercellular bridges. Some 50 h after feeding (at 24°C) the oocytes started to elongate, and shortly afterwards their connections with nurse cells were severed (Fiil, 1974). At 27°C the nurse cells degenerated between 36 and 48 h post-bloodmeal (Judson and de Lumen, 1976).

19.5 FOLLICLE GROWTH AND DEVELOPMENT

19.5.1 Previtellogenic phase

When *Aedes aegypti* are kept at 27°C their primary follicles grow in length from 40 to 110 μm during the first 2.5 days after emergence. At emergence the cells of the follicular epithelium, which may number fewer than 20 per follicle, are already undergoing division and this continues until each primary follicle has 200–250 epithelial cells, about two days after emergence. Females of some other species emerge with their primary follicles at a more advanced stage; in *Mansonia uniformis*, for example, the multiplication of epithelial cells may already be well advanced (Gwadz and Spielman, 1973; Laurence and Simpson, 1974; Fiil, 1976a; Hagedorn *et al.*, 1977). At emergence the follicle cells of *Ae aegypti* appear undifferentiated. However, the follicular epithelium is already permeated by narrow intercellular channels through which horseradish peroxidase can pass (Raikhel and Lea, 1985, 1991). At about the time that the follicles reach a length of 90–100 μm, and the epithelial cells number 200–250, the primary follicles of anautogenous females enter the previtellogenic resting stage, and develop no further until the female takes a blood meal.

In *Ae aegypti* the epithelial cells of the resting stage follicle are cuboidal with large nuclei and little cytoplasm. They are closely applied to each other and joined, at their apical margins, by desmosomes. The nucleus contains a prominent polymorphic nucleolus. The sparse cytoplasm is rich in ribosomes but relatively poor in rough endoplasmic reticulum and Golgi complexes. There is no space between follicle cells and oocyte (Roth and Porter, 1964; Raikhel and Lea, 1982, 1991).

At emergence the oocytes of *Ae aegypti* contain many mitochondria, a little RER, and many ribosomes. When the follicles are in the previtellogenic resting stage their oocytes also contain many lipid droplets, a few much smaller droplets considered to be protein, and dense lysosome-like bodies (Anderson and Spielman, 1971; Tadkowski and Jones, 1978, 1979). In *Culex quinquefasciatus* the staining properties of the lipid droplets suggest that they are rich in phospholipids (Nath *et al.*, 1958). However, the principal lipids of the previtellogenic resting stage ovaries of *Ae aegypti* were identified as 'fatty esters' (possibly esters of long-chain acids and alcohols) and hydrocarbons (Troy *et al.*, 1975).

During the first 12 h after *Ae aegypti* had emerged, the plasma membrane of the follicle cells was closely applied to the oocyte surface. The oocyte plasma membrane was smooth and lacked a glycocalyx, and the cortical region of the ooplasm contained mainly ribosomes. Twenty-four hours after emergence the oocyte plasma membrane bore a few small microvilli, and a few coated and uncoated vesicles were present in the cortical cytoplasm. At 48 h post-emergence the oocytes had well-developed microvilli and a fully developed endocytotic complex (Section 20.4.1). The plasma membrane of the oocyte was covered by a layer of glycocalyx, which was considerably denser between the microvilli where coated pits were numerous.

The cortical layer of ooplasm beneath the plasma membrane, 0.15–0.25 μm thick, was almost free of ribosomes and mitochondria and consisted of a filamentous matrix. Coated vesicles 120–140 nm in diameter, the same size as the coated pits, were the main constituent of the cortical layer, which also contained tubular structures and a few uncoated vesicles or endosomes which varied in size from 0.15 to 0.4 μm. The endosomes were characterized by a fuzzy inner coat and vesicular membrane in the lumen. Horseradish peroxidase injected shortly after emergence reached only the interfollicular and perioocytic space. Injected at 24 h it was seen in a few coated pits and vesicles, but injected at 48 h it was internalized by numerous coated vesicles and appeared within uncoated endosomes (Raikhel and Lea, 1985). Roth *et al.* (1976) found that oocytes in the previtellogenic resting stage were competent to bind vitellogenin *in vitro*.

19.5.2 Initiation phase

The initiation phase starts with the start of blood feeding. It could be considered to end when ovarian ecdysteroidogenic hormone (OEH) is released, which inaugurates the promotion of vitellogenesis and the trophic phase of ovarian development (Section 21.3). In that case, in *Aedes aegypti* the initiation phase lasts for a minimum of three and a maximum of ten hours, and in *Anopheles albimanus* it lasts for some 8–16 h. However, the decapitation of females shortly after blood-feeding is often followed by development of the follicles to stage III (Section 21.2), so it appears that sufficient hormone may be released early in the initiation phase to stimulate vitellogenin synthesis and incorporation and to permit development to the stage III gate. It is interesting that that is often the limit of ovarian development in females that have taken an inadequate blood meal.

Developmental changes can be observed shortly after blood feeding. The rate of protein synthesis in the ovaries of *Ae aegypti* increased 3-fold within 30 min of feeding, and declined between 60 and 120 min post-bloodmeal. The rate of protein synthesis later increased again, peaking at 8 h at a level ten times that of the previtellogenic ovaries, before falling to a somewhat lower rate that was maintained from 12–24 h post-bloodmeal (Figure 21.3A) (Koller and Raikhel, 1991).

Developments occur in the fat body also during the first hours after blood feeding. One hour post-bloodmeal vitellogenin mRNA was detected in females of *Ae aegypti*, and vitellogenin synthesis was observed in the fat body (Racioppi *et al.*, 1986; Raikhel and Lea, 1983). Significant amounts of vitellogenin were extracted from females at three hours post-bloodmeal (Borovsky and Van Handel, 1979).

The capability of the oocytes for vitellogenin uptake had increased in *Ae aegypti* by 1 h post-bloodmeal, judging by the rates of vitellogenin uptake by ovaries *in vitro*. At about 2.5 h post-feeding the capability for vitellogenin uptake increased dramatically, and continued to increase until the end of the first day (Figure 21.3B) (Koller and Raikhel, 1991). In *Ae aegypti* a perioocytic space (0.5–0.7 μm wide in fixed tissues) formed between the follicular epithelium and the oocyte at two hours post-bloodmeal. Thereafter, and until completion of the endochorionic layer, contact between the epithelial cells and oocyte was maintained via the oocyte microvilli, which were joined to the surface of the follicular cells by gap junctions. No space appeared between the nurse cells and the epithelial cells (Raikhel and Lea, 1991).

Within three hours of *Ae aegypti* taking a blood meal the region of the oolemma that faced epithelial cells bore many new microvilli and coated pits, and additional coated vesicles had appeared in the cortical ooplasm. Horseradish peroxidase injected into females appeared in 20-nm-wide channels between follicle cells within an hour of a blood meal, but was not seen in significant quantities in the coated pits and coated vesicles of the oocyte until three hours after feeding (Anderson and Spielman, 1971).

There is probably a marked increase in the patency of the follicles during the first hours after blood feeding, i.e. in the extent of intercellular channels in the follicular epithelium that permit open communication between haemocoele and oocytes. According to some authors, by 4 to 6 hours after *Ae aegypti* had taken a blood meal the cells of the follicular epithelium had separated slightly, allowing the haemolymph to come into contact with the oocyte (Roth and Porter, 1964; Roth, 1966; Matthew and Rai, (1975). This was denied by Anderson and Spielman (1971), who stated that the epithelial cells stayed closely apposed to each other throughout yolk deposition, being separated only by channels approximately 20 nm wide. From observations on the penetration of trypan blue, Yonge and Hagedorn (1977) concluded that intercellular spaces were present in addition to channels, but were relatively rare.

Four hours after *Ae aegypti* had bloodfed, PAS-positive particles, similar to the glycoprotein yolk precursors described in other insects, appeared in the cortical ooplasm where the oocyte faced epithelial cells. None was present opposite the nurse cells. These particles increased in size and number, and six hours after engorgement they stained for both carbohydrate and protein. Eight hours after engorgement yolk particles filled the ooplasm (Laurence and Roshdy, 1963). It is possible that the oocytes also incorporated or synthesized lipid at that time because, in *Culex quinquefasciatus*, the lipid droplets in the ooplasm started to stain for triglyceride as soon as proteinaceous yolk appeared (Nath *et al.*, 1958).

At the time of engorgement in *Ae aegypti* there was little evidence of cell division in the epithelium of the primary ovarian follicles, but a peak of mitosis occurred at four to five hours post-bloodmeal. Cell numbers increased from about 260 at the time of feeding to about 470 ten hours later. By the time the follicles had reached stage IIIb cell division had ceased. When females were injected with [^{3}H]thymidine immediately after feeding, its incorporation into epithelial cell DNA started within two hours, shortly before the peak of mitosis, but incorporation was high from six to ten hours post-bloodmeal, long after the rate of mitosis had declined (Laurence and Simpson, 1974). The significance of this may lie in an observation made with *Anopheles stephensi*, that at 24 h post-bloodmeal the follicular epithelial cells were polyploid (Redfern, 1981a). By 12 h post-bloodmeal the follicle cell cytoplasm contained abundant RER and Golgi complexes (Koller and Raikhel, 1991).

19.5.3 Trophic phase

The trophic phase of ovarian development is taken to begin with the release of ovarian ecdysteroidogenic hormone that stimulates the major burst of 20-hydroxyecdysone secretion. It ends with the cessation of vitellogenin synthesis and the deposition of an intact endochorionic layer between oocyte and epithelial cells. In the first hours of the trophic phase there is a continuation of some of the structural developments initiated in the oocyte shortly after the blood meal, e.g. an increase in the numbers of microvilli and coated pits on the oocyte surface. Indeed, the fact that the capability of the ovaries for vitellogenin uptake increases linearly until 24 h post-bloodmeal (Figure 20.12), indicates that the endocytotic apparatus is continuously synthesized. The presence of the head was required for 16–20 h post-bloodmeal for the normal rates of vitellogenin uptake to be maintained (Koller and Raikhel, 1991). The secretion of the endochorion, which occurs during the trophic phase, is described in Section 19.6.1.

The passage of materials from the haemolymph to the oocyte surface in *Ae aegypti* was investigated by Anderson and Spielman (1971). They found that the basal lamina that surrounds the follicular epithelium (Section 17.2.2) is a coarse mechanical filter which is freely permeable to polysaccharides and proteins of up to 500 kDa molecular mass and with dimension <11 nm. Carbon particles of 30–50 nm were totally excluded. After crossing the basement membrane the exogenous tracers penetrated the narrow channels running between the epithelial cells and accumulated in the enlarged perioocytic space. Vitellogenin in the haemolymph reaches the perioocytic space by the same route. The manner of its internalization and condensation to crystalline yolk protein is described in Section 20.4. Between 36 and 48 hours after the blood meal the intercellular channels between epithelial cells became occluded by desmosomes, coinciding with the cessation of vitellogenesis.

Measurements of vitellogenin accumulation in bloodfed *Ae aegypti* kept at 27°C revealed it increasing from a trace level at 8 h to a maximum at *c.* 40 h post-bloodmeal (Figure 19.2). The mean amount of vitellogenin synthesized by females fed on rabbits was 180 μg (H.H. Hagedorn, personal communication).

The ooplasm of mature mosquito oocytes is packed with vitellin crystals, droplets of lipid yolk and numerous free ribosomes (Tadkowski and Jones, 1979). The oocytes also contain much non-yolk protein, possibly as much as vitellin. These so-called ovary-specific proteins are soluble and additional to the insoluble chorionic proteins. It is possible that some are taken up by endocytosis, but ovaries are capable of protein synthesis *in vitro*. In *Ae aegypti*, *in vitro* synthesis of ovary-specific proteins was maximum about 12 h after the peak of vitellogenin synthesis. In *Cx nigripalpus*, in which ovary development progressed more slowly, the two maxima were separated by about one day (Borovsky and Van Handel, 1980; Ma *et al.*, 1986). The accumulation of substantial quantities of non-yolk protein in oocytes is readily explained by the requirements of the future young embryos for the histones used in chromosome replication and for ribonucleoprotein. As in other organisms with yolky eggs, the abundant ribosomal RNA stored in the ooplasm undoubtedly functions in the synthesis of embryonic proteins during early embryogenesis.

Almost nothing is known about the mechanism

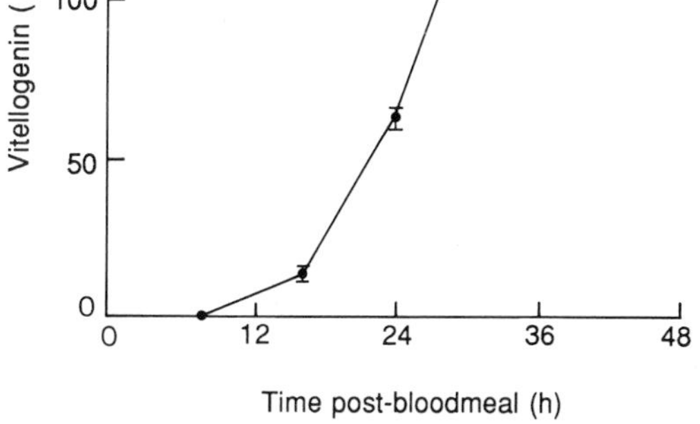

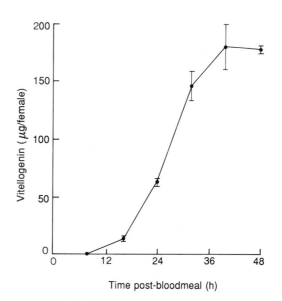

Figure 19.2 The whole-body vitellogenin content of *Aedes aegypti*, kept at 27°C, at different times after feeding on rabbits, measured by rocket immunoelectrophoresis. Means + s.e. (By courtesy of Dr H.H. Hagedorn.)

of lipid synthesis or uptake by mosquito oocytes. Extraction of lipids from the ovaries of *Ae aegypti* at daily intervals after blood feeding showed that the rate of lipid accumulation paralleled the rate of increase in dry weight. In these females, kept at 24–25°C, lipid deposition was greatest between 24 and 48 h after engorging (Figure 20.1) (Troy *et al.*, 1975). Tadkowski and Jones (1978) observed in *Ae aegypti* that lipid droplets did not appear first at the oocyte cortex, and they concluded that the lipid must be synthesized by the oocyte itself. However, the nature of the organelles that they observed in the ooplasm – free ribosomes, short segments of rough ER and many mitochondria – was not consistent with that view.

19.5.4 Post-trophic phase

The post-trophic phase starts when the plaques of endochorionic material, secreted by the epithelial cells, fuse and block the endocytotic activities of the oocyte. The time at which this event occurs is determined by temperature. In females of *Aedes aegypti* kept at 24°C the chorionic plaques obscured most of the oolemma at 48 h post-bloodmeal, and by 54 h the plaques had begun to fuse, separating the oocyte from the epithelial cells (Anderson and Spielman, 1971). When females were kept at 27°C the chorionic plaques fused a little after 36 h post-bloodmeal (Raikhel and Lea, 1982). Endocytosis had stopped by 36 h post-bloodmeal in females kept at 25–28°C (Tadkowski and Jones, 1979). When females were kept at 24°C the length of the ovarian follicles doubled between 50 and 77 h post-bloodmeal, and the dry weight of the whole ovaries increased by one-fifth between 48 and 72 h, one-third of the increase being due to lipid, notably hydrocarbons and 'fatty esters' (Figure 20.1) (Anderson and Spielman, 1971; Troy *el al.*, 1975). There is therefore evidence of continued incorporation of material into ovaries after the trophic phase, at a time when the oocytes are physically separated from the epithelial cells but not from the nurse cells. During the post-trophic phase the oocyte grows to its final size and assumes its final form. Water uptake is possibly involved in the increase in size, and the exochorion is deposited by the follicular epithelium (Section 19.6).

Gravid females of *Culex quinquefasciatus* that were fed only water for seven days after maturing their ovaries laid egg rafts containing only 60% as much glycogen as rafts laid by sugarfed controls. The glycogen content of their egg rafts was completely restored if, after the seven days on water, females were given sugar for 24h. Ovaries containing mature oocytes, when incubated *in vitro* with [^{14}C] glucose for four hours, accumulated substantial quantities of labelled glycogen and much smaller amounts of labelled lipid. The ovaries were incapable of *in vitro* glycogen synthesis during the first two days after blood feeding, the period of vitellogenesis. On the 3rd day they showed a slight capability for glycogen synthesis; this increased to a maximum by the 5th day and remained at that level until the 9th day at least. Freshly laid egg rafts, in which the chorion hardened and darkened within 30 min, showed very little capability for *in vitro* glycogen synthesis. Van Handel (1992) concluded that mature chorionated oocytes of *Cx quinquefasciatus* remain permeable to certain haemolymph constituents, and retain enzymes capable of glycogen synthesis, long after breakdown of the follicular epithelium. Mature ovaries from *Aedes aegypti* incorporated very little [^{14}C] glucose into glycogen, amounts that possibly were within the limits of experimental error.

As in other insects, the anterior–posterior axis of the oocyte conforms to that of the mother, therefore the anterior end of a mosquito oocyte is the end that was adjacent to the nurse cells. Soon after the oocyte has been ovulated and laid the sheath surrounding the germinal nucleus disperses. Entry of sperm stimulates the completion of meiosis (Section 2.1.1) and the female gamete is transformed into an egg.

19.6 CHORION FORMATION

19.6.1 Secretion of the chorion

The chorion is the part of the egg shell that is secreted before oviposition by the follicular epithelium, a maternal tissue. Secretion occurs in two phases and results in formation of the two layers of the chorion, the endochorion and exochorion. Use of the terms 'vitelline membrane' or 'vitelline envelope' for the endochorion is inappropriate (Section 3.1).

(a) *Aedes aegypti*

Two genes, 15a-1 and 15a-2, which share 68% identity, are believed to code for endochorionic proteins. The more fully investigated 15a-1 gene, which was present in single copy, coded for a 9.5 kDa protein that resembled the protein products of four *Drosophila melanogaster* endochorionic genes. The proteins deduced from the sequences of the two 15a genes shared with the *Drosophila* proteins a highly conserved hydrophobic sequence. Northern analysis, with a probe derived from gene 15a-1, of RNA extracted from whole bodies of males and females, and from fat body and ovaries of bloodfed females, revealed the presence of a 650 bp RNA transcript only in the ovaries of bloodfed females. This transcript was detectable at 5 h, abundant at 20–30 h, and declined between 40 and 50 h post-bloodmeal. A tritiated transcript, which was used to probe sections of ovary that had been frozen at 20 h post-bloodmeal, hybridized only to the ovarian follicle cells. It showed that expression of the 15a-1 gene was most pronounced in follicle cells adjacent to the oocyte, and was much less pronounced in those adjacent to nurse cells, suggesting that gene expression was modulated by the local environment within the follicle cell (Lin *et al.*, 1992).

Details of the secretory process have been elucidated by ultrastructural investigations supplemented with radiotracer techniques. The time sequences reported here are from mosquitoes kept at 26–27°C. By 6 h post-bloodmeal the protein synthetic apparatus had started to form in the follicular epithelial cells and it continued to develop until at least 16 h, the cytoplasm becoming packed with RER cisternae and Golgi complexes (Raikhel and Lea, 1982, 1991). With [^{3}H]L-histidine as tracer it was possible to mark the synthesis of chorionic protein in the RER, and then to follow its appearance in membrane-bound granules within Golgi cisternae, the fusion of these granules with the apical cell membrane, and their transfer by exocytosis into the perioocytic space (Anderson and Spielman, 1973). At 6–8 h post-bloodmeal electron-dense granules appeared in the Golgi complexes. At 8–9 h small electron-dense plaques first appeared in the perioocytic space. At 12 h post-bloodmeal many secretory bodies of 0.08–0.1 μm diameter were being released into the perioocytic space, and plaques of chorionic secretion, of 0.3–0.4 μm, were regularly distributed among the oocyte microvilli, separated from the membranes of the microvilli by a layer of glycocalyx. The chorionic material was secreted only by follicle cells adjacent to the oocyte, and not by those adjacent to nurse cells (Raikhel and Lea, 1982, 1991).

By 16 h post-bloodmeal the secretory bodies associated with Golgi complexes were more abundant, and plaques of chorionic secretion, now 0.4–0.6 μm in their longest dimension, were uniformly aligned between adjacent oocyte microvilli. The greatest increase in size of the chorionic plaques occurred between 18 and 30 h, during which time they became columnar in shape and grew to 1.2 μm in their longest dimension, perpendicular to the oocyte surface. By 30–36 h post-bloodmeal the secretory activity of the follicle cells was markedly reduced. The oocyte microvilli subsequently retreated from the region of the plaques which, between 32 and 36 h, fused into a continuous endochorion, 1.5–2.0 μm thick (Raikhel and Lea, 1982, 1991; Powell *et al.*, 1988).

Between about 40 and 60 h post-bloodmeal the elaborately sculptured exochorion was assembled around the smooth endochorion. During this period the rate of incorporation of uridine by epithelial cell nuclei increased again and another cycle of activity took place in the Golgi complexes. The secretory product passed through channels, formed by deep invaginations of the apical plasma membrane, to enter the perioocytic space, and the epithelial cells further altered their morphology, forming cavities which accommodated exochorionic pillars (Anderson and Spielman, 1973; Powell *et al.*, 1988). The exochorion appeared to consist of two thin sheets separated by pillars which resulted from uneven deposition of the secretory material. At their outer ends the pillars were interconnected by struts of similar material. Later a different secretory product, with a fibrous texture, filled the spaces between the pillars and struts. By 60–72 h post-bloodmeal the exochorion was complete and the epithelial cells had degenerated (Anderson and Spielman, 1973; Mathew and Rai, 1975; Powell *et al.*, 1988).

Chorionic secretion appeared only between the epithelial cells and the oocyte, none appeared between the epithelial cells and the nurse cells. [^{3}H]Histidine taken up by epithelial cells and incorporated into chorionic secretion was taken up only by cells in contact with the oocyte; none was taken up by those in contact with nurse cells (Anderson and Spielman, 1973).

Chorions can be isolated for analysis by gently homogenizing ovaries in buffered 1% Triton, followed by washing and centrifuging. The isolated chorions can be examined by scanning electron microscopy and their constituent proteins analysed by SDS gel electrophoresis after treatment with 8M urea + 10% dithiothreitol. Most of the endochorionic protein was secreted between 24 and 48 h post-bloodmeal, the endochorion appearing fully formed at 42 h. The proteins secreted during this early period included at least four of c. 20 kDa, a minor component of 64 kDa and a major one of 90 kDa; none was a glycoprotein. Between 42 and 60 h post-bloodmeal the amount of chorionic protein increased rapidly (Figure 19.4), the exochorionic proteins first appearing at 48 and 54 h. These included a number in the 25–50 kDa range, most of them glycolsylated, and a number of glycosylated proteins of >225 kDa. Material of 110–130 kDa, postulated to be chitin, also appeared (Powell *et al.*, 1986a).

A single chorion was estimated to contain 0.09 μg protein, to which the endochorion contributed 20% and the exochorion 80%. The endochorion was smooth and flexible and was formed of a small number of non-glycosylated proteins. It remained soft until after oviposition when it hardened and darkened (Section 19.6.2). The elaborately sculptured exochorion was more brittle and was formed from a complex array of glycoproteins (Powell *et al.*, 1986b, 1988).

(*b*) *Anopheles atroparvus*

Small globules of chorionic secretion appear in the perioocytic space when the ovarian follicle has reached about one-third of its ultimate size, and not only between the epithelial cells and the oocyte but also between the epithelial cells and the nurse cells. As the oocyte grows in size parts of it extend anteriorly between the layer of chorionic secretion and the nurse cells, and the oocyte eventually envelops the nurse cells save for a narrow isthmus which connects them to the exterior (Figure 19.3A). The chorionic secretion increases in amount and invests the oocyte, forming a gelatinous coat which is complete except at the anterior end. This coat later forms the endochorion. Labelled dopa injected into *Anopheles atroparvus* 24–48 h post-bloodmeal was intensively incorporated into the chorionic secretion.

At the time that chorionic secretion is first seen the epithelial cells adjacent to the funicle become larger than their neighbours and protrude inwards. The central cells of this group secrete material, very like the chorionic secretion, which will later form the micropyle plug, while the peripheral cells migrate further in and assume the form of a rosette (Figure 19.3A, B). Subsequently the remnants of the degenerated nurse cells are extruded from the cavity within the oocyte and the gelatinous coat is extended, narrowing the hole through which the nurse cells passed. The micropyle plug lies immediately under the hole (Figure 19.3C). The endochorion retains its gelatinous consistency until some hours after oviposition when it hardens and forms a thin dark layer (Nicholson, 1921; King, 1964; Laurence, 1977).

In anopheline eggs the floats and micropyle apparatus are parts of the exochorion, and in *An atroparvus* the floats are the first part of the exochorion to be formed. When the follicular epithelial cells divide the cell divisions in two lateral areas become oblique. These oblique divisions continue during the later growth of the follicle until, in the two areas, layers of greatly elongated cells lie at right angles to the long axis of the follicle. A float is formed between the outermost cells and those lying immediately under them, and each corrugation of the float is produced by deposition of material over the outer surface of one of the elongate underlying cells. The main part of the exochorion is first seen as a thin membrane under the epithelium (Figure 19.3D). Local thickenings, which are soon found on this membrane, are the start of the perpendicular processes and tubercles that are constituent parts of the exochorion (Section 3.1.3). The thin membrane does not increase appreciably in thickness but the processes grow far into the cytoplasm of the epithelial cells until they reach their final form and size. The micropyle apparatus is secreted by the group of rosette cells. It is a thick ring surrounding a membrane that is produced into a funnel which passes through the endochorion. Its cavity is the micropyle (Figure 3.2B).

One detail of chorion formation distinguishes the single anopheline species from the two culicine species that have been studied. In *An atroparvus* the secretion that is to form the endochorion appears throughout the perioocytic space except for a small region where the

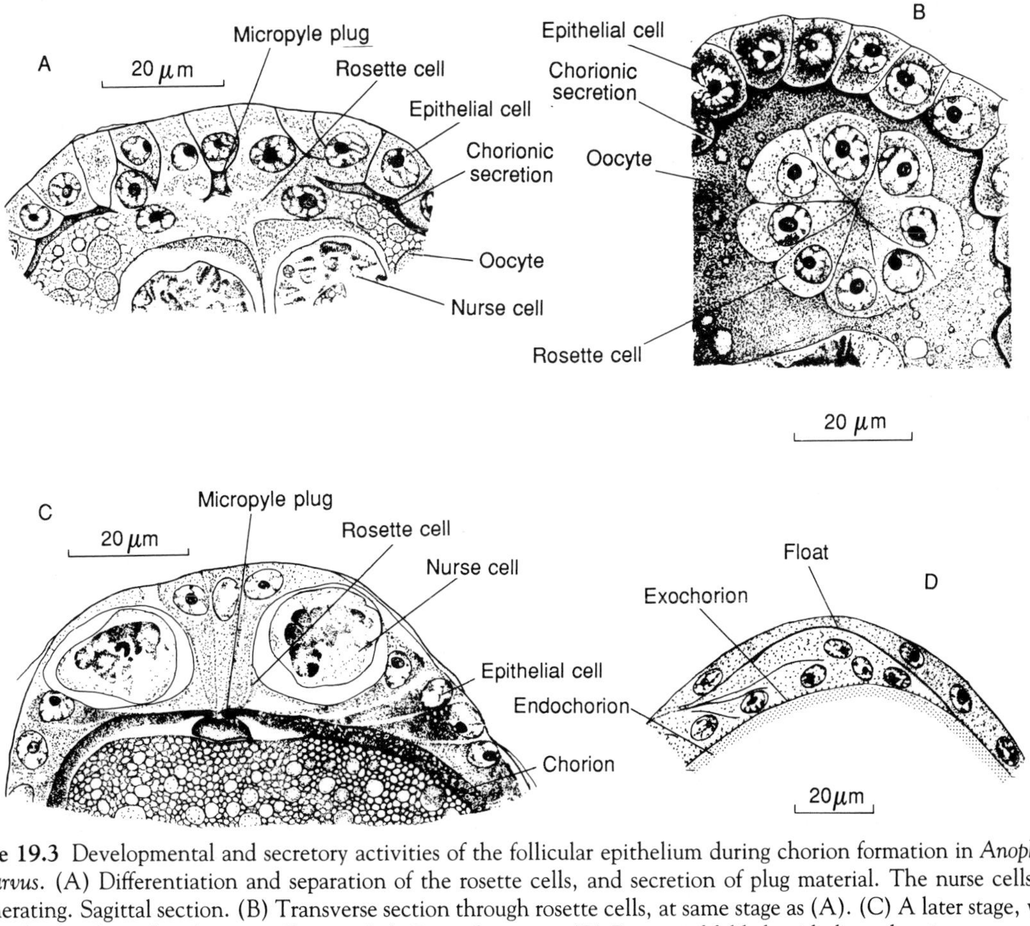

Figure 19.3 Developmental and secretory activities of the follicular epithelium during chorion formation in *Anopheles atroparvus*. (A) Differentiation and separation of the rosette cells, and secretion of plug material. The nurse cells are degenerating. Sagittal section. (B) Transverse section through rosette cells, at same stage as (A). (C) A later stage, with micropyle plug formed and nurse cells extruded. Sagittal section. (D) Region of folded epithelium showing appearance of float material. Transverse section. (After Nicholson (1921.)

micropyle will be formed (Nicholson, 1921). In *Cx pipiens* and *Ae aegypti* this chorionic secretion appears only between the epithelial cells and the oocyte, and not between the epithelial cells and the nurse cells. For a time, therefore, the oocyte is only partly surrounded by the chorionic secretion, and the nurse cells are never enclosed by it (Nath, 1924; Anderson and Spielman, 1973). In *Cx pipiens*, after the transfer of nurse cell cytoplasm into the oocyte and when yolk deposition is almost over, epithelial cells grow inwards between the oocyte and the remnants of the nurse cells and complete the formation of the chorion. Other epithelial cells secrete the complex micropyle apparatus (Nath, 1924).

19.6.2 Hardening and darkening

At oviposition the eggs of *Aedes aegypti* are soft and white but within an hour the chorion begins to darken and harden. Colour development is uniform over the entire surface of the egg, and after four hours the egg shell looks black and is relatively firm. Superficially at least this process resembles sclerotization (Schlaeger and Fuchs, 1974a).

Untanned chorions separated from newly laid eggs were colourless, but when chorions were isolated after they had hardened and darkened naturally, the endochorion appeared brownish-grey and the exochorion colourless. That the undarkened chorion could be dissolved by treatment with urea and dithiothreitol indicated that at that stage there were no other crosslinks than disulphide bonds (Powell *et al.*, 1986b; Smith *et al.*, 1989).

Two enzymes in the synthetic pathway of arylamines involved in sclerotization have been investigated in *Aedes aegypti*. The whole-body titre of one of these, dopa decarboxylase (= aromatic-L-amino-acid decarboxylase), peaked at the times.of larval–pupal and pupal–adult ecdysis. After emergence the titre fell to a very low level, but when blood feeding led to ovary development the titre rose to twice that at pupal–adult ecdysis. The fastest rate of increase occurred 24–48 h post-bloodmeal, and synthesis continued until 96 h (Figure 19.4). About 90% of the enzyme activity was localized in the mature ovaries, and must have been in the oocytes or their chorions because immediately after oviposition the whole-body titre was again very low (Schlaeger and Fuchs, 1974b, c). The site of dopa decarboxylase synthesis is not known. Dopa decarboxylase extracted from gravid females had a molecular mass of 112 kDa and a K_m with L-dopa of 0.24 mM (Kang *et al.*, 1980). The regulation of dopa decarboxylase synthesis by 20-hydroxyecdysone is described in Section 21.5.3.

The product of dopa decarboxylase is dopamine (cf. Figure 7.8). A second enzyme, dopamine *N*-acetyltransferase, the product of which is *N*-acetyldopamine, was detectable in ovaries at 12h post-bloodmeal. The activity level of dopamine *N*-acetyltransferase increased many fold, and almost linearly with time, up to 96h post-bloodmeal (Li and Nappi, 1992).

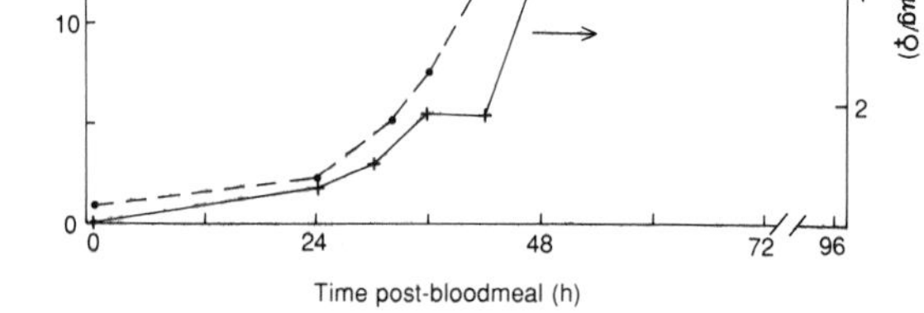

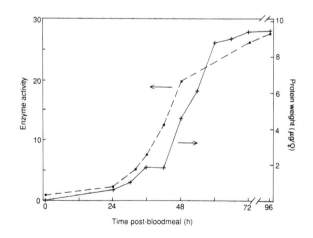

Figure 19.4 Chorion formation in *Aedes aegypti*. Increases in weight of chorionic protein and in whole-body dopa decarboxylase activity, with time after blood feeding. Enzyme activity: μmol × 10^{-4} dopamine formed/min/♀.) (After Powell *et al.*, 1988 and Schlaeger and Fuchs, 1974b.)

After solubilization of isolated untanned chorions with urea and dithiothreitol the constituent proteins could be analysed by SDS gel electrophoresis (Section 19.6.1(a)). When chorions that had been tanned naturally or that had been isolated untanned and then incubated with dopa, dopamine or *N*-acetyldopamine were so treated, only the exochorionic proteins remained apparent. This suggested that all three compounds could be oxidized within the endochorion to a product that would covalently crosslink its proteins.

No endogenous dopamine or *N*-acetyldopamine were detectable in ovarian homogenates (Li and Nappi, 1992). Incubation of isolated untanned endochorion with L-dopa turned it brownish, with dopamine dark grey much like naturally darkened endochorion, and with *N*-acetyldopamine a brilliant yellow (Powell *et al.*, 1986b; Smith *et al.*, 1989). These colour changes may have reflected involvement of the pathway to melanin formation via oxidation products of dopamine, and the blocking effect upon that pathway of acetylation of the amino group of dopamine (Brunet, 1980).

Aedes aegypti eggs failed to darken fully when laid by females that had been injected with α-MDH, an inhibitor of dopa decarboxylase. When injected between 24 and 36 h post-bloodmeal α-MDH prevented darkening in all eggs, and dopa decarboxylase activity was almost completely lacking in 120-h ovaries from females injected at 24 h. From about 40 h the injections started to lose their effectiveness (Schlaeger and Fuchs, 1974a).

The various lines of evidence are consistent with the hypothesis that hardening of the mosquito endochorion is due to sclerotization, and that dopa decarboxylase and dopamine *N*-acetyltransferase are inducible enzymes, constituents of the endochorion, involved in that process (Figure 7.8). The later stages of the pathway, and the role of melanin formation in chorionic darkening, remain to be established.

20

Vitellogenesis

Insect egg yolk contains all nutrients necessary for development of the embryo. Little is known about the synthesis of the lipid droplets present in insect yolk, but the synthesis and incorporation of the crystalline protein constituents have been intensively investigated. The term vitellogenesis embraces all the processes of yolk formation.

The major protein found in the crystalline yolk bodies of most insects is called **vitellin** (Vt); its secretory precursor is called **vitellogenin** (Vg). Vitellogenins are large oligomeric glycolipophosphoproteins, the native proteins usually having a relative molecular mass of 200 000–600 000. They are synthesized in the fat body, first as precursors which undergo proteolytic cleavage into two or more subunits. The subunits are reassembled prior to secretion as the high molecular mass, oligomeric protein vitellogenin. Vitellogenins appear to evolve rapidly since they show little antigenic cross-reaction, except between related species or genera. In some species a small family of genes codes for the vitellogenins. The crystalline yolk bodies may also contain other proteins of lower molecular mass (Raikhel and Dhadialla, 1992).

The so-called 'yolk proteins' of the higher Diptera are very different from the vitellogenins. They consist of three to five discrete peptides that range in molecular mass from 44 000 to 51 000. In *Drosophila*, yolk protein is synthesized by both fat body and follicular epithelium, in *Stomoxys* by the follicular epithelium alone.

Vitellogenin molecules are released by the fat body into the haemolymph, and pass to the oocytes by diffusing through channels between the cells of the follicular epithelium. They bind to vitellogenin receptors located in coated pits on the oolemma, and become internalized when the pits bud-off to form coated vesicles. This phenomenon of 'receptor-mediated endocytosis', which is now recognized as a universal mechanism among animal cells for internalizing functionally important macromolecules, was first reported in an ultrastructural study of vitellogenin uptake by mosquito oocytes (Roth and Porter, 1964).

20.1 THE NATURE OF YOLK

20.1.1 Protein constituents of yolk

(a) Vitellogenin and vitellin

The proteinaceous yolk of newly laid eggs of *Aedes aegypti* was first analysed by Hagedorn and Judson (1972), who reported that it consisted of a single glycophospholipoprotein with a relative molecular mass (M_r) of 270 000. Since that time the yolk protein has been subjected to more advanced analytical and immunological techniques but investigators have been faced with the tendencies of the material to aggregation and proteolysis. It is helpful to approach the analytical data in knowledge of the experimental studies described later which revealed that, within the fat body, provitellogenin (M_r 250 000) is processed by proteolytic cleavage and further modification to yield two polypeptides of 200 000 and 66 000, which later oligomerize to form a native protein of greater molecular mass (Figure 20.5).

Almost all investigators who analysed denatured *Ae aegypti* vitellogenin and vitellin by

SDS-PAGE obtained two subunits of about 200 and 60–70 kDa. Purification of vitellin under non-reducing conditions suggested that the native protein had a molecular mass of 300 kDa or more, possibly of c. 375 kDa (Table 20.1). Denaturation of vitellin produced large and small polypeptides of identical molecular mass to the subunits of vitellogenin.

Borovsky and Whitney (1987) reported a larger number of subunits in vitellogenin and vitellin subjected to SDS-PAGE. In their hands denatured vitellogenin yielded three bands and denatured vitellin yielded six bands. Ramasamy *et al.* (1988) also reported finding numerous bands. The use by these investigators of only one protease inhibitor, PMSF, raises the suspicion that their extracts suffered some proteolysis.

Further investigation of the large and small polypeptides involved use of a library of monoclonal antibodies (mABs) directed against yolk protein from ovaries of *Ae aegypti* 42–48 h post-bloodmeal. All of the mABs reacted only with constituents of fat body, haemolymph, and ovary from vitellogenic females. Immunoblot analysis coupled with SDS-PAGE revealed three groups of antibodies: one group recognized a 200 kDa polypeptide, one recognized a 65 kDa polypeptide, and one recognized both. A small number of antibodies reacted only with yolk proteins at specific stages of synthesis in the trophocytes or of processing in the oocytes. Some recognized the native but not the denatured protein (Raikhel *et al.*, 1986; Raikhel and Bose, 1988).

As described in Section 20.3, use of these mABs showed that the large and small subunits of Vg had a common origin in a 240 kDa precursor, pre-pro-vitellogenin. The subunits of Vg and Vt were found to be indistinguishable. Both large and small subunits were glycosylated. The carbohydrate moieties, which accounted for 10 kDa of the large subunit and 15 kDa of the small subunit, were composed of mannose and *N*-acetylglucosamine. Treatment with the enzyme Endo-H revealed that they were principally asparagine-linked high-mannose oligosaccharides. In both subunits the carbon skeleton was phosphorylated and tyrosine residues were sulphated (Raikhel and Bose, 1988; Dhadialla and Raikhel, 1990).

Table 20.1 Estimates of the relative molecular mass of vitellogenin, vitellin and vitellogenic carboxypeptidase, and of subunits liberated under reducing conditions.

Native protein (kDa)	*Subunits (kDa)*			*References*
Aedes aegypti				
Vt 350	Vg = Vt	170		Harnish and White (1982)
Vt 260	Vg = Vt	200	60	Hagedorn (1985)
–	Vg = Vt	200	66	Ma *et al.* (1986)
Vt 300	Vg [3 subunits, 155 to 62] Vt [6 subunits, 116 to 29]			Borovsky and Whitney (1987)
Vt 375	Vg = Vt	190	70	Thomas *et al.* (1989)
–	Vg	200	66	Dhadialla and Raikhel (1990)
Vcp 53		–		Hays and Raikhel (1990)
Aedes atropalpus				
–	Vg = Vt	200	66–68	Ma *et al.* (1984)
Culex quinquefasciatus				
Vt 380	Vt	160	82	Atlas *et al.* (1978)
Anopheles stephensi				
–	Vg = Vt	175	71	Redfern (1982)
Anopheles albimanus				
–	Vg	165	60	Lu and Hagedorn (1986)

Vg, vitellogenin; Vt vitellin; Vcp, vitellogenic carboxypeptidase.

Polypeptides of 200 and 66-68 kDa were isolated from denatured yolk protein of *Aedes atropalpus* by Ma *et al.* (1984) using mABs coupled with SDS-PAGE. Extracts of *Culex quinquefasciatus* eggs contained a glycophospholipoprotein which under non-denaturing conditions had a relative molecular mass of 380 000, and which when subjected to SDS-PAGE yielded polypeptides of 160 000 and 82 000 (Atlas *et al.*, 1978). Major polypeptides of 175 and 71 kDa were isolated from the ovaries of *Anopheles stephensi* by Redfern (1982). They were sex specific, found in vitellogenic but not previtellogenic ovaries, were present in haemolymph during vitellogenesis, and were secreted by vitellogenic fat body *in vitro*. A minor constituent of 170 kDa was also observed. An extract of mature ovaries of *An albimanus* contained two peptides, with approximate molecular masses of 165 and 60 kDa, which were absent from previtellogenic females and males (Lu and Hagedorn, 1986). Two monoclonals raised against *Aedes aegypti* vitellin showed affinity for vitellin from *Culex annulirostris*, *Cx sitiens* and, more weakly, *Anopheles farauti* (Ramasamy *et al.*, 1988).

(b) *Vitellogenic carboxypeptidase*

The fat bodies of vitellogenic females of *Aedes aegypti* synthesize a protein, immunologically distinct from vitellogenin, which also is secreted into the haemolymph and selectively accumulated by developing oocytes. The ovaries of gravid females contain an immunologically identical protein. This protein was first called 53KP, and later named vitellogenic carboxypeptidase (Vcp). Vcp is female specific, and is synthesized only by the fat bodies of vitellogenic females. It has a relative molecular mass of 53 000 under both reducing and non-reducing conditions. Vcp is glycosylated with mannose at a single position, the carbohydrate moiety accounting for about 2 kDa of the molecular mass. It is not phosphorylated or sulphated. Vcp is internalized by the oocytes without any change in its size. In mature yolk bodies Vcp is located around the crystalline vitelline (Hays and Raikhel, 1990; Cho *et al.*, 1991a; Raikhel and Dhadialla, 1992).

Cell-free translation of Vcp mRNA *in vitro* revealed that the precursor of vitellogenic carboxypeptidase is a 50 kDa polypeptide. Vcp was detectable four hours after blood feeding. The rate of secretion increased progressively until 24 h after which it declined, falling to a very low rate by 36 h. The fat bodies of previtellogenic females synthesized Vcp when exposed to 10^{-6} M 20-hydroxyecdysone *in vitro* (Cho *et al.*, 1991a).

The amino acid sequence of Vcp, which was deduced from the nucleotide sequence of the encoding cDNA, exhibited substantial homology with a family of serine carboxypeptidases. Serine carboxypeptidases (EC 3.4.16.1) are exopeptidases, with optimum activity at acidic pH, that contain a DFP-sensitive serine in their catalytic site, and that cleave single amino acids from the C-terminal end of peptide chains. The amino acid sequence of mosquito Vcp included two conserved domains that are commonly found in serine carboxypeptidases, one of which includes the serine residue present in the catalytic centre of those enzymes.

The molecular mass of Vcp decreased by 0.5–1.0 kDa at the onset of larval embryogenesis. As embryogenesis progressed it underwent further reduction in size, and an immunologically-related protein of 48 kDa appeared. By the end of embryonic development Vcp had been degraded into yet smaller peptides, which disappeared before hatching. The radiolabelled inhibitor [^{3}H]DFP, which binds to the active centre of serine proteases, bound only weakly to the Vcp in oocytes. At the onset of embryogenesis the intensity of its binding to Vcp increased, and reached a maximum in the middle of embryogenesis, when it was shown to bind to the 48 kDa protein. Cho *et al.* (1991a) concluded that Vcp was synthesized by the fat body and internalized by oocytes as a pro-enzyme of 53 kDa. This was activated in eggs at the onset of embryogenesis, upon its conversion to the 48 kDa enzyme. The specific function of the enzyme is uncertain; it might activate hydrolytic enzymes involved in the degradation of yolk proteins, or it

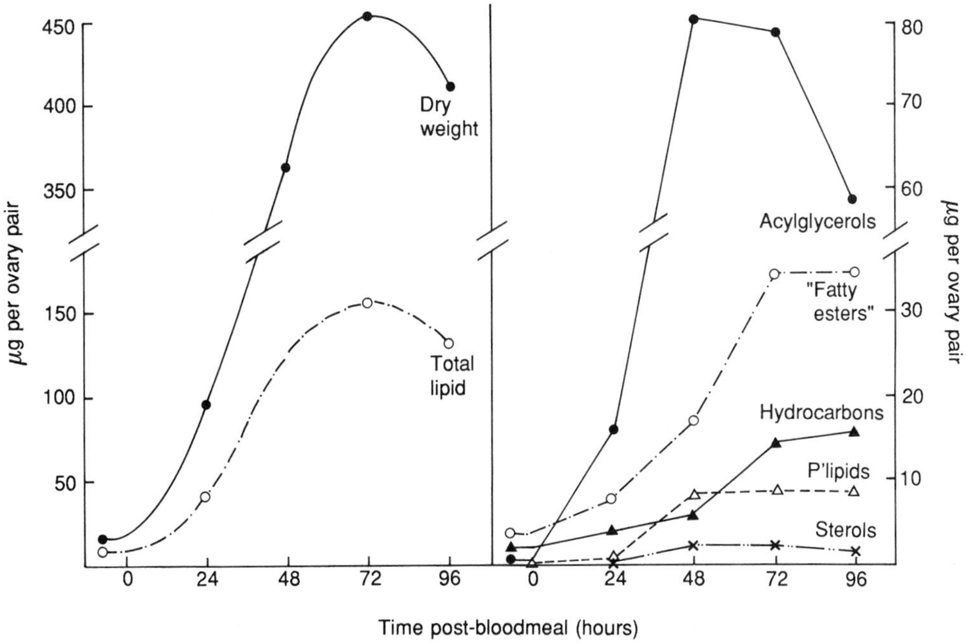

Figure 20.1 The accumulation of lipids in the ovaries of bloodfed *Aedes aegypti* kept at 24–25°C. Lipids were extracted with chloroform–methanol and analysed by TLC. The acylglycerol class includes 2–4% free fatty acids. (After Troy *et al.*, 1975.)

might act as an exopeptidase in the degradation of vitellin.

20.1.2 Lipid constituents of yolk

Histochemical studies on *Culex quinquefasciatus* ovaries showed that triacylglycerols started to accumulate in the oocytes at the time when proteinaceous yolk was first deposited; when vitellogenesis was complete the lipid droplets in the ooplasm consisted almost entirely of saturated triacylglycerols (Nath *et al.*, 1958).

Extraction of lipids from the ovaries of *Aedes aegypti* with chloroform–methanol at 24-h intervals after blood feeding revealed the progressive deposition of acylglycerols, mainly triacylglycerols, and so-called fatty esters (Figure 20.1). Phospholipids appeared in the extracts from 48 h post-bloodmeal, predominantly phosphatidylethanolamine and phosphatidylcholine. It is possible that these extracts included the phospholipid constituents of vitellin. Hydrocarbons appeared in substantial quantity by 72 h post-bloodmeal (Troy *et al.*, 1975).

Lipoproteins of a type known as lipophorins carry lipids to various tissues, and are known to transport the large amounts of lipid that accumulate in insect oocytes. Nothing is known of the lipophorins of mosquitoes.

20.2 EXPRESSION OF THE VITELLOGENIN GENES

20.2.1 Transcription

The *Anopheles gambiae* vitelline gene family consists of closely linked genes located on the right arm of chromosome-2 in zone 18A. Each individual possesses either four or five genes (vg1–vg5). At least four of the genes are in tandem array, with 6.3 kb coding regions separated by approx. 3.0 kb. The coding regions all have extremely well conserved restriction sites, with almost identical nucleotide sequences. The introns and intergenic regions are more variable. The conserved upstream regions contain a sequence that is homologous with a known ovarian enhancer of *Drosophila* yolk protein genes. The upstream regions of vg2 and vg3 contain, within 1 kb of the start site, a two-element motif similar to the fat body enhancer consensus sequences adjacent to *Drosophila* yolk protein genes (P. Romans, personal communication).

The vitellogenin genes of *Aedes aegypti* are present in multiple copies, five having been found in the genome. Four of these genes (A1, A2, B and C) have been cloned and mapped, and they appeared to form a small gene family. A1 and A2 exhibited a high degree of homology. B exhibited a lesser degree of homology with A1 and A2, and C showed only weak homology with the other three genes. Clones of A1, A2 and C all hybridized with mRNA of 6.5 kb which was present in vitellogenic females but absent from non-vitellogenic females and males (Gemmill *et al.*, 1986; Hamblin *et al.*, 1987).

The rate of DNA synthesis in the trophocyte nuclei of females of *Ae aegypti* was low at the time of emergence, but it increased gradually during the first 18 h of adult life and then stabilized at a rate that was about 3-fold higher. The trophocyte nuclei of newly emerged females were diploid. Two days after the peak of DNA synthesis, and of the concomitant increase in total DNA, the trophocyte nuclei started to become polyploid. By the third day 80% of nuclei were tetraploid and 20% were octaploid. The change of ploidy would have increased the copy number of ribosomal and vitellogenin genes, which may have made possible a more rapid response to blood feeding. The change of ploidy was synchronous with post-emergence secretion of juvenile hormone (see Figure 21.2). Ligation of abdomens at emergence prevented over half of the trophocytes from becoming polyploid, but additional polyploidization was induced by topical application of 5 ng juvenile hormone III or 5 pg methoprene (Dittmann *et al.*, 1989; Hagedorn, 1985).

A second burst of DNA synthesis occurred in the fat body of *Ae aegypti* during the first day after a blood meal. The synthesis rate doubled by three hours after feeding, peaked at six hours, and remained elevated up to 12 h (Figure 20.2).

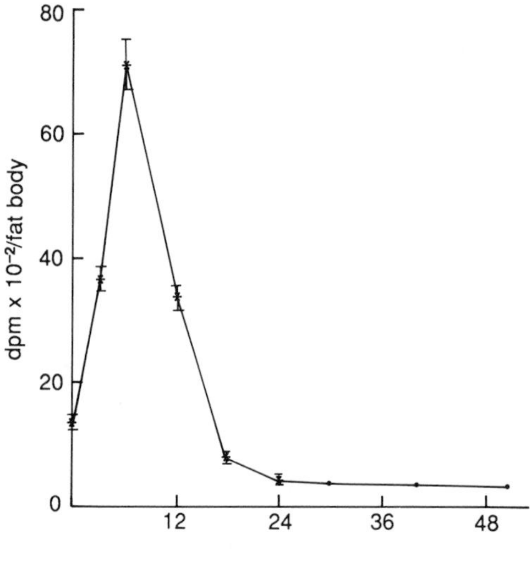

Figure 20.2 DNA synthesis in fat body. [^{3}H]Thymidine was injected into bloodfed females of *Aedes aegypti* at the times indicated by the data points. Two hours later DNA was extracted from abdominal fat body preparations and assayed. Mean synthesis rates therefore correspond more accurately to times 1 h later than the data points. Means ± s.e. (After Hagedorn, 1985.)

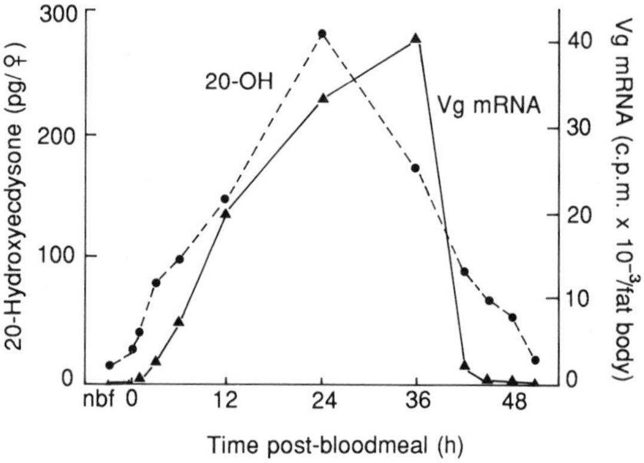

Figure 20.3 Comparison of the accumulation of vitellogenin mRNA in the fat body of *Aedes aegypti* during the first 48 h after a blood meal with changes in whole-body 20-hydroxyecdysone content over the same period (when that was the only ecdysteroid found.) Vitellogenin mRNA was assayed by hybridization to a radiolabelled vitellogenin gene. nbf, non-bloodfed. (From Racioppi *et al.* (1986). Copyright (1986), Pergamon Press plc. Reprinted with permission.)

By 24 h post-bloodmeal the DNA content of abdominal fat body preparations had increased by 20%. During the second and third days post-bloodmeal the rate of DNA synthesis was below that measured prior to feeding. Further polyploidization occurred after blood feeding but it did not start until 48 h after the blood meal. By 72 h 67% of nuclei were octoploid. Since juvenile hormone titres are low during the first 36 h after a blood meal (see Figure 21.9), these observations indicate that juvenile hormone does not stimulate DNA synthesis, although it may regulate polyploidization (Dittmann *et al.*, 1989).

The 6.5 kb mRNA that was believed to encode vitellogenin in *Ae aegypti* was not present before the blood meal but was first detected in fat body one hour after feeding. Its titre in fat body increased steadily and peaked between 24 and 36 h post-bloodmeal, the time varying in different experiments, and then fell rapidly to become undetectable (Figure 20.3). The whole-body titre of free ecdysteroids had also started to rise by 1 h post-bloodmeal, and during the following 24 h the accumulation of vitellogenin message paralleled the rising titre of free ecdysteroids (Racioppi *et al.*, 1986). Comparison of Figures 20.3 and 20.4 shows that that no substantial delay occurred between the synthesis of vitellogenin message and its translation.

Vitellogenic carboxypeptidase cDNA had a single open reading frame that encoded a protein of 441 amino acids, with a deduced molecular weight of 50 153. The size of Vcp mRNA was 1.5 kb. Transcription of vitellogenic carboxypeptidase mRNA in *Ae aegypti* was limited to female fat bodies, and was initiated after a blood meal. The amount of Vcp mRNA paralleled that of vitellogenin mRNA in the fat body, peaking at 24 h post-bloodmeal and then declining (Cho *et al.*, 1991a).

20.2.2 Translation

Abdominal fat body preparations from sugarfed females of *Aedes aegypti* contained c. 4 μg total RNA. The rate of RNA synthesis started to increase shortly after blood feeding, and the total RNA content peaked at 10–12 μg between 8 and 14 h post-bloodmeal. By 40 h it had returned to the prefeeding level (Figure 20.4). Most of this RNA was ribosomal and associated with RER. Measurement of [^{3}H]uridine incorporation into ribosomes showed that the rate of ribosome synthesis continued to increase up to 18 h post-bloodmeal, but that by 24 h it had declined substantially. Although the most striking changes were in the various rRNA species, the tRNAs also underwent a cycle of accumulation and degradation during the vitellogenic cycle (Hagedorn *et al.*, 1973, 1992; Raikhel and Lea, 1983; Hotchkin and Fallon, 1987).

The predominant and specific cell-free translation product of poly(A)$^+$ RNA from fat bodies of vitellogenic *Ae aegypti* was a polypeptide with a relative molecular mass (M_r) of 224 000. Monoclonal antibodies to both large and small Vg subunits precipitated this polypeptide from a pool of translation products. Its size correlated closely with that of the polypeptide that would be derived from a message of 6.5

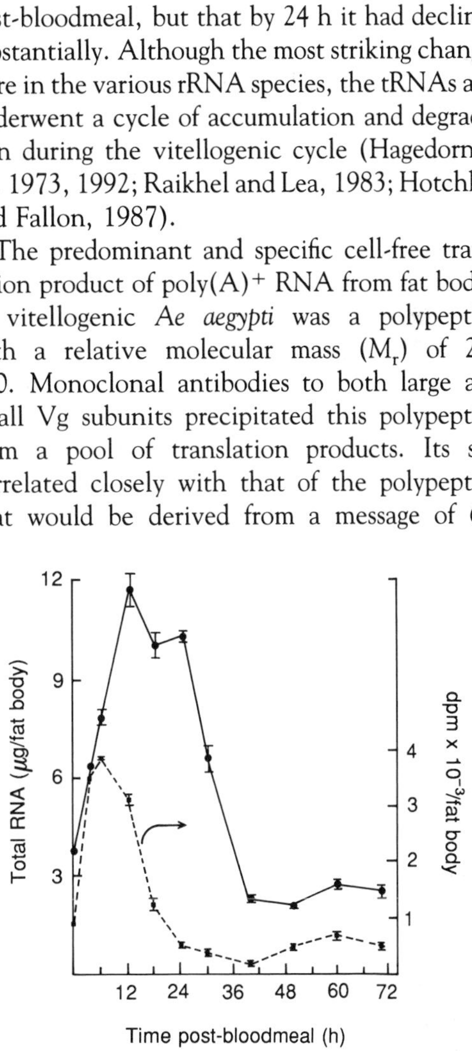

Figure 20.4 RNA synthesis rates and RNA content of *Aedes aegypti* after a blood meal. Females were injected with [^{3}H]uridine at times indicated by the data points. Two hours later RNA was extracted and assayed for ^{3}H label and for mass, therefore mean synthesis rates more accurately correspond to times 1 h later than, and RNA contents to times 2 h later than the data points. Means + s.e. (After Hagedorn, 1985.)

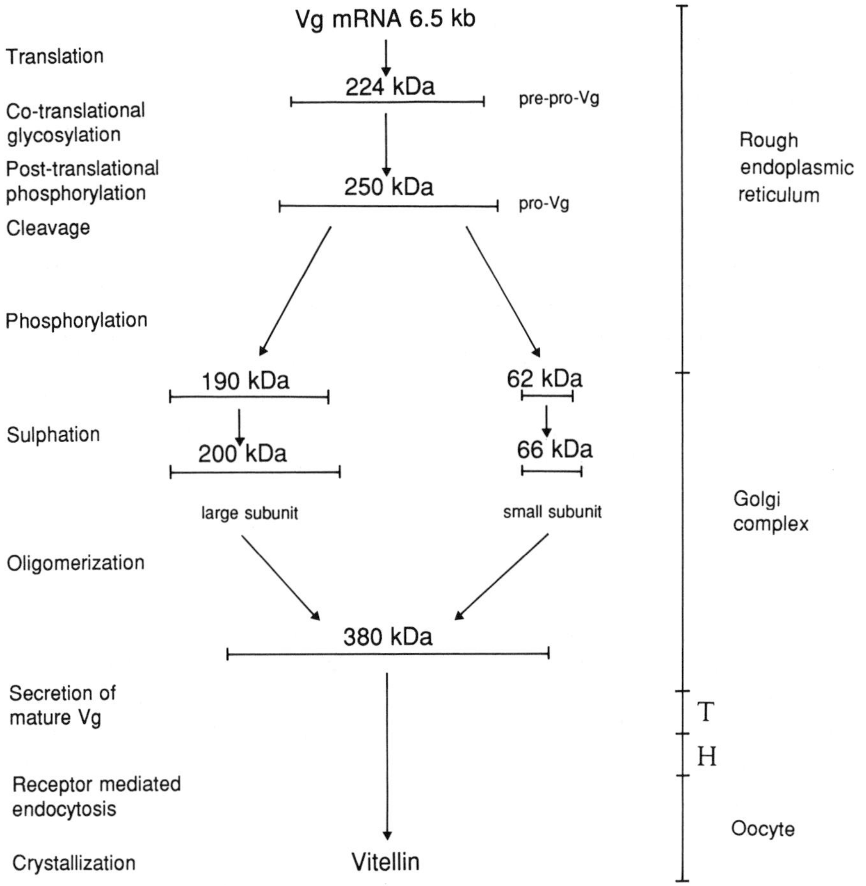

Figure 20.5 The principal stages in the synthesis of vitellogenin and vitellin. Vg, vitellogenin; H, haemolymph; T, transport through cytoplasm. (After Dhadialla and Raikhel, 1990.)

kilobases. When similar poly(A)+ RNA was translated *in vitro* in the presence of microsomal enzymes, the size of the Vg precursor increased to 240 000.

A polypeptide of M_r 250 000 was present in fat bodies dissected from females at 3 h post-bloodmeal and in fat bodies incubated with 20-hydroxyecdysone. It was not readily apparent in fat bodies during the later stages of vitellogenesis, possibly turning over too rapidly for detection at that time. However, analysis of microsomal membranes prepared from vitellogenic fat bodies at 20 h post-bloodmeal revealed its presence within rough endoplasmic reticulum. The 250 kDa polypeptide was immunoprecipitated by mABs specific to both Vg subunits. Dhadialla and Raikhel (1990) named the 224 kDa peptide pre-pro-vitellogenin (pre-pro-Vg) and the 250 kDa peptide pro-vitellogenin (pro-Vg) (Figure 20.5).

Pulse labelling of hormone-stimulated fat bodies with mannose and *N*-acetylglucosamine revealed that pro-Vg was glycosylated. The oligosaccharide moiety on pro-Vg was susceptible to treatment with endo-β-*N*-acetylglucosaminidase (Endo-H) which reduced its mass to 227 kDa. Pulse labelling with $^{32}P_i$ revealed that pro-Vg is phosphorylated. A number of lines of evidence indicated that glycosylation of pre-pro-Vg occurs co-translationally and is completed at this step of processing. First, cell-free translation of Vg mRNA in the presence of microsomal enzymes yielded a product of 240 kDa. Second, a product of 226 kDa was detected only in fat bodies treated with tunicamycin, an inhibitor of co-translational glycosylation. Finally, enzymatic digestion of pro-Vg with Endo-H, which removes high-mannose moieties, reduced its size to 227 kDa. Because tunicamycin blocks phosphorylation as well as glycosylation it appeared that additional phosphorylation must occur at a later stage (Dhadialla and Raikhel, 1990).

The 250 kDa pro-Vg was not secreted into the medium *in vitro* but was further processed during vitellogenin synthesis. Analysis of extracts of vitellogenic fat bodies by immunoprecipitation and SDS-PAGE yielded polypeptide bands of M_r 190 000–200 000 and 62 000–66 000, corresponding to the large and small Vg subunits at various stages of processing. Extracts of fat bodies that had been pulse labelled for just 5 min yielded only two polypeptides, of 190 000 and 62 000. These polypeptides, which were already glycosylated and phosphorylated, could be chased by pulse labelling for >15 min into mature Vg subunits of 200 and 66 kDa. That the two Vg subunits originated from the 224 kDa polypeptide as a common precursor was established by its immunorecognition by mABs directed against either the large or the small Vg subunit and by similarities between peptide maps of the 224 kDa polypeptide and both Vg subunits (Bose and Raikhel, 1988).

The ionophore monensin failed to block the cleavage of pro-Vg, suggesting that cleavage occurred in the endoplasmic reticulum and not the Golgi complex. Monensin totally inhibited the processing of the 190 and 62 kDa polypeptides to the mature Vg subunits, showing that this step takes place in the *trans* compartment of the Golgi complex. Phosphorylation of the subunit precursors was not affected by exposure of fat body to monensin but the incorporation of sulphate was reduced, indicating that approximately 65% of Vg sulphation takes place in the *trans* compartment. The phosphorus moieties of the mature Vg subunits could be removed completely with alkaline phosphatase, which reduced the molecular mass of the large and small subunits by about 7 and 4 kDa respectively. Their phosphorylation was not affected by extraction with chloroform:acetone, nor by treatment with Endo-H, therefore it is the peptide skeleton which is phosphorylated and not the lipid or carbohydrate moieties (Raikhel and Bose, 1988; Dhadialla and Raikhel, 1990).

That the Vg subunits recombine within the trophocytes was shown when proteins labelled with ^{125}I or ^{35}S, extracted from vitellogenic fat bodies and ovaries, were reacted separately with mABs directed to either the large or the small Vg subunit. The products of immunoprecipitation by the two types of mAB proved identical when subjected to SDS-PAGE, comprising in each case the 200 and 66 kDa subunits. Since each mAB individually immunoprecipitated both subunits the subunits must have been combined in oligomeric form, probably bound by hydrogen bonds, within the tissues. Solubilization of the immunoprecipitates in the absence of reducing agents still yielded both polypeptides after SDS-PAGE, showing that they were not bound by covalent sulphydryl bonds (Raikhel and Lea, 1987; Raikhel and Bose, 1988). Gel filtration and HPLC indicated a relative molecular mass for *Ae aegypti* native vitellin of 375 000 ± 25 000 (Thomas *et al.*, 1989). It is not known how the 200 and 66 kDa subunits combine to yield an oligomer of 375 kDa.

Vitellogenin is released from secretion granules into the haemolymph (Section 20.3.2). [^{35}S]Methionine injected into vitellogenic mosquitoes was incorporated into the two Vg subunits in the fat body, but both subunits disappeared from the fat body two hours after the labelling. Extraction of the culture medium in which treated fat bodies had been incubated, and analysis by SDS-PAGE, revealed the two Vg subunits but no other labelled products (Raikhel and Bose, 1988). Vitellogenin extracted from the haemolymph of vitellogenic females and analysed by SDS-PAGE consisted of the large and small Vg subunits (Ma *et al.*, 1986). No changes in the molecular composition of yolk protein subunits were detected by scanning densitometry of SDS-PAGE preparations of extracts of mature eggs, leading Raikhel and Bose (1988) to conclude that vitellogenin had been transformed into vitellin without additional cleavage or further processing.

In summary, the primary product of translation of Vg mRNA is a 224 kDa polypeptide. This is first processed by co-translational glycosylation followed by partial phosphorylation leading to the formation of the 250 kDa pro-vitellogenin. The pro-Vg is proteolytically cleaved and phosphorylated prior to sulphation in the Golgi complex, which yields mature vitellogenin subunits of 200 and 66 kDa. These oligomerize to yield a product of *c.* 375 kDa which is secreted into the haemolymph and taken up by the oocytes within which it is crystallized but not further processed.

20.2.3 Regulation of gene expression

(a) Stimulation of vitellogenin synthesis

During the previtellogenic phase of each ovarian cycle the fat body is made competent for vitellogenin synthesis. Stimulated by juvenile hormone, the trophocytes become polyploid and their nucleoli enlarge, leading to an increase in RNA synthesis and accumulation of non-adenylated RNA (Sections 20.2.1, 20.3.1).

20-Hydroxyecdysone is the one factor that is known to act during the vitellogenic phase to stimulate vitellogenin synthesis, but whether it is the sole factor is not known. At 10^{-7} M, which is the concentration of ecdysteroids in *Aedes aegypti* haemolymph 20 h after blood feeding, 20-hydroxyecdysone can stimulate the *in vitro* synthesis of Vg by fat body from sugarfed females (Hagedorn *et al.*, 1975). The appearance of vitellogenin mRNA in fat body shortly after a blood meal (Section 20.2.1) suggests that the control of vitellogenin synthesis is at the stage of transcription rather than of translation.

The effects of hormonal conditions on vitellogenin gene expression were investigated by measuring mRNA production by means of a subcloned fragment of vitellogenin DNA. Vitellogenin mRNA was normally absent from sugarfed females but was found at very low titre 24 h after injection of 5 μg 20-hydroxyecdysone. Females that had been bloodfed, immediately decapitated and injected with 20-hydroxyecdysone contained, 24 h later, up to 75% as much vitellogenin mRNA as normal bloodfed controls. Females given blood by enema and immediately decapitated did not produce vitellogenin mRNA (Racioppi *et al.*, 1986).

Further evidence that the control of Vg synthesis is at the stage of transcription was obtained in experiments with inhibitors of RNA synthesis. Three hours exposure of fat body *in vitro* to 10^{-6} M α-amanitin, an inhibitor of RNA polymerase II, blocked hormonally stimulated vitellogenin synthesis (Kaczor and Hagedorn,

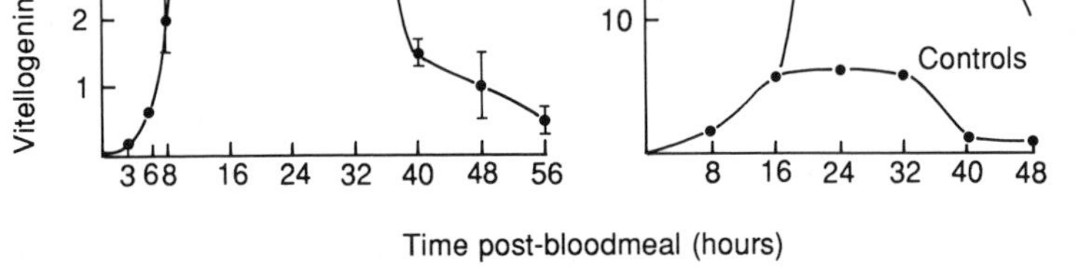

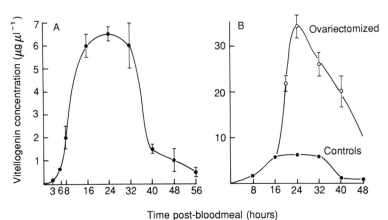

Figure 20.6 Time profile of vitellogenin concentration in the haemolymph of *Aedes aegypti* after a blood meal. (A) Normal females. (B) Normal females (controls) and females ovariectomized at 16 h post-bloodmeal. (From Van Handel and Lea (1984). Copyright (1986), Pergamon Press plc. Reprinted with permission.) The (μg μl^{-1})-values plotted in (A) were similar to the absolute titres of haemolymph vitellogenin measured at different times post-bloodmeal by Ma *et al.* (1986).

1980). Injection into vitellogenic females of actinomycin D, an inhibitor of DNA-primed RNA polymerase, allowed vitellogenin synthesis to proceed at the rate occurring at the time of injection but prevented the normal increase in rate of synthesis (Hagedorn *et al.*, 1973).

(b) *Termination of vitellogenin synthesis*

Both the titre of Vg mRNA in fat body and the amounts of vitellogenin in fat body and haemolymph fall precipitously after approximately 36 h post-bloodmeal (Figures 20.3, 20.4, 20.6). This change occurs several hours after the whole body ecdysteroid titre starts to decline (Figure 21.8A). A number of investigators have attempted to determine the relationship between ecdysteroid titre and the promotion and eventual decline of Vg synthesis. The effect of time of exposure to 20-hydroxyecdysone upon the rate of vitellogenin secretion by fat bodies from sugarfed *Ae aegypti in vitro* was measured by Ma *et al.* (1987). Following an initial period of 6 h exposure of fat bodies to the hormone the medium was sampled for vitellogenin and renewed at 3 h intervals (see Figure 20.7A for details). The rate of secretion was low initially (6–9 h exposure) but started to increase thereafter, reaching a maximum between 21 and 24 h of of 230 ng vitellogenin h^{-1} fat $body^{-1}$. Secretion continued at the same rate until 33 h after which it rapidly declined, becoming negligible by 39 h in the continued presence of the hormone. It is interesting that the curve of vitellogenin secretion in response to 20-hydroxyecdysone (Figure 20.7A) is almost identical with the time profile of vitellogenin concentration in the haemolymph (Figure 20.6A).

The rates of vitellogenin secretion by fat bodies removed from bloodfed females at 6 h post-bloodmeal and then exposed to 10^{-6} M 20-hydroxyecdysone were 2.5 times higher than those by fat bodies from sugarfed females, with a maximum of 590 ng vitellogenin fat $body^{-1}$ h^{-1} (Figure 20.7B). Even so, vitellogenin production *in vitro* was only a small fraction of that *in vivo*, e.g. a total of 9–10μg compared with the 160μg reported from an intact bloodfed mosquito

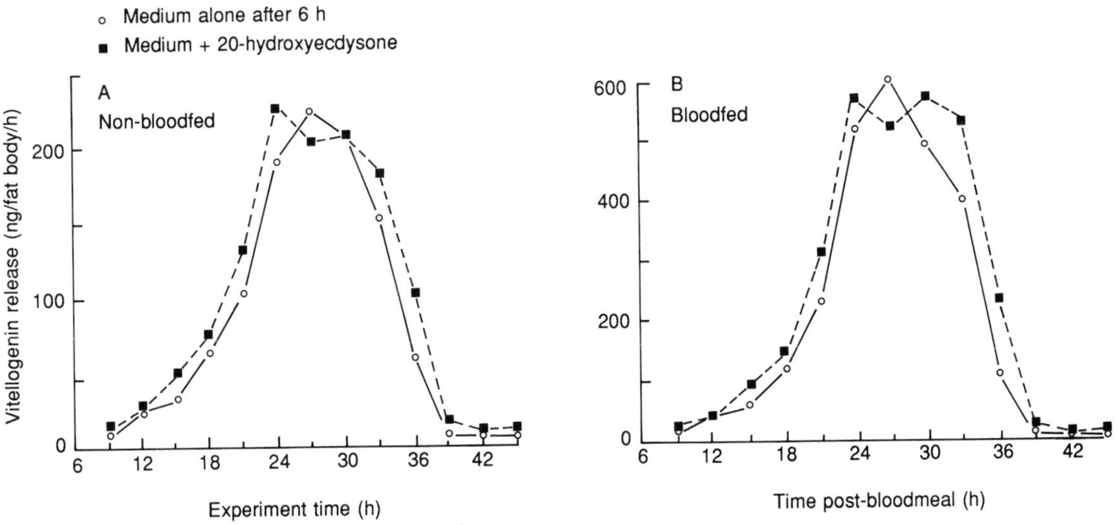

Figure 20.7 Time profile of vitellogenin release *in vitro* by fat bodies from non-bloodfed and bloodfed females of *Aedes aegypti*.

(A) A number of fat bodies from non-bloodfed females were incubated with 10^{-6} M 20-hydroxyecdysone for 6 h and then divided into two series. The fat bodies in one series (■–■) remained exposed to the hormone, the medium being renewed every 3 h and sampled at the end of each 3 h period. Fat bodies in the second series (o–o) were transferred to hormone-free medium after the initial 6 h exposure to hormone, and the hormone-free medium was renewed and sampled on a 3 h schedule. (B) Fat bodies from bloodfed females removed 6 h after the blood meal were treated as in (A) but without the initial 6 h exposure to hormone. (After Ma *et al.*, 1987.)

by Hagedorn (1983). The investigators were surprised to find little difference in the extent or temporal profile of vitellogenin synthesis in fat bodies that had been transferred to hormone-free medium after the initial 6 h exposure to 20-hydroxyecdysone compared with those that were continuously exposed to the hormone (Figure 20.7A, B). They concluded that after a critical period of exposure to the hormone of about six hours the response of fat bodies is a programmed cellular event (Ma *et al.*, 1987).

Observations consistent with that conclusion had been reported earlier by Van Handel and Lea (1984), who ovariectomized females at different times after a blood meal and assayed the haemolymph vitellogenin concentration. When females were ovariectomized at 1, 2, 4, 6 or 8 h after the blood meal the vitellogenin concentration continued to rise for several hours, and the longer the surgery was postponed the higher the concentration rose. The maximum was reached after 8 h, and postponement of the operation to 16 h did not result in a consistently higher vitellogenin concentration. The time course of the rise and fall of haemolymph vitellogenin concentration in mosquitoes ovariectomized at 16 h post-bloodmeal was not different from that in controls, but in the absence of ovaries the maximum haemolymph concentration was several times higher (Figure 20.6). The amounts of vitellogenin that accumulated in the haemolymph after ovariectomy at 8 or 16 h post-bloodmeal were comparable with the amounts deposited in the ovaries of intact controls over the same period.

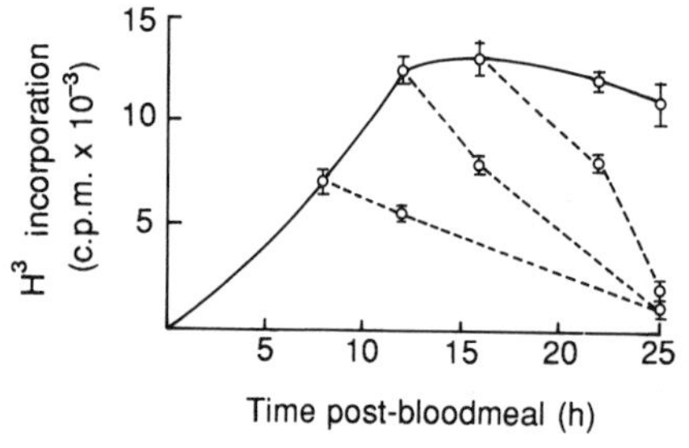

Figure 20.8 Effect of ovariectomy on the rate of vitellogenin synthesis in *Aedes aegypti*. A batch of females was blood-fed at time 0, and sub-batches were ovariectomized after 8, 12 and 16 h. At the times of the data points, operated and unoperated mosquitoes were injected with [^{3}H]valine, and its incorporation into vitellogenin was measured 2 h later. Means + s.e. (From Borovsky (1981a). Copyright (1981), Pergamon Press plc. Reprinted with permission.)

Other investigators had reported a greater dependence of vitellogenin synthesis on the presence of 20-hydroxyecdysone *in vitro* or of the ovaries *in vivo*. Bohm *et al.* (1978) found that the rate of vitellogenin secretion by fat bodies incubated with 10^{-6} M 20-hydroxyecdysone fell rapidly when the hormone was removed after 24 h, but that re-exposure to hormone at 36 h permitted recovery. Borovsky (1981a) reported that ovariectomy of *Ae aegypti* at 8, 12 or 16 h post-bloodmeal caused a decline in the rate of vitellogenin synthesis within 4 h (Figure 20.8). Implantation of an ovary 6 h after ovariectomy led to a resumption of vitellogenin synthesis. Borovsky's (1981a) conclusion that, in the absence of the ovaries, the accumulation of vitellogenin leads to feedback inhibition of synthesis was not supported by the very high haemolymph vitellogenin concentrations he observed (see also Figure 20.6B).

The mechanism regulating termination of vitellogenin synthesis remains to be fully elucidated. Lysosomal degradation of synthetic organelles causes the termination (Section 20.3.3), but whether this is programmed solely by the fat body or whether other factors are involved remains uncertain.

20.3 ULTRASTRUCTURAL CHANGES IN THE TROPHOCYTES

During successive ovarian cycles the fat body passes through a parallel series of cycles of ultrastructural generation and degeneration, during each of which the fat body cells or trophocytes exhibit three stages of activity. (i) A previtellogenic stage during which DNA and RNA synthesis lead to proliferation of biosynthetic organelles. (ii) A vitellogenic stage, after blood feeding, with the formation of more biosynthetic organelles and vitellenin synthesis. (iii) A termination stage during which the biosynthetic apparatus is destroyed. The following account relates to *Aedes aegypti* and is based predominantly on the work of Raikhel (1986a, b, 1987a, b) and Raikhel and Lea (1983).

20.3.1 Previtellogenic development

In newly emerged females the cytoplasm of each trophocyte was rich in reserves, containing lipid droplets, glycogen bodies, and three types of membrane-bounded protein granule. Nucleus and cytoplasm were squeezed between these inclusions. Rough endoplasmic reticulum (RER) and mitochondria were not abundant and the Golgi complexes were undeveloped (Figure 20.9A). The nucleus was relatively small; it contained a compact nucleolus which had a fibrillar centre surrounded by a dense granular component and was associated with a chromosomal nucleolus organizer. Trophocytes of this stage are not capable of vitellogenin synthesis.

During the first three days after emergence a number of changes occurred. The nucleolus grew >3-fold linearly and its enlarging granular component came to contain many maturing pre-ribosomes. Other pre-ribosomes were distributed throughout the nucleus. The number of ribosomes in the cytoplasm increased sharply. Rough endoplasmic reticulum appeared in the form of cisternae, located in rows, concentric circles or around lipid inclusions. The Golgi complexes developed into well-defined organelles consisting of small vesicles, 3–4 short vesicular cisternae, and a few larger vesicles. The plasma membrane became extended into numerous infoldings, increasing its surface area (Figure

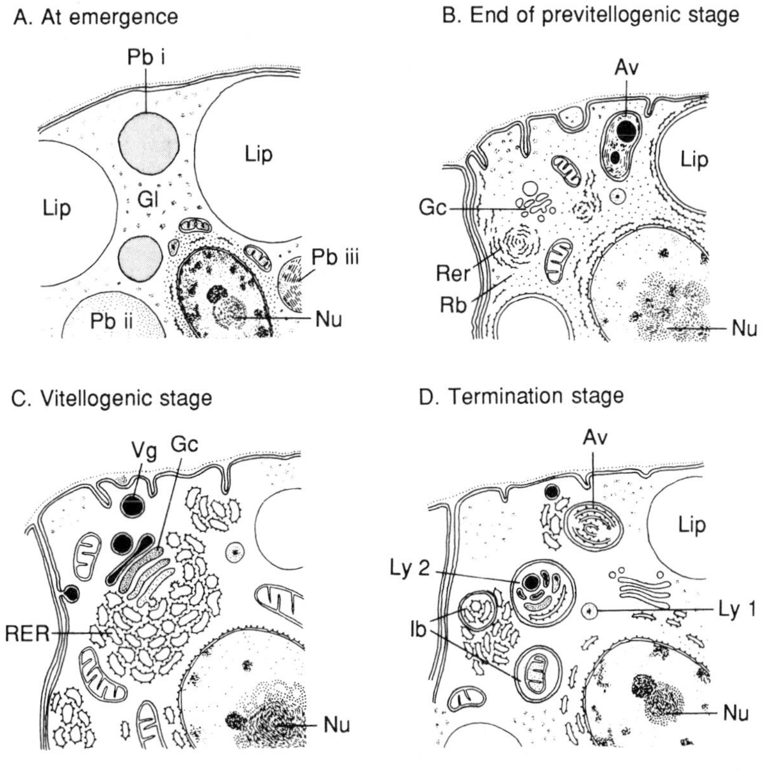

Figure 20.9 Diagram of ultrastructural changes in mosquito trophocytes during the ovarian cycle.
(A) At emergence. (B) At the end of the previtellogenic stage. (C) During the vitellogenic stage. (D) During the termination stage. Av, autophagic vacuole; Gc, Golgi complex; Gl, glycogen; Ib, isolation body; Lip, lipid inclusion; Ly 1, primary lysosome; Ly 2, secondary lysosome; Nu, nucleolus; Pb, protein granule; Rb, ribosomes; RER, rough endoplasmic reticulum; Vg, vitellogenin secretion granule. (After Raikhel and Lea, 1983.)

20.9B). Peripheral trophocytes situated adjacent to the epidermis underwent the least extensive transformation, the synthetic organelles being distributed in islands among extensive lipid and glycogen reserves. Trophocytes closest to the haemocoele underwent most development, massive zones of RER and Golgi complexes occupying almost all the cytoplasm. During the previtellogenic stage most protein bodies disappeared and, in water-fed females, the lipid and glycogen inclusions became somewhat reduced. No vitellogenin was present (Raikhel and Lea, 1983).

Growth of trophocyte nucleoli after emergence was accompanied by an increase in rate of RNA synthesis, which peaked two days after emergence and then declined, and by accumulation of non-polyadenylated RNA, predominantly ribosomal RNA, which reached a maximum after two days and remained high for some days thereafter. Allatectomy at emergence prevented the growth of trophocyte nucleoli. Isolation of the abdomen at emergence had the same effect and also blocked RNA synthesis and the accumulation of poly(A)$^-$ RNA. Implantation of corpora allata into allatectomized females led to normal growth of the nucleoli, and treatment of abdomens isolated at emergence with juvenile hormone III restored the capacity for nucleolar growth, RNA synthesis and accumulation of poly(A)$^-$ RNA (Raikhel and Lea, 1990).

20.3.2 The vitellogenic stage

Ultrastructural changes were detected in the trophocytes three hours after a blood meal: the nucleolus became more multilobed, ribosome proliferation had been renewed, and the rough endoplasmic reticulum was undergoing reorganization. Vitellogenin was first detected by immunofluorescence one hour after feeding, and vitellogenin secretion granules were observed at the cell membrane within three hours.

The nucleolus persisted in its multilobed form until 12 h post-bloodmeal, by which time the RER had proliferated and occupied almost all the cytoplasm. Free ribosomes were rare. This nucleolar activity may account for most of the increase in total RNA that occurs after a blood meal. The number of Golgi complexes increased considerably, becoming maximal between 18 and 24 h post-bloodmeal (Figure 20.9C). As measured by specific immunofluorescence, the vitellogenin content also was maximal between 18 and 24 h.

Between 12 and 36 h after feeding the nucleolus underwent reverse transformation, coming to resemble that in the newly emerged mosquito. Vitellogenin synthesis continued for about another 18 h; it was declining at 30 h post-bloodmeal and had disappeared by 48 h.

High-resolution localization of vitellogenin by use of immunoferritin and immunoperoxidase revealed its presence in the intracisternal spaces of RER, and at higher density in Golgi complexes where its density increased from the *cis* to the *trans* side, reaching a maximum in the membrane-bound secretion granules of 0.2–0.3 μm diameter. Secretion granules were also located near the plasma membrane, and some of them appeared to release their contents into the intercellular space. The paucity of secretion granules outside the Golgi complexes suggested that their release by exocytosis was almost immediate. The sequence of ultrastructural changes that occurs in the trophocytes of *Aedes aegypti* after a blood meal has been described by Behan and Hagedorn (1978), Tadkowski and Jones (1979) and Raikhel and Lea (1983).

Treatment of sections of trophocytes with mABs which recognized either a large (200 kDa) or a small (65 kDa) yolk polypeptide produced the same pattern of immunolabelling. After immunocytochemical treatments all organelles on the secretory pathway were labelled; RER weakly, cisternae of the Golgi complex strongly, and the secretory granules most intensely. By differential immunocytochemical labelling it was shown that the large and small polypeptides were present together in the same Golgi compartments and in the same secretory granules, whether in the cytoplasm or being released. A few mABs from both groups labelled only secretory granules at the *trans* side of the Golgi complex or free in the cytoplasm; it was considered that these mABs probably recognized yolk polypeptides that had completed their post-translational processing in the Golgi complex and were ready for secretion (Raikhel, 1987a).

After a blood meal the lipid content of the fat body rapidly rose to a peak at about 18 h and then equally rapidly declined. There was no deposition of membrane-bounded protein granules in the trophocytes after blood feeding (Behan and Hagedorn, 1978).

20.3.3 The termination stage

Termination of vitellogenin synthesis started between 27 and 30 h post-bloodmeal with a sharp decline in the concentration of vitellogenin in the fat body, which by 48 h was almost devoid of vitellogenin (Figure 20.4). This decline was thought not to result from a failure of transcription, since the rate of transcription continued to increase until about 36 h post-bloodmeal (Figure 20.3), but to be associated with a simultaneous and dramatic rise in lysosomal activity. Ultrastructural studies showed that two processes were involved: (i) autophagy, which involved the sequestration of organelles by isolation membranes, followed by fusion of the so-formed isolation body with primary lysosomes; and (ii) crinophagy, which involved the direct fusion of organelles with primary lysosomes (Raikhel, 1986a, b).

At 24 h post-bloodmeal, during the vitellogenic stage, the trophocyte cytoplasm contained only a few scattered primary lysosomes. Some were in contact with *trans* Golgi cisternae, which also contained lysosomal enzymes, consistent with the concept that lysosomal enzymes undergo post-translational processing in the Golgi complex. There was evidence of some degradation of RER by autophagy.

By 30 h post-bloodmeal, after the start of the termination stage, the number of lysosomes had increased more than 10-fold. The activities of three enzymes representative of the major groups of lysosomal enzymes, acid phosphatase, arylsulphatase A and β-galactosidase, peaked at about 30 h post-bloodmeal while cathepsin D peaked at about 42 h (Raikhel, 1986b). By 30 h numerous isolation membranes were present; these were elongate flat cisternae consisting of paired smooth membranes narrowly separated by an electron translucent space.

Rough endoplasmic reticulum and Golgi complexes were destroyed by a process involving isolation membranes, which were present in large numbers by 30 h. First, isolation membranes wrapped around the organelle forming a double-membraned isolation body. Then primary lysosomes fused with the isolation body producing a secondary lysosome (Figure 20.9D). Secondary lysosomes also sometimes fused with isolation bodies, and there was some evidence of direct sequestration of RER by secondary lysosomes. Vitellogenin secretory granules fused directly with primary or secondary lysosomes. Sequestered Golgi complexes always contained vitellogenin bodies; often only *trans* Golgi compartments were sequestered (Raikhel, 1986a).

Newly formed isolation bodies containing Golgi elements or RER proved negative when tested for acid phosphatase or arylsulphatase A, although free Golgi complexes in close proximity were positive. As the termination phase progressed, secondary lysosomes increasingly fused with each other, forming large lysosomal compartments which contained Golgi elements, RER and vitellogenin secretory granules, and gave strong positive reactions for lysosomal enzymes.

The degradation of organelles during the termination stage was highly selective, causing elimination of protein synthetic machinery and the rapid termination of vitellogenin secretion (Raikhel, 1986a). At this time the fat body was unresponsive to 20-hydroxyecdysone *in vitro*, but by 70 h post-bloodmeal the trophocytes had an ultrastructure like that seen in previtellogenic resting-stage females and responsiveness to 20-hydroxyecdysone had returned (Bohm *et al.*, 1978). The restoration of competence was probably due to re-exposure to juvenile hormone (Section 21.9).

Cathepsin D belongs to the class of aspartic proteinases that have a pH optimum below 5 due to the involvement of an acidic residue in the catalytic process. Characterization of cathepsin D of adult female *Ae aegypti*, derived predominantly from fat body lysosomes, revealed the native protein to have an apparent molecular mass of 80 kDa and to consist of two subunits, both of 40 kDa M_r. The *N*-terminal sequence of the purified enzyme had 74% identity with the homologous sequence of porcine and human cathepsins D. The mosquito enzyme had a pH optimum of 3.0, and its K_m was calculated to be 4.2 μM (Cho *et al.*, 1991b).

20.4 INCORPORATION OF VITELLOGENIN

Developing oocytes accumulate proteinaceous yolk by internalizing large amounts of circulating vitellogenin. Uptake is by selective or receptor-mediated endocytosis. After the vitellogenin has entered the oocyte it is transformed into the final form, crystals of vitellin.

20.4.1 The endocytotic complex

(a) Clathrin

The endocytotic complex is the ultrastructural apparatus that makes receptor-mediated endocytosis possible. A key constituent of it is the protein **clathrin**, which is a hexamer, composed of three copies of the clathrin heavy chain plus three copies of the clathrin light chain. The molecule has the form of a **triskelion**, i.e. of three kinked legs arising from a central point (Figure 20.10). Under suitable conditions, clathrin molecules spontaneously pack together to form polyhedral lattices with one triskelion centre per vertex. The flexibility of the triskelions allows them to form both hexagons and pentagons, which can assemble to make a closed cage. In all animals, clathrin is a constituent of the coat that assembles on the cytoplasmic surface of plasma membranes, over regions that contain certain transmembrane molecules. Each of the latter is composed of an extracellular receptor, with a ligand-binding site, and a smaller intracellular domain to which proteins termed 'adaptors' bind. A clathrin lattice binds to the adaptors. These specialized regions of the plasma membrane invaginate forming 'coated pits'. When the ligand-binding sites of a coated pit are occupied by a specific ligand, the coated pit buds-off forming a coated vesicle; this contains clathrin, adaptors, receptors in the membrane of the vesicle, and their ligand in the interior (Pearse and Crowther, 1987; Pearse and Robinson, 1990). Antibodies against mammalian clathrins were not cross-reactive with mosquito clathrin (Raikhel, 1984a).

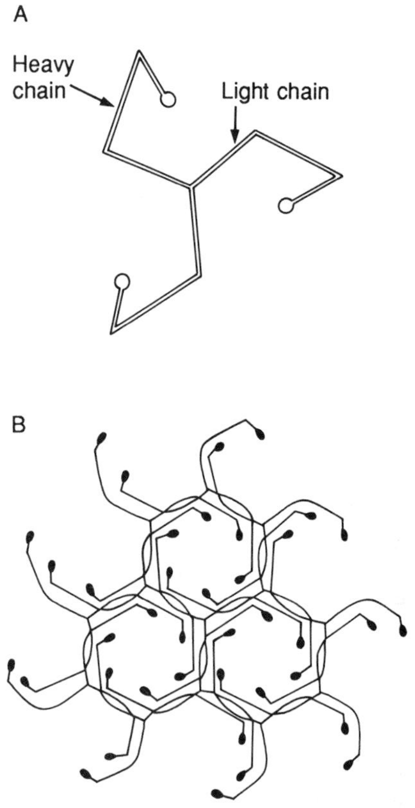

Figure 20.10 (A) Modular structure of a typical clathrin molecule, which has the form of a triskelion. Each leg, about 45 nm long, is composed of a heavy chain of 180 kDa which ends in a globular terminal domain, and of a proximal light chain of 30 or 40 kDa. (B) The packing of clathrin molecules to form a hexagonal lattice. Formation of closed cages requires the introduction of some pentagons into the lattice. (After Pearse and Crowther, 1987.)

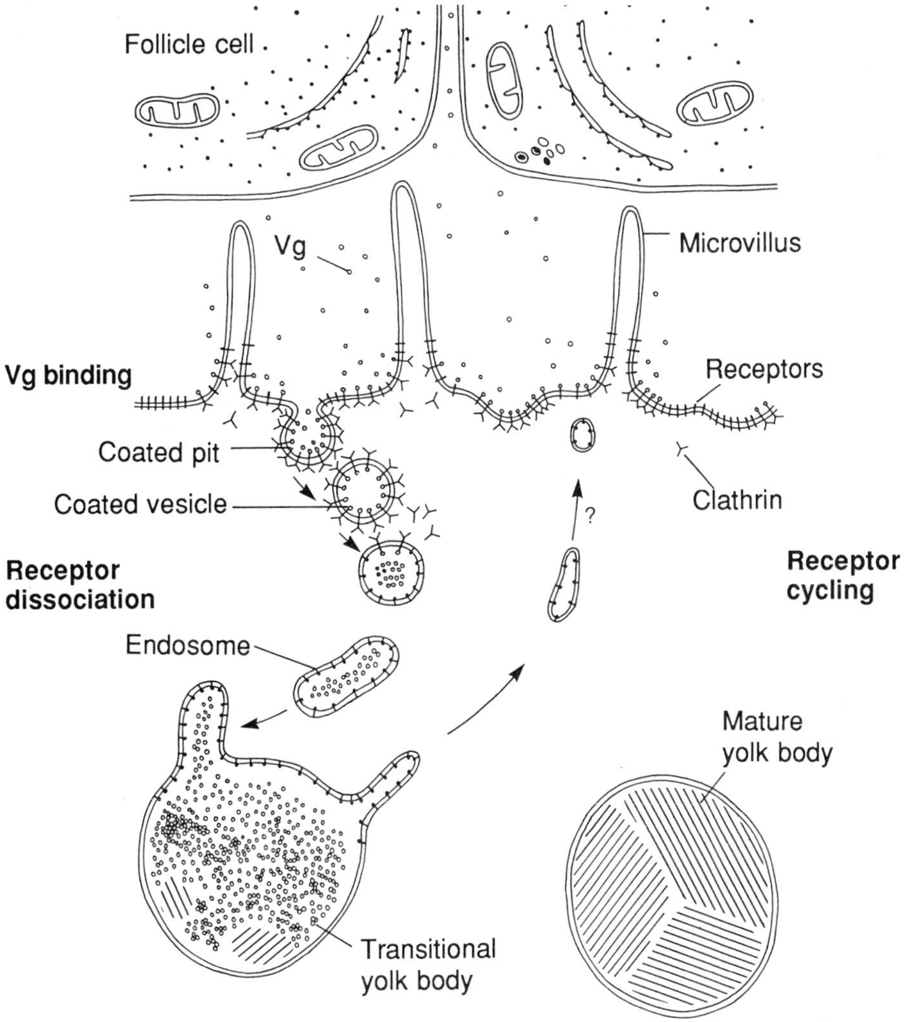

Figure 20.11 The endocytotic cycle. Diagram of the endocytotic apparatus in a mosquito oocyte during vitellogenesis, showing the stages of vitellogenin incorporation. An explanation is given in the text. The adaptor molecules that interconnect the clathrin and receptor molecules are not shown. (After Raikhel, 1984b.)

(b) *Complex formation*

The endocytotic complex is formed in mosquito oocytes during the previtellogenic phase. Aggregates of clathrin-like filamentous material are present in the perinuclear ooplasm at emergence. During the following two days this material becomes increasingly abundant and it is found also in the cortical ooplasm as aggregations of filamentous subunits (clathrin molecules) arranged in networks and as lattice-like cages, 50–60 nm in diameter. The aggregations of clathrin are surrounded by ribosomes, the postulated site of synthesis; most are associated with the *trans* side of a Golgi complex although the significance of this association is not clear (Raikhel, 1984a).

In stage II and vitellogenic oocytes of *Aedes aegypti* and *Culex quinquefasciatus* the bases of the microvilli and the areas of oolemma between them are covered with a thick electron-dense glycocalyx. In addition, a clathrin coat is apparent on the cytoplasmic side of the oolemma in those regions. Between the microvilli the oolemma is invaginated to form many **clathrin-coated pits**, of 120–140 nm diameter, which bear a thick glycocalyx on their outer surface and clathrin-like material on their cytoplasmic surface (Figure 20.11). The cortical ooplasm contains **clathrin-coated vesicles**, derived pinocytotically from coated pits, which are surrounded by a clathrin coat 20 nm thick and lined with a diffuse layer 25 nm thick resembling glycocalyx. Uncoated vesicles of similar size, present in the cortical ooplasm, are believed to have arisen from coated vesicles by loss of the clathrin layer. Structural sequences suggested that fusion of uncoated vesicles is the probable origin of larger uncoated bodies, called endosomes, some of which have the form of tubules of 100–500 nm diameter and variable length. In vitellogenic ovaries immunocytochemical methods show the presence of vitellogenin in all of these organelles. Deeper in the ooplasm uncoated tubules are associated with much larger yolk-containing vesicular organelles called **transitional yolk bodies**. Membrane-bound **mature yolk bodies**, containing crystalline yolk protein, represent the ultimate components of the synthetic apparatus (Roth and Porter, 1964; Anderson and Spielman, 1971; Roth *et al.*, 1976; Raikhel, 1984b; Raikhel and Lea, 1985).

It was postulated that uncoated tubular compartments, found attached to transitional yolk bodies or free in the cortical cytoplasm (Figure 20.11), and which were immunochemically negative for vitellogenin, were involved in the recycling of vitellogenin receptors (Raikhel, 1984b). The fact that endocytosis continued after inhibition of protein synthesis with cycloheximide strongly suggests that the recycling of Vg receptors occurs (Raikhel and Dhadialla, 1992).

The effect of ovarian developmental stage upon the capacity for vitellogenin uptake was investigated by measuring uptake of radiolabelled vitellogenin by ovaries of different ages post-bloodmeal when incubated in a solution of 0.18 μg vitellogenin per μl. The rate of uptake increased steeply and linearly between 6 and 24 h post-bloodmeal (Figure 20.12). The rate was highest between between 24 and 30 h, a period when the haemolymph vitellogenin concentration was maximal (Figure 20.7A), but by 36 h it was virtually zero. At 27°C the chorionic plaques fuse a little after 36 h post-bloodmeal (Section 19.5.4). Surprisingly, the length of the yolk mass almost doubled between 30 and 48 h post-bloodmeal (Koller *et al.*, 1989).

20.4.2 Movement and binding of vitellogenin

Ultrastructural and high-resolution immunocytochemical methods have been used to follow the uptake and incorporation of vitellogenin by oocytes. The later studies included use of polyclonal antibodies raised against vitellin, which also reacted with vitellogenin from fat bodies (Roth and Porter, 1964; Raikhel, 1984b; Raikhel and Lea, 1983, 1985, 1986).

The basal lamina surrounding the follicular epithelium in *Aedes aegypti* was freely permeable to most compounds of molecular mass up to 500 kDa; protein and polysaccharide tracers passed through it with ease but some inert particles penetrated more slowly. Passage of protein from the haemolymph through the narrow intercellular channels between follicular cells and into the perioocytic space

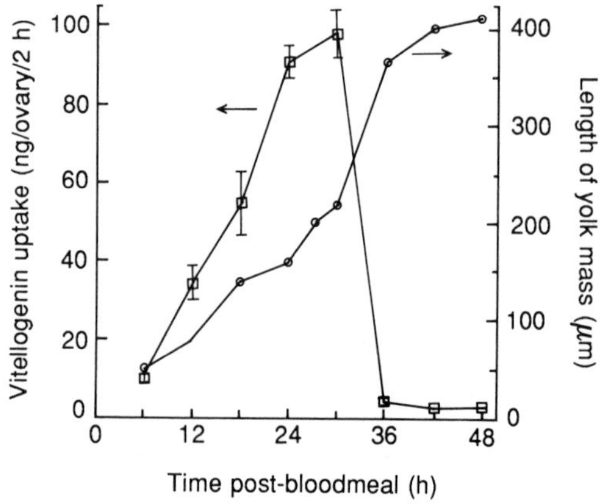

Figure 20.12 Capacity for vitellogenin uptake of ovaries at different developmental stages. Single ovaries of *Aedes aegypti* were incubated for 2 h in 10 μl saline containing 0.18 μg ^{35}S-labelled vitellogenin per μl. Means + s.e. (From Koller *et al.* (1989). Copyright (1989), Pergamon Press plc. Reprinted with permission.)

has been observed with the tracer horseradish peroxidase. Vitellogenin has been demonstrated immunocytochemically in the intercellular channels in vitellogenic ovaries. After fixation with tannic acid, vitellogenin appeared as an electron-dense granular material dispersed throughout the perioocytic space and attached to the region of dense glycocalyx. Vitellogenic carboxypeptidase, the other extraovarian protein, has also been observed in the intercellular channels and perioocytic space (Anderson and Spielman, 1971; Raikhel, 1984b; Hays and Raikhel, 1990).

Ovaries that had not reached the previtellogenic resting stage bound very little yolk protein compared to those in the resting stage. Ovaries from bloodfed females bound even more yolk protein (Roth *et al.*, 1976). Ninety-five percent of vitellogenin bound to the oolemma was located over the region with dense glycocalyx and clathrin lining; 28% was located in coated pits, which occupied only 9% of surface membrane length. It appeared that vitellogenin receptors were located in specific membrane microdomains on the bases of and between the microvilli, and that after binding to these microdomains vitellogenin was concentrated in the coated pits (Raikhel, 1984b; Raikhel and Lea, 1985).

Some characteristics of the binding and internalization processes were determined by measuring the uptake of [^{35}S]methionine-labelled vitellogenin by *Ae aegypti* ovaries *in vitro*. Uptake rates peaked between pH 7.4 and 7.8, corresponding with the pH of mosquito haemolymph, and declined on either side of this optimum. Uptake steadily declined as the pH was reduced from 7.4, almost ceasing at pH 6.3, but it increased somewhat as the pH was reduced further. This activity trough was possibly due to the reduced solubility of vitellogenin at its isoelectric point (which Borovsky and Whitney (1987) had determined to be pH 6.3). Uptake was saturated at a Vg concentration near 8 $\mu g\ \mu l^{-1}$ (= 21 μM) at 27°C. A V_{max} of 3.2 μg ovary^{-1} h^{-1} and a K_{uptake} of 8.6×10^{-6} M were calculated. Uptake was inhibited at 4°C (Koller *et al.*, 1989). In *Ae aegypti*, between 16 and 32 h post-bloodmeal the haemolymph Vg concentration is 6 $\mu g\ \mu l^{-1}$ or more (Figure 20.6A), so the ovaries are in a near-saturating environment. Fat body and Malpighian tubules internalized both Vg and mouse IgG at very low rates; the ovaries, in contrast, showed selectivity, internalizing Vg at a high rate and mouse IgG at a very low rate (Koller *et al.*, 1989). These various experiments showed that the endocytosis of vitellogenin is temperature dependent, tissue specific, saturable and selective; it is therefore receptor mediated.

The presence of specific receptors for vitellogenin in ovary membranes of *Ae aegypti* was confirmed by an *in vitro* binding assay with [^{35}S]vitellogenin and purified membrane fragments representing all three cell types in the ovarian follicles. Binding rates were pH dependent, but the shape of the pH curve varied with Ca^{2+} concentration. At pH 7.0 and in the presence of 5 mM Ca^{2+}, the binding of Vg to its receptor reached equilibrium within 60–90 min at both 4 and 25°C. The binding was specific to ovarian membranes, to which vitellogenin and vitellin bound with equal affinity. Binding under equilibrium conditions demonstrated the existence in the preparation of a single class of binding site with a dissociation constant (K_d) of 0.18 μM. This K_d value was 48-fold lower than the K_{uptake} for intact ovaries, reported above, where the requirement for diffusion of Vg molecules through interfollicular channels would have reduced the rate of Vg uptake by the oocytes. Ligand-blot experiments with solubilized ovary membrane proteins revealed binding of labelled Vg to a single polypeptide with an apparent molecular mass of 205 000 (Dhadialla and Raikhel, 1991).

Uptake of ^{125}I-labelled yolk protein by vitellogenic *Culex quinquefasciatus* ovaries *in vitro* was shown to be selective since the rate of uptake was not affected by competition with other proteins. The transport system was saturable, and had an apparent K_m of $<10^{-7}$ M. Trypan blue, a known uptake inhibitor, competed with yolk protein on an equimolar basis. Yolk protein uptake could be blocked with inhibitors of glycolysis but not with inhibitors of respiration or oxidative phosphorylation (Roth *et al.*, 1976).

20.4.3 Uptake and processing of vitellogenin

Labelled vitellogenin was observed within coated vesicles where it was located in a layer close to the luminal surface of the vesicle membrane. It was also observed in numerous uncoated vesicles and endosomes, in which it was located in the vesicle lumen. After tannic acid fixation vitellogenin could be seen within coated vesicles as granular material lining the inner surface of the membrane and leaving only a small lumen. Some vesicles had only a partial clathrin coat on the cytoplasmic side of the membrane, and in these the granular material was attached only to the coated portion of the membrane. In uncoated vesicular and tubular endosomes the granular material appeared to have detached from the membrane and to have concentrated in the lumen. Raikhel (1984b) concluded that vitellogenin dissociated from its membrane binding sites in the first uncoated compartment.

The uncoated vesicles, which were taken to be the site of ligand–receptor dissociation, had a short life span, appearing to fuse with one another to form small endosomes. These progressively coalesced to form increasingly large transitional yolk bodies, with a maximum diameter of 6 μm. Many of the tubular endosomes were connected to transitional yolk bodies. The granular contents became concentrated to such a degree that individual granules were barely distinguishable, and in some transitional yolk bodies small crystalloids were seen. The contents showed intense immunolabelling (Raikhel, 1984b).

The transitional yolk bodies were characterized (a) by their contact with vitellogenin-containing endosomes, showing the source of their contents, (b) by the even distribution of immunolabelling within the lumen, indicating the presence of vitellogenin uncoupled from its receptors, (c) by contact with unlabelled tubules, indicating that they were the compartment from which vitellogenin receptors were probably recycled, and (d) the presence in some of crystalline yolk.

Mature yolk bodies, which had no tubular or vesicular extensions, were located deeper in the ooplasm. Most of the lumen was occupied by two or four crystalloids separated by a variable quantity of matrix. Both matrix and crystalloid reacted intensively with markers immunospecific to vitellogenin. The crystalloids exhibited lattice patterns of various orientations within a single plane of section, presumably because they had arisen by the fusion of several small units which were already crystalline. Vitellogenic carboxypeptidase was distributed as a rim around the yolk body. All compartments of the vitellogenin accumulative pathway lacked lysosomal enzymes, therefore the mature yolk bodies were endosomal compartments, specialized for long-term storage of yolk protein (Raikhel, 1984b; Hays and Raikhel, 1990).

Individual antibodies from a library of mABs raised against vitellogenin recognized different steps on the accumulative pathway. Certain antibodies labelled all compartments from coated pits to mature yolk bodies, the density of the label being significantly higher over transitional yolk bodies and maximal over mature yolk bodies. Other antibodies recognized different forms of the yolk protein during its accumulative processing. Thus some, apparently recognizing vitellogenin, labelled only the cortex and the rims of mature yolk bodies; others, recognizing vitellin, labelled only the crystalline region of mature yolk bodies (Raikhel, 1987a).

By double immunolabelling with antibodies directed separately against the 200 and 65 kDa yolk polypeptides it was shown that both were present on the dense glycocalyx on the oolemma and in coated pits. Similarly both were present in the majority of coated vesicles as well as in uncoated vesicular and tubular endosomes in the oocyte cortex. There was no evidence of independent routeing. Both labels occurred at high density in transitional and mature yolk bodies (Raikhel, 1987a).

The incorporation of trypan blue by the endocytotic apparatus has been used to observe the start and cessation of endocytosis in *Ae aegypti* maintained at 27.5°C. Oocytes of females injected at 2 h post-bloodmeal took up a small amount of dye, and as the rate of vitellogenin uptake increased so did that of trypan blue. Between 39 and 42 h post-bloodmeal trypan blue was able to penetrate into the perioocytic space of some follicles but not of others within the same ovary. However, even at 39 h, there was no incorporation of trypan blue by any oocyte (Yonge and Hagedorn, 1977).

20.4.4 Internalization and fate of foreign proteins

The internalization of a number of foreign proteins by vitellogenic oocytes after injection into *Aedes aegypti* was reported by Anderson and Spielman (1971). To investigate this phenomenon Raikhel and Lea (1986) injected horseradish peroxidase into sugarfed females with vitellogenically competent ovaries. The peroxidase penetrated the ovarian and ovariolar sheaths, passed through channels between the follicular cells, and entered the perioocytic space where it adhered to the oocyte membrane and accumulated in coated pits. These pits were transformed into coated vesicles, of 110–130 nm diameter, in which the peroxidase lined the inner side of the vesicle membrane. The coated vesicles shed their coats and the uncoated vesicles fused with one another and with uncoated tubules. Within small uncoated vesicles the peroxidase remained attached to the membrane; in larger uncoated vesicles and in uncoated tubules it was located in the lumen. Three to six hours after administration of horseradish peroxidase many large uncoated vesicles filled with peroxidase were present deeper in the ooplasm, in the vicinity of the Golgi complexes and frequently in contact with primary lysosomes. Horseradish peroxidase was also present, with degrading cytoplasmic organelles, in some secondary lysosomes.

Ovaries of *Culex quinquefasciatus* cultured *in vitro* incorporated ferritin into their oocytes when vitellogenin was also present in the culture medium, but not in its absence (Roth *et al.*, 1976). To investigate the incorporation of a foreign protein in the presence of vitellogenin, Raikhel and Lea (1986) implanted competent previtellogenic ovaries into isolated ovariectomized abdomens of bloodfed *Ae aegypti*, in which vitellogenin had accumulated in the haemolymph, and injected horseradish peroxidase. Both proteins entered the oocytes via coated pits, but most coated vesicles and small vesicles contained only one or other protein and only a few contained both. In contrast, all transitional yolk bodies contained both proteins. The horseradish peroxidase was localized at the periphery of mature yolk bodies, often concentrated in vesicular extensions. Peroxidase-containing vesicles of similar size to the extensions were found in the vicinity of the Golgi complexes and lysosomes. It appears that vitellogenin and horseradish peroxidase were internalized in a similar manner, although in separate coated pits, but that after separating from their membrane binding sites in the endosomes they were processed into different pathways. Vitellogenin was then condensed and stored whereas the foreign protein was concentrated in small vesicular extensions which fused with lysosomes.

21

Hormonal regulation of ovarian development in anautogenous mosquitoes

Oogenesis requires integration of the developmental and metabolic activities of separate organs, and this is provided by hormonal mechanisms. Although the molecular structure of insect hormones is highly conserved, at least among the juvenile hormones and ecdysteroids, the regulatory functions of individual hormones vary greatly between different insect taxa. There is, therefore, no general pattern of hormonal regulation of ovarian development in insects with which to compare that of mosquitoes.

In mosquitoes, follicles develop synchronously in all ovarioles during both the previtellogenic and the vitellogenic phases of ovarian development. Mosquitoes are valuable experimental animals for investigations into the regulation of ovarian development because this synchrony of follicular development extends to the ovaries of all females that eclose simultaneously or that are bloodfed simultaneously.

Follicle development may halt at a number of developmental gates (Section 19.2.2), and development beyond most gates is stimulated hormonally. Experiments on mosquitoes have implicated hormones from the brain, corpora allata and the ovaries themselves in the regulation of vitellogenin synthesis and ovarian development, and have provided a conceptual framework involving a number of hormones, of which the actions of two – juvenile hormone and 20-hydroxyecdysone – are generally regarded as fundamental. It is likely that other hormones also have gonadotrophic actions.

Investigations of the past thirty years that were directed at the endocrinology of the vitellogenic phase of mosquito oogenesis were nearly all based on the premiss that the first important step was the release of ovarian ecdysteroidogenic hormone (OEH = EDNH), which occurs three or more hours post-bloodmeal and induces the synthesis and secretion of ecdysone. There was a tendency to overlook earlier evidence that suggested that during or very shortly after blood feeding, releases of hormones occurred that could stimulate limited vitellogenin synthesis and uptake.

Gillett (1956) introduced the terms 'initiation' and 'promotion' for aspects of the endocrine regulation of ovarian development that, respectively, were considered not to involve and to involve a hormone from the head. It was later shown that secretions from the head are involved in both aspects. Nevertheless, it is useful to retain these terms, but redefined. Where used here, as endocrinological concepts, **initiation** and **promotion** respectively concern the mechanisms of regulation of the initiation and trophic phases of ovarian development (Section 19.2.1). The processes of initiation regulate development from the previtellogenic resting stage to the stage III gate; they are insufficient to permit oocytes to develop to full term. The processes of promotion (some of which may be identical with those of initiation) regulate development from the stage III gate to maturation (Section 19.2.2). In some species the initiation and trophic phases of ovarian development are discontinuous, require separate blood meals, and are easily distinguishable. In other species, including *Aedes aegypti*, they are usually continuous, but they may be discontinuous if small blood meals are ingested, and they can be separated experimentally.

Almost all experimentation has been on *Aedes aegypti*, and it is possible that the regulatory systems of mosquitoes of other genera differ to a certain extent. Outlines of the hormonal systems that are believed to regulate previtellogenic and vitellogenic development in *Ae aegypti* are shown in Figures 21.1 and 21.4. Studies of hormonal regulation in autogenous mosquitoes (Section 23.2) have provided additional insights. Basic information on each hormone will be found in Chapter 10.

21.1 THE PREVITELLOGENIC PHASE

21.1.1 Regulation of follicle growth and competence

Two developmental gates control the course of previtellogenic development (Figure 21.1). The first of these is the germarial gate which governs separation of follicles from the germaria. In many species the primary follicles separate from the germaria shortly after emergence, and the later series of follicles separate some hours after the successive blood meals. We have no direct evidence of the mechanism controlling separation of primary follicles from their germaria, but it is known that 20-hydroxyecdysone produced during the first gonotrophic cycle induces separation of the secondary follicles (Section 21.9). It may be that the partial separation of primary follicles from germaria observed in newly emerged *Aedes aegypti* (Gwadz and Spielman, 1973) is a consequence of 20-hydroxyecdysone secretion during the pupal stage (Figure 8.13).

A number of investigations have produced evidence that the development of follicles beyond the stage I gate is stimulated by juvenile hormone. When females of *Ae aegypti* were allatectomized* within one hour of emergence, their primary ovarian follicles made little or no growth and did not develop to the previtellogenic resting stage. If these females were fed blood their primary follicles degenerated. However, full previtellogenic development could be stimulated in females allatectomized at emergence, either by reimplanting corpora allata or by treatment with a juvenile hormone analogue, and these females matured their ovaries after a blood meal (Table 21.1).

When the corpora allata were removed from *Ae aegypti* on the second or third day after emergence, 25 and 50% of the females respectively were able to mature their oocytes after a subsequent blood meal. Therefore in some females the target organs had received sufficient exposure to juvenile hormone during the first two or three days after emergence for completion of the previtellogenic phase. Females allatectomized three days after emergence and bloodfed eight days later could still produce mature oocytes (Lea, 1963; Gwadz and Spielman, 1973). Similar experimental results were obtained with *Culex pipiens* (Spielman, 1974) and *Aedes taeniorhynchus* (Lea, 1963). However, females of *Ae triseriatus* that were allatectomized at emergence and given a blood meal two days later all developed mature eggs; it appeared that sufficient juvenile hormone had been released before emergence to stimulate previtellogenic development. The responses of *Ae sollicitans* were intermediate (Table 21.1) (Lea, 1963).

When *Ae aegypti* were allatectomized within 1 h of emergence and the ovaries were examined 48 or 72 h later the follicles were small and the oocyte cortex remained undifferentiated, i.e. they lacked microvilli and endocytotic organelles. Horseradish peroxidase injected at 48 h penetrated the channels between the follicular cells but was not taken up by the oocytes. Delaying allatectomy until 24 h post-emergence permitted

* Removal of the corpora allata necessarily involves removal of the corpora cardiaca also. The term 'allatectomy' therefore implies removal of both. Decapitation removes the cerebral neurosecretory system, including most of the neurohaemal organ; the corpora allata and corpora cardiaca are left intact but denervated.

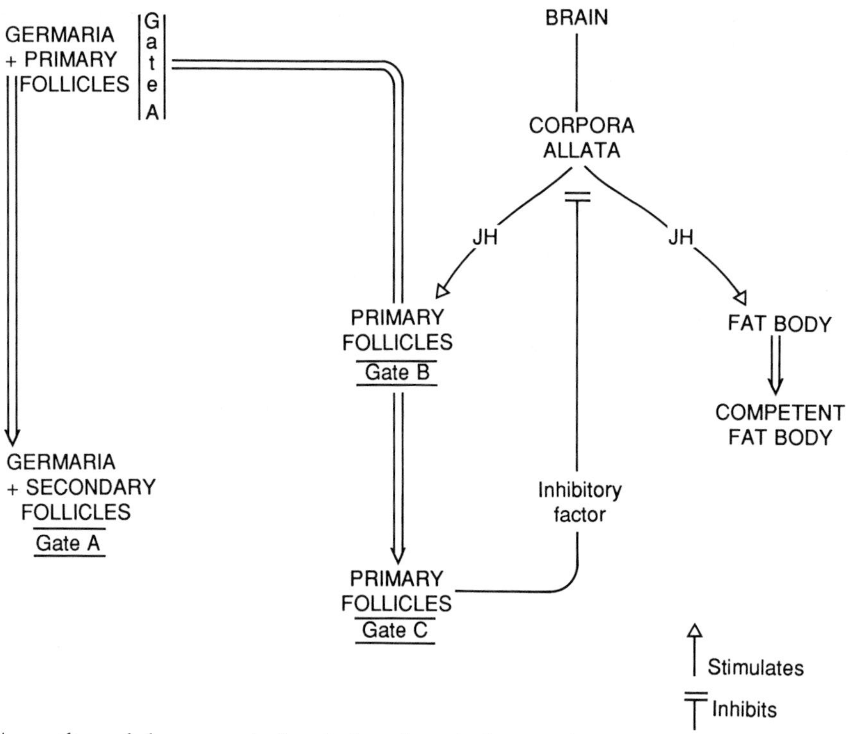

Figure 21.1 An outline of the anatomical and physiological changes that occur during the previtellogenic phase in *Aedes aegypti*, between the time of eclosion and entry of the primary ovarian follicles into the previtellogenic resting stage. Exposure to juvenile hormone induces developmental changes in both primary follicles and fat body that prepare those organs for vitellogenesis. Developmental gates: A, Germarial; B, stage I; C, Previtellogenic (see Section 19.2.2.)

Table 21.1 The effects of prior ablation and implantation of corpora allata upon ovarian maturation in bloodfed anautogenous females of four species of *Aedes*. (From the data of Lea, 1963, 1970; Gwadz and Spielman, 1973.)

Species	*Age at operation*	*Treatment*		*Females maturing oocytes (%)*
		Ablated	*Implanted*	
Ae aegypti	1 hour	CA	–	0*
	1 hour	CA	CA (1 pair)	100
	1 day	CA	–	0
	2 days	CA	–	25
	3 days	CA	–	50
Ae taeniorhynchus	1 hour	CA	–	6
	1 hour	CA	CA (1 pair)	95
	3 days	CA	–	78
Ae triseriatus	1 hour	CA	–	100
Ae sollicitans	1 hour	CA	–	30
	3 days	CA	–	89

* degenerated; CA, corpora allata.

normal formation of the endocytotic complexes within the following 48 h. When females were allatectomized at emergence and, 48 h later, were implanted with one pair of corpora allata, subsequent ovary growth and endocytotic-complex formation were nearly normal, and the oocytes were endocytotically active when tested with horseradish peroxidase. When abdomens were isolated at emergence and treated with 300 pg of JH III, the oocytes later had fully developed endocytotic complexes and could internalize injected horseradish peroxidase (Raikhel and Lea, 1985, 1991).

During the previtellogenic phase juvenile hormone not only stimulates limited growth and development of the ovarian follicles but also renders the follicles competent to respond to ovarian ecdysteroidogenic hormone (OEH) after a blood meal. Ovaries taken from females at 12 h post-emergence did not respond to OEH *in vitro*, but between 24 and 60 h post-emergence the ovaries became responsive and would secrete ecdysone upon exposure to OEH. Treatment with JH-I of abdomens ligated soon after emergence compensated for isolation from the corpora allata (Shapiro and Hagedorn, 1982).

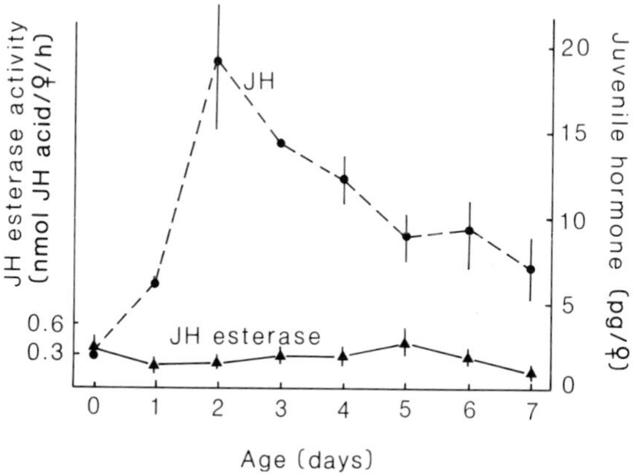

Figure 21.2 Juvenile hormone and juvenile hormone esterase in newly emerged female *Aedes aegypti*. The whole-body JH-III content (recalculated) and the haemolymph juvenile hormone esterase activity of the haemolymph were assayed in sugarfed females during the first seven days of adult life. Means ± s.d. (After Shapiro *et al.*, 1986.)

21.1.2 Control of juvenile hormone secretion

During the first two days after emergence the whole-body juvenile hormone content of *Aedes aegypti* rose from 2 to 20 pg per female. In non-bloodfed females it declined gradually over the next five days. Haemolymph juvenile hormone esterase activity remained low throughout this period (Figure 21.2) (Shapiro *et al.*, 1986). In insects a very high proportion of circulating JH may be inactive due to binding to haemolymph protein (Steel and Davey, 1985).

Corpora allata removed from sugarfed females of *Culex pipiens* on each of the first eight days after emergence, and cultured *in vitro*, secreted JH at the rate of 18–35 fmol h^{-1} per pair. The rate of secretion did not decline after the mosquitoes had entered the previtellogenic resting stage (Readio *et al.*, 1988).

Little is known about the regulation of juvenile hormone secretion in mosquitoes. Decapitation immediately after emergence has the same effect on follicle development as allatectomy (Gwadz and Spielman, 1973; Hagedorn *et al.*, 1977), which suggests that the brain is involved. We may note that in a wide variety of insects the corpora allata are under nervous inhibition via NCC-I and the allatal nerve (Steel and Davey, 1985).

There is evidence from *Ae aegypti* of the ovaries affecting JH secretion. Ovaries removed from pupae before pupal–adult apolysis and implanted into females with previtellogenic resting stage ovaries caused the hosts' ovaries to degenerate while themselves developing to the resting stage. When ovaries from postapolytic pupae were similarly transplanted they did not cause degeneration of their host's ovaries, but nor did they develop further unless the host was treated with methoprene. Ovaries from both pre- and postapolytic pupae developed when implanted into ovariectomized females. Rossignol *et al.* (1981) concluded that the loss of resting stage ovaries, whether by degeneration or ovariectomy, had promoted further JH secretion by removing an inhibitory factor, presumed to have been secreted by resting stage ovaries. They postulated that corpus allatum activity is regulated by ovarian feedback inhibition.

21.1.3 Effects of nutrition

If a female is in a poor state of nourishment at emergence her primary follicles may develop only to stage I and then enter a quiescent state. However, if she is in a satisfactory state of nourishment her primary follicles may undergo full previtellogenic development. The importance of nourishment was shown in experiments with *Aedes aegypti* which had either been crowded as larvae, producing small undernourished adults, or reared without crowding to produce large well-nourished adults. If large females were fed only water after emergence their primary follicles still grew substantially and developed to the previtellogenic resting stage, and after a blood meal their oocytes developed to maturity. Small females contained undersized primary follicles at emergence. If they were given only water their primary follicles failed to grow, but if they were given a blood meal their follicles developed to the previtellogenic resting stage. A second blood meal was needed for vitellogenesis and growth of the oocytes to maturity. When small females were given sucrose after emergence their follicles grew slightly and developed to the previtellogenic resting stage; after a single blood meal the oocytes in two-thirds of these females developed to maturity (Feinsod and Spielman, 1980a). Most females of *Ae aegypti* caught from two wild populations had wing lengths of 2.8–2.9 mm, characteristic of undernourishment (Feinsod and Spielman, 1980a), therefore undernourishment may partly account for the multiple blood feeding recorded among wild females of that species (McClelland and Conway, 1974).

Adult size, sugar-feeding and blood-feeding affect ovary development in *Anopheles sacharovi* very much as described above for *Ae aegypti* (Mer, 1936). High percentages of nulliparous females in wild populations of *An gambiae* s.l. and *An funestus* required two blood meals to develop their primary follicles to maturity, a phenomenon rarely observed in the parous females (Gillies, 1954, 1955).

Experimental evidence points to an interaction between the nutritional condition of newly emerged females and juvenile hormone secretion. Topical treatment of small, undernourished and water-fed females of *Ae aegypti* with 8 ng of the JH analogue methoprene stimulated follicle development to the previtellogenic resting stage in all cases, but treatment with 80 ng was required to enable most of the females to develop their primary follicles to maturity after a single blood meal (Feinsod and Spielman, 1980a). When females of *Ae sollicitans*, derived from well-nourished larvae, were allatectomized at 3 days post-emergence, 89% subsequently matured their ovaries after blood feeding. However, when the females were derived from poorly nourished larvae only 16% allatectomized at 3 days developed their ovaries after blood feeding, although 73% of sham-operated females were able to do so (Lea, 1963).

21.1.4 Fat body competence

During the previtellogenic phase the fat body is rendered competent to synthesize vitellogenin, a capability which is utilized only later, after blood feeding and in response to stimulation by 20-hydroxyecdysone (Section 21.5.2). Females of *Aedes aegypti* that had been allatectomized at emergence failed to synthesize vitellogenin in response to 20-hydroxyecdysone injection unless corpora allata had been implanted (Flanagan and Hagedorn, 1977). Fat bodies removed from female *Ae aegypti* between 0 and 52 h after emergence, and incubated for 3 h with 10^{-6} M 20-hydroxyecdysone, did not secrete vitellogenin into the medium. However, between 56 and 64 h post-emergence the fat bodies developed full competence to respond to the hormone. If the abdomens of newly emerged females were coated with 0.5 μg of the JH analogue methoprene, or if the fat bodies were incubated with that compound for 60 h, vitellogenin was produced by the fat bodies upon incubation with 20-hydroxyecdysone (Ma *et al.*, 1988).

Both of these investigations showed that it is exposure to juvenile hormone that induces fat body to become competent to synthesize vitellogenin. The development of competence involves fat body nuclei becoming polyploid (Section (20.2.1) and synthesis of RNA (Section 20.3.1).

21.2 THE INITIATION OF VITELLOGENIC DEVELOPMENT

Initiation, as an endocrinological concept, is here treated as the regulation of the initiation phase of vitellogenic development. Evidence was presented in Section 19.5.2 that a number of changes occur in bloodfed females well before the major burst of OEH release, *viz* vitellogenin synthesis, an increase in the capability of oocytes for vitellogenin uptake, and an increase in the rate of ovarian protein synthesis. Decapitation experiments have revealed the involvement of hormones from the head in regulating these developments.

When engorged females of *Aedes aegypti* were decapitated within five minutes of starting to blood-feed, and their ovaries were later assayed *in vitro*, it was found that at 30 min post-bloodmeal the rates of protein synthesis in ovaries from bloodfed decapitated females were strikingly higher than those in previtellogenic ovaries (Figure 21.3A). Similarly, at 2 h post-bloodmeal, the capability of ovaries from bloodfed decapitated females for vitellogenin uptake was significantly higher than that of previtellogenic ovaries but below that in ovaries from intact bloodfed controls (Figure 21.3B). In contrast, the accelerated rates of vitellogenin uptake and of ovarian protein synthesis that started at about 4 h post-bloodmeal in the ovaries of intact controls were absent from the ovaries of mosquitoes decapitated immediately after feeding. It appears that these physiological developments were stimulated by hormones from the head, which were released in pulses during and after blood feeding. It is interesting that blood enemas given to intact or decapitated sugar-fed females failed to stimulate ovarian protein synthesis or the rate of vitellogenin uptake, at 4 h post-feeding, whereas substantial stimulation occurred at that time in bloodfed decapitated females (Koller and Raikhel, 1991).

Experimental evidence points to the release of one or more gonadotrophic hormones in *Culex pipiens* during blood feeding. Females decapitated after feeding on quail for four minutes, and

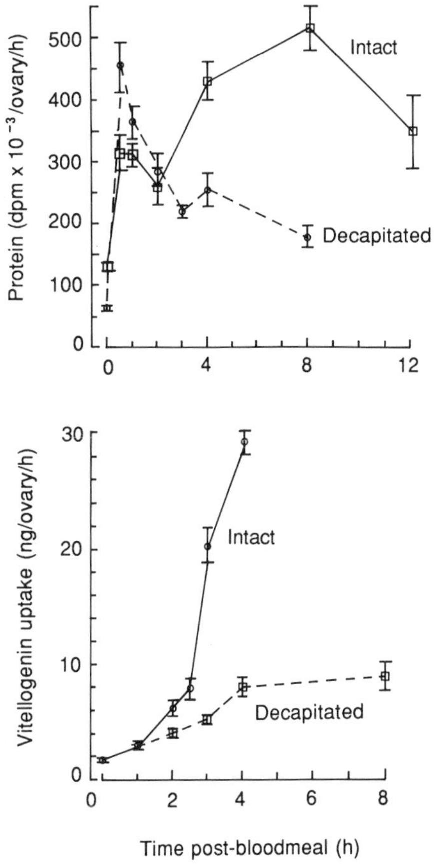

Figure 21.3 Physiological developments in the ovaries of *Aedes aegypti* during the initiation phase, in intact females and in females decapitated within 5 min of starting to blood feed. Kinetic data were obtained *in vitro* with ovaries removed at different times after the females had fed. (A) Rates of protein synthesis. (B) Rates of vitellogenin uptake. (After Koller and Raikhel, 1991.)

Table 21.2 The effects of experimental conditions upon OEH activity in *Aedes aegypti*. (From Hagedorn, 1986.)

Head extract or saline was injected into non-bloodfed females or into bloodfed females that had been decapitated at 2 h post-bloodmeal. Ecdysteroid titres were measured 25 h after the injection. The ovaries were assayed for vitellin after 50 h.

	Ecdysteroid per female (pg)	*Vitellin per female* (μg)	*Females maturing oocytes* (%)
Non-bloodfed:			
Saline injected	22	0	0
Head extract injected	16	0	0
Bloodfed, decapitated:			
Saline injected	37	0	0
Head extract injected	254	75	75
Bloodfed control	452	135	100

which had imbibed sufficient blood for ovarian development, made little ovarian growth, whereas females decapitated after feeding for eight minutes made substantial ovarian growth. Baldridge and Feyereisen (1986) concluded that a gonadotrophic factor had been released from the head between four and eight minutes after the start of blood feeding, which normally lasted for ten minutes.

That hormones are released during the course of blood feeding had earlier been demonstrated by observations on females of *Ae aegypti*. Diuresis started within a minute of the start of engorgement, showing that diuretic hormone could be released and affect the target organ in the abdomen within that time (Section 16.1.2). The decapitation experiments reported above revealed hormonal activity immediately after blood feeding. Evidence that exposure to hormones at that time puts the females into a receptive state for OEH and 20-hydroxyecdysone is found in the heightened responsiveness to injected substances observed in females decapitated shortly after taking a blood meal. Sugarfed females of *Aedes aegypti* did not synthesize ecdysteroids or vitellogenin in response to the injection of head extract containing OEH. In contrast, females that were decapitated two hours after engorging, and immediately injected with head extract, later synthesized ecdysteroids and vitellogenin (Table 21.2) (Wheelock and Hagedorn, 1985; Hagedorn, 1986). Bloodfed decapitated females injected with 5 μg 20-hydroxyecdysone produced much more vitellogenin mRNA than non-bloodfed females injected with the hormone (Gemmill *et al.*, 1986).

Decapitation and ligation experiments on *Ae aegypti* provided further evidence of early developmental changes that apparently were independent of OEH. Females decapitated within 10 min of blood feeding contained significant quantities of vitellogenin mRNA 24 h later (Racioppi *et al.*, 1986). Females that were ligated at the base of the abdomen immediately after a blood meal secreted significant amounts of 20-hydroxyecdysone during the next eight hours (Borovsky and Thomas, 1985). Indeed, in intact females, the ecdysteroid titre was found to have increased measureably by two hours post-bloodmeal (Racioppi *et al.*, 1986), and a small burst of ecdysteroid production at about four hours post-bloodmeal (Figures 20.3, 21.8) was reported to be reproducible (Hagedorn *et al.*, 1975).

Ligation at the base of the abdomen immediately after blood feeding was found to permit ovarian development to stage III in *Ae aegypti* and *Cx pipiens* and to stage IV in *Anopheles atroparvus* (Clements, 1956a). In some experiments a number of females of *Ae aegypti* decapitated within 25 min of feeding developed their ovaries to stages III or IV, whereas in other experiments by the same investigators no vitellogenesis was observed. Nearly all females decapitated within 45–90 min of engorging showed vitellogenesis, and some developed their ovaries to maturity (Clements and Boocock, 1984). Greenplate *et al.* (1985) reported that most females decapitated within four hours of blood feeding developed their follicles beyond the previtellogenic resting stage, follicle development halting at or before stage IIIa.

Little can be said about the regulation of the initiation phase except that hormones from the head are involved which are released during or soon after blood feeding. Although decapitation experiments indicate that in the majority of females of *Ae aegypti* OEH is released between 3 and 11 hours post-bloodmeal, in any individual being released in a short burst (Section 21.3.3, Figure 21.5), the possibility remains that OEH is also one of the hormones first released, being released both then and later. However, it is unlikely that the physiological developments measured shortly after blood feeding are due simply to exceptionally early OEH release in a few individuals causing effects in batch preparations.

21.3 THE PROMOTION OF VITELLOGENIC DEVELOPMENT

Promotion requires continuation of some or all of the activities of the initiation phase, but it is characterized by a massive increase in the concentration of 20-hydroxyecdysone and consequent stimulation of vitellogenin synthesis. The injection into bloodfed decapitated *Aedes aegypti* of head extract believed to contain ovarian ecdysteroidogenic hormone (OEH) led to a substantial increase in ecdysteroid titre and to vitellogenin synthesis, with 75% of females maturing oocytes (Table 21.2). These actions of OEH indicate its important role in the promotion of vitellogenesis and ovarian development. The regulation of OEH release and the nature of OEH action are reviewed in this Section.

21.3.1 OEH synthesis and storage

Experiments on *Aedes sollicitans* implicated the medial neurosecretory cells (MNC), their axons in the nervi corporis cardiaci (NCC), and their release sites in the neurohaemal organ (NHO), in the synthesis, storage and release of ovarian ecdysteroidogenic hormone. OEH activity was demonstrated in both the medial neurosecretory cell perikarya (MNCp) and the NHO. Activity was low in the NHO at emergence but high four days later. Ablation of the MNCp at emergence did not reduce the OEH activity of the NHO four days later, but severance of the NCC at emergence did. If the MNCp were ablated at emergence one blood meal depleted the NHO of most of its OEH. If the MNC were ablated five days after emergence, and the mosquitoes were given two blood meals, most developed eggs after each blood meal (Meola *et al.*, 1970; Meola and Lea, 1971).

The conclusions to be drawn from these experiments on *Ae sollicitans* are that the MNC synthesize OEH, and that at emergence some OEH is already stored in neurosecretory axons within the brain. During the four days after emergence OEH passes to release sites at the axon terminals within the NHO. Blood feeding leads to discharge of OEH from the axon terminals and the hormone is rapidly replaced from stores within the NCC.

21.3.2 OEH-releasing factor

Lea (1972) provided persuasive evidence that the blood meal generates a haemolymph-borne agent that stimulates the release of a promotion-inducing factor from the head. Females of *Aedes sollicitans* that were fed blood and immediately decapitated deposited only traces of yolk in their oocytes. Thus the females underwent the initiation but not the promotion phase of ovarian development. Loss of median neurosecretory cell perikarya and neurohaemal organ would explain this effect of decapitation (Table 21.3). However, if a bloodfed decapitated female was joined in parabiosis to an intact sugarfed female, they both matured their oocytes. Thus a haemolymph-borne agent from the bloodfed decapitated female must have stimulated release from the head of her sugarfed partner of a factor that induced promotion.

Blood feeding by *Ae sollicitans* leads to disappearance of OEH activity from the neurohaemal organ. When females that had been ovariectomized took a blood meal, OEH was not lost from their NHO unless an ovary had been implanted prior to the blood meal (Lea and Van Handel, 1982). Other experiments by these authors with *Ae sollicitans* and *Ae aegypti*, and by Borovsky (1982) with *Ae aegypti*, showed that only if the ovaries and the NHO were present at the same time following a blood meal did OEH release occur, as evidenced by vitellogenesis and ovary development. The conclusion was drawn that following a blood meal a factor secreted by the follicles acted on the axon terminals of the medial neurosecretory cells causing the release of OEH (Figure 21.4). An ovary from *Culex pipiens* could replace *Aedes* ovaries as a source of the releasing factor, but an ovary from *Anopheles quadrimaculatus* could not.

After ovariectomizing and decapitating *Ae aegypti* at hourly intervals following a blood meal, Lea and Van Handel (1982) found that vitellogenin had been secreted by some females operated on as early as 4 h post-bloodmeal; they concluded that secretion of the releasing factor and of OEH must have occurred within that period. Ovaries from sugarfed females implanted into ovariectomized females at 24 h post-bloodmeal were also capable of secreting the releasing factor. Borovsky (1982) reported that the releasing factor was not stored in the ovaries of sugarfed females. The factor was shown not to be 20-hydroxyecdysone and not to stimulate vitellogenin synthesis directly.

The question arises as to the nature of the connection between blood feeding and secretion of the OEH-releasing factor. It is likely that a hormone from the midgut endocrine cells (Section 10.1.2) is released within 6 h of feeding (Section 21.4). The midgut hormone might act

Table 21.3 The effects of prior removal and implantation of the median neurosecretory cells upon ovarian maturation in bloodfed anautogenous females of four species of *Aedes*. (From the data of Lea, 1967, 1970, 1972.)

Species	*Age at removal*	*Treatment*		*Females maturing (%)*
		Removed	*Implanted**	
Ae aegypti	1 hour	MNCp	–	0
	7 days	MNCp	–	0
Ae triseriatus	1 hour	MNCp	–	41
	3 days	MNCp	–	75
Ae taeniorhynchus	1 hour	MNCp	–	4–8
	3 days	MNCp	–	61
	3 days	NHO	–	64
	3 days	MNCp + NHO	–	0
	1 hour	MNCp	MNCp 2 pr	90
	1 hour	MNCp	NHO 4	70
Ae sollicitans	1 hour	MNCp	–	54
	5 min PBM	MBCp	–	most
	5 min PBM	MNCp + NHO	–	0
	Before 2nd BM	MNCp + NHO	–	6
	1 day P2BM	MNCp + NHO	–	93

PBM = post-bloodmeal
NHO = neurohaemal organ
* Implanted 1 day after removal
P2BM = post 2nd blood meal
MNCp = medial neurosecretory cell perikarya

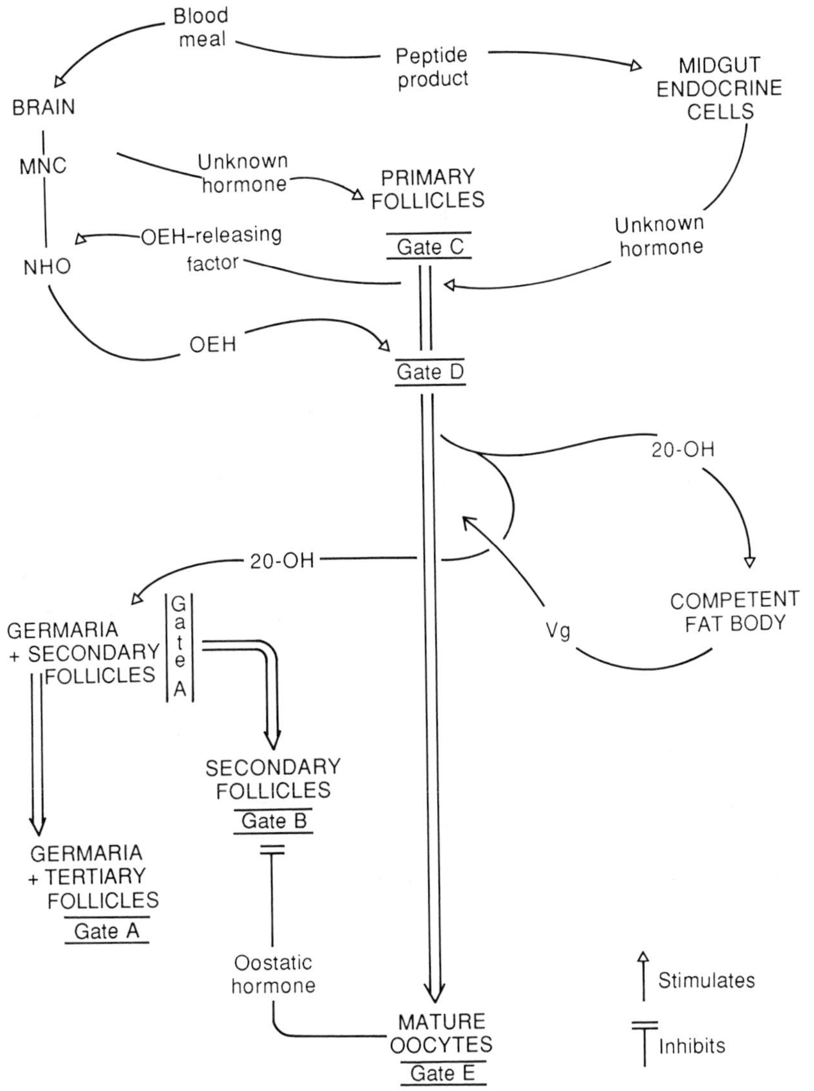

Figure 21.4 An outline of the physiological and anatomical changes that occur in *Aedes aegypti* after blood feeding and during the initiation and promotion phases of ovarian development. For simplicity, the 20-hydroxylation of ecdysone is not shown. OEH, ovarian ecdysteroidogenic hormone; 20-OH, 20-hydroxyecdysone; Vg, vitellogenin. Developmental gates: A, Germarial; B, stage I; C, Previtellogenic; D, stage III; E, Maturation (see Section 19.2.2).

on the ovarian follicles, as schematized in Figure 21.4, or it might act with OEH-releasing factor on the NHO.

21.3.3 OEH secretion

When mosquitoes are decapitated at different times after a blood meal, a plot of percentage of females maturing oocytes against time of decapitation yields a curve which is taken to represent the critical period for OEH release. At 27–28°C the critical period occurs between about 3 and 10 h post-bloodmeal in *Aedes aegypti* and between about 7 and 18 h in *Anopheles albimanus* (Lu and Hagedorn, 1986) (Figures 21.5A, B). The curve distinguishes the percentage of females that have

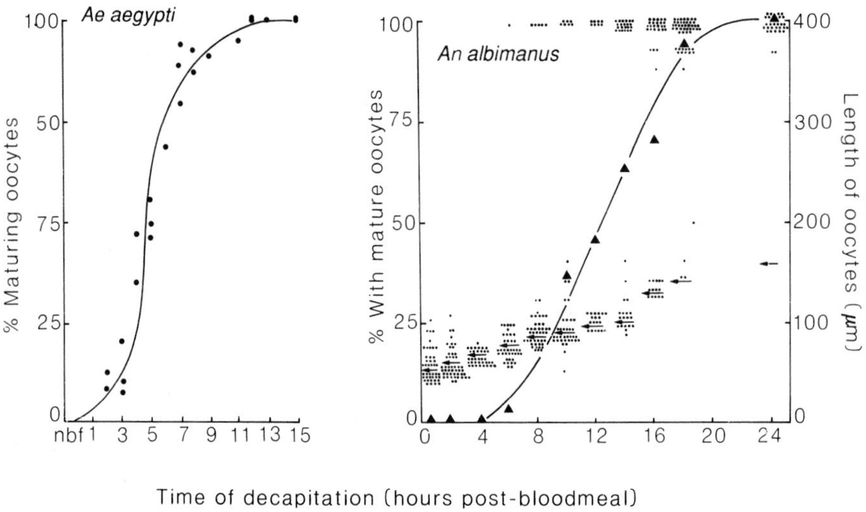

Figure 21.5 Critical periods for OEH release, assessed by the development of oocytes to maturity in females decapitated at different times after blood feeding.

(A) *Aedes aegypti*. Percentages of decapitated females containing maturing oocytes at 20 or more hours post-bloodmeal (from the data of Gillett (1957) and Greenplate *et al.* (1985)). (B) *Anopheles albimanus*. Percentages of decapitated females with mature oocytes at 48 h post-bloodmeal, and distribution of oocyte lengths at 48 h. Arrows, mean oocyte lengths at the times of decapitation; nbf, non-bloodfed. (From Lu and Hagedorn, 1986.)

matured their oocytes from the percentage that have, at most, completed the initiation phase. That there are no intermediates indicates that this is an all-or-none step and that for the individual female OEH release must be a brief event.

The time of OEH release in *Ae aegypti* appears to be affected by a variety of factors. Ingestion of fruit juice on the day before a blood meal delayed the onset of the critical period by two hours and extended its duration by four hours, even though only a trace of fluid remained in the crop at the time of the blood feeding (Gillett, 1957). The critical period occurred some 2–3 h earlier when guinea-pig blood had been ingested than when a similar volume of human blood had been ingested (Judson, 1986), and the onset and duration of the critical period were profoundly affected by the volume of blood given by enema (Figure 21.7) (Klowden, 1987).

There is some evidence for two bursts of OEH release. Females of *Ae aegypti* that were decapitated between 2 and 8 h post-bloodmeal and that developed their ovaries beyond stage III synthesized some 45–65% as much ecdysteroid as the unoperated controls; there was no increase in synthesis if decapitation was delayed through that period. In contrast, females decapitated 12 h or more post-bloodmeal synthesized as much ecdysteroid as unoperated controls. Greenplate *et al.* (1985) concluded that there may be two periods of OEH release, one within about 8 h and the other more than 8 h after feeding. Most females of *An albimanus* decapitated at 18 h post-bloodmeal developed mature oocytes, but these females developed only 60% as many mature oocytes as unoperated females and contained only 50% as much vitellin. Lu and Hagedorn (1986) concluded that there may be a second release of OEH after 18 h post-bloodmeal to permit the development of a full batch of eggs.

When the MNC perikarya (MNCp) were removed from *Ae aegypti*, whether within one hour or seven days of emergence, the females did not mature their ovaries after subsequent blood feeding. However, when the MNCp were removed from females of *Ae taeniorhynchus*, *Ae triseriatus* and *Ae sollicitans* within one hour of emergence, a variable percentage of the females matured their ovaries after blood feeding, and this percentage increased as ablation was delayed for 1, 2 or 3 days after emergence (Table 21.3). It appears that in *Ae aegypti* the MNCp must be present for hormone release, whereas in the other species OEH can be discharged from release sites in the neurohaemal organ after removal of the perikarya.

Removal of the MNCp from *Ae taeniorhynchus* within one hour of emergence blocked subsequent ovarian maturation in all but 8% of females, but fecundity was restored by implanting two pairs of MNCp or four NHO (Table 21.3). If at three days after emergence either the MNCp or the NHO was removed, over 60% of females could still mature their ovaries after blood feeding, but if both MNCp and NHO were removed no females could subsequently mature their ovaries. It appears that OEH-releasing factor can stimulate OEH release equally well from neurosecretory axons and neurohaemal release sites. Possibly, after removal of the neurohaemal organ, axon terminals containing new release sites formed at the severed ends of the cardiacal nerves. Implanting four pairs of MNCp into sugarfed females of *Ae taeniorhynchus* induced oocyte maturation in the absence of a blood meal (Lea, 1967, 1970).

21.3.4 OEH action

Head extracts that stimulate ovaries *in vitro* to synthesize ecdysone are considered to contain OEH. Such extracts can show gonadotrophic activity, but whether this is a direct action or via ecdysone is not known. Several peptides isolated from *Aedes aegypti* heads induced vitellogenesis and ovarian development when injected into females that had been bloodfed and immediately decapitated, and at least one of the peptides possessed ecdysteroidogenic activity *in vitro* (Section 10.3.1(c)). It may be, therefore, that OEH comprises more than one peptide species. It is an unexplained phenomenon that head extracts that exerted a gonadotrophic action when injected into decapitated bloodfed females showed no activity when injected into sugarfed females, since considerable amounts of ecdysone were produced by ovaries from non-bloodfed females which were incubated with whole brain. Midbrain, which contains the medial neurosecretory cells, proved even more stimulatory (Hagedorn *et al.*, 1975, 1979; Wheelock and Hagedorn, 1985).

When ovaries were incubated with head extract, ecdysone secretion started after a lag of 0.5 h and continued at a constant rate for 10 h. After 18 h the medium contained 30 times more ecdysone than was originally present in the ovaries, showing that synthesis as well as release of ecdysone had taken place. These results are consistent with the hypothesis that at some time after blood feeding OEH released into the haemolymph stimulates the ovaries to secrete ecdysone (Figure 21.4) (Hagedorn *et al.*, 1979). Only ovaries that had been exposed to juvenile hormone during previtellogenic development were competent to synthesize ecdysone following exposure to head extract. Incubation of competent ovaries with a partly purified head extract caused a rise in cAMP concentration within 30 s which reached maximum within 2 min. Incubation for 6 h stimulated ecdysone synthesis (Shapiro and Hagedorn, 1982; Shapiro, 1983).

Borovsky (1982) attempted to measure the rate of inactivation of circulating OEH in *Ae aegypti*. Between 6 and 18 h post-bloodmeal only about 30% of OEH activity disappeared, but by 24 h post-bloodmeal no OEH activity remained. He concluded that OEH was stable for about 18 h after secretion.

21.3.5 Insemination factor

There have been reports for a very few anautogenous species that blood feeding stimulates ovarian development in inseminated but not in virgin females. The report for *Anopheles atroparvus* is well documented. Virgin females that took one blood meal 48 h after emergence deposited lipid droplets in the oocyte cytoplasm but showed no oocyte growth beyond that shown by sugarfed females. Virgins that took from three to six blood meals deposited a little more lipid than those that had taken only one meal, but their primary follicles did not develop beyond stage IIb. However, females that had taken two blood meals within two days of emergence and that mated at two or even seven days post-emergence subsequently developed their ovaries to maturity (Marchi *et al.*, 1978; A. Marchi, personal communication). Circumstantial evidence suggesting that insemination is necessary for blood feeding to stimulate ovarian development beyond the previtellogenic resting stage has been provided also for *An subpictus* (Roy, 1940) and *An minimus* (Muirhead-Thomson, 1941).

Mated and unmated sugarfed females of *Aedes aegypti* were equally likely to mature their ovaries after ingesting 2 μl or more of blood. In contrast, waterfed females that ingested 2–3 μl of blood were significantly more likely to mature their ovaries if they had mated. These differences were not observed when the blood meal exceeded 4 μl. The percentage of waterfed virgin females that matured their ovaries after ingesting 2 μl blood was higher among females that had been implanted with one male accessory gland or 0.02 homogenized-gland equivalent. Klowden and Chambers (1991) concluded that a male accessory gland substance altered the metabolic priorities of starved females, rendering them more likely to utilize ingested blood for vitellogenesis than as an energy source.

21.4 STIMULI FROM THE BLOOD MEAL

Experimental evidence presented in Section 21.2 led to the conclusion that the act of blood feeding itself is critically important for the initiation of ovarian development, causing the pulsed release of head factors that stimulate ovarian protein synthesis and increase the capacity for vitellogenin uptake (Koller and Raikhel, 1991). In this Section we shall examine evidence that suggests that stimuli from the ingested blood meal are required for the promotion of ovarian development. Two types of stimulus, physical and chemical, have been proposed to explain the triggering of hormone release by the blood meal, and there is experimental evidence that both abdominal distension and exposure to digestion products are involved.

Administration of 0.1 μl of blood by enema, which produced little distension, led to oocyte maturation in very few females of *Aedes aegypti*, but if the blood enema was extended with saline a higher percentage of females developed mature oocytes, the percentage increasing as the volume of saline was increased up to 3.9 μl (Table 21.4). Fertility was not significantly affected (Klowden, 1987). To determine whether the blood must be digested for stimulation of oogenesis, 0.1 μl of blood was administered simultaneously with 1.9 μl of saline containing trypsin inhibitor. Few of the females receiving this solution matured oocytes (Table 21.4). Van Handel and Lea (1984) had earlier suggested that it was not abdominal distension but a product of digestion that stimulated ovary development after a blood meal because addition of 8 μg trypsin inhibitor to a 2 μl blood enema prevented vitellogenin synthesis.

Transecting the ventral nerve cord of *Ae aegypti* anterior to the 2nd abdominal ganglion prior to administration of 0.5 μl blood + 2.5 μl saline virtually abolished the distension-mediated stimulation of oogenesis (Figure 21.6A). However, distension became a progressively less important stimulus as the proportion of blood was increased

Table 21.4 Effects of saline supplements on the stimulation of oogenesis in *Aedes aegypti* by 0.1 μl blood. (From Klowden, 1987.)

Females were given enemas of blood + saline or of blood or saline alone and the percentages subsequently maturing their oocytes were recorded.

Blood volume (μl)	*Saline volume (μl)*	*% Maturing oocytes*	*Mature oocytes/♀*
0	3.0	0	0
0.1	0	2	–
0.1	0.9	18	17.2
0.1	1.9	75	19.8
0.1	1.9 + TI	2	–
0.1	3.9	80	21.5

TI = Saline containing 8 μg trypsin inhibitor.

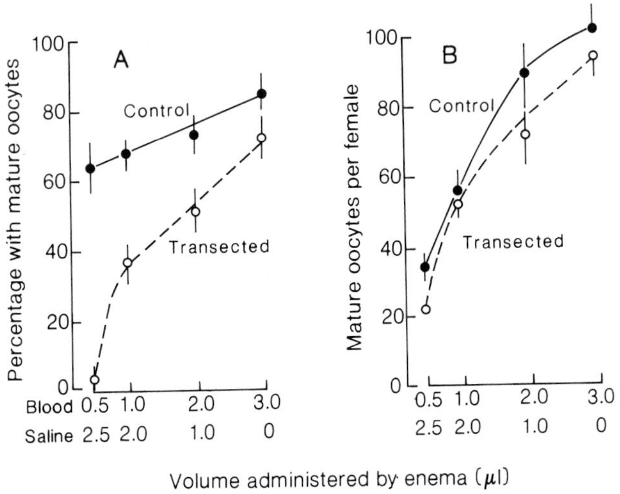

Figure 21.6 Effects of saline supplements and nerve-cord transection on the stimulation of oogenesis by small blood meals in *Aedes aegypti*. The ventral nerve cord was transected anterior to the 2nd abdominal ganglion in experimental females, and the arthrodial membrane was opened in controls. Subsequently the mosquitoes were given enemas of blood + saline or of blood alone to a total volume of 3 µl. Means ± s.e. (From Klowden (1987). Copyright (1987), Pergamon Press plc. Reprinted with permission.)

relative to saline. With 3 µl blood and no saline, transection of the nerve cord caused no significant reduction in the percentage of females maturing their oocytes. Fertility was affected only slightly by transection (Figure 21.6B). The experimental evidence suggests that digestion products from the blood meal provide the principal stimulus

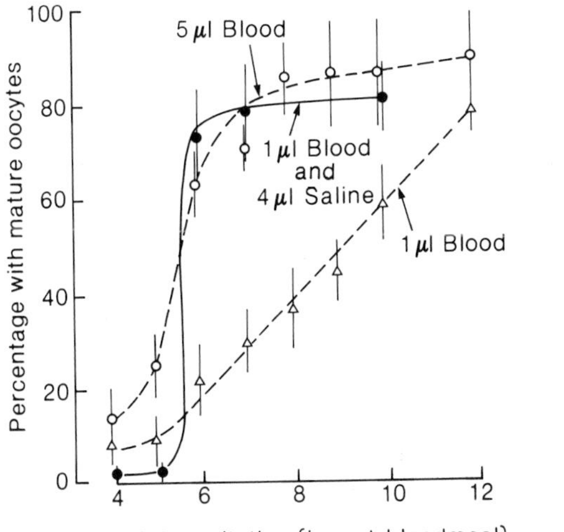

Figure 21.7 Critical periods for presence of the head in *Aedes aegypti* following enemas of 1 µl blood, 5 µl blood, and 1 µl blood + 4 µl saline. Means ± s.e. (After Klowden, 1987.)

for the promotion phase of ovarian development, but that where this stimulus is weak it may be supplemented by abdominal distension. The negative results from 3 µl enemas of saline (Table 21.4) show that distention by itself is ineffective (Klowden, 1987). Females of *Ae sollicitans* still matured oocytes when fed blood after a variety of nerves had been cut, including the ventral nerve cord in the neck, the recurrent nerve anterior to the brain, the NCC-I and -II, the allatal nerves and the oesophageal nerves (Lea, 1972). Klowden (1987) suggested that this was because the females had received full blood meals.

The period for which the head was necessary for promotion has been determined for blood meals of 1 and 5 µl. With 5 µl blood there was a sharp increase in the percentage of females maturing oocytes when decapitation was delayed until 6 h (Figure 21.7). Females receiving 1 µl blood showed no clearly defined critical period. However, when 1 µl blood was administered simultaneously with 4 µl saline, in order to increase abdominal distention, the critical period was similar to that of mosquitoes receiving 5 µl of blood (Klowden, 1987).

Chang and Judson (1977b) designed an experiment to assess the extent of OEH release in

Table 21.5 The stimulation of ovary development in *Aedes aegypti* following ingestion of test solutions. (From the data of Chang and Judson, 1977b.)

Females were fed 2–3 μl of test solution; after 8 h they were fed blood to repletion and immediately decapitated; after 48 h their ovaries were examined. The blood meal provided the possibility of ovary development in cases where the test substance was nutritionally inadequate.

Test solution	Females with ovaries showing: Activation (stage III only) (%)	Females with ovaries showing: Promotion (stages IV–V) (%)
0.15 M L-Isoleucine	0	0
Amino acid mixture, 8.4% w/v	19	0
Amino acid mixture, 8.4% w/v (16 h)*	12	16
Guinea-pig haemoglobin	42	47
Guinea-pig plasma (2×)‡	20	10
Peptides from trypsin digest of guinea pig blood	17	65
Human plasma, dialysable constituents (2×)‡	17	0
Human plasma, non-dialysable constituents (2×)‡	13	75
Peptides from trypsin digest of human blood	26	68
Human γ-globulin, 15%	6	89
Human γ-globulin, 15% + trypsin inhibitor	8	0
Human albumin, 15%	40	40
Bovine serum albumin, 12.5%†	23	59
Bovine serum albumin, 12.5% + trypsin inhibitor†	0	0
Bovine serum albumin, 12.5% + trypsin inhibitor† (not decapitated)	16	64

* Interval between test feed and blood feeding–decapitation step = 16h.
† Interval as above = 12 h. ‡ Concentrated 2-fold.

females of *Ae aegypti* fed different substances. The females were fed 2–3 μl of a test solution; after 8 h they were fed to repletion on human blood and then immediately decapitated; 40 h later they were dissected and the stage of ovarian development was recorded. Their rationale was that a substance that was fed might stimulate OEH release but not be nutritionally adequate to allow maturation to occur. Eight hours was considered a sufficient period for OEH release, and the provision of blood with immediate decapitation at 8 h after the test feed would provide nutrients to allow completion of follicular maturation but would not permit further OEH release. The results of the experiment are summarized in Table 21.5, ovarian development being distinguished according to whether it ceased at stage III or progressed to stages IV–V. The negative response to L-isoleucine confirmed that the blood feeding at 8 h did not itself lead to initiation or promotion. The amino acid mixture caused a modest amount of development. Feeding certain blood proteins caused ovarian development in a high percentage of females, as did feeding peptides derived from trypsin digests of blood. Human γ-globulin and bovine serum albumin were inactive when fed with a trypsin inhibitor, showing that proteins must be digested to cause these effects.

Although many of the experimental diets stimulated ovarian development to stage III, the experimental results do not indicate the nature of the stimulus for the activation phase under normal conditions. The results do, however, suggest that a peptide digestion product is the external stimulus leading to OEH release. Trypsin is not secreted by the midgut until several hours post-bloodmeal (Section 14.3.2) but Graf and Briegel (1982) postulated that aminopeptidase, which is secreted sooner, might hydrolyse small blood peptides and produce a stimulatory factor.

The ingestion of blood plasma or egg albumin by *Culex pipiens pallens* frequently led to oocyte maturation if the meal was deposited in the midgut, but it usually failed to do so if the meal also contained sucrose and was deposited in the diverticula. However, a meal of 17 amino acids and sucrose could stimulate oocyte maturation although it was deposited in the diverticula. An increase in the concentration of threonine produced a striking increase in the percentage of females with developed ovaries (Hosoi *et al.*, 1975).

Within 6 h of blood feeding by *Ae aegypti* the number of stomach endocrine cells that were immunoreactive to the peptide *Aea*-HP-I had been reduced by almost one half, and the remaining cells showed a reduction in immunoreactivity. Brown *et al.* (1986) concluded that a hormone, which might exert a gonadotrophic action, was released from the midgut endocrine cells in response to either secretogogues from the blood meal or stretching of the stomach.

Differences of responsiveness have been reported between batches of females fed blood by mouth or given blood by enema. Females of *Ae aegypti* given blood by enema secreted significantly less ecdysteroid, and synthesized only half as much vitellogenin, as females fed similar quantities of the same blood through a membrane (Greenplate *et al.*, 1985; Hagedorn, 1983, 1986). Females of *An stephensi* that ingested guinea-pig blood by mouth laid many more eggs than females that were given similar volumes by enema. Although the nitrogen content of the egg batch was lower in enema-fed females, the mean N content per egg was significantly higher (Briegel and Rezzonico, 1985). These observations confirm that sensory stimuli associated with blood feeding affect the physiological responses of the female.

21.5 20-HYDROXYECDYSONE

Although there had been occasional reports in the 1950s and 1960s of ecdysteroids in adult insects, ideas that these hormones might function in insect reproductive physiology were blocked by two strongly held beliefs – that no organs capable of ecdysteroid synthesis were present following degeneration of the prothoracic glands soon after emergence, and that juvenile hormone was the key gonadotrophic hormone in insects. The generally held belief that insect gonads did not secrete hormones discouraged investigation of those organs in endocrinological studies. In the early 1970s observations on *Aedes aegypti* provided a breakthrough. It was reported that non-bloodfed females developed eggs when fed or injected with 20-hydroxyecdysone (Spielman *et al.*, 1971; Fallon *et al.*, 1974); that fat body from sugarfed females synthesized vitellogenin when incubated with ovaries from bloodfed females (Hagedorn and Fallon, 1973); and that the ovaries of bloodfed females synthesized ecdysone (Hagedorn *et al.*, 1975). The 1970s and 1980s, which saw intensive research on the putative role of ecdysteroids in the regulation of vitellogenin synthesis, were a period of controversy and debate.

21.5.1 Secretion and titre

Ecdysone is secreted by the ovaries of bloodfed mosquitoes and can be converted to 20-hydroxyecdysone by a monooxygenase present in a number of tissues (Section 10.3.3). Hagedorn *et al.* (1975) reported that ecdysone was the principal and possibly the only ecdysteroid secreted by the ovaries of *Ae aegypti in vitro*, but Smith and Mitchell (1986) found a significant amount of 20-hydroxyecdysone production by the ovaries. Ecdysone added to cultured fat body became 20-hydroxylated, the rate of hydroxylation being twice as fast with fat body from bloodfed females as with fat body from sugarfed females (Hagedorn, 1985). The 20-monooxygenase activity of abdominal homogenates declined between 1 and 24 h post-bloodmeal, but the activity of a fat body preparation peaked at 20–24 h when the activity of the other organs was at a basal rate (Smith and Mitchell, 1986).

Ecdysone and 20-hydroxyecdysone are subject to conjugation and metabolism (Section 10.3.3), consequently mosquitoes may contain a variety of ecdysteroids (Whisenton *et al.*, 1989). A number of investigators used radioimmunoassy to measure the total ecdysteroid content of females and were unable to to specify the amounts of hormone present.

Broadly similar accounts of ecdysteroid production by bloodfed females of *Ae aegypti* have been given by different groups of investigators. Sugarfed females aged 3–5 days contained between 15 and 80 pg ecdysteroid per female. Because mammalian blood contains RIA-active materials (Redfern, 1982), a growing increase in ecdysteroid content must be demonstrated after blood feeding to prove renewed synthesis. The small burst at 4–6 h post-bloodmeal (Figure 21.8A, B), found by some authors to be reproducible (Hagedorn *et al.*, 1975), appears authentic. The principal burst of ecdysteroid synthesis was reported to peak at about 600 pg per female between 18 and 28 h post-bloodmeal, and to decline at different rates in different batches of females (Figure 21.8A, B) (Hagedorn *et al.*, 1975; Greenplate *et al.*, 1985; Racioppi *et al.*, 1986; Borovsky *et al.*, 1986). At 20 h post-bloodmeal the haemolymph contained 55 pg ecdysteroid μl^{-1}, a concentration of 1.1×10^{-7} M (Hanaoka and Hagedorn, 1980).

When measuring the time profile of ecdysteroid production in *Ae aegypti*, Borovsky *et al.* (1986) analysed the ovaries separately from the remainder of the carcase. An early increase in the ecdysteroid content of the carcase was apparent at 4 h and peaked at about 8 h post-bloodmeal (Figure 21.8B). This was followed by a substantially higher peak of >300 pg per carcase between 25 and 40 h. The ecdysteroid content of the ovaries was very low for several hours after the blood meal. It rose from 19 pg at 8 h to a broad peak of 175–250 pg between 15 and 50 h.

Hagedorn *et al.* (1975) reported that 20-hydroxyecdysone was the principal ecdysteroid in females extracted between 14 and 26 h post-bloodmeal. Borovsky *et al.* (1986) reported that ecdysone and 20-hydroxyecdysone constituted 15% of total ecdysteroids at 8 h post-bloodmeal and 80% at 24 h post-bloodmeal, at which time the whole ecdysone content was within the ovaries (Table 21.6). At 40 h post-bloodmeal the ecdysteroid content of the ovaries was only one-tenth of that at 20 h (Hanaoka and Hagedorn, 1980), Unlike some other insects, *Ae aegypti* do not store large quantities of ecdysteroids in their ovaries. Measurements of ecdysteroid concentration in the haemolymph of *Ae aegypti* ranged from undetectable to 2×10^{-8} M for sugarfed females and from 1 to 2×10^{-7} M at 20 h post-bloodmeal (Hanaoka and Hagedorn, 1980; Hagedorn, 1985). The half-life of ecdysteroids circulating outside the ovaries at 24–30 h post-bloodmeal was 3.5 h (Borovsky *et al.*, 1986).

Radioimmunoassay showed the ecdysteroid titre of bloodfed *Anopheles stephensi* starting to rise by 4 h post-bloodmeal in females kept at 25°C, with small peaks at 8 and 12 h. The main peak occurred between 24 and 36 h post-bloodmeal with the titre exceeding 3 ng, some 5-fold higher than in *Ae aegypti* (Figure

Table 21.6 Ecdysteroid titres in *Aedes aegypti* 8 and 24 h after a blood meal. (From the data of Borovsky *et al.*, 1986.)

Eight and 24 h post-bloodmeal (PBM) the ovaries were separated from the carcase. Ecdysteroids were extracted from ovaries and carcases and samples were assayed by radioimmunoassay. The extracts were then purified by TLC and HPLC and their ecdysone and 20-hydroxyecdysone contents assayed by RIA. All measurements are expressed as 20-hydroxyecdysone equivalents.

	Content per female (pg)			
	8 h PBM		*24 h PBM*	
	Ovary	*Carcase*	*Ovary*	*Carcase*
Ecdysone	2.5	0	150	47
20-Hydroxyecdysone	4.2	28	0	82
Other ecdysteroids	12.3	172	0	73
Totals	19.0	200	150	202

21.8C). Materials that co-migrated on TLC with ecdysone and 20-hydroxyecdysone were present at all times post-bloodmeal except time 0. GC analysis revealed that at peak titre, 30–32 h post-bloodmeal, each female contained 300–400 pg 20-hydroxyecdysone and 800–900 pg ecdysone (Redfern, 1982). In *An albimanus* kept at 27.5°C the ecdysteroid titre peaked at 18 h post-bloodmeal at 800 pg per female and declined sharply thereafter. Up to 36 h post-bloodmeal all ecdysteroids were unconjugated; from 42 h onwards all were conjugated (Lu and Hagedorn, 1986).

In an investigation with *Culex pipiens*, Baldridge and Feyereisen (1986) observed a mean (non-conjugated) ecdysteroid titre of 7 pg per female on the day after emergence, which fell below the level of detection (1 pg) by the second day. Ecdysteroid titres rose rapidly after a blood meal, reaching 20 pg per female within 6 h and peaking at 190 pg at 36 h, after which they declined rapidly (Figure 21.8D). Both ecdysone and 20-hydroxyecdysone were present in all positive extracts, ecdysone being found at relatively higher titre than 20-hydroxyecdysone from 6 to 30 h post-bloodmeal and at relatively low titre from 36 to 72 h. A

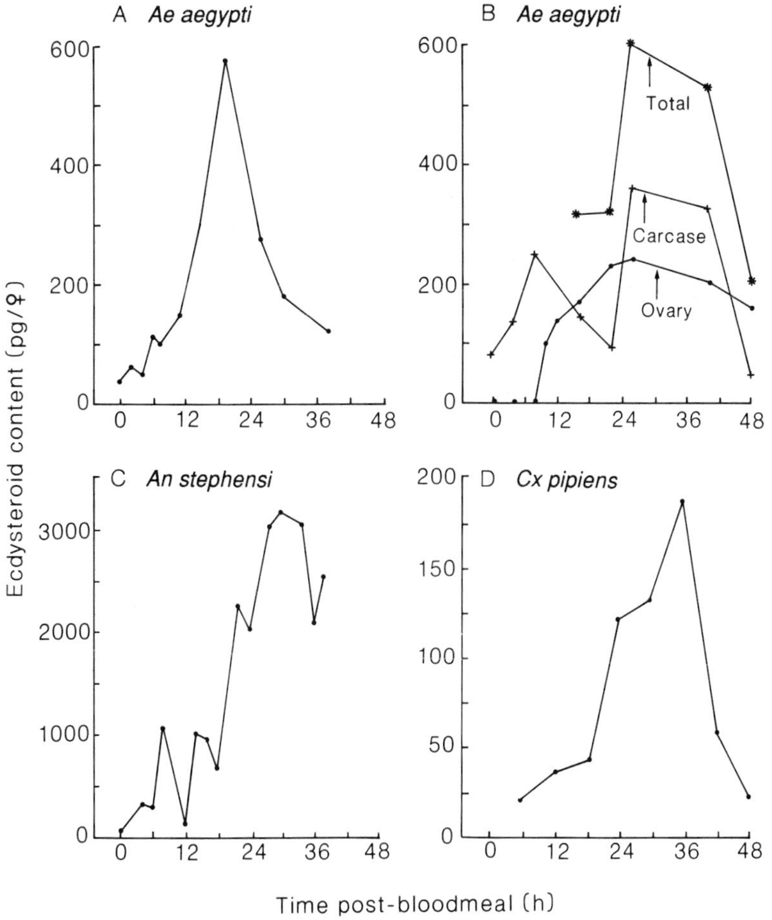

Figure 21.8 Time profiles of the whole-body ecdysteroid contents of bloodfed mosquitoes measured by radio-immunoassay.

(A) *Aedes aegypti* kept at 27°C (after Greenplate *et al.*, 1985). (B) *Aedes aegypti* kept at 26°C (after Borovsky *et al.*, 1986). (C) *Anopheles stephensi* kept at 25°C (after Redfern, 1982). (D) *Culex pipiens* kept at 26°C (after Baldridge and Feyereisen, 1986). Total ecdysteroid titres are expressed as 20-hydroxyecdysone equivalents. Time-0 data points are for non-bloodfed females.

similar burst of ecdysteroid production after blood feeding was reported for *Cx pipiens pallens* (Zhu *et al.*, 1980).

It is not known for any mosquito species whether the burst of ecdysteroid secretion that follows blood feeding includes separate functionally-significant peaks of 20-hydroxyecdysone synthesis.

21.5.2 Regulation of vitellogenin synthesis and ovarian development

Investigations on *Aedes aegypti* undertaken in the 1970s appeared to establish a key role for 20-hydroxyecdysone in the regulation of vitellogenin synthesis. Hagedorn (1983) listed the following observations as evidence.

1. A burst of ecdysteroid secretion occurs some 10–36 h post-bloodmeal (Figure 21.8), which coincides with the period of vitellogenin synthesis.
2. Ovaries taken from females 18 h post-bloodmeal synthesized ecdysone when incubated *in vitro* (Hagedorn *et al.*, 1975).
3. Ovaries taken from sugarfed females synthesized ecdysone when incubated with head extract thought to contain OEH (Hagedorn *et al.*, 1979).
4. Fat body taken from sugarfed females synthesized vitellogenin when incubated with 10^{-7} M 20-hydroxyecdysone, which is the concentration of ecdysteroid in the haemolymph of bloodfed females (Hagedorn *et al.*, 1975; Hagedorn, 1985).
5. Injection of microgram quantities of 20-hydroxyecdysone into sugarfed females stimulated vitellogenin synthesis (Fallon *et al.*, 1974).

On the basis of these observations it was postulated that OEH, released after a blood meal, stimulated the ovaries to secrete ecdysone; that this prohormone was hydroxylated by the fat body, and that the 20-hydroxyecdysone so produced stimulated the fat body to synthesize vitellogenin.

Some investigators remained unpersuaded by the evidence, giving three main reasons.

1. Injection of unphysiological (microgram) quantities of 20-hydroxyecdysone was necessary to stimulate vitellogenesis and ovary development in sugarfed females (Lea, 1982).
2. Transplantation of ovaries from vitellogenic females (12–24 h post-bloodmeal) into sugarfed females failed to stimulate vitellogenin synthesis by fat body (Borovsky and Van Handel, 1979).
3. They had failed to obtain 20-hydroxyecdysone-stimulated vitellogenin synthesis *in vitro* with fat body from sugarfed females (Borovsky and Van Handel, 1979), or at most had obtained synthesis amounting to 5% of that obtained with fat bodies from bloodfed females (Borovsky, 1984).

Addressing the first two of these objections, Hagedorn (1983) suggested that rapid elimination of 20-hydroxyecdysone accounted for the insensitivity of sugarfed females to injected 20-hydroxyecdysone and also for the failure of implanted ovaries to stimulate vitellogenesis. He reported that a dose of 0.5 μg 20-hydroxyecdysone injected into a sugarfed female was reduced to 10 ng of RIA-active ecdysteroid within 6 h, and that a dose of 0.05 μg was reduced to 0.5 ng within the same period (6 h is the time needed for the induction of gene expression). Since 0.5 ng 20-hydroxyecdysone would give a haemolymph concentration of $>10^{-6}$ M, which would stimulate vitellogenin synthesis *in vitro*, Hagedorn's argument also requires that these RIA-active residues are other than 20-hydroxyecdysone, which is not improbable. Hagedorn (1986) suggested that the requirement of sugarfed females for excessive dosages of 20-hydroxyecdysone might be because physiological changes occur shortly after blood feeding, independent of the actions of OEH, which are a necessary precedent for the action of 20-hydroxyecdysone. Evidence for such changes is discussed in Section 21.2.

The question as to whether or not fat body from sugarfed females responds to 20-hydroxyecdysone *in vitro* by vitellogenin synthesis was considered by Fuchs and Kang (1981) who analysed the claims of different groups of investigators (Fallon *et al.*, 1974; Hagedorn *et al.*, 1975;

Hanaoka and Hagedorn, 1980; Borovsky and Van Handel, 1979). They observed that the radioimmunoassay technique that had been used to assay vitellogenin was fraught with experimental pitfalls from which discrepancies between investigators could easily arise.

A fresh approach to this problem made use of a quantitative assay involving ELISA with monoclonal antibodies prepared against vitellogenin and vitellin. Fat bodies from 5–7-day-old sugarfed females synthesized 3.8 μg vitellogenin per fat body when incubated for 45 h with 10^{-6} M 20-hydroxyecdysone. Fat bodies removed from engorged females at 6 h post-bloodmeal and incubated in 10^{-6} M 20-hydroxyecdysone for 39 h synthesized 9.3 μg vitellogenin per fat body (Ma *et al.*, 1984, 1986, 1987). That contrasts poorly with the 8 μg vitellogenin synthesized per hour by intact females fed on rabbits (Figure 19.2), but the incubation medium contained only 2 mmol l^{-1} of the limiting amino acid isoleucine. Human blood contains 7 mmol l^{-1} and rat blood 24 mmol l^{-1} (Briegel, 1985).

It is now generally agreed that 20-hydroxyecdysone is a key factor in vitellogenesis, that it can stimulate some vitellogenin synthesis in fat bodies from non-bloodfed *Ae aegypti in vitro*, and that rapid elimination accounts for at least part of the unphysiological dose of exogenous hormone required to stimulate vitellogenesis in sugarfed females *in vivo*. The action of 20-hydroxyecdysone on the expression of vitellogenin genes is described in Section 20.2.3.

There is only slender evidence for a gonotrophic function for 20-hydroxyecdysone. Injections of physiological doses of 20-hydroxyecdysone increased the capacity for vitellogenin uptake, and the rate of Vg uptake *in vitro*, when injected into bloodfed decapitated females of *Ae aegypti* (Koller and Raikhel, 1991).

21.5.3 Regulation of chorionic-protein and dopa-decarboxylase synthesis

Injection of an unphysiological dose of 20-hydroxyecdysone into *Aedes aegypti* immediately after a blood meal caused premature secretion of chorionic protein by the follicle cells (Raikhel and Lea, 1982). To test whether the 15a-1 gene, which codes for endochorionic protein (Section 19.6.1(a)), is regulated by 20-hydroxyecdysone, females of *Ae aegypti* were bloodfed, immediately decapitated, and injected with 1 μg 20-hydroxyecdysone in saline or with saline alone. After 24 h the ovaries were removed and their total RNA analysed by Northern hybridization. RNA from the ovaries of unoperated bloodfed females, and from the ovaries of decapitated bloodfed females injected with 20-hydroxyecdysone, contained the 650 bp transcript of the 15a-1 gene. In the absence of injected 20-hydroxyecdysone no 15a-1 gene transcript was detected in operated females, suggesting that hormone regulates expression of the gene (Lin *et al.*, 1992).

Females of *Ae aegypti* which were decapitated and given a blood enema failed to develop their ovaries. However, if such females were given two injections of 500 pg 20-hydroxyecdysone, at 4 h and 12 h post-enema, by 36 h the follicles had developed to the stage found in bloodfed controls at 28–30 h post-bloodmeal, and the structure of the endochorion was normal. Raikhel and Lea (1991) concluded that secretion of 20-hydroxyecdysone is necessary for endochorion formation.

Injection of 20-hydroxyecdysone into sugarfed females of *Ae aegypti* stimulated synthesis of dopa decarboxylase; 1.6 μg hormone per female was required to stimulate a half-maximal rate of synthesis. The amount synthesized could be substantially increased by injecting dibutyryl cyclic AMP with the hormone. Injected alone, db-cAMP had no effect on synthesis, so it appeared not to be acting as a second messenger. Males did not synthesize dopa decarboxylase when injected with 20-hydroxyecdysone (Fuchs and Schlaeger, 1973; Schlaeger *et al.*, 1974).

In vitellogenic females 20-hydroxyecdysone is believed to stimulate the synthesis of vitellogenin, endochorionic secretion and dopa decarboxylase, but these secretions are not synchronous. The maximum rate of vitellogenin synthesis occurs between 8 and 24 h post-bloodmeal, of endochorionic plaques between 18 and 30 h, while that of dopa decarboxylase does not start until 24 h. When 20-hydroxyecdysone was injected into sugarfed females, the peak of vitellogenin synthesis occurred at least 24 h earlier than that of dopa decarboxylase (Fallon *et al.*, 1974; Schlaeger *et al.*, 1974). In fact, in bloodfed females the time of fastest increase in dopa decarboxylase activity was a time of low or falling ecdysteroid concentration (Figures 19.4, 21.8A, B). It may be that regulatory genes control the time of response to 20-hydroxyecdysone, or that, in the case of dopa decarboxylase, synthesis is triggered by a falling 20-hydroxyecdysone concentration.

In an attempt to identify the point at which 20-hydroxyecdysone acted to stimulate the synthesis of dopa decarboxylase, Fong and Fuchs (1976) injected *Ae aegypti* with inhibitors of transcription. When females were simultaneously injected with 7.5 ng actinomycin D and 7.5 μg 20-hydroxyecdysone, the hormone-stimulated synthesis of dopa decarboxylase was inhibited at 48 h post injection but had largely recovered by 72 h. However, the simultaneous injection of 3 ng α-amanitin or 0.3 μg cordycepin with 7.5 μg 20-hydroxyecdysone markedly stimulated dopa decarboxylase synthesis in comparison with the injection of 20-hydroxyecdysone alone. The problem clearly needs further investigation.

21.6 JUVENILE HORMONE

Juvenile hormone has important functions in the previtellogenic phase, stimulating development of stage I follicles to the previtellogenic resting stage (Sections 21.1.1, 21.9.1) and inducing the fat body to become competent to synthesize vitellogenin (Section 20.3.1). The present Section is concerned with juvenile hormone in bloodfed females.

When 3-day-old *Aedes aegypti* took a blood meal their juvenile hormone content fell from 14 pg per female before feeding to 6 pg at 3 h, 1.5 pg at 12 h, and 1 pg at 36 h post-bloodmeal. Between 36 and 48 h the titre started to rise again, and by 96 h it had returned to the prefeeding level (Figure 21.9) (Shapiro *et al.*, 1986). Borovsky *et al.* (1985) reported a mean titre of 3 pg per female in a large batch aged 2–10 h post-bloodmeal. It is not known for any of the mosquito juvenile hormone assays how much JH was bound and how much was free.

The capacity of corpora allata dissected from *Culex pipiens* and incubated with [*methyl*^{2}H]methionine to synthesize labelled juvenile hormone III *in vitro* remained stable at between 20 and 30 fmol per corpus allatum pair per hour while the mosquitoes were maintained on a sucrose diet. The capacity for synthesis declined markedly within four hours of blood feeding, and had dropped to 1.4 fmol h^{-1} by 12 h post-bloodmeal. It returned to the pre-bloodmeal rate by 120 h and increased further after oviposition. When females were prevented from ovipositing, their corpora allata showed a gradual decline in ability to synthesize juvenile hormone, but this trend was reversed after the retained eggs were laid (Readio *et al.*, 1988). There is indirect evidence of juvenile hormone secretion during the 24 h after oviposition (Section 21.9), so it may be that egg retention suppresses JH synthesis or release.

The juvenile hormone esterase activity of the haemolymph doubled between 0 and 12 h post-bloodmeal in *Ae aegypti*. The activity subsequently increased sharply, peaking at 36 h before declining to a low level at 48 h (Figure 21.9). Shapiro *et al.* (1986) considered that the initial doubling of esterase activity was insufficient to account for the rapid fall in JH content, and suggested that the sudden fall was due to decreased hormone synthesis plus the actions of JH esterase and other enzymes. They thought that a low JH content might be necessary after a blood meal for normal ovary development, and that the much higher JH esterase activity occurring at 24–42 h post-bloodmeal might serve to scavenge residual JH.

The ability of females allatectomized three days

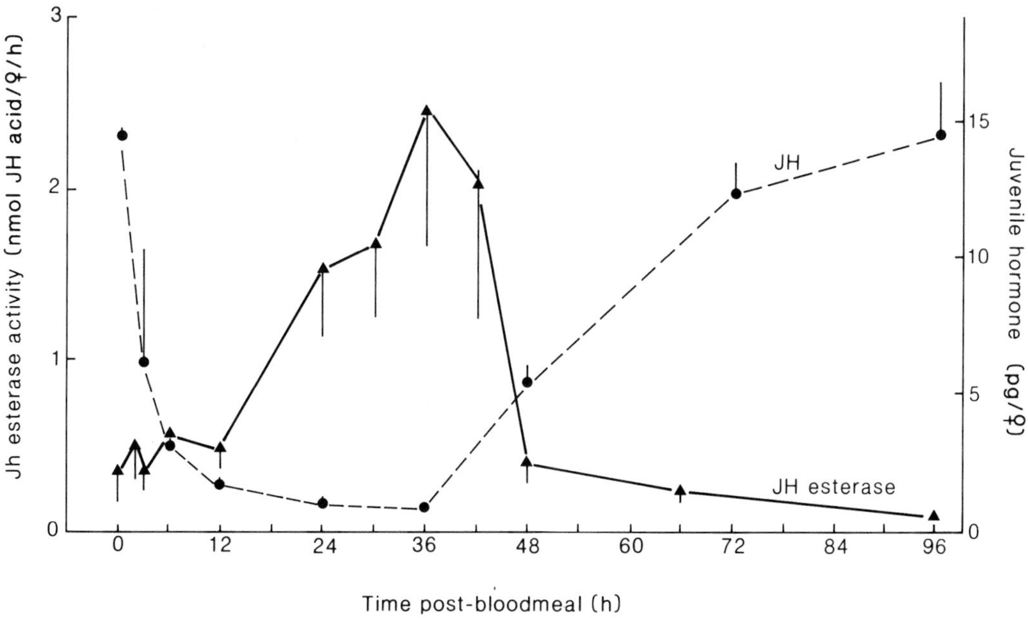

Figure 21.9 Juvenile hormone and juvenile hormone esterase in bloodfed females of *Aedes aegypti*. The JH-III content of whole bodies (means, with range of duplicate samples) and the juvenile hormone esterase activity of the haemolymph (means ± s.d.) assayed during the four days after a blood meal. The blood meal was given on the 3rd day after emergence. (Recalculated from Shapiro *et al.*, 1986.)

after emergence to mature their primary oocytes after a subsequent blood meal (Table 21.3) has been taken by some as proof that JH is not required for post-bloodmeal vitellogenesis and ovarian development. However, the physiological activity of JH applied to bloodfed-female preparations (Section 21.7.2) led Borovsky (1981b) to suggest that JH derived from a source outside the corpora allata could participate in vitellogenesis.

21.7 ANTAGONISTIC AND SYNERGISTIC EFFECTS OF HORMONES

21.7.1 Antagonism

We know from investigations on other insects that simultaneous exposure to 20-hydroxyecdysone and juvenile hormone may enhance a response or may have deleterious effects (Willis and Hollowell, 1976). Observations on *Aedes aegypti* suggest that the exposure of ovarian follicles to one or other of these hormones under inappropriate conditions leads to follicular degeneration or damage. Thus the primary follicles degenerated when allatectomy at emergence, which deprived the follicles of exposure to juvenile hormone, was followed by blood feeding or by injection of 20-hydroxyecdysone. The follicles developed normally if corpora allata were implanted or juvenile hormone analogue was applied 24 h after the allatectomy and before either blood feeding or injection of 20-hydroxyecdysone (Gwadz and Spielman, 1973). When abdomens were isolated by ligation at emergence the primary ovarian follicles would develop to stage II after treatment with juvenile hormone, but simultaneous treatment with juvenile hormone and ecdysone produced no development (Hagedorn *et al.*, 1977). Treatment of intact females with methoprene or a juvenile hormone esterase inhibitor 30 h after a blood meal reduced egg viability (Shapiro *et al.*, 1986).

The periods of secretion of 20-hydroxyecdysone and juvenile hormone are in antiphase, both before and after blood feeding (Figure 21.10), and it has been postulated that this prevents antagonistic actions (Shapiro *et al.*, 1986).

21.7.2 Synergism

An experimental procedure developed by Borovsky (1981b) has proved useful for investigating the interactions of hormones. Females are ligated at the base of the abdomen immediately after taking a blood meal, hormone is applied topically or by injection to the isolated abdomens, and after an appropriate period the ovaries are examined and recorded as undeveloped or maturing. The experimental data in Table 21.7 reflect differences of sensitivity of this preparation in the hands of different investigators. Isolated bloodfed abdomens are much more sensitive to exogenous hormone than are intact sugarfed females, responding to ng or pg instead of μg quantities, due to the actions of hormones released from the head during blood feeding (Section 21.2).

The isolated abdomens increased in sensitivity to 20-hydroxyecdysone with time following the blood meal. If injected immediately after the blood meal and ligation, 700 ng were needed to stimulate ovarian maturation in about 50% of females (Table 21.7). If the injection was delayed for 18 h, which was the time needed for 20-hydroxyecdysone titres to rise in the normal animal, only 150 ng hormone were needed to produce the same effect (Martinez and Hagedorn, 1987). Injection of 50 pg 20-hydroxyecdysone immediately after a blood meal and ligation potentiated the action of the same hormone injected again 18 h later, when only 30 ng were needed (Table 21.7). Martinez and Hagedorn (1987) postulated that the small peak of ecdysone secretion that occurs about 4 h after blood feeding may serve as a primer for the main burst of ecdysone secretion that occurs later.

Juvenile hormone I could stimulate ovary development in isolated bloodfed abdomens, 1250 pg inducing maturation in 50% of cases. Methoprene was much more active (Table 21.7). Application of 25 pg methoprene to isolated abdomens stimulated 20-hydroxyecdysone secretion, the titre reaching 200 pg after 24 h, which was less than half the normal titre, but it failed to stimulate vitellogenin synthesis or ovarian development. However, that treatment had a potentiating effect if followed after 18 h by injection of 20-hydroxyecdysone, when only 5 ng of the second hormone was needed to stimulate ovary maturation, with as much vitellin deposition as in unoperated bloodfed controls (Table 21.7) (Borovsky *et al.*, 1985). Methoprene also potentiated the action of OEH. Injection

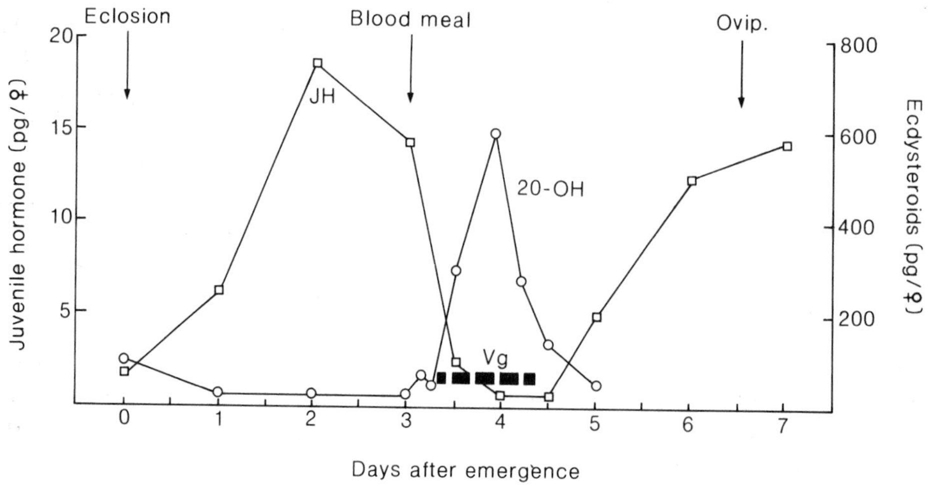

Figure 21.10 A diagrammatic representation of the whole-body content of juvenile hormone and total ecdysteroids in sugarfed then bloodfed females of *Aedes aegypti*. The time of vitellogenin secretion is also shown. The original data are represented in Figures 21.2, 21.8A and 21.9. (After Hagedorn, 1989.)

Table 21.7 The stimulation of ovarian development in *Aedes aegypti* ligated at the base of the abdomen.

Females were bloodfed and immediately ligated, and were treated with hormones at the times and dosages listed. Later the percentages of females containing maturing ovaries were recorded.

Compound	*Treatment*		*Females maturing oocytes (%)*	*Ref.*
	Times PBM (h)	*Dosage*		
Juvenile hormone I	<1	1250pg	50	1
Methoprene	<1	100pg	50	1
20-Hydroxyecdysone	<1	700ng	45	2
20-Hydroxyecdysone	18	150ng	50	2
20-Hydroxyecdysone and 20-Hydroxyecdysone	<1 18	50pg 30ng	50	2
Methoprene	<1	25pg	5	3
Methoprene and 20-Hydroxyecdysone	<1 18	25pg 5ng	60	3
OEH	<1	2h.e.	0	3
Methoprene and OEH	<1 <1	12.5pg 2h.e.	75	3

PBM, post-bloodmeal; h.e., head equivalent.
References: 1. Borovsky (1981b); 2. Martinez and Hagedorn (1987); 3. Borovsky *et al.* (1985).

of 2 head equivalents of OEH was ineffective, but if accompanied by treatment with 12.5 pg methoprene it stimulated ovary development in 75% of abdomens (Table 21.7) (Borovsky *et al.*, 1985).

It is possible that the gonotrophic activity of methoprene, whether applied alone or with other hormones, resulted from 20-hydroxyecdysone secretion, which it has been shown to induce. Bloodfed females injected with [^{14}C]cholesterol were later found to contain ^{14}C-labelled ecdysone and 20-hydroxyecdysone; injection of [^{14}C]cholesterol into bloodfed decapitated females yielded only 5% as much. However, topical application of methoprene and injection of OEH into the ligated abdomens of bloodfed females prior to [^{14}C]cholesterol injection led to as much [^{14}C] ecdysone and [^{14}C]20-hydroxyecdysone synthesis as in unoperated bloodfed females (Borovsky and Thomas, 1985; Borovsky *et al.*, 1986). Incubation of fat bodies with 10^{-6} M 20-hydroxyecdysone stimulated transcription of the vitellogenin gene and vitellogenin synthesis. Addition of methoprene to the incubation medium, in a concentration falling with time from 2 to 0.5 $\times 10^{-8}$ M, coupled with 20-hydroxyecdysone rising from 10^{-7} to 10^{-6} M, doubled the rates of transcription and synthesis. Methoprene alone was inactive (Racioppi *et al.*, 1986).

Insight into one action of juvenile hormone has been provided by Klowden and Chambers (1989). Only 20% of carbohydrate-deprived *Ae aegypti* matured oocytes when given 2 μl rat blood, but the number responding could be raised to 60% or more by treating the females with 0.1 μg JH-III or methoprene 0–8 h after the blood meal. Feeding the females with 1 μl 20% sucrose solution during the same period was equally effective in raising fertility, leading the investigators to postulate that juvenile hormone acted by mobilizing reserves. Their further observations that methoprene treatment of sucrose-fed females caused significant glycogen depletion, and that high mortality rates resulting from treatment with juvenile hormone could be prevented by feeding sucrose were consistent with that hypothesis.

The nature of juvenile hormone action after a blood meal is uncertain. The fall in JH titre at that time (Figure 21.9) may have significance for vitellogenic development, but females allatectomized at three days post-emergence and subsequently bloodfed, sometimes several days later, were able to mature their ovaries (Table 21.1). Some authors have deduced from experiments with isolated abdomens that juvenile hormone may be involved with 20-hydroxyecdysone in the regulation of vitellogenin synthesis after a blood meal (Borovsky, 1984; Kelly *et al.*, 1987), but it is possible that the actions of juvenile hormone in these experimental situations are artefactual.

21.8 OOSTASIS

While the primary follicles develop to maturity after the first blood meal the secondary follicles separate from their germaria and grow slightly, in some species developing to the previtellogenic resting stage. The secondary follicles of non-bloodfed *Aedes aegypti* can be caused to separate from their germaria by treating the females with 20-hydroxyecdysone. If the treated females are later fed blood, yolk deposition occurs simultaneously in the primary and secondary follicles (Beckemeyer and Lea, 1980). Gravid mosquitoes which are unable to oviposit will sometimes continue to take blood meals, especially if unmated, and they might therefore be expected to develop a second series of follicles while retaining mature oocytes from the first. However, experimental evidence suggests that there is a mechanism that can prevent development of another series of follicles in gravid females.

Gravid females of *Ae sollicitans* that were prevented from ovipositing but given a substantial second blood meal failed to deposit yolk in any secondary follicles. This developmental block occurred even in females in which fewer than half of the ovarioles contained mature oocytes (Meola and Lea, 1972b).

Ovary development was blocked after a second blood meal taken by females of *Ae aegypti* (Rockefeller strain) that retained a full complement of mature oocytes. However, when a second blood meal was taken by females which, following a small first blood meal, had developed and retained only 50–60 mature oocytes, most and possibly all secondary follicles started to develop. Secondary follicles that had started to develop in ovarioles that retained mature primary oocytes were resorbed within 24–48 h. About 65 secondary follicles developed to maturity, all in ovarioles not retaining mature oocytes from the first ovarian cycle (Meola and Lea, 1972b). In contrast, 20% of females of the Liverpool strain of *Ae aegypti* retaining a full batch of mature oocytes matured further oocytes in their secondary follicles after a second blood meal. Some females produced over 200 mature oocytes, which was far in excess of the number of ovarioles (Else and Judson, 1972).

In *Culex pipiens* the secondary follicles grew slowly and failed to develop to the previtellogenic resting stage as long as primary follicles remained in the ovarioles, whether as developing or mature oocytes (Readio and Meola, 1985). However, Hosoi (1954a) found that some females of *Culex pipiens* that retained appreciable numbers of mature oocytes from the first ovarian cycle were able to develop additional oocytes to maturity after a second blood meal.

Attempts to analyse the inhibitory effects of retained eggs were made by Meola and Lea (1972b) in experiments with *Ae sollicitans*. Females that had matured and still retained one batch of oocytes each had one ovary removed and replaced with an immature ovary. Half of these operated females were allowed to oviposit and half were forced to retain their mature oocytes; all were fed a second blood meal and dissected five days later. The females that had retained mature oocytes from the first ovarian cycle had not deposited yolk in either their own intact ovary or the transplanted immature ovary. Females without such mature oocytes had developed oocytes to maturity in both their own intact ovary and the transplanted ovary. Thus the females that retained mature oocytes exhibited both intra- and interovarian inhibition of yolk deposition. In further experiments, a third ovary was implanted into females of *Ae sollicitans* that had been fed on blood 24 h earlier. When the implanted ovary came from a donor that had been fed at the same time as the host, oocytes matured in all three ovaries. When the implanted ovary came from an unfed donor the ovaries of the host continued development and produced mature oocytes but the follicles of the implanted ovary did not incorporate any vitellogenin. The results of these experiments suggested that both vitellogenic follicles and mature oocytes produce a diffusible substance which directly or indirectly blocks further development in follicles that are at the previtellogenic gate.

Similar experiments but with *Ae aegypti* were reported by Borovsky *et al.* (1991). A stage II ovary transplanted from a sugarfed donor into a recipient that had been bloodfed 24 h previously incorporated no vitellogenin over the following 48 h, while the host ovaries developed to maturity. In contrast, when ovaries were transplanted from a donor that had been bloodfed 24 h previously into a sugarfed recipient that was fed blood immediately after the transplantation, the transplanted ovaries developed to maturity while the host ovaries remained unchanged. The failure of certain ovaries to develop could not have been due to an absence of vitellogenin, a fact that is pertinent to work described later in this Section.

Borovsky (1985a, 1988) isolated from the ovaries of *Ae aegypti* a peptide, here named trypsin modulating factor (TMF), which when injected into bloodfed females at very high concentration inhibited trypsin synthesis and so affected digestion (Sections 10.3.1(f) and 14.4.3). Injection of 0.24, 1.45 or 2.86 nmol synthetic TMF into bloodfed females caused reductions of 62, 76 and 98% respectively in growth of the yolk mass. Injection of the peptide did not prevent the release of OEH after blood feeding, and Borovsky *et al.* (1990, 1991) concluded that its effect on vitellogenin incorporation was not through actions on the endocrine system or on the ovary but was caused by a reduction in vitellogenin synthesis due to the effects on digestion. TMF was absent from the ovaries of sugarfed females, but its titre within the ovaries increased linearly during the first 48 h after blood feeding (Borovsky, 1985a).

Borovsky (1988) proposed that after a blood meal TMF is synthesized in the ovaries, and that between 24 and 55 h post-bloodmeal it is released into the haemolymph and causes the cessation of digestion by inhibiting midgut trypsin synthesis. The cessation of digestion was considered to stop further vitellogenin synthesis, effectively blocking vitellogenesis. TMF was therefore a putative oostatic hormone.

21.9 THE SECOND OVARIAN CYCLE

Investigations have shown that follicle development in the second ovarian cycle requires a similar series of hormonal stimuli to that in the first.

21.9.1 Previtellogenic development of the secondary follicles

The secondary follicles of *Aedes aegypti* separate from the germaria about 20 h after the first blood meal, during the main period of ecdysteroid secretion. Beckemeyer and Lea (1980) caused premature separation of secondary follicles in sugarfed females by injecting 25 ng 20-hydroxyecdysone. The effect was greater when the females received two injections of 100 and 275 pg 20-hydroxyecdysone 14–16 h apart, which is the interval separating the minor and major peaks of ecdysteroid titre in bloodfed females.

Fat bodies taken from parous females of *Ae aegypti* were incapable of responding to 20-hydroxyecdysone *in vitro* between 48 and 88 h after the blood meal, at which time they were undergoing ultrastructural reorganization (Section 20.3.3). Fat bodies taken at 96 h post-bloodmeal had started to regain competence to respond to the hormone. Fat bodies removed at

48 h and incubated in saline until 96 h could not, at that time, respond to 20-hydroxyecdysone, but fat bodies which had been incubated in 8×10^{-5} M methoprene were able to do so (Ma *et al.*, 1988).

The secondary follicles of *Ae aegypti* grew in length during the 3rd and 4th days after the first blood meal, a period of increasing juvenile hormone titre (Figure 21.9). Decapitation during the first two days after blood feeding prevented this growth but it could be restored by treatment with 8 ng methoprene, suggesting that the decapitated females lacked juvenile hormone and that corpus allatum activity was regulated by nervous or neurosecretory tissues of the head (Feinsod and Spielman, 1980b).

Females of *Ae aegypti* that had been allatectomized at emergence, treated with juvenile hormone, and fed blood developed their primary follicles to maturity. If they were allowed to oviposit and were fed a second blood meal their secondary follicles degenerated. If juvenile hormone was applied before the second blood meal the secondary follicles matured normally but the tertiary follicles failed to develop to the previtellogenic resting stage. It appears that each successive series of follicles needs a new exposure to juvenile hormone if it is to develop to the previtellogenic resting stage (Gwadz and Spielman, 1973).

The secondary follicles of *Culex pipiens* appear to need two periods of exposure to juvenile hormone to complete previtellogenic growth. While the primary follicles developed to maturity following a first blood meal, the secondary follicles grew from 28 to 57 μm in mean length but no further. As long as the primary follicles remained in the ovarioles, whether as developing or retained oocytes, the secondary follicles did not exceed 57 μm. Allatectomy during the first three days after blood feeding stopped the secondary follicles from growing over 44 μm in length, but injection of allatectomized females with JH-III permitted growth to 57 μm. Growth of the secondary follicles was resumed after oviposition, when they grew from 53 to 68 μm within 48 h, reaching the resting stage. The second burst of growth could be prevented by allatectomy performed at 1 h but not at 24 h after oviposition. Thus in *Cx pipiens* the secondary follicles undergo two periods of JH-mediated previtellogenic growth, one before and one after oviposition (Readio and Meola, 1985). The corpora allata exhibit increased synthetic activity after oviposition (Readio *et al.*, 1988).

In some females of *Aedes sollicitans* the secondary follicles may be able to develop without further exposure to juvenile hormone. Lea (1963) allatectomized females three days after emergence and gave them a blood meal; 14 females that oviposited were given a second blood meal and examined some days later. Four of these females that had already laid 108–164 eggs produced another batch after the second blood meal varying from 92 to 104 eggs. Follicular relics present in the ovarioles with mature oocytes confirmed that two egg batches had developed.

21.9.2 Vitellogenic development of the secondary follicles

In a series of experiments Lea (1972) removed the medial neurosecretory cell perikarya, and in some cases the neurohaemal organ also, from parous females of *Aedes sollicitans* before or after a second blood meal. Most females from which the MNCp alone were removed before the second blood meal matured a second batch of oocytes, and Lea concluded that they had released OEH from the neurohaemal organ. Removal of both organs before the second blood meal prevented oocyte maturation in almost all females, indicating the continuing requirement for OEH, but removal one day after the second blood meal did not (Table 21.3).

Gravid females of *Ae aegypti* that were allowed to oviposit and were then decapitated and given a blood meal by enema failed to deposit yolk, confirming that a further release of OEH was required for renewed vitellogenesis. Vitellogenesis and yolk deposition occurred in parous females that had been ovariectomized and implanted with an ovary from a sugarfed donor prior to blood feeding and decapitation. They did not occur in ovariectomized and bloodfed females that received a vitellogenic ovary at the time of decapitation. Lea and Van Handel (1982) concluded that after a second blood meal OEH-releasing factor is secreted by the stage II ovaries.

22

Nutrition and fertility of anautogenous mosquitoes

22.1 FACTORS AFFECTING FECUNDITY AND FERTILITY

The reproductive capacity of female mosquitoes is affected by their nutrition in both the larval and adult stages. Their reproductive potential is established by the end of the larval stage; their subsequent exploitation of that potential depends upon their nutrition during the adult stage. It is necessary to define two terms that are often used rather differently by different investigators. **Fecundity** is the potential reproductive capacity of an organism (or population), measured by the number of gametes formed. **Fertility** is the actual reproductive performance of an organism (or population), measured as the number of viable offspring produced (Lincoln *et al.*, 1982).

The number of ovarioles in the two ovaries of a female governs the number of oocytes present at the start of each ovarian cycle. The mean number of ovarioles in batches of *Culex pipiens* reared under different conditions ranged from 340 to 430 (Hosoi, 1954b). In two studies of *Aedes aegypti*, ovariole numbers were reported to range from 100 to 125 (van den Heuvel, 1963) and from 72 to 158 (Steinwascher, 1984). In *Hodgesia* sp. they ranged from 20 to 60 (Mattingly, 1977), in *Armigeres digitatus* from 33 to 54 (Okazawa *et al.*, 1991), and in *Ae punctor* from 30 to 175 (Packer and Corbet, 1989a). Whenever it was examined, a strong positive correlation was found between female size and fecundity. In the absence of direct observations, fecundity can be estimated from the maximum number of eggs reported in individual egg batches (which is likely to be fewer than the number of ovarioles). The results show substantial differences between species: *Anopheles melanoon* and *An messeae* – 500 (Shannon and Hadjinicalao, 1941; Detinova, 1955), *Culiseta annulata* and *Cs subochrea* – 300 (Marshall, 1938), *Ae vexans* – 300 (Zharov, 1980), *Ae detritus* – 260, *Ae malayensis* – c. 60 (Tesfa-Yohannes, 1982).

Several factors can affect egg-batch size, namely (a) maternal body size, (b) nutritional condition (Section 22.2.3), (c) egg size, (d) physiological age and (e) volume and source of blood meal (Section 22.2.2). Because the conditions of larval development affect adult body size they must also influence ovariole number, which is fecundity. The total number of eggs laid by any female is a function of the number of gonotrophic cycles she completes and the sizes of her egg batches.

22.1.1 Effect of maternal body size on size and utilization of the blood meal

The early investigations into the relationships between maternal body size, blood-meal size and size of egg batch produced variable and inconsistent results, maternal body weight and size of blood meal appearing to be positively correlated in many instances, but not all (Clements, 1963; Hawley, 1985). The inconsistencies possibly arose from inaccuracies in measurement of blood intake due to use of the gravimetric method. Considering all the results it seems likely that, for any species, both blood-meal volume and size of egg batch are positively correlated with maternal size.

Briegel (1990a, b) reinvestigated the relationships between maternal size and fertility, either giving fixed volumes of blood to *Aedes aegypti* by enema, or retrospectively computing the blood intake of naturally fed females of *Aedes* and *Anopheles* from assays of excretory haematin titre and of the haemoglobin concentration of the blood. He also measured the ovarian protein and lipid titres of gravid females and determined the efficiency of blood utilization. Briegel's results are reviewed below.

(a) *Aedes aegypti*

The size of a female mosquito probably determines the number of ovarioles in her ovaries and hence her fecundity. In *Ae aegypti* the number of ovarioles increased by 20 for each 1 mg increase in pupal weight, to a maximum of 158 (Steinwascher, 1984). Increase in ovariole number was more closely correlated with increase in thorax length than with increase in dry weight. Indeed, the number of ovarioles per unit dry weight fell as the mosquitoes increased in weight (Van den Heuvel, 1963) (see Figure 7.4).

Some early investigators, working with a number of species of *Aedes*, obtained a positive correlation between weight of blood ingested and number of eggs produced with smaller blood meals, but found that above a certain volume there was little or no further increase in egg production (Woke *et al.*, 1956; Volozina, 1967; Jalil, 1974; Hien, 1976). Possibly this was because fertility could not exceed fecundity. When *Ae aegypti* were given fixed volumes of blood by enema, the number of eggs produced per μl of blood was substantially reduced at the highest blood intake (see Table 22.4) (Briegel, 1985).

Briegel (1990a) investigated the relationship between maternal size and volume of blood ingested, and the utilization of blood for vitellogenesis. When *Ae aegypti* of different sizes fed to repletion on a human host, blood-meal volumes ranged from 1.3 to 6.6 μl and were positively correlated with female size. Egg production ranged from 18 to 116 eggs per female. Generally, large females ingested more than twice as much blood as small females, and their egg production was almost 4-fold greater in comparison. Data for females of three size classes are presented in Table 22.1 and illustrated in Figure

Table 22.1 The effects of body size in *Aedes aegypti* upon the intake and utilization of blood when fed human blood to repletion. Data for three size classes were computed from regression equations. (From Briegel (1990a) with additional recalculation.)

Wing length (mm)	2.4		2.9		3.4	
Relative body size*	13.8		24.4		39.3	
Blood-meal volume (μl)	2.1		3.2		4.8	
No. of eggs†	26		52		90	
	μg	J	μg	J	μg	J
Blood-meal protein	375.0	6.28	575.0	9.62	825.0	13.81
Ovary protein	20.0	0.33	85.0	1.42	177.5	2.97
Ovary lipid	45.5	1.72	73.3	2.76	113.3	4.27
	65.5	2.05	158.3	4.18	290.8	7.24
Oocyte protein	0.82	13.81‡	1.47	24.68‡	2.0	33.47‡
Oocyte lipid	1.59	59.83	1.28	48.12	1.13	42.68
	2.41	73.64	2.75	72.80	3.13	76.15

* Relative body size is determined as the cube of wing length.
† Corresponding numbers of eggs on guinea-pig blood were 56, 81 and 115.
‡ Joules × 10^{-3}.

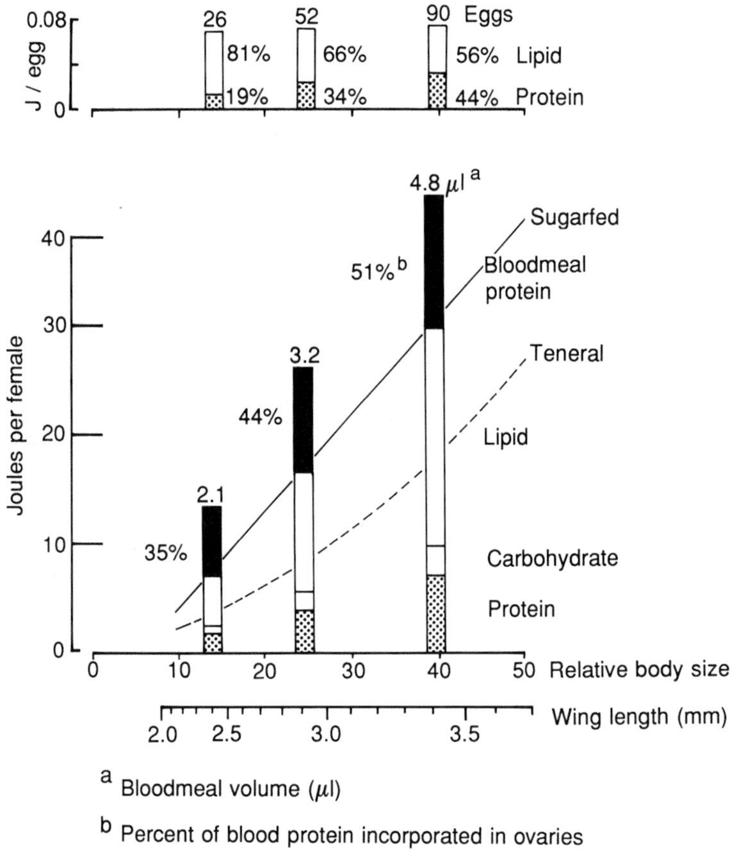

Figure 22.1 The accumulation and utilization of nutritive substances by *Aedes aegypti*. The lower diagram illustrates the accumulation of carbohydrate and lipid by females fed sucrose for 3–5 days *ad lib*, and the size of blood meals imbibed by sucrose-fed females feeding on human blood to repletion. These data are expressed as a function of relative body size, determined as the cube of wing length. The dashed regression line indicates the total energy values of teneral females; the solid regression line those of sucrose-fed females. The heights of the vertical columns indicate the energy values of the protein, carbohydrate and lipid contents of sugarfed females of three relative body sizes (13.8, 24.4 and 39.3). The black segments at the tops of the vertical columns indicate the energy values of the blood meals ingested by the sucrose-fed females, and the figures above them record the volumes of blood ingested. Utilization of the blood meal for oogenesis is depicted in the upper diagram. This details mean egg production, the mean energy value of single eggs (which was relatively constant at *c.* 0.08 J), and the percentages of lipid and protein within them. (After Briegel, 1990a.)

22.1. Utilization of blood protein for synthesis of ovarian protein and lipid amounted to 35, 44 and 51% in the three classes, increasing with female size. The percentage conversion of blood protein to ovarian protein varied markedly according to female size; the percentage conversion to lipid was rather less affected. In the smallest females examined, only about 2% of the dietary protein was converted to yolk protein, while about 98% was deaminated and its nitrogen excreted. In the largest females about 20% was converted to yolk protein and 80% was deaminated.

The deamination products from blood protein are converted to the lipid constituents of yolk, as described below, and also to carbohydrate and lipid reserves which are an important energy source. The rate of oxygen consumption of resting females of *Ae aegypti* increased up to 6-fold after a blood meal and remained high for one to two days. The extra oxygen consumed was positively correlated with the weight of blood ingested, and approximately 30% of the energy value of the blood meal was used in energy production (Heusner and Lavoipierre, 1973; Heusner *et al.*, 1973).

The total energy values (joules) of the protein and lipid in mature ovaries of *Ae aegypti* were linearly correlated with body size, with extremes of from 0.17 to 4.02 J protein (10 to 240 μg) and 1.13 to 5.23 J lipid (30 to 120 μg) per ovary pair. The proportion of ovarian protein to lipid varied; small females had a ratio of 1:5 whereas in large females it approximated 2:3. Under all conditions the total energy value of single mature oocytes was constant at approximately 0.08 J. In contrast to the energy value the mass of single oocytes varied. This was due to a significant positive correlation between protein and maternal body size (Table 22.1) which was slightly offset by a (non-significant) negative correlation of lipid content with female body size. The differences of composition of the mature oocytes of large and small females were shown to have no detectable effects upon the size and metabolism of the F_1 progeny to which they gave rise (Briegel, 1990a).

When large females of *Ae aegypti* (wing length >3 mm) took small blood meals and developed small numbers of oocytes, an investment of maternal reserves was required as a supplement to the blood meal to bring the oocytes to maturity (Figure 22.2). Large females ingesting *c.* 1–2 μl human blood matured fewer than 40 oocytes, and the energy values of the their ovaries were greater than those of their blood meals. Maternal investment was found only in large females; those of intermediate and small size evidently did not carry sufficient protein reserves to be used for oocyte protein synthesis because their ovarian protein always remained below 40% of the blood-meal protein.

Maternal investment was much more prominent when large females of *Ae aegypti* ingested small volumes of guinea-pig blood, when the maternal contribution to oogenesis might exceed that from the blood meal by more than 10-fold (Figure 22.2). Briegel (1990a) postulated that the higher isoleucine content of guinea-pig blood

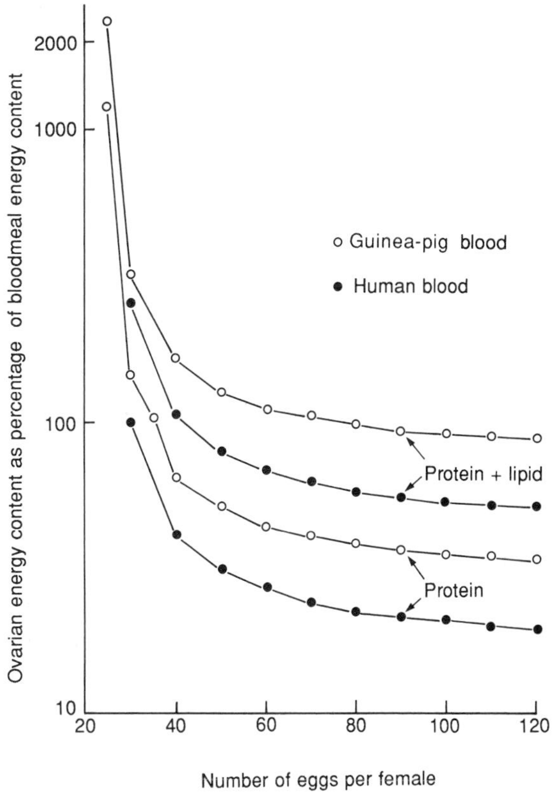

Figure 22.2 Utilization of blood-meal protein and of maternal reserves in production of oocyte protein and lipid by large females of *Aedes aegypti* fed human or guinea-pig blood, expressed as a function of fertility. Females that laid fewer than 40 eggs, due to the nutritional limitations of small blood meals, contributed more from their reserves to egg production than was contained in the blood meals. (After Briegel, 1990a.)

raised the fertility threshold and permitted a greater maternal input, but he suggested that although guinea-pig blood was superior to human blood with respect to fertility it might be considered suboptimal for *Ae aegypti* energetically.

The nutritional properties of the host blood also had a significant effect upon the protein content of the eggs. When a group of females was fed human blood the mean nitrogen content per egg was 0.208 μg (≡ 1.3 μg protein); this was significantly increased to 0.288 μg N (≡ 1.8 μg protein) when females were fed guinea-pig blood. Conversely, the lipid content per egg was 1.84 μg with human blood and significantly less, 1.19 μg per oocyte, with guinea-pig blood.

(*b*) *Anopheles*

When females were allowed to feed to repletion on a human host highly significant linear correlations between relative body size and blood-meal volume (determined from haematin output) were obtained for *An albimanus* and *An gambiae*. There was a highly significant correlation between haematin output and number of oocytes matured for *An albimanus*, *An gambiae* and *An stephensi*. In *An albimanus* the number of oocytes matured with human blood was also significantly correlated with body size, confirming that larger females both ingest larger meals and have a greater fertility (Briegel, 1990b).

A significant correlation was obtained between the energy values (J) of the protein in the blood meal and the protein + lipid content of gravid ovaries in *An albimanus*, *gambiae* and *stephensi*. Some 1–13% of the blood-meal protein was utilized for yolk protein synthesis; an additional 12–20% was utilized for yolk lipid synthesis. When these species fed on guinea-pig a significantly higher percentage of blood-meal nitrogen was recovered from the ovaries.

Ovarian protein ranged between 6 and 120 μg and ovarian lipids ranged between 10 and 90 μg per female. The protein content per egg was significantly correlated with blood-meal volume, but the mean lipid content per egg was constant for each species irrespective of blood-meal volume. Thus as blood-meal volume increased the mature ovary contained the same species-specific quantity of lipid and a greater quantity of protein (Figure 22.3).

The fate of maternal protein and lipid in *An albimanus* was investigated during the course of an ovarian cycle and as a function of body size. Comparison of the protein contents of the ovaries and carcases of gravid females with those of females just before the blood meal, when they were equal to the teneral levels, revealed a net gain of extra-ovarian protein (Figure 22.4).

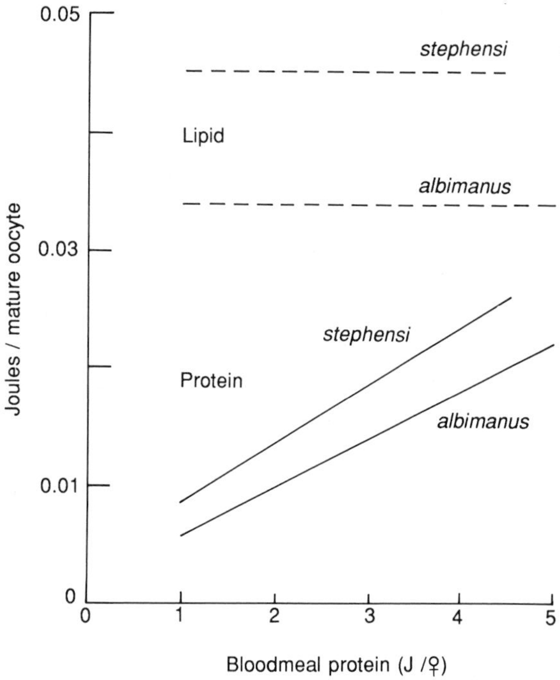

Figure 22.3 The energy values of the protein and lipid in single mature oocytes of *Anopheles stephensi* and *An albimanus* as a function of blood-meal size. (After Briegel, 1990b.)

Additional lipid synthesis occurred when sucrose-fed females of *An albimanus* were given a blood meal (Figure 22.5). Small females incorporated a relatively smaller proportion of the synthesized lipid into yolk and consequently gained a considerable amount of extra-ovarian lipid. Large females contributed part of their lipid reserves, derived from sugar feeding, to yolk lipid. Thus in large females greater proportions of available protein and lipid were contributed to egg formation whereas in smaller females more protein and lipid were allocated to extra-ovarian compartments (Briegel, 1990b).

In *An gambiae* only 19% of the blood-meal protein was incorporated in the ovaries, 4% being converted to oocyte protein and 15% being deaminated and converted to oocyte lipid. An additional 33% of blood-meal protein was incorporated into maternal extra-ovarian deposits (15% protein, 18% lipid). The remaining 49% of blood-meal protein was said to have been 'degraded'. Presumably some of this constituted an energy source during the ovarian cycle (Section 15.2.2). In a group of bloodfed females that failed to initiate vitellogenesis, 36% of the joules were transferred to extra-ovarian deposits (10% protein, 26% lipid); 64% of joules were said to have been degraded (Briegel, 1990b).

In *An stephensi* maternal protein was consistently 3–5% higher after oviposition than before the blood meal. After small blood meals there was a net loss of 1.46 J lipid per female, but after large blood meals there was a net gain of 0.75 J lipid per female. *An quadrimaculatus* showed a stable nitrogen balance before and after a gonotrophic cycle (Briegel, 1990b).

In terms of energy content the efficiency of

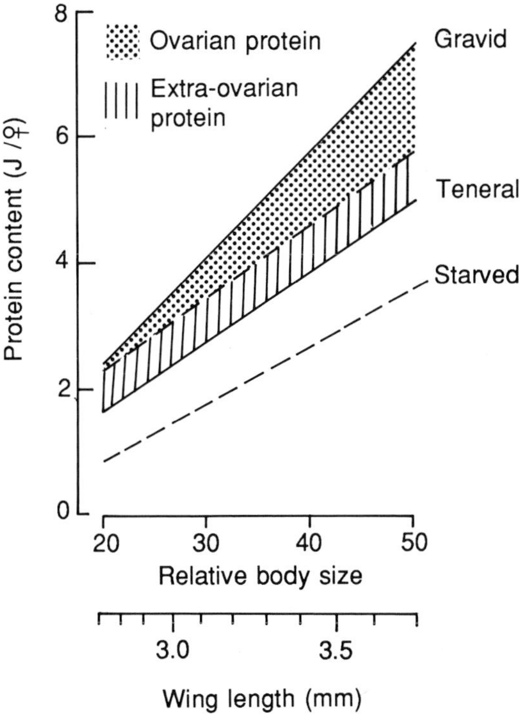

Figure 22.4 Regression lines for the energy values of the whole-body protein content of female *Anopheles albimanus* against relative body size under different physiological conditions, viz. (a) when teneral, (b) when starved to death, (c) when gravid, shortly before oviposition. The dotted and cross-hatched areas between the regression lines for teneral and gravid females indicate how the net protein gain from blood feeding is divided between ovarian and extra-ovarian compartments. (After Briegel, 1990b.)

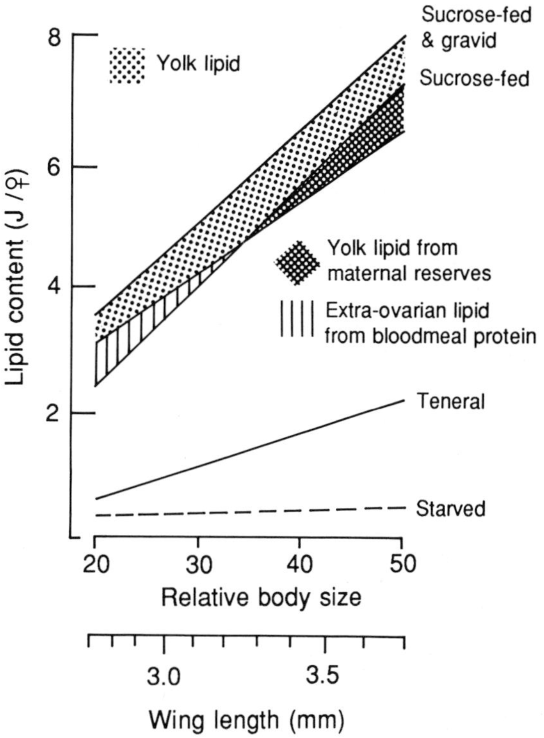

Figure 22.5 Regression lines for the energy values of the whole-body lipid content of female *Anopheles albimanus* as a function of relative body size under different physiological conditions, viz. (a) when teneral, (b) when starved to death, (c) after sugar feeding, (d) after sugar and blood feeding, when gravid, shortly before oviposition. The patterned areas relate to the gravid female only. See text for fuller explanation. (After Briegel, 1990b.)

utilization of blood protein for oogenesis shown by anophelines was significantly lower than that shown by large females of *Aedes aegypti* but comparable with that of smaller *Ae aegypti* females. However, when the conversion of blood protein to maternal extra-ovarian protein and lipid was also considered, the efficiency of utilization of blood protein by anophelines resembled that of large females of *Ae aegypti*. Briegel (1990b) suggested that synthesizing maternal protein and lipid from the blood meal might be an adaptive response to the low reserves of teneral anopheline females, and that the repetitive blood feeding of anophelines might both enhance fertility and build up reserves. He also suggested that human blood, which has a higher protein content than rodent blood, might be a suitable food source for anophelines because, although only a limited portion of the blood protein could be channelled into oogenesis due to its low isoleucine content, an appreciable amount would always be available for the synthesis of extra-ovarian reserves.

22.1.2. Egg size

Egg size varies both between and within species. It appears to affect fertility (egg-batch size) through relationships with both maternal body size and blood-meal volume.

The calculated egg volumes for 52 North American aedine species ranged from 6 to 60 $\times 10^{-3}$ mm^3. In an analysis of his own and published measurements of maternal body size and egg volume in nine anautogenous and autogenous species, Hawley (1985) obtained a highly significant negative correlation between log egg volume and log specific fertility (egg batch size per mg adult weight), i.e. the smaller the eggs the more were produced per mg of adult body weight.

Steinwascher (1984) observed that the eggs of individual females of *Aedes aegypti* varied slightly in length and breadth but that much greater variation occurred between the eggs of different females of a cohort. He obtained a significant regression of mean egg size and number on maternal body mass and blood-meal mass. The regression coefficients showed that body mass and blood-meal mass had about equal effects on the number of eggs produced but that egg size was affected much more by blood mass than by body mass. Briegel (1990a) reported that under all conditions of oogenesis the energy content per egg was constant at about 0.08 J, although the mass varied (Section 22.1.1(a)).

22.1.3 Female longevity

Long-lived females can produce many eggs. For example, three captive females of *Aedes aegypti* that laid 22 times in 87 days produced an average of 1360 eggs (Mathis, 1935). Captive females of *Anopheles atroparvus* that survived to lay more than 10 batches were estimated to have laid 2500 eggs each; one female laid 17 times (Shute, 1936).

Some species may be relatively long-lived in nature, and differences in longevity can cause striking differences of fertility between individual females. Table 22.2 records the percentage distribution among successive gonotrophic cycles of all egg batches produced by five wild populations, viz. of populations of three species of *Anopheles* in Tanzania and of *Mansonia uniformis* and *Culex quinquefasciatus* in Sri Lanka. This reveals that the contribution of long-lived females to the fertility of a population is minor because only a small percentage of females survive for more than three or four gonotrophic cycles.

With most species investigated in the laboratory, the number of eggs laid decreased in successive gonotrophic cycles, although in some cases the decline did not start before the 3rd or 4th

Table 22.2 Percentage distribution, in successive gonotrophic cycles, of the egg batches laid by females in wild populations of five tropical species. (Calculated from life tables derived from the original physiological age data.)

	Percentage distribution of egg batches among successive gonotrophic cycles									
	1st	*2nd*	*3rd*	*4th*	*5th*	*6th*	*7th*	*8th*	*9th*	*10th*
Anopheles funestus[1]	33.4	28.6	17.3	11.0	5.4	2.8	0.8	0.5	0.2	0.04
Anopheles gambiae[1]	37.6	25.2	15.1	9.9	6.1	4.0	1.2	0.4	0.3	0.2
Anopheles arabiensis[1]	45.4	29.3	13.9	7.5	2.7	0.9	0.2	0.1		
Mansonia uniformis[2]	47.8	30.5	14.5	5.2	1.7	0.3				
Culex quinquefasciatus[3]	73.4	23.1	3.0	0.5						

1. Gillies and Wilkes (1965); 2. Samarawickrema (1968); 3. Samarawickrema (1967).

cycle (Roubaud, 1934; Detinova, 1949, 1955; Hien, 1976; Hawley, 1985). For *Aedes aegypti* it was estimated that each successive egg batch contained 15% fewer eggs than the preceding one (Putnam and Shannon, 1934). Different results have been reported for a few species. Individual females of *Ae albopictus* laid variable numbers of eggs in successive gonotrophic cycles, but the mean number per cycle remained fairly constant through the first eight cycles, only starting to decline thereafter (Gubler and Bhattacharya, 1971). Egg batch size did not decline through nine cycles in *An stephensi* (Suleman, 1990). In a wild population of *Ae vexans*, fertility increased with physiological age due to seasonal influences, as described in the next Section.

Detinova (1955) considered that the decline in fertility of *Anopheles messeae* with age was due to the increasing percentage of ovarian follicles that degenerated in each gonotrophic cycle. Volozina (1967) reported that older females of *Ae communis* ingested less blood when fed to repletion than did younger females. She observed that this alone could account for the reduced fertility of older females, but found that even when similar amounts of blood were ingested by younger and older females, the average number of mature follicles was lower in older females.

22.1.4 Seasonal variations

Seasonal variations in size of egg batch have been reported. The temperature prevailing during larval development affects adult size, and a study of wild populations of four species of *Anopheles* showed that seasonal changes in adult size were accompanied by changes in fertility (Shannon and Hadjinicalao, 1941). Because fertility falls with age the occurrence together of females from different generations leads to a population whose members vary widely in reproductive potential (Detinova, 1955).

Such seasonal variations are well illustrated by studies made on *Anopheles messeae* in Moscow Province during 1952. Females that had overwintered laid an average of 195 eggs in their first egg batches in April. In May, when they had passed through two or three gonotrophic cycles, the average batch size had fallen to 172, and it fell to 149 in June. By June, offspring of these overwintering mosquitoes were flying. They had developed at low temperatures, were larger than the overwintered females, and the average size of their first batch was 289 eggs. They continued to lay during July and August but produced progressively smaller egg batches. At the same time other females were emerging which had developed at progressively higher temperatures; their size was progressively smaller as they emerged through July and August, and their first egg batches were smaller than those laid in June, amounting to 263 in July and 256 in August. From the middle of August emerging females started to enter diapause (Detinova, 1955).

A similar but less detailed study of *An messeae* in Bavaria by Kühlhorn (1972) showed a seasonal variation in fertility that was much more pronounced in maximum than in mean egg batch size. Barr *et al.* (1986) described seasonal trends in size of egg raft in *Culiseta inornata* in southern California.

Seasonal influences on the relationship between physiological age and fertility have been described by Zharov (1980). A significant increase in fertility with physiological age was observed in a population of *Aedes vexans* in the Volga delta in 1971, the mean egg batch developed by gravid wild-caught females increasing from 156 in the first gonotrophic cycle to 198 in the fourth. This phenomenon, which occurred in the first generation but not in the second, was ascribed to the conditions under which larvae of the first generation grew – high water temperature and intense crowding (>4000 larvae m^2) – which resulted in most females emerging with primary follicles at stage G and with only traces of reserves in their fat bodies. In later gonotrophic cycles the fat body reserves slowly increased, the previtellogenic ovaries were more advanced, and fertility increased steadily and significantly. The second generation larvae developed in shaded, cooler water and at low density (10–20 m^{-2}); most females emerged with ovaries at stage II and with fat bodies full of reserves, and fertility in the first gonotrophic cycle was high.

22.2 NUTRITIONAL REQUIREMENTS FOR OOGENESIS

In blood-sucking insects, ovary development commonly occurs after the female has taken a single protein meal. Wild females of some mosquito species may require two blood meals to mature their ovaries, very few species require more. This mode of protein intake differs from that of most insects, in which an extended period of feeding permits the gradual accumulation of protein. As a consequence of their blood-feeding habit, the number of ovarian follicles that can be matured by mosquitoes is determined by the volume of blood ingested in the one, or two, blood meals and by the nutritional quality of the blood. Therefore a mechanism is required for equating the number of follicles that mature with the nutritional input. The bloods of different vertebrate hosts can differ sufficiently in composition to affect the number of eggs produced by a particular mosquito species.

Blood-meal protein is the prime nutritional source for egg formation. However, the nutritional condition of a female at emergence profoundly affects her capacity for reproduction. Females derived from poorly nourished larvae require sugar and may have to engorge on blood two or even three times for ovarian maturation (Macdonald, 1956); also, their egg production will be affected throughout life (Mathis, 1935). Probably most of the laboratory studies described here involved females that had received adequate larval nutrition.

22.2.1 Requirements for blood constitutents

The identity of the nutrients required by female mosquitoes for egg production has excited interest for many years. Early experiments in which blood fractions were fed to mosquitoes suggested that protein is the only essential ingredient in blood, and this was confirmed when egg production was stimulated in females of *Aedes aegypti* that had been fed diets of amino acids (Table 22.3) (Dimond *et al.*, 1956; Singh and Brown, 1957). No eggs were produced when any one of eight amino acids was omitted from the diet, viz. arginine, isoleucine, leucine, lysine, phenylalanine, threonine, tryptophan and valine. When histidine or methionine was omitted a few eggs were laid during the first days after the start of the experiment but none later; it was concluded that enough of the deficient amino acid had been carried over from the pupa for the production of a few eggs, but that when this store was depleted it became indispensable. Ten amino acids were therefore essential for egg production, and these were the same amino acids that were essential for larval development (Section 5.3.1). Citrulline could replace arginine in the diet, although fewer eggs were produced, but ornithine could not (Dimond *et al.*, 1956).

Omission of either cystine or glutamic acid led to a fall in the number of eggs developed. Cystine could not be replaced by excess methionine. The stimulating effect of glutamic acid appeared to be due to its contribution of amino groups

Table 22.3 Two diets that permitted egg production in *Aedes aegypti*. (From Dimond *et al.* 1956.)

	Diet A g/100 ml	*Diet B* g/100 ml
D-Fructose	15.00	15.00
D-Glucose	5.00	5.00
Salts	0.15	0.15
*L-Arginine HCl	0.50	0.38
L-Cystine	0.20	0.15
L-Glutamic acid	1.00	1.00
Glycine	0.50	–
*L-Histidine HCl	0.70	0.15
*DL-Isoleucine	1.00	0.50
*L-Leucine	1.00	0.75
*L-Lysine HCl	0.90	0.75
*DL-Methionine	0.20	0.15
*DL-Phenylalanine	0.70	1.20
*DL-Threonine	0.80	0.30
*L-Tryptophan	0.40	0.30
*DL-Valine	1.00	1.00
Total amino acids	8.90 g	6.64 g
Mean fertility	18 eggs/♀	40 eggs/♀

* Essential for egg production.

for the synthesis of non-essential amino acids since it could be replaced with aspartic acid or ammonium acetate. Glutamic and aspartic acids are utilized in transamination in many animals. The D enantiomer of histidine was as effective as the L in promoting egg production. The D enantiomers of methionine, phenylalanine and tryptophan could also be utilized although they were less effective than the corresponding L forms. No eggs were produced when the D enantiomers of isoleucine, leucine, threonine or valine were fed (Dimond *et al.*, 1956).

Attempts to find the optimum concentration of each amino acid led to the production of a number of diets, two of which are detailed in Table 22.3 (Dimond *et al.*, 1956). From the mean egg production of 400 females given continuous access to food for 14 days it can be seen that Diet B was far more satisfactory than Diet A although it contained a lower concentration of amino acids. Haemolysed blood gave an egg yield which was double that on Diet B, but if the blood was diluted to the nitrogen concentration of Diet B egg production dropped to one-tenth of that on the amino acid solution (Dimond *et al.*, 1958).

The omission of salts from a chemically defined diet led to a halving of egg production (Dimond *et al.*, 1958). Sterile females obtained from larvae reared on a chemically defined medium laid eggs when fed only amino acids and sugar (Singh and Brown, 1957), therefore vitamins, nucleic acids and sterols are not essential for egg production, at least in the first gonotrophic cycle. This was confirmed by Dimond *et al.* (1958) who obtained no increase in egg production upon adding those substances. It is characteristic that insects such as the Cimicidae and Pediculidae that feed on vertebrate blood alone throughout their lives require symbionts for normal growth and reproduction, whereas mosquitoes and other insects that feed on blood only in the adult stage do not need symbionts (Brooks, 1964).

Sugars, which were included in the diets to stimulate feeding, have an effect on oogenesis, which is considered in Section 22.2.3.

22.2.2 Host blood

(a) Species differences

For analytical or comparative purposes egg production is often expressed as relative fertility, i.e. number of eggs produced per mg of blood ingested. To discover whether or not relative fertility is affected by the type of blood ingested, mosquitoes have been fed on hosts from a variety of vertebrate taxa (Shelton, 1972; Jalil, 1974; Downe and Archer, 1975; Nayar and Sauerman, 1977). Unfortunately most studies were undertaken before it was known that gravimetric measurement of blood intake can lead to serious underestimation of blood-meal size (Section 11.1.2) and before the technique of blood feeding by enema was established, therefore there is uncertainty over the validity of the results

Table 22.4 Mean distribution of blood-meal nitrogen, at the end of oogenesis, in females of *Aedes aegypti* given different volumes of rat blood by enema. Loss of some excretory ammonia was likely. (From Briegel, 1985.)

Blood meal		*Ovaries*		*Excreta*	*Ovaries + excreta + faeces*
μl	μg N	No. of eggs	μg N	μg N	μg N
0.5	15.1	28	7.6	2.7	10.3
1	30.1	52	13.3	12.2	25.5
2	60.2	86	28.7	29.3	58.0
4	120.4	120	39.1	78.7	117.8

that were obtained. It would be foolish, however, to discount all results of these investigations. The most substantial differences between relative fertilities on the blood of different host species were sufficiently striking to suggest that relative fertility is (a) higher on the blood of certain host species than others, (b) not greatly affected by whether the erythrocytes are nucleate or anucleate, and (c) lower on human blood than on that of most other host species.

Giving meals of rat blood to *Ae aegypti* by enema, Briegel (1985) observed that, with the exception of the smallest meals, almost all dietary nitrogen could be accounted for by nitrogen recovered in the mature ovaries plus that in the excrement (Table 22.4). A substantial proportion of blood-meal nitrogen was excreted, therefore dietary protein was never completely utilized for oogenesis. Indeed, the nitrogen incorporated into mature oocytes rarely exceeded half of the blood-meal nitrogen, even with small meals.

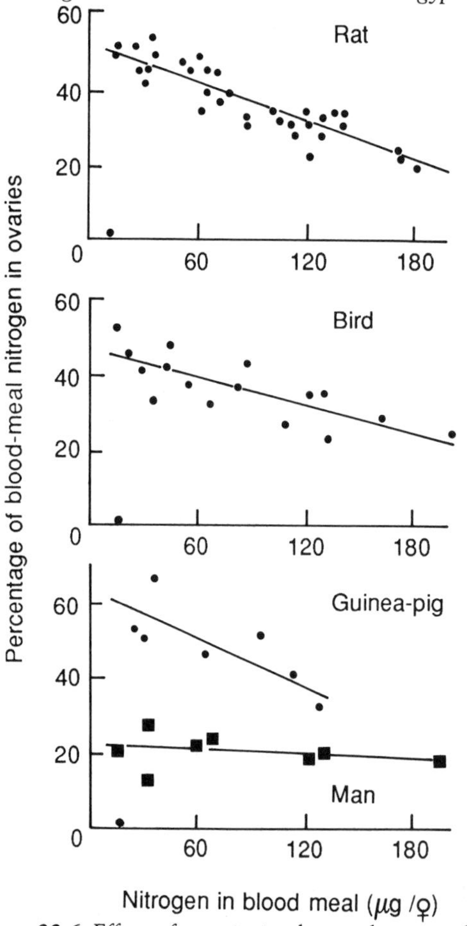

Figure 22.6 Effect of protein intake, as determined by blood enema volume, upon the percent incorporation of blood nitrogen into the ovaries of *Aedes aegypti*. Females were given enemas of different volumes of four different types of blood. Correlation coefficients (r) for the linear regressions were -0.246 for human blood and between -0.767 and -0.899 for the three other types (n = 7–36.) (From Briegel (1985). Copyright (1985), Pergamon Press plc. Reprinted with permission.

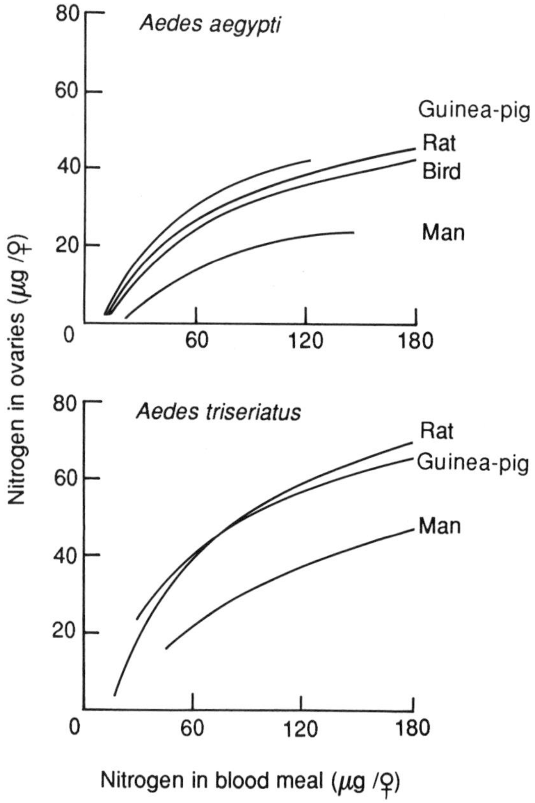

Figure 22.7 Efficacy of incorporation of nitrogen from the blood of different hosts. Females were given four different volumes of blood, from a number of hosts, by enema. Regressions of y upon ln x were computed ($y = a + b \ln x$) and are displayed without data points. All correlations were significant. (From Briegel (1985). Copyright (1985), Pergamon Press plc. Reprinted with permission.)

The degree of incorporation of blood-meal nitrogen into mature oocytes provided a direct measure of the utilization of blood-meal protein. Briegel (1985) combined the enema feeding technique with measurement of nitrogen incorporation in a reinvestigation of the effects of blood-meal volume and host blood type on protein utilization. He obtained negative correlations between amount of nitrogen given in enemas of different sizes and percent incorporation of the nitrogen in mature oocytes, i.e. the bigger the blood meal the less efficiently it was utilized. The correlation coefficients were significant for rat, bird and guinea-pig blood but not for human blood (Figure 22.6). However, it is unlikely that human blood really differs in this respect. In these feeding experiments some 20 to 70% of blood-meal nitrogen appeared in the mature oocytes depending upon blood-meal volume as well as upon host species.

Briegel (1985) observed no significant differences in the incorporation by *Ae aegypti* of nitrogen from the bloods of rat, guinea-pig or bird (chicken or pigeon) into ovarian nitrogen, indicating equal utilization of host blood protein; however, the nitrogen in human blood was less well utilized (Figure 22.7). Females of *Ae triseriatus* incorporated nitrogen from the blood of rat and guinea-pig equally, but again human blood was less efficiently utilized. The percentage of blood-meal nitrogen recovered from the mature ovaries of *Anopheles stephensi* was significantly higher with guinea-pig blood than with human blood (Briegel, 1990b).

(b) Limiting amino acids

A plausible explanation for the differences in egg production obtained with the blood of different hosts is that the amino acid composition of the blood varies and is more suitable for oogenesis in some cases than in others. A seminal study by Lea *et al.* (1958) revealed that availability of L-isoleucine could be limiting for oogenesis. They observed that *Aedes aegypti* laid fewer eggs per mg of human, bovine or sheep blood, in which the isoleucine content of the haemoglobin is low, than per mg of pig or rabbit blood in which the haemoglobin has a much higher isoleucine content. Addition of isoleucine to the different bloods before feeding to *Ae aegypti* reduced the differences in fertility, but addition of nine other essential amino acids had no effect.

Hydrolysates of human blood, guinea-pig blood and ovarian protein from *Ae aegypti* had broadly similar amino acid compositions. Tryptophan, which is destroyed by acid hydrolysis, was not recorded but of the other nine essential amino acids eight occurred in broadly similar proportions in all three sources. Only isoleucine was exceptional. It formed 3.37% of total amino acids in ovarian protein, and 3.41% of total amino acids in guinea-pig blood, but only 1.17% of total amino acids in human blood (Chang and Judson, 1979). Females of *Ae aegypti* laid fewer eggs on human blood than on guinea-pig blood although their protein content was similar. When the isoleucine content of human blood was artificially raised to 80% of that of guinea-pig blood, egg production on the two types of blood became identical (Table 22.5A) (Chang and Judson, 1977a).

Haemoglobin forms a large part of blood protein; for example, it constitutes 80% of human blood protein. In higher vertebrates the haemoglobin molecule consists of four interlocking subunits, each having a coiled polypeptide chain and a haem prosthetic group. Human haemoglobins have five different polypeptides (α, β, γ, etc.) Haemoglobin A, which is the predominant haemoglobin of human adults, consists of two α and two β chains. Haemoglobin A_2, which forms only 2% of the total, consists of two α and two δ chains. The major haemoglobin during later foetal life is haemoglobin F, which has the subunits $\alpha_2\gamma_2$; in the human newborn it still constitutes more than half the total (Stryer, 1988).

Citing a survey by Dayhoff (1969) of the amino acids in haemoglobins from 20 vertebrate species, Briegel (1985) noted that isoleucine is missing from the α and β chains of four hominid primates, four ungulates and one monotreme, whereas in rodents 8–14 isoleucine residues are present in each haemoglobin tetramer. Human haemoglobin F contains four isoleucine residues per tetramer, all in the γ polypeptide chains.

Briegel (1985) analysed the amino acid contents of hydrolysates of human and guinea-pig

Table 22.5 The importance of particular essential amino acids for egg production in *Aedes aegypti*.

A. Egg production by females fed on human and guinea-pig bloods with and without added L-isoleucine. From the data of Chang and Judson, 1977a.

Nutrient	*Protein concn g/100 ml*	*Total isoleucine concn (mM)*	*Mean no. of eggs per mg blood*
Human blood	19.71	18.3	24.47±0.46
Guinea-pig blood	20.98	53.3	35.45±0.85
Human blood with added isoleucine	19.71	43.0	35.47±1.70
Guinea-pig blood with added isoleucine	20.98	123.3	32.26±1.50

B. Incorporation of the isoleucine and phenylalanine in 4 μl human and guinea-pig blood into the ovaries. From the data of Briegel, 1985.

	Human blood			*Guinea-pig blood*		
	nmol in 4 μl blood	*nmol in two ovaries*	*% incorporated in ovaries*	*nmol in 4 μl blood*	*nmol in two ovaries*	*% incorporated in ovaries*
Isoleucine	28	28	100	97	41	42
Phenylalanine	87	46	53	62	68	110

bloods and computed the amino acid titres of 4 μl enemas of those bloods. He also analysed the amino acid contents of mature ovaries of *Ae aegypti* fed human or guinea-pig blood. During oogenesis the 28 nmol of isoleucine present in a 4 μl enema of human blood were transferred in their entirety to the ovaries. In contrast, of the 97 nmol isoleucine in 4 μl guinea-pig blood only 41 nmol (42%) were transferred (Table 22.5B). All phenylalanine in the guinea-pig enema was transferred to the ovaries but only 53% of that in the human blood enema. Utilization for oogenesis of seven other essential amino acids in human and guinea-pig blood varied between 10 and 65%, so they were evidently present in surplus. It appears that, for *Ae aegypti*, isoleucine is the limiting amino acid in human blood and phenylalanine is the limiting amino acid in guinea-pig blood. Human foetal blood, which contained 40 nmol isoleucine per 4 μl, was utilized as efficiently as rodent blood for egg production.

The isoleucine present in ingested plasma proteins is also available for mosquito oogenesis. Enemas of human and guinea-pig plasma were utilized for egg production as efficiently as enemas of whole guinea-pig blood of equal protein content. Analysis showed isoleucine to be present in both plasma enemas in excess of ovarian requirements. While the isoleucine content of cow haemoglobin is lower than that of human haemoglobin the isoleucine content of cow plasma is almost twice that of human plasma. Hydrolysis of a 4 μl sample of whole cow blood yielded 32 nmol isoleucine, compared with 28 nmol for 4 μl human blood, which would account for the slightly greater fertility of *Ae aegypti* on cow blood (Briegel, 1985).

22.2.3 Carbohydrate

Mosquitoes that feed on blood and sugar lay more eggs than those that ingest blood alone. Two reasons probably account for much of this effect of sugar feeding. First, egg production incurs an energy cost, which can be met by sugar intake. Second, sugar intake reduces the utilization of dietary protein for lipid synthesis. Lipids constitute more than 50% of the dry weight of mature oocytes (Figure 22.1), and the distribution of lipids synthesized by gravid females, between ovarian and extra-ovarian compartments, is greatly affected by their nutritional state (Section 22.1.1). The following examples of the protein-sparing effects of sugar feeding are on record.

The amount of human blood required for the maturation of a single egg was 90 μg in sugarfed females and 120 μg in sugar-deprived females of *Aedes diantaeus* (Volozina, 1967). Similar observations on five other species were reported by Nayar and Sauerman (1975c).

Females of *Ae aegypti* which had been provided only with water failed to mature any eggs when fed 1 μl rat blood. About 50% of females fed 2 μl or more of blood developed eggs but the percentage doing so fell progressively as the period on water was increased from three to six days, by which time 80% had died. About 90% of *Ae aegypti* that had been maintained on water were able to develop eggs if given a small sucrose meal one day before or at the time of a 2 μl blood enema, compared with 30% of females denied sucrose or given it one day after the blood enema (Klowden, 1986). Analysing the effect of delay more closely, Klowden and Chambers (1989) found that a meal of 1 μl 20% sucrose supported egg production if given up to 9 h after a blood enema but that by 15 h afterwards it was ineffective.

Wild-caught females of *Ae communis* that were fed human blood once to repletion, and were provided with water but not sugar solution, failed to mature their ovaries. Only when fed sugar solution after the blood meal did up to 15% of females mature their ovaries (Andersson, 1992).

22.3 CORRELATION OF FERTILITY WITH FOOD INTAKE

In nature the nutritional value of blood meals can vary greatly. Some mosquitoes engorge fully while others take only small meals because of the defensive behaviour of their hosts (Edman and Downe, 1964; Edman and Kale, 1971). Mosquitoes may feed upon different species of hosts and ingest blood of different nutritional qualities. There is therefore a conceptual problem of how females match the number of oocytes that mature with the extent and quality of the food intake.

Bellamy and Bracken (1971) suggested that some mechanism adjusts the number of follicles that start upon vitellogenic development to the amount of nutrient available in the blood meal. Volozina (1967) believed that the nutritional inadequacies of a blood meal are compensated for by some follicles remaining dormant, some starting upon vitellogenic development but being resorbed, and the remainder completing development. Lea *et al.* (1978) postulated that if vitellogenic development starts at all after a blood meal it starts in all of the proximate follicles, and that size of egg batch is correlated with food intake by the resorption of developing follicles.

Few investigators have counted the numbers of proximate follicles that start or fail to start upon vitellogenic development after a blood meal. Volozina (1967) reported considerable numbers of dormant follicles in gravid females that had taken small meals, e.g. 20–36% in *Aedes intrudens* and 27–50% in *Ae diantaeus*. In contrast, Lea *et al.* (1978), who gave *Ae aegypti* blood by enema, found that all or nearly all follicles started to deposit yolk whether the female had received 4 μl or 1 μl of blood. That observation was confirmed by Clements and Boocock (1984) who used vital staining to identify early vitellogenic and resorbing follicles. Because of the difficulty of identifying dormant follicles in unstained ovaries that contain mature oocytes, there must be doubt over the validity of Volozina's (1967) observations.

In a series of experiments with *Ae aegypti*, Lea *et al.* (1978) investigated the conditions under which vitellogenic follicles entered arrest or developed further, and under which they subsequently developed to maturity or were resorbed.

When females took a small blood meal their ovaries developed in one of three ways.

1. With a very small meal, e.g. 0.5 μl, all proximate follicles might remain in the previtellogenic resting stage. These follicles could start vitellogenic development after the next blood meal if it was big enough.
2. In some females given a meal of 1 μl, virtually all proximate follicles started vitellogenic development but re-entered arrest in stage III. Follicles that re-entered arrest were not prone to degeneration at that stage and resumed vitellogenic development after another blood meal of sufficient size.
3. In other females given a meal of 1 μl, virtually all proximate follicles started upon vitellogenic development; some produced mature oocytes and the remainder were resorbed. The percentage of follicles resorbed was negatively correlated with the amount of blood ingested.

By giving mosquitoes first a 1 μl and then a 3 μl blood meal, with different intervals between the two, Lea *et al.* (1978) were able to estimate the time after a blood meal at which ovaries that started upon vitellogenic development became programmed to follow one course or another. Females given a large meal 8 h after an earlier small meal utilized both and produced a large batch of eggs, therefore no follicles had been programmed for atresia (resorption) within 8 h of the first meal. Females given the second meal 14 h after the first either utilized only the first meal and produced a small batch of eggs, or utilized both meals and produced a large batch of eggs. It was concluded that by 14 h after the first meal (at 27°C) the ovaries of some females had become committed to particular programmes of development and resorption and were unable to utilize the second blood meal. However, the ovaries of other females were able to utilize a second meal 14 h after the first, possibly having first re-entered arrest. Therefore, before 14 h post-bloodmeal vitellogenic ovaries become programmed to re-enter arrest or to continue development; at about 14 h individual follicles in the developing ovaries become committed to completing development or to atresia. The circumstantial evidence for two bursts of EDNH release, one <8 h and the other >8 h after feeding (Section 21.3.3), may be pertinent to these observations.

The degeneration and resorption of a percentage of vitellogenic follicles occur regularly during ovarian development in species of *Culex, Anopheles* and *Aedes* (Hosoi, 1954a; Detinova, 1962; Volozina, 1967; Lea *et al.*, 1978). Degeneration is first apparent in the epithelial cells. Later the follicle shrinks, loses its cell structure, and becomes reduced to a small pigmented body (Nicholson, 1921; Parks and Larsen, 1965; Clements and Boocock, 1984). The percentage of follicles that mature is determined not only by the volume of blood ingested but also by its nutritional quality (Section 22.2.2). Many more follicles degenerated in females of *Culex pipiens pallens* fed human blood, considered nutritionally inferior, than in females fed bird blood (Hosoi, 1954a). Atresia appears to be the mechanism by which fertility is matched with food intake when post-vitellogenic development goes to full term.

Clements and Boocock (1984) used vital staining to reveal the early stages of yolk deposition and of follicular degeneration in *Ae aegypti* kept at 27°C. By 10 h after females had engorged fully on human blood most follicles had started to deposit proteinaceous yolk, but an average of two follicles per ovary remained at stage II and were apparently resorbed between 10 and 20 h post-bloodmeal. Among the majority of follicles that started to deposit yolk a few lagged in early stage IIIa and they appeared to degenerate between 20 and 25 h post-bloodmeal. A larger number of follicles degenerated in stage IIIb, between 26 and 30 h post-bloodmeal, when the oocytes contained a considerable quantity of yolk. Very few follicles that developed to stage IVa failed to mature. Both bursts of degeneration started long after the critical period, about 14 h post-bloodmeal, when it is postulated that follicles become committed to atresia (Lea *et al.*, 1978). The first burst of degeneration started during the period of high ecdysteroid concentration, and the

Table 22.6 The development and resorption of ovarian follicles in *Aedes aegypti*, recorded at different times after a meal of human blood. (From Clements and Boocock, 1984.)

Elapsed time (h)	*No. of ovaries*	*Principal stage*	*Occurrence of follicle types (mean follicles per single ovary)*			*Mean total follicles per ovary*
			Previtell-ogenic	*Vitellogenic*	*Resorbing*	
Unfed	18	IIa	51.4 ± 2.6	0	0.5	51.9 ± 2.6[a]
10–13	13	IIIa	2.0	51.3 ± 2.3[a]	0.15	53.4 ± 2.0[a]
18–19	21	IIIa/IIIb	0.3	54.0 ± 1.2[a]	1.6	55.9 ± 1.4[a]
33–34	25	IIIb/IVa	0	40.9 ± 1.8[b]	15.4 ± 1.3	56.3 ± 1.8[a]
40–42	22	IVa	0	38.7 ± 1.7[b]	Present*	–
59–61	32	IVb/V	0	36.4 ± 1.1[b]	Present*	–

* Resorbing follicles could not be counted accurately at the later times because some were obscured by the large vitellogenic follicles.

Within a column, figures with the same superscript are not significantly different from one another ($P < 0.05$); where the superscripts differ the means are significantly different.

±, Standard error of the mean.

second when the ecdysteroid concentration was low (cf. Figure 21.8). Both started during the period of active vitellogenin synthesis (Figure 20.6a). The details of one experiment are given in Table 22.6.

Very little is known about the nature of the mechanism that causes some follicles to degenerate while others mature, but there is some evidence that competition between follicles plays a part in the process. Hosoi (1954a) prevented the development of single ovaries in *Culex pipiens* by blocking the 4th and 5th abdominal spiracles on one side. Of the treated females that subsequently fed on blood, five survived. Three of these females developed an average of 108 mature oocytes in their single ovaries, compared with an average of 100 mature oocytes in the two ovaries together of the control females. Thus these three experimental females raised a compensatory number of eggs from follicles that would otherwise have degenerated.

Hormone concentrations may also affect follicular resorption. Schlaeger *et al.* (1974) injected varying but very large amounts of 20-hydroxyecdysone into sugarfed *Ae aegypti*, stimulating yolk deposition. By 48 and 72 hour post-injection it appeared that there had been a higher rate of follicular resorption in females that received less hormone. Greater yolk deposition in the surviving oocytes accompanied the higher resorption rate. Treatment of *Ae aegypti* with Cecropia juvenile hormone or methoprene, 0 or 24 h after engorging, blocked ovarian development beyond stages III to IVa. In treated females, unlike the bloodfed controls, no follicles were resorbed and the blocked follicles retained their cellular integrity throughout the life of the female (Judson and de Lumen, 1976).

23

Autogeny

A number of species have reverted to the use of reserves for production of the first batch of eggs, which has selective advantage under certain conditions. A very few species can produce more than one batch of eggs from their reserves. The change in reproductive strategy necessitates changes in the provision of nutrients for vitellogenesis as well as in the regulation of ovarian development. Autogeny is genetically determined, nevertheless environmental and other non-genetic factors play an important role in its expression.

23.1 THE PHENOMENON OF AUTOGENY

23.1.1 Definitions

Roubaud (1929, 1933) coined the terms **autogeny** for egg production in the absence of all adult feeding, and **anautogeny** for egg production dependent upon blood feeding. Lea (1964) proposed that the definition of autogeny should permit sugar feeding, because no amount of it will enable genetically anautogenous females to produce eggs. We shall therefore define autogeny as the production of eggs (or at least the deposition of some yolk) without ingestion of protein by the adult. However, as will be made clear, sugar feeding can affect egg production by genetically autogenous females.

Females that are genetically autogenous do not invariably initiate vitellogenic ovarian development in the absence of blood feeding; the frequency of initiation in a population is a measure of **penetrance**. Even when the genes for autogeny are penetrant, **expressivity** may be incomplete, as when oocyte development ends before maturity. Penetrance and expressivity probably depend upon both genetic and environmental factors. **Obligatory autogeny** is observed in females that do not engorge during the first ovarian cycle, even when hosts are available. **Facultative autogeny** is observed in genetically autogenous females that will engorge readily during the first ovarian cycle if they meet a host, but which can mature one batch of eggs without a blood meal. Some of these terms are illustrated schematically in Figure 23.1.

23.1.2 Occurrence and characteristics

Autogeny has been reported in many species distributed among at least 15 genera (Clements, 1963; Ellis and Brust, 1973; Rioux *et al.*, 1975; Mogi and Miyagi, 1989). Undoubtedly it will be described in many more species. Laboratory colonies of *Aedes aegypti* are mostly anautogenous, but of 19 wild African populations 14 included some autogenous females (Trpis, 1977).

Autogeny was first described in *Culex pipiens*, and the variability of its expression in that species may exemplify a situation common to many autogenous species. In the northern part of the Palaearctic Region autogeny occurs in small isolated populations of *Cx pipiens* which almost certainly interbreed only rarely with sympatric anautogenous populations. They breed below ground in enclosed spaces with restricted access, such as septic tanks and flooded basements, frequently in water contaminated with organic waste. The autogenous form is stenogamous (requiring little space for mating), feeds readily on

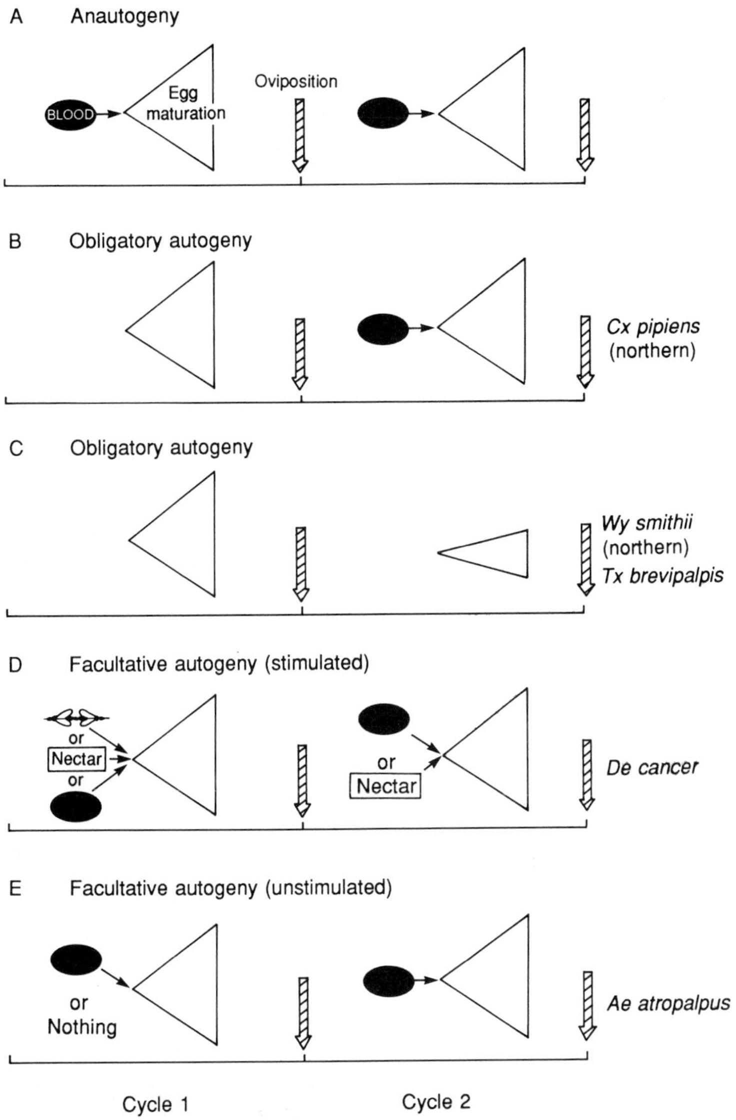

Figure 23.1 Schematic characterizations of anautogeny and of various forms of autogeny, with named species in which the form of autogeny is occasional or invariable. The requirement for mating for fertilization is not indicated. It has been established for all of the named species except *Tx brevipalpis* that sugar is not an essential nutritional requirement for autogenous ovarian development. (After O'Meara, 1985a).

man after the first ovarian cycle, and is incapable of winter diapause (Figure 23.1B). The widespread anautogenous form breeds above ground in unpolluted water, is eurygamous (requiring much space for mating), feeds on birds, and diapauses during the winter (Theobald, 1901; Roubaud, 1929, 1933).

Towards the south of the Palaearctic Region the populations of *Cx pipens* are very different. In Israel, for example, similar percentages of autogenous females were found in open and enclosed sites, in clean and polluted water. No populations were entirely autogenous, and almost certainly interbreeding was frequent between autogenous and anautogenous forms. At sites where a high percentage of females initiated autogenous ovarian development expressivity was also high; where populations were predominantly anautogenous, only incomplete ovarian development was observed in most of the few females that initiated autogenous ovarian development. About one-third of young females accepted blood meals, showing little or no difference in responsiveness to human and bird hosts. Except in the north of Israel, all females were incapable of diapause (Nudelman *et al.*, 1988). North African populations of *Cx pipiens* were similar (Knight and Abdel Malek, 1951).

In the northern part of the Nearctic Region, north of latitude 39°N, all populations are of *Cx pipiens*; south of 36°N all are of *Cx quinquefasciatus*, while in a band across North America, between those two latitudes, both taxa and their hybrids are found. (Opinion remains divided as to whether these taxa are species or subspecies.) In the north, anautogenous diapausing *pipiens* populations breed above ground while autogenous non-diapausing *pipiens* populations breed in man-made underground structures. At a site in Boston which was accessible to autogenous and anautogenous populations, <3% of captured females were heterozygous for autogeny, leading Spielman (1967) to conclude that the populations were largely reproductively isolated. However, if the hypothesis (Section 23.3) is correct that one of the genes coding for autogeny has multiple alleles much hybridization could occur without the manifestation of autogeny. South of latitude 39°N populations are found that include very variable percentages of autogenous females, as in the coastal area south of San Francisco. Interbreeding may occur in locations where autogenous populations are sympatric with anautogenous populations, whether of *pipiens* or *quinquefasciatus*. Autogeny is not found in populations of *quinquefasciatus* (Spielman, 1964, 1971; Barr, 1982).

Populations of *Cx pipiens* that exhibit autogeny have often been classified as an infraspecific form, *Cx pipiens molestus* Forskål. However autogenous and anautogenous strains of *pipiens* are interfertile, unless incompatible due to *Wolbachia pipientis*, and they appear to share a common specific mate recognition system. The autogenous populations are now regarded as physiological variants of *Cx pipiens* L. (Barr, 1982; Harbach *et al.*, 1984); in this volume they are called 'autogenous *Cx pipiens*'.

A relatively high percentage of mosquito species that breed in saline waters are autogenous. They are found both in coastal habitats, such as marshes, mangrove swamps and rock pools, and in inland saline pools, and include *Aedes atropalpus*, *Ae campestris*, *Ae caspius*, *Ae detritus*, *Ae mariae*, *Ae taeniorhynchus*, *Ae togoi*, *Anopheles hilli*, *Deinocerites cancer De pseudes* and *Opifex fuscus*.

It has been noted that within groups of closely related species one may be autogenous and another completely or very largely anautogenous (O'Meara, 1985). Such autogenous:anautogenous pairs include *Cx pipiens* and *quinquefasciatus*; *Ae churchillensis* and *communis*; *Ae taeniorhynchus* and *sollicitans*; *Ae atropalpus* and *epactius*, *Wy vanduzeei* and *mitchellii*. Populations of the species pairs may be extensively sympatric or may be geographically largely separated. Autogeny is much more prevalent in species or populations occurring at higher latitudes. Thus *Cx pipiens* and *Ae atropalpus* occur in more northerly latitudes, *Cx quinquefasciatus* and *Ae epactius* in more southerly. Within a species the characteristics of autogenous populations may show a north–south cline, as in those of *Cx pipiens* described above. At latitudes higher than 40°N females of *Wy smithii* emerge with vitellogenic ovaries, are obligatorily autogenous, and repeatedly mature batches of oocytes without blood feeding. South of 36°N, in a region where the carrying capacity of the pitcher plants is always saturated and density-dependent constraints on larval growth are more severe, the females emerge with undeveloped ovaries. They may develop one egg batch autogenously but all subsequent ovarian development is anautogenous (Smith and Brust, 1971; O'Meara *et al.*, 1981; O'Meara and Lounibos, 1981).

Along the east coast of North America almost all populations of *Ae taeniorhynchus* were found to include some autogenous females. The percentages of autogenous females were high (75–94%)

in south Florida where they bred in mangrove swamps, intermediate in the biotically transitional region of mid- and north Florida, and low (1–8%) further north where breeding occurred in temporary pools in grassy salt marsh. However, among the widely dispersed populations of the Caribbean Region the percentage of autogenous females was highly variable. At a number of locations evidence was obtained of frequent interbreeding between autogenous and anautogenous individuals. The mean autogenous fertility was often <30 eggs per female in populations in which the frequency of autogeny was low, but reached 50–65 eggs per female in populations with a high frequency of autogeny. These differences persisted when larvae were reared in the laboratory on the same highly nutritious diet (O'Meara and Edman, 1975; O'Meara, 1985).

Populations of *Culex tarsalis* exhibit a striking seasonal variation in the expression of autogeny. The earliest females to emerge from pupae collected in March in the northern Sacramento Valley, California, were all anautogenous, but autogeny was recorded at low or moderate levels among females emerging later in the spring. In mid and late summer the extent of autogeny increased greatly, sometimes reaching almost 100%, but it fell in the autumn, and <1% of wild overwintering females were autogenous. A positive correlation was demonstrated between percentage expression of autogeny and autogenous fertility (Spadoni *et al.*, 1974). With material obtained from Washington state, Harwood (1966) showed that the photoperiod and temperature to which the developmental stages were exposed determined the percentage of autogenous females. The F_1 progeny of mosquitoes collected near Bakersfield, California, and reared under simulated summer conditions (LD 16:8 and 25°C) had an autogeny rate of 43%. Other F_1 progeny transferred at the start of the pupal stage to simulated winter conditions (LD 10:14 and 16°C) had a zero autogeny rate and their ovaries remained at stage I, characteristic of diapausing females. Reisen (1986) concluded that cool temperature and short daylength arrested follicular development at stage I, inhibiting the penetrance of autogeny. His field studies suggested that once females had experienced shortening photoperiods and low temperature during the autumn, the alleles for autogeny were not expressed, even under diapause-termination conditions which permitted follicles to develop to the previtellogenic resting stage prior to host finding and anautogenous development. Females from wild-caught pupae which had experienced a short but lengthening photoperiod and a cool temperature exhibited a 16% autogeny rate. Similar observations were made on a population of *Cx tarsalis* in Manitoba (Brust, 1991).

In the Camargue region of France, *Aedes detritus* exhibited autogeny rates of 16–35% between October and March but only of 0.4–4% between April and September. Short days and lower temperatures were shown to increase the expression of autogeny (Guilvard and Rioux, 1986).

In most autogenous species, females that survive after completing one autogenous ovarian cycle then take a blood meal and start upon anautogenous ovarian development. Two species of *Aedes* never take a blood meal and produce all their eggs autogenously in a single batch. They are *Ae churchillensis*, in which the capacity for flight is lost (Hocking, 1954), and the arctic species *Ae rempeli* (Smith and Brust, 1970). North of approximately 40°N females of *Wyeomyia smithii* similarly refuse to blood feed after ovipositing although females occurring south of 36°N will do so (O'Meara and Lounibos, 1981). In genera such as *Toxorhynchites* and *Malaya*, in which the mouthparts are incapable of penetrating skin, the females are invariably autogenous.

23.1.3 Ovarian development

We distinguish autogenous from anautogenous females by the development of the ovaries – but the ovaries are passive, their development regulated by the endocrine system. Ovaries from autogenous females of *Culex pipiens* failed to develop when transplanted into anautogenous females unless the hosts were given a blood meal, whereas ovaries from anautogenous females produced mature oocytes after being transplanted into autogenous females although those females never received a blood meal (Larsen and Bodenstein, 1959).

The females of most autogenous species emerge with their primary ovarian follicles in stage I. During the first days after emergence all of these follicles start to develop. Some develop to maturity without interruption but others are resorbed so that only a percentage of the primary ovarian follicles produce mature oocytes (Larsen and Bodenstein, 1959; Laurence, 1964; Bellamy and Corbet, 1973; Smith and Corbet, 1975; Trpis, 1977). Presumably, as in anautogenous females, follicular resorption provides a means of matching the number of developing follicles with the availability of reserves. When poor larval nutrition restricts the expression of autogeny the ovarian follicles cease development at about stage III (Laurence, 1964).

A few autogenous species exhibit precocious ovarian development. In northern populations of *Wyeomyia smithii*, vitellogenesis starts in the pupal or pharate adult stages and females emerge with their primary follicles in stage III. Development proceeds synchronously in these follicles, and at 20°C it is completed by about two days after emergence (Smith and Brust, 1971). In contrast, the females of a Florida population emerged with previtellogenic follicles (O'Meara and Lounibos, 1981). *Toxorhynchites rutilus* also undergoes precocious ovarian development, and two days before emergence a number of the primary follicles have passed the previtellogenic gate. Development of the primary follicles is irregular; two days after emergence they are found in all stages from I to V, with 20% resorbing. Oviposition is intermittent (Watts and Smith, 1978).

Some species can complete more than one autogenous ovarian cycle. Females of *Wyeomyia smithii* from Massachusetts, which were fed only sugar, laid repeated egg batches (Figure 23.1C). The total number of eggs laid by individual females exceeded the number of ovarioles, and in older females the ovariolar stalks bore up to four dilatations. Oocyte maturation was a cyclic process, with 2–3 days between successive ovipositions. During the first day after oviposition the ovaries contained only previtellogenic or stage III follicles. Egg retention, resulting from denial of access to oviposition sites, inhibited further autogenous egg production. Females of *Wy smithii* from Florida produced only one autogenous egg batch, and subsequent ovarian development was dependent on blood feeding (Lang, 1978; O'Meara *et al.*, 1981).

Females of *Tx brevipalpis* oviposited repeatedly for up to 18 weeks. Egg batch size decreased with age, falling from 20 in the first month to 8 in the fourth (Trpis, 1981). Wild females of *Deinocerites cancer* developed and laid an initial autogenous egg raft, stimulated by insemination and sugar feeding. Some parous females then became facultative blood feeders, producing an anautogenous egg raft after taking a blood meal, but if kept in the laboratory and fed only sugar they produced a second autogenous egg raft with a mean size of 7.4 eggs (Figure 23.1D) (O'Meara and Mook, 1990). Repeated autogenous egg production has been claimed in *Cx pipiens*, based on observation of a second batch of mature oocytes (Sichinava, 1974) and of a second oviposition accompanied by the presence of two dilatations on some ovarioles (Kal'chenko, 1962), but more detailed evidence is needed to authenticate these claims.

Most females of *Ae togoi* that were gravid with autogenously developed oocytes refused to blood feed, but some fed when exposed to a host in the laboratory and developed an additional series of follicles (McGinnis and Brust, 1985). Whether these developed in the same ovarioles as the first batch of oocytes is not clear.

Autogenous egg batches are often rather small and are nearly always smaller than anautogenous egg batches produced by the same species. The egg rafts produced by autogenous females of *Cx pipiens* rarely contained more than 80 eggs; following a blood meal they usually contained from 80 to 100 eggs. This contrasted with rafts of 150–200 eggs laid by anautogenous *Cx pipiens*, presumably fed on bird blood (Christophers, 1945). The mean autogenous batch size of one strain of *Cx tarsalis* was 218 eggs, which was only slightly although significantly less than the mean of

232 eggs in the rafts of an anautogenous strain (Reisen and Milby, 1987). The large autogenous females of *Aedes atropalpus*, given only water, laid batches of 150–200 eggs, fully equal to those of the bloodfed anautogenous females of *Ae epactius*, a closely related but smaller species. The autogenous F_1 hybrids produced only half as many eggs (O'Meara and Krasnick, 1970). The mean size of autogenous egg batch produced by females of *Deinocerites cancer* collected as pupae was 41.8±1.6 eggs, identical with that laid by wild-caught parous females that fed on chicken – 41.5±1.6 eggs (O'Meara and Mook, 1990). Females of *Toxorhynchites brevipalpis* laid twice as many eggs per day as females of *Tx rutilus*, but due to the difference in egg size the weight of eggs produced per day was the same for the two species (Lamb and Smith, 1980).

23.1.4 Effects of larval nutrition

Autogeny shifts the role of gathering resources for egg production from adult to larva. Non-bloodfed females have no exogenous supply of nitrogen and must carry over from the larval stage sufficient protein for vitellogenin production. The lipid reserves are also important, both for yolk formation and as an energy source for the female. Consequently the conditions of food supply and crowding experienced by the larva can affect the expression of autogeny in the adult.

As would be expected there have been no reports of genetically anautogenous females being made phenotypically autogenous through rich larval feeding. However, poor larval nutrition reduced both the penetrance and expressivity of the alleles for autogeny in *Culex pipiens* (Spielman, 1957), *Aedes togoi* (Laurence, 1964) and *Wyeomyia smithii* (Lounibos *et al.*, 1982), i.e. the percentages of females initiating and completing vitellogenic development were both diminished, as was fertility. High larval density and high salinity of the larval medium had adverse effects on *Ae taeniorhynchus*, causing a reduction in adult dry weight, reducing the percentage of autogenous females initiating vitellogenic development, reducing further the percentage completing it, and diminishing fertility (Nayar, 1969b). However, the additional stimulus provided by insemination would very substantially increase the percentage of such females that produced an autogenous egg batch, although without increasing fertility (O'Meara, 1979). Poor larval nutrition did not affect the initiation of autogeny in *Ae atropalpus* but it reduced fertility. However, poor larval diet reduced the penetrance of autogeny in the F_1 progeny of a cross between *Ae atropalpus* and the anautogenous *Ae epactius* (O'Meara and Krasnick, 1970). Severe larval overcrowding affected adult size and viability in *Anopheles hilli*, but all surviving females developed eggs autogenously although with reduced fertility (Russell, 1979).

Comparison of autogenous and anautogenous strains of *Culex pipiens* reared under identical conditions revealed that genetically autogenous individuals had longer larval development times. Newly emerged autogenous females usually had higher fresh and dry weights, more extensive fat body, and greater total lipid, glycogen, protein and nitrogen contents than newly emerged anautogenous females. Expressed in relation to dry weight, autogenous females had a higher percentage composition of lipid and glycogen but a lower percentage composition of nitrogen than anautogenous females (Clements, 1956a; Twohy and Rozeboom, 1957; Lang, 1963; Briegel, 1969). In one investigation, by the time that waterfed autogenous females had completed an ovarian cycle and laid their eggs, waterfed anautogenous females had used up all their lipid and glycogen reserves and had died. Twohy and Rozeboom (1957) concluded that autogenous females require additional energy reserves as well as materials for vitellogenesis. Better survival was observed in an anautogenous strain studied by Briegel (1969).

Protein and lipid are the principal components of mosquito yolk. In waterfed females of *Ae atropalpus* the weights of protein and lipid that appeared in the ovaries equalled the weights lost from other parts of the abdomen (Van Handel, 1976). In waterfed autogenous females the fat body must be the principal source of ovarian lipid, and it may be a source of protein since the ninhydrin-positive droplets disappear after emergence. Another possible protein source is the larval abdominal musculature which is carried over to the adult and histolysed two days after emergence (Roubaud, 1932; Clements, 1956a). In females of *Ae churchillensis* the indirect flight muscles are histolysed during the first two or three weeks after emergence and possibly contribute to autogenous ovarian development (Hocking, 1954; Ellis and Brust, 1973). The fertility of *Wyeomyia smithii* was not significantly increased by adult ingestion of protein or amino acid solutions (Lang, 1978).

23.1.5 Effect of sugar feeding by adults

The necessity for sugar feeding for ovarian development by autogenous females varies greatly between species. Wild-caught females of *Aedes impiger* and *Ae nigripes* did not initiate vitellogenic development unless provided with nectar or sugar solution (Corbet, 1964). In *Ae atropalpus* and *Wyeomyia smithii* the expression of autogeny was not affected by the denial of sugar to females derived from well-nourished larvae, nor was their fertility reduced (Figure 23.1D, E) (Hudson, 1970b; O'Meara and Krasnick, 1970; O'Meara and Lounibos, 1981). Subterranean autogenous populations of *Culex pipiens* reproduce without access to sugar.

Feeding sucrose to females of *Ae taeniorhynchus* derived from well-nourished larvae raised the percentage of phenotypically autogenous females from 60 to 90%. Only 2% of females derived from poorly nourished larvae were phenotypically autogenous; sugar feeding increased this number to 14% (Lea, 1964).

In *Deinocerites cancer* sugar intake is not a nutritional requirement for autogeny since waterfed females will develop their ovaries provided they have been inseminated (Table 23.1). However, in this species sugar intake appears to stimulate autogenous ovarian development in virgin females, in one experiment raising the percentage of vitellogenic females from 2 to 94%. It also increased fertility (O'Meara and Petersen, 1985). Females of this species are active nectar feeders in the field (O'Meara and Mook, 1990).

23.1.6 Effect of insemination

In *Deinocerites cancer* insemination provides a stimulus for autogenous egg production. Autogeny remained unexpressed in waterfed virgins, but 87% of water-fed mated females developed eggs. Sugar feeding by virgins provided an equally effective stimulus and led to higher fertility. Whether sugarfed or not, virgins developed eggs after blood feeding (Table 23.1). Insemination was also important in an autogenous strain of *De pseudes*. Females of these two species have no sexually refractory period and can copulate at the time of emergence (O'Meara and Petersen, 1985). Evidence was obtained for a male factor stimulating autogenous ovarian development in *Wyeomyia vanduzeei* (O'Meara, 1979).

Populations of *Aedes taeniorhynchus* include anautogenous females and two types of autogenous female: (i) those in which autogeny is expressed in virgins, and (ii) those that must have mated to produce eggs autogenously. The requirement for insemination varies with geographical

Table 23.1 Effects of sugar feeding and insemination on the expression of autogeny in two species of *Deinocerites*. (From O'Meara and Petersen, 1985.)

Feeding status	*Mating status*	*Autogeny (%)*	*Eggs per gravid female*
Deinocerites cancer – Vero Beach, Florida			
Water	Virgin	2	16
Sugar	Virgin	94	57
Water	Mated	87	26
Sugar	Mated	100	58
Blood	Virgin	64	21
Blood + sugar	Virgin	100	75
Deinocerites pseudes – Texas			
Sugar	Virgin	13	38
Sugar	Mated	98	54

Females of *De cancer* originated from field-collected 4th instar larvae or pupae. Females of *De pseudes* were from a laboratory strain which originated in Texas. Sugar, when provided, was available continuously from emergence.

location and is further complicated by the influence of larval nutrition. Among laboratory-reared material from Flamingo, Florida, 92% of virgins produced eggs autogenously. In contrast, only 2% of virgins originating from Indian River County, Florida, produced eggs autogenously compared to 66% of mated females. Rearing larvae at high density reduced the percentage of phenotypically autogenous females among virgins from Flamingo; insemination restored the level of autogeny but did not raise fertility. Wild females from Flamingo were of similar size to females reared in the laboratory under crowded conditions, therefore in the field autogeny must be male induced, In all experiments with *Ae taeniorhynchyus* the females were sugarfed (O'Meara and Evans, 1976, 1977; O'Meara, 1979, 1985a). Females of *Ae taeniorhynchus* <36 h old were sexually refractory. After mating there was a lag of 1–2 days before the ovaries started vitellogenic development, consequently the period from emergence to initiation of ovary development lasted about four days, during which time the females would respond to a host and increase their fertility (O'Meara, 1985b).

Injection of an extract of male accessory glands, equivalent to 0.25 of a pair of glands, was as effective as mating in stimulating autogeny in *Ae taeniorhynchus*. An extract of *Ae sollicitans* accessory glands stimulated autogenous ovarian development when injected into females of *Ae taeniorhynchus*, but not when injected into females of *Ae sollicitans* in which autogeny is very rare (O'Meara and Evans, 1977; O'Meara, 1979). Experimental evidence suggests that a factor from male accessory glands stimulates the ovaries to secrete OEH-releasing factor (Section 23.2.2(a)).

23.1.7 Facultative autogeny

In a number of species genetically autogenous females can mature their first batch of eggs with or without blood. Examples of facultative autogeny, both stimulated unstimulated (Figure 23.1D, E), have been discussed in earlier Sections, but here we shall consider the selective advantage of the phenomenon. Females that exhibit facultative autogeny will engorge readily during the first gonotrophic cycle if they meet a host and will produce a large initial egg batch, but in the absence of a host they will develop a small autogenous egg batch.

Most females in arctic populations of *Ae impiger* and *Ae nigripes* studied by Corbet (1964, 1967) showed such facultative autogeny. The small percentage of females exhibiting obligatory autogeny began to mature their ovaries promptly after emergence and were ready to oviposit about ten days later. Females exhibiting facultative autogeny did not immediately mature their ovaries, and engorged readily during the first ten days after emergence if an opportunity occurred; those that engorged matured a larger number of oocytes than the obligatorily autogenous females. If these females were unable to engorge, their primary follicles remained in arrest at stages II or III for about ten days and then matured autogenously, but only if the females had had access to sugar.

Genetically autogenous females of *Ae taeniorhynchus* with undeveloped ovaries will readily engorge and will then develop some 150 eggs, double the number produced by sugarfed females. If autogenous ovarian development has progressed before blood feeding the blood meal does not contribute to fertility (O'Meara and Evans, 1973). Shifting the role of gathering resources for egg production from the adult to the larval stage possibly has survival value in severe environments where opportunities for host finding may be reduced, and in habitats where hosts occur at low or very variable density (Corbet, 1964, 1967). O'Meara (1985b) postulated that facultative autogeny provides flexibility for resource gathering in habitats where food supplies may be limited for both larvae and adults.

23.1.8 Modelling autogeny and anautogeny

A mathematical model has been developed to investigate the relative advantages of autogenous and anautogenous reproduction for mosquito populations (Tsuji, 1989; Tsuji *et al.*, 1990). The model incorporated a number of relevant life history parameters, and the following were shown to affect the advantageousness of autogeny or anautogeny.

1. The searching time to locate a blood meal (T_s).
2. The ratio of the sizes of the autogenous and initial anautogenous egg batches (p), which varies with the quality of larval rearing conditions.
3. The probability of surviving accidental death during blood feeding (s).
4. Duration of the immature stages relative to a designated part of the adult stage (α_0).
5. The difference between the death rates of the preimaginal and adult stages (δ).

Some of these life history parameters have been considered in earlier sections of this chapter, e.g. the longer larval developmental times of autogenous individuals, the effects of larval nutrition upon the expression of autogeny, and the relative size of autogenous and anautogenous egg batches ($0<p<1$). Little information is available for some other parameters, e.g. the death rate during blood feeding.

Employing quantitative data from several species, the model was used to assess the effects of individual parameters upon the intrinsic rate of natural increase (R) of a population, indicating their influence upon the advantageousness or disadvantageousness of autogeny. The strongest influence was exerted by the first three listed parameters, viz. host-searching time (T_s), the ratio of autogenous and anautogenous fertilities (p), and the probability of survival from blood feeding (s). Autogeny was more advantageous when the values of T_s and p were highest and s was lowest. The two other listed parameters were rather less influential, viz. the relative duration of the preimaginal period (α_0) and the difference between the adult and preimaginal death rates (δ).

Figure 23.2A is a worked example from the model in which the relationship between host-searching time (T_s) and the ratio of autogenous and anautogenous fertilities (p) was described through the parameter R. A curve was plotted for the condition in which autogeny and anautogeny resulted in the same intrinsic rate of increase (R), when $p = p_c$, p_c being the critical ratio of autogenous and anautogenous fertilities that resulted in a constant value of R.

The boundary curve (p_c) in the ($T_s \times p$) space moved when one of the other life history parameters changed. Thus the advantage of autogeny or anautogeny was dependent on the relative duration of the immature stages (α_0) (Figure 23.2B), and on the difference between the adult and preimaginal death rates (δ). Larval conditions that resulted in a prolonged larval stage (giving a

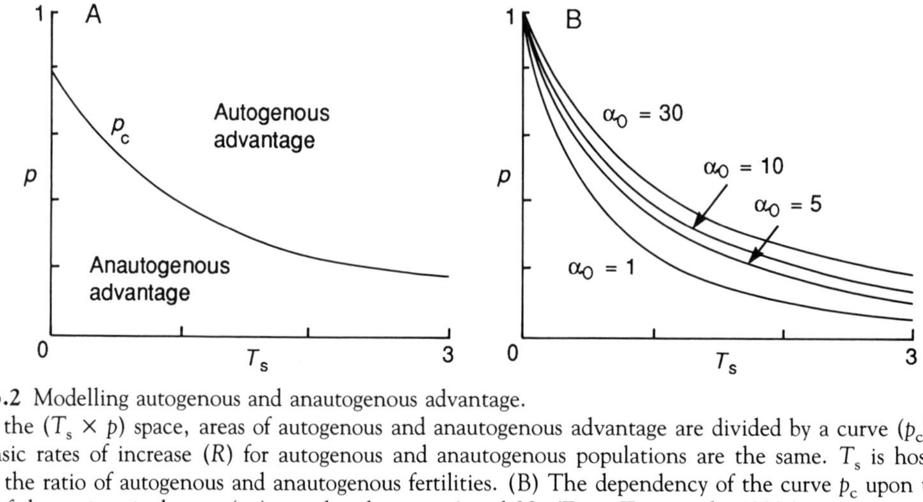

Figure 23.2 Modelling autogenous and anautogenous advantage.
(A) In the ($T_s \times p$) space, areas of autogenous and anautogenous advantage are divided by a curve (p_c) on which the intrinsic rates of increase (R) for autogenous and anautogenous populations are the same. T_s is host-searching time; p is the ratio of autogenous and anautogenous fertilities. (B) The dependency of the curve p_c upon the relative duration of the preimaginal stage (α_0) at values between 1 and 30. (From Tsuji *et al.*, 1990.)

high value of α_0) or high larval mortality (giving a negative value of δ), reduced the advantage of autogeny. High adult mortality reduced the advantage of anautogeny.

23.2 HORMONAL REGULATION OF AUTOGENOUS OVARIAN DEVELOPMENT

We have sufficient knowledge of the regulation of autogenous ovarian development in mosquitoes to believe that, in general, it involves a similar endocrinological framework to that described earlier for anautogenous development and summarized in Figures 21.1 and 21.4, although in autogenous development vitellogenesis proceeds without the stimulus of a blood meal. Certain important endocrinological observations were first made on autogenous mosquitoes, and only later repeated with *Aedes aegypti*, e.g. responsiveness to physiological levels of 20-hydroxyecdysone and the activity of juvenile hormone when applied subsequent to the previtellogenic phase. Most experimental work has been carried out on autogenous strains of *Aedes taeniorhynchus* and on *Ae atropalpus*.

23.2.1 Previtellogenic development

Autogenous females of *Aedes taeniorhynchus* emerge with their primary ovarian follicles at stage I, having passed the germarial gate, and at 27°C the oocytes mature within five days. When the corpora allata (and the adjacent corpora cardiaca) were removed within one hour of emergence, the primary follicles underwent no further development. However, if allatectomy was delayed until two days after emergence, most females developed their primary follicles to maturity (Table 23.2). Reimplantation of one pair of corpora allata into autogenous females of *Ae taeniorhynchus* which had been allatectomized at emergence restored their capacity for autogenous ovarian development (Lea, 1963, 1967, 1970). These results were sufficiently similar to those obtained with anautogenous females of *Ae taeniorhynchus* and other species (Table 21.1) to suggest that

Table 23.2 The effects of removal and implantation of endocrine organs upon oocyte maturation in sugarfed females of autogenous and anautogenous strains of *Aedes taeniorhynchus*. (From the data of Lea, 1963, 1967, 1970.)

Age at operation	*Treatment*		*Females maturing oocytes (%)*
	Removed	*Implanted*	
	Autogenous females		
1 h	CA	–	0
1 h	(sham)	–	85
1 h	CA	CA (1 pr)*	73
1 day	CA	–	20
2 days	CA	–	92
1 h	MNCp	–	0
1 h	MNCp	MNCp (4 pr)*	90
3 days	MNCp + NHO	–	80
	Anautogenous females		
1 day	–	CA (2–4 pr)	8
1 day	–	MNCp (4 pr)	60

* One day after removal.
MNCp = median neurosecretory cell perikarya.
CA = corpora allata.
NHO = neurohaemal organ.

autogenous development also includes a stage I gate as a potential stage of arrest, and that in both anautogenous and autogenous females of *Ae taeniorhynchus* juvenile hormone is released after emergence, stimulating the primary follicles to develop beyond the stage I gate.

At emergence the primary ovarian follicles of the autogenous females of *Ae atropalpus* are about 55 μm long, and are apparently at stage I. The follicles grow after emergence, reaching almost 100 μm length by 12 h about which time they appear to have completed previtellogenic development. Previtellogenic development could proceed in the absence of the head; after decapitation within 0 to 6 h of emergence the follicles grew to about 100 μm length (Kelly and Fuchs, 1980; Masler *et al.*, 1980; Kelly *et al.*, 1981). When females of an autogenous strain of *Culex pipiens* were ligated at the base of the abdomen during the first five hours after emergence, a few did not develop their primary follicles beyond stage I but most developed them to stage IIb, the previtellogenic resting stage (Clements, 1956a). It appears that in *Ae atropalpus* and *Cx pipiens* juvenile hormone is secreted before emergence or that the corpora allata are capable of secretion when separated from the brain.

23.2.2 Ovarian ecdysteroidogenic hormone (OEH)

(a) *OEH-releasing factor*

Circumstantial evidence for the existence of an OEH-releasing factor in *Ae taeniorhynchus* has been provided by Borovsky. Autogenous females of the Vero Beach strain decapitated 24 h after emergence did not synthesize vitellogenin, suggesting a deficiency of OEH. Females ovariectomized 24 h after emergence and 18 h later reimplanted with a *taeniorhynchus* or *aegypti* ovary subsequently underwent vitellogenesis; however, decapitation prior to ovary implantation prevented vitellogenesis, suggesting to Borovsky (1982) that an ovarian signal was necessary for OEH release.

In the Rutgers strain of *Ae taeniorhynchus*, which requires insemination for autogenous ovarian development, females that were mated or injected with a factor from the male accessory glands developed a batch of eggs in three days. Injection of the male factor into decapitated females did not stimulate vitellogenesis although injection of 5 μg 20-hydroxyecdysone did. However, when decapitation followed 12 h after the injection of male factor, vitellogenesis was not prevented. When the heads of intact or ovariectomized females that had been injected with male factor were assayed for OEH activity (by an assay that measured the rate of vitellogenin synthesis in pmol min^{-1}), rates of 9.0 pmol Vg min^{-1} $head^{-1}$ were found in heads from ovariectomized females and only 2.5 pmol Vg min^{-1} $head^{-1}$ in heads from intact females. Borovsky (1985b) concluded that although the male factor and 20-hydroxyecdysone both stimulated vitellogenesis, their roles were not the same. He postulated that after insemination the male factor stimulates the ovary to synthesize and secrete OEH-releasing factor, which acts on the release sites of the median neurosecretory cells. If that is correct, the male factor functions like the hormone from the midgut endocrine cells which, in anautogenous females, is postulated to stimulate the ovaries to secrete OEH-releasing factor (see Figure 21.4). It is not known how this step is controlled in females that are autogenous although virgin.

(b) *OEH release and action*

Secretory material accumulates in certain neurosecretory cells in the pars intercerebralis in young anautogenous females of *Aedes detritus*, but no such accumulation occurs in autogenous females of the same age undergoing the 1st ovarian cycle (Guilvard *et al.*, 1976).

There is clear experimental evidence for the involvement of OEH in autogenous ovarian development. Removal of the median neurosecretory cell perikarya (MNCp) from newly emerged females of *Ae taeniorhynchus* prevented maturation of the primary follicles, but the capability for autogenous development could be restored by

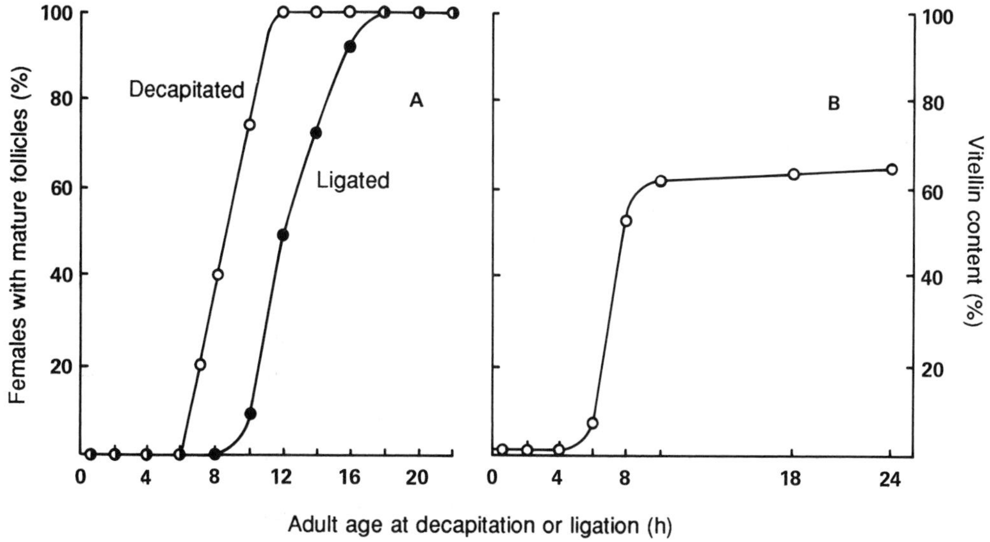

Figure 23.3 Effects of decapitation and ligation at the base of the abdomen, at different ages after emergence, upon vitellogenesis and ovarian development in *Aedes atropalpus*. (A) Percentage of females maturing oocytes. (From Kelly and Fuchs, 1980.) (B) Vitellin content of the ovaries of decapitated females as a percentage of the vitellin content of intact 72 h-old females. (From Masler *et al.*, 1980.)

Table 23.3 Induction of the capacity for ecdysteroid synthesis in *Aedes atropalpus* with exogenous hormone. (From Birnbaum *et al.*, 1984.)

Females were decapitated or ligated at the base of the abdomen within 2 h of emergence and were given either juvenile hormone I by topical application or *Ae aegypti* head extract by injection. Females receiving JH-I were treated within 3 h of emergence and dissected 24 h after emergence. Individuals receiving head extract were treated 24 h after emergence and dissected 24 h later. Operated control females were treated with acetone instead of juvenile hormone or saline instead of head extract. After dissection the ovaries were examined and then incubated for 5 h in either saline or saline containing *Ae aegypti* head extract (0.1 head equivalent μl^{-1}).

Operation	*Treated with*		*Ecdysone synthesis in*		*Primary follicles*
	*Head extract (hd eq.)**	*JH-1 (ng)*	*Saline (pg/ovary)*	*Head extract (pg/ovary)*	
Unoperated	–	–	160.0	–	Vitellogenic
Decapitated	–	–	3.6	–	Previtellogenic
Decapitated	–	–	–	24.6	Previtellogenic
Decapitated	1	–	211.6	–	Vitellogenic
Decapitated	–	500	128.0	–	Vitellogenic
Ligated	–	–	2.1	9.5	Previtellogenic
Ligated	1	–	2.4	–	Previtellogenic
Ligated	–	5	0.1	19.3	Previtellogenic
Ligated	–	500	21.5	17.2	Vitellogenic
Ligated	1	5	56.6	–	Vitellogenic

* hd eq. = head equivalents.

implanting four pairs of MNCp into the operated females (Table 23.2). Removal of the MNCp affected progressively fewer females as their age at removal increased. Removal of both the MNCp and the neurohaemal organ at three days post-emergence did not affect autogenous development (Table 23.2), although that operation blocked ovarian maturation when performed on 3-day-old anautogenous females before blood feeding (Table 21.3). These experiments of Lea (1967, 1970) indicated that OEH had been released in 3-day-old autogenous females of *Ae taeniorhynchus* but not in sugarfed anautogenous *taeniorhynchus* females of that age. That implantation of four pairs of MNCp induced oocyte maturation in 60% of non-bloodfed autogenous females (Table 23.2) provides additional evidence for the role of OEH.

Decapitation of *Ae atropalpus* up to 6 h post-emergence blocked vitellogenic development. As the time of decapitation was delayed between 7 and 10 h, increasing percentages of females developed their primary follicles to maturity (Figure 23.3A, B), revealing the critical period for release of a head factor. Ligation at the base of the abdomen had a similar effect, but the critical period showed a lag of 4 h (Figure 23.3A) (Kelly and Fuchs, 1980). The primary action of OEH is to stimulate ecdysone synthesis. Ovaries from females of *Ae atropalpus* that had been decapitated at emergence were incapable of ecdysone synthesis *in vitro*, but this capacity was restored by prior injection of *Ae aegypti* head extract, the treated females developing primary follicles to maturity (Tables 23.3, 23.4) (Birnbaum *et al.*, 1984; Fuchs *et al.*, 1980). It is interesting that ligated bloodfed *Ae aegypti* did not respond to head extract unless it was supplemented with methoprene (Table 21.7).

Decapitation at emergence had no effect on ovarian development in females from a northern population of *Wyeomyia smithii* which emerge with vitellogenic follicles, suggesting that OEH was released before emergence. Decapitation at emergence blocked vitellogenic development in females of *Wy smithii* from Florida unless they had been exposed to methoprene during the pupal stage (O'Meara and Lounibos, 1981).

It is possible that the endocrinological difference between autogenous and anautogenous females lies in regulation of the secretion of a single hormone. There is no evidence for any important difference during previtellogenic development, but there is a difference in OEH release. In anautogenous females OEH is not released until after a blood meal. In autogenous females OEH is released without the stimulus of a blood meal, although in some the presence of male factor or sugar feeding is required. It may be that the basic endocrinological difference between autogenous and anautogenous females lies in the regulation of OEH release.

23.2.3 20-Hydroxyecdysone and juvenile hormone titres

Changes in ecdysteroid and juvenile hormone titres with age were measured in autogenous females of two species of *Aedes* by Guilvard *et al.* (1984). The total ecdysteroid titre of *Ae*

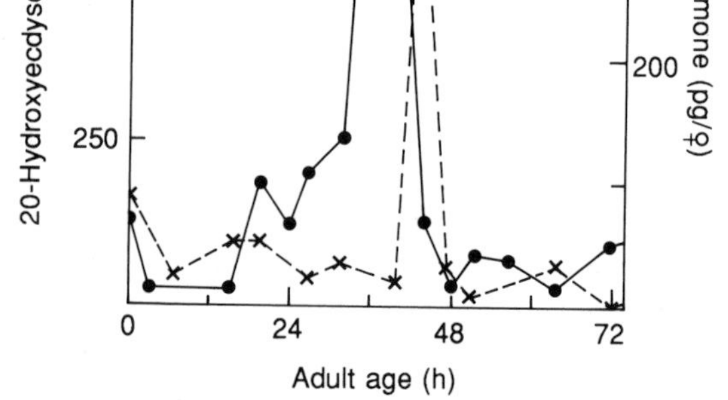

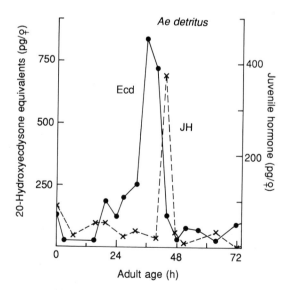

Figure 23.4 Ecdysteroid and juvenile hormone contents of autogenous females of *Aedes detritus* during the first 72 h after emergence. Total RIA-positive ecdysteroids are expressed as 20-hydroxyecdysone equivalents. (After Guilvard *et al.*, 1984.)

Table 23.4 Stimulation of ovarian development in *Aedes atropalpus* with exogenous hormone.

Autogenous females were decapitated or ligatured at the base of the abdomen at various times after emergence, and were subsequently treated with *Aedes aegypti* head extract, juvenile hormone I or 20-hydroxyecdysone. Ovarian development was assessed at 72 h post-emergence.

Age	Substances administered						Females with vitellogenic ovaries	Ref.
	Head extract		JH-I		20-OHecdysone			
(h)	Age (h)	Amount (hd eq.)*	Age (h)	Amount (ng)	Age (h)	Amount (ng)	(%)	
At decapitation								
<1	6	0	–	–	–	–	1	1
<1	6	0.5	–	–	–	–	100	1
1	–	–	3	5	–	–	8	3
1	–	–	3	50	–	–	45	3
1	–	–	3	500	–	–	100	3
<1	–	–	–	–	1	4.8	10	2
<1	–	–	–	–	18	0.48	19	2
<1	–	–	–	–	18	1.5	44	2
<1	–	–	–	–	18	4.8	100	2
At ligation								
<1	–	–	3	5	–	–	10	3
<1	–	–	3	50	–	–	45	3
<1	–	–	3	500	–	–	100	3
2	–	–	–	–	12	4.8	0	3
2	–	–	–	–	12	4.8×10^3	100	2
10	–	–	–	–	10	4.8	90	3
<1	–	–	3	0.05	12	4.8	18	3
<1	–	–	3	0.5	12	4.8	100	3

* hd eq. = head equivalents.
References: 1, Fuchs *et al.* (1980); 2. Kelly and Fuchs (1980); 3. Kelly *et al.* (1981).

detritus declined immediately after emergence, increased slightly and briefly at 20 h, peaked between 35 and 40 h, and declined to a low level by 44 h post-emergence. Subsequently it remained relatively low, with possible minor peaks, at least until after oviposition at 120 h post-emergence (Figure 23.4). The ecdysteroids were not identified further.

The juvenile hormone titre of *Ae detritus* also fell after emergence. It peaked sharply at 43 h post-emergence, some 5.5 h after the mid-point of the major ecdysteroid peak, and subsequently declined almost to zero (Figure 23.4). Thus the juvenile hormone titre rose as the ecdysteroid titre fell. At least 95% of the juvenile hormone titre was of JH-III. The changes of hormone titre with age in *Ae caspius* were similar to those in *Ae detritus*, although the ecdysteroids showed minor late peaks at 72 and 110 h post-emergence.

In these two species the primary follicles were in stage IIb prior to the major ecdysteroid peak. They entered stage IIIa at the time that the ecdysteroid titre started to rise, and reached stage IIIb immediately after the juvenile hormone peak, at which time vitellogenin was being actively incorporated into the oocytes.

Autogenous females of *Ae atropalpus* contained 150–200 pg ecdysteroid at emergence. The ecdysteroid titre increased slowly after 14 h post-emergence, increased considerably between 20 and 24 h, and reached a maximum of 325 pg at 32 h (at 27°C) (Figure 23.5). The ecdysteroid titre fell steadily between 32 and 48 h, stabilizing

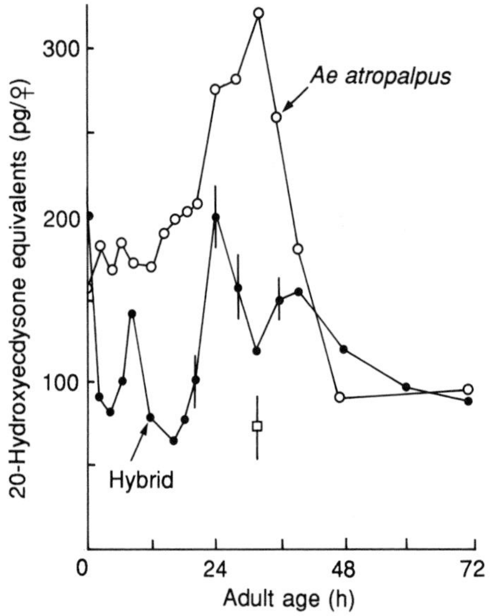

Figure 23.5 Ecdysteroid content of females of *Aedes atropalpus* and of F_1 hybrid females from an *Ae epactius* × *Ae atropalpus* cross during the first 72 h after emergence. Total ecdysteroids are expressed as 20-hydroxyecdysone equivalents. The isolated datum point at 32 h post-emergence is the mean (± range) for *Ae atropalpus* decapitated within 30 min of emergence. (After Masler *et al.*, 1981.)

at a relatively low level. The ovaries incorporated vitellogenin at a high rate from before 24 h until after 48 h post-emergence, while the 20-hydroxyecdysone titre was rising and falling (Masler *et al.*, 1980).

Females of *Ae atropalpus* extracted at 10 h and at 28–36 h post-emergence were reported by Masler *et al.* (1980) to contain 20-hydroxyecdysone but no ecdysone. Kelly *et al.* (1984) reported that at 24 h the females contained 335 pg ecdysone and 82 pg 20-hydroxyecdysone. F_1 hybrids from reciprocal crosses between the autogenous *Ae atropalpus* and anautogenous *Ae epactius* were always autogenous. From 2 h post-emergence the ecdysteroid titre of the hybrid female was much lower than that of *atropalpus*; the hybrids showed a distinct peak at 10 h and a more substantial peak at 24 h (Figure 23.5). The F_1 hybrid females matured only half as many oocytes as either the sugarfed *atropalpus* or the bloodfed *epactius* parents; Masler *et al.* (1981) associated this with the lower ecdysteroid titre.

23.2.4 20-Hydroxyecdysone action

Most recent studies have been on *Aedes atropalpus*, principally investigations of the effects of applications of 20-hydroxyecdysone to females decapitated or ligated at the base of the abdomen at different times after emergence (Fuchs *et al.*, 1980, 1981; Kelly and Fuchs, 1980; Kelly *et al.*, 1981; Masler *et al.*, 1980, 1981; Birnbaum *et al.*, 1984). The most important findings were the following.

(a) Ovaries taken from unoperated females were capable of ecdysteroid synthesis *in vitro* (Table 23.3), producing about half as much 20-hydroxyecdysone as ecdysone (Birnbaum *et al.*, 1984).
(b) Females decapitated 30 min after emergence failed to secrete 20-hydroxyecdysone (Figure 23.5, Table 23.3) and were incapable of vitellogenesis. Vitellogenesis could be prevented by decapitation up to *c.* 6 h post-emergence (Figure 23.3B).
(c) Females decapitated at emergence were unresponsive to injection of 4.8 ng 20-hydroxyecdysone injected at the same time, but if the time of hormone treatment was delayed it stimulated vitellogenesis, at 18 h the response being proportional to dose between 0.48 and 4.8 ng (Table 23.4).
(d) Females ligated at emergence did not respond to low doses of 20-hydroxyecdysone injected at 12 h, only to massive (4 μg) doses. However, as the age at ligation was delayed, an increasing percentage of females responded to 4.8 ng 20-hydroxyecdysone injected at the time of ligation. Thus ligation with simultaneous injection of 4.8 ng hormone at 10 h stimulated vitellogenesis in 90% of cases (Figure 23.6, Table 23.4).
(e) Identical time–response curves were obtained for the injection of 4.8 ng 20-hydroxyecdysone at two hour intervals into females decapitated at emergence and females ligated at the time of injection (Figure 23.6).

An injection of 4.8 ng 20-hydroxyecdysone delivers 15 times more hormone than the

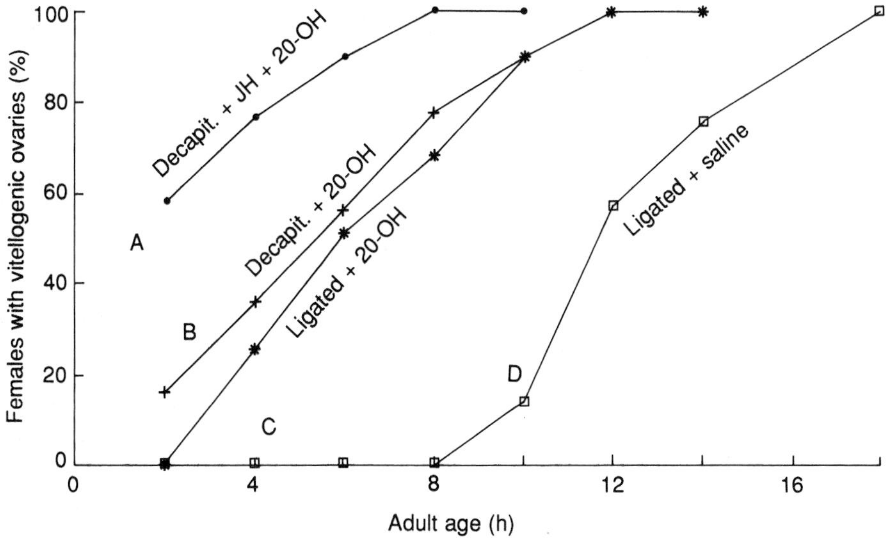

Figure 23.6 Effects of age and prior treatment with juvenile hormone upon the responsiveness of females of *Aedes atropalpus* to exogenous 20-hydroxyecdysone.

Curves A and B: Females were decapitated at emergence, half of them (A) being treated with 0.5 ng JH-I within 15 min. All were later injected with 4.8 ng 20-hydroxyecdysone at ages indicated by the data points. Their ovaries were examined at 72 h. Curves C and D: At ages indicated by the data points, females were ligatured at the base of the abdomen and immediately injected with 4.8 ng 20-hydroxyecdysone (C) or saline (D.). Their ovaries were examined at 48 h. (After Kelly *et al.*, 1981; Fuchs *et al.*, 1981.)

peak endogenous titre of 325 pg found at 32 h post-emergence. Kelly and Fuchs (1980) pointed out that the normal female is exposed to a raised hormone concentration for at least ten hours (Figure 23.5) and that the exogenous hormone is subject to enzymatic degradation, and concluded that the responses to a 4.8 ng dose could be considered physiologically significant.

Comparison of the time–response curves in Figures 23.3 and 23.6 suggests that the requirement for OEH can be satisfied, at least in part, by exogenous 20-hydroxyecdysone. If this is correct, it is likely that the 20-hydroxyecdysone stimulates the ovaries to secrete ecdysone. As in anautogenous mosquitoes the continued presence of ovaries is essential for the maintenance of vitellogenin synthesis. Ovariectomy at 13 h post-emergence, when the rate of vitellogenin synthesis was almost half-maximal, caused a decline in synthesis; in ovariectomized females it had almost ceased by 24 h, the time at which the rate peaked in control females (cf. Figure 20.6) (Borovsky, 1981a).

All females decapitated between 10 and 24 h post-emergence developed some primary follicles to maturity, but whatever the time of decapitation these females synthesized only 60% as much vitellogenin and matured only half as many oocytes as the normal intact females (Figure 23.3) (Masler *et al.*, 1980). Decapitated females receiving 4.8 ng 20-hydroxyecdysone 18 h after emergence produced mature oocytes of normal size, but they produced only half as many oocytes as did the intact controls, containing in total only half as much vitellin. When 4.8 μg 20-hydroxyecdysone was injected, the decapitated females produced the same number of vitellogenic follicles as the controls but with a 35% shorter follicle length and only 25% as much vitellin (Kelly and Fuchs, 1980).

23.2.5 Juvenile hormone action

Females of *Aedes atropalpus* that were decapitated or ligated at the base of the abdomen at emergence matured their ovaries if treated at 3 h post-emergence with some 50–500 ng juvenile hormone I. The dose–response curves were identical for decapitated and ligated females (Table 23.4). JH-III was ten times less active than JH-I on decapitated preparations, and 100 times less active when administered to isolated abdomens in conjunction with 4.8 ng 20-hydroxyecdysone (Kelly *et al.*, 1981).

Treatment of females decapitated or ligated at emergence with 500 ng juvenile hormone I at 3 h induced ecdysteroid synthesis by the ovaries and vitellogenic development (Table 23.3) (Birnbaum *et al.*, 1984), and it produced a peak of whole-body ecdysteroid content that corresponded with the 32 h peak in normal females (Fuchs *et al.*, 1981).

Treatment of females decapitated at emergence with the much lower dose of 0.5 ng JH-I advanced the development of their responsiveness to 20-hydroxyecdysone by 2–3 h (Figure 23.6) (Fuchs *et al.*, 1981). Females ligated soon after emergence were insensitive to a small (4.8 ng) dose of 20-hydroxyecdysone; however, if treated at 3 h with 0.5 ng JH-I they would respond fully to 4.8 ng 20-hydroxyecdysone injected at 12 h (Table 23.4) (Kelly *et al.*, 1981).

No consensus has been reached on the role of juvenile hormone in the later phases of autogenous ovarian development. Ablation experiments with *Ae taeniorhynchus* indicated that the corpora allata are required for previtellogenic development, but the observation that removal of the corpora allata 1–2 days after emergence did not block later ovarian development (Table 23.2) suggests that, as in anautogenous females, their presence may not be necessary for vitellogenesis.

Measurements of juvenile hormone titre are available only for *Ae detritus* (Figure 23.4) and *Ae caspius*. These show a juvenile hormone peak following the ecdysteroid peak, and the burst of juvenile hormone secretion appears as likely to affect the secondary as the primary follicles.

Treatment of decapitated or ligated females of *Ae atropalpus* with 500 ng JH-I stimulated ecdysone synthesis (Table 23.3), and that may account for the gonadotrophic effect of 50–500 ng doses of juvenile hormone (Table 23.4). It is also pertinent, however, that relatively small doses (0.5–5 ng) of juvenile hormone synergized head extract and 20-hydroxyecdysone (Tables 23.3, 23.4). Whether juvenile hormone has a role in normal development after the previtellogenic phase remains to be elucidated.

23.2.6 Oostasis

Extracts from ovaries of *Aedes atropalpus* or *Ae aegypti* that contained mature oocytes yielded a factor which, when injected into females of *Ae atropalpus*, prevented follicle development beyond stage IIb. This ovarian factor, presumably an oostatic hormone, acted on the primary follicles after the phase of JH-dependent previtellogenic growth. Oostatic hormone accumulated in the ovaries between 30 and 48 h post-emergence, which was a period of decline in the rate of ovarian ecdysteroid synthesis. The saline medium in which 'post-vitellogenic' ovaries of *Ae atropalpus* had been incubated contained a factor which inhibited ecdysteroid synthesis by vitellogenic ovaries *in vitro* (Kelly *et al.*, 1986a).

When housefly oostatic hormone was injected into females of *Ae atropalpus* shortly after emergence the primary follicles completed previtellogenic development but remained at stage II; at 24 h post-emergence ecdysteroids were undetectable. Injection of *Ae aegypti* head extract into decapitated females of *Ae atropalpus* led to ecdysteroid synthesis, but this was prevented when housefly oostatic hormone was injected with the head factor. Kelly *et al.* (1984) postulated that the housefly oostatic hormone inhibited a step subsequent to OEH release in *Ae atropalpus*, but they did not rule out the possibility that it also inhibits OEH release.

It appears that the regulation of ovarian development in autogenous mosquitoes involves secretion of an oostatic hormone by post-vitellogenic follicles, which directly or indirectly inhibits ecdysteroid synthesis by less advanced follicles.

23.3 THE GENETIC BASIS OF AUTOGENY

The genetic basis of autogeny has been investigated most intensively in *Culex pipiens*. In the first detailed study, with North American strains, Spielman (1957) showed that autogeny is controlled by two or more non-allelic genes.

In crosses between an autogenous strain of *Cx pipiens* and anautogenous strains of *Cx pipiens* and *Cx quinquefasciatus* he obtained a positive correlation between gene dosage and both the penetrance and expressivity of autogeny.

Subsequently, Aslamkhan and Laven (1970) used anautogenous *quinquefasciatus* from North America and an autogenous strain of *pipiens* from Germany which had dominant marker genes on both autosomes to investigate the inheritance of autogeny. The results of their extensive crosses and backcrosses may be summarized as follows: (a) approximately 75% of F_1 hybrid females were autogenous; (b) all backcrosses to the autogenous stock produced 100% autogenous progeny; (c) there were differences in percentage of autogenous female progeny between reciprocal backcrosses to the anautogenous stock, indicating involvement of a sex-linked gene on chromosome-1; (d) chromosome-2 had little influence on autogeny; and (e) there was a gene with a strong tendency to produce autogeny close to the *Kuf* marker on chromosome-3.

Aslamkhan and Laven (1970) framed the following working hypothesis. The inheritance of autogeny is bifactorial, involving loci on chromosomes 1 and 3. On chromosome-1 the locus has the alleles *D* (autogeny) and *d* (anautogeny), with *D* dominant. The locus on chromosome-3 has multiple alleles, designated a_1, a_2 and a_3, which can occur in six combinations. The genotype *d/d* on chromosome-1 is epistatic over the gene for autogeny on chromosome-3, therefore any female that is *d/d* will be anautogenous whatever the constitution of chromosome-3. Females that are a_2/a_3 or a_3/a_3 will be anautogenous whether *D/D* or *D/d*. Females that are *D/D* or *D/d* and a_1/a_1, a_1/a_2, a_1/a_3 or a_2/a_2 will be autogenous. The existence of some anomalous individuals in the backcrosses was explained by recombination between the marker and autogeny genes, but close linkage (or identity) between *Kuf* and the gene for autogeny on chromosome-3 also seems to be required to explain the experimental results. In summary, females of genotype *d/d* and/or a_2/a_3 or a_3/a_3 will be anautogenous, females of all other genotypes will be autogenous given adequate larval nutrition.

Early studies on autogeny in the *Aedes atropalpus* species group suggested that autogeny is controlled by a single dominant autosomal gene, A. The autogenous species, *Ae atropalpus*, was believed to be homozygous *A/A*, and the anautogenous species *Ae epactius* to be homozygous *a/a*. F_1 hybrid females were almost uniformly autogenous (O'Meara and Craig, 1969). However, certain crosses gave F_1 hybrids that developed only half as many eggs as the autogenous parent, and when the genome of *atropalpus* was progressively replaced by repeated crossing to *epactius*, autogenous fertility decreased in a stepped manner over several generations. It was postulated that the genome of the autogenous species contains modifier genes that enhance the expression of A, whereas the anautogenous species lacks some if not all of these enhancers (O'Meara, 1972).

The *Aedes scutellaris* complex includes species that are anautogenous and 'haematophagous' (biting readily during the 1st ovarian cycle), e.g. *Ae polynesiensis*, and others that are autogenous and that do not bite during the 1st ovarian cycle, e.g. *Ae kesseli* (Trpis, 1978; taxonomy Huang and Hitchcock, 1980). Crosses and backcrosses between these two species showed that both autogeny and haematophagy were controlled by more than one pair of alleles. The offspring that resulted from the crosses and backcrosses could be classified into four physiologically and behaviourally distinct categories: haematophagous anautogenous, haematophagous autogenous, non-haematophagous anautogenous, and non-haematophagous autogenous. Nearly 40% of the F_1 females were autogenous and haematophagous, but only a very small percentage were anautogenous and non-haematophagous. In reciprocal crosses the percentages of haematophagous and of autogenous offspring were both slightly higher when the respective alleles were from the female parents than when they were from the male parents. The inheritance of autogeny in species of the *Ae scutellaris* groups has also been investigated by Hoyer and Rozeboom (1976), and preliminary studies on the inheritance of autogeny have been reported for *Ae togoi* (Thomas and Leng, 1972) and *Ae detritus* (Rioux *et al.*, 1973).

References

Abbitt, B. and Abbitt, L.G. (1981) Fatal exsanguination of cattle attributed to an attack of salt marsh mosquitoes (*Aedes sollicitans*). *J. Am. Vet. Med. Assoc.*, **179**, 1397–1400.

Abdel-Malek, A. and Goulding, R.L. (1948) A study of the rate of growth of two sclerotized regions within larvae of four species of mosquitoes. *Ohio J. Sci.*, **48**, 119–28.

Achundow, I. (1928) Die Modifikation der Anophelen unter den äusseren Bedingungen und kritische Betrachtungen der Rassenfrage. *Arch. Schiffs-u. Tropenhyg.*, **32**, 546–61.

Adak, T., Subbarao, S.K., Sharma, V.P. and Rao, S.R.V. (1988) X-linkage of malic enzyme in *Anopheles culicifacies* species B. *J. Hered.*, **79**, 37–9.

Adam, J.P. (1962) Un anophèle cavernicole nouveau de la République du Congo (Brazzaville) : *Anopheles (Neomyzomyia) hamoni* n.sp. (Diptera–Culicidae). *Bull. Soc. Pathol. Exot.*, **55**, 153–64.

Adam, J.P. and Vattier, G. (1964) Contribution a l'étude biologique d'*Anopheles hamoni* Adam, 1962 (Diptera–Culicidae). *Cahier ORSTOM*, sér. *Entomol. Med.*, (2), 55–71.

Ailus, K., Palosuo, T. and Brummer-Korvenkontio, M. (1985) Demonstration of antibodies to mosquito antigens in man by immunodiffusion and ELISA. *Int. Arch. Allergy Appl. Immunol.*, **78**, 375–9.

Akov, S. (1962a) A qualitative and quantitative study of the nutritional requirements of *Aedes aegypti* L. larvae. *J. Insect. Physiol.*, **8**, 319–35.

Akov, S. (1962b) Antimetabolites in the nutrition of *Aedes aegypti* L. larvae. The substitution of choline by related substances and the effect of choline inhibitors. *J. Insect Physiol.*, **8**, 337–48.

Akov, S. (1972) Protein digestion in hematophagous insects. In *Insect and Mite Nutrition* (ed. J.G. Rodriguez), North-Holland, Amsterdam, pp. 531–40.

Akov, S. and Guggenheim, K. (1963) Antimetabolites in the nutrition of *Aedes aegypti* L. larvae. Nicotinic acid analogues. *Biochem. J.*, **88**, 182–7.

Akstein, E. (1962) The chromosomes of *Aedes aegypti*, and of some other species of mosquitoes. *Bull. Res. Council Israel* B, **11**, 146–55.

Al-Ahdal, M. N., Al-Hussain, K., Thorogood, R. J. *et al.* (1990) Protein constituents of mosquito saliva: studies on *Culex molestus*. *J. Trop. Med. Hyg.*, **93**, 98–105.

Alberts, B., Bray, D., Lewis, J. *et al.* (1989) *Molecular Biology of the Cell*, 2nd edn, Garland Publishing, New York and London.

Allen, J.R. (1966) Passive transfer between experimental animals of hypersensitivity to *Aedes aegypti* bites. *Exp. Parasitol.*, **19**, 132–7.

Allen, J.R. and West, A.S. (1962) Collection of oral secretion from mosquitoes. *Mosq. News*, **22**, 157–9.

Allen, J.R. and West, A.S. (1966) Some properties of oral secretion from *Aedes aegypti*. *Exp. Parasitol.*, **19**, 124–31.

Aly, C. (1983) Feeding behaviour of *Aedes vexans* larvae (Diptera: Culicidae) and its influence on the effectiveness of *Bacillus thuringiensis* var. *israelensis*. *Bull. Soc. Vector Ecol.*, **8**, 94–100.

Aly, C. (1985a) Feeding rate of larval *Aedes vexans* stimulated by food substances. *J. Am. Mosq. Control. Ass.*, **1**, 506–10.

Aly, C. (1985b) Germination of *Bacillus thuringiensis* var. *israelensis* spores in the gut of *Aedes* larvae (Diptera: Culicidae). *J. Invert. Pathol.*, **45**, 1–8.

Aly, C. (1988) Filtration rates of mosquito larvae in suspensions of latex microspheres and yeast cells. *Entomol. Exp. Applic.*, **46**, 55–61.

Aly, C. and Dadd, R.H. (1989) Drinking rate regulation in some fresh-water mosquito larvae. *Physiol. Entomol.*, **14**, 241–56.

Aly, C. and Mulla, M.S. (1986) Orientation and ingestion rates of larval *Anopheles albimanus* in response to floating particles. *Entomol. Exp. Applic.*, **42**, 83–90.

Ameen, M. and Iversen, T.M., (1978) Food of *Aedes* larvae (Diptera, Culicidae) in a temporary forest pool. *Arch. Hydrobiol.*, **83**, 552–64.

Amouriq, L. (1960) Formules hémocytaires de la larve, de la nymphe et de l'adulte de *Culex hortensis* (Dipt. Culicidae). *Bull. Soc. Entomol. France*, **65**, 135–9.

Andersen, S.O. (1989) Enzymatic activities involved in incorporation of N-acetyldopamine into insect cuticle during sclerotization. *Insect Biochem.*, **19**, 375–82.

Anderson, D.T. (1972) The development of holometabolous insects. In *Developmental Systems : Insects* (eds S. J. Counce and C. H. Waddington), Academic Press, London, **1**, 165–242.

Anderson, J.F. (1967) Histopathology of intersexuality in mosquitoes. *J. Exp. Zool.*, **165**, 475–95.

Anderson, J.F. and Horsfall, W.R. (1963) Thermal stress and anomalous development of mosquitoes (Diptera: Culicidae). I. Effect of constant temperature on dimorphism of adults of *Aedes stimulans*. *J. Exp. Zool.*, **154**, 67–107.

Anderson, J.F. and Horsfall, W.R. (1965a) Dimorphic development of transplanted juvenile gonads of mosquitoes. *Science*, **147**, 624–5.

Anderson, J.F. and Horsfall, W.R. (1965b) Thermal stress and anomalous development of mosquitoes (Diptera: Culicidae). V. Effect of temperature on embryogeny of *Aedes stimulans*. *J. Exp. Zool.*, **158**, 211–21.

Anderson, W.A. and Spielman, A. (1971) Permeability of the ovarian follicles of *Aedes aegypti* mosquitoes. *J. Cell Biol.*, **50**, 201–21.

Anderson, W.A. and Spielman, A. (1973) Incorporation of RNA and protein precursors by ovarian follicles of *Aedes aegypti* mosquitoes. *J. Submicroscop. Cytol.*, **5**, 181–98.

Andersson, I.H. (1992) The effect of sugar meals and body size on fecundity and longevity of female *Aedes communis* (Diptera: Culicidae). *Physiol. Entomol.*, **17**, 203–7.

Andreadis, T.G. and Hall, D.W. (1976) *Neoaplectana carpocapsae*: encapsulation in *Aedes aegypti* and changes in host hemocytes and hemolymph proteins. *Exp. Parasitol.*, **39**, 252–61.

Aneshansley, D.J., Marler, C.E. and Beyenbach, K.W. (1988) Transepithelial voltage measurements in isolated Malpighian tubules of *Aedes aegypti*. *J. Insect Physiol.*, **35**, 41–52.

Applebaum, S.W. (1985) Biochemistry of digestion. In *Comprehensive Insect Physiology Biochemistry and Pharmacology* (eds G.A. Kerkkut and L.I. Gilbert), Pergamon Press, Oxford, **4**, 279–311.

Arnal, A. (1950) El aparato digestivo y la digestion en los culicidos hematofagos. *Trabajos Inst. Cien Nat. Jose de Acosta' (S. Biol.)*, **2**, 303–63, 20 pls.

Asakura, K. (1978) Phosphatase activity in the larva of the euryhaline mosquito, *Aedes togoi* Theobald, with special reference to sea-water adaptation. *J. Exp. Marine Biol. Ecol.*, **31**, 325–37.

Asakura, K. (1980) The anal portion as a salt-excreting organ in a seawater mosquito larva, *Aedes togoi* Theobald. *J. Comp. Physiol.*, **138**, 59–65.

Asakura, K. (1982a) A possible role of the gastric caecum in osmoregulation of the seawater mosquito larva, *Aedes togoi* Theobald. *Annotationes Zoologicae Japonenses*, **55**, 1–8.

Asakura, K. (1982b) Ultrastructure and chloride cytochemistry of the hindgut epithelium in the larvae of the seawater mosquito, *Aedes togoi* Theobald. *Archivum Histologicum Japonicum*, **45**, 167–80.

Asakura, M. (1970) Studies on the structural basis for ion- and water-transport by the osmoregulatory organs of mosquitoes. (I). Fine structure of the rectal epithelium of *Aedes albopictus* larva. *Sci. Rep. Kanazawa Univ.*, **15**, 37–55.

Ashburner, M. (1992) Mapping insect genomes. In *Insect Molecular Science* (eds J.M. Crampton and P. Eggleston), Academic Press, London.

Ashida, M., Kinoshita, K. and Brey, P.T. (1990) Studies on prophenoloxidase activation in the mosquito *Aedes aegypti* L. *Eur. J. Biochem.*, **188**, 507–15.

Aslam Khan, M. (1973) Sex-chromosomes and sex-determination in the malaria mosquito, *Anopheles stephensi*. *Pakistan J. Zool.*, **5**, 127–30.

Aslamkhan, M. and Laven, H. (1970) Inheritance of autogeny in the *Culex pipiens* complex. *Pakistan J. Zool.*, **2**, 121–47.

Asman, S.M. (1974) Cytogenetic observations in *Culex tarsalis*: mitosis and meiosis. *J. Med. Entomol.*, **11**, 375–82.

Aspöck, H. (1966) Parasitierung eines im Freiland aufgefundenen Intersexes von *Aedes (Ochlerotatus) communis* De Geer (Insecta, Culicidae) durch einen Pilz der Ordnung Blastocladiales. *Z. Morph. Ökol. Tiere*, **57**, 231–43.

Atlas, S.J., Roth, T.F. and Falcone, A.J. (1978) Purification and partial characterization of *Culex pipiens fatigans* yolk protein. *Insect Biochem.*, **8**, 111–15.

Auclair, J.L. (1963) Aphid feeding and nutrition. *Annual Rev. Entomol.*, **8**, 439–90.

Bacon, J.S.D. and Dickinson, B. (1957) The origin of melezitose: a biochemical relationship between the lime tree (*Tilia* spp.) and an aphis (*Eucallipterus tiliae* L.). *Biochem. J.*, **66**, 289–97.

Baerg, D.C. and Boreham, M.M. (1974) Experimental rearing of *Chagasia bathana* (Dyar) using induced mating, and description of the egg stage (Diptera: Culicidae). *J. Med. Entomol.*, **11**, 631–32.

Baimai, V. (1988) Constitutive heterochromatin differentiation and evolutionary divergence of karyotype in oriental *Anopheles* (*Cellia*). *Pacific Sci.*, **42**, 13–27.

Baker, F.C., Hagedorn, H.H., Schooley, D.A. and Wheelock, G. (1983) Mosquito juvenile hormone: identification and bioassay activity. *J. Insect Physiol.*, **29**, 465–70.

Baker, H.G. and Baker, I. (1983a) Floral nectar sugar constituents in relation to pollinator types. In *Handbook of Experimental Pollination Biology* (eds C.E. Jones and R. J. Little), Van Nostrand Reinhold Company Inc., New York, pp. 117–41.

Baker, H.G. and Baker, I. (1983b) A brief historical review of the chemistry of floral nectar. In *The Biology of Nectaries* (eds B. Bentley and T. Elias), Columbia University Press, New York, pp. 126–52.

Baker, R.H. (1968) The genetics of 'golden', a new sex-linked colour mutant of the mosquito *Culex tritaeniorhynchus* Giles. *Ann. Trop. Med. Parasitol.*, **62**, 193–9.

Baker, R.H., Saifuddin, U.T. and Sakai, R.K. (1977) Variations in the linkage of the sex allele in laboratory colonies of the mosquito *Culex tritaeniorhynchus*. *Japan. J. Genet.*, **52**, 425–30.

Baker, R.H., Sakai, R.K. and Mian, A. (1971) Linkage group-chromosome correlation in *Culex tritaeniorhynchus*. *Science*, **171**, 585–7.

Baker, R.H. and Sakai, R.K. (1973) Genetic studies of two new mutants in linkage group III of the mosquito *Culex tritaeniorhynchus*. *Ann. Trop. Med. Parasitol.*, **67**, 467–73.

Baker, R.H. and Sakai, R.K. (1976) Male determining factor on chromosome 3 in the mosquito, *Culex tritaeniorhynchus*. *J. Hered.*, **67**, 289–94.

Baker, R.H. and Sakai, R.K. (1979) Triploids and male determination in the mosquito, *Anopheles culicifacies*. *J. Hered.*, **70**, 345–6.

Baldridge, G.D. and Fallon, A.M. (1991) Nucleotide sequence of a mosquito 18S ribosomal RNA gene. *Biochim. Biophys. Acta*, **1089**, 396–400.

Baldridge, G.D. and Feyereisen, R. (1986) Ecdysteroid titer and oocyte growth in the northern house mosquito, *Culex pipiens* L. *Comp. Biochem. Physiol.*, **83A**, 325–9.

Bandmann, H.-J. and Bosse, K. (1967) Histologie des Mückenstiches (*Aedes aegypti*). *Arch. Klin. Exp. Dermatol.*, **231**, 59–67.

Barbosa, P. and Peters, T.M. (1969) A comparative study of egg hatching techniques for *Aedes aegypti* (L). *Mosq. News*, **29**, 548–51.

Barillas-Mury. C., Graf, R., Hagedorn, H.H. and Wells, M.A. (1991) cDNA and deduced amino acid sequence of a blood meal-induced trypsin from the mosquito, *Aedes aegypti*. *Insect Biochem.*, **21**, 825–31.

Barkai, A.I. and Williams, R.W. (1983) The exchange of calcium in larvae of the mosquito *Aedes aegypti*. *J. Exp. Biol.*, **104**, 139–48.

Barnard, D.R. and Mulla, M.S. (1977) Effects of photoperiod and temperature on blood feeding, oogenesis and fat body development in the mosquito, *Culex inornata*. *J. Insect Physiol.*, **23**, 1261–6.

Barnard, D.R. and Mulla, M.S. (1978) Seasonal variation of lipid content in the mosquito *Culiseta inornata*. *Ann. Entomol. Soc. Am.*, **71**, 637–9.

Barr, A.R. (1964) Notes on the colonization and biology of *Armigeres subalbatus* (Diptera, Culicidae). *Ann. Trop. Med. Parasitol.*, **58**, 171–9.

Barr, A.R. (1970) Partial compatibility and its effect on eradication by the incompatible male method. *Proc. Pap. Calif. Mosq. Control Ass.*, **37**, 19–24.

Barr, A.R. (1975) A new intersex in *Culex pipiens* L. (Diptera: Culicidae). I. Description. *J. Med. Entomol.*, **12**, 562–7.

Barr, A.R. (1980) Cytoplasmic incompatibility in natural populations of a mosquito, *Culex pipien* L. *Nature*, **283**, 71–2.

Barr, A.R. (1982) Symbiont control of reproduction in *Culex pipiens*. In *Recent Developments in the Genetics of Insect Disease Vectors* (eds W.W.M. Steiner, W.J. Tabachnick, K.S. Rai and S. Narang), Stipes Publishing Co., Champaign, Illinois, pp. 153–8.

Barr, A.R., Fujioka, K.K., Wilmot, T.R. and Cope, S.E. (1986) Seasonal variation in number of eggs laid by *Culiseta incidens* (Diptera: Culicidae). *J. Med. Entomol.*, **23**, 178–81.

Barrow, P.M., McIver, S.B. and Wright, K.A. (1975) Salivary glands of female *Culex pipiens*: morphological changes associated with maturation and blood-feeding. *Can. Entomol.*, **107**, 1153–60.

Bar-Zeev, M. (1958) The effect of temperature on the growth rate and survival of the immature stages of *Aedes aegypti* (L.). *Bull. Entomol. Res.*, **49**, 157–63.

Batchelor, G.K. (1967) *An Introduction to Fluid Dynamics*, Cambridge University Press.

Bates, M. (1939) The use of salt solutions for the demonstration of physiological differences between the larvae of certain European anopheline mosquitoes. *Am. J. Trop. Med.*, **19**, 357–84.

Batzer, D.P. and Sjogren, R.D. (1986) Larval habitat characteristics of *Coquillettidia perturbans* (Diptera: Culicidae) in Minnesota. *Can. Entomol.*, **118**, 1193–8.

Baudoin, R. (1976) Les insectes vivant à la surface et au sein des eaux. In *Traité de Zoologie: Anatomie, Systématique, Biologie* (ed. P.-P. Grassé), Masson, Paris, **8** (Fascicle IV), 843-926.

Bauer, H. (1933) Die wachsenden oocytenkerne einiger Insekten in ihrem Verhalten zur Nuklealfarbung. *Z. Zellforsch.*, **18**, 254–98.

Bauer, P., Rudin, W. and Hecker, H. (1977) Ultrastructural changes in midgut cells of female *Aedes aegypti* L. (Insecta, Diptera) after starvation or sugar diet. *Cell Tissue Res.*, **177**, 215–9.

Beadle, L.C. (1939) Regulation of the haemolymph in the saline water mosquito *Aedes detritus* Edw. *J. Exp. Biol.*, **16**, 346–62.

Beament, J. (1989) Eggs – the neglected insects. *J. R. Army Med. Corps*, **135**, 49–56.

Beament, J. and Corbet, S.A. (1981) Surface properties of *Culex pipiens pipiens* eggs and the behaviour of the female during egg-raft assembly. *Physiol. Entomol.*, **6**, 135–48.

Beckel, W.E. (1958) Investigations of permeability, diapause, and hatching in the eggs of the mosquito *Aedes hexodontus* Dyar. *Can. J. Zool.*, **36**, 541–54.

Beckemeyer, E.F. and Lea, A.O. (1980) Induction of follicle separation in the mosquito by physiological amounts of ecdysterone. *Science*, **209**, 819–21.

Beckett, E.B. (1988) Development and ageing of the salivary glands of adult male *Aedes aegypti* (L.) and *Aedes togoi* (Theobald) mosquitoes (Diptera : Culicidae). *Int. J. Insect Morphol. Embryol*, **17**, 327–33.

Beckett, E.B., Boothroyd, B. and Macdonald, W.W. (1978) A light and electron microscope study of rickettsia-like organisms in the ovaries of mosquitoes of the *Aedes scutellaris* group. *Ann. Trop. Med. Parasitol.*, **72**, 277–83.

Beerntsen, B.T. and Christensen, B.M. (1990) *Dirofilaria immitis*: effect on hemolymph polypeptide synthesis in *Aedes aegypti* during melanotic encapsulation reactions against microfilariae. *Exp. Parasitol.*, **71**, 406–14.

Behan, M. and Hagedorn, H.H. (1978) Ultrastructural changes in the fat body of adult female *Aedes aegypti* in relationship to vitellogenin synthesis. *Cell Tissue Res.*, **18**, 6 499–506.

Beier, J.C., Copeland, R., Oyaro, C. *et al.* (1990) *Anopheles gambiae* complex egg-stage survival in dry soil from larval development sites in western Kenya. *J. Am. Mosq. Control Assoc.*, **6**, 105–9.

Beier, J.C., Perkins, P.V., Wirtz, R.A. *et al.* (1988) Bloodmeal identification by direct enzyme-linked immunosorbent assay (ELISA) tested on *Anopheles* (Diptera: Culicidae) in Kenya. *J. Med. Entomol.*, **25**, 9–16.

Beklemishev, W.N. (1940) Gonotrophic rhythm as a basic principle of the biology of *Anopheles*. *Voprosy Fiziologii i Ekologii Malyariinogo Komara*, **1**, 3–22. (Cited by Detinova, 1962.)

Bekman, A.M. (1935) Interruptions dans la nutrition des larves d'*Anopheles maculipennis*, occasionnées par les mues. *Med. Parasitol., Moscow*, **4**, 389–91.

Belkin, J.N. (1962) *The Mosquitoes of the South Pacific (Diptera, Culicidae)*. Vol. 1, University of California Press, Berkeley and Los Angeles.

Bellamy, R.E. and Bracken, G.K. (1971) Quantitative aspects of ovarian development in mosquitoes. *Can. Entomol.*, **103**, 763–73.

Bellamy, R.E. and Corbet, P.S. (1973) Combined autogenous and anautogenous ovarian development in individual *Culex tarsalis* Coq. (Dipt., Culicidae). *Bull. Entomol. Res.*, **63**, 335–46, 1 pl.

Berger, C.A. (1938a) Cytology of metamorphosis in the Culicinae. *Nature*, **141**, 834–5.

Berger, C.A. (1938b) Multiplication and reduction of somatic chromosome groups as a regular developmental process in the mosquito, *Culex pipiens*. *Contr. Embryol. Carnegie Instn*, **27**, 209–32, 1 pl.

Berner, R., Rudin, W. and Hecker, H. (1983) Peritrophic membranes and protease activity in the midgut of the malaria mosquito, *Anopheles stephensi* (Liston) (Insecta: Diptera) under normal and experimental conditions. *J. Ultrastruct. Res.*, **83**, 195–204.

Berridge, M.J. and Oschman, J.L. (1972) *Transporting Epithelia*, Academic Press, New York and London.

Berry, S. (1985) Insect nucleic acids. In *Comprehensive Insect Physiology Biochemistry and Pharmacology* (eds G.A. Kerkut and L.I. Gilbert), Pergamon Press, Oxford, **10**, 219–53.

Berry, W.O. and Brammer, J.D. (1975) The effect of temperature on oxygen consumption in *Aedes aegypti* pupae. *Ann. Entomol. Soc. Am.*, **68**, 298–300.

Bertram, D.S. and Bird, R.G. (1961) Studies on mosquito-borne viruses in their vectors. I. The normal fine structure of the midgut epithelium of the adult female *Aedes aegypti* (L.) and the functional significance of its modification following a blood meal. *Trans. R. Soc. Trop. Med. Hyg.*, **55**, 404–23, 12 pls.

Besansky, N.J. (1990a) A retrotransposable element from the mosquito *Anopheles gambiae*. *Mol. Cell. Biol.*, **10**, 863–71.

Besansky, N.J. (1990b) Evolution of the T1 retrotransposon family in the *Anopheles gambiae* complex. *Mol. Biol. Evol.*, **7**, 229–46.

Beyenbach, K. and Pannabecker, T. (1991) Evidence of a H^+ATPase in mosquito Malpighian tubules. *FASEB J.*, **5**, A750.

Bhalla, S.C. (1971) A crossover suppressor-enhancer system in the mosquito *Aedes aegypti*. *Can. J. Genet. Cytol.*, **13**, 561–77.

Bhalla, S.C., Santos, J.M. and Barnett, M.F. (1975) Genetics of eye-gap and maroon-eye mutants in the mosquito. *J. Hered.*, **66**, 349–52.

Bhalla, S.C., Cajaiba, A.C.I., Carvalho, W.M.P. and Santos, J.M. (1974) Translocations, inversions, and correlation of linkage groups in the mosquito *Culex pipiens fatigans*. *Can. J. Genet. Cytol.*, **16**, 837–50.

Bhat, U.K.M. and Singh, K.R.P. (1975) The haemocytes of the mosquito *Aedes albopictus* and their comparison with larval cells cultured *in vitro*. *Experientia*, **31**, 1331–2.

Billingsley, P.F. (1990) Blood digestion in the mosquito, *Anopheles stephensi* Liston (Diptera: Culicidae): partial characterization and post-feeding activity of midgut aminopeptidases. *Arch. Insect Biochem. Physiol.*, **15**, 149–63.

Billingsley, P.F. and Hecker, H. (1991) Blood digestion in the mosquito, *Anopheles stephensi* Liston (Diptera: Culicidae): activity and distribution of trypsin, aminopeptidase and α-glucosidase in the midgut. *J. Med. Entomol.*, **28**, 865–71.

Billingsley, P.F. and Rudin, W. (1992) The role of the mosquito peritrophic membrane in bloodmeal digestion and infectivity of *Plasmodium* species. *J. Parasitol.*, **78**, 430–40.

Birnbaum, M.J., Kelly, T.J., Woods, C.W. and Imberski, R.B. (1984) Hormonal regulation of ovarian ecdysteroid production in the autogenous mosquito, *Aedes atropalpus*. *Gen. Comp. Endocrinol.*, **56**, 9–18.

Birtwisle, D. (1971) The concentration of some intermediates of glycolysis in the thorax of the mosquito, *Aedes aegypti*, at rest. *Insect Biochem.*, **1**, 293–8.

Bishop, A. and Gilchrist, B.M. (1946) Experiments upon

the feeding of *Aedes aegypti* through animal membranes with a view to applying this method to the chemotherapy of malaria. *Parasitology*, **37**, 85–100.

Bishop, A. and McConnachie, E.W. (1956) A study of the factors affecting the emergence of the gametocytes of *Plasmodium gallinaceum* from the erythrocytes and the exflagellation of the male gametocytes. *Parasitology*, **46**, 192–215.

Bishopp, F.C. (1933) Mosquitoes kill livestock. *Science*, **77**, 115–16.

Black, W.C., McLain, D.K. and Rai, K.S. (1989) Patterns of variation in the rDNA cistron within and among world populations of a mosquito, *Aedes albopictus* (Skuse). *Genetics*, **121**, 539–50.

Black, W.C. and Rai, K.S. (1988) Genome evolution in mosquitoes: intraspecific and interspecific variation in repetitive DNA amounts and organization. *Genet. Res.*, **51**, 185–96.

Blackmore, M.S. (1989) The efficacy of melanotic encapsulation as a defense against parasite-induced mortality in snowpool mosquitoes (Diptera: Culicidae) infected by mermithid nematodes. *Can. J. Zool.*, **67**, 1725–9.

Blackmore, M.S. and Nielsen, L.T. (1990) Observations on the biology of *Romanomermis* sp. (Nematoda: Mermithidae) parasites of *Aedes* in western Wyoming. *J. Am. Mosq. Control Ass.*, **6**, 229–34.

Bodenstein, D. (1945) A study of the relationship between organ and organic environment in the postembryonic development of the yellow fever mosquito *Aedes aegypti*. *Bull. Connecticut Agric. Exp. Sta.*, **501**, 100–14.

Bohm, M.K., Behan, M. and Hagedorn, H.H. (1978) Termination of vitellogenin synthesis by mosquito fat body, a programmed response to ecdysterone. *Physiol. Entomol.*, **3**, 17–25.

Bonaccorsi, S., Santini, G., Gatti, M. *et al.* (1980) Intraspecific polymorphism of sex chromosome heterochromatin in two species of the *Anopheles gambiae* complex. *Chromosoma*, **76**, 57–64.

Bonne-Webster, J. (1932) A mosquito with larval habits like *Taeniorhynchus*. *Bull. Entomol. Res.*, **23**, 69–72.

Boo, K.S. and Richards, A.G. (1975) Fine structure of scolopidia in Johnston's organ of female *Aedes aegypti* compared with that of the male. *J. Insect Physiol.*, **21**, 1129–39.

Boorman, J. (1987) Induction of salivation in biting midges and mosquitoes, and demonstration of virus in the saliva of infected insects. *Med. Vet. Entomol.*, **1**, 211–14.

Boorman, J.P.T. (1960) Observations on the feeding habits of the mosquito *Aedes (Stegomyia) aegypti* (Linnaeus): the loss of fluid after a blood-meal and the amount of blood taken during feeding. *Ann. Trop. Med. Parasitol.*, **54**, 8–14.

Boreham, M.M. (1970) *Mansonia leberi*, a new mosquito from the Panama Canal zone, with notes on its biology (Diptera: Culicidae). *J. Med. Entomol.*, **7**, 383–90.

Boreham, M.M. and Baerg, D.C. (1974) Description of the larva, pupa and egg of *Anopheles (Lophopodomyia) squamifemur* Antunes with notes on development (Diptera: Culicidae). *J. Med. Entomol.*, **11**, 564–9.

Borovsky, D. (1981a) Feedback regulation of vitellogenin synthesis in *Aedes aegypti* and *Aedes atropalpus*. *Insect Biochem.*, **11**, 207–13.

Borovsky, D. (1981b) *In vivo* stimulation of vitellogenesis in *Aedes aegypti* with juvenile hormone, juvenile hormone analogue (ZR 515) and 20-hydroxyecdysone. *J. Insect Physiol.*, **27**, 371–78.

Borovsky, D. (1982) Release of egg development neurosecretory hormone in *Aedes aegypti* and *Aedes taeniorhynchus* induced by an ovarian factor. *J. Insect Physiol.*, **28**, 311–16.

Borovsky, D. (1984) Control mechanisms for vitellogenin synthesis in mosquitoes. *BioEssays*, **1**, 264–7.

Borovsky, D. (1985a) Isolation and characterization of highly purified mosquito oostatic hormone. *Arch. Insect Biochem. Physiol.*, **2**, 333–49.

Borovsky, D. (1985b) The role of the male accessory gland fluid in stimulating vitellogenesis in *Aedes taeniorhynchus*. *Arch. Insect Biochem. Physiol.*, **2**, 405–13.

Borovsky, D. (1986) Proteolytic enzymes and blood digestion in the mosquito, *Culex nigripalpus*. *Arch. Insect Biochem. Physiol.*, **3**, 147–60.

Borovsky, D. (1988) Oostatic hormone inhibits biosynthesis of midgut proteolytic enzymes and egg development in mosquitoes. *Arch. Insect Biochem. Physiol.*, **7**, 187–210.

Borovsky, D., Carlson, D.A., Griffin, P.R. *et al.* (1990) Mosquito oostatic factor: a novel decapeptide modulating trypsin-like enzyme biosynthesis in the midgut. *J. Fed. Am. Soc. Exp. Biol.*, **4**, 3015–20.

Borovsky, D., Carlson, D.A. and Hunt, D.F. (1991) Mosquito oostatic hormone: a trypsin modulating oostatic factor. In *Insect Neuropeptides: Chemistry, Biology, and Action* (eds J.J. Menn, T.J. Kelly and E.P. Masler), *Am. Chem. Soc. Symp. Ser.*, **453**, 133–42.

Borovsky, D. and Schlein, Y. (1988) Quantitative determination of trypsinlike and chymotrypsinlike enzymes in insects. *Arch. Insect Biochem. Physiol.*, **8**, 249–60.

Borovsky, D. and Thomas, B.R. (1985) Purification and partial characterization of mosquito egg development neurosecretory hormone: evidence for gonadotropic and steroidogenic effects. *Arch. Insect Biochem. Physiol.*, **2**, 265–81.

Borovsky, D., Thomas, B.R., Carlson, D.A. *et al.* (1985) Juvenile hormone and 20-hydroxyecdysone as primary and secondary stimuli of vitellogenesis in *Aedes aegypti*. *Arch. Insect Biochem. Physiol.*, **2**, 75–90.

Borovsky, D., Whisenton, L.R., Thomas, B.R. and Fuchs, M.S. (1986) Biosynthesis and distribution of ecdysone and 20-OH-ecdysone in *Aedes aegypti*. *Arch. Insect Biochem. Physiol.*, **3**, 19–30.

Borovsky, D. and Van Handel, E. (1979) Does ovarian ecdysone stimulate mosquitoes to synthesize vitellogenin? *J. Insect Physiol.*, **25**, 861–865.

Borovsky, D. and Van Handel, E. (1980) Synthesis of ovary-specific proteins in mosquitoes. *Int. J. Invert. Reprod.*, **2**, 153–63.

Borovsky, D. and Whitney, P.L. (1987) Biosynthesis, purification and characterization of *Aedes aegypti* vitellin and vitellogenin. *Arch. Insect Biochem. Physiol.*, **4**, 81–100.

Bose, S.C. and Raikhel, A.S. (1988) Mosquito vitellogenin subunits originate from a common precursor. *Biochem. Biophys. Res. Comm.*, **155**, 436–42.

Boswell, R. E. and Mahowald, A. P. (1985) Cytoplasmic determinants in embryogenesis. In *Comprehensive Insect Physiology Biochemistry and Pharmacology* (eds G.A. Kerkut and L.I. Gilbert), Pergamon Press, Oxford, **1**, 387–405.

Bounias, M., Vivares. C.P. and Nizeyimana, B. (1989) Functional relationships between free amino acids in the hemolymph of fourth instar larvae of the mosquito *Aedes aegypti* (Diptera, Culicidae) as a basis for toxicological studies. *J. Invert. Pathol.*, **54**, 16–22.

Bourassa, J.-P. (1981) La croissance de la capsule céphalique et du siphon respiratoire des formes larvaires d'*Aedes* (*Ochlerotatus*) *atropalpus* (Coquillett) (Diptera: Culicidae). *Can. J. Zool.*, **59**, 1111–14.

Bownes, M. (1992) Molecular aspects of sex determination in insects. In *Insect Molecular Science* (eds J.M. Crampton and P. Eggleston), Academic Press, London.

Bradley, T.J. (1985) The excretory system: structure and physiology. In *Comprehensive Insect Physiology Biochemistry and Pharmacology*, (eds G. A. Kerkut and L. I. Gilbert), Pergamon Press, Oxford, **4**, 421–465.

Bradley, T.J. (1990) Scalariform junctions in the Malpighian tubules of the insects *Rhodnius prolixus* (Hemiptera: Reduviidae) and *Aedes taeniorhynchus* (Diptera: Culicidae). *J. Morphol.*, **206**, 65–9.

Bradley, T.J., Nayar, J.K. and Knight, J.W. (1990) Selection of a strain of *Aedes aegypti* susceptible to *Dirofilaria immitis* and lacking intracellular concretions in the Malpighian tubules. *J. Insect Physiol.*, **36**, 709–17.

Bradley, T.J. and Nayar, J.K. (1985) Intracellular melanization of the larvae of *Dirofilaria immitis* in the Malpighian tubules of the mosquito, *Aedes sollicitans*. *J. Invert. Pathol.*, **45**, 339–45.

Bradley, T.J. and Phillips, J.E. (1975) The secretion of hyperosmotic fluid by the rectum of a saline-water mosquito larva, *Aedes taeniorhynchus*. *J. Exp. Biol.*, **63**, 331–42.

Bradley, T.J. and Phillips, J.E. (1977a) Regulation of rectal secretion in saline-water mosquito larvae living in waters of diverse ionic composition. *J. Exp. Biol.*, **66**, 83–96.

Bradley, T.J. and Phillips, J.E. (1977b) The effect of external salinity on drinking rate and rectal secretion in the larvae of the saline-water mosquito *Aedes taeniorhynchus*. *J. Exp. Biol.*, **66**, 97–110.

Bradley, T.J. and Phillips, J.E. (1977c) The location and mechanism of hyperosmotic fluid secretion in the rectum of the saline water mosquito larvae *Aedes taeniorhynchus*. *J. Exp. Biol.*, **66**, 116–26.

Bradley, T.J. and Snyder, C. (1989) Fluid secretion and microvillar ultrastructure in mosquito Malpighian tubules. *Am. J. Physiol.* (*Reg. Int. Comp. Physiol.*), **257**, R1096–R1102.

Bradley, T.J., Stuart, A.M. and Satir, P. (1982) The ultrastructure of the larval Malpighian tubules of a saline-water mosquito. *Tissue & Cell*, **114**, 759–73.

Bradshaw, W.E. (1980) Thermoperiodism and the thermal environment of the pitcher-plant mosquito, *Wyeomyia smithii*. *Oecologia*, **46**, 13–17.

Bradshaw, W.E. and Lounibos, L.P. (1977) Evolution of dormancy and its photoperiodic control in pitcher-plant mosquitoes. *Evolution*, **31**, 546–67.

Brady, J. (1974) The physiology of insect circadian rhythms. *Adv. Insect Physiol.*, **10**, 1–115.

Brammer, J.D. and White, R.H. (1969) Vitamin A deficiency: effect on mosquito eye ultrastructure. *Science*, **163**, 821–3.

Breeland, S.G. and Pickard, E. (1967) Field observations on twenty-eight broods of floodwater mosquitoes resulting from controlled floodings of a natural habitat in the Tennessee Valley. *Mosq. News*, **27**, 343–58.

Breland, O.P. (1949) The biology and the immature stages of the mosquito, *Megarhinus septentrionalis* Dyar & Knab. *Ann. Entomol. Soc. Am.*, **42**, 38–47.

Breland, O.P., Eddleman, C.D. and Biesele, J.J. (1968) Studies of insect spermatozoa. I. *Entomol. News*, **79**, 197–216.

Breland, O.P., Gassner, G., Riess, R.W. *et al.* (1966) Certain aspects of the centriole adjunct, spermiogenesis, and the mature sperm of insects. *Can. J. Genet. Cytol.*, **8**, 759–73.

Breland, O.P., Gassner, G. and Riemann, J.G. (1964) Studies of meiosis in the male of the mosquito *Culiseta inornata*. *Ann. Entomol. Soc. Am.*, **57**, 472–9.

Briegel, H. (1969) Untersuchungen zum Aminosäuren- und Proteinstoffwechsel während der autogenen und anautogenen Eireifung von *Culex pipiens*. *J. Insect Physiol.*, **15**, 1137–66.

Briegel, H. (1975) Excretion of proteolytic enzymes by *Aedes aegypti* after a blood meal. *J. Insect Physiol.*, **21**, 1681–84.

Briegel, H. (1980a) Determination of uric acid and hematin in a single sample of excreta from blood-fed insects. *Experientia*, **36**, 1428.

Briegel, H. (1980b) Stickstoffexkretion bei Weibchen von *Aedes aegypti* (L.) nicht endokrin reguliert. *Rev. Suisse Zool.*, **87**, 1029–33

Briegel, H. (1983) Manipulation of age-dependent kinetics of the induction of intestinal trypsin in the mosquito *Aedes aegypti* (Diptera: Culicidae). *Entomologia Generalis*, **8**, 217–23.

Briegel, H. (1985) Mosquito reproduction: incomplete utilization of the blood meal protein for for oogenesis. *J. Insect Physiol.*, **31**, 15–21.

Briegel, H. (1986a) Protein catabolism and nitrogen partitioning during oogenesis in the mosquito *Aedes aegypti*. *J. Insect Physiol.*, **32**, 455–62.

Briegel, H. (1986b) Genetic variability of physiological parameters in blood-fed *Aedes aegypti*. In *Host Regulated Developmental Mechanisms in Vector Arthropods* (eds D. Borovsky and A. Spielman), University of Florida, Vero Beach, **1**, 104–7.

Briegel, H. (1990a) Metabolic relationship between female body size, reserves, and fecundity of *Aedes aegypti*. *J. Insect Physiol.*, **36**, 165–72.

Briegel, H. (1990b) Fecundity, metabolism, and body size in *Anopheles* (Diptera: Culicidae), vectors of malaria. *J. Med. Entomol.*, **27**, 839–50.

Briegel, H. and Kaiser, C. (1973) Life-span of mosquitoes (Culicidae, Diptera) under laboratory conditions. *Gerontologia*, **19**, 240–9.

Briegel, H. and Lea, A.O. (1975) Relationship between protein and proteolytic activity in the midgut of mosquitoes. *J. Insect Physiol.*, **21**, 1597–1604.

Briegel, H. and Lea, A.O. (1979) Influence of the endocrine system on tryptic activity in female *Aedes aegypti*. *J. Insect Physiol.*, **25**, 227–30.

Briegel, H., Lea, A.O. and Klowden, M.J. (1979) Hemoglobinometry as a method for measuring blood meal sizes of mosquitoes (Diptera: Culicidae). *J. Med. Entomol.*, **15**, 235–8.

Briegel, H. and Rezzonico, L. (1985) Concentration of host blood protein during feeding by anopheline mosquitoes (Diptera: Culicidae). *J. Med. Entomol.*, **22**, 612–18.

Broadie, K.S. and Bradshaw, W.E. (1991) Mechanisms of interference competition in the western tree-hole mosquito, *Aedes sierrensis*. *Ecol. Entomol.*, **16**, 145–54.

Brocher, F. (1910) Les phénomènes capillaires. Leur importance dans la biologie aquatique. *Ann. Biol. Lacustre*, **4**, 89–138, 1 pl.

Brooks, M.A. (1964) Symbiotes and the nutrition of medically important insects. *Bull. Wld Hlth Org.*, **31**, 555–9.

Brown, A., Griffitts, T.H.D., Eerwin, S. and Dyrenforth, L.Y. (1938) Arthus's phenomenon from mosquito bites. *Southern Med. J.*, **31**, 590–6.

Brown, M.R., Crim, J.W. and Lea, A.O. (1986) FMRFamide- and pancreatic polypeptide-like immunoreactivity of endocrine cells in the midgut of a mosquito. *Tissue & Cell*, **18**, 419–28.

Brown, M.R., Raikhel, A.S. and Lea, A.O. (1985) Ultrastructure of midgut endocrine cells in the adult mosquito, *Aedes aegypti*. *Tissue & Cell*, **17**, 709–21.

Brown, M.R. and Lea, A.O. (1988) FMRFamide- and adipokinetic hormone-like immunoreactivity in the nervous system of the mosquito, *Aedes aegypti*. *J. Comp. Neurol.*, **270**, 606–14.

Brown, M.R. and Lea, A.O. (1989) Neuroendocrine and midgut endocrine systems in the adult mosquito. *Advances Disease Vector Res.*, **6**, 29–58.

Brown, T.M. and Brown, A.W.A. (1980) Accumulation and distribution of methoprene in resistant *Culex pipiens pipiens* larvae. *Entomol. Exp. Applic.*, **27**, 11–22.

Brummer-Korvenkontio, M., Korhonen, P. and Hämeen-Antilla, R. (1971) Ecology and phenology of mosquitoes (Dipt., Culicidae) inhabiting small pools in Finland. *Acta Entomol. Fennica*, **28**, 51–73.

Brummer-Korvenkontio, H., Lappalainen, P., Reunala, T. and Palosuo, T. (1990) Immunization of rabbits with mosquito bites: immunoblot analysis of IgG antimosquito antibodies in rabbit and man. *Int. Arch. Allergy Appl. Immunol.*, **93**, 14–8.

Brumpt, E. (1941) Mécanisme d'éclosion des moustiques. *Ann. Parasitol. Hum. Comp.*, **18**, 75–94.

Brunet, P.C.J. (1980) The metabolism of the aromatic amino acids concerned in the cross-linking of insect cuticle. *Insect Biochem.*, **10**, 467–500.

Brust, R.A. (1967) Weight and development time of different stadia of mosquitoes reared at various constant temperatures. *Can. Entomol.*, **99**, 986–93.

Brust, R.A. (1968a) Effect of starvation on molting and growth in *Aedes aegypti* and *A. vexans*. *J. Econ. Entomol.*, **61**, 1570–2.

Brust, R.A. (1968b) Temperature-induced intersexes in *Aedes* mosquitoes: comparative study of species from Manitoba. *Can. Entomol.*, **100**, 879–91.

Brust, R.A. (1991) Environmetal regulation of autogeny in *Culex tarsalis* (Diptera: Culicidae) from Manitoba, Canada. *J. Med. Entomol.*, **28**, 847–53.

Brust, R.A. and Costello, R.A. (1969) Mosquitoes of Manitoba. II. The effect of storage temperature and relative humidity on hatching of eggs of *Aedes vexans* and *Aedes abserratus* (Diptera: Culicidae). *Can. Entomol.*, **101**, 1285–91.

Brust, R.A. and Horsfall, W.R. (1965) Thermal stress and anomalous development of mosquitoes (Diptera: Culicidae). IV. *Aedes communis*. *Can. J. Zool.*, **43**, 17–53, 8 pls.

Brust, R.A. and Kalpage, K.S. (1967) A rearing method for *Aedes abserratus* (F. and Y.). *Mosq. News*, **27**, 117.

Bryan, J.H. and Coluzzi, M. (1971) Cytogenetic observations on *Anopheles farauti* Laveran. *Bull. Wld Hlth Org.*, **45**, 266–7.

Bryan, J.H., Di Deco, M.A., Petrarca, V. and Coluzzi, M. (1982) Inversion polymorphism and incipient speciation in *Anopheles gambiae* s. str. in The Gambia, West Africa. *Genetica*, **59**, 167–76.

Bryan, J.H., Petrarca, V., Di Deco, M.A. and Coluzzi. M. (1987) Adult behaviour of members of the *Anopheles gambiae* complex in The Gambia with special reference to *An. melas* and its chromosomal variants. *Parassitologia*, **29**, 221–49.

Buffington, J.D. and Zar, J.H. (1968) Changes in fatty acid composition of *Culex pipiens pipiens* during hibernation. *Ann. Entomol. Soc. Am.*, **61**, 774–5.

Burgess, L. (1967) Pycnosis of the nuclei of ecdysial gland cells in *Aedes aegypti* (L.) (Diptera:Culicidae). *Can. J. Zool.*, **45**, 1294–5, 1 pl.

Burgess, L. (1971) Neurosecretory cells and their axon pathways in *Culiseta inornata* (Williston) (Diptera: Culicidae). *Can. J. Zool.*, **49**, 889–901, 1 pl.

Burgess, L. (1973) Axon pathways of the intermediate neurosecretory cells in *Culex tarsalis* Coquillett (Diptera: Culicidae). *Can. J. Zool.*, **51**, 379–82.

Burgess, L. and Rempel, J.G. (1966) The stomodaeal nervous system, the neurosecretory system, and the gland complex in *Aedes aegypti* (L.). *Can. J. Zool.*, **44**, 731–65, 4 pls.

Burkot, T.R., Goodman, W.G. and DeFoliart, G.R. (1981) Identification of mosquito blood meals by enzyme-linked immunosorbent asssay. *Am. J. Trop. Med. Hyg.*, **30**, 1336–41.

Burton, G.J. (1965) Method of attachment of pupae of *Mansonia annulifera* (Theo.) and M. *uniformis* (Theo.) to *Pistia stratiotes*. *Bull. Entomol. Res.*, **55**, 691–6, 1 pl.

Buse, E. and Kuhlow, F. (1979) Rasterelektronenoptische Studien zur Morphologie und Bedeutung der Vorderdarmbewehrung verschiedener Stechmückenarten für die Aufnahme von Mikrofilarien. *Tropenmed. Parasitol.*, **30**, 446–54.

Buxton, P.A. (1935) Changes in the composition of adult *Culex pipiens* during hibernation. *Parasitology*, **27**, 263–5.

Cadeddu, M.G. and Laudani, U. (1974) Testis maturation in *Anopheles atroparvus* van Thiel. *Boll. Zool.*, **41**, 39–42, 5 pls.

Callahan, J.L. and Morris, C.D. (1987) Habitat characteristics of *Coquillettidia perturbans* in central Florida. *J. Am. Mosq. Control Ass.*, **3**, 176–80.

Callan, H.G. and Montalenti, G. (1947) Chiasma interference in mosquitoes. *J. Genet.*, **48**, 119–34.

Candy, D.J. (1985) Intermediary metabolism. In *Comprehensive Insect Biochemistry Physiology and Pharmacology* (eds G.A. Kerkut and L.I. Gilbert), Pergamon Press, Oxford, **10**, 1–41.

Caspari, E. and Watson, G.S. (1959) On the evolutionary importance of cytoplasmic sterility in mosquitoes. *Evolution*, **13**, 568–70.

Causey, O.R., Deane, L.M. and Deane, M.P. (1945) Description of *Chagasia rozeboomi*, an anopheline from Ceara, Brazil. *J. Natl Malaria Soc.*, **4**, 341–50.

Cave, M.D. (1982) Morphological manifestations of ribosomal DNA amplification during insect oogenesis. In *Insect Ultrastructure* (eds R.C. King and H. Akai), Plenum Press, New York and London, **1**, 86–117.

Cerreta, J.M. (1976) Structural microtubules in hind-gut epithelial cells of the mosquito. *Proc. Electron Microscopy Soc. Am.*, **34**, 192–3.

Chambers, G.M. and Klowden, M.J. (1990) Correlation of nutritional reserves with a critical weight for pupation in larval *Aedes aegypti* mosquitoes. *J. Am. Mosq. Control Ass.*, **6**, 394–9.

Chang, Y.H. and Judson, C.L. (1977a) Amino acid composition of human and guinea pig blood proteins, and ovarian proteins of the yellow fever mosquito *Aedes aegypti*; and their effects on the mosquito egg production. *Comp. Biochem. Physiol.*, **62A**, 753–5.

Chang, Y.H. and Judson, C.L. (1977b) Peptides as stimulators of egg development neurosecretory hormone release in the mosquito *Aedes aegypti*. *Comp. Biochem. Physiol.*, **57C**, 147–51.

Chang, Y.H. and Judson, C.L. (1979) Amino acid composition of human and guinea pig blood proteins, and ovarian proteins of the yellow fever mosquito *Aedes aegypti*; and their effects on the mosquito egg production. *Comp. Biochem. Physiol.*, **62A**, 753–55.

Charles, J.-F. (1987) Ultrastructural midgut events in Culicidae larvae fed with *Bacillus sphaericus* 2297 spore/crystal complex. *Ann. Inst. Pasteur/Microbiol.*, **138**, 471–84.

Charles, J.-F. and de Barjac, H. (1981) Variations du pH de l'intestin moyen d'*Aedes aegypti* en relation avec l'intoxication par les cristaux de *Bacillus thuringiensis* var. *israelensis* (sérotype H14). *Bull. Soc. Pathol. Exotique*, **74**, 91–5.

Charles, J.-F. and de Barjac, H. (1983) Action des cristaux de *Bacillus thuringiensis* var. *israelensis* sur l'intestin moyen des larves de *Aedes aegypti* L., en microscopie électronique. *Ann. Microbiol. (Inst. Pasteur)*, **134 A**, 197–218.

Chaudhry, S. (1981) The salivary gland chromosomes of *Culex (Culex) vishnui* (Culicidae: Diptera). *Genetica*, **55**, 171–8.

Chaudhry, S. (1986) Sex chromosome in mosquitoes. I. NOR associated gene activation and asynapsis in the salivary X-chromosome of *Anopheles subpictus* (Culicidae: Diptera). *Caryologia*, **39**, 211–15.

Cheer, A.Y.L. and Koehl, M.A.R. (1987) Paddles and rakes: fluid flow through bristled appendages of small organisms. *J. Theor. Biol.*, **129**, 17–39.

Chen, C.C. (1988) Further evidence of both humoral and cellular encapsulations of sheathed microfilariae of *Brugia pahangi* in *Anopheles quadrimaculatus*. *Int. J. Parasitol.*, **18**, 819–26.

Chen, C.C. and Laurence, B.R. (1985) An ultrastructural study on the encapsulation of microfilariae of *Brugia pahangi* in the haemocoel of *Anopheles quadrimaculatus*. *Int. J. Parasitol.*, **15**, 421–8.

Chen, C.C. and Laurence, B.R. (1987a) *In vitro* study on humoral encapsulation of microfilariae: establishment of technique and description of reaction. *Int. J. Parasitol.*, **17**, 781–7.

Chen, C.C. and Laurence, B.R. (1987b) *In vitro* study on humoral encapsulation of microfilariae: effects of diethyldithiocarbamate and dopachrome on the reaction. *Int. J. Parasitol.*, **17**, 789–94.

Chen, P.S. and Briegel, H. (1965) Studies on the protein metabolism of *Culex pipiens* L. – V. Changes in free amino acids and peptides during embryonic development. *Comp. Biochem. Physiol.*, **14**, 463–73.

Chevone, B.I. and Peters, T.M. (1969) Retardation of larval development of *Aedes aegypti* (L.) by the vital dye, Nile blue sulphate (A). *Mosq. News*, **29**, 243–51.

Chevone, B.I. and Richards, A.G. (1976) Ultrastructure of the atypic muscles associated with terminalial inversion in male *Aedes aegypti* (L.). *Biol. Bull.*, **151**, 283–96.

Chevone, B. I. and Richards, A. G. (1977) Ultrastructural changes in intersegmental cuticle during rotation of the terminal abdominal segments in a mosquito. *Tissue & Cell*, **9**, 241–54.

Cho, W.-L., Deitsch, K.W. and Raikhel, A.S. (1991a) An extraovarian protein accumulated in mosquito oocytes is a carboxypeptidase activated in embryos. *Proc. Natl Acad. Sci. USA*, 88, 10821–4.

Cho, W.-L., Dhadialla, T. S. and Raikhel, A. S. (1991b) Purification and characterization of a lysosomal aspartic protease with cathepsin D activity from the mosquito. *Insect Biochem.*, **21**, 165–76.

Christensen, B.M., Huff, B.M., Miranpuri, G.S. *et al.* (1989) Hemocyte population changes during the immune reponse of *Aedes aegypti* to inoculated microfilariae of *Dirofilaria immitis*. *J. Parasitol.*, **75**, 119–23.

Christensen, B.M. and Forton, K.F. (1986) Hemocyte-mediated melanization of microfilariae in *Aedes aegypti*. *J. Parasitol.*, **72**, 220–5.

Christophers, S.R. (1911) The development of the egg follicle in anophelines. *Paludism*, no. 2, 73–8, 1 pl.

Christophers, S.R. (1945) Structure of the *Culex* egg and egg-raft in relation to function (Diptera). *Trans. R. Entomol. Soc. Lond.*, **95**, 25–34, 4 pls.

Christophers, S.R. (1960) *Aedes aegypt (L.) the Yellow Fever Mosquito*, Cambridge University Press, Cambridge.

Christophers, S.R. and Puri, I.M. (1929) Why do *Anopheles* larvae feed at the surface, and how? *Trans. Far-Eastern Ass. Trop. Med.*, **2**, 736–9, 1 pl.

Clay, M.E., Meola, R.W., Venard, C.E. and Skavaril, R.V. (1973) Changes in weight and concentration of solutions of dextrose in the crops of mosquitoes. *Mosq. News*, **33**, 579–85.

Clay, M.E. and Venard, C.E. (1972) The fine structure of the oesophageal diverticula in the mosquito *Aedes triseriatus*. *Ann. Entomol. Soc. Am.*, **65**, 964–75.

Clements, A.N. (1955) The sources of energy for flight in mosquitoes. *J. Exp. Biol.*, **32**, 547–54.

Clements, A.N. (1956a) Hormonal control of ovary development in mosquitoes. *J. Exp. Biol.*, **33**, 211–23.

Clements, A.N. (1956b) The antennal pulsating organs of mosquitoes and other Diptera. *Quart. J. Microscop. Sci.*, **97**, 429–33.

Clements, A.N. (1963) *The Physiology of Mosquitoes*, Pergamon Press, Oxford.

Clements, A.N. and Boocock, M.R. (1984) Ovarian development in mosquitoes: stages of growth and arrest, and follicular resorption. *Physiol. Entomol.*, **9**, 1–8.

Clements, A.N. and Potter, S.A. (1967) The fine structure of the spermathecae and their ducts in the mosquito *Aedes aegypti*. *J. Insect Physiol.*, **13**, 1825–36, 8 pls.

Clements, A.N., Potter, S.A. and Scales, M.D.C. (1985) The cardiacal neurosecretory system and associated organs of an adult mosquito, *Aedes aegypti*. *J. Insect Physiol.*, **31**, 821–30.

Cockburn, A.F., Mitchell, S.E. and Seawright, J.A. (1990) Cloning of the mitochondrial genome of *Anopheles quadrimaculatus*. *Arch. Insect Biochem. Physiol.*, **14**, 31–6.

Cockburn, A.F. and Mitchell, S.E. (1989) Repetitive DNA interspersion patterns in Diptera. *Arch. Insect Biochem. Physiol.*, **10**, 105–13.

Cockburn, A.F. and Seawright, J.A. (1988) Techniques for mitochondrial and ribosomal DNA analysis of anopheline mosquitoes. *J. Am. Mosq. Control Ass.*, **3**, 261–5.

Cocke, J., Bridges, A.C., Mayer, R.T. and Olson, J.K. (1979) Morphological effects of insect growth regulating compounds on *Aedes aegypti* (Diptera: Culicidae) larvae. *Life Sci.*, **24**, 817–31.

Cole, S.J. and Gillett, J.D. (1978) The influence of the brain hormone on retention of blood in the mid-gut of the mosquito *Aedes aegypti* (L.). II. Early elimination following removal of the medial neurosecretory cells of the brain. *Proc. R. Soc. Lond.* (B), **202**, 307–11.

Cole, S.J. and Gillett, J.D. (1979) The influence of the brain hormone on retention of blood in the mid-gut of the mosquito *Aedes aegypti* (L.). III. The involvement of the ovaries and ecdysone. *Proc. R. Soc. Lond.* (B), **205**, 411–22.

Colless, D.H. (1956) Environmental factors affecting hairiness in mosquito larvae. *Nature*, **177**, 229–30.

Colless, D.H. and Chellapah, W.T. (1960) Effects of body weight and size of blood-meal upon egg production in *Aedes aegypti* (Linnaeuus) (Diptera, Culicidae). *Ann. Trop. Med. Parasit.*, **54**, 475–82.

Collins, F.H., Mendez, M.A., Rasmussen, M.O. *et al.* (1987) . A ribosomal RNA gene probe differentiates member species of the *Anopheles gambiae* complex. *Am. J. Trop. Med. Hyg.*, **37**, 37–41.

Collins, F.H., Paskewitz, S.M. and Finnerty, V. (1989) Ribosomal RNA genes of the *Anopheles gambiae* species complex. *Advances Disease Vector Res.*, **6**, 1–28.

Collins, F.H., Sakai, R.K., Vernick, K.D. *et al.* (1986) Genetic selection of a *Plasmodium*-refractory strain of the malaria vector *Anopheles gambiae*. *Science*, **234**, 607–10.

Coluzzi, M. (1968) Cromosomi politenici delle cellule nutrici ovariche nel complesso gambiae del genere *Anopheles*. *Parassitologia*, **10**, 179–83.

Coluzzi, M. (1982) Spatial distribution of chromosomal inversions and speciation in anopheline mosquitoes. In *Mechanisms of Speciation* (ed. C. Barigozzi), Alan R. Liss, Inc., New York, pp. 143–53.

Coluzzi, M., Cancrini, G. and Di Deco, M. (1970) The polytene chromosomes of *Anopheles superpictus* and relationships with *Anopheles stephensi*. *Parassitologia*, **12**, 101–11, 5 pls.

Coluzzi, M., Concretti, A. and Ascoli, F. (1982) Effect of cibarial armature of mosquitoes (Diptera, Culicidae) on blood-meal haemolysis. *J. Insect Physiol.*, **28**, 885–8.

Coluzzi, M. and Kitzmiller, J.B. (1975) *Anopheline mosquitoes. Handbook of Genetics* (ed. R. C. King), Plenum Press, New York and London, **3**, 285–309.

Coluzzi, M., Petrarca, V. and Di Deco, M.A. (1985) Chromosomal inversion intergradation and incipient speciation in *Anopheles gambiae*. *Boll. Zool.*, **52**, 45–63.

Coluzzi, M., Petrarca, V. and Di Deco, M.A. (1990) Variazioni cromosomiche in *Anopheles gambiae* e caratteristiche ecoclimatiche delle localita di campionamento in Africa occidentale. *Parassitologia*, **32** (Suppl. 1), 66–7.

Coluzzi, M., Sabatini, A., Petrarca, V. and Di Deco, M.A. (1979) Chromosomal differentiation and adaptation to human environments in the *Anopheles gambiae* complex. *Trans. R. Soc. Trop. Med. Hyg.*, **73**, 483–97.

Coluzzi, M. and Trabucchi, R. (1968) Importanza dell'armatura bucco-faringea in *Anopheles* e *Culex* in relazione alle infezioni con *Dirofilaria*. *Parassitologia*, **10**, 47–59, 1 pl.

Consoli,R.A.G.B. and Williams, P. (1978) Laboratory observations on the bionomics of *Aedes fluviatilis* (Lutz) (Diptera: Culicidae). *Bull. Entomol. Res.*, **68**, 123–36, 1 pl.

Constantineaunu, N.J. (1930) Der Aufbau der Sehorgane bei den im Süsswasser lebenden Dipterenlarven und bei Puppen und Imagines von *Culex*. *Zool. Jb. (Anat. Ont.)*, **52**, 253–346, 12 pls.

Cook E.F. (1944) The morphology of the larval heads of certain Culicidae (Diptera). *Microentomology*, **9**, 33–68.

Copeland, E. (1964) A mitochondrial pump in the cells of the anal papillae of mosquito larvae. *J. Cell Biol.*, **23**, 253–63.

Corbet, P.S. (1964) Autogeny and oviposition in arctic mosquitoes. *Nature*, **203**, 668.

Corbet, P.S. (1965) Reproduction in mosquitoes of the high arctic. *Proc. 12th Int. Congr. Entomol.* (1964), pp. 817–18.

Corbet, P.S. (1966) Diel patterns of mosquito activity in a high arctic locality: Hazen Camp, Ellesmere Island, N.W.T. *Can. Entomol.*, **98**, 1238–52.

Corbet, P.S. (1967) Facultative autogeny in arctic mosquitoes. *Nature*, **215**, 662–3.

Corbet, P.S. and Ali, H.J. (1987) Diel patterns of pupation and emergence, and protogyny, in *Toxorhynchites brevipalpis brevipalpis* (Theobald) (Diptera: Culicidae): a laboratory study. *Ann. Trop. Med. Parasitol.*, **81**, 437–43.

Corradetti, A. (1930) Sulle modificazioni delle larve di *Anopheles* in relazione col colore dell'ambiente. *Riv. Malariol.*, **9**, 35–9.

Costello, R.A. and Brust, R.A. (1969) A quantitative study of uptake and loss of water by eggs of *Aedes vexans* (Diptera : Culicidae). *Can. Entomol.*, **101**, 1266–9.

Coulson, R.M.R., Curtis, C.F., Ready, P.D. *et al.* (1990) Amplification and analysis of human DNA present in mosquito bloodmeals. *Med. Vet. Entomol.*, **4**, 357–66.

Craig, G.B. (1965) Genetic control of thermally-induced sex reversal in *Aedes aegypti*. *Proc. 12th Int. Congr. Entomol.* (1964), p. 263.

Craig, G.B. and Hickey, W.A. (1967) Genetics of *Aedes aegypti*. In *Genetics of Insect Vectors of Disease* (eds J.W. Wright and R. Pal), Elsevier, Amsterdam, pp. 67–131.

Crampton, J.M., Morris, A., Lycett, G. *et al.* (1990) Molecular characterization and genome manipulation of the mosquito, *Aedes aegypti*. In *Molecular Insect Science* (eds H.H. Hagedorn, J.G. Hildebrand, M.G. Kidwell and J.H. Law), Plenum Press, New York, pp. 1–11.

Cummins, K.W. (1973) Trophic relations of aquatic insects. *Annual Rev. Entomol.*, **18**, 183–206.

Cummins, K.W. and Klug, M.J. (1979) Feeding ecology of stream invertebrates. *Annual Rev. Ecol. Syst.*, **10**, 147–72.

Cupp, E.W. and Horsfall, W.R. (1970a) Thermal stress and anomalous development of mosquitoes (Diptera, Culicidae). VI. Effect of temperature on embryogeny of *Aedes sierrensis*. *Ann. Zool. Fennici*, **7**, 358–65.

Cupp, E.W. and Horsfall, W.R. (1970b) Thermal stress and anomalous development of mosquitoes (Diptera, Culicidae). VII. Effect of temperature on embryogeny of *Aedes aegypti*. *Ann. Zool. Fennici*, **7**, 370–4.

Curtin, T.J. and Jones, J.C. (1961) The mechanism of ovulation and oviposition in *Aedes aegypti*. *Ann. Entomol. Soc. Am.*, **54**, 298–313.

Curtis, C.F. (1977) Testing systems for the genetic control of mosquitoes. *Proc. XVth Int. Congr. Entomol.* (1976), pp. 106–16.

Curtis, C.F., Brooks, G.D., Ansari, M.A. *et al.* (1982) A field trial on control of *Culex quinquefasciatus* by release of males of a strain integrating cytoplasmic incompatibility and a translocation. *Entomol. Exp. Applic.*, **31**, 181–90.

Curtis, C.F., Grover, K.K., Suguna, S.G., *et al.* (1976) Comparative field cage tests of the population suppressing efficiency of three genetic control systems for *Aedes aegypti*. *Heredity*, **36**, 11–29.

Dadd, R.H. (1968) A method for comparing feeding rates in mosquito larvae. *Mosq. News*, **28**, 226–30.

Dadd, R.H. (1970a) Comparison of rates of ingestion of particulate solids by *Culex pipiens* larvae: phagostimulant effect of water-soluble yeast extract. *Entomol. Exp. Applic.*, **13**, 407–19.

Dadd, R.H. (1970b) Relationship between filtering activity and ingestion of solids by larvae of the mosquito *Culex pipiens*: a method for assessing phagostimulant factors. *J. Med. Entomol.*, **7**, 708–12.

Dadd, R.H. (1971) Effects of size and concentration of particles on rates of ingestion of latex particulates by mosquito larvae. *Ann. Entomol. Soc. Am.*, **64**, 687–92.

Dadd, R.H. (1975a) Ingestion of colloid solutions by filter-feeding mosquito larvae: relationship to viscosity. *J. Exp. Zool.*, **191**, 395–406.

Dadd, R.H. (1975b) Alkalinity within the midgut of mosquito larvae with alkaline-active digestive enzymes. *J. Insect Physiol.*, **21**, 1847–53.

Dadd, R.H. (1976) Loss of alkalinity in chilled or narcotized mosquito larvae. *Ann. Entomol. Soc. Am.*, **69**, 248–254.

Dadd, R.H. (1978) Amino acid requirements of the mosquito *Culex pipiens*: asparagine essential. *J. Insect Physiol.*, **24**, 25–30.

Dadd, R.H. (1979) Nucleotide, nucleoside and base nutritional requirements of the mosquito *Culex pipiens*. *J. Insect Physiol.*, **25**, 353–9.

Dadd, R.H. (1980) Essential fatty acids for the mosquito *Culex pipiens*. *J. Nutrition*, **110**, 1152–60.

Dadd, R.H. (1981) Essential fatty acids for mosquitoes, other insects and vertebrates. In *Current Topics in Insect Endocrinology and Nutrition* (eds G. Bhaskaran, S. Friedman and J. G. Rodriguez), Plenum Publishing Corpn, New York, pp. 189–214.

Dadd, R.H. (1983) Essential fatty acids: insects and vertebrates compared. In *Metabolic Aspects of Lipid Nutrition in Insects* (eds T.E. Mittler and R.H. Dadd), Westview Press, Boulder, Colorado, pp. 107–47.

Dadd, R.H. (1985) Nutrition: organisms. In *Comprehensive Insect Physiology Biochemistry and Pharmacology* (eds G.A. Kerkut and L.I. Gilbert), Pergamon Press, Oxford, **4**, 313–90.

Dadd, R.H., Asman, S.M. and Kleinjan, J.E. (1989) Essential fatty acid status of laboratory-reared mosquitos improved by supplementing crude larval foods with fish oils. *Entomol. Exp. Applic.*, **52**, 149–58.

Dadd, R.H., Friend, W.G. and Kleinjan, J.E. (1980) Arachidonic acid requirement for two species of *Culiseta* reared on synthetic diet. *Can. J. Zool.*, **58**, 184–5.

Dadd, R.H. and Kleinjan, J.E. (1974) Autophagostimulant from *Culex pipiens* larvae: distinction from other mosquito larval factors. *Environ. Entomol.*, **3**, 21–8.

Dadd, R.H. and Kleinjan, J.E. (1976) Chemically defined dietary media for larvae of the mosquito *Culex pipiens* (Diptera: Culicidae): effects of colloid texturizers. *J. Med. Entomol.*, **13**, 285–91.

Dadd, R.H. and Kleinjan, J.E. (1977) Dietary nucleotide requirements of the mosquito, *Culex pipiens*. *J. Insect Physiol.*, **23**, 333–41.

Dadd, R.H. and Kleinjan, J.E. (1978) An essential nutrient for the mosquito *Culex pipiens* associated with certain animal-derived phospholipids. *Ann. Entomol. Soc. Am.*, **71**, 794–800.

Dadd, R.H. and Kleinjan, J.E. (1979a) Essential fatty acid for the mosquito *Culex pipiens*: arachidonic acid. *J. Insect Physiol.*, **25**, 495–502.

Dadd, R.H. and Kleinjan, J.E. (1979b) Vitamin E, ascorbyl palmitate, or propyl gallate protect arachidonic acid in synthetic diets for mosquitoes. *Entomol. Exp. Applic.*, **26**, 222–6.

Dadd, R.H. and Kleinjan, J.E. (1984a) Lecithin-dependent phytosterol utilization by larvae of *Culex pipiens* (Diptera: Culicidae). *Ann. Entomol. Soc. Am.*, **77**, 518–25.

Dadd, R.H. and Kleinjan, J.E. (1984b) Prostaglandin synthetase inhibitors modulate the effect of essential dietary arachidonic acid in the mosquito *Culex pipiens*. *J. Insect Physiol.*, **30**, 721–8.

Dadd, R.H. and Kleinjan, J.E. (1985) Phagostimulation of larval *Culex pipiens* L. by nucleic acid nucleotides, nucleosides and bases. *Physiol. Entomol.*, **10**, 37–44.

Dadd, R.H. and Kleinjan, J.E. (1988) Prostaglandin sparing of dietary arachidonic acid in the mosquito *Culex pipiens*. *J. Insect Physiol.*, **34**, 779–85.

Dadd, R.H., Kleinjan, J.E. and Asman, S.M. (1988) Eicosapentaenoic acid in mosquito tissues: differences between wild and laboratory-reared adults. *Environ. Entomol.*, **17**, 172–80.

Dadd, R.H., Kleinjan, J.E. and Merrill, L.D. (1982) Phagostimulant effects of simple nutrients on larval *Culex pipiens* (Diptera: Culicidae). *Ann. Entomol. Soc. Am.*, **75**, 605–12.

Dadd, R.H., Kleinjan, J.E. and Stanley-Samuelson, D.W. (1987) Polyunsaturated fatty acids of mosquitos reared with single dietary polyunsaturates. *Insect Biochem.*, **17**, 7–16.

Dahl, C. (1973) Emergence and its diel periodicity in *Aedes* (*O.*) *communis* (DeG.), *punctor* (Kirby) and *hexodontus* Dyar in Swedish Lapland. *Aquilo Ser. Zool.*, **14**, 34–45.

Dahl, C., Craig, D.A. and Merritt, R.W. (1990) The

sites of possible mucus-producing glands in the feeding system of mosquito larvae (Culicidae, Diptera). *Ann. Entomol. Soc. Am.*, **83**, 827–33.

Dahl, C., Widahl, L.-E. and Nilsson, C. (1988) Functional analysis of the suspension feeding system in mosquitoes (Diptera: Culicidae). *Ann. Entomol. Soc. Am.*, **81**, 105–27.

Dallai, R. and Afzelius, B.A. (1990) Microtubular diversity in insect spermatozoa: results obtained with a new fixative. *J. Structural Biol.*, **103**, 164–79.

Dallai, R., Baccetti, B., Mazzini, M. and Sabatinelli, G. (1984) The spermatozoon of three species of *Phlebotomus* (Phlebotominae) and the acrosomal evolution in nematoceran dipterans. *Int. J. Insect Morphol. Embryol.*, **13**, 1–10.

Danks H.V. and Corbet, P.S. (1973) Sex ratios at emergence of two species of high-arctic *Aedes* (Diptera : Culicidae). *Can. Entomol.*, **105**, 647–51.

Dapples, C.C. and Lea, A.O. (1974) Inner surface morphology of the alimentary canal in *Aedes aegypti* (L.) (Diptera: Culicidae). *Int. J. Insect Morphol. Embryol.*, **3**, 433–42.

Das, M.K., Mishra, A., Beuria, M.K. and Dash, A.P. (1991) Human natural antibodies to *Culex quinquefasciatus*: age-dependent occurrence. *J. Am. Mosq. Control Ass.*, **7**, 319–21.

Davidson, E.W. (1979) Ultrastructure of midgut events in the pathogenesis of *Bacillus sphaericus* strain SSII-1 infections of *Culex pipiens quinquefasciatus* larvae. *Can. J. Microbiol.*, **25**, 178–84.

Davis, C.W.C. (1967) A comparative study of larval embryogenesis in the mosquito *Culex fatigans* Wiedemann (Diptera: Culicidae) and the sheep-fly *Lucilia sericata* Meigen (Diptera: Calliphoridae). I. Description of embryonic development. *Austral. J. Zool.*, **15**, 547–79, 2 pls.

Davis, C.W.C. (1970) A comparative study of larval embryogenesis in the mosquito *Culex fatigans* Wiedemann (Diptera: Culicidae) and the sheep-fly *Lucilia sericata* Meigen (Diptera Calliphoridae). II. Causal interactions in embryonic development. *Austral. J. Zool.*, **18**, 547–79.

Day, J.F. and Van Handel, E. (1986) Differences between the nutritional reserves of laboratory-maintained and field-collected adult mosquitoes. *J. Am. Mosq. Control Ass.*, **2**, 154–7.

Day, M.F. (1954) The mechanism of food distribution to midgut or diverticula in the mosquito. *Austral. J. Biol. Sci.*, **7**, 515–24, 1 pl.

Day, M.F. and Bennetts, M.J. (1953) Healing of gut wounds in the mosquito *Aedes aegypti* (L.) and the leafhopper *Orosius argentatus* (Ev.). *Austral. J. Biol. Sci.*, **6**, 580–5, 2 pls.

Dayhoff, M.O. (1969) *Atlas of Protein Sequence and Structure*, Vol. 4, Nat. Biomed. Res. Found., Silver Spring, Maryland, USA.

de Boissezon, P. (1930a) Contribution à l'étude de la biologie et de l'histophysiologie de *Culex pipiens*. *Arch. Zool. Exp. Gén.*, **70**, 281–431.

de Boissezon, P. (1930b) Le rôle du corps gras comme rein d'accumulation chez *Culex pipiens* L. et chez *Theobaldia annulata* M. *C. R. Soc., Biol., Paris*, **103**, 1233–5.

De Meillon, B., Sebastian, A. and Khan, Z.H. (1967a) The duration of egg, larval and pupal stages of *Culex pipiens fatigans* in Rangoon, Burma. *Bull. Wld Hlth Org.*, **36**, 7–14.

De Meillon, B., Sebastian, A. and Khan, Z.H. (1967b) Time of arrival of gravid *Culex pipiens fatigans* at an oviposition site, the oviposition cycle and the relationship between time of feeding and time of oviposition. *Bull. Wld Hlth Org.*, **36**, 39–46.

De Meillon, B., Sebastian, A. and Khan, Z.H. (1967c) Exodus from a breeding place and the time of emergence from the pupa of *Culex pipiens fatigans*. *Bull. Wld Hlth Org.*, **36**, 163–7.

De Oliveira, D. and Durand, M. (1978) Head capsule growth in *Culex territans* Walker. *Mosq. News*, **38**, 230–3.

Deane, M.P. and Causey, O.R. (1943) Viability of *Anopheles gambiae* eggs and morphology of unusual types found in Brazil. *Am. J. Trop. Med.*, **23**, 95–102.

DeGuire, D.M. and Fraenkel, G. (1973) The meconium of *Aedes aegypti* (Diptera: Culicidae). *Ann. Entomol. Soc. Am.*, **66**, 475–6.

Dennhöfer, L. (1968) Die Speicheldrüsenchromosomen der Stechmücke *Culex pipiens*. I. Der normale Chromosomenbestand. *Chromosoma*, **25**, 365–76, 2 pls.

Dennhöfer, L. (1974) Die Speicheldrüsenchromosomen der Stechmücke *Culex pipiens* L. II. Ergänzungen zur Kartierung. *Genetica*, **45**, 29–38.

Dennhöfer, L. (1975a) Die Speicheldrüsenchromosomen der Stechmücke *Culex pipiens* L. IV. Der chromosomale Geschlechtsdimorphismus *Genetica*, **45**, 163–75.

Dennhöfer, L. (1975b) Genlokalisation auf den larvalen Speicheldrüsenchromosomen der Stechmücke *Culex pipiens* Theoret. *Appl. Genet.*, **45**, 279–89.

Deslongchamps, P. and Tourneur, J.-C. (1980) Head capsule growth and early sexual dimorphism in *Culex quinquefasciatus* Say (Diptera: Culicidae). *Mosq. News*, **40**, 351–5.

Detinova, T.S. (1944) The relationship between the size of female *Anopheles maculipennis atroparvus* van Thiel and the stage of development of the ovaries on emergence. *Med. Parasitol., Moscow*, **13**, 52–5. [In Russian.]

Detinova, T.S. (1949) Physiological changes in the ovaries of female *Anopheles maculipennis*. *Med. Parasitol., Moscow*, **18**, 410–20. [In Russian.]

Detinova, T.S. (1955) Fertility of the common malarial mosquito *Anopheles maculipennis*. *Med. Parasitol., Moscow*, **24**, 6–11. [In Russian.]

Detinova, T.S. (1962) Age-grouping methods in Diptera

of medical importance with special reference to some vectors of malaria. *Monogr. Ser. Wld Hlth Org.*, No. 47, 216 pp.

Detra, R.L. and Romoser, W.S. (1979a) Permeability of *Aedes aegypti* larval peritrophic membrane to proteolytic enzyme. *Mosq. News*, **39**, 582–5.

Detra, R.L. and Romoser, W.S. (1979b) Proteolytic enzyme concentration, weight change and starvation in larval *Aedes aegypti* (L.) (Diptera: Culicidae). *Mosq. News*, **39**, 594–6.

Devaney, E. (1985) Lectin-binding characteristics of *Brugia pahangi* microfilariae. *Trop. Med. Parasitol.*, **36**, 25–8.

Devine, T.L., Venard, C.E. and Myser, W.C. (1965) Measurement of salivation by *Aedes aegypti* (L.) feeding on a living host. *J. Insect Physiol.*, **11**, 347–53.

Dhadialla, T.S. and Raikhel, A.S. (1990) Biosynthesis of mosquito vitellogenin. *J. Biol. Chem.*, **265**, 9924–33.

Dhadialla, T.S. and Raikhel, A.S. (1991) Binding of vitellogenin to membranes isolated from mosquito ovaries. *Arch. Insect Biochem. Physiol.*, **18**, 55–70.

Diaz, G. and Lewis, K.R. (1975) Interphase chromosome arrangement in *Anopheles atroparvus*. *Chromosoma*, **52**, 27–35.

Dimond, J.B., Lea, A.O. and DeLong, D.M. (1956) The amino acids required for egg production in *Aedes aegypti*. *Can. Entomol.*, **88**, 57–62.

Dimond, J.B., Lea, A.O. and DeLong, D.M. (1958) Nutritional requirements for reproduction in insects. *Proc. 10th Int. Congr. Entomol.*, **2**, 135–7.

Dittmann, F., Kogan, P.H. and Hagedorn, H.H. (1989) Ploidy levels and DNA synthesis in fat body cells of the adult mosquito, *Aedes aegypti*: the role of juvenile hormone. *Arch. Insect Biochem. Physiol.*, **12**, 133–43.

Dixon, M. and Webb, E.C. (1979) *Enzymes*, 3rd edn, Longman, London.

Dobrotworsky, N.V. (1955) The *Culex pipiens* group in south-eastern Australia. IV. Crossbreeding experiments within the *Culex pipiens* group. *Proc. Linnean Soc. New South Wales*, **80**, 33–43.

Dörner, R. and Peters, W. (1988) Localization of sugar components of glycoproteins in peritrophic membranes of larvae of Diptera (Culicidae, Simuliidae). *Entomol. General.*, **14**, 11–24.

Downe, A.E.R. (1975) Internal regulation of rate of digestion of blood meals in the mosquito, *Aedes aegypti*. *J. Insect Physiol.*, **21**, 1835–9.

Downe, A.E.R. and Archer, J.A. (1975) The effects of different blood-meal sources on digestion and egg production in *Culex tarsalis* Coq. (Diptera: Culicidae). *J. Med. Entomol.*, **12**, 431–7.

Downer, R.G.H. (1985) Lipid metabolism. In *Comprehensive Insect Physiology Biochemistry and Pharmacology* (eds G.A. Kerkut and L.I. Gilbert), **10**, 77–113.

Downes, J.A. (1974) The feeding habits of adult Chironomidae. *Entomol. Tidskr.*, **95** (Suppl.), 84–90.

Downs, W.G. (1951) Growth changes of anopheline eggs in water and in saline solutions. *J. Natl Malaria Soc.*, **10**, 17–22.

Drif, L. and Brehélin, M. (1983) The circulating hemocytes of *Culex pipiens* and *Aedes aegypti*: cytology, histochemistry, hemograms and functions. *Dev. Comp. Immunol.*, **7**, 687–90.

Dubin, D.T., HsuChen, C.-C., Cleaves, G.R. and Timko, D. (1984) Sequence and structure of a serine transfer RNA with GCU anticodon from mosquito mitochondria. *J. Molec. Biol.*, **176**, 251–60.

Dubin, D.T., HsuChen, C.-C. and Tillotson, L.E. (1986) Mosquito mitochondrial transfer RNAs for valine, glycine and glutamate: RNA and gene sequences and vicinal genome organization. *Curr. Genet.*, **10**, 701–7.

Dubin, D.T. and HsuChen, C.-C. (1983) The 3′-terminal region of mosquito mitochondrial small ribosomal subunit RNA: sequence and localization of methylated residues. *Plasmid*, **9**, 307–20.

Dubin, D.T. and HsuChen, C.-C. (1985) The 3′ end of large ribosomal subunit RNA from mosquito mitochondria: homogeneity of transcribed moieties. *Plasmid*, **13**, 139–44.

Durbin, J.E., Swerdel, M.R. and Fallon, A.M. (1988) Identification of cDNAs corresponding to mosquito ribosomal protein genes. *Biochim. Biophys. Acta*, **950**, 182–92.

Dyar, H.G. (1903) *Culex atropalpus* Coquillett. *Entomol. News*, **14**, 180–2.

Dye, C. (1984) Competition amongst larval *Aedes aegypti*: the role of interference. *Ecol. Entomol.*, **9**, 355–7.

Eckert, D., Paasonen, M. and Vartiainen, A. (1951) On histamine in the gnat (*Culex pipiens*). *Acta. Pharmacol. Toxicol.*, **7**, 16–22.

Edman, J.D. (1970) Rate of digestion of vertebrate blood in *Aedes aegypti* (L.). Effect of age, mating, and parity. *Am. J. Trop. Med. Hyg.*, **19**, 1031–3.

Edman, J.D. and Downe, A.E.R. (1964) Host-blood sources and multiple-feeding habits of mosquitoes in Kansas. *Mosq. News*, **24**, 154–60.

Edman, J.D. and Kale. H.W. (1971) Host behavior: its influence on the feeding success of mosquitoes. *Ann. Entomol. Soc. Am.*, **64**, 513–6.

Edman, J.D., Webber, L.A. and Schmid, A.A. (1974) Effect of host defenses on the feeding pattern of *Culex nigripalpus* when offered a choice of blood sources. *J. Parasitol.*, **60**, 874–83.

Edwards, F.W. (1919) The larva and pupa of *Taeniorhynchus richiardii* Fic. (Diptera, Culicidae). *Entomol. Mon. Mag.*, **55**, 83–8.

Edwards, H.A. (1982a) *Aedes aegypti*: energetics of osmoregulation. *J. Exp. Biol.*, **101**, 135–41.

Edwards, H.A. (1982b) Ion concentration and activity in the haemolymph of *Aedes aegypti* larvae. *J. Exp. Biol.*, **101**, 143–51.

Edwards, H.A. (1982c) Free amino acids as regulators of

osmotic pressure in aquatic insect larvae. *J. Exp. Biol.*, **101**, 153–60.

Edwards, H.A. (1983) Electrophysiology of mosquito anal papillae. *J. Exp. Biol.*, **102**, 343–6.

Edwards, H.A. and Harrison, J.B. (1983) An osmoregulatory syncytium and associated cells in a freshwater mosquito. *Tissue & Cell*, **15**, 271–80.

Eliason, D.A. (1963) Feeding adult mosquitoes on solid sugars. *Nature*, **200**, 289.

Ellgaard, E.G., Capiola, R.J. and Barber, J.T. (1987) Preferential accumulation of *Culex quinquefasciatus* (Diptera: Culicidae) larvae in response to adenine nucleotides and derivatives. *J. Med. Entomol.*, **24**, 633–6.

Ellis, R.A. and Brust, R.A. (1973) Sibling species delimitation in the *Aedes communis* (Degeer) aggregate (Diptera: Culicidae). *Can. J. Zool.*, **51**, 915–959, 1 pl.

Else, J.G. and Judson, C.L. (1972) Enforced egg-retention and its effects on vitellogenesis in the mosquito, *Aedes aegypti*. *J. Med. Entomol.*, **9**, 527–30.

Elzinga, R.J. (1961) A comparison in time of egg hatching between male and female *Aedes aegypti* (L.). *Mosq. News*, **21**, 307–10.

Falleroni, D. (1926) Fauna anofelica italiana e suo 'habitat' (paludi, risaie, canali). Metodi dilotta contro la malaria. *Riv. Malariol.*, **5**, 553–93, 1 pl.

Fallon, A.M., Blahnik, R.J., Baldridge, G.D. and Park, Y. (1991) Ribosomal DNA structure in *Aedes* mosquitoes (Diptera: Culicidae) and their cell lines. *J. Med. Entomol.*, **28**, 637–44.

Fallon, A.M., Hagedorn, H.H., Wyatt, G.R. and Laufer, H. (1974) Activation of vitellogenin synthesis in the mosquito *Aedes aegypti* by ecdysone. *J. Insect Physiol.*, **20**, 1815–23.

Farci, A., Laudani, U. and Lecis, A.R. (1973) A revision of the banding pattern of salivary chromosomes of *Anopheles atroparvus* (Diptera : Nematocera), *Z. Zool. Systemat. Evolutionforsch.*, **11**, 304–10.

Farnsworth, M.W. (1947) The morphology and musculature of the larval head of *Anopheles quadrimaculatus* Say. *Ann. Entomol. Soc. Am.*, **40**, 137–51.

Farquharson, C.O. (1918) *Harpagomyia* and other Diptera fed by *Cremastogaster* ants in S. Nigeria. *Proc. Entomol. Soc. Lond.*, 1918, xxix–xxxix.

Feinsod, F.M. and Spielman, A. (1980a) Nutrient-mediated juvenile hormone secretion in mosquitoes. *J. Insect Physiol.*, **26**, 113–17.

Feinsod, F.M. and Spielman, A. (1980b) Independently regulated juvenile hormone activity and vitellogenesis in mosquitoes. *J. Insect Physiol.*, **26**, 829–32.

Feir, D., Lengy, J.I. and Owen, W.B. (1961) Contact chemoreception in the mosquito, *Culiseta inornata* (Williston); sensitivity of the tarsi and labella to sucrose and glucose. *J. Insect Physiol.*, **6**, 13–20.

Feldmann, A.M., Billingsley, P.F. and Savelkoul, E. (1991) Blood digestion by strains of *Anopheles stephensi* Liston (Diptera: Culicidae) of differing susceptibility to *Plasmodium falciparum*. *Parasitology*, **101**, 193–200.

Felix, C.R., Betschart, B., Billingsley, P.F. and Freyvogel, T.A. (1991) Post-feeding induction of trypsin in the midgut of *Aedes aegypti* L. (Diptera: Culicidae) is separable into two cellular phases. *Insect Biochem.*, **21**, 197–203.

Fiil, A. (1974) Structural and functional modifications of the nucleus during oogenesis in the mosquito *Aedes aegypti*. *J. Cell Sci.*, **14**, 51–67.

Fiil, A. (1976a) Oogenesis in the malaria mosquito *Anopheles gambiae*. *Cell Tissue Res.*, **167**, 23–35.

Fiil, A. (1976b) Polycomplexes and intranuclear annulate lamellae in mosquito oocytes. *Hereditas*, **84**, 117–20.

Fiil, A. (1978a) Follicle cell bridges in the mosquito ovary: syncytia formation and bridge morphology. *J. Cell Sci.*, **31**, 137–43.

Fiil, A. (1978b) Meiotic chromosome pairing and synaptonemal complex formation in *Culex pipiens quinquefasciatus* oocytes. *Chromosoma*, **69**, 381–95.

Fiil, A. and Moens, P.B. (1973) The development, structure and function of modified synaptonemal complexes in mosquito oocytes. *Chromosoma*, **41**, 37–62.

Fine, P.E.M. (1978) On the dynamics of symbiote-dependent cytoplasmic incompatibility in culicine mosquitoes. *J. Invert. Pathol.*, **30**, 10–18.

Fish, D. (1985) An analysis of adult size variation within natural mosquito populations. In *Ecology of Mosquitoes* (eds L.P. Lounibos, J.R. Rey and J.H. Frank), Florida Medical Entomology Laboratory, Vero Beach, pp. 419–29.

Fish, D. and Carpenter, S.R. (1982) Leaf litter and larval mosquito dynamics in tree-hole ecosystems. *Ecology*, **63**, 283–8.

Fisk, F.W. and Shambaugh, G.F. (1954) Invertase activity in adult *Aedes aegypti* mosquitoes. *Ohio J. Sci.*, **54**, 237–9.

Flanagan, T.R. and Hagedorn, H.H. (1977) Vitellogenin synthesis in the mosquito: the role of juvenile hormone in the development of responsiveness to ecdysone. *Physiol. Entomol.*, **2**, 173–78.

Florey, E. (1982) Excretion in insects: energetics and functional principles. *J. Exp. Biol.*, **99**, 417–24.

Foley, D.A. (1978) Innate cellular defense by mosquito hemocytes. *Comp. Pathobiol.*, **4**, 113–44.

Fong, W.F. and Fuchs, M.S. (1976) Studies on the mode of action of ecdysterone in adult female *Aedes aegypti*. *Mol. Cell. Endocrinol.*, **4**, 341–51.

Ford, E.B. (1961) The theory of genetic polymorphism. *Symp. R. Entomol. Soc. Lond.*, **1**, 11–19.

Forton, K.F., Christensen, B.M. and Sutherland, D.R. (1985) Ultrastructure of the melanization response of *Aedes trivittatus* against inoculated *Dirofilaria immitis* microfilariae. *J. Parasitol.*, **71**, 331–41.

Fraccaro, M., Laudani, U., Marchi, A. and Tiepolo, L. (1976) Karyotype, DNA replication and origin of sex chromosomes in *Anopheles atroparvus*. *Chromosoma*, **55**, 27–36.

Fraccaro, M., Tiepolo, L., Laudani, U. and Jayakar, S. (1977) Y chromosome controls mating behaviour on *Anopheles* mosquitoes. *Nature*, **265**, 326–8.

France, K.R. and Judson, C.L. (1979) Nitrogen partitioning and blood meal utilization by *Aedes aegypti* (Diptera, Culicidae). *J. Insect Physiol.*, **25**, 841–6.

Frank, J.H. and Curtis, G.A. (1977) On the bionomics of bromeliad-inhabiting mosquitoes. III. The probable strategy of larval feeding in *Wyeomyia vanduzeei* and *Wy. medioalbipes*. *Mosq. News*, **37**, 200–6

Frelinger, J.A. and Roth, T.F. (1971) Synthesis and conservation of ribosomal RNA in adult mosquito ovaries. *J. Insect Physiol.* **17**, 1401–10.

French, F.E. (1972) *Aedes aegypti*: histopathology of immediate skin reactions of hypersensitive guinea pigs resulting from bites. *Exp. Parasitol.*, **32**, 175–80.

French, W.L. (1978) Genetic and phenogenetic studies on the dynamic nature of the cytoplasmic inheritance system in *Culex pipiens*. *Genetics*, **88**, 447–55.

French, W.L., Baker, R.H. and Kitzmiller, J.B. (1962) Preparation of mosquito chromosomes. *Mosq. News*, **22**, 377–83.

Freyvogel, T.A., Hunter, R.L. and Smith, E.M. (1968) Non-specific esterases in mosquitoes. *J. Histochem. Cytochem.*, **16**, 765–90.

Freyvogel, T.A. and Jaquet, C. (1965) The prerequisites for the formation of a peritrophic membrane in Culicidae females. *Acta Tropica*, **22**, 148–54.

Freyvogel, T.A. and Stäubli, W. (1965) The formation of the peritrophic membrane in Culicidae. *Acta Tropica*, **22**, 118–47.

Friedman, S. (1985) Carbohydrate metabolism. In *Comprehensive Insect Physiology Biochemistry and Pharmacology* (eds G.A. Kerkut and L.I. Gilbert), Pergamon Press, Oxford, **10**, 43–76.

Friend, W.G. (1978) Physical factors affecting the feeding responses of *Culiseta inornata* to ATP, sucrose, and blood. *Ann. Entomol. Soc. Am.*, **71**, 935–40.

Friend, W.G. (1981) Diet destination in *Culiseta inornata* (Williston): effect of feeding conditions on the response to ATP and sucrose. *Ann. Entomol. Soc. Am.*, **74**, 151–4.

Friend, W.G. (1985) Diet ingestion and destination in *Culiseta inornata* (Diptera: Culicidae): effects of mouthpart deployment and contact of the fascicle and labellum with sucrose, water, saline, or ATP. *Ann. Entomol. Soc. Am.*, **78**, 495–500.

Friend, W.G., Schmidt, J.M., Smith, J.J.B. and Tanner, R.J. (1988) The effect of sugars on ingestion and diet destination in *Culiseta inornata*. *J. Insect Physiol.*, **34**, 955–61.

Friend, W.G. and Smith, J.J.B. (1972) Feeding stimuli and techniques for studying the feeding of haematophagous arthropods under artificial conditions, with special reference to *Rhodnius prolixus*. In *Insect and Mite Nutrition* (ed. J.G. Rodriguez), North-Holland, Amsterdam, pp. 241–56.

Friend, W.G. and Smith, J.J.B. (1977) Factors affecting feeding by bloodsucking insects. *Annual Rev. Entomol.*, **22**, 309–31.

Friend, W.G., Smith, J.J.B., Schmidt, J.M. and Tanner, R.J. (1989) Ingestion and diet destination in *Culiseta inornata*: responses to water, sucrose and cellobiose. *Physiol. Entomol.*, **14**, 137–46.

Fuchs, M.S. and Kang, S.H. (1981) Ecdysone and mosquito vitellogenesis: a critical appraisal. *Insect Biochem.*, **11**, 627–33.

Fuchs, M.S., Kang, S.H., Kelly, T.J. *et al.* (1981) Endocrine control of ovarian development in an autogenous mosquito. In *Regulation of Insect Development and Behaviour International Conference* (eds F. Sehnal, A. Zabza, J.J. Menn and B. Cymborowski), *Sci. Pap. Inst. Org. Phys. Chem. Wroclaw Techn. Univ.*, No. 22, 569–90.

Fuchs, M.S. and Schlaeger, D.A. (1973) The stimulation of DOPA decarboxylase activity by ecdysone and its enhancement by cyclic AMP in adult mosquitoes. *Biochem. Biophys. Res. Comm.*, **54**, 784–9.

Fuchs, M.S., Sunderland, B.R. and Kang, S.-H. (1980) *In vivo* induction of ovarian development in *Aedes atropalpus* by a head extract from *Aedes aegypti*. *Int. J. Invert. Reprod.*, **2**, 121–9.

Fülleborn, F. (1932) Über den Saugakt der Stechmücken. *Arch. Schiffs-u. Tropenhyg.*, **36**, 169–81.

Furumizo, R.T. and Rudnick, A. (1978) Laboratory studies of *Toxorhynchites splendens* (Diptera: Culicidae): biological observations. *Ann. Entomol. Soc. Am.*, **71**, 670–3.

Gale, K. and Crampton, J. (1989) The ribosomal genes of the mosquito, *Aedes aegypti*. *Europ. J. Biochem.*, **185**, 311–17.

Galindo, P., Carpenter, S.J. and Trapido, H. (1951) Ecological observations on forest mosquitoes of an endemic yellow fever area in Panama. *Am. J. Trop. Med.*, **31**, 98–137.

Galindo, P., Carpenter, S.J. and Trapido, H. (1955) A contribution to the ecology and biology of tree hole breeding mosquitoes of Panama. *Ann. Entomol. Soc. Am.*, **48**, 158–64.

Galun, R. (1967) Feeding stimuli and artificial feeding. *Bull. Wld Hlth Org*, **36**, 590–3.

Galun, R. (1987) The evolution of purinergic receptors involved in recognition of a blood meal by hematophagous insects. *Mem. Inst. Oswaldo Cruz*, **82** (Suppl. 3), 5–9.

Galun, R., Friend, W.G. and Nudelman, S. (1988) Purinergic reception by culicine mosquitoes. *J. Comp. Physiol.* (A), **163**, 665–70.

Galun, R., Koontz, L.C. and Gwadz, R.W. (1985a)

Engorgement response of anopheline mosquitoes to blood fractions and artificial solutions. *Physiol. Entomol.*, **10**, 145–9.

Galun, R., Koontz, L.C., Gwadz, R.W. and Ribeiro, J.C. (1985b) Effect of ATP analogues on the gorging response of *Aedes aegypti. Physiol. Entomol.*, **10**, 275–81.

Galun, R., Oren, N. and Zecharia, M. (1984) Effect of plasma components on the feeding response of the mosquito *Aedes aegypti* L. to adenine nucleotides. *Physiol. Entomol.*, **9**, 403–8.

Galun, R. and Rice, M.J. (1971) Role of blood platelets in haematophagy. *Nature, New Biol.*, **233**, 110–11.

Gander, R. (1951) Experimentelle und oekologische Untersuchungen über das Schlupfvermögen der Larven von *Aedes aegypti* L. *Rev. Suisse Zool.*, **58**, 215–78.

Gander, E. (1968) Zur Histochemie und Histologie des Mitteldarmes von *Aedes aegypti* und *Anopheles stephensi* in Zusammenhang mit der Blutverdauung. *Acta Tropica*, **25**, 133–75.

Gander, E.S., Schoenenberger, C. and Freyvogel, T.A. (1980) Ribosome and ribosome-function in the midgut of *Aedes aegypti. Insect Biochem.*, **10**, 441–7.

Garrett, M. and Bradley, T.J. (1984a) The pattern of osmotic regulation in larvae of the mosquito *Culiseta inornata. J. Exp. Biol.*, **113**, 133–41.

Garrett, M. and Bradley, T.J. (1984b) Ultrastructure of osmoregulatory organs in larvae of the brackish-water mosquito, *Culiseta inornata* (Williston). *J. Morphol.*, **182**, 257–77.

Garrett, M.A. and Bradley, T.J. (1987) Extracellular accumulation of proline, serine and trehalose in the haemolymph of osmoconforming brackish-water mosquitoes. *J. Exp. Biol.*, **129**, 231–8.

Gass, R.F. (1977) Influences of blood digestion on the development of *Plasmodium gallinaceum* (Brumpt) in the midgut of *Aedes aegypti* (L.). *Acta Tropica*, **34**, 127–40.

Gatti, M., Bonaccorsi, S., Pimpinelli, S. and Coluzzi, M. (1982) Polymorphism of sex chromosome heterochromatin in the *Anopheles gambiae* complex. In *Recent Developments in the Genetics of Insect Disease Vectors* (eds W.W.M. Steiner, W.J. Tabachnick, K.S. Rai and S. Narang), Stipes Publishing Co., Champaign, Illinois, pp. 32–48.

Geering, K. (1975) Haemolytic activity in the blood clot of *Aedes aegypti. Acta Tropica*, **32**, 145–51.

Geering, K. and Freyvogel, T.A. (1974) The distribution of actylcholine and unspecific esterases in the midgut of female *Aedes aegypti* L. *Comp. Biochem. Physiol.*, **49B**, 775–84.

Geering, K. and Freyvogel, T.A. (1975) Lipase activity and stimulation mechanism of esterases in the midgut of female *Aedes aegypti. J. Insect Physiol.*, **21**, 1251–6.

Gemmill, R.M., Hamblin, M., Glaser, R.L. *et al.* (1986) Isolation of mosquito vitellogenin genes and induction of expression by 20-hydroxyecdysone. *Insect Biochem.* **16**, 761–74.

Giblin, R.M. and Platzer, E.G. (1984) Hemolymph pH of the larvae of three species of mosquitoes, and the effect of *Romanomermis culicivorax* parasitism on the blood pH of *Culex pipiens. J. Invert. Pathol.*, **44**, 63–6.

Giglioli, M.E.C. (1963) The female reproductive system of *Anopheles gambiae melas.* I. The structure and function of the genital ducts and associated organs. *Riv. Malariol.*, **42**, 149–76, 6 pls.

Giglioli, M.E.C. (1964a) Tides, salinity and the breeding of *Anopheles melas* (Theobald, 1903) during the dry season in The Gambia. *Riv. Malariol.*, **43**, 245–63.

Giglioli, M.E.C. (1964b) The female reproductive system of *Anopheles gambiae melas.* II. The ovary. *Riv. Malariol.*, **43**, 265–75, 4 pls.

Giglioli, M.E.C. (1965) The problem of age determination in *Anopheles melas* Theo. 1903, by Polovodova's method. *Cahier ORSTOM, Entomol. Méd.*, 1965 (3,4), 157–77.

Gilchrist, B.M. and Haldane, J.B.S. (1947) Sex linkage and sex determination in a mosquito, *Culex molestus. Hereditas*, **33**, 175–90.

Gillett, J.D. (1955) Behaviour differences in two strains of *Aedes aegypti. Nature*, **176**, 124.

Gillett, J.D. (1956) Initiation and promotion of ovarian development in the mosquito *Aedes (Stegomyia) aegypti* (Linnaeus). *Ann. Trop. Med. Parasitol.*, **50**, 375–80.

Gillett, J.D. (1957) Variation in the time of release of the ovarian development hormone in *Aedes aegypti. Nature*, **180**, 656–7.

Gillett, J.D. (1961) Laboratory observations on the life-history and ethology of *Mansonia* mosquitoes. *Bull. Entomol. Res.*, **52**, 23–30.

Gillett, J.D. (1967) Natural selection and feeding speed in a blood-sucking insect. *Proc. R. Soc. Lond.* (B), **167**, 316–29.

Gillett, J.D. (1981) Blood feeding and the neuroendocrine system in mosquitoes. *Parasitology*, **82**, 97–8.

Gillett, J.D. (1982a) Circulatory and ventilatory movements of the abdomen in mosquitoes. *Proc. R. Soc. Lond.* (B), **215**, 127–34.

Gillett, J.D. (1982b) Diuresis in newly emerged, unfed mosquitoes. I. Fluid loss in normal females and males during the first 20 hours of adult life . *Proc. R. Soc. Lond.* B, **216**, 201–7.

Gillett, J.D. (1983a) Diuresis in newly emerged, unfed mosquitoes. II. The basic pattern in relation to escape from the water, preparation for mature flight, mating and the first blood meal. *Proc. R. Soc. Lond.* (B), **217**, 237–42.

Gillett, J.D. (1983b) Mid-gut air in newly emerged mosquitoes and its elimination. *Proc. Pap. Calif. Mosq. Vector Contr. Ass.*, **50**, 86–7.

Gillett, J.D. (1983c) Abdominal pulses in newly emerged mosquitoes, *Aedes aegypti. Mosq. News*, **43**, 359–61.

Gillett, J.D. (1984) The effects of decapitation and the influence of size and sex on diuresis in newly emerged mosquitoes. *Physiol. Entomol.*, **9**, 139–44.

Gillett, J.D., Cole, S.J. and Reeves, D. (1975) The influence of the brain hormone on retention of blood in the mid-gut of the mosquito *Aedes aegypti*. *Proc. R. Soc. Lond.* (B), **190**, 359–67.

Gillies, M.T. (1953) The duration of the gonotrophic cycle in *Anopheles gambiae* and *Anopheles funestus*, with a note on the efficiency of hand catching. *East African Med. J.*, **30**, 129–35.

Gillies, M.T. (1954) The recognition of age-groups within populations of *Anopheles gambiae* by the pre-gravid rate and the sporozoite rate. *Ann. Trop. Med. Parasitol.*, **48**, 58–74.

Gillies, M.T. (1955) The pre-gravid phase of ovarian development in *Anopheles funestus*. *Ann. Trop. Med. Parasitol.*, **49**, 320–5.

Gillies, M.T. and Wilkes, T.J. (1965) A study of the age-composition of populations of *Anopheles gambiae* Giles and *A. funestus* Giles in north-eastern Tanzania. *Bull. Entomol. Res.*, **56**, 237–62, 1 pl.

Gilpin, M.E. and McClelland, G.A.H. (1979) Systems analysis of the yellow fever mosquito *Aedes aegypti*. *Fortschr. Zool.*, **25**, 355–88.

Gjullin, C.M., Hegarty, C.P. and Bollen, W.B. (1941) The necessity of a low oxygen concentration for the hatching of *Aedes* eggs (Diptera: Culicidae). *J. Cell. Comp. Physiol.*, **17**, 193–202.

Gjullin, C.M., Yates, W.W. and Stage, H.H. (1950) Studies on *Aedes vexans* (Meig.) and *Aedes sticticus* (Meig.), flood-water mosquitoes, in the lower Columbia river valley. *Ann. Entomol. Soc. Am.*, **43**, 262–75.

Gnatzy, W. (1970) Struktur und Entwicklung des Integuments und der Oenocyten von *Culex pipiens* L. (Dipt.). *Z. Zellforsch.*, **110**, 401–43.

Golberg, L. and De Meillon, B. (1948) The nutrition of the larva of *Aedes aegypti* Linnaeus. 4. Protein and amino-acid requirements. *Biochem. J.*, **43**, 379–87.

Goldman, L., Rockwell, E. and Richfield, D.F. (1952) Histopathological studies on cutaneous reactions to the bites of various arthropods. *Am. J. Trop. Med. Hyg.*, **1**, 514–25.

Goma, L.K.H. (1964) The exudation of fluid by the newly emerged adult of the mosquito, *Anopheles gambiae* Giles. *Entomologist*, **97**, 233–9.

Gooding, R.H. (1966a) *In vitro* properties of proteinases in the midgut of adult *Aedes aegypti* L. and *Culex fatigans* (Wiedemann). *Comp. Biochem. Physiol.*, **17**, 115–27.

Gooding, R.H. (1966b) Physiological aspects of digestion of the blood meal by *Aedes aegypti* (Linnaeus) and *Culex fatigans* Wiedemann. *J. Med. Entomol.*, **3**, 53–60.

Gooding, R.H. (1969) Studies on proteinases from some blood-sucking insects. *Proc. Entomol. Soc. Ontario*, **100**, 139–45.

Gooding, R.H., Cheung, A.C. and Rolseth, B.M. (1973) The digestive processes of haematophagous insects. III. Inhibition of trypsin by honey and the possible functions of the oesophageal diverticula of mosquitoes (Diptera). *Can. Entomol.*, **105**, 433–6.

Gordon, R., Finney, J.R., Condon, W.J. and Rusted, T.N. (1979) Lipids in the storage organs of three mermithid nematodes and in the hemolymph of their hosts. *Comp. Biochem. Physiol.*, **64B**, 369–74.

Gordon, R.M. and Lumsden, W.H.R. (1939) A study of the behaviour of the mouthparts of mosquitoes when taking up blood from living tissue ; together with some observations on the ingestion of microfilariae. *Ann. Trop. Med. Parasitol.*, **33**, 259–78.

Götz, P. (1986) Encapsulation in arthropods. In *Immunity in Invertebrates* (ed. M. Bréhelin), Springer-Verlag, Berlin, pp. 153–70.

Graf, R., Binz, H. and Briegel, H. (1988) Monoclonal antibodies as probes for *Aedes aegypti* trypsin. *Insect Biochem.*, **18**, 463–70.

Graf, R., Boehlen, P. and Briegel, H. (1991) Structural diversity of trypsin from different mosquito species feeding on vertebrate blood. *Experientia*, **47**, 603–9.

Graf, R. and Briegel, H. (1982) Comparison between aminopeptidase and trypsin activity in blood-fed females of *Aedes aegypti*. *Rev. Suisse Zool.*, **89**, 845–50.

Graf, R. and Briegel, H. (1985) Isolation of trypsin isozymes from the mosquito *Aedes aegypti* (L.). *Insect Biochem.*, **15**, 611–18.

Graf, R. and Briegel, H. (1989) The synthetic pathway of trypsin in the mosquito *Aedes aegypti* L. (Diptera: Culicidae) and in vitro stimulation in isolated midguts. *Insect Biochem.*, **19**, 129–37.

Graf, R., Raikhel, A.S., Brown, M.R. *et al.* (1986) Mosquito trypsin: immunocytochemical localization in the midgut of blood-fed *Aedes aegypti* (L.). *Cell Tissue Res.*, **245**, 19–27.

Graziosi, C., Sakai, R.K., Romans, P. *et al.* (1990) Method for in situ hybridization to polytene chromosomes from ovarian nurse cells of *Anopheles gambiae* (Diptera: Culicidae). *J. Med. Entomol.*, **27**, 905–12.

Green, C.A. (1972) Cytological maps for the practical identification of females of the three freshwater species of the *Anopheles gambiae* complex. *Ann. Trop. Med. Parasitol.*, **66**, 143–7.

Green, C.A. (1982a) Population genetical studies in the genus *Anopheles*. PhD thesis, University of the Witwatersrand, Johannesburg, South Africa.

Green, C.A. (1982b) Polytene-chromosome relationships of the *Anopheles stephensi* species group from the Afrotropical and Oriental regions. In *Recent Developments in the Genetics of Insect Disease Vectors* (eds W.W.M. Steiner, W.J. Tabachnick, K.S. Rai and S. Narang), Stipes Publishing Co., Champaign, Illinois, pp. 49–61.

Green, C.A. and Hunt, R.H. (1980) Interpretation of variation in ovarian polytene chromosomes of *Anopheles funestus* Giles, *A. parensis* Gillies, and *A. aruni*?. *Genetica*, **51**, 187–95.

Greenplate, J.T., Glaser, R.L. and Hagedorn, H.H. (1985) The role of factors from the head in the regulation of egg development in the mosquito *Aedes aegypti*. *J. Insect Physiol.*, **31**, 323–9.

Greenwood, B., Marsh, K. and Snow, R. (1991) Why do some African children develop severe malaria? *Parasitol. Today*, **7**, 277–81.

Grell, M (1946a) Cytological studies in *Culex*. I. Somatic reduction divisions. *Genetics*, **31**, 60–76, 3 pls.

Grell, M. (1946b) Cytological studies in *Culex*. II. Diploid and meiotic divisions. *Genetics*, **31**, 77–94, 1 pl.

Griffiths, R.B. and Gordon, R.M. (1952) An apparatus which enables the process of feeding by mosquitoes to be observed in the tissues of a live rodent; together with an account of the ejection of saliva and its significance in malaria. *Ann. Trop. Med. Parasitol.*, **46**, 311–9, 3 pl.

Grillot, J.P. (1977) Les organes périsympathiques des Diptères. *Int. J. Insect Morphol. Embryol.*, **6**, 303–43.

Grimstad, P.R. and DeFoliart, G.R. (1974) Nectar sources of Wisconsin mosquitoes. *J. Med. Entomol.*, **11**, 331–41.

Grimstad, P.R. and Walker, E.D. (1991) *Aedes triseriatus* (Diptera: Culicidae) and La Crosse virus. IV. Nutritional deprivation of larvae affects the adult barriers to infection and transmission. *J. Med. Entomol.*, **28**, 378–86.

Grossman. G.L. and Pappas, L.G. (1991) Human skin temperature and mosquito (Diptera: Culicidae) blood feeding rate. *J. Med. Entomol.*, **28**, 456–60.

Grover, K.K., Suguna, S.G., Uppal, D.K. *et al.* (1976) Field experiments on the competitiveness of males carrying genetic control systems for *Aedes aegypti*. *Entomol. Exp. Applic.*, **20**, 8–18.

Gubler, D.J. and Bhattacharya, N.C. (1971) Observations on the reproductive history of *Aedes* (*Stegomyia*) *albopictus* in the laboratory. *Mosq. News*, **31**, 356–9.

Guichard, M. (1971) Etude *in vivo* du développement embryonnaire de *Culex pipiens*. Comparaison avec *Calliphora erythrocephala* (Diptera). *Ann. Soc. Entomol. France*, **7**, 325–41.

Guichard, M. (1973) Recherche des origines embryonnaires de quelques structures ventrales de la tête larvaire chez *Culex pipiens* L. (Diptère Nématocère). *Bull. Soc. Zool. France*, **98**, 23–34.

Guilvard, E., De Reggi, M. and Rioux, J.-A. (1984) Changes in ecdysteroid and juvenile hormone titers correlated to the initiation of vitellogenesis in two *Aedes* species (Diptera, Culicidae). *Gen. Comp. Endocrinol.*, **53**, 218–23.

Guilvard, E., Raabe, M. and Rioux, J.-A. (1976) Autogenèse et neurosécrétion cérébrale chez *Aedes detritus* (Haliday, 1833) (Diptera-Culicidae). *C.R. Acad. Sci., Paris (D)*, **283**, 1217–20.

Guilvard, E. and Rioux, J.-A. (1986) Dynamique de l'autogenèse dans les populations naturelles d'*Aedes* (*O.*) *detritus* (Haliday, 1833) espèce jumelle A [Diptera – Culicidae] en Camargue. Role prééminant de la photopériode. *Ann. Parasitol. Hum. Comp.*, **61**, 109–19.

Gupta, A.P. (1985) Cellular elements in the haemolymph. In *Comprehensive Insect Physiology Biochemistry and Pharmacology* (eds G.A. Kerkut and L.I. Gilbert), Pergamon Press, Oxford, **3**, 401–51.

Guptavanij, P. and Barr, R.A. (1985) Failure of culicine eggs to darken in the field. *J. Med. Entomol.*, **22**, 228-9.

Guptavanij, P. and Venard, C.E. (1965) A radiographic study of the oesophageal diverticula and stomach of *Aedes aegypti* (L.). *Mosq. News*, **25**, 288–93.

Guthrie, M. (1989) *Animals of the Surface Film*, Richmond Publishing Co. Ltd, Slough, UK.

Gwadz, R.W. (1969) Regulation of blood meal size in the mosquito. *J. Insect Physiol.*, **15**, 2039–44.

Gwadz, R.L. and Spielman, A. (1973) Corpus allatum control of ovarian development in *Aedes aegypti*. *J. Insect Physiol.*, **19**, 1441–8.

Haas, G. (1956) Entwicklung des Komplexauges bei *Culex pipiens* und *Aedes aegypti*. *Z. Morphol. Ökol. Tiere*, **45**, 198–216.

Haddow, A.J. (1946) The mosquitoes of Bwamba County, Uganda. IV.–Studies on the genus *Eretmapodites*, Theobald. *Bull. Entomol. Res.*, **37**, 57–82.

Haddow, A.J., Gillett, J.D. and Corbet, P.S. (1959) Laboratory observations on pupation and emergence in the mosquito *Aedes* (*Stegomyia*) *aegypti* (Linnaeus). *Ann. Trop. Med. Parasitol.*, **53**, 123–31.

Haeger, J.S. (1955) The non-blood feeding habits of *Aedes taeniorhynchus* (Diptera, Culicidae) on Sanibel Island, Florida. *Mosq. News*, **15**, 21–6.

Haeger, J.S. and Provost, M.W. (1965) Colonization and biology of *Opifex fuscus*. *Trans. R. Soc. N.Z.*, **6**, 21–31.

Hagedorn, H.H. (1983) The role of ecdysteroids in the adult insect. In *Invertebrate Endocrinology. Vol. 1. Endocrinology of Insects* (eds R.G.H. Downer and H. Laufer), Alan R. Liss Inc., New York, pp. 271–304.

Hagedorn, H.H. (1985) The role of ecdysteroids in reproduction. In *Comprehensive Insect Physiology Biochemistry and Pharmacology* (eds G.A. Kerkut and L.I. Gilbert), Pergamon Press, Oxford, **8**, 205–62.

Hagedorn, H.H. (1986) Simplicity versus complexity: a study of the problems involved in relating in vitro results to the live animal. In *Techniques in the Life Sciences*, Vol. C2, *Techniques in in vitro Invertebrate Hormones and Genes* (eds E. Kurstak and H. Oberlander), Elsevier Scientific Publications Ireland Ltd, **C201**, 1–13.

Hagedorn, H.H. (1989) Physiological roles of hemolymph

ecdysteroids in the adult insect. In *Ecdysone. From Chemistry to Mode of Action* (ed. J. Koolman), Georg Thieme Verlag, Stuttgart, pp. 279–89.

Hagedorn, H.H. and Fallon, A.M. (1973) Ovarian control of vitellogenin synthesis by the fat body in *Aedes aegypti. Nature*, **244**, 103–5.

Hagedorn, H.H., Fallon, A.M. and Laufer, H. (1973) Vitellogenin synthesis by the fat body of the mosquito *Aedes aegypti*: evidence for transcriptional control. *Dev. Biol.*, **31**, 285–94.

Hagedorn, H.H. and Judson, C.L. (1972) Purification and site of synthesis of *Aedes aegypti* yolk proteins. *J. Exp. Zool.*, **182**, 367–77.

Hagedorn, H.H., O'Connor, J.D., Fuchs, M.S. et al. (1975) The ovary as a source of α-ecdysone in an adult insect. *Proc. Natl Acad. Sci. USA*, **72**, 3255–9.

Hagedorn, H.H., Shapiro, J.P. and Hanaoka, K. (1979) Ovarian ecdysone secretion is controlled by a brain hormone in an adult mosquito. *Nature*, **282**, 92–4.

Hagedorn, H.H., Turner, S., Hagedorn, E.A. *et al.* (1977) Postemergence growth of the ovarian follicles of *Aedes aegypti. J. Insect Physiol.*, **23**, 203–6.

Hagstrum, D.W. and Gunstream, S.E. (1971) Salinity, pH, and organic nitrogen of water in relation to presence of mosquito larvae. *Ann. Entomol. Soc. Am.*, **64**, 465–7.

Hagstrum, D.W. and Workman, E.B. (1971) Interaction of temperature and feeding rate in determining the rate of development of larval *Culex tarsalis* (Diptera, Culicidae). *Ann. Entomol. Soc. Am.*, **64**, 668–71.

Hall, D.W. (1987) Gynandromorphism in mosquitoes. *J. Florida Anti-Mosq. Ass.*, **58**, 25-88.

Hall, D.W. and Avery, S.W. (1978) Hemocytes of mosquito larvae. *Florida Entomol.*, **61**, 63–8.

Hall, M.H., Dutro, S.M. and Klowden, M.J. (1990) Determination by near-infrared reflectance spectroscopy of mosquito (Diptera: Culicidae) bloodmeal size. *J. Med. Entomol.*, **27**, 76–9.

Ham, P.J., Phiri, J.S. and Nolan, G.P. (1991) Effect of N-acetyl-D-glucosamine on the migration of *Brugia pahangi* microfilariae into the haemocoel of *Aedes aegypti. Med. Vet. Entomol.*, **5**, 485–93.

Hamblin, M.T., Marx, J.L., Wolfner, M.F. and Hagedorn, H.H. (1987) The vitellogenin gene family of *Aedes aegypti. Mem. Inst. Oswaldo Cruz*, **82** (Suppl. 3), 109–14.

Hanaoka, K. and Hagedorn, H.H. (1980) Brain hormone control of ecdysone secretion by the ovary in a mosquito. In *Progress in Ecdysone Research* (ed. J.A. Hoffmann), Elsevier/North Holland Biomedical Press, pp. 467–80.

Hanec, W. and Brust, R.A. (1967) The effect of temperature on the immature stages of *Culiseta inornata* (Diptera: Culicidae) in the laboratory. *Can. Entomol.*, **99**, 59–64.

Harada, R., Moriya, K. and Yabe, T. (1975) Observations on the survival and longevity of *Culex* and *Aedes* mosquitoes fed on the flowers of nectar plants (IV). *Jap. J. Sanit. Zool.*, **26**, 193–201.

Harada. F., Moriya, K. and Yabe, T. (1976) Observations on the survival and longevity of *Culex* and *Aedes* mosquitoes fed on the flowers of nectar plants (IV supplement). *Jap. J. Sanit. Zool.*, **27**, 307–9.

Harbach, R.E. (1977) Comparative and functional morphology of the mandibles of some fourth stage mosquito larvae (Diptera: Culicidae). *Zoomorphologie*, **87**, 217–36.

Harbach, R.E. (1978) Comparative structure of the labiohypopharynx of fourth stage mosquito larvae (Diptera: Culicidae), with comments on larval morphology, evolution and feeding habits. *Mosq. Systemat.*, **10**, 301–33.

Harbach, R.E., Harrison, B.A. and Gad, A.M. (1984) *Culex (Culex) molestus* Forskål (Diptera: Culicidae): neotype designation, description, variation, and taxonomic status. *Proc. Entomol. Soc. Wash.*, **86**, 521–42.

Harbach, R.E. and Knight, K.L. (1977a) A mosquito taxonomic glossary XII. The larval labiohypopharynx. *Mosq. Systemat.*, **9**, 337–65.

Harbach, R.E. and Knight, K.L. (1977b) A mosquito taxonomic glossary XIII. The larval pharynx. *Mosq. Systemat.*, **9**, 389–401.

Harbach, R.E. and Knight, K.L. (1980) *Taxonomists Glossary of Mosquito Anatomy*. Plexus Publishing Inc., New Jersey.

Harber. P.A. and Mutchmor, J.A. (1970) The early embryonic development of *Culiseta inornata* (Diptera: Culicidae). *Ann. Entomol. Soc. Am.*, **63**, 1609–14.

Hardy, J.L., Houk, E.J., Kramer, L.D. and Reeves, W.C. (1983) Intrinsic factors affecting vector competence of mosquitoes for arboviruses. *Annual Rev. Entomol.*, **28**, 229–62.

Harnish, D.G. and White, B.N. (1982) Insect vitellins: identification, purification, and characterization from eight orders. *J. Exp. Zool.*, **220**, 1–10.

Harnish, D.G., Wyatt, G.R. and White, B.N. (1982) Insect vitellins: identification of primary products of translation. *J. Exp. Zool.*, **220**, 11–19.

Hartberg, W.K. and Johnston, K.W. (1977) Red-eye, a sex-linked mutant in the mosquito *Eretmapodites quinquevittatus* Theobald. *Mosq. News*, **37**, 725–8.

Harwood, R.F. (1958) Development, structure, and function of coverings of eggs of floodwater mosquitoes. II. Postovarian structure. *Ann. Entomol. Soc. Am.*, **51**, 464–71.

Harwood, R.F. (1966) The relationship between photoperiod and autogeny in *Culex tarsalis* (Diptera, Culicidae). *Entomol. Exp. Applic.*, **9**, 327–31.

Harwood, R.F. and Horsfall, W.R. (1959) Development, structure, and function of coverings of eggs of floodwater mosquitoes. III. Functions of coverings. *Ann. Entomol. Soc. Am.*, **52**, 113–16.

Harwood, R.F. and Takata, N. (1965) Effect of photoperiod and temperature on fatty acid composition of the mosquito *Culex tarsalis*. *J. Insect Physiol.*, **11**, 711–16.

Hassett, C.C. and Jenkins, D.W. (1951) The uptake and effect of radiophosphorus in mosquitoes. *Physiol. Zool.*, **24**, 257–66.

Hastings, R.J. and Wood, R.J. (1978) Meiotic drive at the D(MD) locus and fertility in the mosquito, *Aedes aegypti* (L.). *Genetica*, **49**, 159–63.

Hatfield, P.R. (1988) Detection and localization of antibody ingested with a mosquito bloodmeal. *Med. Vet. Entomol.*, **2**, 339–45.

Haufe, W.O. and Burgess, L. (1956) Development of *Aedes* (Diptera: Culicidae) at Fort Churchill, Manitoba, and prediction of dates of emergence. *Ecology*, **37**, 500–19.

Hausermann, W. and Nijhout, H.F. (1975) Permanent loss of male fecundity following sperm depletion in *Aedes aegypti* (L.). *J. Med. Entomol.*, **11**, 707–15.

Hawley, W.A. (1985) A high-fecundity aedine: factors affecting egg production of the western treehole mosquito, *Aedes sierrensis* (Diptera: Culicidae). *J. Med. Entomol.*, **22**, 220–5.

Hayes, T.K., Holman, G.M., Pannabecker, T.L. *et al.* (1992) Culekinin depolarizing peptide: a mosquito leucokinin-like peptide that influences insect Malpighian tubule ion transport. *Regulat. Peptides* (submitted)

Hayes, T.K., Pannabecker, T.L., Hinckley, D.J. *et al.* (1989) Leucokinins, a new family of ion transport stimulators and inhibitors in insect Malpighian tubules. *Life Sci.*, **44**, 1259–66.

Hays, A.R. and Raikhel, A.S. (1990) A novel protein produced by the vitellogenic fat body and accumulated in mosquito oocytes. *Roux's Arch. Dev. Biol.*, **199**, 114–21.

Headlee, T.J. (1942) A continuation of the studies of the relative effects on insect metabolism of temperature derived from constant and varied sources. *J. Econ. Entomol.*, **35**, 785–6.

Hearle, E. (1929) The life history of *Aedes flavescens* Muller – a contribution to the biology of mosquitoes of the Canadian prairies. *Trans. R. Soc. Can.*, **23**, 85–102, 6 pls.

Heatwole, H. and Shine, R. (1976) Mosquitoes feeding on ectothermic vertebrates: a review and new data. *Austral. Zool.*, **19**, 69–75.

Hecker, H. (1977) Structure and function of midgut epithelial cells in Culicidae mosquitoes (Insecta, Diptera). *Cell Tissue Res.*, **184**, 321–41.

Hecker, H. (1978) Intracellular distribution of ribosomes in midgut cells of the malaria mosquito, *Anopheles stephensi* (Liston) (Insecta : Diptera) in response to feeding. *Int. J. Insect Morphol. Embryol.*, **7**, 267–72.

Hecker, H. and Bleiker, S. (1972) Feinstruktur und Funktion des ventralen Darmdivertikels bei *Aedes aegypti*. *Rev. Suisse Zool.*, **79**, 1027–31.

Hecker, H., Brun, R. and Reinhardt, C. (1974) Morphometeric analysis of the midgut of female *Aedes aegypti* (L.) (Insecta, Diptera) under various physiological conditions. *Cell Tissue Res.*, **152**, 31–49.

Hecker, H., Freyvogel, T.A., Briegel, H. and Steiger, R. (1971a) Ultrastructural differentiation of the midgut epithelium in female *Aedes aegypti* (L.) (Insecta, Diptera) imagines. *Acta Tropica*, **28**, 80–104.

Hecker, H., Freyvogel, T.A., Briegel, H. and Steiger, R. (1971b) The ultrastructure of midgut epithelium in *Aedes aegypti* (L.) (Insecta, Diptera) males. *Acta Trop.*, **28**, 275–90.

Hecker, H. and Rudin, W. (1979) Normal versus α-amanitin induced cellular dynamics of the midgut epithelium in female *Aedes aegypti* L. (Insecta, Diptera) in response to blood feeding. *Eur. J. Cell Biol.*, **19**, 160–7.

Hecker, H. and Rudin, W. (1981) Morphometric parameters of the midgut cells of *Aedes aegypti* L. (Insecta, Diptera) under various conditions. *Cell Tissue Res.*, **219**, 619–27.

Hegarty, J.L., Zhang, B., Pannabecker, T.L. *et al.* (1991) cAMP activates bumetanide-sensitive electrolyte transport in Malpighian tubules. *Am. J. Physiol.* **261** (*Cell Physiol.* 30), C521–C529.

Hermansson, M. (1990) The dynamics of dissolved and particulate organic material in surface microlayers. In *The Biology of Particles in Aquatic Systems* (ed. R.S. Wotton), CRC Press, Boca Raton, pp. 145–59.

Hertig, M. (1936) The rickettsia, *Wolbachia pipientis* (gen. et sp.n.) and associated inclusions of the mosquito, *Culex pipiens*. *Parasitology*, **28**, 453–86, 5 pls.

Hervy, J.-P. (1977) Expérience de marquage-lâcher-recapture portant sur *Aedes aegypti* Linné, en zone de savane soudanienne ouest-africaine. I. Le cycle trophogonique. *Cahiers ORSTOM, Entomol. Méd. Parasitol.*, **15**, 353–64.

Heusner, A.A. and Lavoipierre, M.M.J. (1973) Effet énergétique du repas sanguin chez *Aedes aegypti*. *C.R. Acad. Sci., Paris*, **276**, 1725–8.

Heusner, A.A., Lavoipierre, M.M.J. and Bond, D.C. (1973) Etude cinétique de l'effet métabolique d'un repas sanguin chez *Aedes aegypti*. *C.R. Acad. Sci., Paris*, **277**, 2017–20.

Hickey, W.A. (1970) Factors influencing the distortion of sex ratio in *Aedes aegypti*. *J. Med. Entomol.*, **7**, 727–35.

Hickey, W.A. and Craig, G.B. (1966a) Genetic distortion of sex ratio in a mosquito, *Aedes aegypti*. *Genetics*, **53**, 1177–96.

Hickey, W.A. and Craig, G.B. (1966b) Distortion of sex ratio in populations of *Aedes aegypti*. *Can. J. Genet. and Cytol.*, **8**, 260–78.

Hidano, A., Kawakami, M. and Yago, A. (1982) Hypersensitivity to mosquito bite and malignant histiocytosis. *Jap. J. Exp. Med.*, **52**, 303–6.

Hien, Do Si (1976) Biology of *Aedes aegypti* (L., 1762) and *Aedes albopictus* (Skuse, 1895) (Diptera, Culicidae). V. The gonotrophic cycle and oviposition. *Acta Parasitol. Polonica*, **24**, 37–55.

Hilburn, L.R. and Rai, K.S. (1982) Genetic analysis of abnormal male sexual development in *Aedes aegypti* and *Ae. mascarensis* backcross progeny. *J. Hered.*, **73**, 59–63.

Hinke, W. (1961) Das relative postembryonale Wachstum der Hirnteile von *Culex pipiens*, *Drosophila melanogaster* und *Drosophila*-mutanten. *Z. Morphol. Ökol. Tiere*, **50**, 81–118.

Hinman, H.E. (1932) The role of solutes and colloids in the nutrition of anopheline larvae. A preliminary report. *Am. J. Trop. Med.*, **12**, 263–71.

Hinman, H.E. (1933) Enzymes in the digestive tract of mosquito larvae. *Ann. Entomol. Soc. Am.*, **26**, 45–52.

Hinton, H.E. (1947) On the reduction of functional spiracles in the aquatic larvae of the Holometabola, with notes on the moulting process of spiracles. *Trans. R. Entomol. Soc. Lond.*, **98**, 449–73.

Hinton, H.E. (1959) Origin of indirect flight muscles in primitive flies. *Nature*, **183**, 557–8.

Hinton, H.E. (1968a) Structure and protective devices of the egg of the mosquito *Culex pipiens*. *J. Insect Physiol.*, **14**, 145–61, 6 pls.

Hinton, H.E. (1968b) Observations on the biology and taxonomy of the eggs of *Anopheles* mosquitos. *Bull. Entomol. Res.*, **57**, 495–508, 6 pls.

Hinton, H.E. (1973) Neglected phases in metamorphosis: a reply to V.B. Wigglesworth. *J. Entomol.*, (A), **48**, 57–68.

Hinton, H.E. and Service, M.W. (1969) The surface structure of aedine eggs as seen with the scanning electron microscope. *Ann. Trop. Med. Parasitol.*, **63**, 409–11, 2 pls.

Hocking, B. (1953) The intrinsic range and speed of flight of insects. *Trans. R. Entomol. Soc. Lond.*, **104**, 223–346, 6 pls.

Hocking, B. (1954) Flight muscle autolysis in *Aedes communis* (DeGeer). *Mosq. News*, **14**, 121–3.

Hocking, B. (1968) Insect-flower associations in the high Arctic with special reference to nectar. *Oikos*, **19**, 359–87.

Hocking, K.S. and MacInnes, D.G. (1948) Notes on the bionomics of *Anopheles gambiae* and *A. funestus* in East Africa. *Bull. Entomol. Res.*, **39**, 453–65.

Hodapp, C.J. and Jones, J.C. (1961) The anatomy of the adult male reproduction system of *Aedes aegypti* (Linnaeus) (Diptera, Culicidae). *Ann. Entomol. Soc. Am.*, **54**, 832–44.

Hopkins, C.R. (1966) The fine-structural changes observed in the rectal papillae of the mosquito *Aedes aegypti*, L. and their relation to the epithelial transport of water and inorganic ions. *J. R. Microscop. Soc.*, **86**, 235–52.

Hopkins, G.H.E. (1952) *Mosquitoes of the Ethiopian Region. I.-Larval Bionomics of Mosquitoes and Taxonomy of Culicine Larvae*, 2nd edn, British Museum (Natural History), London.

Horsfall, W.R. (1955) *Mosquitoes, their Bionomics and Relation to Disease*. Hafner, New York.

Horsfall, W.R. (1972) Longevity of embryos of *Aedes stimulans*. *J. Econ. Entomol.*, **65**, 891–2.

Horsfall, W.R. (1974) Heteromorphic development of aedine mosquitoes reared at abnormal temperatures. *Ann. Zool. Fennici*, **11**, 224–36.

Horsfall, W.R. and Anderson, J.F. (1961) Suppression of male characteristics of mosquitoes by thermal means. *Science*, **133**, 1830.

Horsfall, W.R. and Anderson, J.F. (1963) Thermally induced genital appendages on mosquitoes. *Science*, **141**, 1183–4.

Horsfall, W.R. and Anderson, J.F. (1964) Thermal stress and anomalous development of mosquitoes (Diptera: Culicidae). II. Effect of alternating temperatures on dimorphism of adults of *Aedes stimulans*. *J. Exp. Zool.*, **156**, 61–89.

Horsfall, W.R., Anderson, J.F. and Brust, R.A. (1964) Thermal stress and anomalous development of mosquitoes (Diptera: Culicidae). III. *Aedes sierrensis*. *Can. Entomol.*, **96**, 1369–72.

Horsfall, W.R., Fowler, H.S., Moretti, L.J. and Larsen, J.R. (1973) *Bionomics and Embryology of the Inland Floodwater Mosquito Aedes vexans*. University of Illinois Press, Urbana.

Horsfall, W.R. and Ronquillo, M.C. (1970) Genesis of the reproductive system of mosquitoes. II. Male of *Aedes stimulans* (Walker). *J. Morphol.*, **131**, 329–57.

Horsfall, W.R., Ronquillo, M.C. and Patterson, W.J. (1972) Genesis of the reproductive system of mosquitoes. IV. Thermal modification of *Aedes stimulans*. *Israel J. Entomol.*, **7**, 73–84.

Hosoi, T. (1954a) Egg production in *Culex pipiens pallens* Coquillett. III. Growth and degeneration of ovarian follicles. *Jap. J. Med. Sci. Biol.*, **7**, 111–27.

Hosoi, T. (1954b) Egg production in *Culex pipiens pallens* Coquillett. IV. Influence of breeding conditions on wing length, body weight and follicle production. *Jap. J. Med. Sci. Biol.*, **7**, 129–34.

Hosoi, T. (1954c) Mechanism enabling the mosquito to ingest blood into the stomach and sugary fluids into the oesophageal diverticula. *Annot. Zool. Jap.*, **27**, 82–90.

Hosoi, T. (1959) Identification of blood components which induce gorging of the mosquito. *J. Insect Physiol.*, **3**, 191–218.

Hosoi, T., Uchida, K., Sato, S. and Matsumura, O. (1975) Initiation of egg development in the mosquito,

Culex pipiens pallens, stimulated by diet amino acids. *J. College Arts Sci., Chiba Univ.*, **B-8**, 73–91. [In Japanese with English abstract.]

Hosselet, C. (1925) Les oenocytes de *Culex annulatus* et l'étude de leur chondriome au cours de la sécrétion. *C.R. Acad. Sci., Paris*, **180**, 399–401.

Hotchkin, P.G. and Fallon, A.M. (1987) Ribosome metabolism during the vitellogenic cycle of the mosquito, *Aedes aegypti. Biochim. Biophys. Acta*, **924**, 352–9.

Houk, E.J. (1977) Midgut ultrastructure of *Culex tarsalis* (Diptera : Culicidae) before and after a bloodmeal. *Tissue & Cell*, **9**, 103–18.

Houk, E.J., Arcus, Y.M. and Hardy, J.L. (1986a) Isolation and characterization of brush border fragments from mosquito mesenterons. *Arch. Insect Biochem. Physiol.*, **3**, 135–46.

Houk, E.J., Chiles, R.E. and Hardy, J.L. (1980) Unique midgut basal lamina in the mosquito, *Aedes dorsalis* (Meigen) (Insecta : Diptera). *Int. J. Insect Morphol. Embryol.*, **9**, 161–4.

Houk, E.J., Cruz, W.O. and Hardy, J.L. (1978) Electrophoretic characterization of the nonspecific esterases of the mosquito, *Culex tarsalis*: conventional and isoelectric focused acrylamide gels. *Comp. Biochem. Physiol.*, **61B**, 291–5.

Houk, E.J., Cruz, W.J. and Hardy, J.L. (1979) Further characterization of the nonspecific esterases of the mosquito, *Culex tarsalis* Coquillett. *Insect Biochem.*, **9**, 429–34.

Houk, E.J. and Hardy, J.L. (1979) *In vivo* negative staining of the midgut continuous junction in the mosquito, *Culex tarsalis* (Diptera; Culicidae). *Acta Tropica*, **36**, 267–75.

Houk, E.J. and Hardy, J.L. (1982) Midgut cellular responses to bloodmeal digestion in the mosquito, *Culex tarsalis* Coquillett (Diptera : Culicidae). *Int. J. Insect Morphol. Embryol.*, **11**, 109–19.

Houk, E.J., Hardy, J.L. and Chiles, R.E. (1981) Permeability of the midgut basal lamina in the mosquito, *Culex tarsalis* Coquillett (Insecta, Diptera). *Acta Tropica*, **38**, 163–71.

Houk, E.J., Hardy, J.L. and Chiles, R.E. (1986b) Mesenteronal epithelial cell surface charge of the mosquito, *Culex tarsalis* Coquillett. Binding of colloidal iron hydroxide, native ferritin and cationized ferritin. *J. Submicroscop. Cytol.*, **18**, 385–96.

Houk, E.J., Hardy, J.L. and Chiles, R.E. (1986c) Histochemical staining of the complex carbohydrates of the midgut of the mosquito, *Culex tarsalis* Coquillett. *Insect Biochem.*, **16**, 667–75.

Houlihan, D.F. (1971) How mosquito pupae escape from the surface. *Nature*, **229**, 489–90.

Houseman, J.G. and Downe, A.E.R. (1986) Methods of measuring blood meal size and proteinase activity for determining effects of mated state on digestive processes of female *Aedes aegypti* (L.) (Diptera: Culicidae). *Can. Entomol.*, **118**, 241–8.

Howard, L.M. (1962) Studies on the mechanism of infection of the mosquito midgut by *Plasmodium gallinaceum*. *Am. J. Hyg.*, **75**, 287–300.

Howard, L.O., Dyar, H.G. and Knab, F. (1912) *The Mosquitoes of North and Central America and the West Indies*, Vol. 2, Carnegie Institution, Washington.

Howland, L.J. (1930a) Bionomical investigation of English mosquito larvae with special reference to their algal food. *J. Ecol.*, **18**, 81–125.

Howland, L.J. (1930b) The nutrition of mosquito larvae, with special reference to their algal food. *Bull. Entomol. Res.*, **21**, 431–40, 1 pl.

Hoyer, L.C. and Rozeboom, L.E. (1976) Inheritance of autogeny in the *Aedes scutellaris* subgroup of mosquitoes. *J. Med. Entomol.*, **13**, 193–7.

HsuChen, C.-C., Kotin, R.M. and Dubin, D.T. (1984) Sequences of the coding and flanking regions of the large ribosomal subunit RNA gene of mosquito mitochondria. *Nucleic Acids Res.*, **12**, 7771–85.

Huang, C.-T. (1971a) Vertebrate serum inhibitors of *Aedes aegypti* trypsin. *Insect Biochem.*, **1**, 27–38.

Huang, C.-T. (1971b) The interactions of *Aedes aegypti* (L.) trypsin with its two inhibitors found in bovine serum. *Insect Biochem.*, **1**, 207–27.

Huang, Y.-M. and Hitchcock, J.C. (1980) A revision of the *Aedes scutellaris* group of Tonga (Diptera: Culicidae). *Contr. Am. Entomol. Inst.*, **17**, 1–107.

Huber, M., Cabib, E. and Miller, L.H. (1991) Malaria parasite chitinase and penetration of the mosquito peritrophic membrane. *Proc. Natl Acad. Sci. USA*, **88**, 2807–10.

Hudson, A. (1964) Some functions of the salivary glands of mosquitoes and other blood-feeding insects. *Can. J. Zool.*, **42**, 113–20.

Hudson, A. (1970a) Notes on the piercing mouthparts of three species of mosquitoes (Diptera: Culicidae) viewed with the scanning electron microscope. *Can. Entomol.*, **102**, 501–9.

Hudson, A. (1970b) Factors affecting egg maturation and oviposition by autogenous *Aedes atropalpus* (Diptera: Culicidae). *Can. Entomol.*, **102**, 939–48.

Hudson, A., Bowman, L. and Orr, C.W.M. (1960) Effects of absence of saliva on blood feeding by mosquitoes. *Science*, **131**, 1730–1.

Huffaker, C.B. (1944) The temperature relations of the immature stages of the malarial mosquito, *Anopheles quadrimaculatus*, Say, with a comparison of the developmental power of constant and variable temperatures in insect metabolism. *Ann. Entomol. Soc. Am.*, **37**, 1–27.

Hulst, F.A. (1906) The histolysis of the musculature of *Culex pungens* during metamorphosis. *Biol. Bull. Woods Hole*, **11**, 277–304, 2 pls.

Hunt, R.H. (1973) A cytological technique for the

study of *Anopheles gambiae* complex. *Parassitologia*, **15**, 137–9.

Hunt, R.H. (1987) Location of genes on chromosome arms in the *Anopheles gambiae* group of species and their correlation to linkage data for other anopheline mosquitoes. *Med. Vet. Entomol.*, **1**, 81–8.

Hurlbut, H.S. (1938) Further notes on the overwintering of the eggs of *Anopheles walkeri* Theobald with a description of the eggs. *J. Parasitol.*, **24**, 521–6.

Hurlbut, H.S. (1966) Mosquito salivation and virus transmission. *Am. J. Trop. Med. Hyg.*, **15**, 989–93.

Hurst, C.H. (1890a) The pupal stage of *Culex*. *Stud. Biol. Lab. Owens Coll.*, **2**, 47–71, 1 pl.

Hurst, C.H. (1890b) The post-embryonic development of a gnat (*Culex*). *Proc Liverpool Biol. Soc.*, **4**, 170–91, 1 pl.

Idris, B.E.M. (1960a) Die Entwicklung im normalen Ei von *Culex pipiens* L. (Diptera). *Z. Morphol. Ökol. Tiere*, **49**, 387–429.

Idris, B.E.M. (1960b) Die Entwicklung im geschnürten Ei von *Culex pipiens* L. (Diptera). *Roux' Arch. Entw.Mech. Organ.*, **152**, 230–62.

Ikeshoji, T. (1966) Bionomics of *Culex* (*Lutzia*) *fuscanus*. *Jap. J. Exp. Med.*, **36**, 321–34.

Ikeshoji, T. (1977) Self-limiting ecomones in the populations of insects and some animals. *J. Pestic. Sci.*, **2**, 77–89.

Ikeshoji, T. and Mulla, M.S. (1970a) Overcrowding factors of mosquito larvae. *J. Econ. Entomol.*, **63**, 90–6.

Ikeshoji, T. and Mulla, M.S. (1970b) Overcrowding factors of mosquito larvae. 2. Growth-retarding and bacteriostatic effects of the overcrowding factors of mosquito larvae. *J. Econ. Entomol.*, **63**, 1737–43.

Ikeshoji, T. and Mulla, M.S. (1974) Overcrowding factors of mosquito larvae: isolation and chemical identification. *Environ. Entomol.*, **3**, 482–6.

Imms, A.D. (1907) On the larval and pupal stages of *Anopheles maculipennis*, Meigen. Part I. The larva. *J. Hygiene*, **7**, 291–318, 2 pls.

Imms, A.D. (1908) On the larval and pupal stages of *Anopheles maculipennis*, Meigen. Part II. The larva (continued). *Parasitology*, **1**, 103–33, 2 pls.

Imms, A.D. (1957) *A General Textbook of Entomology*, Methuen, London.

Ingram, A. and MacFie, J.W.S. (1917) The early stages of certain West African mosquitos. *Bull. Entomol. Res.*, **55**, 691–6, 1 pl.

International Union of Biochemistry (1984) *Enzyme Nomenclature 1984. Recommendations of the Nomenclature Committee of the International Union of Biochemistry on the Nomenclature and Classification of Enzyme-catalysed Reactions*, Academic Press Inc., San Diego.

Inwang, E.E. (1971) Genetic control of enzyme structure in hyperuricemia. *Comp. Biochem. Physiol.*, **39B**, 569–77.

Irby, W.S. and Apperson, C.S. (1989) Immunoblot analysis of digestion of human and rodent blood by *Aedes aegypti* (Diptera: Culicidae). *J. Med. Entomol.*, **26**, 284–93.

Irreverre, F. and Terzian, L.A. (1959) Nitrogen partition in excreta of three species of mosquitoes. *Science*, **129**, 1358–9.

Irving-Bell, R.J. (1974) Cytoplasmic factors in the gonads of *Culex pipiens* complex mosquitoes. *Life Sci.*, **14**, 1149–51.

Irving-Bell, R.J. (1977) Cytoplasmic incompatibility and rickettsial symbiont surveys in members of the *Culex pipiens* complex of mosquitoes. *Adv. Invert. Reprod.*, **1**, 36–48.

Irving-Bell, R.J., Inyang, E.N. and Tamu, G. (1991) Survival of *Aedes vittatus* (Diptera: Culicidae) eggs in hot, dry rockpools. *Trop. Med. Parasitol.*, **42**, 63–6.

Istock, C.A., Wasserman, S.S. and Zimmer, H. (1975) Ecology and evolution of the pitcher-plant mosquito: 1. Population dynamics and laboratory responses to food and population density. *Evolution*, **29**, 296–312.

Ivanova-Kazas, O.M. (1949) Embryonic development of *Anopheles maculipennis* Mg. *Izv. Akad. Nauk. SSR* (ser. *Biol.*), 1949 (2), 140–70. [In Russian.]

Iversen, T.M. (1971) The ecology of a mosquito population (*Aedes communis*) in a temporary pool in a Danish beech wood. *Arch. Hydrobiol. Ichthyol.*, **69**, 309–32.

Iyengar, M.O.T. (1935a) Biology of Indian mosquito larvae that attach themselves to roots of water plants. *Proc. R. Entomol. Soc. Lond.*, **10**, 9–11.

Iyengar, M.O.T. (1935b) Eggs of *Ficalbia minima*, Theo., and notes on breeding habits of three speciesof *Ficalbia*. *Bull. Entomol. Res.*, **26**, 423–5.

Iyengar, M.O.T. and Menon, M.A.U. (1948) Notes on *Harpagomyia genurostris* Leicester (Dipt. Culicidae). *Proc. R. Entomol. Soc. Lond.*(A), **23**, 39–43.

Jacobson, E. (1909) Ein Moskito als Gast und die bischer Schmarotzer der *Cremastogaster difformis* Smith und eine andere schmarotzende Fliege. *Tijdschrift Entomol.*, **52**, 158–64.

Jalil, M. (1974) Observations on the fecundity of *Aedes triseriatus* (Diptera: Culicidae). *Entomol. Exp. Applic.*, **17**, 223–33.

James, A.A., Blackmer, K., Marinotti, O. *et al.* (1991) Isolation and characterization of the gene expressing the major salivary gland protein of the female mosquito, *Aedes aegypti*. *Mol. Biochem. Pharmacol.*, **44**, 245–54.

James, A.A., Blackmer, K. and Racioppi, J.V. (1989) A salivary gland-specific, maltase-like gene of the vector mosquito, *Aedes aegypti*. *Gene*, **75**, 73–83.

Janzen, H.G. and Wright, K.A. (1971) The salivary glands of *Aedes aegypti* (L.): an electron microscope study. *Can. J. Zool.*, **49**, 1343–5, 5 pls.

Jarrett, A. (1973) *The Physiology and Pathophysiology of the Skin*, Vol. 2, Academic Press, London and New York.

Jayakar, S.D., Laudani, U. and Marchi, A. (1982) An

analysis of chiasma frequencies in *Anopheles atroparvus*. *Rend. Sem. Fac. Sci. Univ. Cagliari*, **52**, 127–33.

Jensen, D.V. and Jones, J.C. (1957) The development of the salivary glands in *Anopheles albimanus* Wiedemann (Diptera, Culicidae). *Ann. Entomol. Soc. Am.*, **50**, 464–9.

Jobling, B. (1976) On the fascicle of blood-sucking Diptera. *J. Natural Hist.*, **10**, 457–61.

Jobling, B. (1987) *Anatomical Drawings of Biting Flies*, British Museum (Natural History), London.

Johnson, B.G. and Rowley, W.A. (1972a) Age-related ultrastructual changes in the flight muscle of the mosquito, *Culex tarsalis*. *J. Insect Physiol.*, **18**, 2375–89.

Johnson, B.G. and Rowley, W.A. (1972b) Ultrastructural changes in *Culex tarsalis* flight muscle associated with exhaustive flight. *J. Insect Physiol.* **18**, 2391–9.

Johnston, A.M. and Fallon, A.M. (1985) Characterization of the ribosomal proteins from mosquito (*Aedes albopictus*) cells. *Europ. J. Biochem.*, **150**, 507–15.

Jones, J.C. (1952) Prothoracic aortic sinuses in *Anopheles*, *Culex* and *Aedes*. *Proc. Entomol. Soc. Wash.*, **54**, 244–6.

Jones, J.C. (1953a) On the heart in relation to circulation of hemocytes in insects. *Ann. Entomol. Soc. Am.*, **46**, 366–72.

Jones, J.C. (1953b) Some biometrical constants for *Anopheles quadrimaculatus* Say larvae in relation to age within stadia. *Mosq. News*, **13**, 243–7.

Jones, J.C. (1954) The heart and associated tissues of *Anopheles quadrimaculatus* Say (Diptera: Culicidae). *J. Morphol.*, **94**, 71–123.

Jones, J.C. (1956a) A study of normal heart rates in intact *Anopheles quadrimaculatus* Say larvae. *J. Exp. Zool.*, **131**, 223–33.

Jones, J.C. (1956b) Effects of different gas concentrations on heart rates of intact *Anopheles quadrimaculatus* Say larvae. *J. Exp. Zool.*, **131**, 257–65.

Jones, J.C. (1956c) Effects of drugs on *Anopheles* heart rates. *J. Exp. Zool.*, **133**, 573–88.

Jones, J.C. (1958) Heat-fixation and the blood-cells of *Aedes aegypti* larvae. *Anat. Rec.*, **132**, 461.

Jones, J.C. (1960) The anatomy and rhythmical activities of the alimentary canal of *Anopheles* larvae. *Ann. Entomol. Soc. Am.*, **53**, 459–74.

Jones, J.C. (1967) Spermatocysts in *Aedes aegypti* (Linnaeus). *Biol. Bull.*, **132**, 23–33.

Jones, J.C. (1977) *The Circulatory System of Insects*. C.C. Thomas, Springfield, Illinois.

Jones, J.C. and Brandt, E. (1981) Fluid excretion by adult *Aedes aegypti* mosquitoes. *J. Insect Physiol.*, **27**, 545–9.

Jones, J.C. and Fischman, D.A. (1970) An electron microscopic study of the spermathecal complex of virgin *Aedes aegypti* mosquitoes. *J. Morphol.*, **132**, 293–311.

Jones, J.C. and Pilitt, D.R. (1973) Blood-feeding behaviour of adult *Aedes aegypti* mosquitoes. *Biol. Bull. Woods Hole*, **145**, 127–39.

Jones, J.C. and Madhukar, B.V. (1976) Effects of sucrose on blood avidity in mosquitoes *J. Insect Physiol.*, **22**, 357–60.

Jones, J.C. and Wheeler, R.E. (1965) Studies on spermathecal filling in *Aedes aegypti* (Linnaeus). I. Description. *Biol. Bull. Woods Hole*, **129**, 134–50.

Jones, M.D.R. and Reiter, P. (1975) Entrainment of the pupation and adult activity rhythms during development in the mosquito *Anopheles gambiae*. *Nature*, **254**, 242–4.

Joseph, S.R. (1970) Fruit feeding of mosquitoes in nature. *Proc. New Jersey Mosq. Exterm. Ass.*, **57**, 125–31.

Jost, E. (1970a) Untersuchungen zur Inkompatibilität im *Culex pipiens*-Komplex. *Wilhelm Roux Arch. Entwicklungsmech. Org.*, **166**, 173–88.

Jost, E. (1970b) Genetische Untersuchungen zur Inkompatibilität im *Culex-pipiens-* Komplex. *Theoret. Appl. Genet.*, **40**, 251–6.

Jost, E. (1971) Meiosis in the male of *Culex pipiens* and *Aedes albopictus* and fertilization in the *Culex pipiens*-complex. *Can. J. Genet. Cytol.*, **13**, 237–50.

Jost, E. (1972) Funktion der Spermien in inkompatiblen Kreuzungen der Stechmücke *Culex pipiens* L. *Experientia*, **28**, 1374–5.

Jost, E. and Laven, H. (1971) Meiosis in translocation heterozygotes in the mosquito *Culex pipiens*. *Chromosoma*, **35**, 184–205.

Jost, E. and Mameli, M. (1972) DNA content in nine species of Nematocera with special reference to the sibling species of the *Anopheles maculipennis* group and the *Culex pipiens* group. *Chromosoma*, **37**, 201–8.

Judson, C.L. (1960) The physiology of hatching of aedine mosquito eggs: hatching stimulus. *Ann. Entomol. Soc. Am.*, **53**, 688–91.

Judson, C.L. (1986) Variation in the 'critical period' of *Aedes aegypti* with different hosts. In *Host Regulated Developmental Mechanisms in Vector Arthropods* (eds D. Borovsky and A. Spielman), University of Florida, Vero Beach, **1**, 133–4.

Judson, C.L. and Hokama, Y. (1965) Formation of the line of dehiscence in aedine mosquito. *J. Insect Physiol.*, **11**, 337–45, 2 pls

Judson, C.L., Hokama, Y. and Haydock, I. (1965) The physiology of hatching of aedine mosquito eggs: some larval responses to the hatching stimulus. *J. Insect Physiol.*, **11**, 1169–77.

Judson, C.L., Hokama, Y. and Kliewer, J.W. (1966) Embryogeny and hatching of *Aedes sierrensis* eggs (Diptera: Culicidae). *Ann. Entomol. Soc. Am.*, **59**, 1181–4.

Judson, C.L. and de Lumen, H.Z. (1976) Some effects of juvenile hormone and analogues on ovarian follicles of the mosquito *Aedes aegypti* (Diptera: Culicidae). *J. Med. Entomol.*, **13**, 197–201.

Kaaya, G.P. and Ratcliffe, N.A. (1982) Comparative study of hemocytes and associated cells of some medically important dipterans. *J. Morphol.*, **173**, 351–65.

Kaczor, W.J. and Hagedorn, H.H. (1980) The effects of α-amanitin and cordycepin on vitellogenin synthesis by mosquito fat body. *J. Exp. Zool.*, **214**, 229–33.

Kaiser, P.E. and Seawright, J.A. (1987) The ovarian nurse cell polytene chromosomes of *Anopheles quadrimaculatus*, species A. *J. Am. Mosq. Control Ass.*, **3**, 222–30.

Kaiser, P.E., Seawright, J.A. and Joslyn, D.J. (1979) Cytology of a genetic sexing system in *Anopheles albimanus*. *Can. J. Genet. Cytol.*, **21**, 201–11.

Kal'chenko, Ye.I. (1962) On the biology of the mosquito *Culex pipiens molestus* Forsk. (Diptera, Culicidae). *Entomol. Rev.*, **41**, 52–4.

Kambhampati, S. and Rai, K.S. (1991a) Variation in mitochondrial DNA of *Aedes* species (Diptera: Culicidae). *Evolution*, **45**, 120–9.

Kambhampati, S. and Rai, S.R. (1991b) Mitochondrial DNA variation within and among populations of the mosquito *Aedes albopictus*. *Genome*, **34**, 288–92.

Kamphampati, S. and Rai, K.S. (1991c) Temporal variation in the ribosomal DNA nontranscribed spacer of *Aedes albopictus* (Diptera: Culicidae). *Genome*, **34**, 293–7.

Kanda, T. (1970) The salivary gland chromosomes of *Culex pipiens fatigans* Wiedemann. *Jap. J. Exp. Med.*, **40**, 335–45.

Kang, S.-H., Fuchs, M.S. and Webb, P.M. (1980) Purification and characterization of dopa-decarboxylase from adult gravid *Aedes aegypti*. *Insect Biochem.*, **10**, 501–8.

Kardatzke, J.T. (1979) Hatching of eggs of snow-melt *Aedes* (Diptera: Culicidae). *Ann. Entomol. Soc. Am.*, **72**, 559–62.

Kashin, P. (1966) Electronic recording of the mosquito bite. *J. Insect Physiol.*, **12**, 281–6, 5 pls.

Kashin, P. and Wakeley, H.G. (1965) An insect 'bitometer'. *Nature*, **208**, 462–4.

Kawai, S. (1969) Studies on the follicular development and feeding activity of the females of *Culex tritaeniorhynchus* with special reference to those in autumn. *Trop. Med., Nagasaki*, **11**, 145–69.

Keilin, D. (1932) On the water reservoir of a horse-chestnut tree. *Parasitology*, **24**, 280–2, 1 pl.

Keilin, D. (1944) Respiratory systems and respiratory adaptations in larvae and pupae of Diptera. *Parasitology*, **36**, 1–66, 2 pls.

Keilin, D., Tate, P. and Vincent, M. (1935) The perispiracular glands of mosquito larvae. *Parasitology*, **27**, 257–62.

Kelly, T.J., Adams, T.S., Schwartz, M.B. *et al.* (1987) Juvenile hormone and ovarian maturation in the Diptera: a review of recent results. *Insect Biochem.*, **17**, 1089–93.

Kelly, T.J., Birnbaum, M.J., Woods, C.W. and Borkovec, A.B. (1984) Effects of housefly oostatic hormone on egg development neurosecretory hormone action in *Aedes atropalpus*. *J. Exp. Zool.*, **229**, 491–6.

Kelly, T.J. and Fuchs, M.S. (1980) *In vivo* induction of ovarian development in decapitated *Aedes atropalpus* by physiological levels of 20-hydroxyecdysone. *J. Exp. Zool.*, **213**, 25–32.

Kelly, T.J., Fuchs, M.S. and Kang, S.-H. (1981) Induction of ovarian development in autogenous *Aedes atropalpus* by juvenile hormone and 20-hydroxyecdysone. *Int. J. Invert. Reprod.*, **3**, 101–12.

Kelly, T.J., Masler, E.P., Schwartz, M.B. and Haught, S.B. (1986a) Inhibitory effects of oostatic hormone on ovarian maturation and ecdysteroid production in Diptera. *Insect Biochem.*, **16**, 273–9.

Kelly, T.J., Whisenton, L.R., Katahira, E.J. *et al.* (1986b) Inter-species cross-reactivity of the prothoracicotropic hormone of *Manduca sexta* and egg-development neurosecretory hormone of *Aedes aegypti*. *J. Insect Physiol.*, **32**, 757–62.

Kesavan, S.K. and Reddy, N.P. (1985) On the feeding strategy and the mechanics of blood sucking in insects. *J. Theoret. Biol.*, **113**, 781–3.

Kettle, D.S. (1948) The growth of *Anopheles sergenti* Theobald (Diptera, Culicidae), with special reference to the growth of the anal papillae in varying salinities. *Ann. Trop. Med. Parasitol.*, **42**, 5–29.

Khan, A.A. and Maibach, H.I. (1971) A study of the probing response of *Aedes aegypti*. 2. Effect of desiccation and blood feeding on probing to skin and an artificial target. *J. Econ. Entomol.*, **64**, 439–42.

Khan, A.R. and Ahmed, R. (1975) The duration of the immature stages of *Culex pipiens fatigans* at winter mean temperature of Dacca, Bangladesh. *Bangladesh J. Zool.*, **3**, 111–20.

Kiceniuk, J. and Phillips, J.E. (1974) Magnesium regulation in mosquito larvae (*Aedes campestris*) living in waters of high MgSO4 content. *J. Exp. Biol.*, **61**, 749–60.

King, M. and Pasteur, N. (1985) A rapid technique for obtaining air dried mitotic chromosomes from mosquito egg rafts. *Stain Technol.*, **60**, 119–20.

King, R.C. (1964) Further information concerning the envelopes surrounding dipteran eggs. *Quart. J. Microscop. Sci.*, **105**, 209–11, 2 pls.

King, R.C. and Akai, H. (1982) Insect intercellular junctions: their structure and development. In *Insect Ultrastructure* (eds R.C. King and H. Akai), Plenum Press, New York, **1**, 402–33.

Kitzmiller, J.B. (1967) Mosquito cytogenetics. In *Genetics of Insect Vectors of Disease* (eds J.W. Wright and R. Pal), Elsevier, Amsterdam, pp. 133-50.

Kitzmiller, J.B. (1977) Chromosomal differences among species of *Anopheles* mosquitoes. *Mosq. Systemat.*, **9**, 112–22.

Kleinjan, J.E. and Dadd, R.H. (1977) Vitamin requirements of the larval mosquito, *Culex pipiens*. *Ann. Entomol. Soc. Am.*, **70**, 541–3.

Kliewer, J.W. (1961) Weight and hatchability of *Aedes aegypti* eggs (Diptera : Culicidae). *Ann. Entomol. Soc. Am.*, **54**, 912–17.

Klowden, M.J. (1986) Effects of sugar deprivation on the host-seeking behaviour of gravid *Aedes aegypti* mosquitoes. *J. Insect Physiol.*, **32**, 479–83.

Klowden, M.J. (1987) Distention-mediated egg maturation in the mosquito, *Aedes aegypti*. *J. Insect Physiol.*, **33**, 83–7.

Klowden, M.J. and Chambers, G.M. (1989) Ovarian development and adult mortality in *Aedes aegypti* treated with sucrose, juvenile hormone and methoprene. *J. Insect Physiol.*, **35**, 513–17.

Klowden, M.J. and Chambers, G.M. (1991) Male accessory gland substances activate egg development in nutritionally stressed *Aedes aegypti* mosquitoes. *J. Insect Physiol.*, **37**, 721–26.

Klowden, M.J. and Lea, A.O. (1978) Blood meal size as a factor affecting continued host-seeking by *Aedes aegypti*(L.). *Am. J. Trop. Med. Hyg.*, **27**, 827–31.

Klowden, M.J. and Lea, A.O. (1979) Effect of defensive host behavior on the blood meal size and feeding success of natural populations of mosquitoes (Diptera : Culicidae). *J. Med. Entomol.*, **15**, 514–17.

Knab, F. (1911) The food-habits of *Megarhinus*. *Psyche*, **18**, 80–2.

Knapp, T. and Crampton, J. (1990) Sequences related to immune proteins in the mosquito *Aedes aegypti*. *Trans. R. Soc. Trop. Med. Hyg.*, **84**, 459.

Knight, K.L. (1971) Comparative anatomy of the mandible of the fourth instar mosquito larva (Diptera: Culicidae). *J. Med. Entomol.*, **8**, 189–205.

Knight, K.L. and Abdel Malek, A.A. (1951) A morphological and biological study of *Culex pipiens* in the Cairo area of Egypt [Diptera-Culicidae]. *Bull. Soc. Fouad 1er Entomol.*, **35**, 175–85.

Knight, K.L. and Harbach. R.E. (1977) Maxillae of fourth stage mosquito larvae (Diptera: Culicidae). *Mosq. Systemat.*, **9**, 455–77.

Kobayashi, M., Ogura, N., Tsuruoka, H. *et al.* (1986a) Studies on filariasis VII: histological observation on the encapsulated *Brugia malayi* larvae in the abdominal haemocoel of the mosquitoes, *Armigeres subalbatus*. *Jap. J. Sanit. Zool.*, **37**, 59–65.

Kobayashi, M., Ogura, N. and Yamamoto, H. (1986b) Studies on filariasis VIII: histological observation on the abortive development of *Brugia malayi* larvae in the thoracic muscles of the mosquitoes, *Armigeres subalbatus*. *Jap. J. Sanit. Zool.*, **37**, 127–32.

Kobayashi, M., Yamada, K. and Yamamoto, H. (1988) *In vitro* adhesion of enzyme(s) related to melanin formation in the pupal mosquito haemolymph to the surface of *Brugia pahangi* microfilaria. *Jap. J. Sanit. Zool.*, **39**, 143–6.

Koch, H.J. (1938) The absorption of chloride ions by the anal papillae of Diptera larvae. *J. Exp. Biol.*, **15**, 152–60.

Koehl, M.A.R. and Strickler, J.R. (1981) Copepod feeding currents: food capture at low Reynolds number. *Limnol. Oceanogr.*, **26**, 1062–73.

Koenekoop. R.K. and Livdahl, T.P. (1986) Cannibalism among *Aedes triseriatus* larvae. *Ecol. Entomol.*, **11**, 111–14.

Kogan, P.H. (1990) Substitute blood meal for investigating and maintaining *Aedes aegypti* (Diptera: Culicidae). *J. Med. Entomol.*, **27**, 709–12.

Koller, C.N., Dhadialla, T.S. and Raikhel, A.S. (1989) Selective endocytosis of vitellogenin by oocytes of the mosquito, *Aedes aegypti*: an *in vitro* study. *Insect Biochem.*, **19**, 693–702.

Koller, C.N. and Raikhel, A.S. (1991) Initiation of vitellogenin uptake and protein synthesis in the mosquito (*Aedes aegypti*) ovary in response to a blood meal. *J. Insect Physiol.*, **37**, 703–11.

Konishi, E. (1990) Distribution of immunoglobulin G and E antibody levels to salivary gland extracts of *Aedes albopictus* (Diptera: Culicidae) in several age groups of a Japanese population. *J. Med. Entomol.*, **27**, 519–22.

Konishi, E. and Yamanishi, H. (1984) Estimation of blood meal size of *Aedes albopictus* (Diptera: Culicidae) using enzyme-linked immunosorbent assay. *J. Med. Entomol.*, **21**, 506–13.

Krafsur, E.S. (1977) The bionomics and relative prevalence of *Anopheles* species with respect to the transmission of *Plasmodium* to man in eastern Ethiopia. *J. Med. Entomol.*, **14**, 180–94.

Krafsur, E.S. and Jones, J.C. (1967) Spermiogenesis in *Aedes aegypti* (L.). *Cytologia*, **32**, 450–62.

Kreutzer, R.D. (1970) Activity at the nucleolar organizer locus in nurse cell of *Anopheles atroparvus*. *Am. Zool.*, **10**, 529.

Kreutzer, R.D. (1978) A mosquito with eight chromosomes: *Chagasia bathana*. *Mosq. News*, **38**, 554–8.

Krieg, N.R. and Holt, J.G. (1984) *Bergey's Manual of Systematic Bacteriology*, Vol. 1, Williams & Wilkins, Baltimore, London.

Krogh, A. (1941) *The Comparative Physiology of Respiratory Systems*. University of Pennsylvania Press.

Kroll, M.H. and Schafer, A.I. (1989) Biochemical mechanisms of platelet activation. *Blood*, **74**, 1181–95.

Kühlhorn, F. (1972) Fortpflanzungsbiologische Studien bei *Anopheles messeae messeae* Fall. (Diptera: Culicidae). *Z. Angew. Entomol*,. **70**, 187–203.

Kumar, A. and Rai, K.S. (1990a) Chromosomal localization and copy number of 18S + 28S ribosomal RNA genes in evolutionarily diverse mosquitoes (Diptera, Culicidae). *Hereditas*, **113**, 277–89.

Kumar, A. and Rai, K.S. (1990b) Intraspecific variation in nuclear DNA content among world populations of a mosquito, *Aedes albopictus* (Skuse). *Theor. Appl. Genet.*, **79**, 748–52.

Kunz, P.A (1978) Resolution and properties of the proteinases in the larva of the mosquito, *Aedes aegypti*. *Insect Biochem.*, **8**, 43–51.

Kurihara, Y. (1959) Synecological analysis of the biotic community in a microcosm. VIII. Studies on the limiting factor in determining the distribution of mosquito larvae in the polluted water of a bamboo container, with special reference to relation of larvae to bacteria. *Jap. J. Zool.*, **12**, 391–400.

Kurtti, T.J., Brooks, M.A., Wensman, C. and Lovrien, R. (1979) Direct microcalorimetry of heat generation by individual insects. *J. Thermal Biol.*, **4**, 129–36.

Kuznetsova, L.A. (1981) Cuticular lining of the cardial region of the intestine in *Chironomus plumosus* L. and *Aedes aegypti* L. larvae (Diptera: Chironomidae, Culicidae) and its relationship to the peritrophic membrane. *Moscow Univ. Biol. Sci. Bull.* **36** (2), 9–13. [English language edition.]

Laffoon, J.L. and Knight, K.L. (1973) A mosquito taxonomic glossary IX. The larval cranium. *Mosq. Systemat.*, **5**, 31–96.

Laird, M. (1956) Studies of mosquitoes and freshwater ecology in the South Pacific. *Bull. R. Soc. New Zealand*, No. 6, 1–213, 4 pls.

Laird, M. (1988) *The Natural History of Larval Mosquito Habitats*, Academic Press, London.

Lamb, H. (1911) On the uniform motion of a sphere through a viscous fluid. *Philosophical Magazine*, **21**, 112–21.

Lamb, R.J. and Smith, S.M. (1980) Comparison of egg size and related life-history characteristics for two predaceous tree-hole mosquitoes (*Toxorhynchites*). *Can. J. Zool.*, **58**, 2065–70.

Lane, N.J. (1982) Insect intercellular junctions: their structure and development. In *Insect Ultrastructure*, (eds R.C.King and H.Akai), Plenum Publishing Corpn, **1**, 402–33.

Lang, C.A. (1963) The effect of temperature on the growth and chemical composition of the mosquito. *J. Insect Physiol.*, **9**, 279–86.

Lang, J.T. (1978) Relationship of fecundity to the nutritional quality of larval and adult diets of *Wyeomyia smithii*. *Mosq. News*, **38**, 396–403.

Lang, W.D. (1920) *A Handbook of British Mosquitoes*, British Museum (Natural History), London.

Larsen, J.R. and Bodenstein. D. (1959) The humoral control of egg maturation in the mosquito. *J. Exp. Zool.*, **140**, 343–81.

Larsen, J.R. and Owen, W.B. (1971) Structure and function of the ligula of the mosquito *Culiseta inornata* (Williston). *Trans. Am. Microsop. Soc.*, **90**, 294–308.

Larsson, R. (1983) A Rickettsia-like microorganism similar to *Wolbachia pipientis* and its occurrence in *Culex mosquitoes*. *J. Invert. Path.*, **41**, 387–90.

Laurence, B.R. (1960) The biology of two species of mosquito, *Mansonia africana* (Theobald) and *Mansonia uniformis* (Theobald), belonging to the subgenus *Mansonioides* (Diptera, Culicidae). *Bull. Entomol. Res.*, **51**, 491–517.

Laurence, B.R. (1964) Autogeny in *Aedes* (*Finlaya*) *togoi* Theobald (Diptera, Culicidae). *J. Insect Physiol.*, **10**, 319–31.

Laurence, B.R. (1977) Ovary development in mosquitoes: a review. *Adv. Invert. Reprod.*, **1**, 154–65.

Laurence, B.R., Mori, K., Otsuka, T. *et al.*, (1985) Absolute configuration of mosquito oviposition attractant pheromone, 6-acetoxy-5-hexadecanolide. *J. Chem. Ecology*, **11**, 643–8.

Laurence, B.R. and Pickett, J.A. (1982) *erythro*-6-Acetoxy-5-hexadecanolide, the major component of a mosquito oviposition attractant pheromone. *Chem. Commun. (J. Chem. Soc., D)*, (1), 59–62.

Laurence, B.R. and Pickett, J.A. (1985) An oviposition attractant pheromone in *Culex quinquefasciatus* Say (Diptera: Culicidae). *Bull. Entomol. Res.*, **75**, 283–90.

Laurence, B.R. and Roshdy, M.A. (1963) Ovary development in mosquitoes. *Nature*, **200**, 495–6.

Laurence, B.R. and Simpson, M.G. (1974) Cell replication in the follicular epithelium of the adult mosquito. *J. Insect Physiol.*, **20**, 703–15, 2 pls.

Laven, H. (1957a) Vererbung durch Kerngene und das Problem der ausserkaryotischen Vererbung bei *Culex pipiens*. I. Kernvererbung. *Z. Indukt. Abstammungs-Vererbungslehre*, **88**, 443–77.

Laven, H. (1957b) Vererbung durch Kerngene und das Problem der ausserkaryotischen Vererbung bei *Culex pipiens*. II. Ausserkaryotische Vererbung. *Z. Indukt. Abstammungs-Vererbungslehre*, **88**, 478–516.

Laven, H. (1967a) Formal genetics of *Culex pipiens*. In *Genetics of Insect Vectors of Disease* (eds J.W. Wright and R. Pal), Elsevier, Amsterdam, pp. 17–65.

Laven, H. (1967b) Speciation and evolution in *Culex pipiens*. In *Genetics of Insect Vectors of Disease* (eds J.W. Wright and R. Pal), Elsevier, Amsterdam, pp. 251–75.

Laven, H. (1967c) Eradication of *Culex fatigans* through cytoplasmic incompatibility. *Nature*, **216**, 383–4.

Le Sueur, D. and Sharp, B.L. (1991) Temperature-dependent variation in *Anopheles merus* larval head capsule width and adult wing length: implications for anopheline taxonomy. *Med. Vet. Entomol.*, **5**, 55–62.

Lea, A.O. (1963) Some relationships between environment, corpora allata, and egg maturation in aedine mosquitoes. *J. Insect Physiol.*, **9**, 793–809.

Lea, A.O. (1964) Studies on the dietary and endocrine regulation of autogenous reproduction in *Aedes taeniorhynchus* (Wied.). *J. Med. Entomol.*, **1**, 40–4.

Lea, A.O. (1967) The medial neurosecretory cells and egg maturation in mosquitoes. *J. Insect Physiol.*, **13**, 419–29, 1 pl.

Lea, A.O. (1970) Endocrinology of egg maturation in autogenous and anautogenous *Aedes taeniorhynchus*. *J. Insect Physiol.*, **16**, 1689–96.

Lea, A.O. (1972) Regulation of egg maturation in the mosquito by the neurosecretory system: the role of the corpus cardiacum. *Gen. Comp. Endocrinol.*, Suppl. **3**, 602–8.

Lea, A.O. (1982) Artifactual stimulation of vitellogenesis in *Aedes aegypti* by 20-hydroxyecdysone. *J. Insect Physiol.*, *28*, 173–6.

Lea, A.O., Briegel, H. and Lea, H.M. (1978) Arrest, resorption, or maturation of oocytes in *Aedes aegypti*: a dependence on the quantity of blood and the interval between blood meals. *Physiol. Ent.*, **3**, 309–16.

Lea, A.O. and Brown, M.R. (1990) Neuropeptides of mosquitoes. In *Molecular Insect Science* (eds H.H. Hagedorn, J.G. Hildebrand, M.G. Kidwell and J.H. Law), Plenum Press, New York, pp. 181–8.

Lea, A.O. and DeLong, D.M. (1958) Studies on the nutrition of *Aedes aegypti* larvae. *Proc. 10th Int. Congr. Entomol.* (1956), **2**, 299–302.

Lea, A.O., Dimond, J.B. and DeLong, D.M. (1956) A chemically defined medium for rearing *Aedes aegypti* larvae. *J. Econ. Entomol.*, **49**, 313–5.

Lea, A.O., Dimond, J.B. and DeLong, D.M. (1958) Some nutritional factors in egg production by *Aedes aegypti*. *Proc. 10th Int. Congr. Entomol.* (1956), **3**, 793–6.

Lea, A.O. and Van Handel, E. (1970) Suppression of glycogen synthesis in the mosquito by a hormone from the medial neurosecretory cells. *J. Insect Physiol.*, **16**, 319–21.

Lea, A.O. and Van Handel, E. (1982) A neurosecretory hormone-releasing factor from ovaries of mosquitoes fed blood. *J. Insect Physiol.*, *28*, 503–8.

Lee, C., Barr, A.R. and Chang, P. (1975) A new intersex in *Culex pipiens* L. (Diptera : Culicidae). II. Inheritance. *J. Med. Entomol.*, **12**, 567–70.

Lee, R. (1974) Structure and function of the fascicular stylets, and the labral and cibarial sense organs of male and female *Aedes aegypti* (L.) (Diptera, Culicidae). *Quaest. Entomol.*, **10**, 187–215.

Lee, R.M.K.W. and Craig, D.A. (1983a) Cibarial sensilla and armature in mosquito adults (Diptera: Culicidae). *Can. J. Zool.*, **61**, 633–46.

Lee, R.M.K.W. and Craig, D.A. (1983b) The labrum and labral sensilla of mosquitoes (Diptera: Culicidae): a scanning electron microscope study. *Can. J. Zool.*, **61**, 1568–79.

Lee, R.M.K.W. and Craig, D.A. (1983c) Maxillary, mandibulary, and hypopharyngeal stylets of female mosquitoes (Diptera: Culicidae): a scanning electron microscope study. *Can. Entomol.*, **115**, 1503–12.

Lee, R.M.K.W. and Davies, D.M. (1978) Cibarial sensilla of *Toxorhynchites* mosquitoes (Diptera: Culicidae). *Int. J. Insect Morphol. Embryol.*, **7**, 189–94.

Lehane, M.J. and Laurence, B.R. (1978) Development of the calyx and lateral oviduct during oogenesis in *Aedes aegypti*. *Cell Tissue Res.*, *193*, 125–37.

Lengy, J. and Gold, D (1966) Studies on *Culex pipiens molestus* in Israel. 1. Clinical observations on the bite reaction in man. *Int. Arch. Allergy*, **29**, 404–14.

Leung, M.E. and Romoser, W.S. (1979) Functions of the respiratory trumpets and first abdominal spiracles in *Aedes triseriatus* (Say) pupae (Diptera; Culicidae). *Mosq. News*, **39**, 483–9.

Lewis, D.J. (1949) Tracheal gills in some African culicine mosquito larvae. *Proc. R. Entomol. Soc. Lond.* (A), **24**, 60–6.

Li, J. and Christensen, B.M. (1990) Immune competence of *Aedes trivittatus* hemocytes as assessed by lectin binding. *J. Parasitol.*, **76**, 276–8.

Li, J. and Nappi, A.J. (1992) *N*-Acetyltransferase activity during ovarian development in the mosquito *Aedes aegypti* following blood feeding. *Insect Biochem. Molec. Biol.*, **22**, 49–54.

Li, J., Tracy, J.W. and Christensen, B.M. (1989) Hemocyte monophenol oxidase activity in mosquitoes exposed to microfilariae of *Dirofilaria immitis*. *J. Parasitol.*, **75**, 1–5.

Lin, Y., Hamblin, M.T., Edwards, M.J. *et al.* (1992) Structure, expression and hormonal control of genes from the mosquito *Aedes aegypti*, which encode proteins similar to the vitelline membrane proteins of *Drosophila melanogaster*. *Developmental Biology*, [Submitted]

Lincoln, D.C.R. (1965) Structure of the egg-shell of *Culex pipiens* and *Mansonia africana* (Culicidae, Diptera). *Proc. Zool. Soc. Lond.*, **145**, 9–17.

Lincoln, R.J., Boxshall, G.A. and Clark, P.F. (1982) *A Dictionary of Ecology, Evolution and Systematics*, Cambridge University Press.

Linley, J.R. (1989) Egg of *Mansonia dyari* described and compared with egg of *Mansonia titillans* (Diptera: Culicidae). *J. Med. Entomol.*, **26**, 41–5.

Linley, J.R., Linley, P.A. and Lounibos, L.P. (1986) Light and scanning electron microscopy of the egg of *Mansonia titillans* (Diptera: Culicidae). *J. Med. Entomol.*, **23**, 99–104.

Linley, J.R., Lounibos, L.P. and Linley, P.A. (1990) Fine structure of the egg of *Trichoprosopon digitatum* (Diptera: Culicidae) and its relationship to egg raft formation. *J. Med. Entomol.*, **17**, 578–85.

Linley, J.R. and Simmons, K.R. (1981) Sperm motility and spermathecal filling in lower Diptera. *Int. J. Invert. Reprod.*, **4**, 137–46.

Lock, M.A. (1990) The dynamics of dissolved and particulate organic material over the substratum of water bodies. In *The Biology of Particles in Aquatic Systems* (ed. R.S. Wotton), CRC Press, Boca Raton, pp. 117–44.

Logan, J.A. (1953) *The Sardinian Project. An Experiment in the Eradication of an Indigenous Malarious Vector*, Johns Hopkins Press, Baltimore.

Lomen, F. (1914) Der Hoden von *Culex pipiens* L. (Spermatogenese, Hodenwandungen und Degenerationen). *Jenaische Z. Naturwiss.*, **52**, 567–628.

Lounibos, L.P. (1980) The bionomics of three sympatric *Eretmapodites* (Diptera: Culicidae) at the Kenya coast. *Bull. Entomol. Res.*, **70**, 309–20.

Lounibos, L.P., Van Dover, C. and O'Meara G.F. (1982) Fecundity, autogeny, and the larval environment of the pitcher-plant mosquito, *Wyeomyia smithii*. *Oecologia*, **55**, 160–4.

Lu, Y.H. and Hagedorn, H.H. (1986) Egg development in the mosquito *Anopheles albimanus*. *Int. J. Invert. Reprod. Dev.*, **9**, 79–94.

Lum, P.T. (1961a) The reproductive system of some Florida mosquitoes. I. The male reproductive tract. *Ann. Entomol. Soc. Am.*, **54**, 397–401.

Lum, P.T. (1961b) The reproductive system of some Florida mosquitoes. II. The male accessory glands and their role. *Ann. Entomol. Soc. Am.*, **54**, 430–3.

Lum, P.T.M., Nayar, J.K. and Provost, M.W. (1968) The pupation rhythm in *Aedes taeniorhynchus*. III. Factors in developmental synchrony. *Ann. Entomol. Soc. Am.*, **61**, 889–99.

Lund, J.W.G. (1957) Chemical analysis in ecology illustrated from Lake District tarns and lakes. 2. Algal differences. *Proc. Linnean Soc. Lond.*, **167**, 165–71.

Ma, M., Gong, H., Newton, P.B. and Borkovec, A.B. (1986) Monitoring *Aedes aegypti* vitellogenin production and uptake with hybridoma antibodies. *J. Insect Physiol.*, **32**, 207–13.

Ma, M., Gong, H., Zhang, J.Z. and Gwadz, R. (1987) Response of cultured *Aedes aegypti* fat bodies to 20-hydroxyecdysone. *J. Insect Physiol.*, **33**, 89–93.

Ma, M., Newton, P.B., Gong, H. *et al.* (1984) Development of monoclonal antibodies for monitoring *Aedes atropalpus* vitellogenesis. *J. Insect Physiol.*, **30**, 529–36.

Ma, M., Zhang, J.Z., Gong, H. and Gwadz, R. (1988) Permissive action of juvenile hormone on vitellogenin production by the mosquito, *Aedes aegypti*. *J. Insect Physiol.*, *34*, 593–6.

Macdonald, W.W. (1956) *Aedes aegypti* in Malaya. II. Larval and adult biology. *Ann. trop. Med. Parasitol.*, **50**, 399–414.

Macdonald, W.W. (1976) Mosquito genetics in relation to filarial infections. In *Genetic Aspects of Host-Parasite Relationships*, *Symp. Brit. Soc. Parasitol.*, (eds A.E.R. Taylor and R. Muller), Blackwell Scientific Publications, Oxford, **14**, 1–24.

Macdonald, W.W. and Sheppard, P.M. (1965) Cross-over values in the sex chromosomes of the mosquito *Aedes aegypti* and evidence of the presence of inversions. *Ann. Trop. Med. Parasitol.*, **59**, 74–87.

Macdonald, W.W. and Traub, R. (1960) An introduction to the ecology of the mosquitoes of the lowland dipterocarp forest of Selangor, Malaya. *Stud. Inst. Med. Res. F.M.S.*, No. 29, 79–109.

Macfie, J.W.S (1917) The limitations of kerosene as a larvicide, with some observations on the cutaneous respiration of the mosquito larvae. *Bull. Entomol. Res.*, **7**, 277–95.

MacGregor, M.E. (1921) The influence of the hydrogen-ion concentration in the development of mosquito larvae (preliminary communication). *Parasitology*, **13**, 348–51.

MacGregor, M.E. (1927) *Mosquito Surveys. A Handbook for Anti-Malarial and Anti-Mosquito Field Workers*, Ballière, Tindall & Cox, London.

MacGregor, M.E. (1930) The artificial feeding of mosquitoes by a new method which demonstrates certain functions of the diverticula. *Trans. R. Soc. Trop. Med. Hyg.*, **23**, 329–31.

Mack, S.R., Foley, D.A. and Vanderberg, J.P. (1979a) Hemolymph volume of noninfected and *Plasmodium berghei*-infected *Anopheles stephensi*. *J. Invert. Pathol.*, **34**, 105–9.

Mack, S.R., Samuels, S. and Vanderberg, J.P. (1979b) Hemolymph of *Anopheles stephensi* from noninfected and *Plasmodium berghei*-infected mosquitoes. 3. Carbohydrates. *J. Parasitol.*, **65**, 217–21.

Mack, S.R. and Vanderberg, J.P. (1978) Hemolymph of *Anopheles stephensi* from noninfected and *Plasmodium berghei*-infected mosquitoes. 1. Collection procedure and physical characteristics. *J. Parasitol.*, **64**, 918–23.

Mackerras, M.J. and Lemerle, T.H. (1949) Laboratory breeding of *Anopheles punctulatus punctulatus*, Donitz. *Bull. Entomol. Res.*, **40**, 27–41.

Maddrell, S.H.P. (1977) Insect Malpighian tubules. In *Transport of Ions and Water in Animals* (eds B. L. Gupta, R.B. Moreton, J.L. Oschman and B.J. Wall), Academic Press, New York, London, pp. 541–69.

Maddrell, S.H.P. (1980) Characteristics of epithelial transport in insect Malpighian tubules. *Curr. Topics Membranes Transport*, **14**, 427–63.

Maddrell, S.H.P. (1991) The fastest fluid-secreting cell known: the upper Malpighian tubule cell of *Rhodnius*. *BioEssays*, **13**, 357–62.

Maddrell, S.H.P and Gardiner, B.O.C. (1976) Diuretic hormone in adult Rhodnius prolixus: total store and speed of release. *Physiol. Entomol.* **1**, 265–69.

Maddrell, S.H.P. and Phillips, J.E. (1975) Active transport of sulphate ions by the Malpighian tubules of larvae of the mosquito *Aedes campestris*. *J. Exp. Biol.*, **62**, 367–78.

Maddrell, S.H.P. and Phillips, J.E. (1978) Induction of sulphate transport and hormonal control of fluid secretion by Malpighian tubules of larvae of the mosquito *Aedes taeniorhynchus*. *J. Exp. Biol.*, **72**, 181–202.

Maddrell, S.H.P., Whittembury, G. Mooney, R.L. *et al.*, (1991) The fate of calcium in the diet of *Rhodnius prolixus*: storage of concretion bodies in the Malpighian tubules. *J. Exp. Biol.*, **57**, 483–502.

Magnarelli, L.A. (1979a) Diurnal nectar-feeding of *Aedes*

cantator and *A. sollicitans* (Diptera: Culicidae). *Environ. Entomol.*, **8**, 949–5.

Magnarelli, L.A. (1979b) Feeding behavior of mosquitoes (Diptera: Culicidae) on man, raccoons and white-footed mice. *Ann. Entomol. Soc. Am.*, **72**, 162–6.

Magnarelli, L.A. (1983) Nectar sugars and caloric reserves in natural populations of *Aedes canadensis* and *Aedes stimulans* (Diptera: Culicidae). *Environ. Entomol.*, **12**, 1482–6.

Magnin, M. and Pasteur, N. (1987a) Phénomènes d'incompatibilité cytoplasmique chez le moustique *Culex pipiens* L.(Diptera : Culicidae) du sud de la France. Effet de la tétracycline. *Cahier ORSTOM, Entomol. Méd. Parasitol.*, **25**, 21–5.

Magnin, M. and Pasteur, N. (1987b) Incompatibilités cytoplasmiques dans le complexe *Culex pipiens*. Une revue. *Cahier ORSTOM, Entomol. Méd. Parasitol.*, **25**, 45–53.

Magnin, M., Pasteur, N. and Raymond, M. (1987) Multiple incompatibilities within populations of *Culex pipiens* L. in southern France. *Genetica*, **74**, 125–30.

Malcolm, C.A. and Hall, L.M.C. (1990) Cloning and characterisation of a mosquito acetylcholinesterase gene. In *Molecular Insect Science* (eds H.H. Hagedorn, J.G. Hildebrand, M.G. Kidwell and J.H. Law), Plenum Press, New York, pp. 57–65.

Manjra, A.A. (1971) Regulation of threshold to sucrose in a mosquito, *Culiseta inornata* (Williston). *Mosq. News*, **31**, 387–90.

Manning, D.L. (1978) Mouthparts of larvae of *Opifex fuscus* and *Aedes australis* (Diptera: Culicidae); a scanning electron microscope study. *New Zealand J. Zool.*, **5**, 801–6.

Mara, D. (1976) *Sewage Treatment in Hot Climates*, Wiley, London and New York.

Marchi, A. and Mezzanotte, R. (1990) Inter- and intraspecific heterochromatin variation detected by endonuclease digestion in two sibling species of the *Anopheles maculipennis* complex. *Heredity*, **65**, 135–42.

Marchi, A., Mezzanotte, R. and Ferrucci, L. (1980) Characterization of the metaphase chromosomes in *Anopheles stephensi* (Liston,1901) by Q-, G- and C-banding. *Cytologia*, **45**, 549–53.

Marchi, A., Mezzanotte, R. and Ferrucci, L. (1981) Identification of chromosome markers in interphase nuclei. *Basic Appl. Histochem.*, **25**, 105–11.

Marchi, A. and Rai, K.S. (1986) Chromosome banding homologies in three species of *Aedes* (*Stegomyia*). *Can. J. Genet. Cytol.*, **28**, 198–202.

Marchi, A., Sirigu, G. and Laudani, U. (1978) Influenza dell copulazione sulla maturazione del follicolo in *Anopheles atroparvus* Van Thiel. *Parassitologia*, **20** 29–37.

Marcovitch, S. (1960) Experiments on prolongation of the life of mosquito larvae by underfeeding. *J. Econ. Entomol.*, **53**, 169.

Margulis, L., Corliss, J.O., Melkonian, M. and Chapman, D.J. (1989) *Handbook of Protoctista*. Jones and Bartlett, Boston.

Marinotti, O. and James, A.A. (1990) An α-glucosidase in the salivary glands of the vector mosquito, *Aedes aegypti*. *Insect Biochem.*, **20**, 619–23.

Marinotti, O., James, A.A. and Ribeiro, J.M.C. (1990) Diet and salivation in female *Aedes aegypti* mosquitoes. *J. Insect Physiol.*, **36**, 545–8.

Marks, E.N. (1954) A review of the *Aedes scutellaris* subgroup with a study of variation in *Aedes pseudoscutellaris* (Theobald) (Diptera: Culicidae). *Bull. Brit. Mus. (Nat. Hist.) Entomol.*, **3**, 349–414, 1 pl.

Marshall, J.F. (1938) *The British Mosquitoes*, British Museum (Natural History), London.

Marshall, J.F. and Staley, J. (1932) On the distribution of air in the oesophageal diverticula and intestine of mosquitoes. Its relation to emergence, feeding and hypopygial rotation. *Parasitology*, **24**, 368–81.

Marshall, J.F. and Staley, J. (1935) Generic and subgeneric differences in the mouthparts of male mosquitos. *Bull. Entomol. Res.*, **26**, 531–2.

Marten. G.G. (1986) Mosquito control by plankton management: the potential of indigestible green algae. *J. Trop. Med. Hyg.*, **89**, 213–22.

Martin, M.M. (1987) *Invertebrate-microbial Interactions*, Cornell University Press, Ithaca and London.

Martin, M.M., Martin, J.S., Kukor, J.J. and Merrit, R.W. (1980) The digestion of protein and carbohydrate by the stream detritivore, *Tipula abdominalis* (Diptera, Tipulidae). *Oecologia*, **46**, 360–4.

Martinez, T. and Hagedorn, H.H. (1987) Development of responsiveness to hormones after a blood meal in the mosquito *Aedes aegypti*. *Insect Biochem.*, **17**, 1095–98.

Mashiko, K. and Asakura, K. (1968) An electron microscopic study of the anal papillae of the mosquito larvae, *Culex pipiens pallens*. *Ann. Rep. Noto Marine Lab., Kanazawa Univ.*, **8**, 19–27.

Masler, E.P., Fuchs, M.S., Sage, B. and O'Connor, J.D. (1980) Endocrine regulation of ovarian development in the autogenous mosquito, *Aedes atropalpus*. *Gen. Comp. Endocrinol.*, **41**, 250–9.

Masler, E.P., Fuchs, M.S., Sage, B. and O'Connor, J.D. (1981) A positive correlation between oocyte production and ecdysteroid levels in adult *Aedes*. *Physiol. Entomol.*, **6**, 45–9.

Masler, E.P., Kelly, T.J., Kochansky, J.P. *et al.* (1991) Egg development neurosecretory hormone activity in the mosquito *Aedes aegypti*. In *Insect Neuropeptides: Chemistry, Biology and Action* (eds J.J. Menn, T.J. Kelly and E.P. Masler), *Am. Chem. Soc. Symp. Ser.*, **453**, 124–32.

Masler, E.P., Whisenton, L.R., Schlaeger, D.A. *et al.* (1983) Chymotrypsin and trypsin levels in adult *Aedes atropalpus* and *Toxorhynchites brevipalpis* (Theobald). *Comp. Biochem. Physiol.*, **75B**, 435–40.

Mason, A.Z. and Simkiss, K. (1982) Sites of mineral deposition in metal-accumulating cells. *Exp. Cell Res.*, **139**, 383–91.

Mason, G.F. (1967) Genetic studies on mutations in species A and B of the *Anopheles gambiae* complex. *Genet. Res.*, **10**, 205–17.

Mason, G.F. (1980) A gynandromorph in *Anopheles gambiae*. *Mosq. News*, **40**, 104–6.

Mathew, G. and Rai, K.S. (1975) Structure and formation of egg membranes in *Aedes aegypti* (L.) (Diptera: Culicidae). *Int. J. Insect Morphol. Embryol.*, **4**, 369–80.

Mathew, G. and Rai, K.S. (1976a) Fine structure of the Malpighian tubule in *Aedes aegypti*. *Ann. Entomol. Soc. Am.*, **69**, 659–61.

Mathew, G. and Rai, K.S. (1976b) Ring canals in the ovarian follicles of *Aedes aegypti*. *Ann. Entomol. Soc. Am.*, **69**, 662–5.

Mathis, M. (1935) Sur la nutrition sanguine et la fécondité de Stegomyia: *Aedes aegypti*. *Bull. Soc. Pathol. Exotique*, **28**, 231–4.

Matsuda, R. (1965) Morphology and evolution of the insect head. *Mem. Am. Entomol. Inst.*, No. 4, 1–334.

Matsumoto, S., Brown, M.R., Crim, J.W. *et al.* (1989a) Isolation and primary structure of neuropeptides from the mosquito, *Aedes aegypti*, immunoreactive to FMRF-amide antiserum. *Insect Biochem.*, **19**, 277–83.

Matsumoto, S., Brown, M.R., Suzuki, A. and Lea, A.O. (1989b) Isolation and characterization of ovarian ecdysteroidogenic hormones from the mosquito, *Aedes aegypti*. *Insect Biochem.*, **19**, 651–6.

Matsuo, K., Yoshida, Y. and Lien, J.C. (1974) Scanning electron microscopy of mosquitoes. II. The egg surface structure of 13 species from Taiwan. *J. Med. Entomol.*, **11**, 179–88.

Matthews, T.C. and Munstermann, L.E. (1990) Linkage maps for 20 enzyme loci in *Aedes triseriatus*. *J. Heredity*, **81**, 101–6.

Mattingly, P.F. (1969) Mosquito eggs III. Tribe Anophelini. *Mosq. Syst. Newslett.*, **1**, 41–50.

Mattingly, P.F. (1971) Ecological aspects of mosquito evolution. *Parassitologia*, **13**, 31–65.

Mattingly, P.F. (1973) Mosquito eggs XXIII. Eggs of *Toxorhynchites amboinensis* containing two-headed monsters. *Mosq. Systemat.*, **5**, 197–9.

Mattingly, P.F. (1977) Mosquito eggs XXIX. Genus *Hodgesia* Theobald. *Mosq. Systemat.*, **9**, 333–6.

Matutani, K., Matsumoto, A., Sekoguti, Y. and Yashika, K. (1983) Studies on the mechanisms of ionic regulation in mosquito larvae. 3. Temperature dependent ionic regulation. *Jap. J. Sanit. Zool.*, **34**, 89–94. [In Japanese, English summary.]

Mayer, M.S. (1966) Decline in male/female sex ratio of *Aedes aegypti* (L.) during hatching. *Mosq. News*, **26**, 82–4.

Mayne, B. (1928) The influence of relative humidity on the presence of parasites in the insect carrier and the initial seasonal appearance of malaria in a selected area in India. *Indian J. Med. Res.*, **15**, 1073–84.

McClelland, G.A.H. (1966) Sex-linkage at two loci affecting eye pigment in the mosquito *Aedes aegypti* (Diptera : Culicidae). *Can. J. Genet. Cytol.*, **8**, 192–8.

McClelland, G.A.H. and Conway, G.R. (1974) Frequency of blood feeding in the mosquito *Aedes aegypti*. *Nature*, **232**, 485–6.

McClelland, G.A.H. and Green, C.A. (1970) Subtle periodicity of pupation in rapidly developing mosquitoes. With particular reference to *Aedes vittatus* and *Aedes aegypti*. *Bull. Wld Hlth Org.*, **42**, 951–5.

McCrae, A.W.R. (1968) [Communication] *Proc. R. Entomol. Soc. Lond.*, C, **33**, 27.

McCrae, A.W.R., Boreham, P.F.L. and Ssenkubuge, Y. (1976) The behavioural ecology of host selection in *Anopheles implexus* (Theobald) (Diptera, Culicidae). *Bull. Entomol. Res.*, **66**, 587–631, 2 pls.

McCrae, A.W.R., Ssenkubuge, Y., Mawejji, C. and Kitama, A. (1969) Mosquito activity at nectar sources. *Rep. E. Afr. Virus Res. Inst.*, **17**, 64–5.

McDonald, P.T., Asman, S.M. and Terwedow, H.A. (1978) Sex-linked translocations in *Culex tarsalis*: chromosome-linkage group correlation and segregation patterns. *J. Hered.*, **69**, 304–10.

McDonald, P.T. and Rai, K.S. (1970) Correlation of linkage groups with chromosomes in the mosquito, *Aedes aegypti*. *Genetics*, **66**, 475–85.

McGeachin, R.L., Willis, T.G., Roulston, E.F. and Lang, C.A. (1972) Variations in alpha-amylase during the life span of the mosquito. *Comp. Biochem. Physiol.*, **43B**, 185–91.

McGinnis, K.M. and Brust, R.A. (1983) Effect of different sea salt concentrations and temperatures on larval development of *Aedes togoi* (Diptera: Culicidae) from British Columbia. *Environ. Entomol.*, **12**, 1406–11.

McGinnis, K.M. and Brust, R.A. (1985) Oogenesis in North American population of *Aedes* (*Finlaya*) *togoi* (Theobald) (Diptera: Culicidae). *Can. J. Zool.*, **63**, 2168–71.

McGrane, V., Carlson, J.O., Miller, B.R. and Beaty, B.J. (1988) Microinjection of DNA into *Aedes triseriatus* ova and detection of integration. *Am. J. Trop. Med. Hyg.*, **39**, 502–10.

McGreevy, P.B., Bryan, J.H., Oothuman, P. and Kolstrup, N. (1978) The lethal effects of the cibarial and pharyngeal armatures of mosquitoes on microfilariae. *Trans. R. Soc. Trop. Med. Hyg.*, **72**, 361–8.

McGregor, D.D. (1965) Aspects of the biology of *Opifex fuscus* Hutton (Diptera : Culicidae). *Proc. R. Entomol. Soc. Lond.* (A), **40**, 9–14.

McIver, S. (1971) Comparative studies on the sense organs on the antennae and maxillary palps of selected male culicine mosquitoes. *Can. J. Zool.*, **49**, 235–39, 1 pl.

McIver, S.B. (1972) Fine structure of pegs on the palps of female culicine mosquitoes. *Can. J. Zool.*, **50**, 571–6, 6 pls.

McIver, S.B. (1982) Sensilla of mosquitoes (Diptera: Culicidae). *J. Med. Entomol.*, **19**, 489–535.

McIver, S. and Charlton, C. (1970) Studies on the sense organs on the palps of selected culicine mosquitoes. *Can. J. Zool.*, **48**, 293–5, 1 pl.

McIver, S. and Siemicki, R. (1975a) Campaniform sensilla on the palps of *Anopheles stephensi* Liston (Diptera: Culicidae). *Int. J. Insect Morphol. Embryol.*, **4**, 127–30.

McIver, S. and Siemicki, R. (1975b) Palpal sensilla of selected anopheline mosquitoes. *J. Parasitol.*, **61**, 535–8.

McIver, S. and Siemicki, R. (1977) Observations on *Aedes aegypti* (L.) as scavengers. *Mosq. News*, **37**, 519–21.

McIver, S. and Siemicki, R. (1978) Fine structure of tarsal sensilla of *Aedes aegypti* (L.) (Diptera: Culicidae). *J. Morphol.*, **155**, 137–55.

McIver, S. and Siemicki, R. (1981) Innervation of cibarial sensilla of *Aedes aegypti* (L.) (Diptera: Culicidae). *Int. J. Insect Morphol. Embryol.*, **10**, 355–7.

McKean, T.A. (1973) The latencies of several component parts of the labellar response of the mosquito *Culiseta inornata*. *Ann. Entomol. Soc. Am.*, **66**, 404–6.

McKiel, J.A. (1959) Sensitization to mosquito bites. *Can. J. Zool.*, **37**, 341–51.

McKiel, J.A. and West, A.S. (1961) Nature and causation of insect bite reactions. *The Pediatric Clinics of North America*, **8**, 795–816.

McLain, D.K. and Collins, F.H. (1989) Structure of rDNA in the mosquito *Anopheles gambiae* and rDNA sequence variation within and between species of the *A. gambiae* complex. *Heredity*, **62**, 233–42.

McLain, D.K., Collins, F.H., Brandling-Bennett, A. D. and Were, J.B.O. (1988) Microgeographic variation in rDNA intergenic spacers of *Anopheles gambiae* in western Kenya. *Heredity*, **62**, 257–64.

McLain, D.K., Rai, K.S. and Fraser, M.J. (1986) Interspecific variation in the abundance of highly repeated DNA sequences in the *Aedes scutellaris* (Diptera: Culicidae) subgroup. *Ann. Entomol. Soc. Am.*, **79**, 784–91.

McLain, D.K., Rai, K.S. and Fraser, M.J. (1987) Intraspecific and interspecific variation in the sequence and abundance of highly repeated DNA among mosquitoes of the *Aedes albopictus* subgroup. *Heredity*, **58**, 373–81.

McMullen, A.I., Reiter, P. and Phillips, M.C. (1977) Mode of action of insoluble monolayers on mosquito pupal respiration. *Nature*, **267**, 244–5.

Meek, S.R. (1984) Occurrence of rickettsia-like symbionts among species of the *Aedes scutellaris* group (Diptera: Culicidae). *Ann. Trop. Med. Parasitol.*, **78**, 377–81.

Meek, S..R. and Macdonald, W.W. (1984) Crossing relationships among seven members of the group of *Aedes scutellaris* (Walker) (Diptera: Culicidae). *Bull. Entomol. Res.*, **74**, 65–78.

de Meijere, J.C.H. (1909) Drei myrmecophile Dipteren aus Java. *Tijdschr. Entomol.*, **52**, 165–74, 1 pl.

de Meijere, J.C.H. (1911) Zur Metamorphose der myrmecophilen Culicide *Harpagomyia splendens* de Meij. *Tijdschr. Entomol.*, **54**, 162–7, 1 pl.

Mellanby, K. (1946) Man's reaction to mosquito bites. *Nature*, **158**, 554.

Mellink, J.J. (1980) Mosquito bites, host responses and virus transmission. PhD thesis, University of Amsterdam, xx + 170 pp.

Mellink, J.J., Poppe, D.M.C. and Van Duin, G.J.T. (1982) Factors affecting the blood- feeding process of a laboratory strain of *Aedes aegypti* on rodents. *Entomol. Exp. Applic.*, **31**, 229–38.

Mellink, J.J. and van den Bovenkamp, W. (1981) Functional aspects of mosquito salivation in blood feeding of *Aedes aegypti*. *Mosq. News*, **41**, 115–19.

Mellink, J.J. and van Zeben, M.S. (1976) Age related differences of saliva composition in *Aedes aegypti*. *Mosq. News*, **36**, 247–50.

Meola, R. (1964) The influence of temperature and humidity on embryonic longevity in *Aedes aegypti*. *Ann. Entomol. Soc. Am.*, **57**, 468–72.

Meola, S.M. (1970) Sensitive paraldehyde-fuchsin technique for neurosecretory system of mosquitoes. *Trans. Am. Microscop. Soc.*, **89**, 66–71.

Meola, R. and Lea, A.O. (1971) Independence of paraldehyde-fuchsin staining of the corpus cardiacum and the presence of the neurosecretory hormone required for egg development in the mosquito. *Gen. Comp. Endocrinol.*, **16**, 105–11.

Meola, S.M. and Lea, A.O. (1972a) The ultrastructure of the corpus cardiacum of *Aedes sollicitans* and the histology of the cerebral neurosecretory system of mosquitoes. *Gen. Comp. Endocrinol.*, **18**, 210–34.

Meola, R. and Lea, A.O. (1972b) Humoral inhibition of egg development in mosquitoes. *J. Med. Entomol.*, **9**, 99–103.

Meola, S.M., Lea, A.O. and Meola, R. (1970) Corpus cardiacum: induced fluctuation in paraldehyde-fuchsin material in *Aedes sollicitans*. *Trans. Am. Microscop. Soc.*, **89**, 418–23.

Meola, S.M., Mollenhauer, H.H. and Thompson, J.M. (1977) Cytoplasmic bridges within the follicular epithelium of the ovarioles of two Diptera, *Aedes aegypti* and *Stomoxys calcitrans*. *J. Morphol.*, **153**, 81–5.

Meola, R. and Readio, J. (1988) Juvenile hormone regulation of biting behavior and egg development in mosquitoes. *Adv. Disease Vector Res.*, **5**, 1-24.

Mer, G. (1931) Notes on the bionomics of *Anopheles*

elutus, Edw. (Dipt., Culic.). *Bull. Entomol. Res.*, **22** 137–45.

Mer, G. (1936) Experimental study on the development of the ovary in *Anopheles elutus*, Edw. (Dipt. Culic.). *Bull. Entomol. Res.*, **27**, 351–9.

Mer, G. (1937) Variations saisonnières des caractères de *Anopheles elutus* en Palestine. *Bull. Soc. Pathol. Exotique*, **30**, 38–42.

Meredith, J. and Phillips, J.E. (1973a) Ultrastructure of the anal papillae of a salt-water mosquito larva, *Aedes campestris*. *J. Insect Physiol.*, **19**, 1157–72.

Meredith, J. and Phillips, J.E. (1973b) Ultrastructure of anal papillae from a seawater mosquito larva (*Aedes togoi* Theobald). *Can. J. Zool.*, **51**, 349–53, 1 pl.

Meredith, J. and Phillips, J.E. (1973c) Rectal ultrastructure in salt- and freshwater mosquito larvae in relation to physiological state. *Z. Zellforsch. Mikroscop. Anat.*, **138**, 1–22.

Merritt, R.W. (1987) Do different instars of *Aedes triseriatus* feed on particles of the same size? *J. Am. Mosq. Control Ass.*, **3**, 94–6.

Merritt, R.W. and Craig, D.A. (1987) Larval mosquito (Diptera: Culicidae) feeding mechanisms: mucosubstance production for capture of fine particles. *J. Med. Entomol.*, *24*, 275–8.

Merritt, R.W., Craig, D.A., Walker, E.D. *et al.* (1992a) Interfacial feeding behavior and particle flow patterns of *Anopheles quadrimaculatus* (Diptera: Culicidae). *J. Insect Behavior*, **5** (in press).

Merritt, R.W. and Cummins, K.W. (1984) *An Introduction to the Aquatic Insects of North America.* 2nd edn, Kendall/HuntPublishing Co., Dubuque, Iowa.

Merritt, R.W., Dadd, R.H. and Walker, E.D. (1992b) Feeding behavior, natural food, and nutritional relationships of larval mosquitoes. *Annual Rev. Entomol.*, **37**, 349–76.

Merritt, R.W., Mortland, M.M., Gersabeck, E.F. and Ross, D.H. (1978) X-ray diffraction analysis of particles ingested by filter-feeding animals. *Entomol. Exp. Applic.*, **24**, 27-34.

Merritt, R.W., Olds, E.J. and Walker, E.D. (1990) Natural food and feeding behavior of *Coquillettidia perturbans* larvae. *J. Am. Mosq. Control. Ass.*, **5**, 35–42.

Mescher, A.L. and Rai, K.S. (1966) Spermatogenesis in *Aedes aegypti*. *Mosq. News*, **26**, 45–51.

Metcalf, R.L. (1945) The physiology of the salivary glands of *Anopheles quadrimaculatus*. *J. Natl Malaria Soc.*, **4**, 271–8.

Mezzanotte, R., Ferrucci, L. and Contini, C. (1979) Identification of sex chromosomes and characterization of the heterochromatinin in *Culiseta longiareolata* (Macquart, 1838). *Genetica*, **50**, 135–9.

Micks, D.W., de Caires, P.F. and Franco, L.B. (1948) The relationship of exflagellation in avian plasmodia to pH and immunity in the mosquito. *Am. J. Hyg.*, **48**, 182–90.

Micks, D.W. and Rougeau, D. (1976) Entry and movement of petroleum derivatives in the tracheal system of mosquito larvae. *Mosq. News*, **36**, 449–54.

Micks, D.W., Starr, C.F. and Partridge, M.H. (1959) The vitamin B content of mosquitoes. *Ann. Entomol. Soc. Am.*, **52**, 26–8.

Milby, M.M. and Meyer, R.P. (1986) The influence of constant versus fluctuating water temperatures on the preimaginal development of *Culex tarsalis*. *J. Am. Mosq. Control Ass.*, **2**, 7–10.

Mill, P.J. (1974) Respiration: aquatic insects. In *The Physiology of Insecta* (ed. M. Rockstein), 2nd edn, Academic Press, New York and London, **6**, 403-467.

Mitchell, C.J. and Briegel, H. (1989a) Inability of diapausing *Culex pipiens* (Diptera: Culicidae) to use blood for producing lipid reserves for overwinter survival. *J. Med. Entomol.*, **26**, 318–26.

Mitchell, C.J. and Briegel, H. (1989b) Fate of the blood meal in force-fed, diapausing *Culex pipiens* (Diptera: Culicidae). *J. Med. Entomol.*, **26**, 332–41.

Mitchell, M.J. and Wood, R.J. (1984) Genetic variation in tolerance of ammonium chloride in *Aedes aegypti*. *Mosq. News*, **44**, 498–501.

Moens, P.B. (1978) Ultrastructural studies of chiasma distribution. *Annual Rev. Genet.*, **12**, 433–50.

Moeur, J.E. and Istock, C.A. (1982) Chromosomal polymorphisms in the pitcher-plant mosquito, *Wyeomyia smithii*. *Chromosoma*, **84**, 623–51.

Moffett, A.A. (1936) The origin and behaviour of chiasmata. XIII. Diploid and tetraploid *Culex pipiens*. *Cytologia*, **7**, 184–97.

Mogi, M. (1978) Intra- and interspecific predation in filter feeding mosquito larvae. *Trop. Med.*, **20**, 15–27.

Mogi, M. and Miyagi, I. (1989) Sugar feeding of *Topomyia pseudobarbus* (Diptera: Culicidae) in nature. *J. Med. Entomol.*, **26**, 370–1.

Mohamed, H.A., Ingram, G.A., Molyneux, D.H. and Sawyer, B.V. (1991) Use of fluorescein-labelled lectin binding of salivary glands to distinguish between *Anopheles stephensi* and *An. albimanus* species and strains. *Insect Biochem.*, **21**, 767–73.

Mohrig, W. (1964) Faunistisch-ökologische Untersuchungen an Culiciden der Umgebung von Greifswald. *Deutsche Entomol. Z.*, **11**, 327–52.

Möllring, F.K. (1956) Autogene und anautogene Eibildung bei *Culex* L. Zugleich ein Beitrag zur Frage der Unterscheidung autogener und anautogener Weibchen an Hand von Eiröhrenzahl und Flügellänge. *Z. Tropenmed. Parasitol.*, **7**, 15–48.

Molyneux, D.H., Okolo, C.J., Lines, J.D. and Kamhawi, M. (1990) Variation in fluorescein-labelled lectin staining of salivary glands in the *Anopheles gambiae* complex. *Med. Vet. Entomol.*, **4**, 459–62.

Moore, C.G. and Fisher, B.R. (1969) Competition in mosquitoes. Density and species ratio effects on growth, mortality, fecundity, and production of growth retardant. *Ann. Entomol. Soc. Am.*, **62**, 1325–31.

Moore, C.G. and Whitacre, D.M. (1972) Competition in mosquitoes. 2. Production of *Aedes aegypti* larval growth retardant at various densities and nutrition levels. *Ann. Entomol. Soc. Am.*, **65**, 915–18.

Mori, A. (1979) Effects of larval density and nutrition on some attributes of immature and adult *Aedes albopictus*. *Trop. Med.*, **21**, 85–103.

Morris, A. C., Eggleston, P. and Crampton, J.M. (1989) Genetic transformation of the mosquito *Aedes aegypti* by micro-injection of DNA. *Med. Vet. Entomol.*, **3**, 1–7.

Mortenson, E.W. (1950) The use of sodium hypochlorite to study *Aedes nigromaculis* (Ludlow) embryos (Diptera: Culicidae). *Mosq. News*, **10**, 211–12.

Mosha, F.W. and Mutero, C.M. (1982) The influence of salinity on larval development and population dynamics of *Anopheles merus* Dönitz (Diptera: Culicidae). *Bull. Entomol. Res.*, **72**, 119–28.

Mossé, E. and Hartman, M.J. (1980) The effects of thermal stress upon gonads of *Culex tarsalis* (Diptera: Culicidae). *J. Med. Entomol.*, **17**, 487–8.

Motara, M.A. (1982) Giemsa C-banding in four species of mosquitoes. *Chromosoma*, **86**, 319–23.

Motara, M.A., Pathak, S., Satya-Prakash, K.L. and Hsu, T.C. (1985) Argentophilic structures of spermatogenesis in the yellow fever mosquito. *J. Hered.*, **76**, 295–300.

Motara, M.A. and Rai, K.S. (1977) Chromosomal differentiation in two species of *Aedes* and their hybrids revealed by Giemsa C-banding. *Chromosoma*, **64**, 125–32.

Motara, M.A. and Rai, K.S. (1978) Giemsa C-banding patterns in *Aedes* (*Stegomyia*) mosquitoes. *Chromosoma*, **70**, 51–8.

Muirhead-Thomson, R.C. (1941) Studies on the behaviour of *Anopheles minimus*. Part IV. The composition of the water and the influence of organic pollution and silt. *J. Malaria Inst. India*, **4**, 63–102, 4 pls.

Muirhead-Thomson, R.C. (1951) *Mosquito Behaviour in Relation to Malaria Transmission and Control in the Tropics*, Arnold, London.

Müller, W. (1968) Die Distanz- und Kontakt-Orientierung der Stechmücken (*Aedes aegypti*) (Wirtsfindung, Stechverhalten und Blutmahlzeit). *Z. Vergl. Physiol.*, **58**, 241–303.

Munkirs, D.D., Christensen, B.M. and Tracy, J.W. (1990) High-pressure liquid chromatographic analysis of hemolymph plasma catecholamines in immune-reactive *Aedes aegypti*. *J. Invert. Pathol.*, **56**, 267–79.

Munstermann, L.E. (1990a) Gene map of the yellow fever mosquito (Aedes (Stegomyia) aegypti) (2N=6). In *Genetic Maps: Locus Maps of Complex Genomes* (ed. S.J. O'Brien), 5th edn, Cold Spring Harbor Laboratory Press, pp. 3.179–3.183.

Munstermann, L.E. (1990b) Gene map of the eastern north American tree hole mosquito, *Aedes* (*Protomacleaya*) *triseriatus* (2N=6). In *Genetic Maps: Locus Maps of Complex Genomes* (ed. S.J. O'Brien), 5th edn, Cold Spring Harbor Laboratory Press, pp. 3.184–3.187.

Munstermann, L.E. and Craig, G.B. (1979) Genetics of *Aedes aegypti*. Updating the linkage map. *J. Hered.*, **70**, 291–6.

Munstermann, L.E. and Marchi, A. (1986) Cytogenetic and isozyme profile of *Sabethes cyaneus*. *J. Hered.*, **77**, 241–8.

Munstermann, L.E., Marchi, A., Sabatini, A. and Coluzzi, M. (1985) Polytene chromosomes of *Orthopodomyia pulcripalpis* (Diptera, Culicidae). *Parassitologia*, **27**, 267–77.

Nagl, W. (1978) *Endopolyploidy and Polyteny in Differentiation and Evolution*, North-Holland Publishing Co., Amsterdam.

Nakayama, Y., Kawamoto, F., Suto, C. *et al.* (1985) Histamine and esterases in the salivary gland of the mosquito, *Culex pipiens pallens*. *Jap. J. Sanit. Zool.*, **36**, 315–26.

Nappi, A.J. and Christensen, B.M. (1986) Hemocyte cell surface changes in *Aedes aegypti* in response to microfilariae of *Dirofilaria immitis*. *J. Parasitol.*, **72**, 875–9.

Narang, N., Narang, S. and Kitzmiller, J.B. (1972) Karyological studies on four species of *Anopheles* subgenus *Cellia*. *Caryologia*, **25**, 259–74.

Narang, S. and Seawright, J.A. (1982) Linkage relationships and genetic mapping in *Culex* and *Anopheles*. In *Recent Developments in the Genetics of Insect Disease Vectors* (eds W.W.M. Steiner, W.J. Tabachnick, K.S. Rai and S. Narang), Stipes Publishing Co., Champaign, Illinois, pp. 231–89.

Narang, S.K. and Seawright, J.A. (1990) Linkage map of the mosquito (*Anopheles albimanus*) (2N=6). In *Genetic Maps: Locus Maps of Complex Genomes* (ed. S.J. O'Brien), Cold Spring Harbor Laboratory Press, 5th edn, pp. 3.190–3.193.

Nasci, R.S. (1990) Relationship of wing length to adult dry weight in several mosquito species (Diptera: Culicidae). *J. Med. Entomol.*, **27**, 716–19.

Nath, V. (1924) Egg-follicle of *Culex*. *Quart. J. Microscop. Sci.*, **69**, 151–75, 2 pls.

Nath, V., Gupta, B.L. and Bains, G.S. (1958) Histochemical and morphological studies of the lipids in oogenesis. V. The egg-follicle of *Culex fatigans*. *Res. Bull. Panjab Univ.*, No. 148, 135–48a, 2 pls.

Nayar, J.K. (1966) A method of rearing salt-marsh mosquito larvae in a defined sterile medium. *Ann. Entomol. Soc. Am.*, **59**, 1283–5.

Nayar, J.K. (1967a) Endogenous diurnal rhythm of pupation in a mosquito population. *Nature*, **214**, 828–9.

Nayar, J.K. (1967b) The pupation rhythm in *Aedes taeniorhynchus* (Diptera: Culicidae). *Ann. Entomol. Soc. Am.*, **60**, 946–71.

Nayar, J.K. (1968a) The pupation rhythm in *Aedes taeniorhynchus*. IV. Further studies of the endogenous diurnal (circadian) rhythm of pupation. *Ann. Entomol. Soc. Am.*, **61**, 1408–17.

Nayar, J.K. (1968b) Biology of *Culex nigripalpus* Theobald (Diptera: Culicidae). Part 1: Effects of rearing conditions on growth and the diurnal rhythm of pupation and emergence. *J. Med. Entomol.*, **5**, 39–46.

Nayar, J.K. (1968c) The biology of *Culex nigripalpus* Theobald (Diptera: Culicidae). Part 2. Adult characteristics at emergence and adult survival without nourishment. *J. Med. Entomol.*, **5**, 203–10.

Nayar, J.K. (1969a) The pupation rhythm in *Aedes taeniorhynchus* (Diptera: Culicidae). V. Physiology of growth and endogenous diurnal rhythm of pupation. *Ann. Entomol. Soc. Am.*, **62**, 1079–87.

Nayar, J.K. (1969b) Effects of larval and pupal environmental factors on biological status of adults at emergence in *Aedes taeniorhynchus* (Wied.). *Bull. Entomol. Res.*, **58**, 811–27.

Nayar, J.K. (1978) The detection of nectar sugars in field-collected *Culex nigripalpus* and its application. *Ann. Entomol. Soc. Am.*, **71**, 55–9.

Nayar, J.K. (1982) Bionomics and physiology of *Culex nigripalpus* (Diptera: Culicidae) of Florida: an important vector of diseases. *Bull. Florida Agric. Exp. Sta.*, **82**, 71–3.

Nayar, J.K. (1985) Bionomics and physiology of *Aedes taeniorhynchus* and *Aedes sollicitans*, the salt marsh mosquitoes of Florida. *Bull. Univ. Florida Agric. Exp. Sta.*, **852**, 1–148.

Nayar, J.K. and Bradley, T.J. (1987) Effects of infection with *Dirofilaria immitis* on diuresis and oocyte development in *Aedes taeniorhynchus* and *Anopheles quadrimaculatus* (Diptera, Culicidae) *J. Med. Entomol.*, **24**, 617–22.

Nayar, J.K., Crossman, R.A. and Pierce, P. (1978) Circadian rhythm of emergence in the mosquito *Wyomyia mitchellii* and the effects of light cycles on the pupation rhythm of *Culex nigripalpus*. *Ann. Entomol. Soc. Am.*, **71**, 257–63.

Nayar, J.K., Knight, J.W. and Bradley, T.J. (1988) Further characterization of refractoriness in *Aedes aegypti* (L.) to infection by *Dirofilaria immitis* (Leidy). *Exp. Parasitol.*, **66**, 124–31.

Nayar, J.K., Knight, J.W. and Vickery, A.C. (1989) Intracellular melanization in the mosquito *Anopheles quadrimaculatus* (Diptera: Culicidae) against the filarial nematodes, *Brugia* spp. (Nematoda: Filarioidea). *J. Med. Entomol.*, **26**, 159–66.

Nayar, J.K. and Pierce, P.A. (1977) Utilization of energy reserves during survival after emergence in Florida mosquitoes. *J. Med. Entomol.*, **14**, 54–9.

Nayar, J.K., Samarawickrema, W.A. and Sauerman, D.L. (1973) Photoperiodic control of egg hatching in the mosquito *Mansonia titillans*. *Ann. Entomol. Soc. Am.*, **66**, 831–5.

Nayar, J.K. and Sauerman, D.L. (1970a) A comparative study of growth and development in Florida mosquitoes. Part 1: Effects of environmental factors on ontogenetic timings, endogenous diurnal rhythm and synchrony of pupation and emergence. *J. Med. Entomol.*, **7**, 163–74.

Nayar, J.K. and Sauerman, D.L. (1970b) A comparative study of growth and development in Florida mosquitoes. Part 2: Effects of larval nurture on adult characteristics at emergence. *J. Med. Entomol.*, **7**, 235–41.

Nayar, J.K. and Sauerman, D.M. (1971a) The effects of diet on life-span, fecundity and flight potential of *Aedes taeniorhynchus* adults. *J. Med. Entomol.*, **8**, 506–13.

Nayar, J.K. and Sauerman, D.M. (1971b) Physiological effects of carbohydrates on survival, metabolism, and flight potential of female *Aedes taeniorhynchus*. *J. Insect Physiol.*, **17**, 2221–33.

Nayar, J.K. and Sauerman, D.M. (1972) Flight performance and fuel utilization as a function of age in female *Aedes taeniorhynchus*. *Israel J. Entomol.*, **7**, 27–35.

Nayar, J.K. and Sauerman, D.M. (1973) A comparative study of flight performance and fuel utilization as a function of age in females of Florida mosquitoes. *J. Insect Physiol.*, **19**, 1977–88.

Nayar, J.K. and Sauerman, D.M. (1974a) Osmoregulation in larvae of the salt-marsh mosquito, *Aedes taeniorhynchus* Wiedemann. *Entomol. Exp. Applic.*, **17**, 367–80.

Nayar, J.K. and Sauerman, D.M. (1974b) Long-term regulation of sucrose intake by the female mosquito, *Aedes taeniorhynchus*. *J. Insect Physiol.*, **20**, 1203–8.

Nayar, J.K. and Sauerman, D.M. (1975a) The effects of nutrition on survival and fecundity in Florida mosquitoes. Part 1. Utilization of sugar for survival. *J. Med. Entomol.*, **12**, 92–8.

Nayar, J.K. and Sauerman, D.M. (1975b) The effects of nutrition on survival and fecundity in Florida mosquitoes. Part 2. Utilization of a blood meal for survival. *J. Med. Entomol.*, **12**, 99–103.

Nayar, J.K. and Sauerman, D.M. (1975c) The effects of nutrition on survival and fecundity in Florida mosquitoes. Part 3. Utilization of blood and sugar for fecundity. *J. Med. Entomol.*, **12**, 220–5.

Nayar, J.K. and Sauerman, D.M. (1977) The effects of nutrition on survival and fecundity in Florida mosquitoes. Part 4. Effects of blood source on oocyte development. *J. Med. Entomol.*, **14**, 167–74.

Nayar, J.K. and Van Handel, E. (1971) The fuel for sustained mosquito flight. *J. Insect Physiol.*, **17**, 471–81, addendum 1391–2.

Nehman, B.F. (1968) An electron microscope study of the distal portion of the hypopharynx of female *Aedes aegypti*. *Ann. Entomol. Soc. Am.*, **61**, 1274–8.

Nemjo, J. and Slaff, M. (1984) Head capsule width as a tool for instar and species identification of *Mansonia dyari*, *M. titillans*, and *Coquillettidia perturbans* (Diptera: Culicidae). *Ann. Entomol. Soc. Am.*, **77**, 633–5.

Neville, A.C. (1983) Daily cuticular growth layers and the teneral stage in adult insects: a review. *J. Insect Physiol.*, **29**, 211–19.

Newkirk, M.R. (1955) On the eggs of some man-biting mosquitoes. *Ann. Entomol. Soc. Am.*, **48**, 60–6.

Newton, M.E., Southern, D.I. and Wood, R.J. (1974) X and Y chromosomes of *Aedes aegypti* (L.) distinguished by Giemsa C-banding. *Chromosoma*, **49**, 41–9.

Newton, M.E., Southern, D.I. and Wood, R.J. (1978a) Relative DNA content of normal and sex-ratio distorting spermatozoa of the mosquito, *Aedes aegypti*. *Chromosoma*, **67**, 253–61.

Newton, M.E., Wood, R.J. and Southern, D.I. (1976) A cytogenetic analysis of meiotic drive in the mosquito, *Aedes aegypti* (L.). *Genetica*, **46**, 297–318.

Newton, M.E., Wood, R.J. and Southern, D.I. (1978b) Cytological mapping of the M and D loci in the mosquito, *Aedes aegypti* (L.). *Genetica*, **48**, 137–43.

Nicholson, A.J. (1921) The development of the ovary and ovarian egg of a mosquito, *Anopheles maculipennis*, Meig. *Quart. J. Microscop. Sci.*, **65**, 395–448, 4 pl.

Nicolson, S.W. (1972) Osmoregulation in larvae of the New Zealand salt-water *mosquito Opifex fuscus* Hutton. *J. Entomol.* (A), **47**, 101–8.

Nicolson, S.W. and Leader, J.P. (1974) The permeability to water of the cuticle of the larva of *Opifex fuscus* (Hutton) (Diptera, Culicidae). *J. Exp. Biol.*, **60**, 593–603.

Nielsen, E.T. and Haeger, J.S. (1954) Pupation and emergence in *Aedes taeniorhynchus* (Wied.). *Bull. Entomol. Res.*, **45**, 757–68.

Nijhout, H.F. and Carrow, G.M. (1978) Diuresis after a bloodmeal in female *Anopheles freeborni*. *J. Insect Physiol.*, **24**, 293–8.

Nilsson, C. (1986) Energetics of the suspension feeding mosquito larva *Culex torrentium*. *Holarctic Ecol.*, **9**, 267–71.

Nizeyimana, B., Bounias, M. and Vivares, C.P. (1986) Manifestations biochimiques de l'intoxication des larves de *Aedes aegypti* (Insecte, Diptère), par la δ-endotoxine de *Bacillus thuringiensis israelensis*. I. Les glucides de l'hémolymphe. *C.R. Soc. Biol.*, *180*, 551–63.

Noirot-Timothée, C. and Noirot, C. (1980) Septate and scalariform junctions in arthropods. *Int. Rev. Cytol.*, **63**, 97–140.

Norkrans, B. (1980) Surface microlayers in aquatic environments. *Advances in Microbial Ecology*, **4**, 51–85.

Normann, T.C. (1983) Cephalic neurohemal organs in adult Diptera. In *Neurohemal Organs of Arthropods*, (ed. A.P. Gupta), C.C. Thomas, Springfield, Illinois, pp. 454–80.

Nudelman, S., Galun, R., Kitron, U. and Spielman, A. (1988) Physiological characteristics of *Culex pipiens* populations in the Middle East. *Med. Vet. Entomol.*, **2**, 161–9.

Núñez, J.A. (1963) Über den Feinbau des Spermienschwanzes von *Culex pipiens*. *Biol. Zentralbl.*, **82**, 1–7.

Nuttall,, G.H.F. and Shipley, A.E. (1901, 1902 and 1903) Studies in relation to malaria. II. The structure and biology of *Anopheles* (*Anopheles maculipennis*). *J. Hyg.*, **1**, 45–77, 2 pls; **1**, 269–76; **1**, 451–84, 3 pls; **2**, 58–84; **3**, 166–215, 4 pls.

Oberlander, H. (1985) The imaginal discs. In *Comprehensive Insect Physiology Biochemistry and Pharmacology*, (eds G.A. Kerkut and L.I. Gilbert), Pergamon Press, Oxford, **2**, 151–82.

O'Brien, J.F. (1966a) Development of the muscular network of the midgut in the larval stages of the mosquito, *Aedes aegypti* Linnaeus. *J. NY Entomol. Soc.*, **73**, 226–31.

O'Brien, J.F. (1966b) Origin and structural function of the basal cells of the larval midgut in the mosquito, *Aedes aegypti* Linnaeus. *J. NY Entomol. Soc.*, **74**, 59–63.

Oda, T. (1968) Studies of the follicular development and overwintering of the house mosquito, *Culex pipiens pallens* in Nagasaki area. *Trop. Med., Nagasaki*, **10**, 195–216.

Odland, G.D. and Jones, J.C. (1975) Contractions of the hindgut of adult *Aedes aegypti* with special reference to the development of a physiological saline. *Ann. Entomol. Soc. Am.*, **68**, 613–16.

O'Donnell, M.J. and Maddrell, S.H.P. (1983) Paracellular and transcellular routes for water and solute movements across insect epithelia. *J. Exp. Biol.*, **106**, 231–53.

O'Donnell, M.J., Maddrell, S.H.P. and Gardiner, B. (1984) O.C. Passage of solutes through walls of Malpighian tubules of *Rhodnius* by paracellular and transcellular routes. *Am. J. Physiol.*, *246*, R759–69.

Oelhafen, F. (1961) Zur Embryogenese von *Culex pipiens*: Markierungen und Exstirpationen mit UV-Strahlenstich. *Roux' Arch. Entw.Mech. Organ.*, **153**, 120–57.

O'Gower, A.K. (1956) The rate of digestion of human blood by certain species of mosquitoes. *Austral. J. Biol. Sci.*, **9**, 125–9.

Oguma, Y. and Kanda, T. (1976) Laboratory colonization of *Anopheles sinensis* Wiedemann, 1828. *Jap. J. Sanit. Zool.*, **27**, 319–24.

Ogura, N. (1986) Haemagglutinating activity and melanin deposition on microfilariae of *Brugia pahangi* and *B. malayi* in the mosquito, *Armigeres subalbatus*. *Jap. J. Parasitol.*, **35**, 542–9.

Ogura, N. (1987a) *In vitro* melanin deposition on microfilariae of *Brugia pahangi* and *B. malayi* in haemolymph of the mosquito, *Armigeres subalbatus*. *Jap. J. Parasitol.*, **36**, 242–7.

Ogura, N. (1987b) The effect of exogenous haemagglutinin on *in vitro* melanin deposition on microfilariae of *Brugia pahangi* in haemolymph of the mosquito, *Armigeres subalbatus*. *Jap. J. Parasitol.*, **36**, 291–7.

Ogura, N. (1988) Origin of the various-sized components with melanin in haemolymph of the mosquito, *Armigeres subalbatus*. *Jap J. Parasitol.*, **37**, 365–7.

Oka, H. (1955) A morphological study on the growth of the mosquito larvae. *Bull. Yamaguchi Med. School*, **2**, 137–47.

Oka, K. (1989) Correlation of *Aedes albopictus* bite reaction with IgE antibody assay and lymphocyte transformation test to mosquito salivary antigens. *J. Dermatol.*, **16**, 341–7.

Oka, K., Ohtaki, N., Yasuhara, T. and Nakajima, T. (1989) A study of mosquito salivary gland components and their effects on man. *J. Dermatol.*, **16**, 469–74.

Okazawa. T., Miyagi, M., Toma, T. *et al.*, (1991) Egg morphology and observations on the laboratory biology of *Armigeres (Leicesteria) digitatus* (Diptera: Culicidae) from Sarawak. *J. Med. Entomol.*, **28**, 606–10.

Olifan, V.I. (1949) Some regularities in changes of gas exchange in *Anopheles* pupae. *Dokl. Akad. Nauk SSSR*, **65**, 577–80. [In Russian.]

Olson, J.K. and Horsfall, W.R. (1972) Thermal stress and anomalous development of mosquitoes (Diptera, Culicidae). VIII. Gonadal response to fluctuating temperature. *Ann. Zool. Fennici*, **9**, 98–110.

O'Meara, G.F. (1972) Polygenic regulation of fecundity in autogenous *Aedes atropalpus*. *Entomol. Exp. Applic.*, **15**, 81–9.

O'Meara, G.F. (1976) Saltmarsh mosquitoes. In *Marine Insects* (ed. L. Cheng), North-Holland, Amsterdam, pp. 303–33.

O'Meara, G.F. (1979) Variable expressions of autogeny in three mosquito species. *Int. J. Invert. Rep.*, **1**, 253–61.

O'Meara, G.F. (1985a) Gonotrophic interactions in mosquitoes: kicking the blood-feeding habit. *Florida Entomol.*, **68**, 122–33.

O'Meara, G.F. (1985b) Ecology of autogeny in mosquitoes. In *Ecology of Mosquitoes: Proceedings of a Workshop* (eds L.P. Lounibos, J.R. Rey and J.H. Frank), Florida Medical Entomology Laboratory, Vero Beach, pp. 459–71.

O'Meara, G.F. and Craig, G.B. (1969) Monofactorial inheritance of autogeny in *Aedes atropalpus*. *Mosq. News*, **29**, 14–22.

O'Meara, G.F. and Edman, J.D. (1975) Autogenous egg production in the salt-marsh mosquito, *Aedes taeniorhynchus*. *Biol. Bull.*, **149**, 384–96.

O'Meara, G.F. and Evans, D.G. (1973) Blood-feeding requirements of the mosquito: geographical variation in *Aedes taeniorhynchus*. *Science*, **180**, 1291–3.

O'Meara, G.F. and Evans, D.G. (1976) The influence of mating on autogenous egg development in the mosquito, *Aedes taeniorhynchus*. *J. Insect Physiol.*, **22**, 613–17.

O'Meara, G.F. and Evans, D.G. (1977) Autogeny in saltmarsh mosquitoes induced by a substance from the male accessory gland. *Nature*, **267**, 342–4.

O'Meara, G.F. and Krasnick, G.J. (1970) Dietary and genetic control of the expression of autogenous reproduction in *Aedes atropalpus* (Coq.) (Diptera: Culicidae). *J. Med. Entomol.*, **7**, 328–34.

O'Meara, G.F. and Lounibos, L.P. (1981) Reproductive maturation in the pitcher-plant mosquito, *Wyeomyia smithii*. *Physiol. Entomol.*, **6**, 437–43.

O'Meara, G.F., Lounibos, L.P. and Brust, R.A. (1981) Repeated egg clutches without blood in the pitcher-plant mosquito. *Ann. Entomol. Soc. Am.*, **74**, 68–72.

O'Meara, G.F. and Mook, D.H. (1990) Facultative blood-feeding in the crabhole mosquito, *Deinocerites cancer*. *Med. Vet. Entomol.*, **4**, 117–23.

O'Meara, G.F. and Petersen, J.L. (1985) Effects of mating and sugar feeding on the expression of autogeny in crabhole mosquitoes of the genus *Deinocerites* (Diptera: Culicidae). *J. Med. Entomol.*, **22**, 485–90.

O'Meara, G.F. and Van Handel, E. (1971) Triglyceride metabolism in thermally-feminized males of *Aedes aegypti*. *J. Insect Physiol.*, **17**, 1411–3.

Onyeka, J.O.A. and Boreham, P.F.L. (1987) Population studies, physiological state and mortality factors of overwintering adult populations of females of *Culex pipiens* L. (Diptera: Culicidae). *Bull. Entomol. Res.*, **77**, 99–112.

Orr, C.W.M., Hudson, A. and West, A.S. (1961) The salivary glands of *Aedes aegypti*. Histological-histochemical studies. *Can. J. Zool.*, **39**, 265–72.

Owen, A.R.G. (1949) A possible interpretation of the apparent interference across the centromere found by Callan and Montalenti in *Culex pipiens*. *Heredity*, **3**, 357–67.

Owen, W.B. (1963) The contact chemoreceptor organs of the mosquito and their function in feeding behaviour. *J. Insect Physiol.*, **9**, 73–87.

Owen, W.B. (1965) Structure and function of the gustatory organs of the mosquito. *Proc. 12th Int. Congr. Entomol.*, p. 793.

Owen, W.B. (1971) Taste receptors of the mosquito *Anopheles atroparvus* van Thiel. *J. Med. Entomol.*, **8**, 491–4.

Owen, W.B. (1985) Morphology of the head skeleton and muscles of the mosquito, *Culiseta inornata* (Williston) (Diptera: Culicidae). *J. Morphol.*, **183**, 51–85.

Owen, W.B., Larsen, J.R. and Pappas, L.G. (1974) Functional units in the labellar chemosensory hairs of the mosquito *Culiseta inornata* (Williston). *J. Exp. Zool.*, **188**, 235–47.

Owen, W.B. and McClain, E. (1981) Hyperphagia and the control of ingestion in the female mosquito, *Culiseta inornata* (Williston) (Diptera: Culicidae). *J. Exp. Zool.*, **217**, 179–83.

Packer, M.J. and Corbet, P.S. (1989a) Size variation and

reproductive success of female *Aedes punctor* (Diptera: Culicidae). *Ecol. Entomol.*, **14**, 297–309.

Packer, M.J. and Corbet, P.S. (1989b) Seasonal emergence, host-seeking activity, age composition and reproductive biology of the mosquito *Aedes punctor*. *Ecol. Entomol.*, **14**, 433–42.

Padmaja, K. and Rajulu, G.S. (1981) Chemical nature of the chorionic pad of the egg of *Aedes aegypti*. *Mosq. News*, **41**, 674–6.

Pajot, F.X. (1964) Contribution a l'étude de la biologie d'*Anopheles caroni* Adam, 1961. *Bull. Soc. Pathol. Exotique*, **57**, 1290–306.

Pajot, F.X., Le Pont, F. and Molez, J.F. (1975) Donées sur l'alimentation non sanguine chez *Anopheles* (*Nyssorhynchus*) *darlingi* Root, 1926 (Diptera, Culicidae) en Guayane française. *Cahier ORSTOM, Entomol. Méd. Parasitol.*, **13**, 131–4.

Pal, R. (1944) Nephrocytes in some Culicidae-Diptera. *Indian J. Entomol.*, **6**, 143–8, 2 pls.

Palatroni, P., Gabrielli, M.G. and Scattolini, B. (1981) Histochemical localization of carbonic anhydrase in Malpighian tubules of *Culex pipiens*. *Experientia*, **37**, 409–11.

Pannabecker, T.L., Aneshansley, D.J. and Beyenbach, K.W. (1992) Unique electrophysiological effects of dinitrophenol in Malpighian tubules. *Am. J. Physiol.* (in press).

Pappas, L.G. (1988) Stimulation and sequence operation of cibarial and pharyngeal pumps during sugar feeding by mosquitoes (Diptera: Culicidae). *Ann. Entomol. Soc. Am.*, **81**, 274–7.

Pappas, L.G., Pappas, C.D. and Grossman, G.L. (1986) Hemodynamics of human skin during mosquito (Diptera: Culicidae) blood feeding. *J. Med. Entomol.*, **23**, 581–7.

Pappas, L.G. and Larsen, J.R. (1976a) Gustatory hairs on the mosquito, *Culiseta inornata*. *J. Exp. Zool.*, **196**, 351–60.

Pappas, L.G. and Larsen, J.R. (1976b) Labellar chordotonal organs of the mosquito, *Culiseta inornata* (Williston) (Diptera : Culicidae). *Int J. Insect Morphol. Embryol.*, **5**, 145–50.

Pappas, L.G. and Larsen, J.R. (1978) Gustatory mechanisms and sugar-feeding in the mosquito, *Culiseta inornata*. *Physiol. Entomol.*, **3**, 115–19.

Park, Y.-J. and Fallon, A.M. (1990) Mosquito ribosomal RNA genes: characterization of gene structure and evidence for changes in copy number during development. *Insect Biochem.*, **20**, 1–11.

Parkes, A.S. (1960) *Marshall's Physiology of Reproduction*, Vol. II, Longmans, Green and Co., London, New York, Toronto.

Parks, J.J. and Larsen, J.R. (1965) A morphological study of the female reproductive system and follicular development in the mosquito *Aedes aegypti* (L.). *Trans. Am. Microscop. Soc.*, **84**, 88–98.

Paskewitz, S.M., Brown, M.R., Collins, F.H. and Lea, A.O. (1989) Ultrastructural localization of phenoloxidase in the midgut of refractory *Anopheles gambiae* and association of the enzyme with encapsulated *Plasmodium cynomolgi*. *J. Parasitol.*, **75**, 594–600.

Paskewitz, S.M., Brown, M.R., Lea, A.O. and Collins F.H. (1988) Ultrastructure of the encapsulation of *Plasmodium cynomolgi* (B strain) on the midgut of a refractory strain of *Anopheles gambiae*. *J. Parasitol.*, **74**, 432–9.

Paskewitz, S.M. and Collins, F.H. (1989) Site-specific ribosomal DNA insertion elements in *Anopheles gambiae* and *A. arabiensis*: nucleotide sequence of gene-element boundaries. *Nucleic Acids Res.*, **17**, 8125–33.

Paskewitz, S.M. and Collins, F.H. (1990) Use of the polymerase chain reaction to identify mosquito species of the *Anopheles gambiae* complex. *Med. Vet. Entomol.*, **4**, 367–73.

Pearse, B.M.F. and Crowther, R.A. (1987) Structure and assembly of coated vesicles. *Annu. Rev. Biophysics Biophys. Chem.*, **16**, 49–68.

Pearse, B.M.F. and Robinson, M.S. (1990) Clathrin, adaptors, and sorting. *Annual Rev. Cell Biol.*, **6**, 151–71.

Penneys, N.S., Nayar, J.K., Bernstein, H. et al. (1989) Mosquito salivary gland antigens identified by circulating human antibodies. *Arch. Dermatol.*, **125**, 219–22.

Penzlin, H. (1985) Stomatogastric nervous system. In *Insect Physiology Biochemistry and Pharmacology* (eds G.A. Kerkut and L.I. Gilbert), Pergamon Press, Oxford, **5**, 371–406.

Perrone, J.B., DeMaio, J. and Spielman, A. (1986) Regions of mosquito salivary glands distinguished by surface lectin-binding characteristics. *Insect Biochem.*, **16**, 313–18.

Perrone, J.B. and Spielman, A. (1988) Time and site of assembly of the peritrophic membrane of the mosquito *Aedes aegypti*. *Cell Tissue Res.*, **252**, 473–8.

Peters, W. (1979) The fine structure of peritrophic membranes of mosquito and blackfly larvae of the genera *Aedes*, *Anopheles*, *Culex*, and *Odagmia* (Diptera: Culicidae/Simuliidae). *Entomol. Gen.*, **5**, 289–99.

Peters, W. and Wiese, B. (1986) Permeability of the peritrophic membranes of some Diptera to labelled dextrans. *J. Insect Physiol.*, **32**, 43–9.

Petersen, J.J. and Chapman, H.C. (1969) Chemical factors of water in tree holes and related breeding of mosquitoes. *Mosq. News*, **29**, 29–36.

Petersen, J.J. and Chapman, H.C. (1970) Chemical characteristics of habitats producing larvae of *Aedes sollicitans*, *Aedes taeniorhynchus*, and *Psorophora confinnis* in Louisiana. *Mosq. News*, **30**, 156–61.

Petzel, D.H., Berg, M.M. and Beyenbach, K.W. (1987) Hormone-controlled cAMP-mediated fluid secretion in yellow-fever mosquito. *Am. J. Physiol.*, **253** (*Reg. Int. Comp. Physiol.* 22), R701–11.

Petzel, D. and Conlon, J.M. (1991) Evidence for an antidiuretic factor affecting fluid secretion in mosquito Malpighian tubules. *FASEB J.*, **5**, A1059.

Petzel, D.H., Hagedorn, H.H. and Beyenbach, K.W. (1985) Preliminary isolation of mosquito natriuretic factor *Am. J. Physiol.* **249** (*Reg. Int. Comp. Physiol.* 18), R379–86.

Petzel, D.H., Hagedorn, H.H. and Beyenbach, K.W. (1986) Peptide nature of two mosquito natriuretic factors. *Am. J. Physiol.*, **250** (*Reg. Int. Comp. Physiol.* 19), R328–32.

Petzel, D.H. and Stanley-Samuelson, D.W. (1992) Inhibition of eicosanoid biosynthesis modulates basal fluid secretion in the Malpighian tubules of the yellow fever mosquito (*Aedes aegypti*). *J. Insect Physiol.*, **38**, 1–8.

Pflugfelder, O. (1937) Die Entwicklung der optischen Ganglien von *Culex pipiens*. *Zool. Anz.*, **117**, 31–6.

Phillips, D.M. (1969) Exceptions to the prevailing pattern of tubules (9 + 9 + 2) in the sperm flagella of certain insect species. *J. Cell Biol.*, **40**, 28–43.

Phillips, D.M. (1970a) Insect flagellar tubule patterns, theme and variations. In *Comparative Spermatology* (ed. B. Baccetti), Academic Press, New York, pp. 263–73, 4 pls.

Phillips, D.M. (1970b) Insect sperm: their structure and morphogenesis. *J. Cell Biol.*, **44**, 243–77.

Phillips, D.M. (1974) Structural variants in invertebrate sperm flagella and their relationship to motility. In *Cilia and Flagella* (ed. M.A. Sleigh), Academic Press, London, pp. 379–402.

Phillips, J.E. and Bradley, T.J. (1977) Osmotic and ionic regulation in saline-water mosquito larvae. In *Transport of Ions and Water in Animals* (eds B.L. Gupta, R.B. Moreton, J.L. Oschman and B.L. Wall), Academic Press, London, pp. 709–34.

Phillips, J.E., Bradley, T.J. and Maddrell, S.H.P. (1978) Mechanisms of ionic and osmotic regulation in saline-water mosquito larvae. In *Comparative Physiology: Water, Ions and Fluid Mechanics* (eds K. Schmidt-Nielsen, L. Bolis and S.H.P. Maddrell), Cambridge University Press, pp. 151–71.

Phillips, J.E. and Maddrell, S.H.P. (1974) Active transport of magnesium by the Malpighian tubules of the larvae of the mosquito, *Aedes campestris*. *J. Exp. Biol.*, **61**, 761–71.

Phillips, J.E. and Meredith, J. (1969) Active sodium and chloride transport by anal papillae of a salt water mosquito larva (Aedes campestris). *Nature*, **222**, 168–9.

Pichon, G. and Rivière, F. (1979) Observations sur la biologie préimaginale du moustique prédateur *Toxorhynchites amboinensis* (Diptera, Culicidae). *Cahier ORSTOM, Entomol. Méd. Parasitol.*, **17**, 221–4.

Pimentel, G.E. and Rossignol, P.A. (1990) Age dependence of salivary bacteriolytic activity in adult mosquitoes. *Comp. Biochem. Physiol.*, **96B**, 540–51.

Plawner, L., Pannabecker, T.L., Laufer, S. *et al.* (1991) Control of diuresis in the yellow fever mosquito *Aedes aegypti*: evidence for similar mechanisms in the male and female. *J. Insect Physiol.*, **37**, 119–28.

Poehling, H.-M. (1979) Distribution of specific proteins in the salivary gland lobes of Culicidae and their relation to age and blood sucking. *J. Insect Physiol.*, **25**, 3–8.

Poehling, H.-M. and Meyer, W. (1980) Esterases and glycoproteins in the salivary glands of *Anopheles stephensi*. *Insect Biochem.*, **10**, 189–98.

Poinar, G.O. and Leutenegger, R. (1971) Ultrastructural investigation of the melanization process in *Culex pipiens* (Culicidae) in response to a nematode. *J. Ultrastructural Res.*, **36**, 149–58.

Pollard, S.R., Motara, M.A. and Cross, R.H.M. (1986) The ultrastructure of oogenesis in *Culex theileri*. *S. African J. Zool.*, **21**, 217–23.

Ponnudurai, T., Billingsley, P.F. and Rudin, W. (1988) Differential infectivity of *Plasmodium* for mosquitoes. *Parasitol. Today*, **4**, 319–21.

Portaro, J.K. and Barr, A.R. (1975) 'Curing' Wolbachia infections in *Culex pipiens*. *J. Med. Entomol.*, **12**, 265.

Powell, J.R., Hollander, A.L. and Fuchs, M.S. (1986a) A preparation method for *Aedes aegypti* and *Aedes atropalpus* chorions for protein characterization. *Insect Biochem.*, **16**, 835–42.

Powell, J.R., Hollander, A.L. and Fuchs, M.S. (1986b) Chorion synthesis: tanning and sclerotization in *Aedes aegypti* mosquitoes. In *Host Regulated Developmental Mechanisms in Vector Arthropods* (eds D. Borovsky and A. Spielman), University of Florida, **1**, 3–7.

Powell, J.R., Hollander, A.L. and Fuchs, M.S. (1988) Development of the *Aedes aegypti* chorion: proteins and ultrastructure. *Int. J. Invert. Reprod. Dev.*, **13**, 39–54.

Powers, K.S. and Platzer, E.G. (1984) Oxygen consumption in mosquito larvae parasitized by *Romanomermis culicivorax* (Nematoda). *Comp. Biochem. Physiol.*, **78A**, 119–22.

Powers, K.S., Platzer, E.G. and Bradley, T.J. (1984) The effect of nematode parasitism on the osmolality and major cation concentration in the haemolymph of three larval mosquito species. *J. Insect Physiol.*, **30**, 547–50.

Prashad, B. (1918) The development of the dorsal series of thoracic imaginal buds of the mosquito, and certain observations on the phylogeny of the insects. *Indian J. Med. Res.*, **5**, 641–55, 5 pls.

Price, R.D. (1958) Observations on a unique monster embryo of *Wyeomyia smithii* (Coquillett) (Diptera : Culicidae). *Ann. Entomol. Soc. Am.*, **51**, 600–4.

Provost, M.W. (1969) The natural history of *Culex nigripalpus*. *Florida State Bd Hlth Monogr. Ser.*, **12**, 46–62.

Provost, M.W. and Lum, P.T.M. (1967) The pupation rhythm in *Aedes taeniorhynchus* (Diptera: Culicidae). I. Introduction. *Ann. Entomol. Soc. Am.*, **60**, 138–49.

Pryor, S.C. and Ferrel, R.E. (1981) Biochemical genetics of the *Culex pipiens* complex – III. α-Glycerophosphate dehydrogenase. *Comp. Biochem. Physiol.*, **69B**, 23–8.

Pucat, A.M. (1965) The functional morphology of the mouthparts of some mosquito larvae. *Quaest. Entomol.*, **1**, 41–86.

Puchkova, L.V. (1976) The inner surface of mosquito labella (Culicidae). *Dopov. Akad. Nauk Ukr. RSR*, ser. B, *Heol. Khim. Biol. Nauky*, (6), 588–60. [In Ukrainian.]

Puchkova, L.V. (1977) Morpho-functional characteristics of the mouthparts of blood sucking mosquitoes (Culicidae). *Vest. Zool.*, (6), 61–6. [In Russian.]

Putnam, P. and Shannon, R.C. (1934) The biology of *Stegomyia* under laboratory conditions. *Proc. Entomol. Soc. Wash.*, **36**, 217–42.

Qutubuddin, M. (1953) The emergence and sex ratio of *Culex fatigans* Wied. (Diptera, Culicidae) in laboratory experiments. *Bull. Entomol. Res.*, **43**, 549–65.

Raabe, M., Baudry, N., Grillot, J.P. and Provansal, A. (1974) The perisympathetic organs of insects. In *Neurosecretion – the Final Common Neuroendocrine Pathway* (eds F. Knowles and L. Vollrath), Springer-Verlag, Heidelberg and New York, pp. 59–71.

Rabbani, M.G. and Kitzmiller, J.B. (1975) Studies on X-ray induced chromosomal translocations in *Anopheles albimanus*. I. Chromosomal translocations and genetic control. *Am. J. Trop. Med. Hyg.*, **24**, 1019–26.

Racioppi, J.V., Gemmill, R.M., Kogan, P.H. *et al.* (1986) Expression and regulation of vitellogenin messenger RNA in the mosquito, *Aedes aegypti*. *Insect Biochem.*, **16**, 255–62.

Racioppi, J.V. and Spielman, A. (1987) Secretory proteins from the salivary glands of adult *Aedes aegypti* mosquitoes. *Insect Biochem.*, **17**, 503–11.

Raikhel, A.S. (1984a) Accumulations of membrane-free clathrin-like lattices in the mosquito oocyte. *Eur. J. Cell Biol.*, **35**, 279–83.

Raikhel, A.S. (1984b) The accumulative pathway of vitellogenin in the mosquito oocyte: a high-resolution immuno- and cytochemical study. *J. Ultrastruct. Res.*, **87**, 285–302.

Raikhel, A.S. (1986a) Lysosomes in the cessation of vitellogenin secretion by the mosquito fat body; selective degradation of Golgi complexes and secretory granules. *Tissue & Cell*, **18**, 125–42.

Raikhel, A.S. (1986b) Role of lysosomes in regulating of vitellogenin secretion in the mosquito fat body. *J. Insect Physiol.*, **32**, 597–604.

Raikhel, A.S. (1987a) Monoclonal antibodies as probes for processing of the mosquito yolk protein; a high-resolution immunolocalization of secretory and accumulative pathways. *Tissue & Cell*, **19**, 515–29.

Raikhel, A.S. (1987b) The cell biology of mosquito vitellogenesis. *Mem. Inst. Oswaldo Cruz*, **82** (Suppl. 3), 93–101.

Raikhel, A.S. and Bose, S.G. (1988) Properties of the mosquito yolk protein: a study using monoclonal antibodies. *Insect Biochem.*, **18**, 565–75.

Raikhel, A.S. and Dhadialla, T.S. (1992) Accumulation of yolk proteins in insect oocytes. *Annual Rev. Entomol.*, **37**, 217–51.

Raikhel, A.S. and Lea, A.O. (1982) Abnormal vitelline envelope induced by unphysiological doses of ecdysterone in *Aedes aegypti*. *Physiol. Entomol.*, **7**, 55–64.

Raikhel, A.S. and Lea, A.O. (1983) Previtellogenic development and vitellogenin synthesis in the fat body of a mosquito: an ultrastructural and immunocytochemical study. *Tissue & Cell*, **15**, 281–300.

Raikhel, A.S. and Lea, A.O. (1985) Hormone-mediated formation of the endocytic complex in mosquito oocytes. *Gen. Comp. Endocrinol.*, **57**, 422–33.

Raikhel, A.S. and Lea, A.O. (1986) Internalized proteins directed into accumulative compartments of mosquito oocytes by the specific ligand, vitellogenin. *Tissue & Cell*, **18**, 559–74.

Raikhel, A.S. and Lea, A.O. (1987) Analysis of mosquito yolk protein by monoclonal antibodies. *Molecular Entomology* (ed. J. Law), Alan R. Liss Inc., pp. 403–13.

Raikhel, A.S. and Lea, A.O. (1990) Juvenile hormone controls previtellogenic proliferation of ribosomal RNA in the mosquito fat body. *Gen. Comp. Endocrinol.*, **77**, 423–34.

Raikhel, A.S. and Lea, A.O. (1991) Control of follicular epithelium development and vitelline envelope formation in the mosquito; role of juvenile hormone and 20-hydroxyecdysone. *Tissue and Cell*, **23**, 577–91.

Raikhel, A.S., Pratt, L.H. and Lea, A.O. (1986) Monoclonal antibodies as probes for processing of yolk protein in the mosquito; production and characterization. *J. Insect Physiol.*, **32**, 879–90.

Ramasamy, M.S., Sands, M., Gale, J. and Ramasamy, R. (1988) Mosquito vitellin: structural and functional studies with monoclonal antibodies. *Insect Sci. Applic.*, **9**, 499–504.

Raminani, L.N. and Cupp, E.W. (1975) Early embryology of *Aedes aegypti* (L.) (Diptera: Culicidae). *Int. J. Insect Morphol. Embryol.*, **4**, 517–28.

Raminani, L.N. and Cupp, E.W. (1977) The histopathology of thermally induced sterility in *Aedes aegypti* (Diptera: Culicidae). *J. Med. Entomol.*, **13**, 603–9.

Raminani, L.N. and Cupp, E.W. (1978) Embryology of *Aedes aegypti* (L.) (Diptera: Culicidae); organogenesis. *Int. J. Insect Morphol. Embryol.*, **7**, 273–96.

Ramsay, J.A. (1950) Osmotic regulation in mosquito larvae. *J. Exp. Biol.*, **27**, 145–57, 1 pl.

Ramsay, J.A. (1951) Osmotic regulation in mosquito larvae: the role of the Malpighian tubules. *J. Exp. Biol.*, **28**, 62–73.

Ramsay, J.A. (1953a) Exchanges of sodium and potassium in mosquito larvae. *J. Exp. Biol.*, **30**, 79–89.
Ramsay, J.A. (1953b) Active transport of potassium by the Malpighian tubules of insects. *J. Exp. Biol.*, **30**, 358–69.
Rao, P.N. and Rai, K.S. (1987a) Inter and intraspecific variation in nuclear DNA content in *Aedes* mosquitoes. *Heredity*, **59**, 253–8.
Rao, P.N. and Rai, K.S. (1987b) Comparative karyotypes and chromosomal evolution in some genera of nematocereous (Diptera: Nematocera) families. *Ann. Entomol. Soc. Am.*, **80**, 321–32.
Rao, P.N. and Rai, K.S. (1990) Genome evolution in the mosquitoes and other closely related members of superfamily Culicoidea. *Hereditas*, **113**, 139–44.
Rashed, S.S. and Mulla, M.S. (1989) Factors influencing ingestion of particulate materials by mosquito larvae (Diptera: Culicidae). *J. Med. Entomol.*, **26**, 210–6.
Rashed, S.S. and Mulla, M.S. (1990) Comparative functional morphology of the mouth brushes of mosquito larvae (Diptera: Culicidae). *Ann. Entomol. Soc. Am.*, **27**, 429–39.
Rasnitsyn, S.P. and Yasyukevich, V.V. (1989a) Feeding peculiarities of malaria mosquito larvae. *Zool. Zh.*, **68** (7), 155–8. [In Russian.]
Rasnitsyn, S.P. and Yasyukevich, V.V. (1989b) On the ability of mosquito larvae (Diptera, Culicidae) to endure starvation. *Entomol. Rev.*, **68**, 143–51.
Readio, J. and Meola, R. (1985) Two stages of juvenile hormone-mediated growth of secondary follicles in *Culex pipiens*. *J. Insect Physiol.*, **31**, 559–62.
Readio, J., Peck, K., Meola, R. and Dahm, K.H. (1988) Corpus allatum activity (*in vitro*) in female *Culex pipiens* during adult life cycle. *J. Insect Physiol.*, **34**, 131–5.
Réaumur, R.A.F. de (1750) *Mémoires pour servir à l'histoire des insectes*. Vol. 4, Pt 2, Mém. 13. *Histoires des cousins*, Pierre Mortier, Amsterdam, pp. 372–455, 6 pls.
Redecker, B. and Zebe, E. (1988) Aerobic metabolism in aquatic insect larvae: studies on *Chironomus thummi* and *Culex pipiens*. *J. Comp. Physiol. B*, **158**, 307–15.
Redfern, C.P.F. (1981a) Satellite DNA of *Anopheles stephensi* Liston (Diptera : Culicidae). Chromosomal location and under-replication in polytene nuclei. *Chromosoma*, **82**, 561–81.
Redfern, C.P.F. (1981b) Homologous banding patterns in the polytene chromosomes from the larval salivary glands and ovarian nurse cells of *Anopheles stephensi* Liston (Culicidae). *Chromosoma*, **83**, 221–40.
Redfern, C.P.F. (1981c) DNA replication in polytene chromosomes. Similarity of termination patterns in somatic and germ-line derived polytene chromosomes of *Anopheles stephensi* Liston (Diptera: Culicidae). *Chromosoma*, **84**, 33–47.
Redfern, C.P.F. (1982) 20-Hydroxy-ecdysone and ovarian development in *Anopheles stephensi*. *J. Insect Physiol.*, **28**, 97–109.
Redington, B.C. and Hockmeyer, W.T. (1976) A method for estimating blood meal volume in *Aedes aegypti* using a radioisotope. *J. Insect Physiol.*, **22**, 961–6.
Reid, J.A. (1963) A note on the structure of the pupal trumpet in some mosquitoes. *Proc. R. Entomol. Soc. Lond.* (A), **38**, 32–8.
Reinhardt, C. and Hecker, H. (1973) Structure and function of the basal lamina and of the cell junctions in the midgut epithelium (stomach) of female *Aedes aegypti* L. (Insecta, Diptera). *Acta Tropica*, **30**, 213–36.
Reisen, W.K. (1975) Intraspecific competition in *Anopheles stephensi* Liston. *Mosq. News*, **35**, 473–82.
Reisen, W.K. (1986) Studies on autogeny in *Culex tarsalis*: 2. Simulated diapause induction and termination in genetically autogenous females. *J. Am. Mosq. Control Ass.*, **2**, 44–7.
Reisen, W.K., Meyer, R.P. and Milby, M.M. (1986) Patterns of fructose feeding by *Culex tarsalis* (Diptera: Culicidae). *J. Med. Entomol.*, **23**, 366–73.
Reisen, W.K. and Emory, R.W. (1977) Intraspecific competition in *Anopheles stephensi* (Diptera: Culicidae). II. The effects of more crowded densities and the addition of antibiotics. *Can. Entomol.*, **109**, 1475–80.
Reisen, W.K. and Milby, M.M. (1987) Studies on autogeny in *Culex tarsalis*: 3. Life table attributes of autogenous and anautogenous strains under laboratory conditions. *J. Am. Mosq. Control Ass.*, **3**, 619–25.
Reiter, P. (1978a) The influence of dissolved oxygen content on the survival of submerged mosquito larvae. *Mosq. News*, **38**, 334–7.
Reiter, P. (1978b) The action of lecithin monolayers on mosquitoes. II. Action on the respiratory structures. *Ann. Trop. Med. Parasitol.*, **72**, 169–76.
Reiter, P. (1980) The action of lecithin monolayers on mosquitoes. III. Studies in irrigated rice-fields in Kenya. *Ann. Trop. Med. Parasitol.*, **74**, 541–57.
Reiter, P. and Jones, M.D.R. (1975) An eclosion timing mechanism in the mosquito *Anopheles gambiae*. *J. Entomol.* (A), **50**, 161–8.
Rempel, J.G. and Rueffel, P.G. (1964) The retrocerebral glands of mosquito larvae. *Can. J. Zool.*, **42**, 39–51, 2 pls.
Renn, C.E. (1941) The food economy of *Anopheles quadrimaculatus* and *A. crucians* larvae. Relationships of the air-water interface and the surface-feeding mechanism. In *A Symposium on Hydrobiology*, University of Wisconsin Press, Madison, pp. 329–42.
Ribeiro, J.M.C. (1987) Role of saliva in blood-feeding by arthropods. *Annual Rev. Entomol.*, **32**, 463–78.
Ribeiro, J.M.C., Rossignol, P.A. and Spielman, A. (1984a) Role of mosquito saliva in blood vessel location. *J. Exp. Biol.*, **108**, 1–7.
Ribeiro, J.M.C., Sarkis, J.J.F., Rossignol, P.A. and Spielman, A. (1984b) Salivary apyrase of *Aedes aegypti*: characterization and secretory fate. *Comp. Biochem. Physiol.*, **79B**, 81–6.

Ribeiro, J.M.C., Rossignol, P.A. and Spielman, A. (1985a) *Aedes aegypti*: model for blood finding strategy and prediction of parasite manipulation. *Exp. Parasitol.*, **60**, 118–32.

Ribeiro, J.M.C., Rossignol, P.A. and Spielman, A. (1985b) Salivary gland apyrase determines probing time in anopheline mosquitoes. *J. Insect Physiol.*, **31**, 689–92.

Richards, A.G. and Meier, T.J. (1974) The osmolarity of the blood of a mosquito larva (*Aedes aegypti* L.) reared under several different culture conditions. *Ann. Entomol. Soc. Am.*, **67**, 424–6.

Richards, A.G. and Richards, P.A. (1971) Origin and composition of the peritrophic membrane of the mosquito, *Aedes aegypti*. *J. Insect Physiol.*, **17**, 2253–75.

Richards, A.G. and Seilheimer, S.H. (1977) The proventricular flanges in mosquito larvae. *Entomol. News*, **88**, 1–9.

Richards, G. (1985) Polytene chromosomes. In *Comprehensive Insect Physiology Biochemistry and Pharmacology* (eds G.A. Kekut and L.I. Gilbert), Pergamon Press, Oxford, **2**, 255–300.

Richards, P.A., Richards, A.G. and Hodson, A.C. (1977) Development and variation in flanges on the proventriculus of larvae of *Aedes aegypti* (Diptera : Culicidae). *J. Med. Entomol.*, **13**, 517–23.

Richardson, M.W. and Romoser, W.S. (1972) The formation of the peritrophic membrane in adult *Aedes triseriatus* (Say) (Diptera: Culicidae). *J. Med. Entomol.*, **9**, 495–500.

Richins, C.A. (1938) The metamorphosis of the digestive tract of *Aedes dorsalis* Meigen. *Ann. Entomol. Soc. Am.*, **31**, 74–87.

Richins, C.A. (1945) The development of the midgut in the larva of *Aedes dorsalis* Meigen. *Ann. Entomol. Soc. Am.*, **38**, 314–20.

Rioux, J.-A., Croset, H., Gabinaud, A. *et al.* (1973) Hérédité monofactorielle de l'autogenèse chez *Aedes (Ochlerotatus) detritus* (Haliday, 1833) (Diptera-Culicidae). *C.R. Acad. Sci. Paris (D)*, **276**, 991–4.

Rioux, J.-A., Croset, H., Pech-Périères, J. *et al.* (1975) L'autogenèse chez les diptères culicides. Tableau synoptique des espèces autogènes (1). *Ann. Parasitol. Hum. Comp.*, **50**, 134–40.

Rishikesh, N. (1959) Chromosome behaviour during spermatogenesis of *Anopheles stephensi* sensu stricto. *Cytologia*, **24**, 447–58.

Risler, H. (1959) Polyploidie und somatische Reduktion in der Larvenepidermis von *Aedes aegypti* (L.). *Chromosoma*, **10**, 184–209.

Risler, H. (1961) Untersuchungen zur somatischen Reduktion in der Metamorphose des Stechmücken-darms. *Biol. Zentralbl.*, **80**, 413–28.

Rivière, F. (1985) Effects of two predators on community composition and biological control of *Aedes aegypti* and *Aedes polynesiensis*. In *Ecology of Mosquitoes* (eds L.P. Lounibos, J.R. Rey and J.H. Frank), Florida Medical Entomology Laboratory, Vero Beach, pp. 121-35.

Robinson, G.G. (1939) The mouthparts and their function in the female mosquito, *Anopheles maculipennis*. *Parasitology*, **31**, 212–42.

Rockwell, E.M. and Johnson, P. (1952) The insect bite reaction. II. Evaluation of the allergic reaction. *J. Invest. Dermatol.*, **19**, 137–55.

Romans, P., Seeley, D.C., Kew Y. and Gwadz, R.W. (1991) Use of a restriction fragment length polymorphism (RFLP) as a genetic marker in crosses of *Anopheles gambiae* (Diptera: Culicidae): independent assortment of a diphenol oxidase RFLP and an esterase locus. *J. Med. Entomol.*, **28**, 147–51.

Romoser, W.S. (1974) Peritrophic membranes in the midgut of pupal and pre-blood meal adult mosquitoes. *J. Med. Entomol.*, **11**, 397–402.

Romoser, W.S. and Cody, E. (1975) The formation and fate of the peritrophic membrane in adult *Culex nigripalpus* (Diptera: Culicidae). *J. Med. Entomol.*, **12**, 371–8.

Romoser, W.S., Edman, J.D., Lorenz, L.H. and Scott, T.W. (1989) Histological parameters useful in the identification of multiple bloodmeals in mosquitoes. *Am. J. Trop. Med. Hyg.*, **41**, 737–42.

Romoser, W.S. and Nasci, R.S. (1979) Functions of the ventral air space and first abdominal spiracles in *Aedes aegypti* pupae (Diptera: Culicidae). *J. Med. Entomol.*, **15**, 109–14.

Romoser, W.S. and Rothman, M.E. (1973) The presence of a peritrophic membrane in pupal mosquitoes (Diptera: Culicidae). *J. Med. Entomol.*, **10**, 312–14.

Romoser, W.S. and Venard, C.E. (1966) The development of the ventral oesophageal diverticulum in *Aedes triseriatus* (Diptera: Culicidae). *Ann. Entomol. Soc. Am.*, **59**, 484–9.

Romoser, W.S. and Venard, C.E. (1967) Development of the dorsal oesophageal diverticula in *Aedes triseriatus* (Diptera: Culicidae). *Ann. Entomol. Soc. Am.*, **60**, 617–23.

Romoser, W.S. and Venard, C.E. (1969) The myoblastic masses and neural tissue associated with the proventriculus of larvae of *Aedes triseriatus*, *A. aegypti* and *Anopheles quadrimaculatus*. *Ann. Entomol. Soc. Am.*, **62**, 345–8.

Ronquillo, M.C. and Horsfall, W.R. (1969) Genesis of the reproductive systgem of mosquitoes. I. Female of *Aedes stimulans* (Walker). *J. Morphol.*, **129**, 249–80.

Rosay, B. (1961) Anatomical indicators for assessing the age of mosquitoes: the teneral adult (Diptera: Culicidae) *Ann. Entomol. Soc. Am.*, **54**, 526–9.

Rosen, L. and Rozeboom, L.E. (1954) Morphologic variations of larvae of the scutellaris group of *Aedes* (Diptera, Culicidae) in Polynesia. *Am. J. Trop. Med. Hyg.*, **3**, 529–38.

Rosenberg, R. (1980) Ovarian control of blood meal

retention in the mosquito *Anopheles freeborni*. *J. Insect Physiol.*, **26**, 477–80

Ross, R. (1898) Report on the cultivation of *Proteosoma*, Labbé, in grey mosquitos. Office of the Superintendent of Government Printing, Calcutta, pp. 1–21, 5 pls.

Rossignol, P.A., Feinsod, F.M. and Spielman, A. (1981) Inhibitory regulation of corpus allatum activity in mosquitoes. *J. Insect Physiol.*, **27**, 651–4.

Rossignol, P.A. and Lueders, A.M. (1986) Bacteriolytic factor in the salivary glands of *Aedes aegypti*. *Comp. Biochem. Physiol.*, **83B**, 819–22.

Rossignol, P.A., Ribeiro, J.M.C., Jungery, M. *et al.* (1985) Enhanced mosquito blood-finding success on parasitemic hosts: evidence for vector-parasite mutualism. *Proc. Natl Acad. Sci. USA*, **82**, 7725–7.

Rossignol, P.A., Ribeiro, J.M.C and Spielman, A. (1984) Increased intradermal probing time in sporozoite-infected mosquitoes. *Am. J. Trop. Med. Hyg.*, **33**, 17–20.

Rossignol, P.A. and Spielman, A. (1982) Fluid transport across the ducts of the salivary glands of a mosquito. *J. Insect Physiol.*, **28**, 579–83.

Rossignol, P.A., Spielman, A. and Jacobs, M.S. (1982) Rough endoplasmic reticulum in midgut cells of mosquitoes (Diptera: Culicidae): aggregation stimulated by juvenile hormone. *J. Med. Entomol.*, **19**, 719–21.

Roth, L.M. (1948) A study of mosquito behavior. An experimental study of the sexual behavior of *Aedes aegypti* (Linnaeus). *Am. Midland Nat.*, **40**, 265–352.

Roth, T.F. (1966) Changes in the synaptinemal complex during meiotic prophase in mosquito oocytes. *Protoplasma*, **61**, 346–86.

Roth, T.F., Cutting, J.A. and Atlas, S.B. (1976) Protein transport: a selective membrane mechanism. *J. Supramolec. Res.*, **4**, 527–48.

Roth, T.F., Gord, D. and Dodson, J. (1973) Meiotic crossing-over in synchronous oocytes uniquely temperature sensitive at early pachynema. *J. Cell Biol.*, **59**, 293a.

Roth, T.F. and Porter, K.R. (1964) Yolk protein uptake in the oocyte of the mosquito *Aedes aegypti* L. *J. Cell Biol.*, **20**, 313–32.

Roubaud, E. (1929) Cycle autogene d'attente et générations hivernales suractives inapparentes chez le moustique commun, *Culex pipiens*. *C.R. Acad. Sci., Paris*, **188**, 735–8.

Roubaud, E. (1932) Des phénomènes d'histolyse larvaire post-nymphale et d'alimentation imaginal autotrophe chez le moustique commun (*Culex pipiens*). *C.R. Acad. Sci., Paris*, **194**, 389–91.

Roubaud, E. (1933) Essai synthétique sur la vie du moustique commun (*Culex pipiens*). L'évolution humaine et les adaptations biologiques du moustique. *Ann. Sci. Nat. (Zool.)*, **16**, 5–168, 8 pls.

Roubaud, E. (1934) Observations sur la fécondité des Anophelines. *Bull. Soc. Pathol. Exotique*, **27**, 853–4.

Rousset, F., Raymond, M. and Kjellberg, F. (1991) Cytoplasmic incompatibilities in the mosquito *Culex pipiens*: how to explain acytotype polymorphism? *J. Evolutionary Biol.*, **4**, 69–81.

Rousset, F. and Raymond, M. (1991) Cytoplasmic incompatibility in insects: why sterilize females? *Trends in Ecology and Evolution*, **6**, 54–7.

Roy, D.N. (1940) Influence of spermathecal stimulation on the physiological activities of *Anopheles subpictus*. *Nature*, **145**, 747–8.

Rubenstein, D.I. and Koehl, M.A.R. (1977) The mechanisms of filter feeding: some theoretical considerations. *Am. Nat.*, **111**, 981–94.

Rudin, W., Billingsley, P.F. and Saladin, S. (1991) The fate of *Plasmodium gallinaceum* in *Anopheles stephensi* Liston and possible barriers to transmission. *Ann. Soc. Belg. Mèd. Trop.*, **71**, 167–77.

Rudin, W. and Hecker, H. (1976) Morphometric comparison of the midgut epithelial cells in male and female *Aedes aegypti* L. (Insecta, Diptera). *Tissue & Cell*, **8**, 459–70.

Rudin, W. and Hecker, H. (1979) Functional morphology of the midgut of *Aedes aegypti* L. (Insecta, Diptera) during blood digestion. *Cell Tissue Res.*, **200**, 193–203.

Rudin, W. and Hecker, H. (1989) Lectin-binding sites in the midgut of the mosquitoes *Anopheles stephensi* Liston and *Aedes aegypti* L. (Diptera: Culicidae). *Parasitol. Res.*, **75**, 268–79.

Russell, P.F. and Mohan, B.N. (1939) Experimental infections in *Anopheles stephensi* (type) from contrasting larva environments. *Am. J. Hyg.*, **30**, 73–9.

Russell, R.C. (1979) A study of the influence of some environmental factors on the development of *Anopheles annulipes* Walker and *Anopheles amictus hilli* Woodhill and Lee (Diptera: Culicidae). Part 2: Influence of salinity, temperature and larval density during the development of the immature stages on adult fecundity. *Gen. Appl. Entomol.*, **11**, 42–5.

Russo, R. (1986) Comparison of predatory behavior in five species of *Toxorhynchites* (Diptera:Culicidae). *Ann. Entomol. Soc. Am.*, **79**, 715–22.

Russo, R.J. and Westbrook, A.L. (1986) Ecdysteroid levels in eggs and larvae of *Toxorhynchites amboinensis*. In *Insect Neurochemistry and Neurophysiology*, (eds A.B. Borkovec and D.B. Gelman), Humana Press, Clifton, New Jersey, USA, pp. 415–20.

Sahlén, G. (1990) Egg raft adhesion and chorion structure in *Culex pipiens* L. (Diptera : Culicidae). *Int. J. Insect Morphol. Embryol.*, **19**, 307–14.

Sakai, R.K., Ainsley, R.W. and Baker, R.H. (1977) The inheritance of rose eye, a sex linked mutant in the malaria vector *Anopheles culicifacies*. *Can. J. Genet. Cytol.*, **19**, 633–6.

Sakai, R.K., Baker, R.H., Raana, K. and Hassan, M. (1979) Crossing-over in the long arm of the X and Y chromsomes in *Anopheles culicifacies*. *Chromosoma*, **74**, 209–18.

Sakurai, H. and Makiya, K. (1981) Body composition of *Culex pipiens pallens*, with special reference to body size of overwintering females. *Jap. J. Sanit. Zool.*, **32**, 72–4.

Salama, H.S. (1966) The function of mosquito taste receptors. *J. Insect Physiol.*, **12**, 1051–60.

Salama, H.S. (1967) Nutritive values and taste sensitivity to carbohydrates for mosquitoes. *Mosq. News*, **27**, 32–5.

Samarawickrema, W.A. (1967) A study of the age-composition of natural populations of *Culex pipiens fatigans* Wiedemann in relation to the transmission of filariasis due to *Wuchereria bancrofti* (Cobbold) in Ceylon. *Bull. Wld Hlth Org.*, **37**, 117–37.

Samarawickrema, W.A. (1968) Biting cycles and parity of the mosquito *Mansonia* (*Mansonioides*) *uniformis* (Theo.) in Ceylon. *Bull. Entomol. Res.*, **58**, 299–314.

Samish, M. and Akov, S. (1972) Influence of feeding on midgut protease activity in *Aedes aegypti*. *Israel J. Entomol.*, **7**, 41–8.

Samtleben, B. (1929) Zur Kenntnis der Histologie und Metamorphose des Mitteldarms der Stechmückenlarven. *Zool. Anz.*, **81**, 97–109.

Sanburg, L.L. and Larsen, J.R. (1973) Effect of photoperiod and temperature on ovarian development in *Culex pipiens pipiens*. *J. Insect Physiol.*, **19**, 1173–90.

Satmary, W.M. and Bradley, T.J. (1984) The distribution of cell types in the Malpighian tubules of *Aedes taeniorhynchus* (Wiedemann) (Diptera : Culicidae). *Int. J. Insect Morphol. Embryol.*, **13**, 209–14.

Satô, S. (1951) Development of the compound eye of *Culex pipiens* var. *pallens* Coquillett. (Morphological studies on the compound eye in the mosquito, No. II.) *Sci. Rep. Tôhoku Univ.* (*Biol.*), **19**, 23–8.

Satô, S. (1953) Structure and development of the compound eye of *Anopheles hyrcanus sinensis* Wiedemann. (Morphological studies of the compound eye in the mosquito, No. IV.) *Sci. Rep. Tôhoku Univ.* (*Biol.*), **20**, 43–53.

Satô, S. (1960) Structure and development of the compound eye of *Armigeres* (*Armigeres*) *subalbatus* (Coquillett). (Morphological studies on the compound eye in the mosquito, No. VII.) *Sci. Rep. Tôhoku Univ.* (*Biol.*), **26**, 227–38, 2 pls.

Sautet, J. and Audibert, Y. (1946) Etudes biologiques et morphologiques sur certaines larves de moustiques en vue d'applications pratiques pour leur destruction. *Bull. Soc. Pathol. Exotique*, **39**, 43–61.

Sawyer, D.B. and Beyenbach, K.W. (1985) Dibutyryl-cAMP increases basolateral sodium conductance of mosquito Malpighian tubules. *Am. J. Physiol.*, (*Reg. Int. Comp. Physiol.* 17), **248**, R339–45.

Schaefer, C.H., Miura, T. and Washino, R.K. (1971) Studies on the overwintering biology of natural populations of *Anopheles freeborni* and *Culex tarsalis* in California. *Mosq. News*, **31**, 153–57.

Schaefer, C.H. and Miura, T. (1972) Sources of energy utilized by natural populations of the mosquito, *Culex tarsalis*, for overwintering. *J. Insect Physiol.*, **18**, 797–805.

Schaefer, C.H. and Washino, R.K. (1969) Changes in the composition of lipids and fatty acids in adult *Culex tarsalis* and *Anopheles freeborni* during the overwintering period. *J. Insect Physiol.*, **15**, 395–402.

Schaefer, C.H. and Washino, R.K. (1970) Synthesis of energy for overwintering in natural populations of the mosquito *Culex tarsalis*. *Comp. Biochem. Physiol.*, **35**, 503–6.

Schaefer, C.H. and Washino, R.K. (1974) Lipid contents of some overwintering adult mosquitoes collected from different parts of northern California. *Mosq. News*, **34**, 207–10.

Schiemenz, H. (1957) Vergleichende funktionell-anatomische Untersuchungen der Kopfmuskulatur von *Theobaldia* und *Eristalis* (Dipt. Culicid. und Syrphid.). *Deutsch Entomol. Z.*, **44**, 168–221.

Schildmacher, H. (1950) Darmkanal und Verdauung bei Stechmückenlarven. *Biol. Zentralblatt*, **69**, 390–438.

Schlaeger, D.A. and Fuchs, M.S. (1974a) Effect of DOPA-decarboxylase inhibition on *Aedes aegypti* eggs: evidence for sclerotization. *J. Insect Physiol.*, **20**, 349–57.

Schlaeger, D.A. and Fuchs, M.S. (1974b) DOPA decarboxylase activity in *Aedes aegypti*: a preadult profile and its subsequent correlation with ovarian development. *Dev. Biol.*, **38**, 209–19.

Schlaeger, D.A. and Fuchs, M.S. (1974c) Localization of DOPA decarboxylase in adult *Aedes aegypti* females. *J. Exp. Zool.*, **187**, 217–21.

Schlaeger, D.A., Fuchs, M.S. and Kang, S.H. (1974) Ecdysone-mediated stimulation of dopa decarboxylase activity and its relationship to ovarian development in *Aedes aegypti*. *J. Cell Biol.*, **61**, 454–65.

Schlein, Y. (1979) Age grouping of anopheline malaria vectors (Diptera: Culicidae) by the cuticular growth lines. *J. Med. Entomol.*, **16**, 502–6.

Schlein, J. and Gratz, N.G. (1972) Age determination of some flies and mosquitos by daily growth layers of skeletal apodemes. *Bull. Wld Hlth Org.*, **47**, 71–4, 1 pl.

Schlein, Y. and Gratz, N.G. (1973) Determination of the age of some anopheline mosquitos by daily growth layers of skeletal apodemes. *Bull. Wld Hlth Org.*, **49**, 371–5.

Schmidt, J.M. and Friend, W.G. (1991) Ingestion and diet destination in the mosquito *Culiseta inornata*: effects of carbohydrate configuration. *J. Insect Physiol.*, **37**, 817–28.

Schmidt, S.P. and Platzer, E.G. (1980) Changes in body

tissues and hemolymph composition of *Culex pipiens* in response to infection by *Romanomermis culicivorax*. *J. Invert. Pathol.*, **36**, 240–4.

Schneider, M. Rudin, W. and Hecker, H. (1987) Absorption and transport of radioactive tracers in the midgut of the malaria mosquito, *Anopheles stephensi*. *J. Ultrastruct. Mol. Struct. Res.*, **97**, 50–63.

Schooley, D.A. and Baker, F.C. (1985) Juvenile hormone biosynthesis. In *Comprehensive Insect Physiology Biochemistry and Pharmacology* (eds G.A. Kerkut and L.I. Gilbert), Pergamon Press, Oxford, **7**, 363–89.

Schremmer, F. (1949) Morphologische und funktionelle Analyse der Mundteile und des Pharynx der Larve von *Anopheles maculipennis*. *Österreich. Zool. Z.*, **2**, 173–222.

Schuh, J.E. (1951) Some effects of colchicine on the metamorphosis of *Culex pipiens*. *Chromosoma*, **4**, 456–69.

Scudder, G.C.E. (1969) The fauna of saline lakes on the Fraser Plateau in British Columbia. *Verh. Int. Verein. Limnol.*, **17**, 430–9.

Seawright, J.A., Benedict, M.Q. and Narang, S. (1985) Studies of the X chromosome of *Anopheles albimanus*. *Can. J. Genet. Cytol.*, **27**, 74–82.

Seifert, R.P. and Barrera, R. (1981) Cohort studies on mosquito (Diptera: Culicidae) larvae living in the water-filled floral bracts of *Heliconia aurea* (Zingiberales: Musaceae). *Ecol. Entomol.*, **6**, 191–7.

Senberg, R. (1980) Ovarian control of blood meal retention in the mosquito *Anopheles freeborni*. *J. Insect Physiol.*, **26**, 477–80.

Senior-White, R. (1928) Algae and the food of anopheline larvae. *Indian J. Med. Res.*, **15**, 969–88, 2 pls.

Service, M.W. (1968a) Observations on feeding and oviposition in some British mosquitoes. *Entomol. Exp. Applic.*, **11**, 277–85.

Service, M.W. (1968b) Some environmental effects on blood-fed hibernating *Culiseta annulata* (Diptera: Culicidae). *Entomol. Exp. Applic.*, **11**, 286–90.

Service, M.W. (1971) Feeding behaviour and host preferences of British mosquitoes. *Bull. Entomol. Res.*, **60**, 653–61.

Service, M.W., Voller, A. and Bidwell, D.E. (1986) The enzyme-linked immunosorbent asssay (ELISA) test for the identification of blood-meals of haematophagous insects. *Bull. Entomol. Res.*, **76,** 321–30.

Shahid, A.A., Parveen, T. and Reisen, W.K. (1980) Changes in weight, calories and triglyceride content of sucrose fed *Culex tritaeniorhynchus* Giles and *Anopheles stephensi* Liston. *Pakistan J. Zool.*, **12**, 163–9.

Shalaby, A.M. (1959) Forced retention of eggs in *Culex (Barraudius) pusillus* Macq. (Diptera : Culicidae). *Indian J. Malariol.*, **13**, 199–208.

Shalaby, A.M. (1971) Changes in the ovaries of *Anopheles multicolor* and *Anopheles pharoensis* (Diptera: Culicidae) following oviposition. *Z. Angew. Entomol.*, **69**, 187–97.

Shambaugh, G.F. (1954) Protease stimulation by foods in adult *Aedes aegypti Ohio J. Sci.*, **54**, 151–60.

Shampengtong, L. and Wong, K.P. (1989) An *in vitro* assay of 20-hydroxyecdysone sulfotransferase in the mosquito, *Aedes togoi*. *Insect Biochem.*, **19**. 191–6.

Shampengtong, L., Wong, K.P. and Ho, B.C. (1987a) N-Acetylation of dopamine and tyramine by mosquito pupae (*Aedes togoi*). *Insect Biochem.*, **17**, 111–16.

Shampengtong, L., Wong, K.P. and Ho, B.C. (1987b) N-Acetylation of dopamine as determined by liquid chromatography with electrochemical detection (LCEC). *Biogenic Amines*, **4**, 179–88.

Shampengtong, L., Wong, K.P. and Ho, B.C. (1988) A study of dopa decarboxylase activity in mosquito, *Aedes togoi*, using high-pressure liquid chromatography with electrochemical detection (HPLC-ECD). *Biogenic Amines*, **5**, 7–15.

Shannon, R.C. and Hadjinicalao, J. (1941) Egg production of Greek anophelines in nature. *J. Econ. Entomol.*, **34**, 300–5.

Shapiro, J.P. (1983) Ovarian cyclic AMP and response to a brain hormone from the mosquito *Aedes aegypti*. *Insect Biochem.*, **13**, 273–9.

Shapiro, J.P. and Hagedorn, H.H. (1982) Juvenile hormone and the development of ovarian responsiveness to a brain hormone in the mosquito, *Aedes aegypti*. *Gen. Comp. Endocrinol*,. **46**, 176–83.

Shapiro, A.B., Wheelock, G.D., Hagedorn, H.H. *et al.* (1986) Juvenile hormone and juvenile hormone esterase in adult females of the mosquito *Aedes aegypti*. *J. Insect Physiol.*, **32**, 867–77.

Sharma, G.P., Chaudhry, S. and Safaya, A. (1986) Polytene chromsomes in *Aedes vittatus* Bigot (Culicidae: Diptera). *Microbios Letters*, **33**, 153–6.

Sharma, V.P., Hollingworth, R.M. and Paschke, J.D. (1970) Incorporation of tritiated thymidine in male and female mosquitoes, *Culex pipiens* with particular reference to spermatogenesis. *J. Insect Physiol.*, **16**, 429–36.

Sharma, G.P., Mittal, O.P., Chaudhry, S. and Pal, V. (1979) A preliminary map of the salivary gland chromosomes of *Aedes (Stegomyia) aegypti* (Culicidae, Diptera). *Cytobios*, **22**, 169–78.

Sharma, G.P., Parshad, R., Narang, S.L. and Kitzmiller, J.B. (1969) The salivary gland chromosomes of *Anopheles stephensi*. *J. Med. Entomol.*, **6**, 68–71.

Shaw, J. and Stobbart, R.H. (1963) Osmotic and ionic regulation in insects. *Advances Insect Physiol.*, **1**, 315–99.

Shelton, R.M. (1972) The effects of blood source and quantity on production of eggs by *Culex salinarius* Coquillett (Diptera: Culicidae). *Mosq. News*, **32**, 32–7.

Sheplay, A.W. and Bradley, T.J. (1982) A comparative study of magnesium sulphate tolerance in saline-water mosquito larvae. *J. Insect Physiol.*, **28**, 641–6.

Shipitzina, N.K. (1941) The influence of the density of the powdery pellicle on the filtration of food by the *Anopheles* larva. *Med. Parasitol. Moscow*, **10**, 396–401. [In Russian. Cited from *Rev. Appl. Entomol.* (*B*), (1943), **31**, 87–8.]

Shishliaeva-Matova, Z.S. (1942) Comparative study of salivary glands of Culicinae of the Samarkand District. Report I. Histology and comparative morphology of mosquito salivary glands. *Medskaya Parazit.*, **11**, 61–6. [In Russian.]

Shlenova, M.F. (1938) Vitesse de la digestion du sang par la femelle de l'*Anopheles maculipennis messeae* aux températures effectives constantes. *Med. Parasitol., Moscow*, **7**, 716–35. [In Russian, with French summary.]

Shroyer, D.A. and Craig, G.B. (1981) Seasonal variation in sex ratio of *Aedes triseriatus* (Diptera: Culicidae) and its dependence on egg hatching behavior. *Environ. Entomol.*, **10**, 147–52.

Shute, P.G. (1936) A study of laboratory-bred *Anopheles maculipennis* var. *atroparvus*, with special reference to egg laying. *Ann. Trop. Med. Parasitol.*, **30**, 11–6.

Sichinava, Sh.G. (1974) The capacity of female *Culex pipiens molestus* Forsk. for repeated autogenous oogenesis. *Akad. Nauk Gruz SSR Soobshch.*, **75**, 193–6. [In Russian.]

Siddiqui, T.F., Aslam, Y. and Reisen, W.K. (1976) The effects of larval density on selected immature and adult attributes in *Culex tritaeniorhynchus* Giles. *Trop. Med.*, **18**, 195–202.

Sieber, K.P., Huber, M., Kaslow, D. *et al.* (1991) The peritrophic membrane as a barrier: its penetration by *Plasmodium gallinaceum* and the effect of a monoclonal antibody to ookinetes. *Exp. Parasitol.*, **72**, 145–56.

Singh, K.R.P. and Brown, A.W.A. (1957) Nutritional requirements of *Aedes aegypti* L. *J. Insect Physiol.*, **1**, 199–220.

Singh, K.R.P., Curtis, C.F. and Krishnamurthy, B.S. (1976) Partial loss of cytoplasmic incompatibility with age in males of *Culex fatigans*. *Ann. Trop. Med. Parasitol.*, **70**, 463–6.

Singh, K.R.P. and Micks, D.W. (1957) Synthesis of amino acids in *Aedes aegypti*. *Mosq. News*, **17**, 248–51.

Sinton, J.A. and Covell, G. (1927) The relation of the morphology of the buccal cavity to the classification of anopheline mosquitoes. *Indian J. Med. Res.*, **15**, 301–9, 5 pls.

Sinitsyna, Y.Y. (1971) Bioelectric activity of the contact chemoreceptive sensilla of *Aedes aegypti* L. (Diptera, Culicidae) and food reactions to two-component solutions. *Entomol. Rev.*, **50**, 151–55.

Sirivanakarn, S. (1976) A revision of the subgenus *Culex* in the Oriental Region (Diptera: Culicidae). *Contr. Am. Entomol. Inst.*, **12**, 1–272.

Sirivanakarn, S. (1978) The female cibarial armature of New World *Culex*, subgenus *Melanconion* and related subgenera with notes on this character in subgenera *Culex*, *Lutzia* and *Neoculex* and genera *Galindomyia* and *Deinocerites* (Diptera: Culicidae). *Mosq. Systemat.*, **10**, 474–92.

Slater, J.D. and Pritchard, G. (1979) A stepwise computer program for estimating development time and survival of *Aedes vexans* (Diptera: Culicidae) larvae and pupae in field populations in southern Alberta. *Can. Entomol.*, **111**, 1241–53.

Smith, R.P. and Hartberg, W.K. (1974) Spermatogenesis in *Aedes albopictus* (Skuse). *Mosq. News*, **34**, 42–7.

Smith, S.L. and Mitchell, M.J. (1986) Ecdysone 20-monooxygenase systems in a larval and an adult dipteran. An overview of their biochemistry, physiology and pharmacology. *Insect Biochem.*, **16**, 49–55.

Smith, S.M. and Brust, R.A. (1970) Autogeny and stenogamy of *Aedes rempeli* (Diptera: Culicidae) in arctic Cana. *Can. Entomol.*, **102**, 253–6.

Smith, S.M. and Brust, R.A. (1971) Photoperiodic control of the maintenance and termination of larval diapause in *Wyeomyia smithii* (Coq.) (Diptera: Culicidae) with notes on oogenesis in the adult female. *Can. J. Zool.*, **49**, 1065–73, 1 pl.

Smith, S.M. and Corbet, P.S. (1975) Observations on the reproductive biology of *Aedes* (*Ochlerotatus*) *fryeri* (Theo.) (Diptera, Culicidae). *Bull. Entomol. Res.*, **65**, 285–93, 1 pl.

Smith, T.J., Powell, J.R., Hollander, A.L. and Fuchs, M.S. (1989) Characterization of dopamine N-acetyltransferase from *Aedes aegypti*: probable role in cuticular sclerotization, but not ovarian development. *Comp. Biochem. Physiol.*, **93B**, 721–5.

Sneller, V.-P. and Dadd, R.H. (1977) Requirement for sugar in a chemically defined diet for larval *Aedes aegypti* (Diptera: Culicidae). *J. Med. Entomol.*, **14**, 387–92.

Sneller, V.-P. and Dadd, R.H. (1981) Interaction of amino acids and glucose on growth of *Aedes aegypti* (Diptera: Culicidae) in a synthetic rearing medium. *J. Med. Entomol.*, **18**, 235–9.

Snodgrass, R.E. (1935) *Principles of Insect Morphology*, 4th edn, McGraw-Hill Book Company, New York and London.

Snodgrass, R.E. (1944) The feeding apparatus of biting and sucking insects affecting man and animals. *Smithsonian Misc. Coll.*, **104** (7), 1–113.

Snodgrass, R.E. (1959) The anatomical life of the mosquito. *Smithsonian Misc. Coll.*, **139** (8), 1–87.

Snow, K.R. (1987) Seasonal emergence patterns and sex ratios of *Aedes punctor* and *Aedes cantans* (Diptera; Culicidae) in southern England. *Entomol. Gaz.*, **38**, 253–62.

Sohal, R.S. and Copeland, E. (1966) Ultrastructural variations in the anal papillae of *Aedes aegypti* (L.) at different environmental salinities. *J. Insect Physiol.*, **12**, 429–34, 5 pls.

Spadoni, R.D., Nelson, R.L. and Reeves, W.C. (1974)

Seasonal occurrence, egg production, and bloodfeeding activity of autogenous *Culex tarsalis*. *Ann. Entomol. Soc. Am.*, **67**, 895–902.

Spielman, A. (1957) The inheritance of autogeny in the *Culex pipiens* complex of mosquitoes. *Am. J. Hyg.*, **65**, 404–25.

Spielman, A. (1964) Studies on autogeny in *Culex pipiens* populations in nature. I. Reproductive isolation between autogenous and anautogenous populations. *Am. J. Hyg.*, **80**, 175–83.

Spielman, A. (1967) Population structure in the *Culex pipiens* complex of mosquitos. *Bull. Wld Hlth Org.*, **37**, 271–6.

Spielman, A. (1971) Studies on autogeny in natural populations of *Culex pipiens*. II. Seasonal abundance of autogenous and anautogenous populations. *J. Med. Entomol.*, **8**, 555–61.

Spielman, A. (1974) Effect of synthetic juvenile hormone on ovarian diapause of *Culex pipiens* mosquitoes. *J. Med. Entomol.*, **11**, 223–5.

Spielman, A., Gwadz, R.W. and Anderson, W.A. (1971) Ecdysone-initiated ovarian development in mosquitoes. *J. Insect Physiol.*, **17**, 1807-14.

Spielman, A., Ribeiro, J.M.C., Rossignol, P.A. and Perrone, J.R. (1986) Food-associated regulation of salivary production by mosquitoes. In *Host Regulated Developmental Mechanisms in Vector Arthropods* (eds D. Borovsky and A. Spielman), University of Florida, Vero Beach, **1**, 100–3.

Spielman, A. and Wong, J. (1973) Studies on autogeny in natural populations of *Culex pipiens*. III. Midsummer preparation for hibernation in anautogenous populations. *J. Med. Entomol.*, **10**, 319–24.

Spinner, W. (1969) Transplantionsversuche zur Blastemgliederung, Regenerations- und Differenzierungsleistung der Beinanlagen von *Culex pipiens* (L.). *Wilhelm Roux' Arch. Entw.Mech. Organ.*, **163**, 259–86.

Spiro-Kern, A. (1974) Untersuchungen über die Proteasen bei *Culex pipiens*. *J. Comp. Physiol.*, **90**, 53–70.

Spiro-Kern, A. and Chen, P.S. (1972) Über die Proteasen der Stechmücke *Culex pipiens*. *Rev. Suisse Zool.*, **79**, 1151–9.

Spring, J.H. (1990) Endocrine regulation of diuresis in insects. *J. Insect Physiol.*, **36**, 13–22.

Standfast, H.A. and Dyce, A.L. (1968) Attacks on cattle by mosquitoes and biting midges. *Austral. Vet. J.*, **44**, 585–6.

Stanley-Samuelson, D.W. (1991) Comparative physiology of eicosanoids in invertebrate animals. *Am. J. Physiol.*, **260** (*Reg. Int. Comp. Physiol.* 29), R849–R853.

Stanley-Samuelson, D.W., Mackay, M.E. and Blomquist, G.J. (1989) Arachidonic acid and prostaglandins in the mosquito, *Aedes aegypti*. In *Host Regulated Mechanisms in Vector Arthropods* (eds D. Borovsky and A. Spielman), University of Florida, Vero Beach, **2**, 179–88.

Starratt, A.N. and Osgood, C.E. (1972) An oviposition pheromone of the mosquito *Culex tarsalis*: diglyceride composition of the active fraction. *Biochim. Biophys. Acta*, **280**, 187–93.

Stäubli, W., Freyvogel, T.A. and Suter, J. (1966) Structural modification of the endoplasmic reticulum of midgut epithelial cells of mosquitoes in relation to blood intake. *J. Microscopie*, **5**, 189–204, 6 pls.

Steel, C.G.H. and Davey, K.G. (1985) Integration in the insect endocrine system. In *Comprehensive Insect Physiology Biochemistry and Pharmacology* (eds G.A. Kerkut and L.I. Gilbert), Pergamon Press, Oxford, **8**, 1–35.

Steelman, C.D., White, T.W. and Schilling, P.E. (1973) Effects of mosquitoes on the average daily gain of Hereford and Brahman breed steers in southern Louisiana. *J. Econ. Entomol.*, **66**, 1081–3.

Steelman, C.D., White, T.W. and Schilling, P.E. (1976) Efficacy of Brahman characters in reducing weight loss of steers exposed to mosquito attack. *J. Econ. Entomol.*, **69**, 499–502.

Steffan, W.A. and Evenhuis, N.L. (1981) Biology of *Toxorhynchites*. *Annual Rev. Entomol.*, **26**, 159–81.

Steinwascher, K. (1984) Egg size variation in *Aedes aegypti*: relationship to body size and other variables. *Am. Midland Nat.*, **112**, 76–84.

Steward, C.C. and Atwood, C.A. (1963) The sensory organs of the mosquito antenna. *Can. J. Zool.*, **41**, 577–94.

Stewart, G.L., Soifer, K.F. and Stewart, J.B. (1979) The effects of various carbohydrate diets on *Aedes aegygpti* infected with *Dirofilaria immitis*. *J. Invert. Pathol.*, **33**, 75–80.

Stewart, W.W.A (1974) The rate of larval development of *Aedes implicatus* Vockeroth in field and laboratory. *Mosq. News*, **34**, 283–5.

Stich, H. and Grell, M. (1955) Incorporation of phosphorus-32 into the Malpighian tubes during the metamorphosis of *Culex pipiens*. *Nature*, **176**, 930–1.

Stidham, J.D. and Liles, J.N. (1969a) Some aspects of the metabolic fate of ^{14}C-labeled alanine and asparatic acid in the aging female mosquito, *Aedes aegypti* (L.). *Comp. Biochem. Physiol.*, **31**, 513–21.

Stidham, J.D. and Liles, J.N. (1969b) Free amino acid composition of the ageing female mosquito *Aedes aegypti* as determined by automatic ion-exchange chromatography. *J. Insect Physiol.*, **15**, 1969–80.

Stiles, B. and Paschke, J.D. (1980) Midgut pH in different instars of three *Aedes* mosquito species and the relation between pH and susceptibility of larvae to a nuclear polyhedrosis virus. *J. Invert. Pathol.*, **35**, 58–64.

Stobbart, R.H. (1959) Studies on the exchange and regulation of sodium in the larva of *Aedes aegypti* (L.). I. The steady-state exchange. *J. Exp. Biol.*, **36**, 641–53.

Stobbart, R.H. (1960) Studies on the exchange and

regulation of sodium in the larva of *Aedes aegypti* (L.). II. The net transport and the fluxes associated with it. *J. Exp. Biol.*, **37**, 594–608.

Stobbart, R.H. (1965) The effect of some anions and cations upon the fluxes and net uptake of sodium in the larva of *Aedes aegypti*. *J. Exp. Biol.*, **42**, 29–43.

Stobbart, R.H. (1967) The effect of some anions and cations upon fluxes and net uptake of chloride in the larva of *Aedes aegypti* (L.), and the nature of the uptake mechanisms for sodium and chloride. *J. Exp. Biol.*, **47**, 35–57.

Stobbart, R.H. (1971a) Evidence of Na^+/H^+ and Cl^-/HCO_3^- exchanges during independent sodium and chloride uptake by the larva of the mosquito *Aedes aegypti* (L.). *J. Exp. Biol.*, **54**, 19–27.

Stobbart, R.H. (1971b) The control of sodium uptake by the larva of the mosquito *Aedes aegypti* (L.). *J. Exp. Biol.*, **54**, 29–66.

Stobbart, R.H. (1971c) Factors affecting the control of body volume in the larvae of the mosquitoes *Aedes aegypti* (L.) and *Aedes detritus* Edw. *J. Exp. Biol.*, **54**, 67–82.

Stobbart, R.H. (1974) Electrical potential differences and ionic transport in the larva of the mosquito *Aedes aegypti* (L.). *J. Exp. Biol.*, **60**, 493–533.

Stobbart, R.H. (1977) The control of the diuresis following a blood meal in females of the yellow fever mosquito *Aedes aegypti* *J. Exp. Biol.*, **69**, 53–85.

Stobbart, R.H. (1992) Selection of the yellow fever mosquito *Aedes aegypti* for cheap and easy maintenance without bloodmeals. *Med. Vet. Entomol.*, **6**, 87–9.

Stohler, H. (1957) Analyse des Infektionsverlaufes von *Plasmodium gallinaceum* im Darme von *Aedes aegypti*. *Acta Tropica*, **14**, 302–52.

Stokes, R.H. and Mills, R. (1965) *Viscosity of Electrolytes and Related Properties*. Pergamon Press, Oxford.

Strange, K. and Phillips, J.E. (1984) Mechanisms of CO_2 transport in rectal salt gland of *Aedes*. I. Ionic requirements of CO_2 secretion. *Am. J. Physiol.* (*Reg. Int. Comp. Physiol.* 15), **246**, R727–34.

Strange, K. and Phillips, J.E. (1985) Cellular mechanism of HCO_3^- and Cl^- transport in insect salt gland. *J. Membrane Biol.*, **83**, 25–37.

Strange, K., Phillips, J.E. and Quamme, G.A. (1982) Active HCO_3^- secretion in the rectal salt gland of a mosquito larva inhabiting $NaHCO_3$–CO_3 lakes. *J. Exp. Biol.*, **101**, 171–86.

Strange, K. Phillips, J.E. and Quamme, G.A. (1984) Mechanisms of CO_2 transport in rectal salt gland of *Aedes*. II. Site of Cl^-–HCO_3^- exchange. *Am. J. Physiol.* (*Reg. Int. Comp. Physiol.* 15), **246**, R735–40.

Stryer, L. (1988) *Biochemistry*, 3rd edn, Freeman, New York.

Stueben, E.B. (1978) A new fast fluorescent dye technique to detect parasite infection in mosquitoes. *Mosq. News*, **38**, 586–7.

Subbarao, S.K., Krishnamurthy, B.S., Curtis, C.F. *et al.* (1977a) Further studies on variation of cytoplasmic incompatibility in the *Culex pipiens* complex. *Indian J. Med. Res.*, **65** (Suppl.), 21–33.

Subbarao, S.K., Krishnamurthy, B.S., Curtis, C.F. *et al.* (1977b) Segregation of cytoplasmic incompatibility properties in *Culex pipiens fatigans*. *Genetics*, **87**, 381–90.

Suenega, O. (1982) Treatment of *Wolbachia pipientis* infection with tetracycline hydrochloride and the change of compatibility in a strain of *Culex pipiens* complex. *Trop. Med.*, **24**, 9–15. [In Japanese, with English Abstract.]

Suenaga, Y. (1987) Mosquito bites – especially on mosquito hypersensitivity and malignant histiocytosis. *Nishinihon Hifuka*, **49**, 252–9. [In Japanese.]

Sugumaran, M. and Semensi, V. (1987) Sclerotization of mosquito cuticle. *Experientia*, **43**, 172–4.

Suguna, S.G., Wood, R.J., Curtis, C.F. *et al.* (1977) Resistance to meiotic drive at the MD locus in an Indian wild population of *Aedes aegypti*. *Genet. Res.*, **29**, 123–32.

Suguri, S., Tongu, Y., Itano, K. *et al.* (1969) The ultrastructure of mosquitoes. 2. Malpighian tubule of *Culex pipiens pallens*. *Jap. J. Sanit. Zool.*, **20**, 1–6.

Suguri, S., Tongu, Y., Sakumoto, D. *et al.* (1972) The ultrastructure of mosquitoes. 5. Salivary gland of *Culex tritaeniorhynchus*. *Res. Filar. Schistosom.*, **2**, 51–65.

Sulaiman, I. (1985) Recombination frequencies and the order of the genes *re* (red eye colour), *fi* (filarial susceptibility – *Dirofilaria immitis*) and M (sex) in the sex chromosomes of *Aedes aegypti*. *Trop. Biomed.*, **47**, 47–53.

Suleman, M. (1982) The effects of intraspecific competition for food and space on the larval development of *Culex quinquefasciatus*. *Mosq. News*, **42**, 347–56.

Suleman, M. (1990) Intraspecific variation in the reproductive capacity of *Anopheles stephensi* (Diptera: Culicidae). *J. Med. Entomol.*, **27**, 819–28.

Surtees,G. (1959) Functional and morphological adaptations of the larval mouthparts in the sub-family Culicinae (Diptera) with a review of some related studies by Montschadsky. *Proc. R. Entomol. Soc. Lond.* (A), **34**, 7–16.

Sutton, E. (1942) Salivary gland type chromosomes in mosquitoes. *Proc. Nat. Acad. Sci., Wash.*, **28**, 268–72.

Suzuki, S., Negishi, K. Tomizawa, S. *et al.* (1976) A case of mosquito allergy. Immunological studies. *Acta Allergol.*, **31**, 428–41.

Sverdrup, H.U., Johnson, M.W. and Fleming, R.H. (1942) *The Oceans, their Physics, Chemistry and General Biology*. Prentice-Hall, New York.

Svoboda, J.A., Thompson, M.J. Herbert, E.L. *et al.* (1982) Utilization and metabolism of dietary sterols in the honey bee and the yellow fever mosquito. *Lipids*, **17**, 220–5.

Svoboda, J.A. and Thompson, M.J. (1985) Steroids. In *Comprehensive Insect Physiology Biochemistry Pharmacology* (eds G.A. Kerkut and L.I. Gilbert), Pergamon Press, Oxford, **10**, 137–75.

Swan, M.A. (1981) The generation and propagation of double waves in mosquito (Aedes notoscriptus) sperm-tails. *Gamete Res.*, **4**, 241–50.

Sweeney, T.L. and Barr, A.R. (1978) Sex ratio distortion caused by meiotic drive in a mosquito, *Culex pipiens* L. *Genetics*, **88**, 427–66.

Sweeny, T.L., Guptavanij, P. and Barr, A.R. (1987) Abnormal salivary gland puff associated with meiotic drive in mosquitoes (Diptera: Culicidae). *J. Med. Entomol.*, **24**, 623–7.

Swellengrebel, N.H. (1929) La dissociation des fonctions sexuelles et nutritives (dissociation gonotrophique) d'*Anopheles maculipennis* comme cause du paludisme dans le pays-bas et ses rapports avec l'infection domicilaire. *Ann. Inst. Pasteur, Paris*, **43**, 1370–89.

Sybenga, J. (1975) *Meiotic Configurations: a Source of Information for Estimating Genetic Parameters*. Springer-Verlag, Berlin.

Tadano, T. and Mogi, M. (1987) Inheritance of orange pupa and phosphoglucomutase in the mosquito *Armigeres subalbatus*. *J. Am. Mosq. Control Ass.*, **3**, 642–3.

Tadkowski, T.M. and Jones, J.C. (1978) Endogenous synthesis of lipid yolk in mosquito oocytes. *Experientia*, **34**, 627.

Tadkowski, T.M. and Jones, J.C. (1979) Changes in the fat body and oocysts during starvation and vitellogenesis in a mosquito. *J. Morphol.*, **159**, 185–203.

Takahashi, C. and Harwood, R.F. (1964) Glycogen levels of adult *Culex tarsalis* in response to photoperiod. *Ann. Entomol. Soc. Am.*, **57**, 621–3.

Taketomi, M. (1967) Ovariole and age changes in *Anopheles sinensis* Wiedemann, with special reference to the relation to temperature and season. *Endemic Dis. Bull.*, **8**, 170–90.

Taylor, E.W. (1958) *Examination of Water and Water Supplies*, Churchill, London.

Telfer, W.H. (1975) Development and physiology of the oocyte-nurse cell syncytium. *Adv. Insect Physiol.*, **11**, 223–319.

Tempelis, C.H. (1975) Host-feeding patterns of mosquitoes, with a review of advances in analysis of bloodmeals by serology. *J. Med. Entomol.*, **11**, 635–53.

Terra, W.R. (1990) Evolution of digestive systems of insects. *Annual Rev. Entomol.*, **35**, 181–200.

Terzakis, J.A. (1967) Substructure in an epithelial basal lamina (basement membrane). *J. Cell Biol.*, **35**, 273–8.

Terzian, L.A., Irreverre, F. and Stahler, N. (1957) A study of nitrogen and uric acid patterns in the excreta and body tissues of adult *Aedes aegypti*. *J. Insect Physiol.*, **1**, 221–8.

Tesfa-Yohannes, T.-M. (1982) Reproductive biology of *Aedes* (*S.*) *malayensis* (Diptera: Culicidae). *J. Med. Entomol.*, **19**, 29–33.

Thayer, D.W. and Terzian, L.A. (1971) Amino acid partition in excreta of ageing female *Aedes aegypti* mosquitoes. *J. Insect Physiol.*, **17**, 1731–4.

Thayer, D.W., Terzian, L.A. and Price, P.A. (1971a) Digestion of the avian blood-meal by the mosquito, *Aedes aegypti*. *J. Insect Physiol.*, **17**, 2193–204.

Thayer, D.W., Terzian, L.A. and Price, P.A. (1971b) Digestion of the human blood-meal by the mosquito, *Aedes aegypti*. *J. Insect Physiol.*, **17**, 2469–73.

Theobald, F.V. (1901) *A Monograph of the Culicidae or Mosquitoes*, Vol. 1, British Museum (Natural History), London.

Thiery, I., Nicolas, L., Rippka, R. and de Marsac, N.T. (1991) Selection of cyanobacteria isolated from mosquito breeding sites as a potential food source for mosquito larvae. *Appl. Environ. Microbiol.*, **57**, 1354–9.

Thomas, B.R., Fuchs, M.S., Bonavaliker, S. *et al.* (1986) Juvenile hormone binding proteins of *Aedes atropalpus* and *Aedes aegypti*. In *Host Regulated Developmental Mechanisms in Vector Arthropods* (eds D. Borovsky and A. Spielman), University of Florida, Vero Beach, **1**, 51–9.

Thomas, B.R., Lee, S.K. and Fuchs, M.S. (1989) Vitellogenin, vitellin, and nonvitellin ovarian proteins of *Aedes aegypti*. In *Host Regulated Developmental Mechanisms in Vector Arthropods* (eds D. Borovsky and A. Spielman), University of Florida, Vero Beach, **2**, 64–71.

Thomas, V. (1971) Studies on cytoplasmic incompatibility in southeast asian *Culex pipiens fatigans*. *S.E. Asian J. Trop. Med. Publ. Hlth*, **2**, 469–73.

Thomas, V. and Leng, Y.P. (1972) The inheritance of autogeny in *Aedes* (*Finlaya*) *togoi* (Theobald) from Malaysia and some aspects of its biology. *S.E. Asian J. Trop. Med. Publ. Hlth*, **3**, 163–74.

Thompson, M.T. (1905) Alimentary canal of the mosquito. *Proc. Boston Soc. Natural Hist.*, **32**, 145–202, 6 pls.

Tiepolo, L., Fraccaro, M., Laudani, U. and Diaz, G. (1975) Homologous bands on the long arms of the X and Y chromosomes of *Anopheles atroparvus*. *Chromosoma*, **49**, 371–4.

Tiepolo, L. and Laudani, U. (1972) DNA synthesis in polytenic chromosomes of *Anopheles atroparvus*. *Chromosoma*, **36**, 305–12.

Tokura, Y., Tamura, Y., Takigawa, M. *et al.* (1990) Severe hypersensitivity to mosquito bites associated with natural killer lymphocytosis. *Arch. Dermatol.*, **126**, 362–8.

Tongu, Y., Suguri, S., Itano, K. *et al.* (1968) The ultrastructure of mosquitoes. 1. Spermatozoa in *Culex pipiens pallens*. *Jap. J. Sanit. Zool.*, **19**, 215–22. [In Japanese, English summary.]

Tongu, Y., Suguri, S., Sakumoto, D. *et al.* (1969) The

ultrastructure of mosquitoes. 3. Hindgut of *Aedes aegypti*. *Jap. J. Sanit. Zool.*, **20**, 168–76.

Townson, H. and Chaithong, U. (1991) Mosquito host influences on development of filariae. *Ann. Trop. Med. Parasitol.*, **85**, 149–63.

Trager, W. (1937) Cell size in relation to the growth and metamorphosis of the mosquito, *Aedes aegypti*. *J. Exp. Zool.*, **76**, 467–89.

Treherne, J.E. (1954) The exchange of labelled sodium in the larva of *Aedes aegypti* L. *J. Exp. Biol.*, **31**, 386–401.

Trembley, H.L. (1951) Pyloric spines in mosquitoes. *J. Natl Malaria Soc.*, **10**, 213–5.

Trembley, H.L. (1952) The distribution of certain liquids in the esophageal diverticula and stomach of mosquitoes. *Am. J. Trop. Med. Hyg.*, **1**, 693–710.

Trimble, R.M. and Smith, S.M. (1978) Geographic variation in development time and predation in the tree-hole mosquito, *Toxorhynchites rutilus septentrionalis* (Diptera: Culicidae). *Can. J. Zool.*, **56**, 2156–65.

Tritton, D.J. (1988) *Physical Fluid Dynamics*, 2nd edn, Clarendon Press, Oxford.

Troy, S., Anderson, W.A. and Spielman, A. (1975) Lipid content of maturing ovaries of *Aedes aegypti* mosquitoes. *Comp. Biochem. Physiol.*, **50B**, 457–61.

Trpis, M. (1970) A new bleaching and decalcifying method for general use in zoology. *Can. J. Zool.*, **48**, 892–3, 1 pl.

Trpis, M. (1972a) Seasonal changes in the larval populations of *Aedes aegypti* in two biotopes in Dar es Salaam, Tanzania. *Bull. Wld Hlth Org.*, **47**, 245–55.

Trpis, M. (1972b) Dry season survival of *Aedes aegypti* eggs in various breeding sites in the Dar es Salaam area, Tanzania. *Bull. Wld Hlth Org.*, **47**, 433–7.

Trpis, M. (1972c) Development and predatory behavior of *Toxorhynchites brevipalpis* (Diptera: Culicidae) in relation to temperature. *Environ. Entomol.*, **1**, 537–46.

Trpis, M. (1977) Autogeny in diverse populations of *Aedes aegypti* from East Africa. *Tropenmed. Parasitol.*, **28**, 77–82.

Trpis, M. (1978) Genetics of hematophagy and autogeny in the *Aedes scutellaris* complex (Diptera: Culicidae). *J. Med. Entomol.*, **15**, 73–80.

Trpis, M. (1981) Survivorship and age-specific fertility of *Toxorhynchites brevipalpis* females (Diptera: Culicidae). *J. Med. Entomol.*, **18**, 481–6.

Trpis, M., Haufe, W.O. and Shemanchuk, J.A. (1973) Embryonic development of *Aedes* (*O.*) *sticticus* (Diptera: Culicidae) in relation to different constant temperatures. *Can. Entomol.*, **105**, 43–50.

Trpis, M. and Horsfall, W.R. (1969) Development of *Aedes sticticus* (Meigen) in relation to temperature, diet, density and depth. *Ann. Zool. Fennici*, **6**, 156–60.

Trpis, M., Perrone, J.B., Reissig, M. and Parker, K.L. (1981) Control of cytoplasmic incompatibility in the *Aedes scutellaris* complex. *J. Hered.*, **72**, 313–17.

Trpis, M. and Shemanchuk, J.A. (1969) The effect of temperature on pre-adult development of *Aedes flavescens* (Diptera: Culicidae). *Can. Entomol.*, **101**, 128–32.

Tsuji, N. (1989) Autogenous and anautogenous mosquitoes: the effect of survival rate during blood feeding. *Acta Eruditiorum*, **8**, 1–14.

Tsuji, N., Okazawa, T. and Yamamura, N. (1990) Autogenous and anautogenous mosquitoes: a mathematical analysis of reproductive strategies. *J. Med. Entomol.*, **27**, 446–53.

Tu, Z. and Hagedorn, H.H. (1992) Purification and characterization of pyruvate carboxylase from the honey bee and some properties of related bioin-containing proteins in other insects. *Arch. Insect Biochem. Physiol.*, **19**, 53–66.

Twohy, D.W. and Rozeboom. L.E. (1957) A comparison of food reserves in autogenous and anautogenous *Culex pipiens* populations. *Am. J. Hyg.*, **65**, 316–24.

Uchida, K. (1979) Cibarial sensilla and pharyngeal valves in *Aedes albopictus* (Skuse) and *Culex pipiens pallens* Coquillett (Diptera: Culicidae). *Int. J. Insect Morphol. Embryol.*, **8**, 159–67.

Uchida, K., Ohmori, D., Yamakura, F. and Suzuki, K (1990) . Changes in free amino acid concentration in the hemolymph of the female *Culex pipiens pallens* (Diptera: Culicidae), after a blood meal. *J. Med. Entomol.*, **27**, 302–8.

Uchida, K. and Suzuki, K. (1981) Elimination of protein-food ingested into the crop, and failure of ovarian development in female mosquitoes, *Culex pipiens pallens*. *Physiol. Entomol.*, **6**, 445–50.

Vachereau, A. and Ribeiro, J.M.C. (1989) Immunoreactivity of salivary gland apyrase of *Aedes aegypti* with antibodies against a similar hydrolase present in the pancreas of mammals. *Insect Biochem.*, **19**, 527–34.

Van den Assem, J. (1958) Some experimental evidence for the survival value of the rootpiercing habits of *Mansonia* larvae (Culicidae) to predators. *Entomol. Exp. Applic.*, **1**, 125–9.

Van den Assem, J. (1959) Notes on New Guinean species of Tripteroides, subgenus *Rachisoura* (Diptera, Culicidae), with descriptions of two new species. *Tijdschr. Entomol.*, **102**, 35–55, 1 pl.

Van den Heuvel, M.J. (1963) The effect of rearing temperature on the wing length, thorax length, leg length and ovariole number of the adult mosquito, *Aedes aegypti* (L.). *Trans. R. Entomol. Soc. Lond.*, **115**, 197–216.

Van der Linde, T.C. de K., Hewitt, P.H., Nel, A. and Van der Westhuizen, M.C. (1990) Development rates and percentage hatching of *Culex* (*Culex*) *theileri* Theobald (Diptera: Culicidae) eggs at various constant temperatures. *J. Entomol. Soc. Southern Africa*, **53**, 17–26.

Van Handel, E. (1965a) Microseparation of glycogen, sugars, and lipids. *Anal. Biochem.*, **11**, 266–71.

Van Handel, E. (1965b) The obese mosquito. *J. Physiol.*, **181**, 478–86.

Van Handel, E. (1966) Temperature independence of the composition of triglyceride fatty acids synthesized de novo by the mosquito. *J. Lipid Res.*, **7**, 112–15.

Van Handel, E. (1967a) Determination of fructose and fructose-yielding carbohydrates with cold anthrone. *Analyt. Biochem.*, **19**, 193–4.

Van Handel, E. (1967b) Non-dependence of the saturation of depot fat on temperature and photoperiod in a hibernating mosquito. *J. Exp. Biol.*, **46**, 487–90.

Van Handel, E. (1969a) Metabolism of hexoses in the intact mosquito: exclusion of glucose and trehalose as intermediates. *Comp. Biochem. Physiol.*, **29**, 413–21.

Van Handel, E. (1969b) The equilibrium reaction sorbitol $\rightleftharpoons$ fructose in the intact mosquito. *Comp. Biochem. Physiol.*, **29**, 1023–30.

Van Handel, E. (1972a) The detection of nectar in mosquitoes. *Mosq. News*, **32**, 458.

Van Handel, E. (1972b) Simple biological and chemical methods to determine the caloric reserves of mosquitoes. *Mosq. News*, **32**, 589–91.

Van Handel, E. (1973) Temperature dependence of caloric expenditure and mortality in the starving mosquito. *Comp. Biochem. Physiol.*, **44A**, 1321–3.

Van Handel, E. (1976) The chemistry of egg maturation in the unfed mosquito *Aedes atropalpus*. *J. Insect Physiol.*, **22**, 521–2.

Van Handel, E. (1985a) Rapid determination of glycogen and sugars in mosquitoes. *J. Am. Mosq. Control Ass.*, **1**, 299–301.

Van Handel, E. (1985b) Rapid determination of total lipids in mosquitoes. *J. Am. Mosq. Control Ass.*, **1**, 302–4.

Van Handel, E. (1986) Determination and significance of suspended protein in wastewater. *J. Am. Mosq. Control Ass.*, **2**, 146–9.

Van Handel, E. (1992) Postvitellogenic metabolism of the mosquito (*Culex quinquefasciatus*) ovary. *J. Insect Physiol.*, **38**, 75–9.

Van Handel, E. and Day, J.F. (1988) Assay of lipids, glycogen and sugars in individual mosquitoes: correlations with wing length in field-collected *Aedes vexans*. *J. Am. Mosq. Control Ass.*, **4**, 549–50.

Van Handel, E., Haeger, J.S. and Hansen, C.W. (1972) The sugars of some Florida nectars. *Am J. Bot.*, **59**, 1030–2.

Van Handel, E. and Lea, A.O. (1965) Medial neurosecretory cells as regulators of glycogen and triglyceride synthesis. *Science*, **149**, 298–300.

Van Handel, E. and Lea, A.O. (1970) Control of glycogen and fat metabolism in the mosquito. *Gen. Comp. Endocrinol.*, **14**, 381–4.

Van Handel, E. and Lea, A.O. (1984) Vitellogenin synthesis in blood-fed *Aedes aegypti* in the absence of the head, thorax and ovaries. *J. Insect Physiol.*, **30**, 871–5.

Van Handel, E. and Lum, P.T.M. (1961) Sex as regulator of triglyceride metabolism in the mosquito. *Science*, **134**, 1979–80.

Van Handel, E. and Romoser, W.S. (1987) Proteolytic activity in the ectoperitrophic fluid of blood-fed *Culex nigripalpus*. *Med. Vet. Entomol.*, **1**, 251–5.

Van Pletezen, R. and van der Linde. T.C.deK. (1981) Studies on the biology of *Culiseta longiareolata* (Macquart) (Diptera: Culicidae). *Bull. Entomol. Res.*, **71**, 71–9, 2 pls.

Vargo, A.M. and Foster, W.A. (1984) Gonotrophic state and parity of nectar-feeding mosquitoes. *Mosq. News*, **44**, 6–10.

Vaughan J.A. and Azad, A.F. (1988) Passage of host immunoglobulin G from blood meal into hemolymph of selected mosquito species (Diptera: Culicidae). *J. Med. Entomol.*, **25**, 472–4.

Vaughan, J.A., Noden, B.H. and Beier, J.C. (1991) Concentration of human erythrocytes by anopheline mosquitoes (Diptera: Culicidae) during feeding. *J. Med. Entomol.*, **28**, 780–6.

Ved Brat, S. and Rai, K.S. (1973) An analysis of chiasma frequencies in *Aedes aegypti*. *Nucleus*, **16**, 184–93.

Ved Brat, S.S. and Whitt, G.S. (1974) Lactate dehydrogenase and glycerol-3-phosphate dehydrogenase gene expression during ontogeny of the mosquito (*Anopheles albimanus*). *J. Exp. Zool.*, **187**, 135–40.

Veenstra, J.A. (1988) Effects of 5-hydroxtryptamine on the Malpighian tubules of *Aedes aegypti*. *J. Insect Physiol.*, **34**, 299–304.

Venard, C.E. and Guptavanij, P. (1966) Inflation of the oesophageal diverticula and elimination of air from the stomach in newly emerged mosquitoes. *Mosq. News*, **26**, 65–9.

Vizzi, F.F. (1953) The mouthparts of the male mosquito *Anopheles quadrimaculatus* Say (Diptera: Culicidae). *Ann. Entomol. Soc. Am.*, **46**, 496–504.

Vogel, R. (1921) Kritische und ergänzende Mitteilungen zur Anatomie des Stechsapparats der Culiciden und Tabaniden. *Zool. Jb.*, **42**, 259–82, 1 pl.

Vogel, S. (1981) *Life in Moving Fluids. The Physical Biology of Flow*, Willard Grant Press, Boston, Mass.

Volkmann, A. and Peters, W. (1989a) Investigations on the midgut caeca of mosquito larvae-I. Fine structure. *Tissue & Cell*, **21**, 243–51.

Volkmann, A. and Peters, W. (1989b) Investigations on the midgut caeca of mosquito larvae-II. Functional aspects. *Tissue & Cell*, **21**, 253–61.

Volozina, N.V. (1967) The effect of the amount of blood taken and additional carbohydrate nutrition on oogenesis in females of blood-sucking mosquitoes of the genus *Aedes* (Diptera, Culicidae) of various weights and ages. *Entomol. Rev.*, **46**, 27–32.

Von Gernet, G. and Buerger, G. (1966) Labral and cibarial sense organs of some mosquitoes. *Quaest. Entomol.*, **2**, 259–70.

Voorhees, F.R. and Horsfall, W.R. (1971) Genesis of the reproductive system of mosquitoes. III. Supernumerary male genitalia. *J. Morphol.*, **133**, 399–407.

Vrtiska, L.A. and Pappas, L.G. (1984) Chemical analysis of mosquito larval habitats in southeastern Nebraska. *Mosq. News*, **44**, 506–9.

Wade, J.O. and Macdonald, W.W. (1977) Compatible and incompatible crosses within the *Aedes scutellaris* group. *Trans. R. Soc. Trop. Med. Hyg.*, **71**, 109.

Waldbauer, G.P. (1962) The mouth parts of female *Psorophora ciliata* (Diptera, Culicidae) with a new interpretation of the functions of the labral muscles. *J. Morphol.*, **111**, 201–15.

Walker, E.D. and Merritt, R.W. (1988) The significance of leaf detritus to mosquito (Diptera: Culicidae) productivity from treeholes. *Environ. Entomol.*, **17**, 199–206.

Walker, E.D. and Merritt, R.W. (1991) Behavior of larval *Aedes triseriatus* (Diptera: Culicidae). *J. Med. Entomol.*, **28**, 581–9.

Walker, E.D., Olds, E.J. and Merritt, R.W. (1988) Gut content analysis of mosquito larvae (Diptera: Culicidae) using DAPI stain and epifluorescence microscopy. *J. Med. Entomol.*, **25**, 551–4.

Walker, M.C. and Romoser, W.S. (1987) The origin and movement of gas during adult emergence in *Aedes aegypti*: an hypothesis. *J. Am. Mosq. Control Ass.*, **3**, 429–32.

Wallace, A.J. and Newton, M.E. (1987) Heterochromatin diversity and cyclic responses to selective silver staining in *Aedes aegypti* (L.). *Chromosoma*, **95**, 89–93.

Wandall, A. and Svendsen, A. (1983) The synaptonemal complex karyotype from spread spermatocytes of a dipteran (*Aedes aegypti*). *Can. J. Genet. Cytol.*, **25**, 361–9.

Wandall, A. and Svendsen, A. (1985) Transition from somatic to meiotic pairing and progressional changes of the synaptonemal complex in spermatocytes of *Aedes aegypti*. *Chromosoma*, **92**, 254–64.

Warren, A.M. and Crampton, J.M. (1991) The *Aedes aegypti* genome: complexity and organisation. *Genetical Res.*, **58**, 225–32.

Warren, M.E. and Breland, O.P. (1963) Studies on the gonads of some immature mosquitoes. *Ann. Entomol. Soc. Am.*, **56**, 619–24.

Washino, R.K. (1970) Physiological condition of overwintering female *Anopheles freeborni* in California (Diptera: Culicidae). *Ann. Entomol. Soc. Am.*, **63**, 210–16.

Washino, R.K., Gieke, P.A. and Schaefer, C.H. (1971) Physiological changes in the overwintering females of *Anopheles freeborni* (Diptera: Culicidae) in California. *J. Med. Entomol.*, **8**, 279–82.

Washino, R.K. and Tempelis, C.H. (1983) Mosquito host bloodmeal indentification: methodology and data analysis. *Annual Rev. Entomol.*, **28**, 179–201.

Waterhouse, D.F. (1953) The occurrence and significance of the peritrophic membrane, with special reference to adult Lepidoptera. *Austral. J. Zool.*, **1**, 299–318, 2 pls.

Watts, R.B. and Smith, S.M. (1978) Oogenesis in *Toxorhynchites rutilus* (Diptera: Culicidae). *Can. J. Zool.*, **56**, 136–39.

Weathersby, A.B. and Noblet, R. (1973) *Plasmodium gallinaceum*: development in *Aedes aegypti* maintained on various carbohydrate diets. *Exp. Parasitol.*, **34**, 426–31.

Weaver, S.C. and Scott, T.W. (1990a) Peritrophic membrane formation and cellular turnover in the midgut of *Culiseta melanura* (Diptera: Culicidae). *J. Med. Entomol.*, **27**, 864–73.

Weaver, S.C. and Scott, T.W. (1990b) Ultrastructural changes in the abdominal midgut of the mosquito, *Culiseta malanura*, during the gonotrophic cycle. *Tissue & Cell*, **22**, 895–909.

Weed, R.I. (1965) Exaggerated delayed hypersensitivity to mosquito bites in chronic lymphocytic leukemia. *Blood*, **26**, 257–68.

Wenk, P. (1961) Die Muskulatur der Mandibel einiger blutsaugender Culiciden. *Zool. Anz.*, **167**, 254–9.

Wesenberg-Lund, C. (1918) Anatomical description of the larva of *Mansonia richardii* (Ficalbi) found in Danish freshwaters. *Vidensk. Meddr Dansk Naturh. Foren.*, **69**, 277–328.

Wesenberg-Lund, C. (1921) Contributions to the biology of the Danish Culicidae. *K. Danske Vidensk. Selsk.*, **7**, 1–210, 21 pls.

Westbrook, A. and Russo, R. (1985) Ecdysone titers during the fourth larval instar of three species of *Toxorhynchites*. *Proc. New Jersey Mosq. Control Ass.*, **72**, 63–70.

Weyer, F. (1935) Die Variabilität der Grosse bei den Rassen von *Anopheles maculipennis* unter naturlichen Bedingungen und im Experiment. *Arch. Schiffs-u. Tropenhyg.*, **39**, 399–408.

Wheater, P.R., Burkitt, H.G. and Daniels, V.G. (1979) *Functional Histology*, Churchill Livingstone, Edinburgh, London and New York.

Wheelock, G.D., Petzel, D.H., Gillett, J.D. *et al.* (1988) Evidence for hormonal control of diuresis after a blood meal in the mosquito *Aedes aegypti*. *Arch. Insect Biochem. Physiol.*, **7**, 75–89.

Wheelock, G.D. and Hagedorn, H.H. (1985) Egg maturation and ecdysiotropic activity in extracts of mosquito (*Aedes aegypti*) heads. *Gen. Comp. Endocrinol.*, **60**, 196–203.

Whisenton, L.R. and Bollenbacher, W.E. (1986) Presence of gonadotropic and prothoracicotropic factors in pupal and adult heads of mosquitoes. In *Insect*

Neurochemistry and Neurophysiology (eds A.R. Borkovec and D.B. Gelman), Humana Press, Clifton, New Jersey, pp. 339–42.

Whisenton, L.R., Kelly, T.J. and Bollenbacher, W.E. (1987) Multiple forms of cerebral peptides with steroidogenic functions in pupal and adult brains of the yellow fever mosquito, *Aedes aegypti*. *Mol. Cell. Endocrinol.*, **50**, 3–14.

Whisenton, L.R., Warren, J.T., Manning, M. and Bollenbacher, W.E. (1989) Ecdysteroid titres during pupal-adult development of *Aedes aegypti*: basis for a sexual dimorphism in the rate of development. *J. Insect Physiol.*, **35**, 67–73.

White, G.B. (1980) Academic and applied aspects of mosquito cytogenetics. *Symp. R. Entomol. Soc. Lond.*, **10**, 245–74.

White, M.J.D. (1973) *Animal Cytology and Evolution*, 3rd edn, Cambridge University Press, Cambridge.

White, R.H. (1961) Analysis of the development of the compound eye in the mosquito, *Aedes aegypti*. *J. Exp. Zool.*, **148**, 223–39.

White, R.H. (1963) Evidence for the existence of a differentiation center in the developing eye of the mosquito. *J. Exp. Zool.*, **152**, 139–47.

Whiting, P.W. (1917) The chromosomes of the common house mosquito, *Culex pipiens* L. *J. Morphol.*, **28**, 523–77.

Widahl, L.-E. (1988) Some morphometric differences between container and pool breeding Culicidae. *J. Am. Mosq. Control Ass.*, **4**, 76–81.

Widahl, L.-E. (1991) Flow patterns around suspension feeding mosquito larvae (Diptera: Culicidae). *Ann. Entomol. Soc. Am.*, **84**, 91–5.

Wigglesworth, V.B. (1929) Delayed metamorphosis in a predaceous mosquito larva and a possible practical application. *Nature*, **123**, 17.

Wigglesworth, V.B. (1930) The formation of the peritrophic membrane in insects, with special reference to the larvae of mosquitoes. *Quart. J. Microscop. Sci.*, **73**, 593–616.

Wigglesworth, V.B. (1932) On the function of the so-called 'rectal glands' of insects. *Quart. J. Microscop. Sci.*, **75**, 131–50.

Wigglesworth, V.B. (1933a) The function of the anal gills of the mosquito larva. *J. Exp. Biol.*, **10**, 16–26.

Wigglesworth, V.B. (1933b) The adaptation of mosquito larvae to salt water. *J. Exp. Biol.*, **10**, 27–37.

Wigglesworth, V.B. (1938a) The regulation of osmotic pressure and chloride concentration in the haemolymph of mosquito larvae. *J. Exp. Biol.*, **15**, 235–47.

Wigglesworth, V.B. (1938b) The absorption of fluid from the tracheal system of mosquito larvae at hatching and moulting. *J. Exp. Biol.*, **15**, 248–54.

Wigglesworth, V.B. (1942) The storage of protein, fat, glycogen and uric acid in the fat body and other tissues of mosquito larvae. *J. Exp. Biol.*, **19**, 56–77.

Wigglesworth, V.B. (1943) The fate of haemoglobin in *Rhodnius prolixus* (Hemiptera) and other blood-sucking arthropods. *Proc. R. Soc. Lond.* (*B*), **131**, 313–39.

Wigglesworth, V.B. (1950) *The Principles of Insect Physiology*, 4th edn, Methuen, London.

Wigglesworth, V.B. (1973) The significance of 'apolysis' in the moulting of insects. *J. Entomol.* (*A*), **47**, 141–9.

Wigglesworth, V.B. (1981) The natural history of insect tracheoles. *Physiol. Entomol.*, **6**, 121–8.

Wigglesworth, V.B. (1987) Histochemical studies of uric acid in some insects. 2. Uric acid and polyphenols in the fat body. *Tissue & Cell*, **19**, 93–100.

Wilkes, T.J. and Charlwood, J.D. (1979) A rapid gonotrophic cycle in *Chagasia bonneae* from Brazil. *Mosq. News*, **39**, 137–9.

Wilkinson, R.N., Gould, D.J., Boonyakanist, P. and Segal, H.E. (1978) Observations on *Anopheles balabacensis* (Diptera: Culicidae) in Thailand. *J. Med. Entomol.*, **14**, 666–71.

Williams, J.C. and Beyenbach, K.W. (1983) Differential effects of secretagogues on Na and K secretion in the Malpighian tubules of *Aedes aegypti*. *J. Comp. Physiol.* (*B*), **149**, 511–17.

Williams, J.C. and Beyenbach, K.W. (1984) Differential effects of secretagogues on the electrophysiology of the Malpighian tubules of the yellow fever mosquito. *J. Comp. Physiol.* (*B*), **154**, 301–9.

Williams, J.C., Hagedorn, H.H. and Beyenbach, K.W. (1983) Dynamic changes in flow rate and composition of urine during the post-bloodmeal diuresis in *Aedes aegypti*. *J. Comp. Physiol.* (*B*), **153**, 257–65.

Willis, D.F. and Hollowell, M.P. (1976) The interaction between juvenile hormone and ecdysone: antagonistic, synergistic, or permissive? In *The Juvenile Hormones* (ed. L.I. Gilbert), Plenum Press, New York, pp. 1–35.

Wilson, A.B. and Clements, A.N. (1965) The nature of the skin reaction to mosquito bites in laboratory animals. *Int. Arch. Allergy*, **26**, 294–314.

Woke, P.A. (1955) Deferred oviposition in *Aedes aegypti* (Linnaeus) (Diptera: Culicidae). *Ann. Entomol. Soc. Am.*, **48**, 39–46.

Woke, P.A., Ally, M.S. and Rosenberger, C.R. (1956) The numbers of eggs developed related to the quantities of human blood ingested in *Aedes aegypti* (L.) (Diptera: Culicidae). *Ann. Entomol. Soc. Am.*, **49**, 435–41.

Womersley, C. and Platzer, E.G. (1982) The effect of parasitism by the mermithid *Romanomermis culicivorax* on the dry weight and hemolymph soluble protein content of three species of mosquitoes. *J. Invert. Pathol.*, **40**, 406–12.

Womersley, C. and Platzer, E.G. (1984) A comparison of tricarboxylic acid cycle intermediates in the haemolymph of healthy *Culex pipiens*, *Aedes taeniorhynchus* and *Anopheles quadrimaculatus* larvae and larvae parasitised by the mermithid *Romanomermis culicivorax*. *Insect Biochem.*, **14**, 401–6.

Wood, R.J. (1961) Biological and genetical studies on sex ratio in DDT resistant and susceptible strains of *Aedes aegypti*. *Genetica Agraria*, **13**, 287–307.

Wood, R.J. (1976a) Lethal genes on the sex chromosomes concealed in a population of the mosquito *Aedes aegypti* L. *Genetica*, **46**, 49–66.

Wood, R.J. (1976b) Between-family variation in sex ratio in the Trinidad (T-30) strain of *Aedes aegypti* (L.) indicating differences in sensitivity to the meiotic drive gene. *Genetica*, **46**, 345–61.

Wood, R.J. (1977) Meiotic drive and sex ratio distortion in the mosquito *Aedes aegypti*. *Proc. 15th Int. Congr. Entomol.* (1976), pp. 97–105.

Wood, R.J. and Newton, M.E. (1991) Sex-ratio distortion caused by meiotic drive in mosquitoes. *Am. Naturalist*, **137**, 379–91.

Wood, R.J. and Ouda, N.A. (1987) The genetic basis of resistance and sensitivity to the meiotic drive gene D in the mosquito *Aedes aegypti* L. *Genetica*, **72**, 69–79.

Wooley, T.A. (1943) The metamorphosis of the nervous system of *Aedes dorsalis* Meigen (Diptera-Culicidae). *Ann. Entomol. Soc. Am.*, **36**, 432–47.

Wotton, R.S. (1990a) The classification of particulate and dissolved material. In *The Biology of Particles in Aquatic Systems* (ed. R.S. Wotton), CRC Press, Boca Raton, pp. 1–7.

Wotton, R.S. (1990b) Particulate and dissolved organic material as food. In *The Biology of Particles in Aquatic Systems* (ed. R.S. Wotton), CRC Press, Boca Raton, pp. 213–61.

Wright, J.D. and Barr, A.R. (1980) The ultrastructure and symbiotic relationships of *Wolbachia* of mosquitoes of the *Aedes scutellaris* group. *J. Ultrastruc. Res.*, **72**, 52–64.

Wright, J.D. and Barr, A.R. (1981) *Wolbachia* and the normal and incompatible eggs of *Aedes polynesiensis* (Diptera: Culicidae). *J. Invert. Pathol.*, **38**, 409–18.

Wright, J.D., Sjöstrand, F.S., Portaro, J.K., and Barr, A.R. (1978) The ultrastructure of the rickettsia-like microorganism *Wolbachia pipientis* and associated virus-like bodies in the mosquito *Culex pipiens*. *J. Ultrastruc. Res.*, **63**, 79–85.

Wright, J.D. and Wang, B.-T. (1980) Observations on Wolbachiae in mosquitoes. *J. Invert. Pathol.* **35**, 200–8.

Wright, J.M. and Beyenbach, K.W. (1987) Chloride channels in apical membranes of mosquito Malpighian tubules. *Fed. Proc.*,**46**, 270.

Wright, K.A. (1969) The anatomy of salivary glands of *Anopheles stephensi* Liston. *Can. J. Zool.*, **47**, 579–87, 12 pls.

Wülker, W. (1961) Untersuchungen über die Intersexualität der Chironomiden (Dipt.) nach *Paramermis*-Infektion. *Arch. Hydrobiol.*, **25** (Suppl.), 128–81, 5 pls.

Yaguzhinskaya, L.V. (1954) New data on the physiology and anatomy of the dipteran heart. (Structure and function of the heart of *Anopheles maculipennis* Mgn.). *Byull. Mosk. Obshch. Ispyatelej Pirody, Otd. Biol.*, **59**, 41–50. [In Russian.]

Yajima, T. (1973) Ecological studies on the gonotrophic cycle of the mosquito, *Culex tritaeniorhynchus summorosus* Dyar, in relation to the ovarian condition. *Sci. Rep. Tôhoku Univ.* (*Biol.*), **36**, 241–53.

Yang, Y.J. and Davies, D.M. (1968) Amylase activity in black-flies and mosquitoes (Diptera). *J. Med. Entomol.*, **5**, 9–13.

Yang, Y.J. and Davies, D.M. (1971a) Trypsin and chymotrypsin during metamorphosis in *Aedes aegypti* and properties of the chymotrypsin. *J. Insect Physiol.*, **17**, 117–31.

Yang, Y.J. and Davies, D.M. (1971b) Digestive enzymes in the excreta of *Aedes aegypti* larvae. *J. Insect Physiol.*, **17**, 2119–23.

Yang, Y.J. and Davies, D.M. (1972a) The effect of cations on chymotrypsin from *Aedes aegypti* larvae. *J. Insect Physiol.*, **18**, 747–55.

Yap, H.H., Cutkomp, L.K. and Halberg, F. (1974) Circadian rhythms in rate of oxygen consumption by larvae of the mosquito, *Aedes aegypti* (L). *Chronobiologia*, **1**, 54–61.

Yasuno, M. and Tonn, R.J. (1970) A study of biting habits of *Aedes aegypti* in Bangkok, Thailand. *Bull. Wld Hlth Org.*, **43**, 319–25.

Yeates, R.A. (1980) The mosquito *Aedes aegypti* (L.): evidence for three new proteinases. *Z. Parasit.*, **61**, 277–86.

Yen, J.H. and Barr, A.R. (1973) The etiological agent of cytoplasmic incompatibility in *Culex pipiens*. *J. Invert. Pathol.*, **22**, 242–50.

Yen, J.H. and Barr, A.R. (1974) Incompatibility in *Culex pipiens*. In *The Use of Genetics in Insect Control* (eds R. Pal and M.J. Whitten), Elsevier/North Holland, Amsterdam, pp. 97–118.

Yen, Y.H. (1975) Transovarial transmission of rickettsia-like microorganisms in mosquitoes. *Ann. NY Acad. Sci.*, **266**, 152–61.

Yonge, C. and Hagedorn, H.H. (1977) Dynamics of vitellogenin uptake in *Aedes aegypti* as demonstrated by trypan blue. *J. Insect Physiol.*, **23**, 1199–1203.

Youngson, J.H.A.M., Welch, H.M. and Wood, R.J. (1981) Meiotic drive at the D(MD) locus and fertility in the mosquito, *Aedes aegypti* (L.). *Genetica*, **54**, 335–40.

Zavortink, T.J. (1986) The occurrence of *Runchomyia frontosa* in carnivorous bromeliads in Venezuela, with notes on the biology of its immatures (Diptera, Culicidae, Sabethini). *Wasmann J. Biol.*, **44**, 127–9.

Zharov, A.A. (1980) A method for determining the actual fertility of bloodsucking female mosquitoes as exemplified by *Aedes vexans* Meigen (Diptera,

Culicidae). *Med. Parazitol. Parazit. Bolezni*, **49**, 19-24. [In Russian.]

Zheng, L., Saunders, R.D.C., Fortini, D. *et al.* (1991) Low resolution genome map of the malaria mosquito, *Anopheles gambiae*. *Proc. Natl Acad. Sci. USA*, **88**, 11187–91.

Zhu, X., Chen, Z. and Cao, M. (1980) Endogenous molting hormone level and vitellogenin synthesis during the adult stage of the mosquito, *Culex pipiens pallens* Coq. *Contr. Shanghai Inst. Entomol.*, **1**, 63–8. [In Chinese.]

Zhuzhikov, D.P. (1970) Permeability of the peritrophic membrane in the larvae of *Aedes aegypti*. *J. Insect Physiol.*, **16**, 1193–1202.

Zhuzhikov, D.P. and Dubrovin, N.N. (1969) pH of the midgut contents in larvae of bloodsucking mosquitoes (Diptera, Culicidae). *Entomol. Rev.*, **48**, 293–6.

Zhuzhikov, D.P., Kuznetsova, L.A. and Kondratjeva, V.A. (1971) Ultrastructure of peritrophic membranes of some Diptera. *Nauch. Dokl.Vyssh. Shkoly, Biol. Nauk.*, **0** (9), 5–8. [In Russian.]

Zhuzhikov, D.P., Kuznetsova, L.A. and Kurch, T.N. (1970) The effect of food on the formation of the peritrophic membrane in female mosquitoes, *Aedes aegypti* L. *Med. Parazitol. Parazitar. Bolezni*, **39**, 690–4. [In Russian.]

Zomer, E. and Lipke, H. (1980) Time-course of tyrosine metabolism in *Aedes aegypti*. Micro-organisms as cariers of labelled amino acids. *Insect Biochem.*, **10**, 595–605.

Zomer, E. and Lipke, H. (1981) Tyrosine metabolism in *Aedes aegypti*. II. Arrest of sclerotization by MON 0585 and diflubenzuron. *Pestic. Biochem. Physiol.*, **16**, 28–37.

Zomer, E. and Lipke, H. (1983a) Arylated proteins from the yellow-fever mosquito, *Aedes aegypti*. *Biochem. Soc. Trans.*, **11**, 788.

Zomer, E. and Lipke, H. (1983b) Tyrosine metabolism in *Aedes aegypti* – III Covalently bound aromatic components of the pupal cuticle. *Insect Biochem.*, **13**, 577–83.

Species index

Aedeomyia africana Neveu-Lemaire
  larval respiration 122
Aedes abserratus (Felt and Young)
  larval growth 154
Aedes aegypti (L.)
  allergens 260–2
  anal papillae 149
  apyrase 256–7
  autogeny 424
  blood-feeding mechanism
    236–40
  blood-feeding regulation 246–50
  blood meal, stimuli from 392–5
  blood-meal utilization 410–12
  blood-meal volume 223, 409
  cathepsin D 374–5
  chiasmata 29, 33
  chorion deposition 354–6, 358–9
  chorion deposition, regulation of
    399–400
  compound eye development
    184–8
  corpora allata 210–11, 213
  corpora cardiaca 209, 213–14
  crop 263
  cuticle deposition 162–5
  defaecation 325–6
  digestion
    adult 281–8
    larval 106–9
  digestive enzymes 273–5, 277–8
  diuresis after emergence 306
  diuresis after feeding 306–9
  diuresis, regulation of 320–2
  diuretic hormones 217
  dorsal vessel 195–6, 198
  ecdysis 165–6, 169
  ecdysteroid titre 192–3, 395–402
  eclosion 166–7, 169
  egg shell 67, 69–71
  egg size 414
  endopolyploidy 159, 188
  excretion
    adult 322–6
    larval 117–18
  fat body 116–18, 371–4
  fertility 408–14 421–3
  follicle formation 341
  follicular development
    343, 345–6
  genome organization 15–16
  genome size 14–15
  gut
    adult 264–70
    larval 101–3
  haemocytes 200–1
  haemolymph 198–200
  haemolymph regulation
    126–8, 131
  hatching 72–3
  heterochromatic DNA 16
  20-hydroxyecdysone 219,
    395–403
  imaginal disks 173
  immune responses 202–5
  insemination factor 391–2
  ion uptake 135–7
  ionic regulation 138–9
  juvenile hormone 218–19,
    400–4
  larval drinking rate 78–9
  larval feeding mode 96
  larval growth 150–1, 153–7, 191
  larval ingestion rate 98–9
  larval pigmentation 164
  lectins 204
  linkage groups 11–13
  Malpighian tubule pharmacology
    314–18
  Malpighian tubules 270–1
  meiosis 23–7
  midgut endocrine cells 210,
    215, 269
  mitotic karyotype 2–3, 33
  mouthparts, adult 225–9
  neuropeptides 215–17, 316
  nitrogen balance 417–19
  nurse cells 350
  nutrition
    larval 109–10, 114–15
    adult 416–21
  OEH 217, 389–91
  OEH-releasing factor 386–9
  oenocytes 191
  oostasis 404–5
  peritrophic membrane
    adult 279–80
    larval 104–6
    secretion 278–9
  phagostimulants 246–7
  polytene chromosomes 7
  previtellogenic development
    381–5, 405–6
  prothoracic glands 212–13
  prothoracicotrophic hormone
    217
  rectal papillae 270
  reserves 116–17, 293–8, 300
  respiration 123
  respiratory trumpet 119
  retrotransposons 21–2
  rRNA genes 18
  saliva 255–60
  salivary glands 251–4
  sclerotization 163
  sex chromosomes 33
  sex ratio regulation 35–7
  sexual differentiation 33–4, 180
  spermatogenesis 334, 336
  spermatozoa 337–9
  sugar-feeding regulation 241–2,
    244, 249
  synaptonemal complexes 24–8
  testes and gonoducts 327–8
  total vitellogenin content 353
  trypsin modulating factor 217

trypsin synthesis 276–7, 288–91
tubular fluid production 310–14
vitellogenic carboxy-peptidase 362–3
vitellogenin 353, 360–2
vitellogenin genes 364–5
vitellogenin incorporation 374–9
vitellogenin synthesis 365–70
vitellogensis, regulation of 385–407
water flux 132–3
yolk lipids 363
Aedes africanus (Theobald)
blood feeding 236
Aedes albopictus (Skuse)
allergens 259, 261
chiasmata 29
cibarial armature 233
destination of diet 248
fertility 415
genome organization 15
genome size 14–16
larval rectum 147
mitochondrial genome 20–1
rRNA genes 18
salivary glands 256
Aedes apicoargenteus (Theobald)
larval feeding mode 96
Aedes argenteopunctatus (Theobald)
cutaneous respiration 121
Aedes atropalpus (Coquillett)
autogeny 425–6, 429
chymotrypsin 274
ecdysteroid actions 438
ecdysteroid titre 437
fertility 429
genetics of autogeny 441
JH-binding protein 219
juvenile hormone actions 439–40
larval feeding mode 77, 96
larval growth 150
larval gut pH 107
metabolism 300
midgut endocrine cells 269
OEH action 436
oostatic factor 440
regulation of autogeny 433–5
reserves 429
sugar requirement 430
vitellogenin 361–2
Aedes bahamensis Berlin
genome size 14
Aedes campestris Dyar and Knab
anal papillae 143, 148–9
autogeny 426
ionic regulation 137–40
larval drinking 134
larval habitat 125
larval rectum 146
osmoregulation 131
Aedes canadensis (Theobald)
genome size 14
metabolic reserves 299
Aedes cantans (Meigen)
blood feeding 236
blood meal volume 223
exochorion 67
Aedes caspius (Pallas)
autogeny 426
juvenile hormone titre 437
phagostimulants 246–7
Aedes churchillensis Ellis and Brust
autogeny 426–7
flight capacity 427
flight muscle histolysis 430
Aedes cinereus Meigen
blood feeding 236
blood meal volume 223
Aedes communis (De Geer)
allergenicity 259–61
anautogeny 426
crop capacity 221
eclosion 169
intersexes 182
larval feeding mechanism 92–4
larval food 76
larval habitat 124
sugar utilization 421
thermal constants 153
edes detritus (Haliday)
autogeny 426–7
ecdysteroid titre 436–7
fecundity 408
genetics of autogeny 441
ionic regulation 140
juvenile hormone 218, 436–7
larval habitat 125–6
neurosecretion 434
osmoregulation 130
Aedes diantaeus Howard, Dyar and Knab
blood meal utilization 421
dormant follicles 421
Aedes dorsalis (Meigen)
basal lamina 267
haemolymph 199
ionic regulation 141–3
larval drinking 134
larval habitat 125
metamorphosis of CNS 184
permeability of gut 281-2
regenerative cells 189
Aedes epactius Dyar and Knab
anautogeny 426, 429
fertility 429
larval gut pH 107
trypsins 273
Aedes excrucians (Walker)
genome size 14
larval food 76
thermal constant 153
Aedes fitchii (Felt and Young)
intersexes 182
Aedes flavescens (M13ller)
cutaneous respiration 121
larval growth 153
Aedes flavopictus Yamada
genome size 14
pH tolerance 126
Aedes geniculatus (Olivier)
larval habitat 126
Aedes hexodontus Dyar
eclosion 169
egg survival 71
larval food 76
serosal cuticle 67
thermal constant 153
Aedes impiger (Walker)
eclosion 169
facultative autogeny 431
gonotrophic cycle 347
larval food 76
larval growth 153
sex ratio 35
sugar requirement 430
Aedes intrudens Dyar
dormant follicles 421
Aedes kesseli Belkin
autogeny 441
Aedes malayensis Colless
fecundity 408
Aedes mariae (Sergent and Sergent)
autogeny 426
spermatozoa 337
Aedes mascarensis MacGregor
sexual differentiation 34
Aedes mediovittatus (Coquillett)
rDNA 19
Aedes multicolor Cambouliu
larval habitat 125
Aedes natronius Edwards
larval habitat 125
Aedes nigripes (Zetterstedt)
facultative autogeny 431
gonotrophic cycle 347
sex ratio 35
sugar requirement 430
Aedes nigromaculis (Ludlow)
larval growth 151
Aedes notoscriptus (Skuse)
sperm motility 339
Aedes pembaensis Theobald
larval habitat 125
Aedes polynesiensis Marks
anautogeny 441
aposymbiotic adults 39

larval food 75
Aedes pseudoscutellaris (Theobald)
genome size 14
Aedes punctor (Kirby)
bacteriolytic factor 255
crop capacity 221
fecundity 408
intersexes 182
larval growth 153
sex ratio 35
Aedes rempeli Vockeroth
autogeny 427
larval growth 153
Aedes scutellaris (Walker)
cytoplasmotypes 43
genetics of autogeny 441
genome organization 16
genome size 14
larval gut pH 107
mitochondrial DNA 21
Rickettsial symbiont 39, 42
species group 43
Aedes sierrensis (Ludlow)
temperature and development 62
growth-retarding factors 155
hatching 72
intersexes 182
Aedes sollicitans (Walker)
anautogeny 426
blood intake 223
damage to cattle 224
flight metabolism 302
gonadotrophic factor 393
juvenile hormone actions 381, 384, 406
larval habitat 125
larval–pupal ecdysis 169
lipid metabolism 297–300
neurosecretory cells 207–8
OEH activity 387
OEH secretion 391, 406
OEH-releasing factor 387
oostasis 404
semen factor 431
sugar metabolism 295–300
trehalose 294
Aedes sticticus (Meigen)
larval growth 155
oviposition site 70
Aedes stimulans (Walker)
intersexes 180–2
metabolic reserves 299
metamorphosis of gut 188
sexual differentiation 175–80
spermatogenesis 334
vasa deferentia 328
Aedes taeniorhynchus (Wiedemann)
adult size 156–7
autogeny 426, 429
blood intake 223
diuresis 306
diuretic hormone 217
eclosion rhythm 169
effect of insemination 430
facultative autogeny 431
feeding on honeydew 220
flight metabolism 302
fluid secretion 310
haemolymph 199, 200
hatching 167
ionic regulation 139–43
juvenile hormone actions 381
larval drinking 79, 134
larval growth 154–5
larval habitat 125–6
larval respiration 123
larval urine 134
larval–pupal ecdysis 167–9
lipid metabolism 296–7, 300–1
Malpighian tubules 103, 145, 190, 314
metabolic factor 218
OEH action 434
OEH secretion 390–1
OEH-releasing factor 434
osmoregulation 129–31
regulation of autogeny 433–4
semen factor 431
sugar metabolism 296–8, 301
sugar requirement 430
survival 222
Aedes togoi (Theobald)
anal papillae 148
autogeny 426, 428, 429
dopamine-*N*-acetyltransferase 358
gastric caecae 145
genetics of autogeny 441
ionic regulation 141
larval
drinking 79, 134
growth 155
habitat 125
rectum 147
osmoregulation 130, 134
ovarian follicles 341
Aedes triseriatus (Say)
adult peritrophic membrane 279
adult size 156
basal lamina 267
blood intake 223
crop permeability 264
egg shell 69
genome organization 15
genome size 14
juvenile hormone actions 381
larval feeding 97
larval feeding mode 77–8, 96
larval gut 102
larval–pupal ecdysis 169
linage groups 12
linkage map 11
metamorphosis of gut 189
OEH secretion 391
rDNA 19
sex ratio 72
trypsins 273
Aedes trivittatus (Coquillett)
encapsulation 203–4
haemocytes 201, 205
Aedes vexans (Meigen)
destination of diet 248
egg survival 70
fecundity 408
fertility 415
karyogamy 46
larval feeding 99
larval growth 151, 153–4
larval habitat 125
larval–pupal ecdysis 169
nectar feeding 222
oviposition site 70
phagostimulants 98
primordial germ cells 333
stomatogastric nervous system 54
sugar feeding 242
Aedes vittatus (Bigot)
egg survival 70
larval growth 153
polytene chromosomes 7
Aedes zoosophus Dyar and Knab
genome size 14
Anopheles albimanus Wiedemann
blood intake 223
body composition 296
destination of diet 248
diuresis 223
ecdysteroid titre 396–7
fertility 412
follicular development 343
genome organization 15
hydration of eggs 71
larval drinking 79
larval feeding 98
larval salivary glands 106
linkage map 11
lipid content 413
lipid synthesis 412
nitrogen budget 325
nucleolus 8
OEH secretion 389–90
ovarian development 351
permeability of gut 281
polytene chromosomes 7
protein content 413
protein metabolism 412
trypsins 273

Anopheles albimanus (continued)
vitellogenin 361
Anopheles arabiensis Patton
blood intake 223
fertility 414
paracentric inversions 10
rDNA 18–9
retrotransposons 21
salivary glands 255
sex chromosomes 4
Anopheles atroparvus Van Thiel
chiasmata 29–30
chorion deposition 356–7
cibarial armature 233
DNA replication 9
eggshell 68
fertility 414
follicle structure 330
genetics of stenogamy 5
genome size 14
gonadotrophic factor 386
insemination factor 391
karyotype 4
larval habitat 125
metabolic reserves 299
nucleolus organizer 8, 341
nurse cells 341, 350
ovarian development 347
polytene chromosomes 7–8
salinity tolerance 124
sex chromosomes 4, 17
spermatogenesis 334
testis 327
Anopheles balabacensis Baisas
egg survival 71
karyotype 4
polytene chromosomes 6
Anopheles bradleyi King
larval–pupal ecdysis 169
Anopheles bwambae White
paracentric inversions 10
retrotransposons 22
Anopheles caroni Adam
ovarian development 346
Anopheles culicifacies Giles
chiasmata 29
linkage groups 14
sex chromosomes 4
triploidy 32
Anopheles darlingi Root
crop contents 249
Anopheles dirus Peyton and Harrison
blood intake 223
phagostimulants 247
Anopheles farauti Laveran
cibarial pump 234
larval feeding mode 78
polytene chromosomes 6
vitellin 362
Anopheles freeborni Aitken
anticoagulant 256
defaecation 326
diuresis 310, 320–1
diuretic hormone 217
genome organization 15
genome size 14
lipids 293
metabolic reserves 299
permeability of gut 281
phagostimulants 247
salivary glands 253
Anopheles funestus Giles
fertility 414
ovarian development 346–7, 384
Anopheles gambiae Giles
blood meal volume 223
body composition 296
chromosome map 9
cuticle deposition 194
diuresis 306
eclosion rhythm 170
egg shell 68
egg survival 71
encapsulation 204
fertility 412, 414
genome size 14
gynandromorph 32
karyosphere formation 28
larval–pupal ecdysis 168–9
linkage groups 14
linkage map 11
midgut ultrastructure 266
nucleolus 349
nurse cells 350
oocyte cells 341
ovarian development 347, 384
ovarian stages 345
paracentric inversions 10–11
peritrophic membrane 278, 280
permeability of gut 281
phagostimulants 247
phragmata 194
polytene chromosomes 6–7
protein metabolism 412
rate of digestion 75
rDNA 18–19
regenerative cells 268
retrotransposons 21–2
salivary glands 255
sex chromosomes 4
sex linkage 31–2
vitellogenin genes 363
Anopheles hamoni Adam
ovarian development 346
Anopheles hilli Woodhill and Lee
autogeny 426, 429
Anopheles labranchiae Falleroni
genetics of eurygamy 5
genome size 14
karyotype 4
larval habitat 125
polytene chromosomes 8
sex chromosomes 17
Anopheles listeri De Meillon
polytene chromosomes 6
Anopheles maculipennis Meigen
adult mouthparts 229
cleavage 47
crop permeability 264
gonad formation 53
heart 196
imaginal disks 173
larval feeding rate 109
larval pharynx 95–6
midgut ultrastructure 266
oesophageal invagination 54
ovarian development 347
paracentric inversions 10
polytene chromosomes 8
salinity tolerance 124
saliva 258
sex chromosomes 4
visceral nervous system 210
Anopheles melanoon Hackett
fecundity 408
polytene chromosomes 8
Anopheles melas Theobald
calyx 332
egg survival 71
hatching 72
larval habitat 125
ovary structure 329
oviduct 332
paracentric inversions 10–11
rDNA 18–19
Anopheles merus Dönitz
larval habitat 125
paracentric inversions 10
rDNA 18–19
retrotransposons 22
Anopheles messeae Falleroni
fecundity 408
fertility 415
polytene chromosomes 8
rate of digestion 285
Anopheles minimus Theobald
insemination factor 392
larval habitat 126
Anopheles plumbeus Stephens
blood feeding 236
larval gut 102
larval habitat 126
speed of feeding 237
Anopheles punctimacula Dyar and Knab
egg survival 70
Anopheles punctulatus Dönitz

egg shell 69
Anopheles quadriannulatus Theobald
paracentric inversions 10
rDNA 18–19
Anopheles quadrimaculatus Say
agglutinins 256
anticoagulant 256
bacteriolytic factor 255
blood intake 223
body composition 296
crop permeability 264
destination of diet 248
encapsulation 202–3
genome organization 15
genome size 14
gut peristalsis 108–9
haemocytes 201
haemolymph 199
heart beat 195, 197, 198
larval feeding 96, 98
larval-feeding mechanism 89, 91–2
larval growth 150, 153–4
larval mouthparts 89
larval pharynx 96
larval respiration 123
larval–pupal ecdysis 169
midgut pH 273
mitochondrial DNA 19
nitrogen balance 413
nucleolus organizer 8
ovarian follicles 342
polytene chromosomes 6, 8
rDNA 19
salivary glands 238, 251
survival 222
termination of feeding 250
trypsins 273
vasa deferentia 328
Anopheles sacharovi Favre
larval habitat 125
metabolism 300
ovarian development 347, 384
salivary glands 253
starvation 117
Anopheles sergentii (Theobald)
egg shell 68
larval growth 150
Anopheles sinensis Wiedemann
egg survival 71
ovarian development 346
ovarian follicles 341
Anopheles squamifemur Antunes
hatching 72
Anopheles stephensi Liston
acetylcholinesterase gene 8
adult labrum 226
adult size 157
aminopeptidase 274–5, 283
anticoagulant 256
blood meal volume 223
body composition 296
cibarial armature 233
cutaneous respiration 120
DNA replication 9
ecdysteroid titre 396–7
ecdysteroids 219
fertility 395, 412, 415
genome size 14
α-glucosidase 277, 284
growth-retarding factors 155
haemocytes 201
haemolymph 200
haemolymph volume 198
larval digestion 107–108
larval feeding mode 78
larval gut 101, 103
larval gut pH 107
larval habitat 126
linkage groups 14
metabolic reserves 300
midgut ultrastructure 265–7
nitrogen balance 419
nitrogen budget 325
nurse cells 350
ovarian development 353
ovarian follicles 342
pericardial cells 197
peritrophic membrane 104, 106, 278–80, 283, 287
permeability of gut 281–2
phagostimulants 247
polytene chromosomes 6–8
protein metabolism 413
rate of digestion 285
retrotransposons 22
salivary glands 238, 252–4
satellite DNA 16–17
sex chromosomes 4
sex linkage 32
testis 327
trypsins 273–4, 283
vitellogenin 361–2
Anopheles subpictus Grassi
insemination factor 392
larval feeding 98
polytene chromosomes 6
Anopheles sundaicus (Rodenwaldt)
larval habitat 125
Anopheles superpictus Grassi
larval feeding rate 109
nurse cells 350
polytene chromosomes 6, 8
salinity tolerance 124
Anopheles vagus Dönitz
larval habitat 126
Anopheles walkeri Theobald
egg shell 68
Armigeres subalbatus (Coquillett)
adult hypopharynx 229
adult maxilla 230
encapsulation 203–5
genome size 14
haemolymph 204
larval habitat 126
lectins 204
pH tolerance 126
rDNA 19
sex determination 32
spermatogenesis 334
termination of feeding 250

Bacillus thuringiensis israelensis (de Barjac)
δ-endotoxin 107
Brugia malayi (Brug)
immune responses to 203
Brugia pahangi (Buckley and Edeson)
immune responses to 202, 204

Chagasia bathana (Dyar)
egg shell 69
karyotype 5, 7
Chagasia bonneae Root
gonotrophic cycle 347
Chagasia rozeboomi Causey, Deane and Deane
egg shell 69
Coquillettidia fuscopennata (Theobald)
gonotrophic cycle 347
Coquillettidia perturbans (Walker)
larval food 76
larval respiration 122
Coquillettidia richiardii (Ficalbi)
blood feeding 236
blood meal volume 223
larval air sacs 118
speed of feeding 237
Culex annulirostris Skuse
damage to cattle 224
larval food 75
vitellin 362
Culex australicus Dobrotworsky and Drummond
aposymbiosis 42
taxonomic status 42
Culex bitaeniorhynchus Giles
larval feeding mode 77, 96
larval food 76
Culex fuscanus Wiedemann
larval feeding rate 109
Culex globocoxitus Dobrotworsky
aposymbiosis 40, 42
taxonomic status 42
Culex halifaxii Theobald
larval pharynx 87

Culex hortensis Ficalbi
haemocytes 200–1
Culex nigripalpus Theobald
adult peritrophic membrane 280, 287
adult size 156–7
chymotrypsin 274
digestion 286
eclosion rhythm 169–70
larval growth 151, 154–5
larval–pupal ecdysis 169
metabolic reserves 300
OEH-releasing factor 217
ovarian development 353
trypsins 274
Culex pipiens L.
agglutinins 256
amino acid requirements 110
anal papillae 148
anautogeny 425–6
antennal pulsating organ 196
anticoagulant 256
arachidonic acid requirement 112
autogeny 424–9
bacteriolytic factor 255
blastoderm formation 47
blood as energy source 222
body weight 429
carboxypeptidase 275
chemically-defined diet 109–10
chiasmata 29
chorion
deposition 358
sclerotization 69
structure 64, 71
cibarial armature 233
cleavage 46
cuticle
deposition 161–2
structure 161
cytoplasmic incompatibility 37–8, 40–1
cytoplasmotypes 37, 42–3
destination of diet 249
development of CNS 183–4
development of compound eyes 186–7
digestion 285
ecdysteroid titre 397
egg raft 66
egg survival 71
endomitosis 55
endopolyploidy 159
experimental embryology 56–9
fatty acid requirements 111, 113–14
fecundity 408
fertility 428
fertilization 65
flight metabolism 299, 302
follicle degeneration 422–3
gastric caecae 144
gastrulation 48
genetics of autogeny 440–1
genome organization 15
genome size 14
gonadotrophic factor 385–6, 395
growth rate 429
haemocytes 200–1
haemolymph 126, 199–200
haemolymph amino acids 288
hydration of eggs 71
imaginal disks 172–3
intersex mutant 34
juvenile hormone actions 381, 406
juvenile hormone secretion 383
juvenile hormone synthesis 218, 400
larval digestion 75, 107–8
larval feeding 97–9
larval feeding rate 109
larval growth 154
larval gut pH 106–7
larval pharynx 95
larval respiration 123
larval salivary glands 106
linkage groups 12
lipids 293
Malpighian tubules 190
meiosis 46
meiotic drive 36
metabolic reserves 299, 301
metamorphosis of gut 188, 190
micropyle apparatus 65
midgut pH 273
nucleotide requirements 115–16
OEH-releasing factor 217, 388
oenocytes 163
ovarian diapause 347
ovarian follicles 341–2, 404
ovarian stages 345
permeability of gut 281
phagostimulants 97–8, 246–7
polyspermy 46
polytene chromosomes 32, 36
regulation of autogeny 434
respiratory siphon 119
Rickettsial symbiont 39, 42
saliva 259
salivary gland protein 238
salivary glands 253–4, 258
segmentation 50, 53
sex determination 32
sex linkage 31
sex locus 32, 36
sex ratio regulation 35–6
somatic reduction 159, 190
spermatogenesis 334
spermatozoa 336–7, 339
sterol requirements 114
stylets 230
sugar feeding 221, 243
sugar requirement 430
synaptonemal complexes 28
taxonomic status 42
testis 327
thoracic endocrine complexes 213
uric acid 322
vitamin requirements 115
Culex poicilipes (Theobald)
larval respiration 122
Culex quinquefasciatus Say
allergens 259
anal papillae 148
anautogeny 426
clathrin-coated pits 375
cleavage 47
cytoplasmic incompatibility 42
cytoplasmotypes 43
destination of diet 249
eclosion rhythm 169
egg survival 71
excretion 323
experimental embryology 56
fertility 414
follicular epithelium 331
gastrulation 49–50
genome organization 15
genome size 14
gonotrophic cycle 347
growth-retarding factors 155
haemocytes 200–1
karyogamy 46
larval drinking 79
larval feeding 98
larval food 75
larval growth 150–1, 154
larval gut 101, 103
larval habitat 126
linkage groups 12
midgut enzymes 273
nurse cells 349
oenocytoids 201
oocyte 341, 350, 352, 354
oviposition pheromone 67
peritrophic membrane 104, 106, 279
rate of digestion 285
rDNA 19
sex-ratio regulation 35
spermatogensis 334
spermatozoa 336
survival 221
synaptonemal complexes 26–7
termination of feeding 250

trypsins 274, 286
vitellogenin 361–2
vitellogenin uptake 377, 379
yolk lipids 363
Culex restuans Theobald
genome size 14
nectar feeding 222
Culex salinarius Coquillett
culekinins 216
culetachykinins 216
larval–pupal ecdysis 169
Culex sinaiticus Kirkpatrick
cutaneous respiration 121
Culex sitiens Wiedemann
vitellin 362
Culex tarsalis Coquillett
adult peritrophic membrane 278, 283
aminopeptidase 274
anticoagulant 256
autogeny 427
basal lamina 267
chiasmata 29
crop contents 255
fatty acid requirements 111–13
fertility 428
flight metabolism 302
flight muscle 303
glycerophospholipids 112
haemocytes 201
larval feeding 97–9
larval food 76
larval growth 154
linkage groups 12
lipids 293
Malpighian tubules 103
melezitose in crop 220
metabolic reserves 299
midgut esterases 277
midgut ultrastructure 266–7
nectar feeding 221
osmoregulation 128–9
oviposition pheromone 66
permeability of gut 281–2
sugar feeding 221
testis 327
Culex territans Walker
larval growth 150
Culex thalassius Theobald
cutaneous respiration 120
Culex torrentium Martini
larval feeding rate 109
larval respiration 123
Rickettsial symbiont 42
Culex tritaeniorhynchus Giles
larval growth 150
larval habitat 125
linkage groups 12
metabolic reserves 300
ovarian development 346
ovarian follicles 341
salivary duct 251
salivary glands 253, 254
sex determination 32
sex linkage 29
survival 221
Culiseta annulata (Schrank)
adult head structure 227–8, 232
blood meal volume 223
fecundity 408
gonotrophic cycle 347
oenocytes 191
rate of digestion 285
Culiseta incidens (Thomson)
asparagine requirement 110
egg survival 71
sugar feeding 221
Culiseta inornata (Williston)
adult maxilla 230
anal papillae 148
chemosensilla 241
cibarial pump 234
cibarial sensilla 243
destination of diet 248
fertility 415
hatching 72
labellar receptors 231
larval feeding mode 77, 96
larval growth 151
larval habitat 125
larval rectum 147
Malpighian tubules 103, 145
nectar feeding 235
neurosecretory cells 207
osmoregulation 128–9, 139
phagostimulants 246–7
polyspermy 46
spermatogenesis 334
spermatozoa 336
stylets 230
sugar feeding 221–4
sugar metabolism 301
sugar receptors 243–5
termination of feeding 250
Culiseta litorea (Shute)
genome size 14
Culiseta longiareolata (Macquart)
chiasmata 29
karyotype 3
Culiseta melanura (Coquillett)
adult peritrophic membrane 279
genome size 14
larval–pupal ecdysis 169
midgut 268
Culiseta morsitans (Theobald)
cutaneous respiration 121
genome size 14
larval feeding mechanism 92–4
Culiseta subochrea (Edwards)
fecundity 408

Deinocerites cancer Theobald
autogeny 426, 428, 430
effect of insemination 430
fertility 429
larval habitat 125
larval–pupal ecdysis 169
Deinocerites pseudes Dyar and Knab
autogeny 426, 430
effect of insemination 430
Dirofilaria immitis (Leidy)
immune responses to 201, 203, 205

Eretmapodites chrysogaster Graham
crop permeability 264
larval feeding mode 96
visceral nervous system 210
Eretmapodites quinquevittatus Theobald
sex determination 32

Haemagogus equinus Theobald
genome size 14
rDNA 19

Malaya genurostris Leicester
cutaneous respiration 121
Malaya jacobsoni (Edwards)
feeding from ants 221
Mansonia africana (Theobald)
hatching rhythm 167
Mansonia annulifera (Theobald)
respiratory trumpets 122
Mansonia titillans (Walker)
egg shell 69
hatching rhythm 167
larval growth 150
metabolic reserves 298
Mansonia uniformis (Theobald)
fertility 414
hatching rhythm 167
larval respiration 122
ovarian development 350
ovarian follicles 341
respiratory trumpets 122
Mimomyia hybrida (Leicester)
respiratory siphon 122
respiratory trumpets 122
Mimomyia modesta (King and Hoogstraal)
respiratory siphon 122
Mimomyia splendens Theobald
larval respiration 122

Opifex fuscus Hutton
autogeny 426
haemolymph 199
larval habitat 125
osmoregulation 130
Orthopodomyia pulcripalpis (Rondani)
larval air sacs 118
polytene chromosomes 7

Plasmodium gallinaceum Brumpt
penetration of peritrophic membrane 280
disruption of haemostasis 257
Psorophora ciliata (Fabricius)
adult food canal 227
egg survival 70
haemocytes 201
Psorophora columbiae (Dyar and Knab)
damage to cattle 224
Psorophora confinnis (Lynch Arribalzaga)
larval habitat 126
larval–pupal ecdysis 169
lipid metabolism 297
sugar metabolism 297
Psorophora cyanescens (Coquillett)
egg survival 70
spermatozoa 336
Psorophora discolor (Coquillett)
larval feeding 98
Psorophora ferox (Von Humboldt)
larval–pupal ecdysis 169
midgut 268
Psorophora howardii Coquillett
vasa deferentia 328

Runchomyia frontosa (Theobald)
larval feeding mode 97

Sabethes cyaneus (Fabricius)
genome size 14
polytene chromosomes 7
rDNA 19

Toxorhynchites amboinensis (Doleschall)
aberrant embryo 60–1
ecdysteroid content 192
haemolymph 200
Toxorhynchites brevipalpis Theobald
autogeny 425, 428
chymotrypsin 274
eclosion rhythm 169
fertility 428–9
larval growth 151, 153–4
larval–pupal ecdysis 169
polytene chromosomes 7
prothoracicotrophic hormone 217
salivary glands 252
visceral nervous system 210
Toxorhynchites rutilus (Coquillett)
autogeny 428
ecdysteroid content 192
fertility 429
larval growth 151, 154
larval pharynx 87
salivary glands 252
Toxorhynchites splendens (Wiedemann)
genome size 14
Trichoprosopon digitatum (Rondani)
egg shell 69
Tripteroides bambusa (Yamada)
rDNA 19

Wolbachia pipientis Hertig
symbiosis 39, 40, 42–3
Wyeomyia medioalbipes Lutz
larval growth 154
Wyeomyia mitchellii (Theobald)
anautogeny 426
eclosion rhythm 170
larval–pupal ecdysis 169
Wyeomyia smithii (Coquillett)
aberrant embryo 61
autogeny 425, 427–9
bicaudal larva 61
fertility 430
genome size 14
larval growth 154, 156–7
maxillary palps 231
OEH release 436
paracentric inversions 11
polytene chromosomes 7
rDNA 19
sugar requirement 430
Wyeomyia vanduzeei Dyar and Knab
autogeny 426
larval growth 154
larval–pupal ecdysis 169
semen factor 431

Subject index

Accessory glands 328
Accessory pulsatile organs 196–7
Adenyl nucleotides 246–7
Agglutinin 256
Air/water interface 74, 78, 89, 92
Alimentary canal
  adult 263–70, 282–3
  cytogenetics 158–9
  larva 104
  larval 100–4
  metamorphosis 188–90
Allergen 259
Allometric growth 183
Amino acids
  essential for ovary development 416, 419–20
  essential for growth 109–10
  in excreta 322–3
  in haemolymph 199
Aminopeptidase 108, 274–5, 283, 286, 324
Amnioserosal membrane 49
Amylase 108
Anal papillae
  embryology 53
  in ion uptake 135–8
  ultrastructure 148–9
Anautogeny
  defined 424
Anterior imaginal ring 101, 188
Anticoagulant 256
Antidiuretic factor 317
Ants 221
Apolysis 160, 174, 191
Apyrase 256–8
Arachidonic acid 112–14
Autogeny
  defined 424
  effect of insemination 430–1
  effect of larval nutrition 429
  effect of sugar feeding 430
  facultative 424–5, 431
  genetics 440–1
  hormonal regulation of 433–7
  modelled 431–2
  occurrence 424–7
  ovarian development 427–8
Autolysis 191

Bicaudal larvae 59–62
Blastoderm 46–8, 51, 56–7
Blastokinesis 51, 63
Blood feeding
  mechanism 236–40, 247–8
  regulation 246–50
Blood loss in cattle 224
Blood meal
  discharge 223–4
  electrolytes 309
  nitrogen 223–4, 417–18
  nutritional properties 222, 411–12, 418, 420
  stimulus for ovarian development 392–5
  utilization 412–13, 418–19
  volume 222–3, 409, 411–12
Boundary layer 88

Cathepsin D 373–4
Central nervous system 183–4
Cerebral neurosecretory system 206–8, 215–16
Chorion
  defined 63
  genes 355
  permeability 71
  sclerotization 69–70, 358–9
  secretion 354–8
  structure 64–71
Chromosomes
  chiasmata 11, 25–6, 29
  crossing over 11, 26, 33
  domains 23, 25
  endopolyploid 5
  euchromatic, *see* DNA, euchromatic
  heterochromatic, *see* DNA, heterochromatic
  idiograms 2, 33
  inversions 10–11
  karyotypes 1–5, 10–11, 32
  linkage groups 11–14
  maps 8–9, 11
  mitotic 1, 4, 12, 16–17
  polycomplexes 27–8
  polytene 1, 5–12, 17, 19, 32, 101, 160, 350
  puffs 7–8, 10, 32
  replication 26
  sex 3–4, 8, 12, 17, 29, 33, 35–6
  somatic pairing 1, 7, 22–3, 26
  translocation 12
Chymotrypsin 108, 274
Cibarial armature 233–4
Cibarial pump 228, 231–4
Circadian rhythms 71, 167–70
Clathrin 374–5
Cleavage 46, 56
Cleavage energids 46, 56, 58
Colloidal particles 99
Compound eyes, development 184–8
Corpora allata 54, 210–12
Corpora cardiaca 54, 208–9, 212–16
Crop
  capacity 221
  function 166, 248
  structure 263
Culekinins 216, 317
Cuticle
  deposition 160–3, 193
  sclerotization 163–4
  ultrastructure 161–3
Cystocytes 341
Cytoplasmic incompatibility 37–44

Cytoplasmic inheritance 37–9, 44
Cytoplasmotypes 37–9, 42–4

Defaecation 325–6
Detritus 74, 76
Developmental gates 345–6, 380–1
Diapause 299–301
Diet
 for egg production 416–17
 for larva 109–12, 115–16
Digestion
 adult 281–6, 288
 larval 106–8
 regulation 288–91
Dissolved organic matter 75, 79
Diuresis
 in larva 133–4
 post-emergence 166, 306
 post-feeding 223, 306–10
 regulation 320–2
Diuretic hormones 133, 217, 317
DNA
 euchromatic 1, 3, 4, 7–8, 16–17, 29
 heterochromatic 1–4, 7–8, 16–17, 24
 mitochondrial 19–21
 repetitive 15–16
 replication 9, 46
 satellite 16–17
 transcription 9
Dopa decarboxylase 164–5, 358–9
Dopamine *N*-acetyltransferase 164–5, 358–9
Dorsal closure 51–2
Dorsal vessel 195–6
Double cephalon 61–2
Dyar's rule 150

Ecdysis 160, 165, 167–70, 191
Ecdysteroids
 and metamorphosis 192–3
 metabolism 219
 secretion and titre 395–8, 437
 synthesis 219
Eclosion 166, 169–70
Ectoderm 48, 50, 55, 61
Ectoperitrophic space 133, 281, 286–7
EDNH, *see* Ovarian ecdysteroidogenic hormone
Egg
 dormancy 71
 hatching 70–3, 167
 organization 45, 63, 354
 water loss 70
 water uptake 71
Egg burster 73
Egg membranes 63
Egg raft 64, 66, 69, 73
Egg shell
 sclerotization 69, 358, 359
 terminology 63
Egg size 414
Eggs
 protein content 411
 summer 68
 winter 68–9
Eicosanoids 111, 317–18
Embryonic cells 101
Emergence 166, 169–70
Encapsulation 202–3
Endocytotic complex 374–5
Endogenous rhythms, *see* Circadian rhythms
Endomitosis 55
Endopolyploidy 158–9, 188, 190
Endoreduplication 5, 7
Energids 56, 61
Epidermis, metamorphosis 171, 191
Esterases 277
Euchromatin, *see* DNA, euchromatic
Excretion
 in adult 322-5
 in larva 117–18
 regulation of 325–6

Fat body
 competence 384–5
 in excretion 118
 metamorphosis 191
 polyploidy 159
 reserves 116–17
 structure 116
 vitellogenesis 371–2
Fecundity
 defined 408
 effect of body size 409
 see also Fertility
Feeding modes
 adult 235
 larval 76–8
Fertility
 defined 408
 effect of age 414–15
 effect of blood meal 409–11, 417
 effect of body size 408–9
 of autogenous females 428–9
 relative 417
 seasonal variation 415
Fertilization 22, 36, 40, 45
Filter feeding 77
Flight metabolism 302–3
Food
 adult 220, 222
 larval 74–6
Frontal ganglion 54, 212–13

Gastric caeca, ultrastructure 144–5
Gastrulation 48–50, 59–60
Gates, *see* Developmental gates
Genetic mapping 1
Genetic maps 8
Genitalia, rotation 194
Genome
 cytoplasmic 40
 maps 1
 mass 14–16
 organization 15
α-Glucosidase 255–6, 258, 277, 284
Glycogen 108, 116–17, 292–3, 295–303, 371–2
Gonadal primordia 176–81
Gonotrophic cycle
 and fertility 414
 defined 342
Growth rate 152, 154
Growth-retarding factors 155
Gynandromorphs 30, 32–3

Haematin 324
Haemocytes 200–5
Haemoglobin
 chemistry 419
 digestion 273, 281, 284, 288
Haemolymph
 amino acids 199, 288
 constituents 199–200, 308
 lectins 204
 osmotic pressure 127–31
 pH 199–200
 phenoloxidase 205
 regulation 126–31, 138–9
 volume 198
Haemolysis 233, 255, 288
Hatching, *see* Egg hatching
Hatching spine 55
Heart
 beat 197–8
 structure 195–6
Heterochromatin, *see* DNA, heterochromatic
Histamine 256, 260–1
Histolysis 117, 191, 193
Honeydew 220–1
Humidity, effect on digestion 285
Hydrodynamics 87–8, 92, 95
20-Hydroxyecdysone
 action in autogenous females 438–9
 defaecation 326
 chorionic protein synthesis 399
 dopa decarboxylase synthesis 399–400

previtellogenic development 389, 405
secretion and titre 395–8, 402, 436–8
vitellogenin synthesis 389, 398–9
Hypersensitivity to mosquito bites 259–62
Hypocerebral ganglion 54, 212–13, 215

Imaginal cells 160
Imaginal disks 171–6, 178–81
Immune response of mosquitoes 202–5
Initiation of ovarian development 380, 385–7
Insemination factor 391–2
Intersexes 34, 180–2
Isoenzymes 11

Juvenile hormone
actions 267, 372, 401, 439–40
control of secretion 383–4
fat body competence 384–5
metabolism 218–19, 400
mobilization of reserves 403
previtellogenic development 381–3, 406
structure 218
synergism 402–3
synthesis 400
titre 400–2, 436–7
Juvenile hormone esterase 383, 401

Karyogamy 41, 46
Karyosphere 27–8, 45–6, 348–9

Larva
alimentary canal 100–4
current generation 89–94
drinking rate 79, 134
feeding modes 76–8, 96–7
food 74–6
growth 150–1
gut contents 75–6
habitats 124–6
head structure 80–1
ingestion rate 97–9
ion uptake 135–8
mass 151
mouthparts 81–5, 89, 92, 94
particle capture 94–6
phagostimulants 97–8
pharate 70, 72
pharynx 85–7, 94–6, 100
water balance 131–4
Larval food
particle size 97–8
Lectins 204–5, 255, 267
Leucokinins 215, 317
Lipids, ovarian 363

Malpighian tubules
concretion bodies 270
fluid secretion 311–14, 318–20
in ion regulation 140
metamorphosis 190
pharmacology 316–18
phosphorus accumulation 190
preparations 310–11
response to cAMP 314–16, 318
structure 103
ultrastructure 145–6, 270–1
Matrone, effect on digestion 286
Maturation 193–4
Meconium 166, 305, 322
Meiosis 22–6, 28–9, 36–7, 41–2, 45–7
Meiotic drive 36–7
Melanin 164, 202, 359
Mesenchyme cells 171
Mesoderm 48, 50, 53, 171
Metamorphosis 151, 158, 160, 171, 183–4, 191
Micropyle 45, 63, 65, 68, 357
Micropyle apparatus 65, 341, 356
Midgut
endocrine cells 210, 215, 269, 389
rudiments 49–51, 53, 58–60
Mitochondrial DNA 19
Mobile genetic elements 21–2
Moulting 160–2, 165–6, 192
Mouth 228, 231
Mouthparts
adult 224–31
larval 80–5, 89, 92, 94

Natriuretic hormone 317
Nectar 220–1, 244, 248, 301
Nephrocytes 197
Nervous system
peripheral 254
ventral visceral 196
Neuroblasts 54
Neurohaemal organs 208–11, 213, 216
Neurosecretory systems 206–7
Nitrogen balance 324–5, 417
Nucleolus 8, 348–9
Nucleolus organizer 8, 19, 341, 348–9
Nucleotides
larval requirements 115–16
phagostimulants 97, 246–7
Nurse cells 6, 39, 328, 330–1, 341, 349–50
Nurse cell chromosomes 5, 8–9
Nutrition
effects on body size 157
effects on growth rate 154
effects on ovarian development 384
requirements of larva 109

OEH-releasing factor 217, 387–9, 434
Oenocytes 55, 163, 191
Oenocytoids 201
Oesophageal invagination 53–4
Oesophageal nerves 54
Oesophagus 53
Oocyte
glycogen synthesis 354
organization 330, 354
rRNA synthesis 347–9
Oogenesis, nutritional requirements 416, 419–21
Oostasis 389, 404–5, 440
Optic lobes 184
Osmolality
defined 127
Ovarian cycle 342
Ovarian development
external factors 347
phases 342, 344
previtellogenic 381–3
stages 344–5
use of reserves 412
Ovarian ecdysteroidogenic hormone
action 326, 386, 388–9, 391
definition 216
follicle separation 406
secretion 387–91, 434, 436
synthesis and storage 387
Ovarian follicles
development 340–3, 348, 350–8
resorption 421–3
structure 330–1
Ovariole
number 408–9
structure 328–31
Ovary
development 177–8
energy values 409–13
structure 328–9
Oviducts 331–2
Oviposition pheromone 66–7
Ovulation 65

Palatum 54
Particulate organic matter 74
Peptide hormones 215–17

Pericardial cells 197
Perispiracular gland 119
Peristalsis 108–9, 288
Perisympathetic organs 210
Peritrophic membrane
  adult 279–80
  function 280, 287
  larval 104–6
  permeability 105–6, 287
  secretion 104, 265, 278–9
pH
  of adult gut 273
  of haemolymph 199–200
  of larval gut 106–7
  of larval habitats 125–6
  optima for digestion 107–8, 273
Phagocytes 191
Phagocytosis 191, 201
Phagostimulants 97, 246–7
Pharyngeal pump 232, 234
Pheromone 66–7
Physical gill 66, 68
Pigment 67, 163–5
Plastron 66, 68
Polar bodies 28, 34, 46
Polar granules 45, 53, 61
Pole cells 47, 49, 53, 58–9, 61
Polyspermy 46
Polyunsaturated fatty acids 76, 111–14, 293
Pool feeding 239–41
Posterior imaginal ring 103, 190
Primary germ cells 58, 61, 334
Primordial germ cells 53
Proctodaeum 49, 51, 53, 190
Promotion of ovarian development 380, 387
Prostaglandin synthesis 114
Prothoracic glands 212
Prothoracicotrophic hormone 217
Protocerebrum 54
Pyloric bristles 103, 224, 269
Pyloric chamber 53

rDNA, *see* rRNA genes
Receptor-mediated endocytosis 360, 375–6, 378–9
Rectum
  embryonic development 53
  in ion regulation 139, 141–3
  ultrastructure 146–7
Recurrent nerve 54, 208, 213
Regenerative cells 101–2, 268
Reserves
  adult 293–5, 297, 299–300, 410, 413
  larval 116–17
  of autogenous females 429
  synthesis 295–6, 410–11
  utilization 300–3, 411
Respiration
  larval 118–23
  pupal 119–20, 122
Respiratory siphon 55
Retrotransposons 21–2
Reynolds number 87–8, 92, 94, 240
Rickettsial symbionts 39–42
rRNA
  genes 16–19
  mitochondrial 20
  ovarian 347–8

Saliva
  allergens 259
  constituents 255
  function 256
  in sugar feeding 221
  secretion 258–9
  synthesis 258
Salivary glands
  adult 251
  chromosomes 5, 8
  embryonic development 54–5
  larval 100, 106
  metamorphosis 191, 193
  structure 251–2
  ultrastructure 253–4
Salivary valve 232, 235
Segmentation 50, 53–4, 56
Seminal vesicles 327–8
Sensilla on mouthparts 241
Serosa 50–2, 64, 67
Serosal cuticle 51, 64, 67, 71, 73
Sex determination 30–2
Sex linkage 29, 31–2
Sex locus 32–3, 36
Sex mosaics 33
Sex ratio 30–1, 35–7, 72
Sex ratio distortion 35–6
Sexual differentiation 30, 33, 175, 180–1
Skin reactions 259–62
Small intestine 53
Somatic reduction 158–9, 188
Spermatids 46
Spermatocytes 23–5, 35, 39, 327, 333–4
Spermatogenesis 333–4
Spermatozoa
  motility 334, 339
  non-functional 36
  number 334
  storage 328
  ultrastructure 336–9
Spermiogenesis 335–6
Spiracles 118–19, 191
Stenogamy, inheritance 5
Sterols, larval requirements 114
Stomatogastric nervous system 54, 212
Stomodaeum 50–3, 188
Sugar feeding
  behaviour 235–6
  regulation 241–4, 248–50
Sugar metabolism 295–8
Sugars, structure–activity relationships 244–5
Surface membrane 92, 118
Surface microlayer 74, 89
Surface tension 65–6, 89, 92
Suspension feeding 77
Symbionts 39, 417
Synaptonemal complexes 22–8

Temperature
  effects on body mass 151
  effects on body size 156–7
  effects on digestion 285
  effects on growth rate 150, 152–3
  effects on respiration 123
  effects on sexual differentiation 180–3
Testis
  development 176
  structure 327
Tracheae, moulting of 165
Trehalose 200, 292, 295, 300–2
Triacylglycerols 292–3, 295, 297–300, 302, 363
Trophocytes
  defined 293
  previtellogenic development 364, 368, 371–2
  ultrastructure 371
  vitellogenesis 361, 367, 370, 372–3
Trypsin inhibitors 286, 394
Trypsin modulating factor 217, 291, 405
Trypsin synthesis 276–7, 282–4
Trypsins 108, 273–4, 286, 324
Tubular fluid
  defined 305
  formation 311–14, 318–20
  larval 132–3, 135, 138–40
Tyrosine, and pigmentation 165

Vas deferens 327–8
Vas efferens 327–8
Ventral neurosecretory system 210
Vessel feeding 239–40
Vitamins
  adult requirements 417
  larval requirements 115

Vitellin 360–1
Vitelline envelope 63
Vitelline membrane 63
Vitellogenic carboxypeptidase
  362, 365, 377
Vitellogenin
  composition 360–2
  incorporation 375–8
  receptors 377
  synthesis 365–9
  synthesis termination 369–70,
    373
  whole-body content 353
Vitellogenin genes
  regulation 368
  structure 363–4
  transcription 363–5
  translation 365–7
Vitellophages 47, 50, 52

Water balance 131
Water surface 89
white-eye 31–2

Yolk lipids 363
Yolk proteins
  of higher Diptera 360